Handbook of Research on Developing Smart Cities Based on Digital Twins

Matteo Del Giudice
Politecnico di Torino, Italy

Anna Osello
Politecnico di Torino, Italy

IGI Global
PUBLISHER of TIMELY KNOWLEDGE

A volume in the Advances in Civil and Industrial Engineering (ACIE) Book Series

Published in the United States of America by
IGI Global
Engineering Science Reference (an imprint of IGI Global)
701 E. Chocolate Avenue
Hershey PA, USA 17033
Tel: 717-533-8845
Fax: 717-533-8661
E-mail: cust@igi-global.com
Web site: http://www.igi-global.com

Library of Congress Cataloging-in-Publication Data

Names: Del Giudice, Matteo, 1987- editor. | Osello, Anna, editor.
Title: Handbook of research on developing smart cities based on digital twins / Matteo Del Giudice, and Anna Osello, editors.
Description: Hershey, PA : Engineering Science Reference, an imprint of IGI Global, [2021] | Includes bibliographical references and index. | Summary: "This book contains different contributions focused on the description of methods, processes and tools that can be adopted to achieve smart city goals, discussing how the advent of connected, smart technologies for the built environment can provide significant value to developing digital city models"-- Provided by publisher.
Identifiers: LCCN 2020036922 (print) | LCCN 2020036923 (ebook) | ISBN 9781799870913 (hardcover) | ISBN 9781799870937 (ebook)
Subjects: LCSH: Smart cities. | Cities and towns--Computer simulation. | Information modeling.
Classification: LCC TD159.4 .H3648 2021 (print) | LCC TD159.4 (ebook) | DDC 307.730285--dc23
LC record available at https://lccn.loc.gov/2020036922
LC ebook record available at https://lccn.loc.gov/2020036923

This book is published in the IGI Global book series Advances in Civil and Industrial Engineering (ACIE) (ISSN: 2326-6139; eISSN: 2326-6155)

British Cataloguing in Publication Data
A Cataloguing in Publication record for this book is available from the British Library.

All work contributed to this book is new, previously-unpublished material. The views expressed in this book are those of the authors, but not necessarily of the publisher.

For electronic access to this publication, please contact: eresources@igi-global.com.

Advances in Civil and Industrial Engineering (ACIE) Book Series

Ioan Constantin Dima
University Valahia of Târgovişte, Romania

ISSN:2326-6139
EISSN:2326-6155

Mission

Private and public sector infrastructures begin to age, or require change in the face of developing technologies, the fields of civil and industrial engineering have become increasingly important as a method to mitigate and manage these changes. As governments and the public at large begin to grapple with climate change and growing populations, civil engineering has become more interdisciplinary and the need for publications that discuss the rapid changes and advancements in the field have become more in-demand. Additionally, private corporations and companies are facing similar changes and challenges, with the pressure for new and innovative methods being placed on those involved in industrial engineering.

The **Advances in Civil and Industrial Engineering (ACIE) Book Series** aims to present research and methodology that will provide solutions and discussions to meet such needs. The latest methodologies, applications, tools, and analysis will be published through the books included in **ACIE** in order to keep the available research in civil and industrial engineering as current and timely as possible.

Coverage

- Productivity
- Hydraulic Engineering
- Urban Engineering
- Structural Engineering
- Transportation Engineering
- Materials Management
- Production Planning and Control
- Optimization Techniques
- Coastal Engineering
- Operations Research

IGI Global is currently accepting manuscripts for publication within this series. To submit a proposal for a volume in this series, please contact our Acquisition Editors at Acquisitions@igi-global.com or visit: http://www.igi-global.com/publish/.

The Advances in Civil and Industrial Engineering (ACIE) Book Series (ISSN 2326-6139) is published by IGI Global, 701 E. Chocolate Avenue, Hershey, PA 17033-1240, USA, www.igi-global.com. This series is composed of titles available for purchase individually; each title is edited to be contextually exclusive from any other title within the series. For pricing and ordering information please visit http://www.igi-global.com/book-series/advances-civil-industrial-engineering/73673. Postmaster: Send all address changes to above address.

Titles in this Series

For a list of additional titles in this series, please visit:
http://www.igi-global.com/book-series/advances-civil-industrial-engineering/73673

Structural Dynamics and Static Nonlinear Analysis From Theory to Application
Mourad Belgasmia (Setif 1 University, Algeria)
Engineering Science Reference • © 2021 • 347pp • H/C (ISBN: 9781799843993) • US $195.00

Data-Driven Optimization of Manufacturing Processes
Kanak Kalita (Vel Tech Rangarajan Dr. Sagunthala R&D Institute of Science and Technology, India) Ranjan Kumar Ghadai (Sikkim Manipal Institute of Technology, Sikkim Manipal University, Majhitar, India) and Xiao-Zhi Gao (University of Eastern Finland, Finland)
Engineering Science Reference • © 2021 • 298pp • H/C (ISBN: 9781799872061) • US $225.00

Developing and Monitoring Smart Environments for Intelligent Cities
Zaigham Mahmood (University of Northampton, UK & Shijiazhuang Tiedao University, China)
Engineering Science Reference • © 2021 • 367pp • H/C (ISBN: 9781799850625) • US $215.00

Applications and Challenges of Maintenance and Safety Engineering in Industry 4.0
Alberto Martinetti (University of Twente, The Netherlands) Micaela Demichela (Politecnico di Torino, Italy) and Sarbjeet Singh (Luleå University of Technology, Sweden)
Engineering Science Reference • © 2020 • 321pp • H/C (ISBN: 9781799839040) • US $265.00

Reconstructing Urban Ambiance in Smart Public Places
Hisham Abusaada (Housing and Building National Research Center (HBRC), Egypt) Ashraf M. Salama (University of Strathclyde, Glasgow, UK) and Abeer Elshater (Ain Shams University, Egypt)
Engineering Science Reference • © 2020 • 280pp • H/C (ISBN: 9781799838562) • US $225.00

Additive Manufacturing Applications for Metals and Composites
K.R. Balasubramanian (National Institute of Technology, Tiruchirappalli, India) and V. Senthilkumar (National Institute of Technology, Trichy, India)
Engineering Science Reference • © 2020 • 348pp • H/C (ISBN: 9781799840541) • US $205.00

Multi-Objective Optimization of Industrial Power Generation Systems Emerging Research and Opportunities
Timothy Ganesan (Royal Bank of Canada, Canada)
Engineering Science Reference • © 2020 • 233pp • H/C (ISBN: 9781799817109) • US $195.00

IGI Global
PUBLISHER of TIMELY KNOWLEDGE

701 East Chocolate Avenue, Hershey, PA 17033, USA
Tel: 717-533-8845 x100 • Fax: 717-533-8661
E-Mail: cust@igi-global.com • www.igi-global.com

Editorial Advisory Board

List of Contributors

Table of Contents

Detailed Table of Contents

The term smart city is often synonymous with a sustainable city. The word smart implies the use of digital technology that serves to make processes and services more efficient and to connect the different actors on the urban scene. However, this is no guarantee of sustainability. A city can become sustainable if it changes its metabolism and from linear to circular as in nature's ecosystems. For this to happen, it is necessary to overcome the paradigm of quantitative economic growth based on the infinite substitutability between natural and economic capital. If smart city governance stakeholders primarily pursue profit according to the logic of the free market, the city may be smarter and efficient in the use of energy and resources, but it is not sustainable, often not even inclusive. The challenge of sustainability implies a paradigm shift and the use of digital technologies at the service of the collective good. In this context, after a general analysis of the characteristics of smart cities, the chapter focuses on an Italian case study, Turin Smart City.

For the near future, forecasts predict an uncontrolled growth of urbanization in the world, in which cities are fragmented and uneven systems in relation to fast evolving environmental, economic, and social phenomena. The traditional urban planning approach, essentially theoretical-predictive, adapts poorly to face future challenges. Hence, the need to rethink how to govern the transformations of cities, which can be described by models of urban metabolism. The city sensing has changed the way cities are explored and used. With the transition from digitalization to datafication, through the computational approach, georeferenced big data can be analysed and exploited by algorithms. They originate a generative computational urban planning process, which can achieve a higher quality of the project and provide cities with adaptive capability. This process exploits data provided by public administrations, companies, and citizens who take part in an inclusive and adaptive urban planning.

Digital twin has gained a great interest during the last few years from academia and industry because of the development in IT technology, communication field, and sensor technology. The general vision of the DT is to provide a detailed physical and functional description of a component, product, and systems. Nevertheless, the digital twin is a highly dynamic concept growing in complexity during the product life cycle, which leads to an enormous amount of data and information. Motivated by these, this chapter investigates the concepts and architecture of DT to cover its challenges and explore its applications in various fields such as smart cities, smart manufacturing and factories, and healthcare sectors. In the end, the challenges and research areas will be presented.

This chapter describes the results of a sociological investigation carried out within an EU-funded project. The project was aimed at creating a tool to visualise and compute energy data at an urban district level, with the broader aim to optimise the local district heating (DH) network's distribution policies. This chapter identifies the features of the main categories of actors (from the DH operator to final users) having a role within that network. Special attention is paid to the identification of the barriers and frictions preventing a stronger collaboration and communication among these actors to happen. It is argued that the identification and resolution, in situated and complex socio-technical systems, of these non-strictly-technical problems may be, at least in some cases, a pre-requisite for any ICT-based solution to deploy its full potential.

More than two thirds of the European population live in urban areas. Cities are places where both problems emerge and solutions are found. They are fertile ground for science and technology, for culture and innovation, for individual and collective creativity, and for mitigating the impact of climate change. Cities are communities where to study environmental, social, economic impact of new energy technologies. A tool to unlock the planning potential of EU cities has been developed thanks to the collaboration of different EU research centers, SMEs and consultancy firms led by Rina Consulting involved in Planheat EU Project (funded by the European Union's H2020 Programme under grant agreement 723757).

PLANHEAT Consortium developed and validated an integrated and easy-to-use, GIS-based, and open source tool to support local authorities in selecting, simulating and comparing alternative low carbon and economically sustainable scenarios for heating and cooling.

The architecture, engineering, and construction (AEC) sector is facing the digital transformation by promoting the building information modelling (BIM) as a standard methodology for the digital managing of information along the whole lifecycle of a construction work. Although small-medium enterprises (SMEs) and public administrations (PAs) are aware of the BIM benefit, they ask for pilot actions and tools for testing BIM applications in their daily activities in order to measure benefits and difficulties in detail. This chapter discusses the BIM Simulation Lab initiative which aims at establishing a laboratory for an effective and efficient BIM implementation, by promoting a physical space and specific services for supporting the territory in the DT. The authors describe the concept of the lab and they introduce an assessment method that adopts an indirect approach of the BIM benefit assessment leveraging principles from construction cost estimation and probabilistic risk management.

The post-shock scenario is outlined: an uncertain future with a "new normality." The embryos of the new paradigm are alongside the powerful discontinuity generated by COVID-19. With examples and anecdotes from Shanghai and China, a transformation already underway is portrayed. No more perfect shock could be thought to reconsider the role of humans on this planet, on our cities. There is a crucial need for resilience of local systems, for short chains, for autonomous energy and food self-sufficiency, for decentralizing essential products. Security and contingency plans are needed and must operate on a global scale, but at the same time at the country and the city level. The future must be reinvented, acting in depth, for shifting from "exploitation" to "cooperation" with natural systems. Topics are education, work, services, transport, food safety.

In the era of connections and information and communication technologies, the building industry is facing the challenge of digitization at the building and urban scale. Several researches have been carried out to generate virtual city models to manage and represent a variety of data to reach the smart city concept. Therefore, the development of building/urban digital twins is directly linked to the definition of innovative

digital models useful information related to the analyzed object, such as life cycle, geometry, mechanical properties, and parameters related to management and maintenance. This contribution aims to evaluate building information modeling methodology in the industrial context, as a cyber-physical system, developing flexible 3D parametric models as a data set, where information can be visualized and optimized management, using different visualization tools. The research has underlined the importance to share information between virtual and real worlds through virtual and augmented reality (VAR) systems.

The chapter proposed aims at facing the various implications underlying the smart city concept based on digital twins. The structure of the text is articulated in three main themes: the use of the term "smart city" and the role that technologies had in its definition; the "3D city model" meaning and the integration procedures between BIM (building information modeling) and GIS (geographic information system); the classification of 3D city models by use cases. The chapter can provide researchers with a detailed dissertation aimed at clarifying both the theoretical and technical features belonging to smart city and its related innovative technologies.

A groundswell of opinion exist about the present and future use of smart cities and digital twin technologies and processes and, despite increasing use of information modelling, artificial intelligence, and internet of things, many challenges remain in designing and implement integrated smart systems in large scale contexts. Often, the big picture is shadowed by fragmented processes, and there is a disconnect between the problem and the solution. This chapter aims to address this inverted approach, based on a solution looking for a problem by focusing on the problems of developing integrated solutions for smart cities based on digital twins. The narrative in this chapter is informed by a research project exploring the digitalisation of facilities management processes in Bath, UK. The conclusion is that the development of digital twins goes far beyond linking digital models to sensors.

This work presents the developments of representing a part of the city districts of Manchester, UK and Turin, IT initiated within the FP7 DIMMER project completed in 2016 and continued in the last years by the Center of Excellence GIS of CSI Piemonte. The DIMMER system integrates BIM (building information modelling) and district level 3D models with real-time data from sensors and user feedback to analyze and correlate buildings utilization and provide real-time feedback about energy-related behaviors. The emerging standard 3D Tiles endorsed by the OGC was applied to represent CityGML data of two districts of Turin, Italy and Manchester, UK. 3D Tiles allows for a high level of detail (LOD) visualization that displays increasing detail of geometric features and their alphanumeric properties. Currently, the OGC 3D Tiles technology is mature, and the OGC CityGML specification will be soon updated to version three, with the adoption of exciting innovations like the support of time-dependent properties defined Dynamizers.

Industry 4.0 is encouraging the introduction of pioneering technologies even in the construction industry. Along with the development of high technology, such as augmented reality, virtual reality, and cloud computing, the development of digital twin has been growing. This contribution aims to present the potential of digital twin in the construction field, suggesting a framework that outlines the many different possible applications in construction, with reference to school buildings. First, it summarizes the current overview of digital twin applications in building construction. Then it shows that significant steps that are being taken beyond the digital model, even if the implementation of the digital twin concept in its full meaning is still a long way off. The research is moving in this direction and the evolution of the current state of the art, combined with the experience gained in the industrial sector, will soon bring a new revolution in the construction industry.

Technological progress and the evolution of the directives concerning the construction sector have led to a significant digitalization of information concerning buildings. One of the challenges that have arisen concerns the use of data in the phases of the life cycle following the construction. How can this information be exploited? Is it possible to use it directly in a facility management tool that is within the reach of expert users and not? The aim of this chapter is reporting the developed framework to support the management and maintenance operations in buildings, defined within the project AR4FM - Augmented Reality for Facility Management. In particular, AR4FM aims to create an innovative software ecosystem that uses the most modern technologies to support facility management (FM) operations through a set of bespoke applications. The first tool is the web-based ICT platform usable through web-browser, while the second tool refers to a mobile App, called AR4FM mobile app, which will enable visualization for both mobile and wearable devices, such as smart-glasses.

This chapter presents a novel distributed software infrastructure to enable energy management and simulation of novel control strategies in smart cities. In this context, the following heterogeneous information, describing the different entities in a city, needs to be taken into account to form a unified district information model: internet-of-things (IoT) devices, building information model, system information model, and georeferenced information system. IoT devices are crucial to monitor in (near-) real-time both building energy trends and environmental data. Thus, the proposed solution fulfills the integration and interoperability of such data sources providing also a correlation among them. Such correlation is the key feature to unlock management and simulation of novel energy policies aimed at optimizing the energy usage accounting also for its impact on building comfort. The platform has been deployed in a real-world district and a novel control policy for the heating distribution network has been developed and tested. Finally, experimental results are presented and discussed.

Building information models (BIMs) make it possible to share structured data in the planning and building process and during the live cycle of the building. The models can be enrichened with information in all stages from planning to end-of-life. Predefined information requirements, data structures, and file formats allow the model to be used for design, cost planning, and time scheduling as well as analysis of building performance, and ideally also for facilities management. This chapter describes the possibilities of using application programming interfaces (APIs) to enhance the usefulness of BIM models. This provides the user with the possibility of creating own applications with real time bidirectional data exchange. Of particular interest is the exchange of data with web-based data sources. The applicability is illustrated with examples of environmental analysis based on web-based data and the use of web-based forms to enrichen BIM models with data input from building users.

The chapter describes the approach for the South Tyrolean city of Meran in the creation of use cases and the implementation of an urban agenda (roadmap) for the development of the city from a smart city perspective, with the involvement of citizens, experts, and local administrators. A list of key services, based on a technical and economic pre-feasibility study and social impact assessment, has been developed and will be able to be provided through a digital platform. In particular, the example of the concrete development of a use case about public lighting is presented in detail: starting from the identification of the use case to the execution of the installation of 100 intelligent streetlights and 5 test-sites for the monitoring of water consumption up to the visualization of the collected data.

The existing building stock presents a high potential of energy savings and CO2 emissions reductions. To this purpose, literature provides novel city-scale building-oriented studies, aimed at developing suitable tools for stakeholders, city planners, and decision-makers. To achieve an effective urban energy planning, urban energy systems (UES) models are developed; they employ a multi-domain approach, embracing the complex interactions in urban areas, such as energy flows, environmental indicators, social and economic factors. To perform an advanced modelling and to simulate the complexity of the UES, ICT (information and communications technology) represents nowadays the right answer to the needs of integration of data, tools, and actors in different domains. The chapter investigates the current studies in the field of building stock energy modeling and the application of advanced technologies to develop UES models. As an exemplification, the technological approach followed in the SEMANCO project to support urban scale energy modelling is presented.

The assessment of building energy performance requires data from multiple domains (energy, architecture, planning, economy) and scales (building, district, city) to be processed with a diversity of applications used by experts from various fields. In order to properly assess the performance of the building stock, and to develop and apply the most effective energy efficiency measures, it is necessary to adopt a comprehensive, holistic approach. In this chapter, three research projects are presented which apply Semantic Web technologies to create energy data models from multiple data sources and domains in order

Preface

INTRODUCTION

In recent years, humanity has witnessed climate change causing exceptional events such as the global warming that are also due to the behavior of human beings themselves. Since the beginning of the Industrial Revolution at the end of 18th century, people have burned more and more fossil fuels and changed vast areas of land from forests to farmland. Moreover, the world's leading climate scientists think human activities are almost certainly the main cause of the warming observed since the middle of the 20th century. Population growth is having an increasing influence on climate change also. Due to the constant increase in energy and CO_2 emissions, we are witnessing an ecosystem reaction characterized by a holistic process, which is difficult to control with cause-effect mechanisms.

The scientific community considers that in 2050, two thirds of the world population will live in towns, consuming over 70% of energy and emitting just as much greenhouse gases. As city populations grow, the demand for services but also pressure on resources increases. This demand puts a strain on energy, water, waste, mobility and any other services that would be essential to a city's prosperity and sustainability(https://ec.europa.eu/digital-single-market/en/smart-cities-smart-living). In this sense, the scientific community is debating how to manage the coexistence between the natural environment and the artificial environment built by human beings who are increasingly concentrated in cities.

The advent of the fourth industrial revolution offers the opportunity to rethink in an innovative way the management of resources, the design of the built environment through new methods and tools using Information and Communication Technologies (ICTs). The advent of connected, smart technologies for the built environment may promise a significant value that have to be reached to develop digital city models. Of course, as at international level the role of digital twin is strictly related to massive amount of data that needs to be collected and then processed. This topic proposes several challenges in terms of digital technologies capability, computing, interoperability, simulation, calibration and representation. Thus, the building industry is crossing a digital transition from traditional design processes to information modeling.

In this context, the integration of several data domains can lead to the creation of an extensive repository able to provide a series of useful pointers towards energy saving and reduction of CO_2 emissions, enhancing the livability of the city. The above is reflected in the concepts of smart cities, smart buildings and smart living. Over time, the scientific community has given a series of definitions linked to the concept of the smart city, including many subsystems (economy, environment, governance, living, mobility, people). Certainly the issue of governance has a priority in strategic terms to improve the coexistence of the natural and built environment through different strategies that can be summarized in three actions: knowing, understanding and interpreting (Testoni, 2016).

Public and private actors will be able to learn about and understand emerging problems related to the environment, technology and well-being in order to govern and solve the needs that the urban reality poses on a daily basis. In this context, the amount of data that networks make available constitutes the disruptive technology, typical of the 4th industrial revolution, which offers the opportunity for public administrations to take decisions, strategies and courses of action with greater definition and faster than in the past. The topic of data therefore becomes central in terms of volume, speed, reliability and variety. The classification of data and the relationship between them generates information that must be put into a system for the development of smart cities. For this reason, the construction industry must go through the digital transition in order to innovate its supply chain by applying new technologies for the economic, social and cultural purposes of the community. People need to become active participants in the smart city development process, which becomes a new way of life to increase awareness of being a community using resources that need to be managed in a system to ensure sustainable development.

As mentioned above, the basis of this digital transformation is the value assigned to data, which, when related to each other, becomes information that needs to be used to increase knowledge and awareness in people, thus achieving the objectives of smart cities.

The book provides a valuable benefit to researchers and developers in various fields such as:

- Building Information Modelling (BIM) that aims at improving the building process and, as innovative methodology. It is focused on the information management of all technical resources that are able/essential to represent the complexity of a building. It is based on a development of a 3D parametric model of buildings, enriched with semantic information such as measures, materials and costs;
- Geographic Information System (GIS) that provides geographical location of buildings, energy distribution networks and/or other elements and it is used for data automation and compilation, management, analysis and modelling of advanced cartography;
- Building Energy model (BEM), that focuses on building performance management. BEM models are able to develop energy performance behavior by characterizing the building and system components with essential information for the building envelope, the heating/cooling systems, the occupancy schedule and so on;
- Building Automation Systems (BAS), that consists of a temperature and humidity sensor framework aiming to collect data, improve occupant comfort and efficient operation of building systems, reduce energy consumption and operating costs.

Clearly, Internet of Things (IoT) platforms are fundamental considering of information sharing concerning the smart city topic, providing seamless access to data independently from the hardware devices and making possible the interoperability with other data sources (IGI Global, 2019).

the book has a multidisciplinary approach since the smart city and the digital twin topics are a crosscutting issues that are investigated by many actors who address this issues with different and complementary points of view. For this reason, the book is addressed to:

- Researchers carrying out research on building industry, energy and computer science fields
- Public administrators, managing cadastral and building information using 3D visualization of "static characteristics" (e.g.: construction period, heating system etc.) and buildings selection;

- Building managers who wish to optimize the operation and maintenance phase of the building creating refurbishment scenarios
- Energy professionals keen on improving energy management with innovative strategy for energy optimization

The book can be considered a valid medium for disseminating innovative data management methods related to smart city topics.

THE CHALLENGES

The idea of smart city has been investigated in recent years by the international community to meet the requirements of sustainable development. It consists on a motivator for the development of policies that contribute to a better society, and consequently to improvements in citizens' quality of life (De Guimaraes, 2020).

In order to define new policies to be applied in the Architecture, Engineer and Construction (AEC) Industry, it is necessary to describe reality through one or more models that are able to: i) gather data related to various topics; ii) organize the information in a logical way; iii) visualize the information through interfaces; iv) simulate alternative scenarios; v) provide answers based on data processing.

These characteristics are typical of a digital twin that can be defined as a realistic digital representation of assets, processes or systems in the built or natural environment. Many people see it as a simple digital replica of a real thing – a 'twin', but "what distinguishes a digital twin from any other digital model or replica is its connection to its physical twin ", with 'connection' meaning there is some type of relationship and association between the physical and digital. The concept of digital twins is not new: Nasa has run complex simulations of spacecraft for decades. However, the advance of the 'fourth industrial revolution' is making cyber-physical systems in the built environment practical, useful and affordable, and digital twins are a key part of this. (Bolton A, 2018). According to the definition of Dr. Grieves in 2002, a Digital twin has to be composed by three main parts: a) physical products in Real Space, b) virtual products in Virtual Space, and c) the connections of data and information that ties the virtual and real products together. (Grieves, 2014).

Considering the Built environment field, a digital twin has to represent physical reality at a level of accuracy suited to its purpose. The extent of realism depends on three essentials: i) Data – the quality of the data on which the twin is based. ii) Model – the fidelity of the algorithms, the validity of the assumptions and the competence of the code at the heart of the digital representation; iii) Visualization – the quality of presentation of the output. Digital twins may be developed for a range of purposes, operate at different scales or adopt different approaches to modelling. A number of digital twins have already begun to appear within the built environment, serving a variety of purposes. However, few digital twins at present are connected or share data across organizations, sectors or geographies. Lack of interoperability is a key constraint. (Bolton A, 2018).

Thus, the need to develop a digital representation of a physical environment for optimizing a city's data management is in line with the objectives of energy saving and technological innovation. For this reason, the book focuses on the role of digital twin for the development of smart city, referring to the built environment. The book aims to enhance the value of interdisciplinarity and interoperability for the development of smart city concepts, addressing the sectors of Architecture, Engineer and Construction

(AEC) Industry and in particular to the representation field. In recent years these fields have been innovating the way to represent data which is collected in heterogeneous databases such as Building Information Models (BIMs), Geographic Information Systems (GISs) and Building Management systems (BMSs).

These are connected in an innovative way, providing information at district level to achieve smart city objectives. For this reason, the book proposes a methodology based on the development of a digital twin for the building industry able to facilitate the use of data for connected cities and communities. Starting from the outcomes of some European projects (e.g. DIMMER, INDICATE, SEMANCO), several research fields are investigated from building to district level, paying particular attention to software engineering. Thanks to the synergy of the different skills involved a series of results are presented to suggest new policies at both building and district level, focusing on the development of strategies to create a smart city based on digital twins.

The challenges faced by the authors of the various contributions have therefore focused on the elaboration of virtual models with different purposes and uses, on the interaction of various heterogeneous data domains, evaluating the effectiveness of collaboration between the stakeholders involved in the construction industry.

Thus, the volume emphasizes among the most important challenges:

- Digital transformation oriented to data management;
- The development of a new generation of design tools based on parametric modeling and interoperability between software applications required by the construction industry;
- Interoperability at the technological, decision-making and process levels;
- The development of digital twin models aimed at energy savings;
- The enhancement of citizens' awareness as an active part of the smart city development process.

Clearly, the connection between graphical and alphanumerical databases plays a key role through the adoption of ICT hardware and software infrastructures.

For the above reasons, this publication contains different contributions focused on the description of methods, processes and tools that can be adopted to achieve smart city goals. It aims at collecting all pertinent project knowledge that supports their adoption and utilization, related to the development of a web-based ICT platform where different sets of static and dynamic data are collected, processed and displayed through various visualization tools as well as energy efficiency and cost analysis engines. As representation language is always considered a mean to communicate and share a design idea among the actors involved, this book enhances the value of data visualization to support the management of heterogeneous information. The interaction of different data from building to district scale allows the definition of an information system able to share data in relation to the scale of visualization, taking into account the user needs. As the data sharing methods considered have been investigated by three EU research projects, the book presents the effectiveness of dissemination activities that gathered feedback on the results.

This book can also be used as a valid example in the field of local community. In fact, one of the objectives of the projects was to raise awareness about eco-sustainability among the young generations through a gamification approach.

SEARCHING FOR A SOLUTION

The research presented in this volume showed that the challenges listed above were addressed with different approaches, adopting different procedures, models and standardizations that led to the analysis of the objectives identified by the smart city idea through the scientific method.

In line with the directions specified by the European Commission, the various contributions have shown aspects related to the smart city such as a place where traditional networks and services are made more efficient with the use of digital and telecommunication technologies for the benefit of its inhabitants and business (https://ec.europa.eu/digital-single-market/en/smart-cities-smart-living).

The results shown by the authors in the various manuscripts refer to European Commission guidelines and can be outlined as follows:

- Contribute to the opening of a market for ICT-based customized solutions integrating numerous products from different vendors and offering services from design of integrated systems to the operation and maintenance phases;
- Establish a collaboration framework between the ICT and the building industry aimed at exploiting opportunities for the development of ICT-based systems in compliance with the Energy Performance of Buildings Directive;
- Reduce energy consumption and CO2 emissions, in line with the policy framework for facilitating the transition to an energy-efficient and low-carbon economy through ICT.

It is obvious that, for a complex system such as digital twin, standardization takes on an essential role in the communication of data that need to be exchanged, while interoperability, in its wide meaning of capability of two or more systems or components to exchange information and to use the exchanged information, represents the key element for problem solution. (Osello,2012).

In addition, the ability to develop a new generation of design tools based on data sharing and user collaboration will require the construction industry to increase its maturity by acquiring critical evaluation, detailed description of data and processes, and development of advanced prototypes.

The development of smart cities based on the elaboration of digital twin will require in the near future the necessary training of people with new skills and abilities related to the many aspects of the world of construction, ICTs, energy saving.

In conclusion, the solutions proposed in the book suggest the way for the construction industry to resiliently adapt to the challenges thrown up by the digital age.

ORGANIZATION OF THE BOOK

The book is organized into 25 chapters and they are grouped by topic into four main areas: general, drawing, ICT, environment. A brief description of each of the chapters follows:

Chapter 1 describes the concept of smart city by associating it with sustainability issues. The city needs to be considered in continuous mutation, overcoming the paradigm of quantitative economic growth based on the infinite substitutability between natural and economic capital, adopting digital technologies at the service of the collective welfare.

Chapter 2 shows how effectively Big Data Analysis, as one of the most crucial contemporary tools employable in enquiring and updating, virtually and continuously, both the urban metabolism and its complexity, could "improve the art of urban planning", as Rogers stated.

Chapter 3 investigates the concepts and architecture of Digital Twin to cover its challenges and explore its applications in various fields such as smart cities, smart manufacturing and factories and healthcare sectors.

Chapter 4 describes the results of a sociological survey conducted as part of an EU-funded project to visualize and calculate energy data at the urban district level in order to optimize the distribution policies of the local District Heating (DH) network. The chapter identifies the characteristics of the main categories of actors (from DH operator to end users) that play a role within the network, identifying the barriers and frictions that prevent greater collaboration and communication between these actors.

Chapter 5 presents an integrated GIS-based and open source urban planning tool to support local authorities in selecting, simulating and comparing alternative low-carbon and economically sustainable heating and cooling scenarios. The achieved results are obtained thanks to the collaboration with several EU research centers, SMEs and consulting companies led by Rina Consulting involved in the Planheat EU project.

Chapter 6 discusses the BIM Simulation Lab initiative which aims at establishing a laboratory for an effective and efficient BIM implementation, by promoting a physical space and specific services for supporting the territory in the DT. The authors describe the concept of the lab and they introduce an assessment method that adopts an indirect approach of the BIM benefit assessment leveraging principles from construction cost estimation and probabilistic risk management.

Chapter 7 addresses the post-pandemic issue from Covid 19, with examples and anecdotes from Shanghai and China. The analysis of the situation generated by the health crisis is addressed in the study, highlighting a crucial need for resilience of local systems, short chains, autonomous energy and food self-sufficiency, and decentralization of essential products. The authors point to a future that needs to be reinvented, acting in depth, to move from "exploitation" to "cooperation" with natural systems.

Chapter 8 aims at presenting the District Information Modelling methodology that is strictly related to the Digital Twin concept, starting from heterogeneous data domains, arriving at the various tools developed to reach the users' needs. The chapter considers the digital transition phase of the building industry at the building and urban scale. Through collaboration with several research units within the EU DIMMER project, the chapter extends the BIM methodology at the urban scale to manage and represent a variety of data to reach the smart city concept.

Chapter 9 presents several different approaches towards the assessment of cities' performance, discussing advantages and disadvantages of such methods to get more farsighted results in terms of decision-making strategies. The chapter analyses the different context in which KPIs are applied, defining the fields of application and describing the characteristics of such indicators. Two examples of European projects using indicators for evaluation purposes are studied at city level in the first project (INDICATE) and at a district/building level in the second one (DIMMER).

Chapter 10 presents the result of the study carried out by Politecnico di Torino within the Energy Efficient Buildings (EEB) project. Collected data on urban and building scale are managed in specialized, independent and heterogeneous domains such as GIS, BIM and IoT devices for energy and electrical monitoring. Possible relationships among these datasets in the perspective of system integration have been carried out according to a rich matrix of experimentations.

Chapter 11 aims to evaluate Building Information Modeling methodology in the industrial context, as a cyber-physical system, developing flexible 3D parametric models as a data set, where information can be visualized and optimized management, using different visualization tools. The research has underlined the importance to share information between virtual and real worlds through Virtual and Augmented Reality (VAR) systems in the era of the fourth industrial revolution.

Chapter 12 aims at facing the various implications underlying the Smart City concept based on Digital Twins. The structure of the text is articulated in three main themes: the use of the term "Smart city" and the role that technologies had in its definition; the "3D city model" meaning and the integration procedures between BIM (Building Information Modeling) and GIS (Geographic Information System); the classification of 3D city models by use cases.

Chapter 13 addresses the approach, based on a solution looking for a problem by focusing on the problems of developing integrated solutions for smart cities based on digital twins. The narrative is based on a research project exploring the digitalization of facilities management processes in Bath, the UK The conclusion is that the development of digital twins goes far beyond linking digital models to sensors.

Chapter 14 presents the developments of representing a part of the city districts of Manchester, UK and Turin, IT initiated within the FP7 DIMMER project completed in 2016 and continued in the last years by the Center of Excellence GIS of CSI Piemonte. The use of BIM and GIS at district level have been investigated together with real-time data gathered from sensors and user feedback to analyze and correlate buildings utilization and provide real-time feedback about energy-related behaviors. The emerging standard 3D Tiles endorsed by the OGC was applied to represent CityGML data of two districts of Turin, Italy and Manchester, UK.

Chapter 15 aims to present the potential of Digital Twin in the construction field, suggesting a framework that outlines the many different possible applications in construction, with reference to school buildings. First, it summarizes the current overview of Digital Twin applications in building construction. Then it shows that significant steps that are being taken beyond the digital model.

Chapter 16 reports the developed framework to support the management and maintenance operations in buildings, defined within the project AR4FM - Augmented Reality for Facility Management. It focuses on an innovative software ecosystem that uses the most modern technologies to support Facility Management (FM) operations through a set of bespoke applications. The first tool is the Web-based ICT Platform usable through web-browser, while the second tool refers to a mobile App, called AR4FM Mobile App, which will enable visualization for both mobile and wearable devices, such as Smart-Glasses.

Chapter 17 opens up discussions upon the relevance of interaction of representations and data visualization modes for smart cities design, planning, and development that occur beyond paper and computer drawing. The chapter deals with the study of concepts that relate to both physical and virtual prototyping, evaluating the frontier technology of 4D/5D printing which underlines an interdisciplinary approach to design and the impact of integrating biologically inspired principles from different backgrounds to the field of smart cities design.

Chapter 18 emphasizes the integration of residents' perceptions and objective measures to understand the impact of environmental features on the behavior of neighborhood residents, using survey data and assuming relief given by geo-referenced interviews. The study focuses on a specific district of Turin involved in complex processes of transformation and regeneration, through the use of some indicators and graphic codes. The results broaden the understanding of the hierarchy and criteria of the needs for autonomous mobility on foot of people in relation to the urban environment, assessing their safety and welfare

Chapter 19 presents a novel distributed software infrastructure to enable energy management and simulation of novel control strategies in Smart Cities. In this context, the following heterogeneous information, describing the different entities in a city, needs to be taken into account to form a unified District Information Model: Internet-of-Things (IoT) devices, Building Information Model, System Information Model and Georeferenced Information System. The proposed solution fulfills the integration and interoperability of such data sources providing a correlation among them. The platform has been deployed in a real-world district.

Chapter 20 describes the possibilities of using Application Programming Interfaces (APIs) to enhance the usefulness of BIM models. This provides the user with the possibility of creating own applications with real time bidirectional data exchange. Of particular interest is the exchange of data with web based data sources. The applicability is illustrated with examples of environmental analysis based on web-based data and the use of web-based forms to enrichen BIM models with data input from building users

Chapter 21 describes the approach for the South Tyrolean city of Meran in the creation of use cases and the implementation of an urban agenda (roadmap) for the development of the city from a smart city perspective, with the involvement of citizens, experts and local administrators. A list of key services, based on a technical and economic pre-feasibility study and social impact assessment, has been developed for a subsequent digital platform. The example of the concrete development of a use case about public lighting is presented in detail: starting from the identification of the use case, to the execution of the installation of streetlights and sensors for the monitoring of water consumption, up to the visualization of the collected data.

Chapter 22 investigates the current studies in the field of building stock energy modeling and the application of advanced technologies to develop UES models. As an exemplification, the technological approach followed in the SEMANCO project to support urban scale energy modelling is presented. To achieve an effective urban energy planning, Urban Energy Systems (UES) models are developed; they employ a multi-domain approach, embracing the complex interactions in urban areas, such as energy flows, environmental indicators, social and economic factors.

Chapter 23 presents three research projects which apply Semantic Web technologies to create energy data models from multiple data sources and domains in order to support decision making in energy efficient building renovation projects: SEMANCO, OptEEmAL, and OPTIMUS. A final reflection on the results achieved in these projects and their links to ongoing research on digital twins is presented. The assessment of building energy performance required data from multiple domains (energy, architecture, planning, economy) and scales (building, district, city) to be processed with a diversity of applications used by experts from various fields.

Chapter 24 describes the obtained results within the INDICATE project that focused on the creation of an innovative interactive software capable of providing designers, urban planners and companies a "decision support system" in all urban development phases of a city: from the construction of a single building to the design of a master plan. Galliera test site experience was developed following the idea of using the INDICATE tool on Genoa site, to understand how the tool can work in the planning stages, as if the preliminary design of the new hospital were not already defined, to understand the optimal solution to select from the energy point of view.

Chapter 25 assesses the main perceived obstacles of the applicability of a BIM model for Facility Management, starting from an Italian framework overview. The tool chosen by the authors as a preliminary approach to the problem, the SWOT analysis, allows an effective synthesis of the strengths and weaknesses resulting from such implementation. Various aspects are described in the contribution such as the need to attract the private sector to ensure the digitalization of the AEC sector, where more than 50% of investments are invested in the era of digitalization.

REFERENCES

Astleithner, F., Hamedinger, A., Holman, N., & Rydin, Y. (2004). Institutions and indicators – The discourse about indicators in the context of sustainability. *Journal of Housing and the Built Environment*, *19*(1), 7–24. doi:10.1023/B:JOHO.0000017704.49593.00

Bolton, A., Enzer, M., & Schooling, J. (2018). *The Gemini Principles: Guiding values for the national digital twin and information management framework*. Academic Press.

Grieves, M. (2014). *Digital Twin: Manufacturing Excellence through Virtual Factory Replication*. Whitepaper.

Kumar, B. (2018). *Contemporary Strategies and Approaches in 3-D Information Modeling*. IGI Global. doi:10.4018/978-1-5225-5625-1

Voghera, A., & La Riccia, L. (2019). *Spatial Planning in the Big Data Revolution*. IGI Global. doi:10.4018/978-1-5225-7927-4

Acknowledgment

Each chapter has been reviewed by two blind peers, scientific experts that composed the book's editorial advisory board. For this reason, we must express our acknowledgement to the members of editorial advisory board and colleagues that inspired this book: Ricardo Codinhoto (The University of Bath), Kjartan Gudmundsson (KTH Royal Institute of Technology in Stockholm), Enrico Macii (Politecnico di Torino), Andrea Acquaviva (Università di Bologna), Edoardo Patti (Politecnico di Torino), Lorenzo Bottaccioli (Politecnico di Torino), Nicolas Urbina, (Istituto Giannina Gaslini), Vincenzo Corrado (Politecnico di Torino), Ilaria Ballarini (Politecnico di Torino), Maurizio Marco Bocconcino (Politecnico di Torino), Valentina Villa (Politecnico di Torino), Manuela Rebaudengo (Politecnico di Torino), Leandro Madrazo (ARC, Engineering and Architecture La Salle, Ramon Llull University), Álvaro Sicilia (ARC, Engineering and Architecture La Salle, Ramon Llull University), Giuliano Pairone (Aviation Projects), Francesca Abastante (Politecnico di Torino), Marco Piovano (Fraunhofer Italia Scarl).

Chapter 1
Smart Cities and Sustainability: A Complex and Strategic Issue – The Case of Torino Smart City

Caterina Mele
Politecnico di Torino, Italy

ABSTRACT

The term smart city is often synonymous with a sustainable city. The word smart implies the use of digital technology that serves to make processes and services more efficient and to connect the different actors on the urban scene. However, this is no guarantee of sustainability. A city can become sustainable if it changes its metabolism and from linear to circular as in nature's ecosystems. For this to happen, it is necessary to overcome the paradigm of quantitative economic growth based on the infinite substitutability between natural and economic capital. If smart city governance stakeholders primarily pursue profit according to the logic of the free market, the city may be smarter and efficient in the use of energy and resources, but it is not sustainable, often not even inclusive. The challenge of sustainability implies a paradigm shift and the use of digital technologies at the service of the collective good. In this context, after a general analysis of the characteristics of smart cities, the chapter focuses on an Italian case study, Turin Smart City.

SMART CITY: ONE CONCEPT MANY FACETS

The term Smart City has considerable ambiguity. The more you investigate it, the more discordant elements emerge, as can be observed in particular in the plan of Torino Smart City, which will be analyzed later. On the one hand, the term smart means intelligent. A smart city, in this regard, is a city that, thanks to the use of digital platforms, makes urban processes more efficient and allows for greater interconnection between the different actors on the urban scene (Giffinger et al., 2007). The smart city is therefore an infrastructure/platform of exchange and connection between physical and virtual elements of the city (Murgante & Borruso, 2013). This definition puts the emphasis on the digital infrastructure and ICT technologies that become the predominant feature of the city and subordinate to them the

DOI: 10.4018/978-1-7998-7091-3.ch001

different functions and actions that take place on the urban scene, that seems to have almost no real physical dimension. On the other hand, the term smart city is often synonymous with sustainable city or a city whose vocation is to achieve sustainability. However, the intersection between the objectives of sustainability and the smart dimension is not obvious and it is not even easy to evaluate, since evaluation is an operation deriving from multiple elements that depend on who makes the evaluation, the choices of how to evaluate and what to evaluate. The interest aroused by the issue of the smart city with all its ambiguities has resulted in the last decade in a very extensive scientific literature accompanied by countless generalist and popular articles available online. So popular is the term that, following the example of Borelli (Borelli, 2020), if you type the term smart city on the Google search engine, in less than a second just under 4 billion results appear. Without pretending to be exhaustive in retracing the birth of the smart city concept and leaving the most in-depth articulations to the social urban science studies, it is nevertheless necessary to identify some chronological and descriptive elements of the concept. Following the American New Urbanism of the 80s of the last century the term smart city began to appear more frequently in scientific studies (Cuppini, 2020) of the following decade, with reference to some newly built cities in Malaysia and Australia equipped with an ICT infrastructure (Arun,1999). In the same years several multinational information technology companies such as IBM, Cisco, Siemens began to propose smart solutions and ideas for cities. New terms like Virtual City spread. The spread of the internet and the use of Virtual Reality Modeling Language (VRML) had in fact allowed the three-dimensional representation of the city on the net, modeling urban development projects through virtual laboratories. A little later other adjectives related to the aspects of digitization and connection are used with reference to cities, among them Computable City (Batty, 1995), Digital City or Wired City, or Ubiquitous city which emphasizes the possibility of obtaining any kind of service, at any time with any device (Lee et al.,2008). In the following years the concept was analyzed and then articulated in several aspects (Cocchia, 2014). Six main elements that generally characterize the smart city arise: economy, (smart economy), governance (smart governance), population (smart people), lifestyle (smart living), mobility, (smart mobility) and environment (smart environment). In the IBM website "Building a smarter planet" at the question: "What does it mean to become Smarter?" the answer is: "Measuring, Monitoring, Modeling and Managing" and in relation to sustainability, the document explains that: "… you can't make a product greener, whether it's a car, a refrigerator or a traffic system, without making it smarter - smarter materials, smarter software or smarter design."- moreover: "ICT accounts for 2% of global carbon emissions: ICT can significantly contribute to control and reduce the 98% of CO2 emissions caused by other activities and industries." Moreover "Instrumentation increasingly captures more data… now to make it into real intelligence to enable smarter decisions for a greener society". In the document "Deliver Smarter Cities Agenda - Networks for more competitive, effective and efficient cities" (2011), IBM as a response to global challenges identifies the achievement of the smarter planet through 6 actions: "accelerating globalization, evolving societal relationships, expanding impact of technology, changing demographics, rising environmental concerns, growing threats to social, stability and order." How do we do that? Through a smarter city. In fact a smarter city is "…one that integrates technology into a strategic approach to sustainability, citizen well-being, and economic development … but also stressing the importance of integration and interaction across multiple domains." Moreover "The main role of technology in urban systems is to sense the activity to be able to monitor, manage, visualize and optimize" (IBM,2009). In all these definitions the degree of monitoring, digitization, interconnection of data from different economic and social sectors characterize smartness. The physical city with all its problems and vulnerabilities remains in the background.

Smartness and Sustainability, an Unresolved Issue

It seems, therefore, that the utopian business suggestions of the big technological companies have very soon conveyed the adjective smart, focused on process efficiency, in the urban debate. The issue of urban sustainability is immediately using the term smart almost as a synonym for a sustainable city. This narrative is still supported by a rhetoric that sees cities as the main drivers of development, driving elements of the economy, capable of generating technological innovation like centers of attraction for capital (Vanolo, 2015), from which resources for the sustainable transformation of cities are almost automatically drawn. However, this axiom smart equals greater energy efficiency equals sustainable, greatly simplifies the problem and raises the issue of sustainability in a superficial way. Following the debate triggered by the attempts to implement the Kyoto Protocol on sustainable development and the subsequent COP (Conferences of the Parties), global and local Agenda 21, in the first decade of the current century the countries of the European Union identify the smart city as one of the main tools to combat the challenges of sustainability. The Smart Europe 2020 Strategy's initiatives to fight climate change follow the 2008 Covenant of Majors. The European Commission subsequently launches the Strategic Energy Technology Plan (SET Plan) which, through the smart city, aims to "advance by 2020 towards a 40% reduction in greenhouse gas emissions through the use and sustainable production of energy", through "systemic approaches and organizational innovation, energy efficiency, low emission technologies and intelligent management of supply and demand" (European Commission, 2007). In accordance with these European strategies, a study entitled "Città intelligenti e sostenibili (Smart and Sustainable Cities)" was published by the Italian Foundation for Sustainable Development in 2012. The study aims to identify a unified strategic model for cities, consisting of shared priorities, objectives and targets, thus creating a master plan or strategic plan within which to place the individual sectorial plans. The smart city is the city of the future capable of facing the challenges of fighting climate change, relaunching the economy, social equity, sustainability and well-being. This study and a subsequent in-depth study (Italian Foundation for Sustainable Development, 2020) identifies a set of indicators through which to monitor progress towards the objectives of the research targets. The indicators were calibrated for a medium-sized city (under 500,000 inhabitants) like Piacenza used as a sample. Underlying the system of indicators is a framework, which constitutes the organized set of objectives and priorities identified in the first phase of the research. The scheme adopted starts from the strategic objective of the European Smart Cities initiative, building low-carbon cities that focus on economic development combined with quality of life, in the direction of overcoming GDP (Fitoussi et al., 2010). Based on these strategic objectives, a series of sub-objectives have been identified, in line with the indications of the European Smart Cities strategy (European Commission, 2020). Performance indicators measure performance against the objectives. At the basis of the approach adopted is the idea that a system of indicators for Smart Cities should be built from a series of strategic objectives, then declined into specific objectives and individual targets (like the EU 2020, now 2030 Agenda, and the 17 SDGs). The time window adopted is a decade (2010-2020). Twenty-four indicators chosen according to a multidimensional approach measure the degree of intelligence of the city on different sectors. An index developed for each of the three macro-objectives provides a summary reading of the trend of the indicators. The Figure 1 shows the system of indicators by sector and according to the strategic macro-objectives of the city of Piacenza used as a model. The 24 indicators are divided into 9 strategic areas (Energy, buildings, mobility, employment, income and consumption, housing conditions, health and environmental quality, services and social inclusion, culture and innovation) in relation to 3 strategic goals (a low-carbon city; a city of economic well-being; a city of quality of life). Unfortunately, the model developed by the Foundation for Sustainable Development using the city of Piacenza has remained essentially a theoretical study, and there are no aggregated and updated data available on the web to assess whether the strategies of the European 2020 smart cities set plan have achieved significant results in relation to the stated objectives.

Figure 1. The table shows the system of 24 indicators according to strategic areas and the three main goals of the city of Piacenza
Source: Italian Foundation for Sustainable Development, 2012

GOALS	STRATEGIC AREAS	INDICATORS
A low carbon city	Energy	1.CO2 Emissions 2.Aggregated energy consumption 3.Photovoltaic deployment
	Building	4.Household energy consumption 5.Population served by district heating
	Mobility	6.local public transport demand 7.Motorization rate 8.Cycle-pedestrian index
A economic well-being city	Employment	9.Employment rate 10.Employee index
	Income & consumption	11.Median income per capita 12.Equity in income distribution
	Living conditions	13.Accommodation expenditure 14.Eviction orders
A quality of life city	Health & environmental quality	15.Air quality 16.Urban waste management 17.Road and workplace accidents
	Services, social inclusion, participation	18.Availability of places in kindergartens 19.Involvement in public life 20.Crime rate 21.Entrepreneurship differentials of foreigners 22.Female entrepreneurship differentials
	Culture & innovation	23.Ageing population 24.Access to culture

In general, it can be said that the objectives of the 2020 strategy for a smart, inclusive and sustainable Europe were almost immediately overshadowed by the 2008 sovereign debt crisis. The austerity policies launched by the Junker Commission until the renewal of the European Parliament in 2019 have obscured the vision of economic growth not aimed on GDP. European policy coordination objectives have focused on macroeconomic aspects that have marginalized the environmental issue and the demands for inclusiveness and social justice.

Smart City: a City Without Spatial Dimension

In the debate and programs on the smart city, its spatial, physical dimension seems almost completely absent, In the past "Architecture used to be about the creation of community and making the best effort at symbolizing that community" (Koolhaas, 2024). Its physical characteristics are secondary to the centrality of the digital infrastructure: "In the Smart city the network metaphor overlaps with the urban metaphor, in this sense acting as a new, different infrastructure capable of conveying relationships and interactions, as well as being influenced and shaped by these processes, in a similar way to a public transport network that develops in a city to connect and serve places, and then evolves to create 'new' places" (Murgante & Borruso, 2013) Yet the most relevant global phenomenon of our time is undoubtedly the spatial and demographic growth of urbanized areas. The phenomenon is so significant that today 55% of the world's population lives in cities and UN projections say that by 2050 almost 70% of the world's inhabitants will live in cities (United Nation, 2018). The issue concerns both emerging and developed countries, albeit with different dynamics. In fact, in the so-called Brics (Brazil, Russia, India, China and South Africa) urban areas are growing with impressive numbers in size and number of inhabitants. One example for all: the megalopolis along the axis of the Pearl River in China occupies an area of about 26 million square meters for a population of over 40 million inhabitants. In the countries of older urbanization, on the other hand, cities grow not in number of inhabitants but expand into the surrounding territory, which becomes a non-place, where the border between town and country no longer exists, with significant ecosystem impacts, such as soil consumption. In Italy, with the current trend and without a strong regulation, according to data from the IPSRA Observatory, more than 8000 square kilometers of soil could be lost by 2050 (ISPRA, 2018). These dynamics of growth of urban areas are inevitably followed by a strong increase in built areas, the forecast is that without stringent policies and corrective measures in 2050 will be built another 230 billion square meters of new buildings (ISPRA,2018). Furthermore, energy efficiency, however desirable, can be achieved by drawing on both fossil and renewable sources. According to data from the GABC Dossier 2020, buildings and constructions are responsible for 39% of total CO2 emissions (GABC, 2020). The dossier highlights the rapid growth of construction on a global scale: without corrective measures, the surface area of buildings in the world will double by 2060. Although the efficiency of the energy sector has improved in recent years, it is still not sufficient to reduce the increase in energy demand. CO2 emissions have continued to grow by one percent per year since 2010, and more than four million deaths per year are attributable to diseases caused by pollution from buildings. A study (De Pascali et al.,2015) on residential energy consumption in Italy in the period 1970-2005 has shown that while average consumption per home has decreased (as a result of tax exemption policies), overall residential energy consumption has increased with an average annual increase of 1.3%. From this it can be deduced that in order to achieve significant energy savings, it is necessary to go beyond the scale of the single housing unit and the single building to consider all the buildings, the urban layout and the most efficient technological solutions on the urban scale. But since cities are mostly made up of buildings built in the second half of the twentieth century, and are inadequate and inefficient from an energy point of view, the most important game from the point of view of sustainability is played by the redevelopment of the existing heritage, which must be consistent with the European objectives that provide for an 80% reduction by 2050 in carbon dioxide emissions. Currently, each year an area equal to 1% of the total building is redeveloped, with energy improvements of the order of 15-25%, generally on individual apartments. This is a positive result but totally inadequate in relation to the decarbonisation objectives of the European Union. With these data it seems very difficult to reach the objectives of the Paris Accords of 2015 and to contain the global warming of this century within 2 degrees. The impact on the Earth's biosphere (the ecological footprint) of the growth of urban systems is therefore central both to the amount of resources directly consumed and to the effects caused in terms of emissions and pollution. The flows of energy and matter passing through cities only proceed in one direction, i.e. towards the city itself. In systemic terms, it can therefore be said that, unlike natural ecosystems, which are mainly closed and powered by solar energy, the city behaves like an open thermodynamic system dependent on the environment in which it is located and on the use of fossil fuels. In other words, its system of energy flows is not sustainable. In order to become sustainable, the urban organism should establish a relationship with its territory (or ecosystem area) of reduction of input (material and energy) and output (entropy and waste) flows. The achievement of this objective implies a transformation of the urban metabolism from linear to circular. Sustainability therefore requires the transformation of the current economic model based on consumption and waste into a model in which, as in natural cycles, the city reaches a condition of self-reliant. The issue cannot therefore be reduced to a mere question of energy flow efficiency, as seems to be implicit in the smart city model. Moreover, the challenge of sustainability applied to urban settlements implies an inclusive socio-cultural model, a healthy and ecologically complex environment, with a good quality of life not only for the present but also for future generations. The challenge of sustainability requires much more radical choices. We need to overcome the model of infinite substitutability between natural and economic capital and redesign relationships and functions that profoundly and circularly change the metabolism of today's cities. There is still one aspect to underline, that of the inclusion or exclusion of a significant part of the population from the consequences of technological smartness. In a country like Italy where a significant part of the population is elderly and where the diffusion of digital infrastructures is scarce or even absent in entire areas of the country, the smartness of services for citizens can be a further element of inequality and social unsustainability.

Smart Governance: What are the Priorities?

In the rhetoric of smart cities a relevant point is that of city government. Assuming that the path of sustainability can be practiced through the efficiency and digitization of processes, which technologies are preferable? With what capitals? If the construction of infrastructures is left to the market, it is clear that profit becomes the main objective to be pursued and not the common good. According to the presentation on ICT and Smart City made by D. Mosso at the Politecnico di Torino on 10 June 2016 the priorities for cities that decide to become smart are "to provide more efficient services to citizens, optimizing resources and capitalizing on investments; to attract and maintain qualified human capital; to promote industrial growth and economic growth of the territory; to compete with other territories with innovative territorial marketing tools; to respond to regulatory obligations of transparency and public evidence. Technologies for the future development of cities thanks to digital systems tend to integrate complex systems, through the diffusion of sensor networks, cloud computing and environmental technologies". The smart city government seems to be mainly explicit through a "government of information" about what happens within the urban system, aimed at economic growth. In other words, the goverment of the smart city seems to take place mainly through governance, i.e. a non-hierarchical model of government in which a plurality of public and private actors participate in the strategic policies of the city. Thanks to the network of sensors and digital components placed on the territory, the city is potentially able to provide real-time data on the status of its own components on which it can act quickly in order to progressively reduce the time that elapses between the analysis of the urban layouts and the preparation of intervention actions. This can make democratic spatial planning tools marginal. Citizens all become user-mappers but monitoring, sensor equipment and smart tools like apps for citizens are something else from the real government of the physical city: "when we look at the visual language through which the smart city is represented, it is typically with simplistic, child-like rounded edges and bright colors. The citizens the smart city claims to serve are treated like infants" (Koolhaas, 2014). In Rio de Janerio, the city that won the World Smart City Awards in 2013, the possibility of using apps on the smartphone to move quickly and safely in the city is denied by the dramatic situations of social and physical degradation in which the real city is living (Cuppini, 2015). The many European funding opportunities on smart cities, which have followed one another in the last decade, have favored the construction of private public consortia, because " "solving societal problems is not merely a question of developing good policies but much more a managerial question of organizing strong collaboration between government and other stakeholders" (Meyer & Bolivar, 2015). The global market for smart city technologies is worth many billions of dollars a year. The consequence of all this is a tendency to privatize the decision-making process of city communities with the risk of an exercise of power isolated from democratic responsibilities (Swyngedouw, 2005). As further evidence, in relation to the Smart Cities Initiative involving 24 Italian cities, in the European Regional Development Fund (ERDF) 2014-2020 the share of direct responsibility for the implementation of the program belongs for the most part to private companies and consortia (Grossi & Pianezzi, 2017). In this vision, citizens become consumers and therefore services and common goods are subject to the laws of the free market. Public governance dissolves in favor of the governance of entities related to corporate interests. Finally, the rhetoric of smart cities oriented towards process efficiency and the deployment of technological solutions to promote urban sustainability has the effect of distracting attention from the real causes of urban problems by focusing on the effects. By transforming citizens into consumers/users, the smart city omits political responses to the real problems of communities in favour of only technological/technocratic responses. "Smart cities and politics have been diverging, growing in separate worlds. It is absolutely critical that the two converge again" (Koolhaas, 2024).

The Case of Torino Smart City

The case of Torino Smart City is paradigmatic of what has been said so far. The Torino Smart City project has had a rather long elaboration. The first step was taken with the establishment of the Torino Smart City Foundation for Sustainable Development in 2002. The founding members are the City Council, Environment Park (a publicly owned S.P.A.) with a share of 13.29% and the Province of Turin (share 1.558). The aim that the foundation intends to pursue, as reported in the County of Torino document, is "to design, promote and disseminate projects, studies and research aimed at educating the general public, institutions and actors in the economic-productive fabric to a culture of respect for the environment and energy. The Foundation, in particular, intends to research and promote the rationalization of the management of local and environmental energy resources, through the improvement of energy efficiency and the promotion of renewable energy" (Provincia di Torino, 2002). The project towards the Smart City started in 2009 after joining the Covenant of Mayors and after 2010 with the approval of the Sustainable Energy Action Plan. In 2011 the European Commission launches the program Smart Cities & Communities, the Torino Smart City objective is promoted through the Torino Smart City Foundation for Sustainable Development through participation in European and national calls for proposals on smart city issues. The City of Turin participates in the calls as a partner in projects of the Community programs

CIP (Competitiveness and Innovation Program), VII Framework Program, Interreg and Urbact II. The city collaborates in various forms in 18 research projects (6 funded) submitted to the 2012 MIUR call " Smart Cities and Communities and Social Innovation", finally in 2013 the master plan of Torino Smart City called SMILE, is presented. SMILE means: Smart mobility, Inclusion, Life&Health, Energy, and correspond to the 5 areas of intervention in which to make the city smart. The ideas and areas identified are the result of a 6-month work in which 350 people representing public institutions, private subjects, companies mainly in the ICT sector, universities and research bodies, trade associations were involved. Mobility, Inclusion, Life&Health, Energy plus Integration are the application domains on which the Torino Smart City project is based. 45 actions are proposed, 7 transversal, 7 related to mobility, 9 inclusion, 12 Life and Health. The goal of the master plan is: "an efficient and non-polluting mobility, a rational use of energy from renewable sources, a society open to health and people's needs, also with the help of new technologies, a higher quality of life and attractiveness of tourists and investments and an increasingly efficient public administration, thanks to digital services". All the proposals "will be accompanied in their evolution also thanks to the new actors that can be part of this path and to the identification of public and private resources". The coordination of the master plan is committed to the Torino Wireless Foundation (a public-private partnership company). The SMILE plan is the instrument with which the City of Turin "has decided to set up the model of the smart city, which will guide it in the identification of development trajectories, objectives and priorities, through strategic actions and key projects, beyond the logic of isolated experimentation and promoting sustainable and replicable models and solutions... able to enhance the local specificities to make Torino the first city in Italy that has a smart city action plan and that, above all, makes it ready for the challenges of European calls for proposals... which will require that cities (and in particular those with a metropolitan vocation) have a clear and shared strategy, with all development partners, primarily research institutions and businesses" (Greenreport, 2013).

As you can see, the Torino Smart City project is the result of a process perfectly consistent with the conceptual approach of smart cities at an international level. The city becomes safe, sustainable and inclusive thanks to the new digital infrastructure that makes products and services more efficient, interfacing smartly with citizens and the various players on the urban scene. The process of transformation of the city into a "smart city" lacks references to the planning of the physical dimension of the urban and building framework of the city.

To what extent do Torino Smart City's objectives interface with Agendas 21 and 2030, which should involve city administrations at local level? Have the 17 SDGs (Sustainable Development Goals) of Sustainable Development been included? Since the launch of the Torino Smart City project in 2011 and the publication of the SMILE master plan in 2013, almost a decade has passed, what has happened in the meantime to the real city?

There is actually a significant distance between what really happened to the physical city and the smart, digital city of the SMILE master plan. There is no doubt that action to make public administration services more efficient and to improve the connectivity of citizens will allow progress. As well as some initiatives linked to the provision of shared mobility and micro-mobility of private managers have been launched with encouraging results, although not as successful as desired. The monitoring of air pollution and other data related to the biophysical components of the city are certainly useful, but very little has been done in terms of action on the causes of air pollution. As far as mobility is concerned, despite the offer of shared and alternative mobility, road traffic and cars continue to be predominant. The energy efficiency of the city and the transition to renewables is still insufficient. The energy efficiency of the city and the transition to renewables is still insufficient. The lower spending capacity of the municipal administrations, thanks to the consequences of the austerity policies due to the 2008 crisis, has resulted in a poor capacity of the municipality of Torino to undertake actions of wider impact, such as the extension of the city's metro system. The economic indicators of the city (Eurostat data in January 2020) of the last twenty years show a negative trend and are essentially linked to the phenomena of deindustrialization that have been very strong in Turin. At the same time, the loss of activities directly or indirectly linked to Fiat company has not been accompanied by the rise of other businesses linked to services or trade. The international financial crisis of 2008 damaged the economic capacity of local public administrations, making the situation even more critical in Torino due to the high level of previous debt (4200 million euros) due to the 2006 Winter Olympics. The revival of the city as an urban cultural, tourist and innovative mix has not yet had time to establish itself. The economic crisis has had the consequence at territorial level of increasing the gap between center and periphery, with serious social and economic repercussions on the weaker population groups and on the dynamics of social inclusion and integration. It is no coincidence that one of the main sociological reports on Torino in 2019 was entitled by its authors "Futuro rinviato" (Postponed future) (Davico et al., 2019). Some of the main critical elements emerge from the 2019 Rota Report. The urban transformations started by the PRGC (municipal masterplan) in 1995 were almost exhausted in 2006, the year of the Winter Olympics, leaving two of the most important interventions of the plan unfinished. The first concerns the western area centered on "Corso Marche", and the second the "Variante 200" in the northern area of the city, an area that has long been at the center of the city's debate on the redevelopment of the suburbs (Consiglio Comunale di Torino, 2009). Of the intense debate on the urban transformations of these areas of the city, there remain massive speculative residential building interventions that are not followed by important interventions of urban and environmental quality redevelopment. On the economic side, in the last twenty years small businesses with less than 10 employees have increased, but bigger ones have decreased by 12%, causing a decrease in the number of employees (unlike what happened in other Italian metropolitan areas). Moreover, since 2010 the balance of companies that open and close in the Torino metropolitan area is worse than the Italian average, so the wealth produced by the local economic system has declined to the penultimate place among the cities in the center-north. A growing economic sector, which is still weak, however, is tourism. Torino is at the 11th place for tourist attraction among Italian cities. As far as external mobility is concerned, the intensity of connections on high-speed rail lines, which are now the main axis of national development, has left the city in a marginal position. Caselle airport, although it has increased its activity by 45%, is also in fourth place in terms of number of passengers among Italian airports. As far as internal mobility was concerned, the first metro line that the city lacked was built. The metropolitan railway service was launched with 8 lines, and the urban cycle network was implemented with results, although growing, which are still not very significant, as private road mobility is still the most used. As far as telematics infrastructure is concerned, Torino is in an intermediate position among the Italian metropolitan areas. The main sector of the metropolitan area in terms of attractiveness and quantity of investments is today research, 80% of which is financed by private companies. As regards air pollution, the increased detection efficiency has not been associated with a substantial improvement in air quality. In fact, the metropolitan area of Torino is affected by a phenomenon of serious air pollution with frequent exceeding of the limits of contaminants provided for by current legislation, with serious effects on the health of citizens. The Torino Smart City objective has generated the launch of many strategic development plans that have involved many public and private actors and raised many expectations. All this however, has not been supported by a strong and structured implementation capacity, also due to the increasingly scarce public resources. Moreover, most of the plans concerned only the capital and not the entire metropolitan area. Therefore, despite the many projects of Torino Smart City, the perception by citizens about the quality of urban life in the last twenty years is that it has not improved overall. The phenomenon of deindustrialization that has hit the local reality hard with the progressive divestment of activities linked to automotive production has not been counterbalanced by the birth of a strong alternative identity. Turin, a tourist city and a university city, are objectives in progress, not yet consolidated and the current pandemic crisis has frustrated the expectations of the various territorial actors. The attractions of Turin as a university city and as a tourist city can find good allies in the various smart projects of digital services to citizens or in smart mobility. However, it is in the urban and building framework of the physical city that a concrete, tangible spin-off of smart city projects is missing.

IN CONCLUSION, IS THE SMART CITY A SUSTAINABLE CITY?

Is the smart city a sustainable city? No, a smart city is not automatically sustainable, despite the emphasis on this aspect declared by many different advocates of the smart city. A smart city, thanks to the integration of digital technologies in one or more infrastructure networks, can allow a better efficiency in the use of energy and available resources and can be the platform within which to deploy strategies and actions based on sustainability. Moreover, the challenges posed by climate change, with the intensification of meteorological phenomena and the increase in vulnerability and risks, can be better understood and monitored thanks to the implementation of the data of the different environmental, physical and human components (also replicated in digital twin models). However, it is clear that the challenge of sustainability implies a radical rethinking of our society's development model. The physical city with its contradictions and its environmental, social and economic problems is the most relevant phenomenon of the unsustainability of a development model based on consumption and waste. The ecosystem surface of a contemporary city is on average 200 times larger than its spatial dimension. It is therefore an extremely complex and vulnerable organism at the same time, in which the enormous flow of resources and energy moves in a single direction - towards the city itself. The contemporary city is the consequence of this economic model, not the cause. Only if the urban metabolism, in imitation of nature's cycles, becomes circular can the city become sustainable, but for this to happen a drastic change in the current development model is necessary. In terms of energy, on the other hand, to achieve an effective reduction in consumption and to reduce greenhouse gas emissions, current policies of tax exemption for measures to improve the energy performance of buildings are not enough, but it is necessary to identify the most efficient technological solutions at the urban building scale. On the energy and environmental side, actions and strategies are needed to promote urban self-sufficiency, reducing the flow of materials and energy in and out, and emissions and wastewater, bringing the functioning of urban metabolism closer to that of natural ecosystems. The integration and use of renewable energy sources must be encouraged by promoting the creation of smart grids, in close relation with the specific functional needs and the different environmental conditions of the city (smart cities). It is also necessary to integrate into global and local policies the positive ideas coming from participatory and "bottom-up" cooperative experiences of changing lifestyles in relation to more sustainable housing and urban models. The use of digital technologies to monitor the billions of data concerning different urban structures/sectors and their management is a useful cognitive tool. The possibility of interconnecting in real time the different actors acting on the urban scene is certainly relevant, but the change in the functioning of the urban organism and its transformation in the sense of sustainability require a change of economic and social paradigm. Transformation requires management independent of the mechanisms that regulate the global market. The objective of sustainability needs a public government, based on democratic choices shared by citizens that must be involved in achieving common goals, as citizens and not as consumer users. The scenarios of the model World3 of the 1972 Limits to Growth report (Meadows et al. 1972) and subsequent updates have come true as we see in Figure 2. The nine planetary boundaries are suffering and global warming is evidence confirmed by the facts, to which is added the very recent pandemic crisis that has shown in all its drama the vulnerability of our society and our lives.

Figure 2. The Limits to Growth baseline scenario revisited in 2012. The graph confirms the predictions of exceedances of the limits of the 1972 and subsequent studies
Source: Deborah McKenzie, New Scientist, 2012

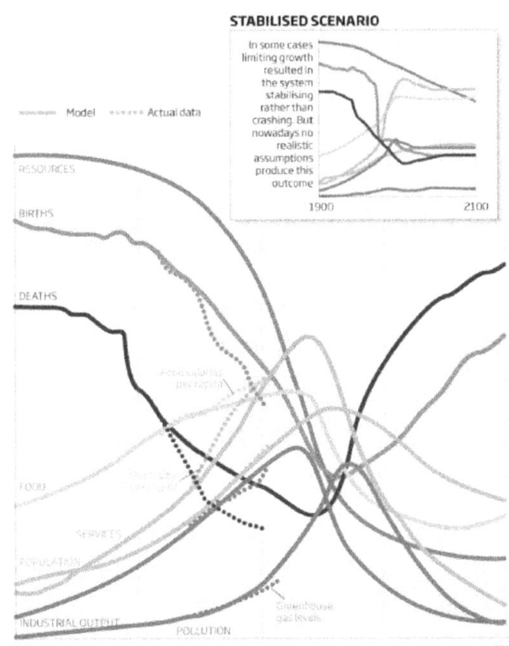

An urban smartness detached from the rest and the spatial dimension of urban phenomena is not enough to solve the strategic and epochal issues of sustainability. Parallel to this, a sustainable society needs a radical redesign of the existing city. There are many visions and proposals but the redesign implies some strategic axes on which to set the design, such as, for example, the creation of self-sufficient neighborhoods not only from the point of view of energy, but from the point of view of economic and productive activities. Mobility must be based on routes suitable for a plurality of means, integrated by a mix of collective and individual public services. From the environmental point of view, the districts and individual blocks should be equipped with large green and wooded areas to restore the main ecosystem functions such as soil permeability and normal drainage of surface water, and more. The redesign of the city in its spatial and physical dimension and the transformation from linear to circular of the flows of matter and energy entering the city are strategic for the goal of sustainability. The use of smart digital platforms and technologies on the urban scene can allow to manage more efficiently resources, processes and services. The increasingly sophisticated digital tools have enormous potential. Spatial modeling of an urban agglomeration fully replicated in the digital twin solutions available today can offer an even more advanced tool to optimize and guide urban planning. Monitoring and simulation of the functioning of the urban organism can descend to the districts, block and single building level. Smartness can therefore be an extremely useful tool for the pursuit of urban sustainability goals, but in itself, like all tools, it is not enough.

REFERENCES

Albino, V., Berardi, U., & Dangelico, R. M. (2015). Smart Cities: Definitions, Dimensions, Performance and Initiative. *Journal of Urban Technology*, *22*(1), 1, 3–21. doi:10.1080/10630732.2014.942092

Arun, M. (1999). Smart cities. The Singapore case. *Cities (London, England)*, *16*(1), 13–18. doi:10.1016/S0264-2751(98)00050-X

Associazione Torino Internazionale. (2015). *Il terzo Piano Strategico dell'area metropolitana di Torino* [The third Strategic Plan for the Turin metropolitan area]. Author.

Batty, M. (1995, July). The computable city. In *Fourth international Conference on Computers in Urban Planning and Urban Management*. School of Environmental Studies. University of Melbourne.

Batty, M., Axhausen, K. W., Giannotti, F., Pozdnoukhov, A., Bazzani, A., Wachowicz, M., Ouzounis, G., & Portugali, Y. (2012). Smart Cities of the Future. *The European Physical Journal. Special Topics*, *214*(1), 481–518. doi:10.1140/epjst/e2012-01703-3

Borelli, G. (2020). La Smart city e la fine della città [The Smart city and the end of the city]. *Sociologia Urbana e Rurale*, *122*(122), 96–122. doi:10.3280/SUR2020-122007

Cocchia, A. (2014). Smart and Digital City: A Systematic Literature Review. In R. P. Dameri & C. Rosenthal-Sabroux (Eds.), *Smart City: How to Create Public and Economic Value with High Technology in Urban Space* (pp. 13–43). Springer.

Commissione Europea. (2020). *EUROSTAT Statistiche europee* [EUROSTAT Statistics]. https://ec.europa.eu/info/departments/eurostat-european-statistics_it

Consiglio Comunale di Torino. (2009). Deliberazione 2008 09659/009. *Documento Programmatico finalizzato alla redazione della variante strutturale n. 200 al P.R.G. ai sensi della LR n.56 del 1977 e s.m.i. e della L.R. n.1 del 2007 s.m.i.* [Programme Document for the drafting of structural variant no. 200 to the P.R.G. in accordance with LR no. 56 of 1977 and subsequent amendments and additions and Regional Law no. 1 of 2007 and subsequent amendments and additions]. Author.

Cuppini, N. (2020). Cosa si muove sotto lo skyline digitale? Storia e politica della smart city [What moves under the digital skyline? History and politics of the smart city]. *Sociologia Urbana e Rurale*, *122*(122), 12–26. doi:10.3280/SUR2020-122002

Davico, L., Cabodi, C., Guiati, F., Gullino, V., Staricco, L., & Vitale Brovarone, E. (2019). Futuro Rinviato [Posponed Future]. XX RGR. Centro di Ricerca e Documentazione Luigi Einaudi. Torino.

De Ioris, D., Di Pasqua, G., Reginaldi, M., De Pascali, P., Alberti, V., (2015). *L'energia nelle trasformazioni del territorio. Ricerche su governance ed energia nelle trasformazioni del territorio* [Energy in the transformations of the land. Research on governance and energy in the transformation of the land]. Franco Angeli.

European Commission. (2007). *A European strategic energy technology plan (SET-plan) - Towards a low carbon future*, COM (2007)723.

European Commission. (2020). *Smart Cities*. https://ec.europa.eu/info/eu-regional-and-urban-development/topics/cities-and-urban-development/city-initiatives/smart-cities_en

Fitoussi, J. P., Sen, A., & Stiglitz, J. (2010). *Mismeasuring our lives: why GDP doesn't add up*. New Press.

Fondazione per lo Sviluppo Sostenibile. (2012). *Città intelligenti e sostenibili* [Smart and sustainable cities]. Author.

Fondazione per lo Sviluppo Sostenibile. (2012). *Piacenza Smart City 2020. Il modello smart city per la sostenibilità delle città medie Piacenza Smart City 2020* [The smart city model for the sustainability of medium-sized cities]. Author.

Fondazione Torino Smart City. (2013). *Deliberazione della Giunta Comunale 2013 07373/068. SMILE - Smart mobility*. Inclusion, Life & Health, Energy.

Giffinger, R., Fertner, C., Kramar, H., Meijers, E., & Pichler-Milhanovic, N. (2007). *Smart Cities: Ranking of European medium-sized cities*. Centre of Regional Science.

Global Alliance for Buildings and Construction (GABC). (2020). *GlobalABC for Building and Construction 2020-2050 Toward a zero emission, efficient and resilient buildings and construction sector*. United Nation Environment Programme. IEA.

Greenreport Review. (2013). *Ecco Torino Smart City: Smile, la città intelligente e sostenibile* [Here is Torino Smart City: Smile, the intelligent and sustainable city]. https://www.greenreport.it/news/comunicazione/ecco-torino-smart-city-smile-la-citta-intelligente-e-sostenibile/

Grossi, G., & Pianezzi, D. (2017). Smart cities: Utopia or neoliberal ideologism? *Cities (London, England)*, *69*(9), 79–85. doi:10.1016/j.cities.2017.07.012

Hollands, R. G. (2008). Will the real smart city please stand up? Intelligent, progressive or entrepreneurial? *City*, *12*(3), 303–320. doi:10.1080/13604810802479126

IBM. (2009). *Smarter Thinking for a Smarter Planet*. https://www.ibm.com/smarterplanet/us/en/

Istituto Superiore per la Protezione e la Ricerca Ambientale. (2018). *Consumo di suolo, dinamiche territoriali e servizi ecosistemici* [Soil consumption, territorial dynamics and ecosystem services] (Rapporti ISPRA 288). ISPRA.

Koolhaas, R. (2014). *Are Smart Cities Condemned to Be Stupid?* https://www.archdaily.com/576480/rem-koolhaas-asks-are-smart-cities-condemned-to-be-stupid

Lee, S. H., Han, J. H., Leem, Y. T., & Yigitcanlar, T. (2008). Towards ubiquitous city: concept, planning, and experiences in the Republic of Korea. In T. Yigitcanlar, K. Velibeyoglu, & S. Baum (Eds.), Knowledge-Based Urban Development: Planning and Applications in the Information Era (pp. 148-169). IGI Global. doi:10.4018/978-1-59904-720-1.ch009

Manville, C., Cochrane, G., Cave, J., Millard, J., Pederson, J. K., Thaarup, R. K., Liebe, A., Wissner, M., Massink, R., & Kotterink, B. (2014), Mapping smart cities in the EU. European Parliament. Policy Department A: Economic and Scientific Policy, Bruxelles.

Meadows, D. H., Meadows, D. L., & Randers, J. (2004). *Limits to Growth. The 30-Year Update*. Earthscan.

Meadows, D.H., Meadows, D.L., Renders, J., & Berhens, W.W. (1972). *The Limits to Growth*. Universe Book.

Meijer, A., & Bolivar, M. P. R. (2015). Governing the smart city: A review of the literature on smart urban governance. *International Review of Administrative Sciences*, *82*(2), 392–408. doi:10.1177/0020852314564308

Murgante, B., & Borruso, G. (2013). Smart cities: Un'analisi critica delle opportunità e dei rischi [Smart cities: a critical analysis of opportunities and risks]. *GEOmedia*, *3*, 6–10.

Provincia di Torino, (2002). *Scheda della Fondazione di Torino Smart City per lo Sviluppo Sostenibile* [Torino Smart City Foundation for Sustainable Development sheet]. Relazione Annuale, Allegato1.

Provincia di Torino. (2011). *Relazione illustrativa del Piano Territoriale di Coordinamento Provinciale – Ptc 2* [Illustrative report of the Provincial Coordination Territorial Plan - Ptc2]. Città Metropolitana di Torino.

Renda, A., & Cellai, C. (2016). *Strategia Europa 2020: cronaca di una revisione annunciata* [Europe Strategy 2020: chronicle of an announced revision]. Seminario CINSEDO-Centro Interregionale Studi e Documentazione.

Swyngedouw, E. (2005). Governance innovation and the citizen: The Janus face of governance beyond the state. *Urban Studies (Edinburgh, Scotland)*, *42*(11), 1991–2006. doi:10.1080/00420980500279869

Tocci, G. (2020). La Governance delle città intelligenti. Tecnologia, innovazione, smartness [Governance of smart cities. Technology, innovation, smartness]. *Sociologia Urbana e Rurale*, *122*(122), 44–61. doi:10.3280/SUR2020-122004

Vanolo, A. (2015). Smart city e sviluppo urbano: alcune note per un'agenda critica [Smart city and urban development: some notes for a critical agenda]. *Scienze del Territorio 3. Ricostruire le città,* 111-118. . doi:10.13128/Scienze_Territorio-16256

KEY TERMS AND DEFINITIONS

Governance: A non-hierarchical model of government in which a plurality of public and private actors participate in the public strategic policies.

Rota Report (RGR): Annual sociological study on the social, economic, and environmental situation of the city of Torino, carried out by Fondazione Einaudi.

Smart City: One city that uses digital technologies to make processes and relations between different urban actors more efficient.

SMILE: Torino Smart City masterplan. The acronym means Smart mobility, Inclusion, Life & Health, Energy.

Sustainability: A pivotal concept for sustainable development, involving the fair and intergenerational use of environmental resources.

Sustainable Development Goals: Sustainable development objectives according to the United Nations Agenda 2030 signed in 2015 by 150 states.

Urban Metabolism: Flow of matter and energy derived from natural resources and human transformations that defines the functioning of cities assimilated to living organisms.

Chapter 2
Generative Computational Urban Planning Through Big Data Analysis

Luca Saverio Valzano
Politecnico di Torino, Italy

Carlo Caldera
Politecnico di Torino, Italy

Carlo Luigi Ostorero
Politecnico di Torino, Italy

Valentino Manni
Politecnico di Torino, Italy

Andrea Galli
Accurat, Italy

ABSTRACT

For the near future, forecasts predict an uncontrolled growth of urbanization in the world, in which cities are fragmented and uneven systems in relation to fast evolving environmental, economic, and social phenomena. The traditional urban planning approach, essentially theoretical-predictive, adapts poorly to face future challenges. Hence, the need to rethink how to govern the transformations of cities, which can be described by models of urban metabolism. The city sensing has changed the way cities are explored and used. With the transition from digitalization to datafication, through the computational approach, georeferenced big data can be analysed and exploited by algorithms. They originate a generative computational urban planning process, which can achieve a higher quality of the project and provide cities with adaptive capability. This process exploits data provided by public administrations, companies, and citizens who take part in an inclusive and adaptive urban planning.

DOI: 10.4018/978-1-7998-7091-3.ch002

INTRODUCTION

Nowadays about 55% of the world's population lives in urban areas, a percentage that is expected to increase. In 1990 there were just 10 megacities, classed as places with populations of 10 million or more. Globally, the number of megacities is expected to rise from 33 in 2018 to 43 in 2030. By 2030, the number of cities with one to 5 million inhabitants is expected to grow to 597 (United Nations Organization, 2018). In relation to what can be deduced from the data shown, cities represent one of the major problems to face in our contemporary era, both on the side of the environmental issues (energy consumption, pollution, etc.) and on the side of the economic and social ones (Figures 1, 2).

Some decades ago, Davis (1965) already wrote that urbanized societies, in which most people lived in crowded cities, were a new and basic step in man's social evolution. It was clear that modern urbanization could be best understood in terms of its connection with economic growth and availability of resources. Moreover, in those days, there was hardly any widespread perception of the problems that urbanization could imply and only a few, non profit, international associations, such as the "Club of Rome", founded in 1968, started to study solutions.

In relation to the increasing urbanization of contemporary society, Kevin Lynch's last work, "Wasting away" (1990), is not optimistic. The satisfaction of voracious urban metabolism transforms incoming resources into waste that is not compatible with the environment. This pessimistic urban scenario of the near future is not very dissimilar from dystopic Orwell's and Kafka's visions earlier in the century. In fact, future scenarios foreshadow cities as extremely fragmented and uneven systems in relation to fast evolving environmental, economic, energy and social phenomena. According to Lynch, the basic assumption seemed, that far, to be that technology and advanced urban planning could solve every problem related to urban metabolism. Under this point of view the movies of the "Qatsi Trilogy", an ecological eschatology, directed by Godfrey Reggio, Koyaanisquatsi (Coppola, 1982), Powaqqatsi (Coppola, 1988), Naquoyqatzi (Coppola, 2002), were visionary prophecies becoming little by little real in our time.

In his book, "Cities for a Small Planet", Richard Rogers stated:

It is ironic that mankind's habitat - our cities - are the major destroyer of the ecosystem and the greatest threat to mankind's survival on the planet...While the need for cities and the inevitability of their continued growth will not diminish, city living per se need not lead to civilization's self-destruction.

But he also stated:

I passionately believe that the arts of architecture and city planning could be evolved to provide crucial tools for safeguarding our future, creating cities that provide sustainable and civilising environments. ... My cause for optimism is derived from three factors: the spread of ecological awareness, of communications technology and of automated production. All are contributory conditions for the development of an environmentally aware and socially responsible post-industrial urban culture (Rogers, 1998, pp.4-5).

If we observe the latest transformations that have taken place in our cities, we realize that probably, at some point in history, we have lost the ability to control the evolutionary processes of urban systems and the environment (Borga, 2013).

Otherwise, according to John Wilmoth, director of the UN's Population Division, progressive urbanization, if ruled, could prove a positive factor both for economy and quality of life. Furthermore, the concentration of the population in a large inhabited centre can help to minimize our environmental impact on the planet, optimizing the location of resources, provided that administrations develop policies and practices to prepare for a huge influx of people (Meredith, 2018).

Figure 1. World's urbanization over the past 500 years (Ritchie & Roser, 2018)[1]

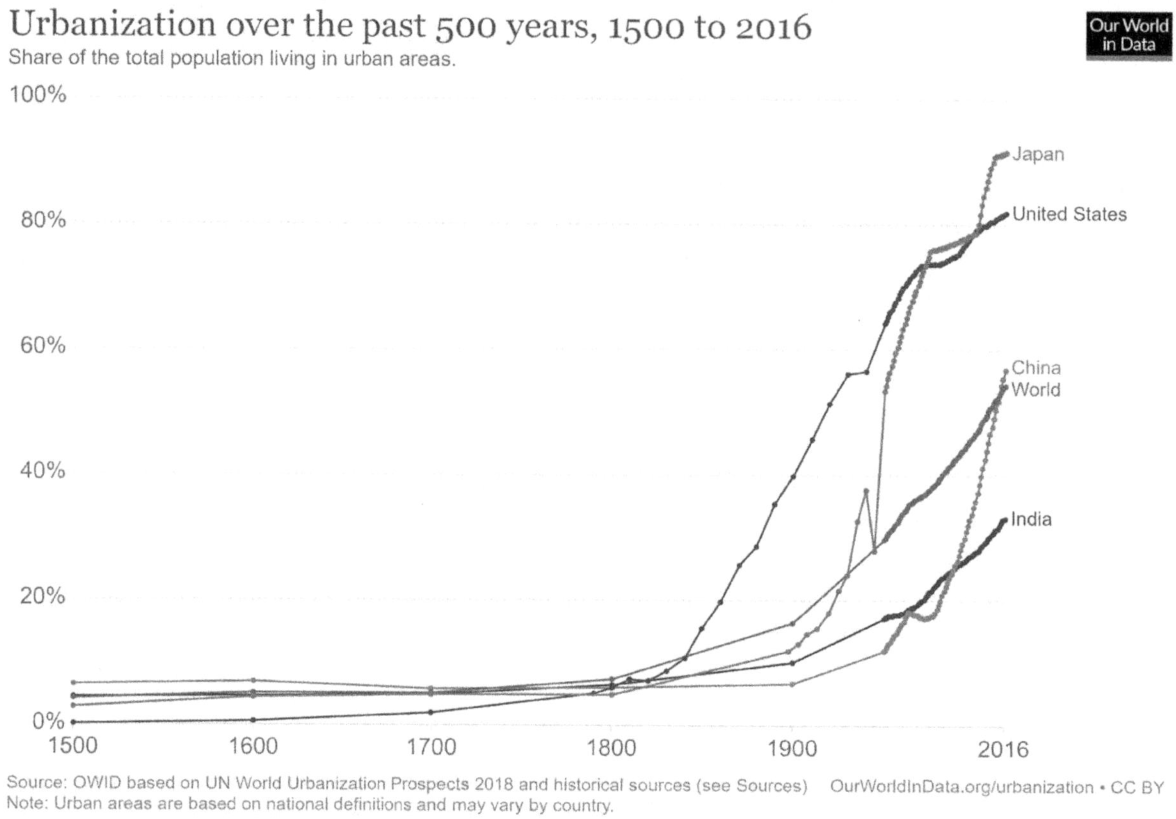

Whether urbanization could be a positive or negative phenomenon, the application of consolidated urban planning tools, based on a theoretical-predictive approach, adapts in a hardly effective way to the above-mentioned aspects due to the ever-changing stresses that act on the anthropic environment. A city can be defined as a constantly evolving metabolic organism, which self-regenerates by optimizing its configuration over time depending on the needs and availability of resources (Wolman, 1965). Lynch's basic assumption is that technology and advanced urban planning can solve every problem related to urban metabolism. The need will follow to rethink of ways of managing the transformations of a city.

Big Data Analysis is one of the most crucial contemporary tools in enquiring and updating, virtually and continuously, both the urban metabolism and its complexity so as to manage such transformations. In consideration of such premises, the aim of this paper is to show how effectively this analysis, result of datafication, could "*improve the art of urban planning*", as Rogers stated (1998).

If the historical consolidated experiences in urban planning were based on the recovery of past legacies, today the phenomenon of digitization is revolutionizing the concept of city, starting from an

Figure 2. Left: World's population growth since 1800. Source: UN. Adapted from source (Rosslyn, 2018); Right: World Urban and Rural Population projections from 2010 to 2100. Source: UN Department of Economics and Social Affairs – Population Division (2014)

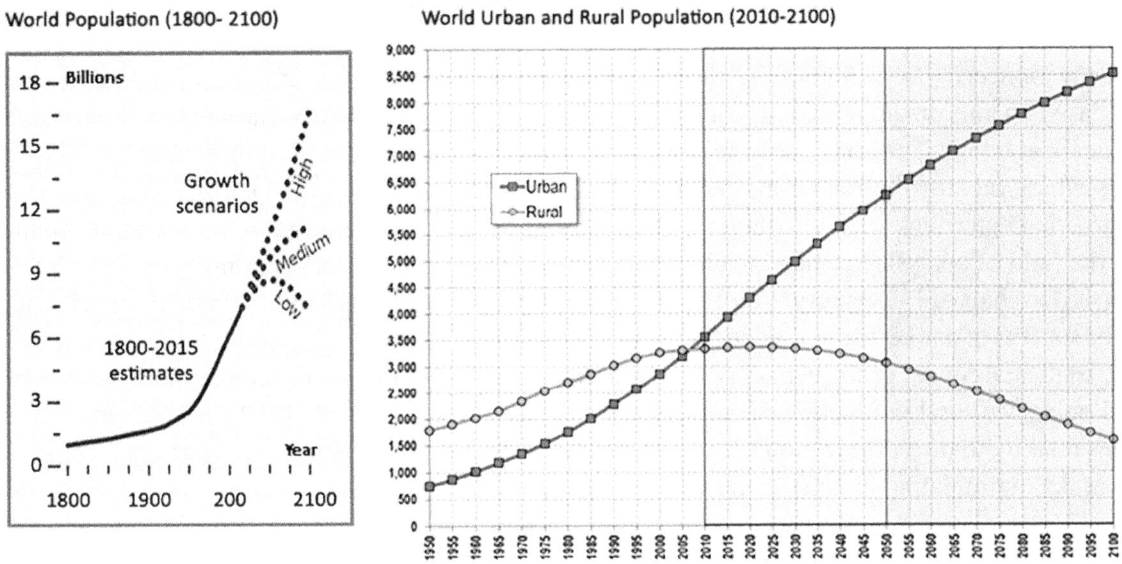

entirely new opposite process. In fact, digitization is global, accessible to everyone and from everywhere. It makes new space available, it stimulates sharing and connects things and people. The static image of a monolithic city, handed down through centuries in western urban history, is subverted in the digital age by the places of the sharing society, where public space becomes a protagonist again. The main European cities are creating new neighbourhoods where public space plays a central role taking inspiration from principles of shared models in a holistic form (Smart City, resilient city, green city, etc.). This is pursued with the common goal of creating cities more resilient, more efficient, healthier, safer and, consequently, more liveable. The potential offered by Computational Design, and Big Data Analysis can exploit theoretical models of urban metabolism elaborated through the last two centuries so as to achieve a deeper knowledge of phenomena and develop new tools able to integrate multiple levels of stakeholders and interests. This research describes a new approach to urban planning, based on the exploitation of Big Data Analysis to support inclusive generative and adaptive processes that create and modify the urban form of a city.

BACKGROUND

Cultural background in computational urban planning was born in the 19th century with the early theories about urban metabolism, that tried to interpret the distribution of economic forces during the First Industrial Revolution. During the decades those studies have been developed and refined. In the 20th century, these studies had a new impetus and were applied to the urban planning praxis. The advent of parametric urban design opened new addresses to research. Recently, the transition from digitization to datafication empowered the computational approach, providing huge amounts of data to the algorithmic computation which can be regulated by the theoretical models of urban metabolism.

Theoretical Models of Urban Metabolism Since Von Thünen

Since the pioneering contributions of the beginning of the 19th century, theories on localization, which were based on interpretative schemes of human activities, have followed one another in a conceptual and technical evolution. They have been followed and preceded by technological innovation, with the aim of describing the distribution of objects or activities in space, as well as the identification and study of the laws that determine such distribution. Originally, they found application in the economic field, but soon the field of analysis extended to all production sectors and to urban planning.

In the first half of 19th century, Von Thünen, witness of the second agricultural revolution in Germany, saw the success of market economy and the formation of prices of agricultural products (Torricelli, 2003a). The first spatial model of the economy was born in this context of first or proto-capitalism, essentially based on agricultural production. In his Concentric Model (von Thünen, 1842), he deepened the first studies relating to the optimization of the distribution of economic resources in a territory around a city (Figure 3.a), stimulating interest in the analysis of resource location (Parr, 2015).

Von Thünen's world was still essentially agricultural. Instead, Alfred Weber's world, between the end of the 19th and the beginning of the 20th century, had already undergone many transformations, mainly with the transition from an essentially agricultural economy and animal-drawn means of transportation to industrialization, railways, urbanization and industrial concentrations (Torricelli, 2003b).

By introducing variables relating to time and costs, Weber's model (Figure 3.b) tries to solve the problem of determining, in an isotropic space, the point where a production source must be located in order to minimize costs (transport cost, labour costs, cost of energy, etc.) and optimize the distribution of the produced goods. In its formulation, the Weber model defines three fundamental aspects: the minimum point of the transport cost, attracting workforce and the agglomeration force of production factors. These elements represent the legacy of the Weber model that will become the founding element of economic geography as well as of regional and urban economy (Torricelli, 2003b).

Weber's analysis was very abstract because it considered the location behaviour of individual companies regardless of the relationships they have with other economic entities. In addition it did not consider the strong influence of geomorphology. The problem of the spatial distribution of the interaction between supply and demand for goods and services in a territory remained open. Christaller's model (Figure 4.a) was based on the assumption that urban centres exist for the exchange of goods and services which they must produce or offer to the population, spatially dispersed over a homogeneous and isotropic area. The model's commitment is to understand how products or services, and in particular the functions of the services sector, are organized in an area, creating an urban hierarchy. The theory formulated by Christaller tried to explain the distribution of urban centres, their shape and their hierarchical structure on the territory, to provide a rational basis for their accommodation, contradicting the previous thesis that married the idea of a random location of urban centres, depending on gravitational areas. Christaller presented the concepts of threshold and range which expressed in geographical terms the usual economic forces that organize activities in space, the transport costs and agglomeration economies, specifically the economies of scale.

Figure 3. a. Von Thünen's Concentric Model (Torricelli, 2003a); b. Weber's model (Torricelli, 2003b)

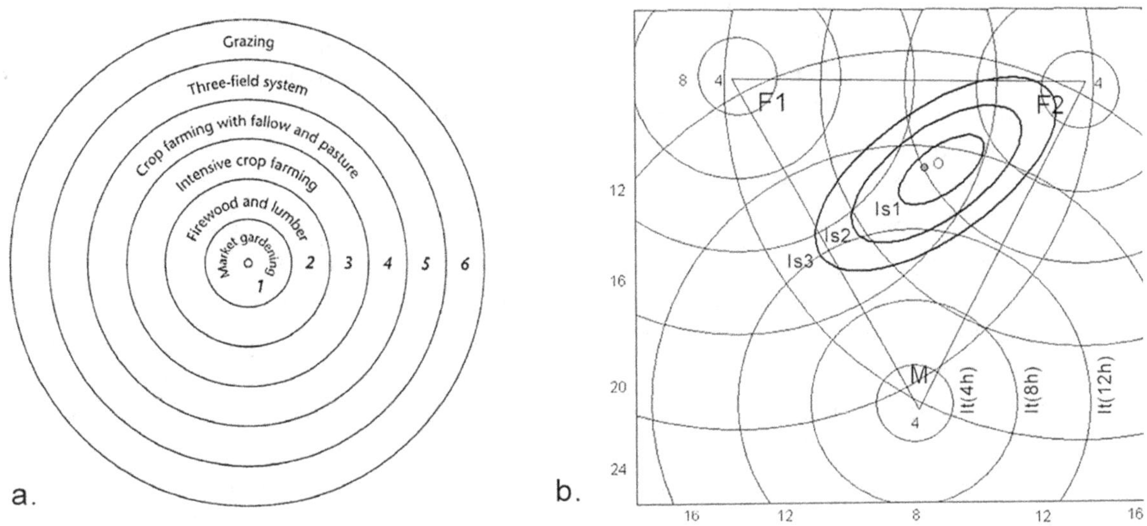

Agglomeration economies indicate a trend towards a large urban dimension, however, they do not explain why in reality, within apparently balanced urban systems, cities of different sizes coexist. Generally, smaller centres accommodate functions that find their way into local demand as a result of the limited economies of scale and their production size, while major centres host rarer and more specialized functions in which more relevant economies of scale appear. So as to represent different hierarchical layouts, Christaller introduced a factor of proportionality *K* which shows how much the Sphere of Influence of the central places extends.

Starting in the 1930s, August Lösch began to build his own theory upon Christaller's model (Lösch, 1954). His model was less narrow than Christaller's (Figure 4.b), in that it treated the range, threshold, and hexagonal hinterland of each function separately. The resulting pattern of central places is much more complex than Christaller's, and yields a continuous, rather than stepped, distribution of population sizes. Lösch modified Christaller's model because he noticed that the variation in K parameter was very relevant in shaping the organizations of centres and the numbers of centres at each level in a hierarchy. While Christaller arbitrarily chose K-values, on the contrary Lösch argued that no particular K-value could be considered axiomatic, suggesting that a large number of proportionality values can be used, still maintaining the hexagonal pattern model.

John Parr's Comparative Statics Model (1978) attempted the geometric exploration of the ways of change in the structure of an urban hierarchy, using a taxonomy of the types of change that can be associated with an urban structure (with variable K), maintaining the hypothesis of the presence of all the lower functions in the higher order centres. Parr highlights how one of the fundamental forces of structural change is to be found in the technical progress and innovative processes that cause the mix of functions in a centre and the change of the size of its market area, until the disappearance of a well-defined hierarchical level (Camagni, 1993).

John Parr (2017) criticized Christaller's model. In fact, on the scale of a small agricultural region with no major metropolitan centres it may be possible to identify something resembling a pure central place system. This is very unlikely, however, in the case of the typical region within a developed economy, where economic activity is not exclusively of the central place type. In this setting, the central place structure is only one part of a wider urban system. As a consequence, Central Place Theory can only represent a partial approach to the analysis and understanding of an urban system. As demonstrated, the typical urban system contains an element based on central place activity as well as one based on a specialized function activity. The two elements are inextricably interwoven and cannot be regarded as independent or separately functioning subsystems in any meaningful sense.

Generally, this theory provides an appropriate framework for many types of urban analysis, especially because it is able to draw attention to the manner in which centres are interrelated in competitive as well as complementary terms. This aspect is not emphasized in most analyses of urban systems, where the concern is typically with cross-sectional studies, involving all centres without regard either to functional composition or hierarchical level. Mention may be made of studies on the size distribution of urban centres, such as those by Gabaix (1999) and Eeckhout (2004). While work of this kind is valuable in drawing attention to the form of the urban system under various conditions of change, nothing is said about the spatial distribution of centres or the interaction among these. The central place theory must still be regarded as incomplete, as there are several important areas that require further development and elaboration. Despite its incompleteness, the central place theory is able to cast considerable light on the functioning of an urban system as well as on some of the policy issues relating to this.

Figure 4. a. Christaller's model (1933); b. Lösch's model (1954)

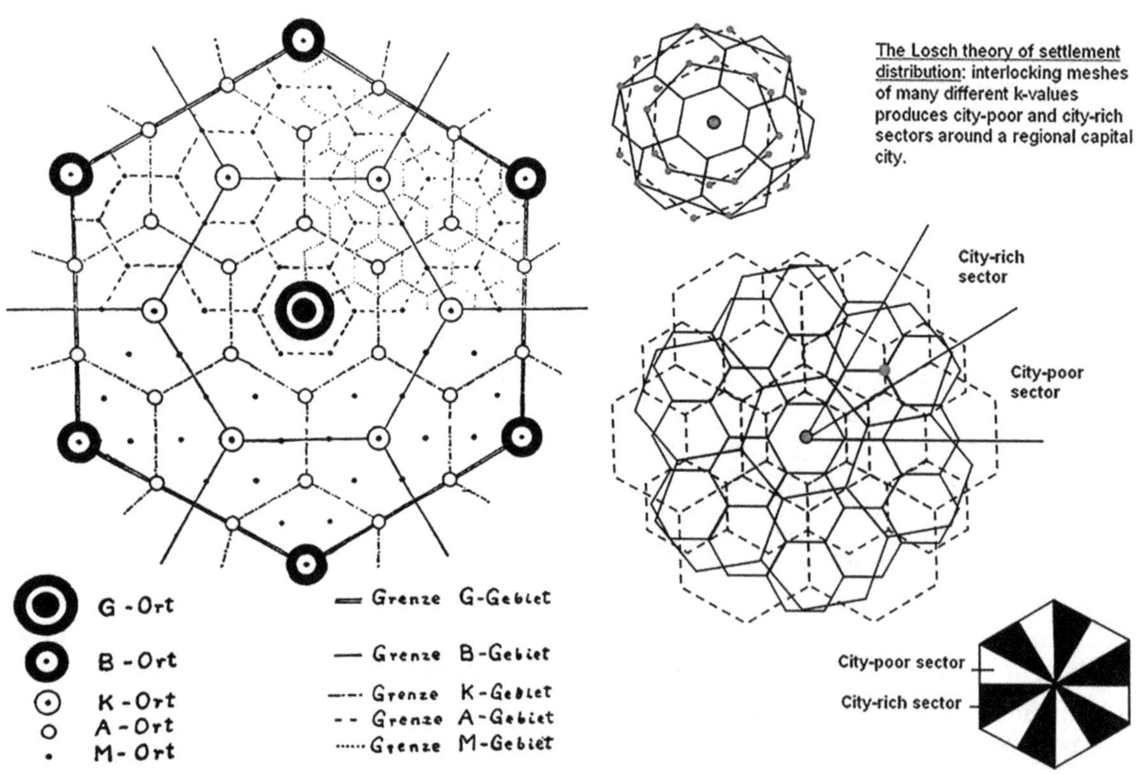

a. b.

Contemporary urbanization has stimulated new research that has linked the phenomenon to a qualitative change in hierarchical relationships. Structures of centres characterized by strong interdependence and accentuated production specialization, are gradually consolidating with the lack of evident internal hierarchical relationships and with a tightly interconnected network of intermediate centres. Space and economy are organized on network relationships: networks of local or international sub-supply relationships and global financial networks. The traditional urban structure inherited from the previous agricultural and industrial societies, characterized by a Christallerian hierarchy, organized directly on the hierarchical structure of the agricultural and industrial markets and on the structure of the bureaucratic and public administrative functions, was replaced by a mixed Crystallerian-Löschian structure in K variable. In conclusion, the new paradigm can be defined as follows: city networks are sets of horizontal and non-hierarchical relationships between similar or complementary centres. It creates synergies and the formation of economies or externalities of specialization.

Otto Frei (2011) elaborated some relevant thoughts on territories and spheres of influence with particular reference to urban settlements. Otto's work is particularly relevant today with the assertion of an eco-culture of the city and focuses on the analysis of the existing processes for the occupation of space and interconnections. It identified a striking analogy between the aggregation processes that occur in the physical, animal and vegetable world and the spontaneous forms and logic behind human settlements (Figure 5). The two are characterized by two opposing trends, one of attraction and the other of distance, and by the dominance of triangular and hexagonal forms, as opposed to the orthogonal logic typical of rational and artificial human desire. Otto Frei demonstrated that town planning, based on this artificial model, as it has developed over time, is contrary to an economic and ecological use of the land. Cities, settlements and road networks change and evolve with their own rules, following natural world's logic and are essentially impossible to plan. Although a city can be forced into a certain scheme, it is certainly more important to encourage spontaneous development and optimization processes. To do so, however, these processes must be understood and stimulated. Otto didn't suggest a mere zoomorphic imitation of natural forms, but the application of natural conformations to technical solutions. Many forms spontaneously created by nature are optimal models, as nature expresses itself and develops its phenomena with the minimum expenditure of energy. A new model of town planning based on sustainable development cannot, in terms of ecology and energy saving, disregard the knowledge of nature and spontaneous processes or, equally, the potential of new technology (Sola, 2009).

Significant studies by Otto & Rasch (1996) on the research of minimal-path patterns in complex systems, were based on empirical evidence, as a special form of analog computing. Analysis of complexity has produced many diagrams which have often been dropped onto architectural structures or typologies. All objects, however, should be understood as part of a process of emergence (Spuybroek, 2005). Richard Rogers (1998) also deepened such studies, focusing on pedestrian minimal paths and metabolism of a city (Figure 6). Parametric methodology applied to urban planning provides new tools to understand, describe and drive such processes.

Parametricism and Computational Urban Planning

As illustrated, the evolutionary processes of a city can be described by mathematical models that express organizational forms aimed at optimizing complex processes of interaction between individual, community and environment.

Figure 5. a. Frei Otto's comparative study on connection patterns in anthropic and natural environment; 1. Thread models 2. Paths trampled by hoofed animals 3. Dragonfly's wing (section) 4. Maple leaf 5. Stoneware glaze 6. Instanbul, road network 7. Village in the Serengeti 8. Crack pattern, gelatine 9. Soap-bubble raft 10. Minimal-path network (Frei 2011); b. Frei Otto et al., schemes of the empirical strands of the wool-thread machine which generates an optimized path system between targets. From left to right: Step 1. System, Step 2. Over-lengthened, Step 3. After wetting and self-organizing (Spuybroek, 2005).

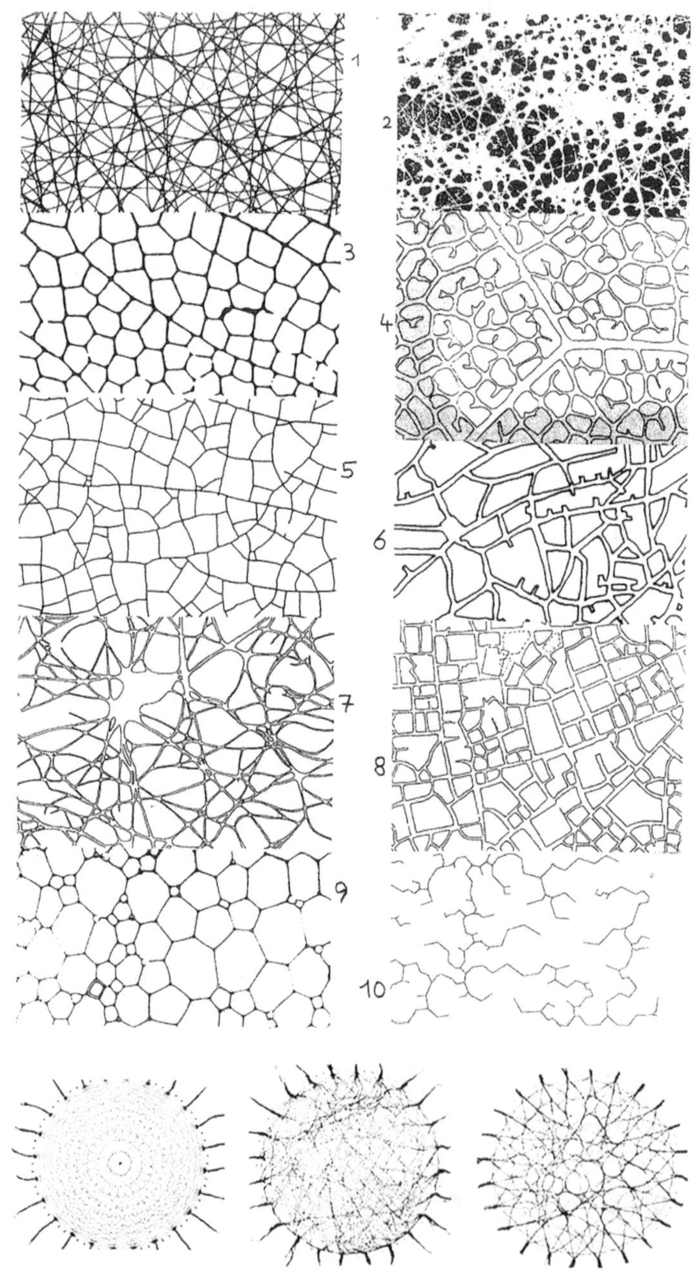

a.

b.

Figure 6. Rogers' studies on pedestrian minimal paths and urban metabolism (Rogers, 1998)

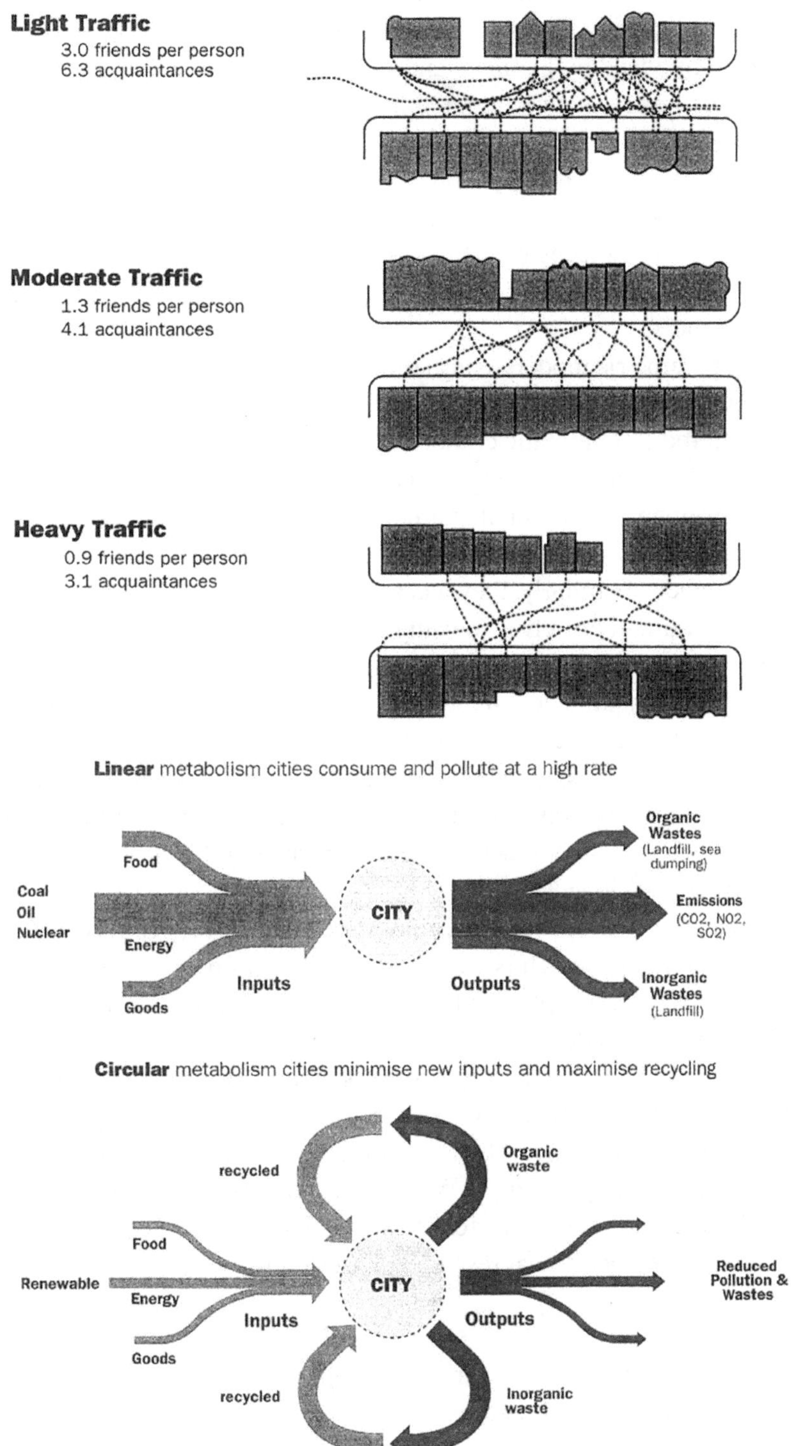

The formal models by von Thünen, Weber and Christaller had demonstrated the opportunity to study geography, on a regional and urban scale, starting from a model built on postulates and axioms, it was, therefore, an abstract and deductive model and in some way, predictive. Those new models, however, went, at the beginning, unnoticed by planners and only later, after the WW2, Christaller's work began to be fully appreciated and effectively integrated into the conceptual heritage of economic geography and urban planning.

In fact, only since the 1950s, models by Von Thünen, Weber and others, especially in Great Britain and the United States have been reinterpreted and reworked within the framework of broader theories, in particular that of the central places by Christaller and subsequent works by August Lösch. These theories appear able to describe reality (especially geographic-economic), with the support of more advanced mathematical tools.

The development of parametric models using computational tools in design has its cultural roots in the avant-garde of the 1960s. In that context Luigi Moretti inaugurated the studies on algorithmic urbanism, trusting the potential offered by automatic computing. In his work "Structure Comme Forme" (Moretti, 1954), he wrote about architecture based on parameters, referring to the theory of Galois groups. In his view, the modernist movement and rationalist architecture had not been able to respond effectively to the needs of functionality, limiting themselves to proposing empirical and essentially formal solutions for merely utilitarian instances. Similarly, they had not grasped the possibilities offered by new materials and new technologies, limiting themselves to using traditional construction methods.

Moretti defines Parametric Architecture an architecture that assigns precise functions to mathematical values by relating them as parameters. Once defined, the field of these relationships constitutes an area in which freedom of formal decisions is exercised. The parameters and their interrelationships, expressed by mathematical formulas, thus become the expression of a new design methodology: forms and their relationships create a structure responding to given practical, ethical and social functions. The contradictions inherent in this approach consist of entrusting the choice of parameters to personal intuition, compromising scientific validity.

In the late 50s, Moretti argued the possibility, innovative for the time, of applying Operational Research to urban planning. Operational Research is a methodology that aims to identify, with procedures based on mathematical and statistical concepts, the best course of action to achieve an a priori assigned goal (Esposito de Vita, 2009). Complex decision problems are analysed and solved by means of mathematical models and advanced quantitative methods (optimization, simulation, etc.) as a support for the decisions themselves. The use of operational research dates back to the late 18th century, when the mathematician Monge addressed the problem of transport by examining its economic aspects. The development of operational research had great impetus during WW2, for the resolution of problems in the field of military tactics and logistics. In the years in which Moretti was active, it was still very little used and certainly not used in city planning. The operational research phases consist of:

1. Formulation of the problem;
2. Examination of the real situation and data collection;
3. Construction of the mathematical model;
4. Solving the mathematical model;
5. Analysis and verification of the solutions obtained;
6. Implementation of the solutions obtained.

An urban plan is a configuration that develops over time. It can be seen as a function of probabilistically variable parameters. Thus, only operational research, according to Moretti, allows to identify the number, the measures and the reciprocal relationships.

Thanks to this methodology and to cybernetics, Moretti believed that it was possible to evaluate and control the social, financial, economic and productive complex phenomena that urban planning is called upon to interpret and regulate. The method suggested for urban planning by Moretti consists of defining the themes of each research with the help of researchers from different scientific fields. It consists, then, of establishing elements of judgment and data processing methods, in order to arrive at defining probable groups of solutions in quantitative and non-descriptive terms, so that they can be mathematically interpreted.

For this specific purpose Moretti brought together, in the 1950s, an interdisciplinary study group, with which in 1957 he founded the IRMOU (National Institute of Mathematical and Operational Research for Urban Planning) based in Rome. First of its kind, the institute brought together scholars from different fields (mathematics, geometry, topology, statistics, financial mathematics, electronics, cybernetics, architecture, engineering, biology, psychology, sociology and economics) in order to promote and increase applications of operational research to town planning and related problems. In 1960 the IRMOU set up the "Parametric Architecture and Mathematical and Operational Research Exhibition for Urban Planning" at the "XII Triennale di Milano" (Figure 7). On that occasion, studies were presented on the form of football stadiums, swimming-pools and cinemas, which were based on the so called "equi-appetibilità visive", i.e. an equal visual appeal from any viewpoint on the surfaces, with double-valve or tulip shapes. In addition, examples of economic geography, such as the distribution of public services (schools, offices, telephone exchanges) in relation to the structure of inhabited centres were illustrated as a demonstration of the applications of operational research and cybernetics in urban planning, exploiting the scheme by Lösch.

Figure 7. Luigi Moretti design studies for the XII Triennale in Milan, 1960. a. stadium; b. swimming-pool centre. (Esposito de Vita, 2009)

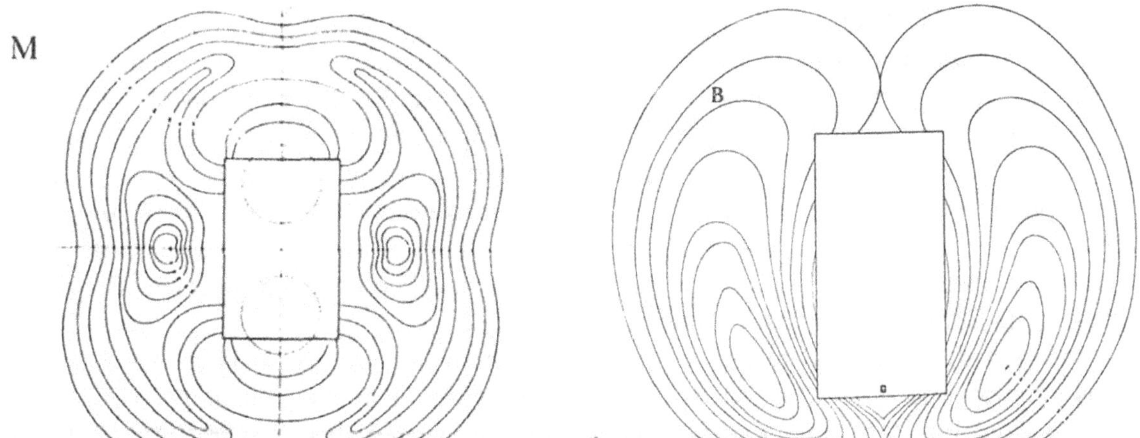

Ferguson (1975), a forerunner of a mathematical methodology, in his book "Architecture, City and the Systems Approach", presents a general survey of the ways in which systems analysis can be applied to decision making in the design and planning professions.

In "An Evolutionary Architecture", Frazer (1995), considering architecture as a form of artificial life, attempted to make form and structure evolve in emulation of natural evolutionary processes. This approach exploiting generative computational techniques to implement the evolutionary reiterative process, led to the conception of Shape Grammars, specific class of production systems that generate geometric shapes driven by preset rules.

A Shape Grammar with parametric features, improved with the capability of an algorithmic optimisation, might provide the flexibility demanded in urban design in order to achieve preset design objectives. It produces analytical and adaptive solutions in large-scale urban projects (Aydin and Schnabel, 2013).

From Digitization to Datafication

In the last 50 years, contemporary culture has been enriched with new terms referable to the translation of entities and processes from the physical world to the digital one.

The term data comes from Latin *datum*, whose meaning is 'something given' (Casi, 2019). Today this word recovers its original meaning, which it had lost over time.

Digitization is the process that converts data from the analog field of continuous values, to the digital one of discrete values, readable by digital devices. According to Negroponte, digitization could be seen as the shift from atoms to bits (Negroponte, 1995). Digitization is essential for data processing, storage and transmission. Analogical data, however, is normally more stable, digital data can more easily be accessed and shared indefinitely. The beginning of the digital revolution dates back to the late 50s with the transition from analog to digital technology. It has continued up to the present day, within the framework of the third and fourth Industrial Revolutions.

Datafication is the phenomenon introduced and consolidated by the continuous development of ICT. It turns aspects of everyday life into data and valuable information, going significantly beyond digitization. This term was introduced in 2013 by Mayer-Schöenberger and Cukier (2013), although datafication had its roots in the remote past of human civilization, since the birth of writing, as a solution to the needs of registration and calculation. Nowadays digitization has given a new impetus to datafication. The match between datafication and digitization has given rise to Big Data.

The term Big Data commonly refers to a collection of data so huge in terms of volume, speed and variability to request specific technologies and methods for its analysis. Big data refers to a further step in the paradigm of the interpretation of economic and social phenomena through analytical techniques. Thanks to Data Mining, i.e. the process of extracting from extensive amounts of data the most relevant ones in relation to preset purposes, data 'varieties' can be identified, grouped, stored and processed at high speed, even in real time (Agicom, 2018). The acquisition of Big Data has been made possible through the progress of digital devices and data transmission networks that have reached capabilities and levels of performance unimaginable until a few years ago.

The diffusion of connected digital devices has changed the contemporary lifestyle, changing individual behaviours due to the pervasive and obsessive use of such devices, which either voluntarily or involuntarily generates information. ICTs of the Fourth Industrial Revolution strengthened the relationship between physical reality and digital world, attributing their own digital tracks to real phenomena and to human behaviours. New ethical problems, therefore, have arisen. Big Data ethics tries to respond to them by introducing the principles of right and wrong conduct in relation to data management, in particular the exploitation of personal data by third parties.

Big Data Analysis is the evolution of traditional analytical processes. It extrapolates, collects, analyses and correlates huge amounts of heterogeneous, structured and unstructured data provided by digital devices, with the purpose to extract latent information. When huge amounts of historical data are available, Big Data Analysis allows to perform predictive analyses on statistical bases. For instance, the use of Big Data Analysis allows businesses to observe customers' patterns and trends. Observing customers' behaviour is important to foster their loyalty and provide customized services or goods. Through a proper Big Data Analysis, a business will have the capability to derive critical customers' behavioural insights to be acted on in order to retain their customer base.

Big Data Analysis consists of 6 C's: Connection (achieved through sensors and networks), Cloud (which means computing and data on demand), Cyber (which consists of a digital model & memory), Content/Context (i.e. meaning and correlation), Community (whose features are sharing & collaboration) and Customization (i.e. personalization and value) (Lee, Bagheri, & Kao 2014). The 6 C's of Big Data Analysis make it possible to interpret and exploit the exceptional information content of Big Data. The potential of this new analytical tool can be successfully used to drive adaptive and inclusive territorial policies, making stakeholders' interests converge towards a shared objective.

Contemporary Experiences in Generative Computational Urban Planning That Exploit Datafication

The design methodology introduced by Luigi Moretti was antithetical to the practice of his time, which was inspired by a rigid descriptive formalism, unable to provide effective tools for managing the complexity of the modern world. The systemic approach, intuited by Moretti, allows to go beyond the urban structure to focus attention on the relationships between components that become part of the model.

The adoption of parametric tools in urban planning allows the definition of a new methodology which consists of the theoretical development of systemic models whose formulation is based on the comparison between urbanism and autopoietic, evolutionary, self-regulating systems. Pragmatically, the transition from the concept of arbitrary topological and typological description of the urban form to the concept of self-regulating process is obtained. This involves overcoming the logic of the typological composition and zoning of urban planning and adopting a vision based on multiple interconnected elements. Through the propagation of its effects, the transformation of a single element, as the parameters vary, can involve the modification of an entire architectural or urban organism. Hence, the final form is an output generated by a procedure. Design becomes the definition of intelligent rules (Fusero & Massimiano, 2013).

Three different approaches can be identified in Generative Computational Urban Planning that exploits Big Data. The first approach, starting from the data relating to the anthropic and environmental context, generates, through the parametric calculation, the most responsive solution to the set objectives. The second approach consists of the definition of a "master algorithm" which, as the Big Data changes, depending on the conditions imposed, generates multiple alternative scenarios. These scenarios, then, are screened by the stakeholders involved. The third approach, once the generative rules are defined, produces the urban form and also attributes an adaptive ability to the city, through the definition of threshold values, whose overstepping triggers self-regulation processes (Galli & Massimiano, 2015). Many contemporary examples show the effectiveness of the Generative Computational Urban Planning.

Examples of the above-mentioned first approach are Kartal-Pendik and King Abdullah City for Atomic and Renewable Energy (Ka-Care) Masterplans, in which data related to the context generate an optimized solution.

Kartal-Pendik Masterplan by Zaha Hadid Architects (2006) is aimed at the redevelopment of an abandoned industrial site through the planning of a new central business district, residential areas and cultural and tourist facilities (Figure 8). The site lies at the confluence of many crucial infrastructural links. The masterplan is inspired by Frei Otto's studies on path optimization and links together the basic infrastructural and urban context of the surrounding site. In the west and in the east sides of the area, lateral lines link together the major road connections. The integration of these lateral connections with the main longitudinal axis creates a soft grid that draws the underlying framework for the project. At a local level, this net can be grouped together to form, programmatically, areas of higher density through a vertical growth of the urban fabric. An algorithmic script, then, generates different typologies of buildings that respond to the different demands of each district. The urban form, therefore, is generated by a dynamic and iterative process that adapts to preset requirements, while achieving a recognizable image and a smart integration of the new area with the existing surroundings (Kang, 2012).

King Abdullah City for Atomic and Renewable Energy (Ka-Care) Masterplan, developed in 2010 by Carlo Ratti Associati (2010), foreshadows an eco-city in the desert near Riyadh, in Saudi Arabia (Figure 9).

Exploiting algorithmic urban design, a script has been developed to guide the evolutionary growth of the city, in accordance with site-specific environmental parameters. The algorithm, defined by a list of instructions, runs a calculation that, starting from an initial state, proceeds through series of recursive runs, finally achieving an optimized final solution. The rules were established in advance, in order to incorporate the information extracted from the environment and architectural prototypes into a coherent and unitary result.

Data about landscape morphology, water proximity, solar radiation and wind direction is taken into account by the algorithm, in order to highlight plots with different degrees of attractiveness to host the first nuclei of the city. By integrating these virtual spaces with information relating to economy and technological conditions, as well as the building typologies and degree of interconnection within the city, different evolutionary scenarios emerge for evaluation and comparison. Planners, stakeholders and future citizens can use the parametric model to read the actual situation of the system and to lead its evolution achieving a self-regulating urban organism (EcoLogic Studio, 2011).

Water is collected to its very maximum on site, storing annual rainfall in natural aquifers that minimize evaporation. The water and energy systems are controlled by a smart grid that enables optimum control of urban processes. The display of ambient data throughout the city promotes public engagement and prompts behavioural change in its citizenry, who can monitor the city through their hand-held devices. These systems offer the dynamic support system for a real-time, rapid-response city, leveraging parametric digital design to evolve organically, optimizing the provision of social, economic and physical resources. (Atmos Studio, 2010).

Overlapped City research by Remixstudio (2012) is an example of the second approach in Generative Computational Urban Planning by the above-mentioned data exploitation. It deepens, through the algorithmic computation, urban energy efficiency. This research consists of a large-scale, regional, masterplan developed through three levels: redefinition of urban boundaries and clusters, optimization of the framework of energy infrastructure and definition of a new set of urban codes. The testing site is in West Houston, the fastest growing urban area in the US, with a rich potential for renewable energy production.

Figure 8. Kartal-Pendik Masterplan design process (Hristov, 2015): a. direct links network; b. minimized detour network – lake not included; c. minimized detour network – lake included; Images edited by Zaha Hadid Architects (2006): d. minimal paths infrastructural grid; e. urban tissue densification; f. parametric masterplanning outcome

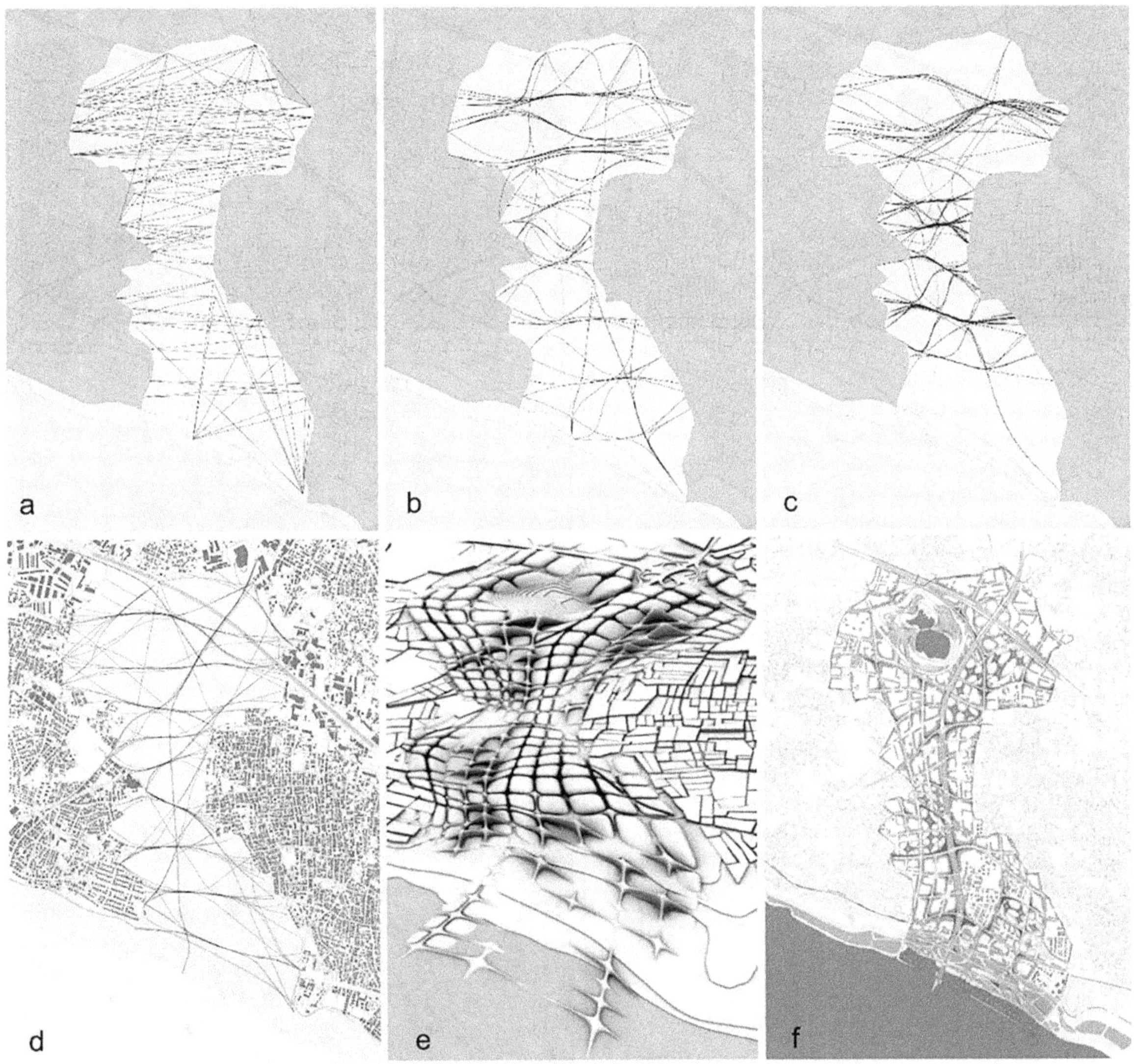

Parametric tools help to recognize the different energy potentials associated with various massing strategies. The design of urban typologies is articulated following the concept of an 'verlapped city': active surfaces are maximized while multiple functions are overlapped in compact urban areas. A policy of inter-block energy balance incentivizes the citizens to maximize their local energy production and reduce their reliance on the centralized energy supply, turning the new city work into a resilient ecosystem. In such context, different degrees of flexibility can be allowed in order to adapt to the market economy. New patterns of energy specialization arise, defining the blocks that are more prone to production and the ones that are more prone to consumption. Through a system of energy trading a cluster meets its

Figure 9. Generative computational master planning of King Abdullah City for Atomic and Renewable Energy (Ka-Care) by Carlo Ratti Associati (Atmos Studio, 2010)

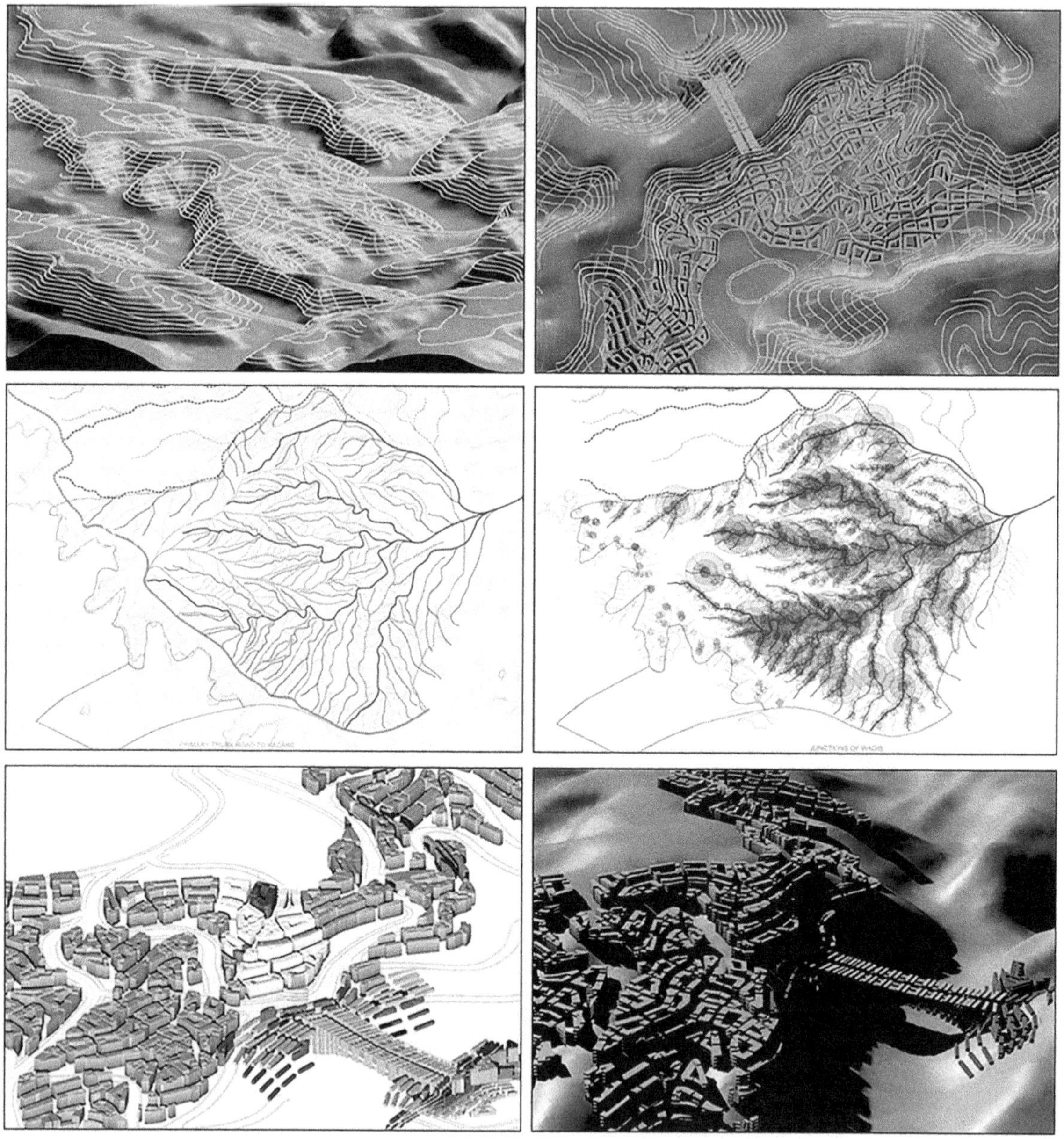

overall energy production target through compensation among the different blocks and the creation of additional energy sources within the green corridors.

As result of the new urban code, some new interesting and highly diverse spatial conditions and relationships arise, disclosing the potential of this approach to create new paradigms of urbanization. Thus, a variety of radically different scenarios are generated, tested and evaluated (Figure 10).

Figure 10. Generative computational master planning of the west Houston area. Adapted from source (Remixstudio, 2012). Some relevant steps in the process are: 1. Definition of the clusters in the West Houston area; 2. Urban cell prototype, building typologies and energy potentials; 3.Catalogue of energy-related building typologies; 4. Inter-block energy balancing; 5. Energy resilient urban cell prototype; 6. Urban cell prototype adaptability to different local energy productivity targets; 7. Regional masterplan implementation; 8: Vision to the year 2050

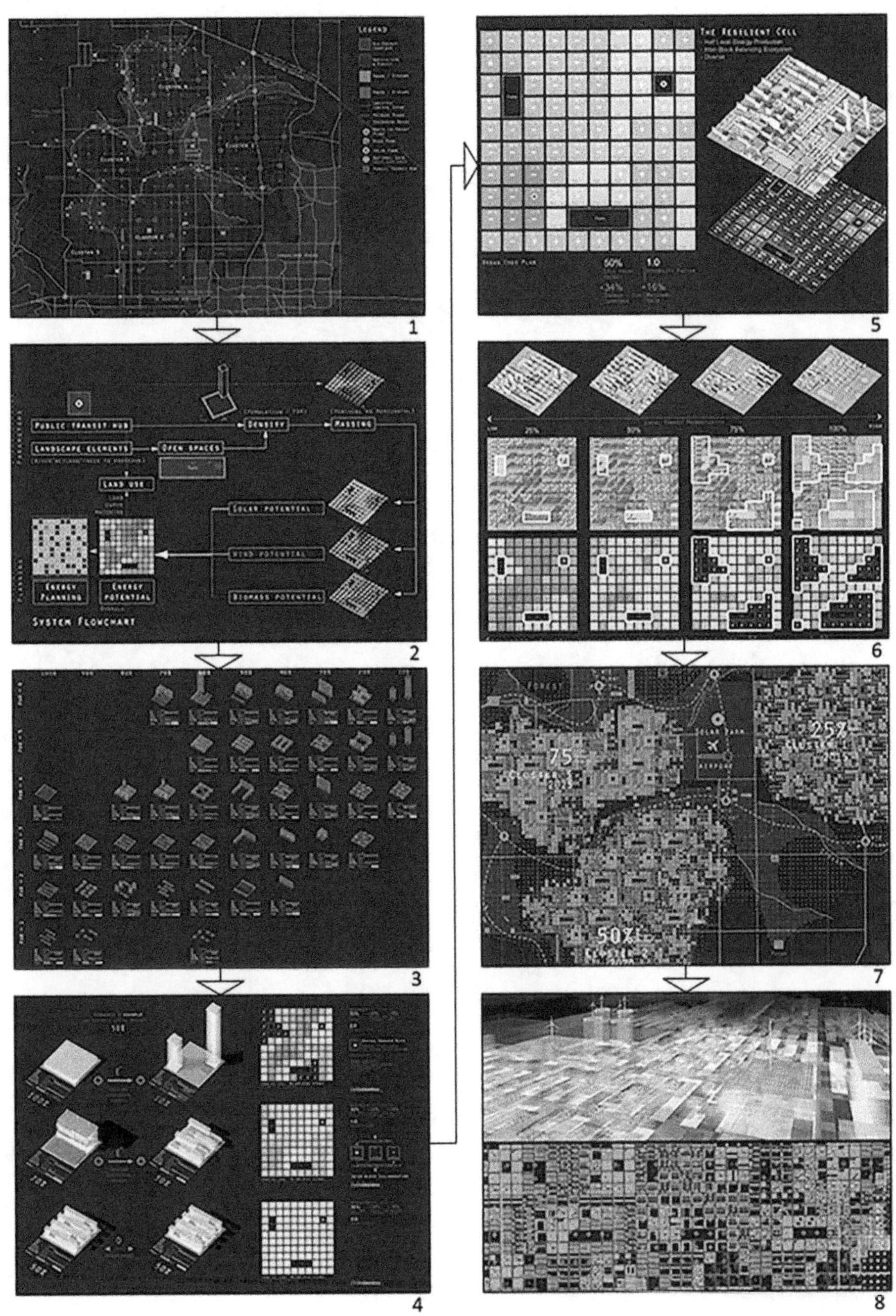

Figure 11. The IDEA District and eastern waterfront geography (Sidewalk Labs 2019a, p.259)

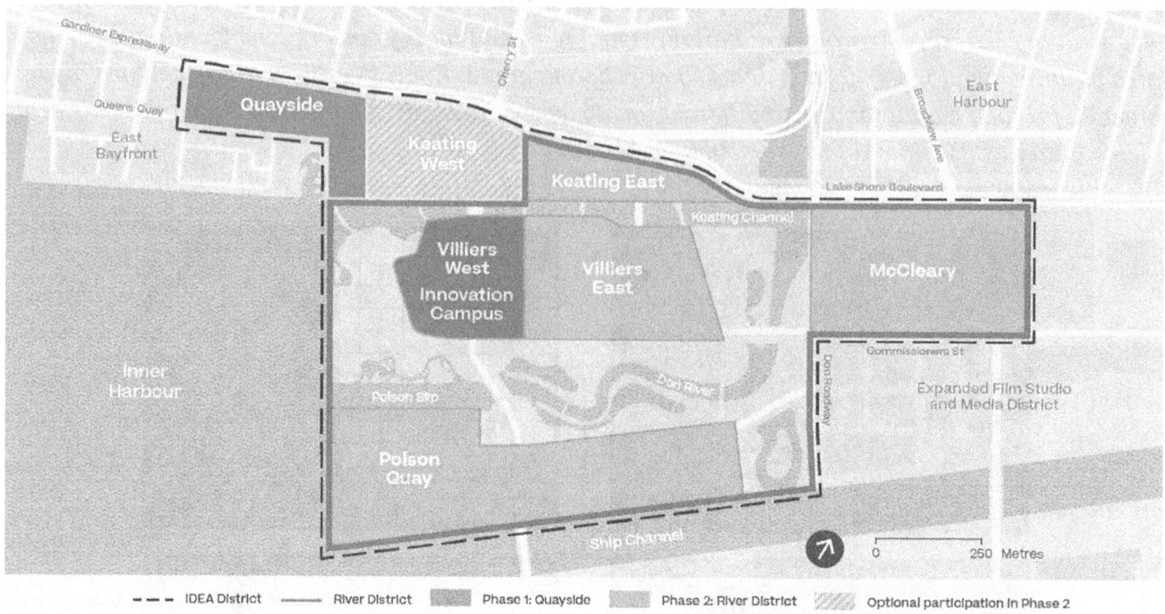

Figure 12. Generative master planning of Villier East neighbourhood in Toronto's waterfront. Adapted from source (Sidewalk Labs, 2019a, pp.324-325)

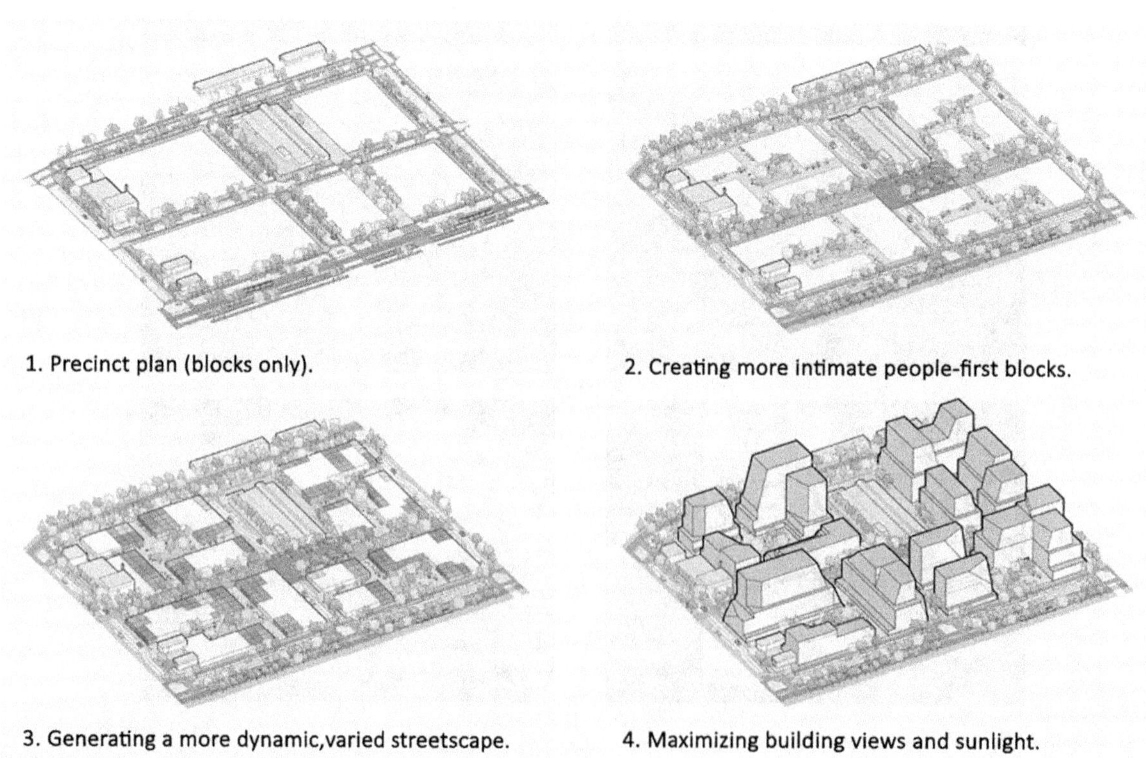

Figure 13. Generative master planning of Villier East neighbourhood in Toronto. Using digital tools it has been possible to assess thousands of options. Adapted from source (Sidewalk Labs, 2019a, pp.327-329)

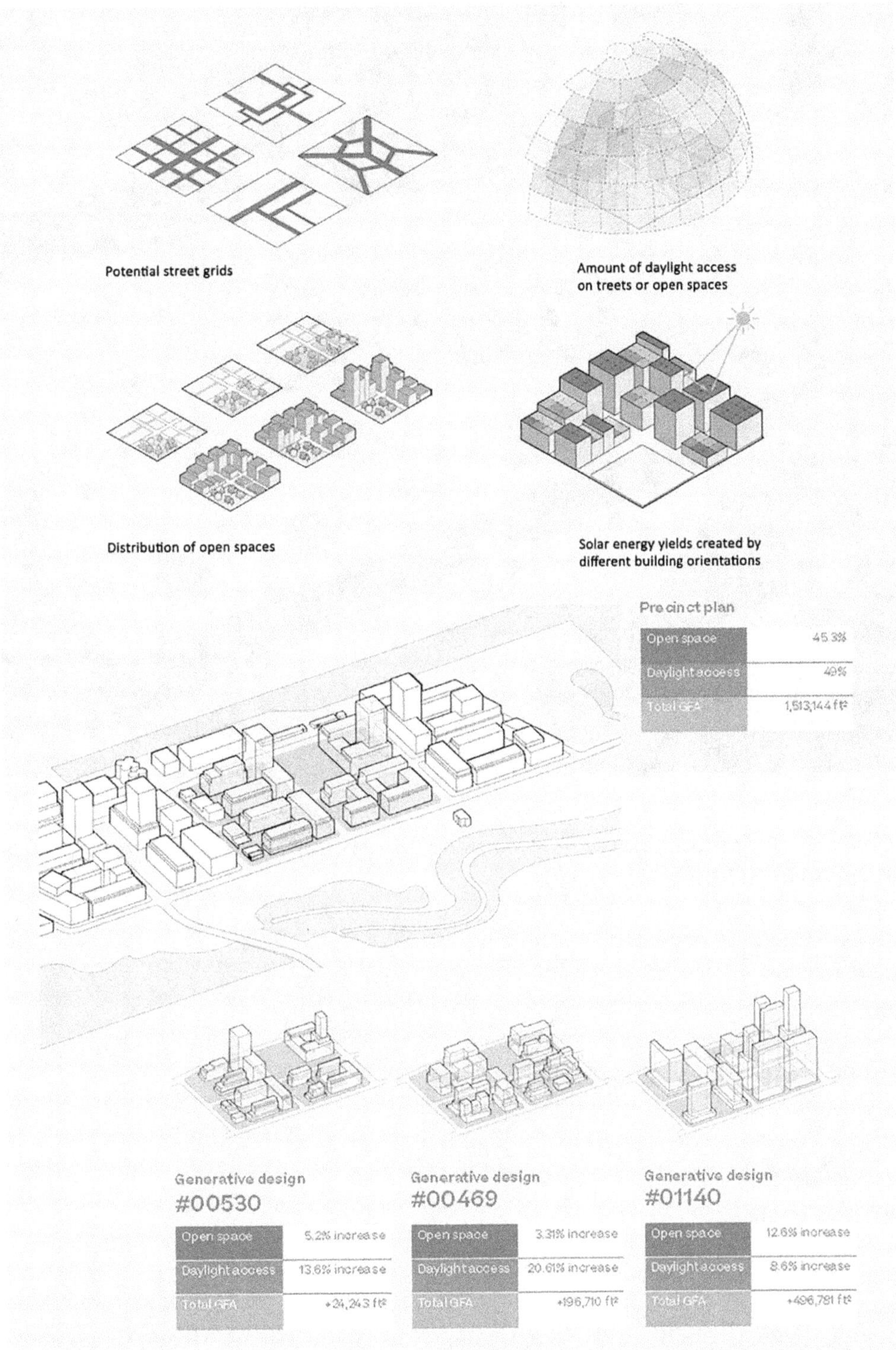

Figure 14. Interactive interface for Sidewalk Toronto (Wilson et al., 2019)

Figure 15. Comparison of traditional master planning process (top) with Computational Urban Design methodology (bottom) (Wilson et al., 2019)

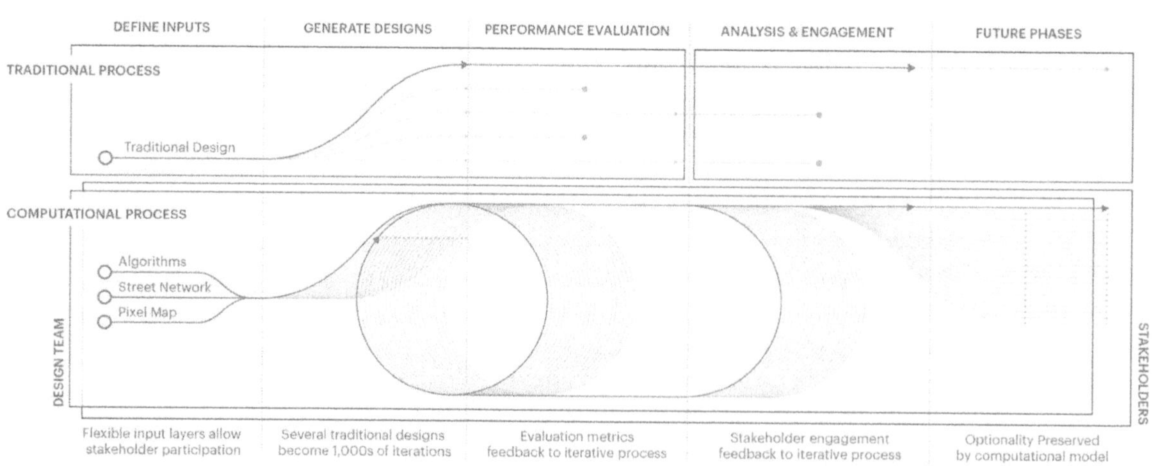

The project of the IDEA (Innovative Design and Economic Acceleration) district is an example of the third approach in Generative Computational Urban Planning through Big Data Analysis (Figure 11). This study was developed within the framework of Toronto's Waterfront redevelopment. It was conceived by Sidewalk Labs, a subsidiary company of Alphabet Inc., a sister company of Google. This project, whose goal was to renovate a portion of a disused industrial area, was first initiated in 2017. The Master Innovation Development Plan (MIDP) was, then, created in 2019 through interviews with thousands of citizens and aimed to be an innovative reinvention of Toronto's eastern waterfront. The plan consisted of 7 new neighbourhoods: Quayside, Keating West, Keating East, Villier West, Villier East, McCleary and Polson Quay. The project was cancelled in 2020 due to the uncertainty of the economy owing to the COVID-19 pandemic.

Sidewalk Toronto IDEA had aimed to exploit technology to create a smart urban area so as to improve the quality of life of residents, also using it as a testing ground for future technology and urban design projects. The high-tech plan was criticized, mainly in relation to data privacy issues regarding the massive citizens' personal data collection. The project goes beyond the computational master planning, adding to a city adaptive capabilities in real-time through the exploitation of Remote Sensing, City Sensing and Big Data Analysis. In fact, not only the generative computational urban planning produced IDEA's urban form, but also the urban form of the district can modify itself. The IDEA district is a self-regulating systemic organism which continuously adapts and re-generates its morphology, ruled by computational scripts and Big Data Analysis. Such capabilities can be based on users' needs, on the action of economic forces and on the changing environmental conditions.

Within the framework of the IDEA masterplan, the layout of Villiers East has been generated using a dedicated computational tool (Sidewalk Labs, 2019a, pp. 323-329). In Villiers East, the existing City of Toronto precinct plan calls for high-density housing to support the economy. That objective required finding a balance between the required density and the need for high-quality open spaces to support residents, visitors, and workers. As planning proceeded, the generative design tool supported planners in evaluating the performance of different options by running thousands of simulations that assessed key-factors like building massing, access to natural light and wind. Among various strategies to achieve these objectives, one involved breaking down the blocks into a series of small buildings with pedestrian courtyards. This solution generated more intimate spaces for residents. As a test, designers used the generative design tool to conduct a preliminary study of possible courtyard configurations, aiming to optimize the percentage of open space, sunlight access in the courtyard and density (Figure 12). In an initial iteration, the tool analysed thousands of permutations and generated hundreds of design concepts that created more open space and allowed more sunlight to reach the streets than the precinct plan base-lines, while also adding more density (Figure 13).

A team of Kohn Pedersen Fox Associates (Wilson, Danforth, Cerezo Davila, & Harvey, 2019), working with Sidewalk Labs, developed a computational urban design model to assist with the master planning of the IDEA district. As part of a public exhibition they developed a digital model of an abstract urban site with inputs that included a representative sampling of options under consideration for the waterfront development, such as building density, street grids, and green spaces. The digital models were made accessible through a physical interface that allowed the public to engage with the computational urban design process (Figure 14). Visitors experimented combinations of density, public space and street grids by a physical control device. A reiterative computational process allowed users to create and optimize the type of neighbourhood they wanted and to understand how those design decisions impact the functioning of a complex system like a city (Figure 15).

Figure 16. Short-term urban adaptation in Bonnycastle Street in Quayside sub-plot of IDEA District in Toronto by Sidewalk Labs. Images adapted from source (Sidewalk Labs, 2019a, pp. 122-123). A. Street lighting. Street lights would adjust to real-time conditions. B. Adaptive traffic signals. Adaptive signals can make real-time adjustments to balance traffic needs. C. Traffic coordination. A proposed mobility management system would monitor space availability and parking areas, (direct) directing drivers to open spaces and adjusting fees in real time. D. Dynamic curb. Pick-up and drop-off spaces would expand or shrink based on demand. E. Retractable facades. Facades would open during the summer months, but close during rain and snow storms. F. Heating. A number of modular pavers will pilot heating capabilities to clear snow and ice, improving safety and eliminating the need for salting. G. Streetscape installations. Pavers would be designed to enable easy installation and removal of street infrastructure, such as signs, traffic lights, etc. H. Paver maintenance. Minor repairs can be completed quickly with hand-held tools, reducing cost and neighbourhood disruption. I. Hexagonal design. Each paver's shape distributes vehicle weight more evenly than traditional rectangles, helping to minimize cracks and potholes

Figure 17. Long-term urban infrastructural adaptation in Queens Quay sub-plot of IDEA District in Toronto. Project by Sidewalk Labs (Sidewalk Labs, 2019a, pp. 121). Above: Queens Quay in 2025. The initial plans for Queens Quay feature wide sidewalks and bike lanes, as well as plentiful dynamic curb spaces along the north side of the street. Below: Queens Quay in 2035. In the future, when self-driving vehicles can replace traditional cars and share space with light rail transit lanes without impeding operations, the two traffic lanes will be converted to pedestrian spaces

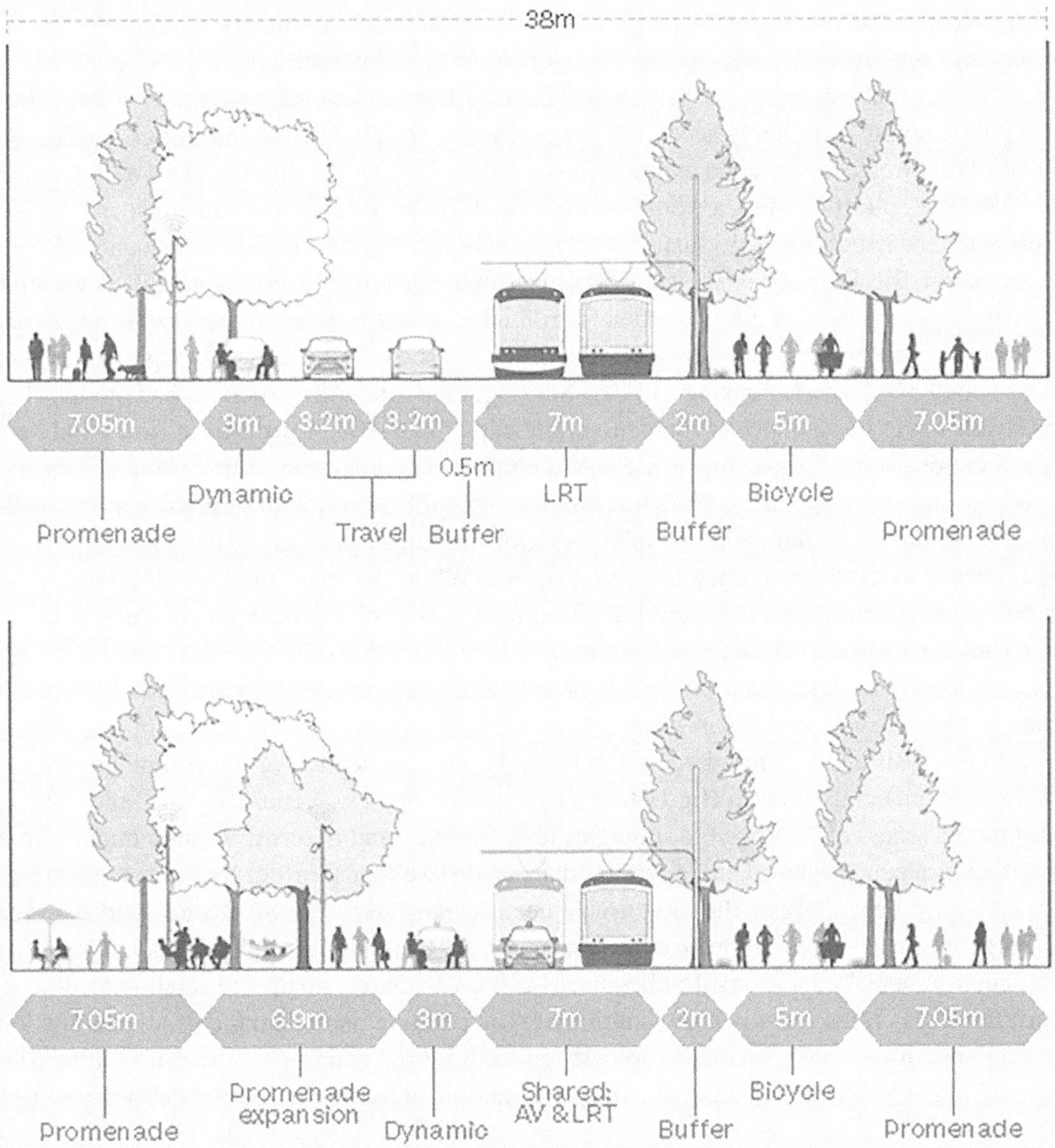

The mentioned examples demonstrate how the Generative Computational Urban Planning can exploit huge amount of data such as urban codes, environmental data and citizens' feedback to generate the initial urban form of a city. A city can also, however, adapt and reconfigure itself based on data analysis.

Sidewalk Labs' vision consists of catalysing digital innovations that help to face urban challenges and establish a new standard for the responsible collection and use of data in cities (Sidewalk Labs, 2019b, pp. 376-463). Digital infrastructure creates connectivity that helps residents, companies and organizations to use data to launch new services that improve urban life. It is also the catalyst for new services or businesses and the cornerstone of a digital economy. Sidewalk Labs propose deployment of two primary types of digital infrastructure: ubiquitous connectivity and standardized mounts. Ubiquitous connectivity can become the sub-layer of a widespread and atomized digital technology, made of personal devices and urban sensors that can collect Urban Data. The term Urban Data refers to information gathered in the city's public realm, its publicly accessible spaces, and even some private buildings. An independent Urban Data Trust would oversee the exploitation of Urban Data in the district. (Sidewalk Labs, 2019a, pp. 323-329). The collection of data would be stored and shared as Open Data in an Open City Network contributing to face issues related to sustainability, mobility, public realm and performance of buildings.

The results of Big Data Analysis can drive measures capable of giving the built environment an adaptive capacity, thereby changing its urban form. In this case, the generative process can change continuously the morphology of a city in response to different types of stress. For example, based on traffic flow data provided by Remote Sensing and City Sensing, roads and infrastructures can modify their own section, favouring, as needed, private vehicular traffic, public traffic or pedestrian flows. Public spaces can expand in non-working time slots in view of the indications inferable from the analysis of Big Data related to how citizens use the city. The same happens with the detection of climatic conditions. In fact, depending on rainfall, snowfall, wind or solar radiation, buildings can change the configuration of their envelope. Facade systems are retractable and adaptive. While the traditional urban planning is based on baselines that statically govern the built environment, this new approach reconfigures and balances the urban standards based on needs. The paradigm of the monolithic city is undermined by the adaptive capacity conferred to the built environment by the technologies of the digital age and datafication, which contribute to make a city a real Smart City.

Big Data Analysis allows a multi-spectrum analysis. The application of the computational tool as a processor of an apparently uneven Big Data Analysis opens up new scenarios. This holds true both in terms of predictiveness of the possible strategies to be adopted and in terms of on-demand responsiveness. It will in turn generate an adaptive process that can drive either short- or long-term transformations of a city, e.g. daily, depending on the most urgent needs (Figures 16, 17). Streets can widen and modify themselves adapting to vehicular traffic or pedestrian flows. Building Envelope can adapt to climate and changing environmental conditions. Public spaces can get wider or smaller in relation to time slots or the day of the week. Hubs can stock and share energy depending on demand, e.g. supporting domestic rather than public use. Transformations projected in the long run can improve mobility infrastructures, building massing, the establishment or shifting of economic activities or resource allocation.

Generative Urban Planning Through Big Data Analysis

The atomized technology of Industry 4.0 sensors allows to gather information all over a city. Remote Sensing provides information acquired by satellite or aircraft-based sensor technologies to detect and classify objects on the territory. City Sensing, then, exploits small, miniaturized, portable or low-cost personal technological detection devices to penetrate deeply into complex contexts, such as urban ones, thus allowing to analyse them from within. The dissemination of numerous connected points of acquisition throughout the territory has changed the way cities are explored and used, i.e. inhabited. Inhabiting multiple spaces virtually, simultaneously and in real-time, thanks to the development of ICT, allows to qualify and connote the places in relation to one's habits and needs and to share the experience. The physical space of contemporary society is configured as a real substrate of a virtual and digital space of relationships, located on the web and characterized by its own rules and dynamics. This virtual space is the *medium* which real individuals use to interact (Caldera, C., Ostorero, C.L., Manni, V., Galli, A. & Valzano, L.S., 2020).

Hence, interconnected urban space, but not only urban, is crossed by continuous data flows coming from a myriad of personal technological devices. By sharing georeferenced information collected in person, the users of digital devices become themselves 'sensors' of ongoing phenomena. This in turn allows to monitor complex phenomena that otherwise could not be observed (Bolognesi & Galli 2017). Data Visualizations try to visualize such data flows contributing to the interpretation of urban phenomena. The City Model, a virtual entity consisting of digital data of a city, is a multi-resolution object accessible in real-time. City Sensing and City Model mutually confer meaning and effectiveness, supporting multi-actor interactions and territorial governance processes.

As Borga states in his book "City Sensing" (2013), the level reached by New Technologies in the field of acquisition and processing of territorial data is significant, but in public administrations technical and cultural evolution did not take place so as to fully exploit those innovative tools of knowledge. The Open Government doctrine aims to provide open information to citizens for the sake of participation in the decision-making process. The use of ICTs allows to acquire and share Open Data, i.e. information collected and freely provided to citizens by public administrations. Open Data could also be considered as a movement based on a set of principles to make data accessible and intelligently open: its goals differ from the ones of Big Data. In fact, Big Data is mainly collected by private companies for business purposes (Arcella, 2018). If Big Data were shared by virtue of benefit agreements between private companies and public administrations, it could contribute considerably to territorial governance. The huge amount of Big Data is, however, unmanageable when applied to traditional decision-making, and particularly to the consolidated urban planning methods still used by public administrations.

Hence, Big Data Analysis can support information management in innovative generative computational multi-layer planning methods, going beyond the limits of the traditional approach. Generative Computational Urban Planning through Big Data Analysis, based on algorithmic calculation, can exploit the acquisition and analysis of Big Data to generate multiple scenarios that can be corrected and validated by human decision-makers to meet stakeholders' interests. The real-time comparison of alternative design scenarios obtained by varying the mutual influence of system parameters allows to evaluate the real needs while maintaining consistency with the preset goals and optimizing the results.

This process is data-driven, i.e. it uses the information derived from data analysis to guide the development of new and more efficient services, products, organizational processes and methods. In data-driven approaches, the selection of data is intrinsically a real project operation, since it influences the result directly. So, in the urban planning of a city, one will first have to decide what information the city should provide and how it ought to be used.

A possible set-up of a process of Generative Computational Urban Planning through Big Data Analysis is given (Figure 18). It updates a previous study and consists of the following main steps (Caldera et al., 2020):

Figure 18. Generative Computational Urban Planning flowchart. This scheme recalls and updates a previous work (Caldera et al., 2020)

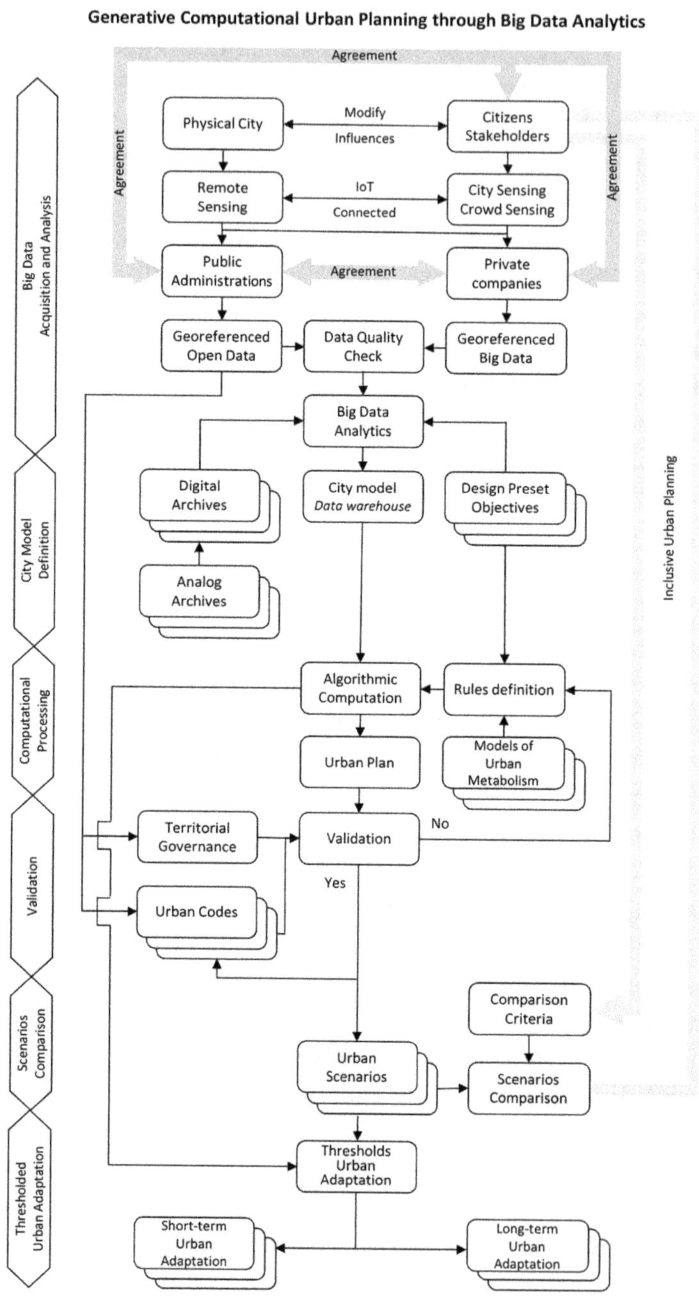

1) Big Data Acquisition and Analysis;
2) City Model Definition;
3) Computational Processing of the City Model (distribution of functions, density and massing);
4) Urban Plan Validation;
5) Comparison of Urban Scenarios;
6) Thresholded Urban Adaptation.

CONCLUSIONS

As mentioned before, cities represent one of the major problems our contemporary era needs to face from the viewpoint of both environmental issues (energy consumption, soil consumption, pollution, waste stocking and recycling etc.) and economic and social ones.

At some point in history, the ability to control the evolutionary processes of urban systems and the environment was probably lost (Borga, 2013). Loss of control of urban growth is the result of the acceleration of technological development since the First Industrial Revolution. In the past, the planning of our cities could not foresee either the technological progress or the growth and functioning of contemporary urbanization. Likewise it is difficult, today, to predict the complexity of future cities, as they are the product of phenomena not yet in progress or just in the making. Despite the degree of uncertainty of the forecasts for the city of the future, nevertheless, we cannot but consider the incipient or not yet completely manifest actions. The entire urban organism, therefore, must be given the capability to adapt, in response to its own metabolism, so that it can survive itself, as in the case of an autoimmune disease.

In the 60s, Wolman (1965) thought that a city could self-regulate. In fact, he defined a city as a constantly evolving metabolic organism which self-regenerates by optimizing its configuration over time depending on its needs and the availability of resources.

Lynch's basic assumption (1990) is that technology and advanced urban planning can solve every problem related to urban metabolism.

Though he stated that cities are the major destroyers of the ecosystem and the greatest threat to the survival of mankind on the planet, Richard Rogers (1998) believed that architecture and city planning could evolve to provide new crucial technological tools for safeguarding the future.

The need will arise to rethink city planning by exploiting new technologies. Generative Computational Urban Planning through Big Data Analysis, as Toronto's case study demonstrates, can exploit and evolve theoretical models of urban metabolism. Its aim is not only to generate an optimized initial urban form, but also to provide a capability as an adaptive city and generate otherwise hardly predictable design scenarios. The described adaptive capability of a city can be interpreted in analogy with the autopoietic processes typical of biological systems. The Generative Computational Urban Planning through Big Data Analysis can drive a city in analogy with self-regulating natural processes. It can be a new step in the development of a real on-demand design tool that can face the complexity of the metabolism of a city of the future.

In a traditional consolidated urban planning process (linear process) the availability of Big Data collections is useless, because they can't be reiteratively processed. In the contemporary multidisciplinary scenario the availability of data, the diffusion of IoT devices and computing tools are the drivers of a new approach. This applies to both urban planning and the continuous adaptation into the development of a city, as previously shown. In such a way, a city generates and extends its Digital Twin, a multi-dimensional and multi-layered digital model of the urban environment.

Possible scenarios may be foreseen, regarding both technical features and social and ethical issues. In fact, the exploitation of ICT gives voice (both in terms of expression of needs and in terms of information contributions) to local stakeholders initially excluded from top-down organizational models. The result is a sharing and democratization of bottom-up decision-making processes related to city planning that involve wider and wider social strata.

Nevertheless, in the exploitation of the potential of the new ICT lie, as limits, latent contradictions between the role and value attributed to, and held by, datafication and data management and the role of individual freedoms. The mutual acceptance and positioning of the boundary between citizens, public administrations, private companies and ICT Trust can balance mutual interests and inalienable rights.

These issues, therefore, encourage reflection towards both a holistic approach and a multidisciplinary involvement that relates not only to the technical-scientific aspects but, also as much, to the ethical-philosophical ones.

REFERENCES

Agicom. (2018). *Big Data - Interim Report Nell'ambito Dell'Indagine Conoscitiva Di Cui Alla Delibera n. 217/17/CONS*. Rome, Italy: Agicom. Retrieved July 30, 2020, from https://www.agcom.it/documents/10179/10875949/Studio-Ricerca+08-06-2018/c72b5230-354d-444f-9e3f-5467ca450714?version=1.0

Arcella, G. (2018, May 23). *Open Data e Big Data Pubblici, Cosa Sono e Perché Sono Connessi*. Retrieved July 30, 2020, from https://www.agendadigitale.eu/dati/open-data-e-big-data/

Aydin, S., & Schnabel, M. A. (2013). Coding Shape Grammars. Hints for generating a parametric design tool for large-scale urban renewal projects. In *Cutting Edge: 47th International Conference of the Architectural Science Association*, (pp. 177-186). The Architectural Science Association (ANZAScA), Australia.

Bolognesi, C., & Galli, A. (2017). The Representation of Socials in Maps: Map of a Great Event in Monza Park. In La Città Creativa; Spazi Pubblici e Luoghi Della Creatività, (pp. 627–35). Rome, Italy: CNAPPC.

Borga, G. (2013). *City Sensing. Approcci, Metodi e Tecnologie Innovative per La Città Intelligente*. Franco Angeli.

Caldera, C., Ostorero, C. L., Manni, V., Galli, A., & Valzano, L. S. (2020). *Cities in Transformation. Computational Urban Planning through Big Data Analytics. Techne, Special Issue, 2*. Firenze University Press.

Camagni, R. (1993). *Principi Di Economia Urbana e Territoriale*. Carocci Editore.

Carlo Ratti Associati. (2010). *Ka Care*. Retrieved February 30, 2020, from https://carloratti.com/project/ka-care/

Casi, F. (2019 January 28). *Una Nuova Frontiera: La Datizzazione*. Retrieved July 30, 2020, from https://www.consultadibioetica.org/una-nuova-frontiera-la-datizzazione-di-fiorello-casi/

Christaller, W. (1933). *Die Zentralen Orte in Süddeutschland.* Gustav Fischer Verlag.

Coppola, F. F. (Producer), & Reggio, G. (Director). (1982). *Koyaanisqatsi: Life Out of Balance* [Motion Picture]. USA: Island Alive.

Coppola, F. F. (Producer), & Reggio, G. (Director). (1988). *Powaqqatsi: Life in Trasformation.* [Motion Picture]. USA: Island Alive.

Coppola, F. F. (Producer), & Reggio, G. (Director). (2002). *Naqoyqatsi: Life as War.* [Motion Picture]. USA: QATSI Productions, LLC and Miramax Film Corp.

Davis, K. (1965). The Urbanization of the Human Population. *Scientific American, 213*(3), 41–53. doi :10.1038cientificamerican0965-40

EcoLogic Studio. (2011 January 10). *Ka-Care.* Retrieved March 12, 2020, from http://ecologicstudio.com/v2/project.php?idcat=3&idsubcat=66&idproj=121

Eeckhout, J. (2004). Gibrat's Law for (All) Cities. *The American Economic Review, 94*(5), 1429–1451. doi:10.1257/0002828043052303

Esposito de Vita, G. (2009). Le Fonti per l'Urbanistica del XXI Secolo: Luigi Moretti (1907-1973). In Luigi Moretti e La Fondazione Della Rocca Urbanistica e Ricerca Operativa, (pp. 15-76). Rome, Italy: Fondazione Aldo della Rocca.

Ferguson, F. (1975). *Architecture, Cities and the Systems Approach.* Braziller.

Frazer, J. (1995). *Themes VII: An Evolutionary Architecture.* Architectural Association.

Fusero, P., Massimiano, L., Tedeschi, A., & Lepidi, S. (2013). *Parametric Urbanism: a New Frontier for Smart Cities. Planum – The Journal of Urbanism, 2(27).*

Gabaix, X. (1999). Zipf's Law for Cities: An Explanation. *The Quarterly Journal of Economics, 114*(3), 739–767. doi:10.1162/003355399556133

Galli, A., & Massimiano, L. (2015). Progettazione Parametrica della Città Attraverso gli Open Data. *Urbanistica Informazioni, Proceedings of IX Giornata di Studi INU, Special Issue September-October 2015.* Rome, Italy: INU Edizioni.

Hristov, T. (2015 May 8). *Zha Kartal Pendik Masterplan Breakdown.* Retrieved February 4, 2020, from https://tshristov.wordpress.com/2015/05/08/kartal-pendik-masterplan-by-zaha-hadid-architects-work-in-progress/

Kang, D. W. (2012 August 13). *Kartal – Pendik Masterplan.* Retrieved February 4, 2020, from https://arcspace.com/feature/kartal-pendik- masterplan/

Labs, S. (2019a). *Toronto Tomorrow - A new approach for inclusive growth – Vol. 1 The Plans.* New York: Sidewalk Labs. Retrieved August 20, 2020, from https://sidewalk-toronto-ca.storage.googleapis.com/wp-content/uploads/2019/06/23135619/MIDP_Volume1.pdf

Labs, S. (2019b). Toronto Tomorrow - A new approach for inclusive growth: Vol. 2. *Urban Innovations*. Sidewalk Labs. https://sidewalk-toronto-ca.storage.googleapis.com/wp-content/uploads/2019/06/23135715/MIDP_Volume2.pdf

Lee, J., Bagheri, B., & Kao, H. A. (2014). Recent Advances and Trends of Cyber-Physical Systems and Big Data Analysis in Industrial Informatics. *International Conference on Industrial Informatics (INDIN)*.

Lösch, A. (1954). The Economics of Location: A Pioneer Book in the Relations Between Economic Goods and Geography. Yale University Press.

Lynch, K. (1990). Wasting Away. Sierra Club Books.

Mayer-Schöenberger, V., & Cukier, K. (2013). *Big data: a revolution that will transform how we live, work and think*. John Murray.

Meredith, S. (2018 May 17). *Two-Thirds of World Population Will Live in Cities by 2050 - UN Says*. Retrieved February 4, 2020, from https://www.theguardian.com/world/2018/may/17/two-thirds-of-world-population-will-live-in-cities-by-2050-says-un

Moretti, L. (1954). *Structure Comme Forme*. United States Lines Paris Review.

Negroponte, N. (1995). *Being Digital*. Alfred A. Knopf Inc.

Otto, F. (2011). Occupyind and Connecting - Thoughs on Territories and Spheres of Influence with Particular Reference to Human Settlement. Edition Axel Menges GmbH.

Otto, F., & Rasch, B. (1996). *Gestalt Finden. Auf Dem Weg Zu Einer Baukunst Des Minimalen*. Edition Axel Menges GmbH.

Parr, J. B. (1978). Models of the Central Place System: A More General Approach. *Urban Studies (Edinburgh, Scotland)*, *15*(1), 33–50. doi:10.1080/713702286

Parr, J. B. (2015). Exploring the Urban System of von Thünen's Isolated State. *Papers in Regional Science*, *94*(1), 1161–1175.

Parr, J. B. (2017). Central Place Theory: An Evolution. *Review of Urban and Regional Development Studies*, *29*(3), 151–162.

Remixstudio. (2012). *Overlapped City*. Retrieved February 30, 2020, from http://www.remixstudio.org/research/design/overlapped-city

Ritchie, H., & Roser, M. (2018). *Urbanization, Our World in Data*. Oxford, UK: Oxford University Press. Retrieved July 30, 2020, from https://ourworldindata.org/urbanization

Rogers, R. (1998). Cities for a Small Planet. Westview Press.

Rosslyn, I. (2018 July 24). Action on Overpopulation Still Urgent. *Victorian Humanist, 57*(6). Retrieved February 4, 2020, from https://vichumanist.org.au/environment/action-overpopulation-urgent/

Sola, G. (2009). *Natural Forms and Optimum Models*. Retrieved February 30, 2020, from https://www.domusweb.it/en/reviews/2009/03/05/natural-form-and-optimum-models.html

Spuybroek, L. (2005). *The Structure of Vagueness. Textile The Journal of Cloth and Culture, 3(1).*

Studio, A. (2010). *Ka Care*. Retrieved July 30, 2020, from http://www.atmosstudio.com/KA-CARE

Torricelli, G. P. (2003a). *Introduzione Alla Geografia Economica - Lezione 2*. Retrieved February 30, 2020, from http://www.gpt.adhoc.ch/Geoeconomica/Modulo2/Lezione2.pdf

Torricelli, G. P. (2003b). *Introduzione Alla Geografia Economica - Lezione 3*. Retrieved February 30, 2020, from http://www.gpt.adhoc.ch/Geoeconomica/Modulo2/Lezione3.pdf

United Nations Organization. (2018). *The World's Cities in 2018.* Retrieved February 10, 2020, from https://www.un.org/en/events/citiesday/assets/pdf/the_worlds_cities_in_2018_data_booklet.pdf

United Nations Organization DESA. (2014). *World Urbanization Prospects*. Retrieved February 10, 2020, from https://population.un.org/wpp/

von Thünen, J. H. (1842). Der Isolierte Staat in Beziehung Auf Landwirtschaft Und Nationalökonomie, 2018th ed. London, UK: Forgotten Books.

Wilson, L., Danforth, J., Cerezo Davila, C., & Harvey, D. (2019). How to Generate a Thousand Master Plans: A Framework for Computational Urban Design. In *Proceedings of SimAUD 2019, Society for Modeling & Simulation International (SCS)*. Society for Modeling & Simulation International (SCS).

Wolman, A. (1965). The Metabolism of Cities. *Scientific American, 213*(3), 179–190. doi:10.1038cien tificamerican0965-178 PMID:14338520

Zaha Hadid Architects. (2006). *Kartal Masterplan*. Retrieved February 30, 2020, from https://www.zaha-hadid.com/masterplans/kartal-pendik-masterplan/

ENDNOTE

1 Urbanization has largely been confined to the past 200 years. By 1800, still over 90 percent of the global population lived in rural areas. World's urbanization began to increase rapidly through the 19th century with the First Industrial Revolution.

Chapter 3
Digital Twins Concepts, Challenges, and Future Trends

Amged Sayed A. Mahmoud
Department of Industrial Electronics and Control Engineering, Faculty of Electronic Engineering, Menoufia University, Egypt

Ezz El-Din Hemdan
Department of Computer Science and Engineering, Faculty of Electronic Engineering, Menoufia University, Egypt

ABSTRACT

Digital twin has gained a great interest during the last few years from academia and industry because of the development in IT technology, communication field, and sensor technology. The general vision of the DT is to provide a detailed physical and functional description of a component, product, and systems. Nevertheless, the digital twin is a highly dynamic concept growing in complexity during the product life cycle, which leads to an enormous amount of data and information. Motivated by these, this chapter investigates the concepts and architecture of DT to cover its challenges and explore its applications in various fields such as smart cities, smart manufacturing and factories, and healthcare sectors. In the end, the challenges and research areas will be presented.

INTRODUCTION

During the last decade, the growth of the Internet of Things (IoT), Cyber-Physical System and communication technologies are shifting the way to build a system and way to transfer information between different sources. Indeed, digitalization leads to a new revolution in all aspect of life. Owing to this development and revolution in Information Technology and Industry 4.0, Digital Twins (DT) has become one of the promising concepts nowadays (Barricelli et al., 2019; Fuller et al., 2020; Jones et al., 2020; Negri et al., 2017; Qi & Tao, 2018; Rasheed et al., 2020; Tao, Zhang, et al., 2019; Uhlemann et al., 2017). The Digital Twins (DT) is a virtual representation of the physical products in the context of Cyber-Physical Systems. Hence, DT is a digital counterpart of a certain product that simulates its real behavior. The data from the physical system is represented by the digital information come from the virtual system which is

DOI: 10.4018/978-1-7998-7091-3.ch003

connected to a real system throughout entire lifecycle process. Therefore, the integration of digital with its physical twins creates an effective way to manage, regulate and improve the decision making while the real system is running (Rosen et al., 2015). The Digital Twin stores all information gathered from the sensors and the output of the execution system. Thus, the main significance of DT is to prognosis and diagnosis the behavior of the real system to predict any fault or error and provide data back to the system to provide the optimal maintenance (Boschert & Rosen, 2016).

The Digital Twin 's general vision addresses an extensive physical and functional definition of a device, assets, or a system. Its main purpose is to introduce a virtual representing of the system to increase productivity, make the operation more flexible, fault isolation and reduce the cost and time of the maintenance. Then, DT can be implemented based on the application either on the system or available on a cloud or edge computing that can tackle Industry 4.0 's challenges which occur during creating connected elements. On the other hand, DT is providing both an explanation of the behavior for the practical system and optimal solutions for the real model. Hence, a simplified modelling system, simulation processes, and transparent simulation are included in DT to improve control action, support decision making, and predict the future behavior of the system (Alam & El Saddik, 2017; El Saddik, 2018; M. Grieves et al., 2017; Schleich et al., 2017).

The concept of digital twins was first presented by Grieves in 2002 (M. Grieves, 2014) and its first application was done by NASA for building virtual models of its space machines. However, in only a few years DT technology is applied in many research fields and popular in many industries and business environments. In Literature, it is showed that existing DT work and implementation is still in its early stage and need a lot of efforts. However, the Digital Twins technology is successfully integrated into many applications such as manufacturing, Smart Cities, Healthcare, aerospace industry, agriculture, weather forecasting and many real-time monitoring systems (Jones et al., 2020; Rasheed et al., 2020).

The digital twin is integrating new technology such as IoT, Artificial Intelligence (AI), Cyber-Physical Systems (CPS), Big data, and sensor data fusion to create a digital information structure that is updated and altered as their physical counterpart change (Alam & El Saddik, 2017; Huang et al., 2020; Lee et al., 2020; Negri et al., 2017; Tao et al., 2018). Then to build a Digital twin for a physical product, a team of engineers and computer scientists are needed. The engineers who will design the prototype of the required product or system, whereas the computer scientists are responsible to generate the precise description of the virtual system. The physical product and the digital shadow must be identical in their features and characteristics. Then Digital twins are programmed to collect data from sensors during the manufacturing of the product to provide real-time information and the feedback to the process that explain the behavior of the a simulated model. For example, Grieves says, "we can collect the torque readings of every bolt that attaches a fuel pump to an engine in order to ensure that each engine/fuel pump attachment is successfully performed.

This chapter is organized as follows. The concepts and characteristics of a digital twin are introduced in Section 3. Then in section 4, the construction and benefits of Digital Twins are given. In Section 5, the applications in manufacturing, smart cities and healthcare are presented and the challenges and Future directions and gaps in research are examined in section 6. followed by conclusions.

Concepts and Characteristic of Digital Twins

The first approach of Digital Twins is back to NASA's Apollo program in the 1970s, there was two identical space, one on earth for training the flight and the other one was launched to simulate the behavior with one on earth. However, the basic concept of digital twin model was firstly presented by Grieves in 2002 for product life cycle management (PLM) (M. W. Grieves, 2003). Its basic idea is that the digital simulation model is constructed for the physical system and the digital information from this model is Twined with the data from the real system. This digital model is linked to the system during its entire life. Therefore, the Digital Twin is continuously monitoring the behavior and performance of the product/system and then determine when and where failure is expected to happen thus to give the optimal solution. Hence the digital Twins is considered a dynamic concept to cope up with the variations in the real systems. Relevantly, the digitalization of product lifecycle management (PLM) increase the effectiveness of the physical product and ensure reliable and fast operation.

Practically, the DT is consisting of three main component that must be taken into consideration: the real system, the virtual model and the communication link that feed the information from the virtual model to the real system and the data from the real system to the virtual model. The general frame of Digital Twins is depicted in figure.1. A group of engineering and computer science are creating the Digital Twins for a physical object by first build a prototype and a mathematical model. Then they build a software simulation that is optimally duplicate and identical for the object. This digital representation is received measured data from sensors of the real-time system using a various technique of communication and visualize it. Then this mimic representation produces a feed signal and digital information back to a physical object to improve the performance and to describe how the object would act in a real model. Therefore, a DT is considered an advanced system that can simulate, observe, analyze, adjust, control, and optimize the system behavior and process.

Digital Twins have certain characteristics that distinguish them from other technologies. One of the main characteristics of DT is the reflection of the real-time behavior of the object with the digital process that make it smarter and reprogrammable(Negri et al., 2017). As the virtual representation is developing

Figure 1. Concept of Digital Twins

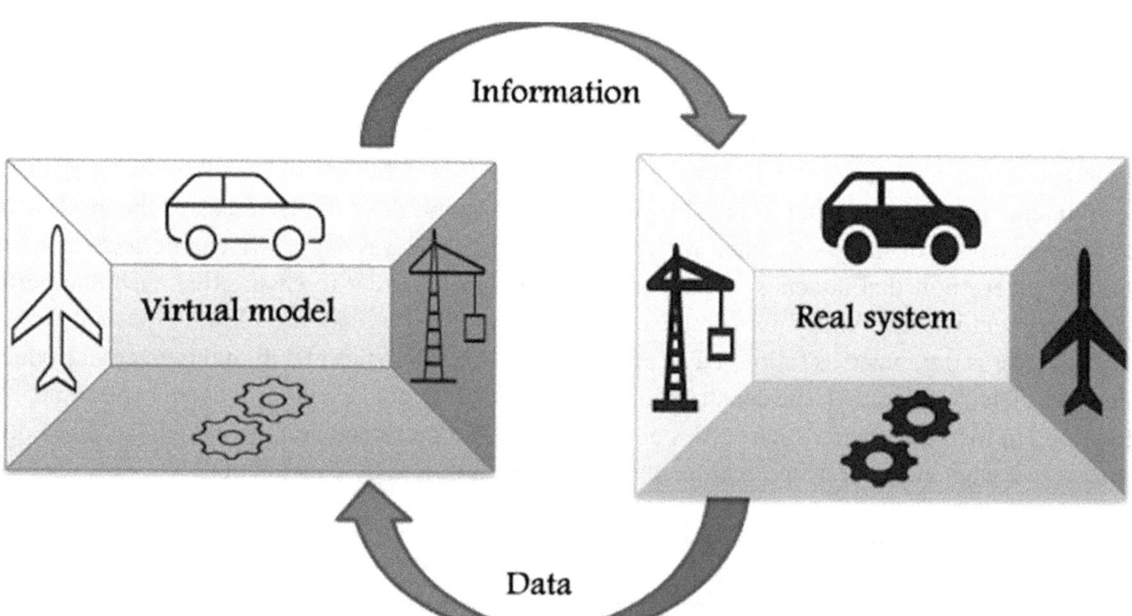

simultaneously with the real system during the running process using AI and predictive analysis and synchronize the new information with the physical environment to improve the functionalities(Tao et al., 2018). Another important characteristic is the homogenization of data within a Digital Twins. The generated data produced in various spaces of the physical environment can interact with each other. Also, the dynamic data and historical data are interacting with each other to provide the required action. Therefore, due to huge amount of data available, DT must be integrated with analysis technique, Big data technique and data fusion algorithms to handle these data and to provide accurate and useful information (Qi & Tao, 2018). Hence, the Digital Twin allow smooth connection between the virtual and physical system and make more information on physical objects to be stored digitally and become derived from the object itself.

Another important characteristic of The DT is the digital traces for diagnosis and prognosis (Cai et al., 2017). The usage of sensor data fusion and feature extraction is integrated to determine where the fault occurred and use this diagnosis to improve the design of the system so that the same error and problem will occur less in future. Also, DT has an ability of self-adaptation and self-parametrization capabilities (Schleich et al., 2017), this characteristic allows to tweak and improve the physical model. With this modularity ability of DT, the changes in one module will not affect other modules in the process and it is easy to replace the poorly performed module with a better component to improve the whole process. Hence, the modularity ensures an easy modification and self-evaluation of the physical system. Besides, the optimization algorithms are applied within DT technology to achieve the best output by using a set of objectives, criteria, and constraints to determine a set of high-value alternative actions or decisions. These prescriptive analytics are one of DT's characteristic and is responsible to take an action relevant to its destiny without the need of operator for handling the processing .

DT technology encounters emerging technology such as IoT and AI (Rasheed et al., 2020). Implementation of IoT with DT enables good visibility of physical Twins and the status of their function. Also, it facilitates the connection and documented the function of the system to understand and explain the performance of the physical devices. AI that refers to supervised / non-supervised learning algorithms, which constantly improves its ability by manipulating data received from the physical twin and surrounding environment. This leads to detect the patterns, classify the information, and identify the model through data analysis and prediction. Finally, the DT can interact with the model and simulate the application realistically and differentiate between the real and physical object using what-if analysis. Recently, the integration of new technology like Deep Learning and Blockchain has facilitated the usage of Digital Twins in practical application (Huang et al., 2020; Lee et al., 2020). Deep Learning will provide the required functionality such as classification, regression, clustering, and pattern recognition of the huge data and information generated from the DT to produce a required action.

Construction and Benefits of Digital Twins

Digital twin (DT) is emerging technology nowadays due to its significant influences on linking the gap between the physical, digital environment, data and services (Figure2). According to Gartner Research (Costello & Omale, 2019), DT is expected to be used by two-third of companies which implementing IoT already in their operation in the next few years. Many companies provide Digital Twins solutions such as Microsoft Azure IoT, GE predix, IBM, Cisco Systems, Oracle Cloud SAP Leonardo (Figure.3) and other industrial vendors. In the Literature review, work and development on DTs are expected to undergo a further boom over the next 3–5 years.

Figure 2. Vision of Digital Twins in PLM (M. Grieves et al., 2017)

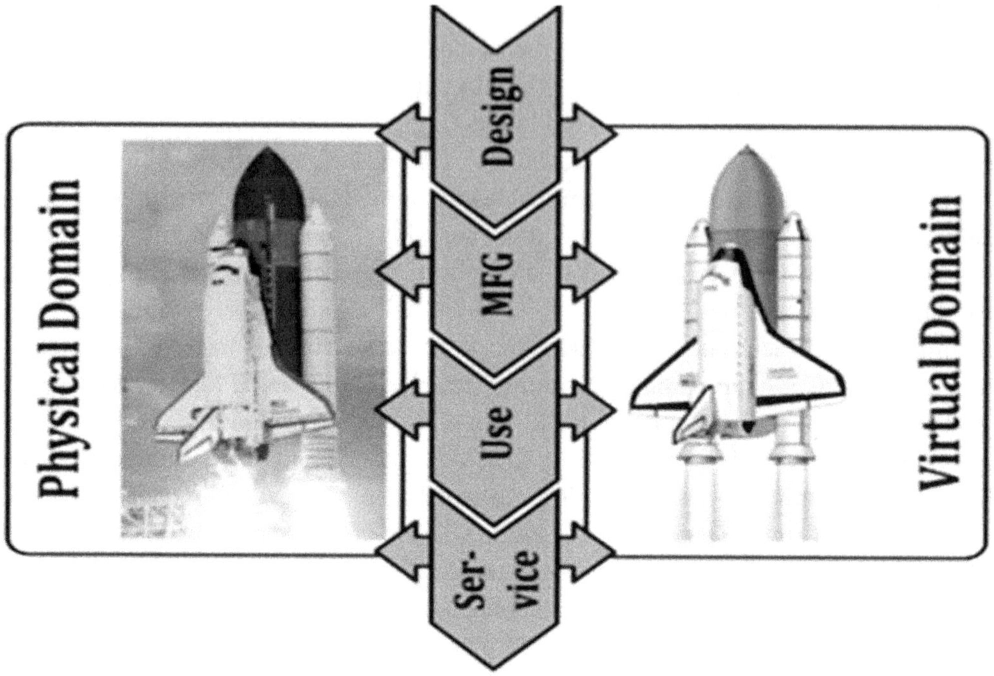

The Digital Twins is a combination of different such as computer science, engineering and communication scientist production engineering. The proposed Digital Twin model is composed of major parts including digital simulation, data fusion and knowledgebase (Tao, Zhang, et al., 2019). The Digital simulation is digitally replicating the physical objects and its components, modelling the performance and abnormalities of the system, and modelling the communication. It explains the configuration of subsystems, subassemblies, and modules and generates a virtual representation of each device with the obtained sensing measurements from production, activities, and operation. To increase the system's fidelity, a virtual sensor can be created inside the digital model (Wang et al., 2019). By identifying and classifying the main component of physical objects, digital modelling should be a reflection of the real objects. Besides, the communication techniques, data storage, information transmission, data format conversion are essential for digital simulation and facilitate data processing in DT. The mathematical modelling and simulation are required for operation analysis and validate the virtual model of the process.

The second important part of DT is Data fusion is used to diagnose, predict, and prescribe the behaviour of the physical system by preprocessing, and optimizing the data. It performs processing for a massive amount of data and information to analyse the performance of the physical objects. This analyzing data is passed through knowledge rule-based for the diagnosis and prognosis troubleshooting in the physical systems. Also, this helps in determining the appropriate action, setting the ideal maintenance's instructions and improve behavior of the physical system. During the life cycle of the process, digital modelling is dynamically updating and adapting based on analyzed data.

Interactions and services are the last important part of DT to tackle complex system. Collaboration between the parts allows us to carry out a complicated function that can not be carried out by any particular unit, to exchange crucial knowledge between the parts and continuously refining the virtual

Figure 3. An Example of Vendor for Digital Twins [SAP Leonardo portfolio]

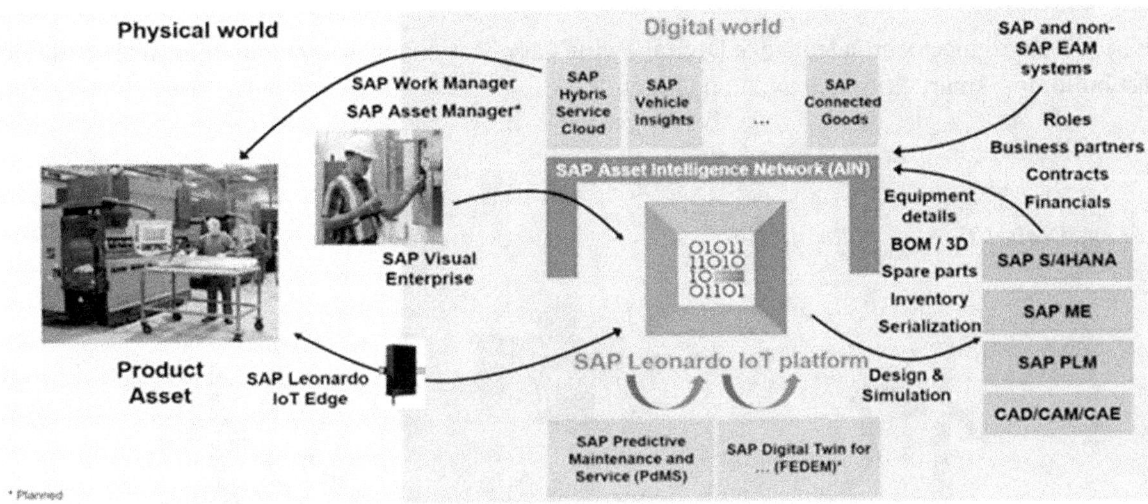

model. Services include modelling, structure monitoring, fault-tolerant and service optimization. With service principles, Digital Twins describe a physical object and identify the failure information then recommend the suitable repair for it. As a result of the foregoing, an important question appears: Why the Digital Twins will be the backbone of industry in the future? The fundamental principle of Digital Twins is to replica the physical sources with information and data. Hence, the main merit of using DT is to increase productivity and reduce the cost of the operation. One of DT's characteristic is self-adaptation and self-learning. This help to modify the process and give a future vision of the improvement for the system to be designed. So, the information and data from the DT will be used when prepared to design the new product, update the revision, or offering the new system to decrease the problem encountered during the operation or product life cycle. Another important advantage is the real-time remote control of the physical system. As the operator can access and monitor the operation of the process at any time and anywhere. Besides the accessibility, this will improve the performance and increase the safety as the operator will remotely manage and monitor the process without the need of existence in the physical field.

Reducing the time of repairing, improving maintenance process, and predicting the failure are another merits of employing DT. As Digital Twins collects the data from sensors and analysis it to diagnosis the behavior of the physical object and predict the optimal fixing. Then it classifies the data to prognosis and schedules the necessary maintenance in the field. Also, the Digital Twin continuously estimates the wellbeing of the object, the lasting valuable cycle, and the possibility of operation ability. So, it can diagnosis and tolerate the fault to prevent predicted and unpredicted catastrophic operations in the future and the spare part will also be easily estimated. Likewise, the simulation models and operating history was given by the Digital Twin are also the foundation for more dynamic service development alongside the data from the online measurement. Likewise, DT provides better communication and documentation of the process so that the stockholders can easily see the real-time report and the available information.

Applications of Digital Twins

This section introduces a broad range of Digital Twins's applications such as manufacturing, Health-care, smart buildings, smart cities and aviation (Figure 4)

Figure 4. Digital Twin's Applications

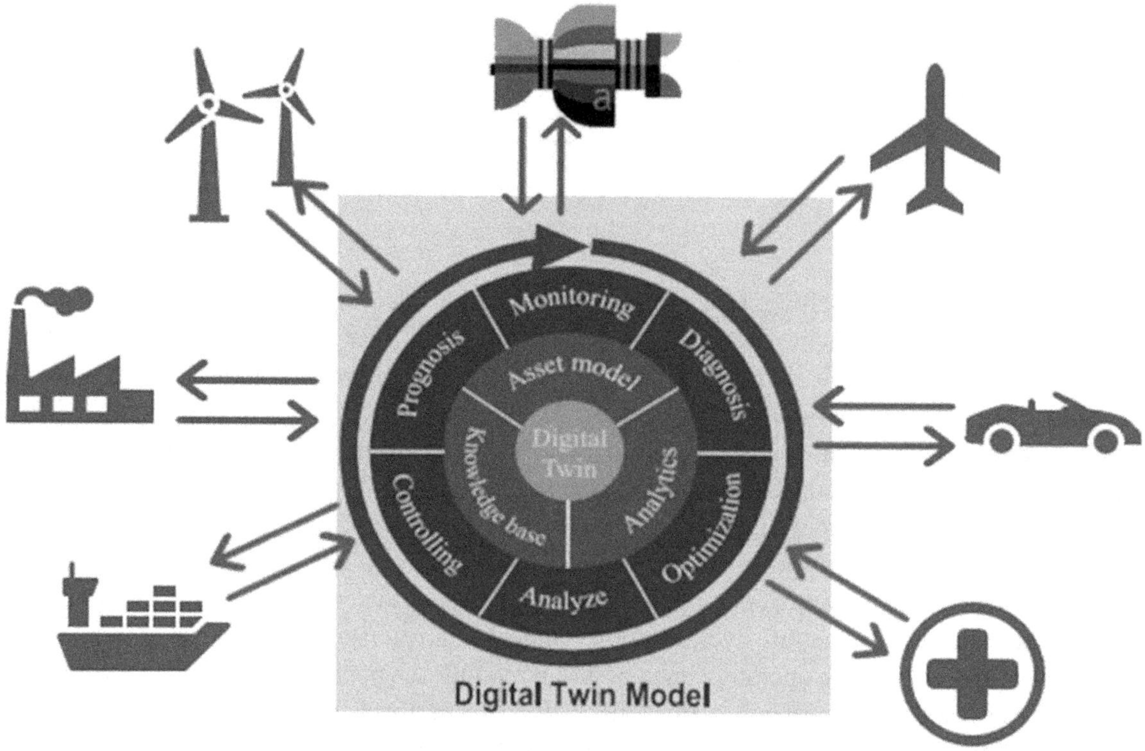

Smart Manufacturing

The industry and smart manufacturing is the main sector where the Digital Twins is applied. In terms of manufacturing, the application of DTs is concentrated on the field of Product lifecycle systems (PLM) such as plan and manufacture. It is also applied for prognostics and well-being monitoring during operating life. So, in a lifecycle approach, the DT's technique is to collect and analyze data from a complete life cycle product/process and then using this data to upgrade the same product/process (Hinchy et al., 2019; Negri et al., 2017). In (Qi & Tao, 2018), the details overview of the application of Digital Twins for Big data and smart Manufacturing is presented. It provides the difference between emerging technologies and Digital Twins then it gives the importance of DT in developing smart manufacturing. The historical perspective of DT in product manufacturing and the product life cycle is highlighted in (Tao, Zhang, et al., 2019). The authors presented the main fields DT is applied in the production and manufacturing of the product. The solution of data management problems for digital twin of a product based on blockchain technology is given in (Huang et al., 2020). The proposed product data management is based on a constructed network to connect all component involved in the product life cycle and then Blockchain is used to ensure the authorized person to access the stored data. (Tao, Sui, et al., 2019) presents a new framework to design a product called Digital Twins product design. It promotes the manufacturing process by guiding companies to design a digital twin and then use the data provided by the digital twin.

The emerging AI technology such as Deep learning with Cyber-Physical System (CPS) is applied accompanying with digital twins to enhance the performance and improve the production of the manufacturing in (Lee et al., 2020). It provides a vision towards smart manufacturing and Industry 4.0 by using the new architecture utilizing the new technologies. (Booyse et al., 2020) proposed new framework using Deep learning and Digital Twins to fault diagnosis and fault prognosis. By monitoring the deterioration of the physical object and comparing between the normal and failure, the training algorithm using Deep learning can easily detect the failure in the process. In (Wang et al., 2019), the fault diagnosis for a rotating machine using Digital Twins is introduced. The experimental work is given to prove the ability of DT to detect the fault and correct the unbalanced model. Design and production can be synchronized through DTs. (Cheng et al., 2020) proposed a new DT with the Industrial Internet to enhance the vision towards smart manufacturing. They considered that the application of digital twins is the core for developing the Industrial Internet and this will improve the interaction between users and the industry chain.

In (Zhou et al., 2019), DT is proposed to be applied in the power grids. With the application of DT, the time required for analyzing of the large power grid is reduced and the decision-making for the operation is improved which will reduce the time-cost value as compared to other technologies such as CPSs. Recently, the developing of DT in power electronics in introduced in (Milton et al., 2020). The layer of the converter is simulated and visualized to diagnostic analysis with integrated the DT in FPGA. The comparative analysis with the real-time action is provided to detect abnormalities in the power electronics. The Experimental Digital Twins (EDT) proposed in (Schluse et al., 2018) to achieve a seamless connection between physical and virtual spaces and to develop representation field. The control algorithm, intelligent system and simulation-based technique are realized by the application of EDT. The conceptual work for 3D printing using Digital Twin with machine learning and Big Data is presented in (Mukherjee & DebRoy, 2019). In this model, computational demanding time for trial and error is shortened and the time-to-production of the product is fall.

Smart Cities

The digital twin technologies turn out to be more intelligent to present effective tools in smart cities. The research in this field has been significantly grown as the development of IoT and communications which lead to motivation in the digitalization of our lives. (Austin et al., 2020) explores the trend technologies to provide the Digital Twins for Smart Cities. A Knowledge-base approach with machine learning is presented in the proposed architecture to developed an identification and decision-making activities for the management of the infrastructure and the energy in a metropolitan area. In (Laamarti et al., 2020) a framework using Digital Twins is used to develop smart healthcare services of the individual in the smart cities. An experimental result shows the efficiency of the proposed technique by twining the data gathered then analyzing and storing the information.

Energy management using Digital Twins in smart cities, (Francisco et al., 2020) proposed a benchmark from analyzed the data of smart meters to develop an energy building guideline of daily used. The guideline will decide how the new assumption vary from traditional and annual energy assessment methods and to examine how these steps can contribute to energy management in approximately real-time. (Fan et al., 2019) introduces a vision for disaster management using DT technology. The proposed paradigm incorporated the ICT technology with DT to provide a powerful pledge for boosting the functioning of disaster management activities. Application of Digital Twins in livestock farm is presented by (Jo et al., 2018) to produce a feasibility study to make a smart farm to improve the welfare. It is mentioned the significance of Industry 4.0 for the introduction of this initiative, as it allows the delivery of Digital Twin a problem in diverse systems.

Healthcare

One of the winning fields from applying the Digital Twins' concepts is healthcare. The emerging healthcare IoT, wearable devices, and Telemedicine technologies have underpinned the usage for integration Digital Twins in medical and e-health sectors. Digital Twins can used in the medical sector to predict the maintenance of medical devises, diagnosis and suggest the treatment, identify the lifestyle's changes, optimize the hospital functions, remote surgery facility and support the governments for the patient's healthcare. A DT virtualized biomedical model will help physicians to make expectation in vitro methods about how the actual organ will function in any given situation. (Bruynseels et al., 2018) explore how the emerging implication of Digital Twins in the healthcare paradigm treatment, diagnosis, and wellbeing. For the clinical field, it emphasizes that the building of human DT and the data from it can be used in physiotherapy, enhancement, lifestyle, nutritional habits and expect the occurrence of a disease. El-Saddik in his paper (El Saddik, 2018) presented that the usage of DT in the medical field would facilitate the monitor and diagnosis to enhance healthcare and wellbeing of the living thing like nonliving objects.

Saracco et al. explored in the white paper, the usage of the Digital Twins in Epidemics and case study Covid-19 (Saracco et al., 2020) and defined the concepet of Personal Digital Twins (PDT). Many companies like General Electric, Siemens, Philips, enter this field due to the importance and merit of the control of pandemic and its effect on the economy. The usage of PDT has a great role to real-time monitor of health signals, trace the physical location of the person, and specifying the healthcare resources available for infected people. For medical surgeries, DT with trend technology such as Industry 4.0, Virtual reality (VR) and IoT have been used to make a prototype in autonomous medical surgery (Laaki et al., 2019). The prototype uses a robotic arm, VR connected via a 4G mobile communication, to proceed a remote surgery for patients. The authors discuss the complexities and difficulties of integration Digital Twins with a prototype and how new technologies ease these challenges. (Liu et al., 2019) introduced a framework for providing healthcare services by integration Digital Twins with cloud technology to build a frame that supports to observe, examine, and predict a patient's condition. The IoT wearable sensor and in-home sensor were used to expect the illness and the difficulty of the elderly people with the help of DT and Cloud technology.

The Challenges and Open Research in Digital Twins

Despite the advantages of using digital twins in many aspects of our life, there are many obstacles and challenges that still open for the researchers and scientists (Barricelli et al., 2019; Fuller et al., 2020). One of the major problems of applying is the high cost problem for large companies, as the requirement of various software layers besides layers of hardware components to build a DT. Therefore, a lot of research has been conducted to design a low cost or develop a common DT platform in manufacturing and medical field.

Security and data privacy are other major challenges in Digital Twins. Because the visualization of information and data from the process, securing this data against viruses and hacking is one of the vital issues in DT as it can damage all information in digital environments. Although new technologies such as Blockchain has been applied to give a solution for securing problem in DT, this field is still open research area. Data management and storage are other challenges, as there is a huge amount of data and information from/to physical objects. Big data cloud technologies, Artificial intelligence and sensor data fusion are techniques used to pave the way to get tools that help in solving this issue.

Real-time communication and the data latency are technical problem DT can be encountered when working in physical environments. New technologies are been utilized to deal with this problem such as 5G communication, IoT technology and data compression. Also, another problem arises when applying DT is the need for continuous upgrading and updating of the digital environments. Machine Learning technique such as Deep Learning algorithms could potentially have the accomplishment to enrich and enhance the Digital Twins technology in this aspect. Standardization and ethical issue are major challenges in DT. Till now, there is no standardization for the use of DT whether in industries or health sectors. The ethical issue is research area as it is very important, especially in healthcare and medicine, as access to high-fidelity data that containing crucial information of the patient and his lifestyle needs ethical rules to deal with it.

CONCLUSION

Digital Twin is a highly dynamic concept and it is growing complexity during the product life cycle that leads to an enormous amount of data and information. This chapter presented the concept and architecture of the Digital Twins and the values of its utilizing. It is concluded that the DT is emerging technology nowadays due to its significant influences on linking the gap between the physical, digital environment, data and services. So due to its merits, it has been developed and used in many fields and effectiveness to digitalize the physical environments, such as manufacturing,smart cities and healthcare sector. It is also emphasize that despite the advantages of using digital twins in many aspects of our life, there are many obstacles and challenges that still open for the researchers and scientists..

References

Alam, K. M., & El Saddik, A. (2017). C2PS: A digital twin architecture reference model for the cloud-based cyber-physical systems. *IEEE Access: Practical Innovations, Open Solutions*, *5*, 2050–2062. doi:10.1109/ACCESS.2017.2657006

Austin, M., Delgoshaei, P., Coelho, M., & Heidarinejad, M. (2020). Architecting Smart City Digital Twins: Combined Semantic Model and Machine Learning Approach. *Journal of Management Engineering*, *36*(4), 1–14. doi:10.1061/(ASCE)ME.1943-5479.0000774

Barricelli, B. R., Casiraghi, E., & Fogli, D. (2019). A survey on digital twin: Definitions, characteristics, applications, and design implications. *IEEE Access, 7*(MI), 167653–167671. doi:10.1109/ACCESS.2019.2953499

Booyse, W., Wilke, D. N., & Heyns, S. (2020). Deep digital twins for detection, diagnostics and prognostics. *Mechanical Systems and Signal Processing, 140*, 106612. doi:10.1016/j.ymssp.2019.106612

Boschert, S., & Rosen, R. (2016). Digital twin—the simulation aspect. In *Mechatronic futures* (pp. 59–74). Springer International Publishing. doi:10.1007/978-3-319-32156-1_1

Bruynseels, K., de Sio, F. S., & van den Hoven, J. (2018). Digital Twins in health care: Ethical implications of an emerging engineering paradigm. *Frontiers in Genetics*, *9*(FEB), 1–11. doi:10.3389/fgene.2018.00031 PMID:29487613

Cai, Y., Starly, B., Cohen, P., & Lee, Y. S. (2017). Sensor Data and Information Fusion to Construct Digital-twins Virtual Machine Tools for Cyber-physical Manufacturing. *Procedia Manufacturing*, *10*, 1031–1042. doi:10.1016/j.promfg.2017.07.094

Cheng, J., Zhang, H., Tao, F., & Juang, C. F. (2020). DT-II:Digital twin enhanced Industrial Internet reference framework towards smart manufacturing. *Robotics and Computer-integrated Manufacturing*, *62*, 101881. doi:10.1016/j.rcim.2019.101881

Costello, K., & Omale, G. (2019). *Gartner Survey Reveals Digital Twins Are Entering Mainstream Use*. https://www.gartner.com/en/newsroom/press-releases/2019-02-20-gartner-survey-reveals-digital-twins-are-entering-mai

El Saddik, A. (2018). Digital Twins: The Convergence of Multimedia Technologies. *IEEE MultiMedia*, *25*(2), 87–92. doi:10.1109/MMUL.2018.023121167

Fan, C., Zhang, C., Yahja, A., & Mostafavi, A. (2019). Disaster City Digital Twin: A vision for integrating artificial and human intelligence for disaster management. *International Journal of Information Management*, *102049*(November). doi:10.1016/j.ijinfomgt.2019.102049

Francisco, A., Mohammadi, N., & Taylor, J. E. (2020). Smart City Digital Twin-Enabled Energy Management: Toward Real-Time Urban Building Energy Benchmarking. *Journal of Management Engineering*, *36*(2), 04019045. Advance online publication. doi:10.1061/(ASCE)ME.1943-5479.0000741

Fuller, A., Fan, Z., Day, C., & Barlow, C. (2020). Digital Twin: Enabling Technologies, Challenges and Open Research. *IEEE Access: Practical Innovations, Open Solutions*, *8*, 1–1. doi:10.1109/ACCESS.2020.2998358

Grieves, M. (2014). *Digital Twin : Manufacturing Excellence through Virtual Factory Replication.* White Paper, March. https://www.researchgate.net/publication/275211047_Digital_Twin_Manufacturing_Excellence_through_Virtual_Factory_Replication

Grieves, M., Vickers, J., Kahlen, F. J., Flumerfelt, S., & Alves, A. (2017). Digital Twin: Mitigating Unpredictable, Undesirable Emergent Behavior in Complex Systems. In F.-J. Kahlen, S. Flumerfelt, & A. Alves (Eds.), Transdisciplinary Perspectives on Complex Systems: New Findings and Approaches (pp. 85–113). Springer International Publishing. doi:10.1007/978-3-319-38756-7_4

Grieves, M. W. (2003). PLM-Beyond lean manufacturing. SOC Manufacturing Engineers.

Hinchy, E. P., O'Dowd, N. P., & McCarthy, C. T. (2019). Using open-source microcontrollers to enable digital twin communication for smart manufacturing. *Procedia Manufacturing*, *38*, 1213–1219. doi:10.1016/j.promfg.2020.01.212

Huang, S., Wang, G., Yan, Y., & Fang, X. (2020). Blockchain-based data management for digital twin of product. *Journal of Manufacturing Systems, 54*(August), 361–371. doi:10.1016/j.jmsy.2020.01.009

Jo, S. K., Park, D. H., Park, H., & Kim, S. H. (2018). Smart Livestock Farms Using Digital Twin: Feasibility Study. *9th International Conference on Information and Communication Technology Convergence: ICT Convergence Powered by Smart Intelligence, ICTC 2018*, 1461–1463. 10.1109/ICTC.2018.8539516

Jones, D., Snider, C., Nassehi, A., Yon, J., & Hicks, B. (2020). Characterising the Digital Twin: A systematic literature review. *CIRP Journal of Manufacturing Science and Technology*, *2019*, 36–52. Advance online publication. doi:10.1016/j.cirpj.2020.02.002

Laaki, H., Miche, Y., & Tammi, K. (2019). Prototyping a Digital Twin for Real Time Remote Control over Mobile Networks: Application of Remote Surgery. *IEEE Access: Practical Innovations, Open Solutions*, *7*, 20235–20336. doi:10.1109/ACCESS.2019.2897018

Laamarti, F., Badawi, H. F., Ding, Y., Arafsha, F., Hafidh, B., & El Saddik, A. (2020). An ISO/IEEE 11073 Standardized Digital Twin Framework for Health and Well-Being in Smart Cities. *IEEE Access: Practical Innovations, Open Solutions*, *8*, 105950–105961. doi:10.1109/ACCESS.2020.2999871

Lee, J., Azamfar, M., Singh, J., & Siahpour, S. (2020). Integration of digital twin and deep learning in cyber-physical systems: Towards smart manufacturing. *IET Collaborative Intelligent Manufacturing*, *2*(1), 34–36. doi:10.1049/iet-cim.2020.0009

Liu, Y., Zhang, L., Yang, Y., Zhou, L., Ren, L., Wang, F., Liu, R., Pang, Z., & Deen, M. J. (2019). A Novel Cloud-Based Framework for the Elderly Healthcare Services Using Digital Twin. *IEEE Access: Practical Innovations, Open Solutions*, *7*, 49088–49101. doi:10.1109/ACCESS.2019.2909828

Milton, M., De La, C. O., Ginn, H. L., & Benigni, A. (2020). Controller-Embeddable Probabilistic Real-Time Digital Twins for Power Electronic Converter Diagnostics. *IEEE Transactions on Power Electronics*, *35*(9), 9852–9866. doi:10.1109/TPEL.2020.2971775

Mukherjee, T., & DebRoy, T. (2019). A digital twin for rapid qualification of 3D printed metallic components. *Applied Materials Today*, *14*, 59–65. doi:10.1016/j.apmt.2018.11.003

Negri, E., Fumagalli, L., & Macchi, M. (2017). A Review of the Roles of Digital Twin in CPS-based Production Systems. *Procedia Manufacturing*, *11*(June), 939–948. doi:10.1016/j.promfg.2017.07.198

Qi, Q., & Tao, F. (2018). Digital Twin and Big Data Towards Smart Manufacturing and Industry 4.0: 360 Degree Comparison. *IEEE Access: Practical Innovations, Open Solutions*, *6*, 3585–3593. doi:10.1109/ACCESS.2018.2793265

Rasheed, A., San, O., & Kvamsdal, T. (2020). Digital twin: Values, challenges and enablers from a modeling perspective. *IEEE Access: Practical Innovations, Open Solutions*, *8*, 21980–22012. doi:10.1109/ACCESS.2020.2970143

Rosen, R., Von Wichert, G., Lo, G., & Bettenhausen, K. D. (2015). About the importance of autonomy and digital twins for the future of manufacturing. *IFAC-PapersOnLine*, *28*(3), 567–572. doi:10.1016/j.ifacol.2015.06.141

Saracco, R., De Kerckhove, D., & Nisiotis, L. (2020). *The Role of Personal Digital Twins in Control of Epidemics.* An IEEE Digital Reality White Paper.

Schleich, B., Anwer, N., Mathieu, L., & Wartzack, S. (2017). Shaping the digital twin for design and production engineering. *CIRP Annals - Manufacturing Technology, 66*(1), 141–144. doi:10.1016/j.cirp.2017.04.040

Schluse, M., Priggemeyer, M., Atorf, L., & Rossmann, J. (2018). Experimentable Digital Twins-Streamlining Simulation-Based Systems Engineering for Industry 4.0. *IEEE Transactions on Industrial Informatics*, *14*(4), 1722–1731. doi:10.1109/TII.2018.2804917

Tao, F., Cheng, J., Qi, Q., Zhang, M., Zhang, H., & Sui, F. (2018). Digital twin-driven product design, manufacturing and service with big data. *International Journal of Advanced Manufacturing Technology*, *94*(9–12), 3563–3576. doi:10.100700170-017-0233-1

Tao, F., Sui, F., Liu, A., Qi, Q., Zhang, M., Song, B., Guo, Z., Lu, S. C. Y., & Nee, A. Y. C. (2019). Digital twin-driven product design framework. *International Journal of Production Research*, *57*(12), 3935–3953. doi:10.1080/00207543.2018.1443229

Tao, F., Zhang, H., Liu, A., & Nee, A. Y. C. (2019). Digital Twin in Industry: State-of-the-Art. *IEEE Transactions on Industrial Informatics*, *15*(4), 2405–2415. doi:10.1109/TII.2018.2873186

Uhlemann, T. H. J., Lehmann, C., & Steinhilper, R. (2017). The Digital Twin: Realizing the Cyber-Physical Production System for Industry 4.0. *Procedia CIRP*, *61*, 335–340. doi:10.1016/j.procir.2016.11.152

Wang, J., Ye, L., Gao, R. X., Li, C., & Zhang, L. (2019). Digital Twin for rotating machinery fault diagnosis in smart manufacturing. *International Journal of Production Research*, *57*(12), 3920–3934. doi:10.1080/00207543.2018.1552032

Zhou, M., Yan, J., & Feng, D. (2019). Digital twin and its application to power grid online analysis. *CSEE Journal of Power and Energy Systems*, *5*(3), 391–398. doi:10.17775/CSEEJPES.2018.01460

Chapter 4
A District Heating Socio-Technical System Approaching the Energy Transition:
Issues of Energy Data Flows at the Urban Level

Osman Arrobbio
Università di Torino, Italy

Dario Padovan
https://orcid.org/0000-0002-5280-1120
Università di Torino, Italy

Alessandro Sciullo
Università di Torino, Italy

ABSTRACT

This chapter describes the results of a sociological investigation carried out within an EU-funded project. The project was aimed at creating a tool to visualise and compute energy data at an urban district level, with the broader aim to optimise the local district heating (DH) network's distribution policies. This chapter identifies the features of the main categories of actors (from the DH operator to final users) having a role within that network. Special attention is paid to the identification of the barriers and frictions preventing a stronger collaboration and communication among these actors to happen. It is argued that the identification and resolution, in situated and complex socio-technical systems, of these non-strictly-technical problems may be, at least in some cases, a pre-requisite for any ICT-based solution to deploy its full potential.

DOI: 10.4018/978-1-7998-7091-3.ch004

INTRODUCTION

Significant efforts are delivered to make urban areas a supporting element of the ecological transition (Broto, 2017; Frantzeskaki et al., 2016). One of the paramount fields concerning this issue relates to resource use, among which energy. The ecological transition entails the so-called energy transition. In its merely physical aspect, 'energy transition' is used to mean the passage from a certain energy mix to another energy mix (Smil, 2010). In the field of environmental and energy policies, 'energy transition' is mostly used to mean the transition from an energy mix mainly based on non-renewable sources, to one which is mainly, or more, based on renewable sources. However, the energy transition is not just a set of physical, chemical, and mechanical processes (Markard, 2018; Millet et al., 2013; Sovacool, 2016).

The energy transition is aimed at answering to two broader urgencies. Firstly, satisfying energy demand in a moment when the exhaustion of relatively low-cost energy sources is approaching. Secondly, reducing the negative consequences – emissions of greenhouse gases and pollutants being among the most renown - deriving from energy consumption. Two main strategies can be identified: reducing energy production and consumption (conservation/sufficiency strategies), and; reducing the emissions per energy unit (efficiency/optimisation strategies) (Alcott, 2008; Boulanger, 2010; Lorek & Spangenberg, 2019; Lutzenhiser, 2014; Princen, 2005). The latter comprises the following: increasing the reliance on lower-emission and less-pollutant energy sources (i.e., renewables); reducing energy wastages (at the transmission, distribution, and consumption stages) and; increasing efficiency (in energy conversion processes, including the energy services consumption stage).

What has been done so far towards the energy transition is far from being satisfactory. Depending on the geographical area one refers to, energy consumption has been increasing, or only slightly decreasing. This may be because improvements may take time to fully become 'operational', and that rebound effects may intervene to reduce or counteract their potential beneficial effects (Saunders, 2015; Stern, 2017; Wei et al., 2019). It may also depend on the fact that efforts may have been few. The efforts needed to pursue the ecological transition in a way that is in line with objectives such as those deriving from the Paris Climate Agreement are monumental. That means that any sector and everyone, at any level, must give a contribution to it and that no single type of action could be sufficient if taken alone.

High expectations have been put on ICTs' potentialities and on citizen's involvement for the energy transition. One of the fields where the two can be found together is that of smart grids (Clastres, 2011; Ramchurn et al., 2012). When graphics is used to explain what a smart grid is, energy grids (i.e., the not-yet-smart ones) may be represented as composed of a few elements: a limited number of energy production sites, the transmission and distribution lines, and the multiple places where energy is consumed. More detailed graphical representations would show the different modes of energy production (e.g., thermoelectric, nuclear, hydroelectric), there may be transmission lines having different capacities, and there may be the places of energy consumption, usually represented according to a well-established categorisation: industry, tertiary, residential. There may be arrows indicating the direction of energy flows. Thus, energy moves in the direction indicated by the same order in which the elements composing an energy grid have been just presented: from production sites, through power lines or pipes, to the places where energy is consumed. As in many other cases where the adjective 'smart' is added to something that before did not need it, current (or conventional) energy networks are all but "dumb". Indeed, they are more and more reliable systems and are for most of the time invisible, silent, and unnoticed. They had to pass many 'tests' to become what they are. However, the changes that occurred during the few previous decades were minor, at least according to a certain standpoint: energy networks had to increase their capacity and extension, while end-users were becoming accustomed to higher and higher levels of energy services. In a graphical representation of a smart grid, none of the previously present elements would need to be removed, while others would need to be added. For example, an increased number of energy production sites (usually from renewables), as well as quite or completely new elements such as electric vehicles, and energy accumulators. It will be shown that energy now flows in multiple directions. Most importantly, there may be many lines connecting all or almost all elements and flowing to and from something representing things like computers or data centres. New flows are therefore added to the energy ones: information and data. These flows need new elements that would require an even higher-resolution graphical representation to be detected, such as sensors, meters, displays, applications, and actuators. And, of course, humans. Not to mention other intangible, albeit very important, elements such as normative frameworks and users' attitudes (Blumsack & Fernandez, 2012; Gangale et al., 2013; Mengolini & Vasiljevska, 2013). Anyway, more and more information and data are expected to be collected, in part they will be eventually refined through elaborations, and finally they will be sent to someone and / or to something. But what data will be sent to whom and to what? What are data and information expected to make do to those humans or machines that will receive them? To what extent will recipients 'understand' the data they receive? How to improve data understanding and effectiveness in bringing about the desired changes? And what are the actual desires and needs of the actors operating in situated energy systems? These are some of the questions this chapter aims at answering to.

The next section will focus on some issues related to the prospected more incisive presence of end-users in energy networks. Section 2 will describe a case study about thermal energy (i.e., district heating) from an EU-funded project aimed at realising a visualisation tool to optimise energy distribution at the urban district level. Section 3 will provide some recommendations.

THE GROUND FOR DATA FLOWS AND USERS' ENGAGEMENT IN THE ENERGY TRANSITION

Room for New Data

New data, together with the new tools that are needed to produce them and to make them flow, will need to find a place in a well-established and coherent assemblage, or actor-network (Callon, 1987; Latour, 2005; Schick & Winthereik, 2013). This is reliable (despite occasional blackouts and failures) (Silvast, 2013) and characterised by incumbent positions, path-dependencies, and routines. Personal and professional lives have evolved in this system, in a densely packed spatio-temporal organisation that leaves few ways of manoeuvring differently (Ellegård & Vilhelmson, 2004; Powells et al., 2014).

Although closely intertwined, data production and use refer to two different perspectives of observation. The first relates to the material objects (e.g., sensors, meters, and screens) that need to be properly designed and adopted in order data to be produced and circulated. This perspective raises the issue of the acceptance, or suitable embracement, of these devices into professionals' and final users' practices. The second is related to the actual exploitation of the information contained in data, that is to take into account how data can be properly processed, to what aim and what final uses and users they are supposed to address (Geelenn et al., 2013).

A More Central Role for Non-Energy-Sector Actors

Currently, the role of citizens and, more broadly, of non-energy-sector actors, in the 'management' of energy systems is very limited. It has not always been so. Procuring fuel, storing it, and disposing of combustion residues, required final users more direct monetary, time and physical efforts than today. This may have also resulted in a greater awareness of the limits to resource consumption, an aspect which is now almost completely lost: of course, there may be awareness of it, but much less experience of it. Not differently from other fields in technically advanced society, what was gained in energy systems was instead security, easiness-of-use, and reliability which came out because of labour division, of having left the management of the 'energy life cycle' to specialised actors.

The energy transition requires citizens and the actors not belonging to the energy economic sector, to play a greater role than today. This is not an easy endeavour and one may be tempted to think that it would be preferable to continue leaving this task in the hands of those – individuals and organisations – that deals with energy professionally (Viitanen & Kingston, 2014). Energy systems performances can be improved in 'invisible' ways, through new organisational and technical arrangements. Data can flow and serve energy systems without common users barely or at all noticing it. Grids may become more reliable, energy costs, energy intensity and emission intensity may become lower than would have been without such data flows, but these facts may be unknown to most people, and the effects on the demand for energy services may be not clearly detectable. In any case, all this would not solve the issue of the social acceptance of the energy transition, whereby by 'social acceptance' it is meant either the willingness of final users to give a contribution to it and their acceptance of the changes 'imposed' on them (Wolsink, 2018; Wüstenhagen et al., 2007).

Accepting the Energy Transition

More and more citizens agree ecological problems are severe and dangerous. However, it is questionable whether they are prone to transform awareness into consistent behaviours that would bring massive changes to their ways of living and consuming, and whether they are aware how far we are from the required reduction in greenhouse gas concentration.

Non-energy-sector actors can actively participate in the energy transition in many ways: by producing energy from renewable sources; by increasing the efficiency of their energy consumption practices and operations (e.g. reducing wastages, using more efficient devices, favouring demand-response mechanisms, exchanging energy with their neighbours); by reducing energy demand; and by participating to collective actions and manifesting their wills and desires through votes, complaints and protests. These possibilities of action are not being pursued with comparable intensities.

A greater role in the energy transition is expected to be given to citizens and non-energy sector actors by energy communities, as defined in EU Directives 2018/2001 and 2019/944, soon to be transposed into EU member states legislation. With energy communities, the number of subjects operating in the energy field will increase. Most of these will not be operators formally belonging to the energy, or energy-related, economic sectors. This will somehow lead to the need of bridging the difference in the level of skills and knowledge compared to those who deal with energy professionally and regularly. Energy skills will also need to be more widespread. Not only to create new professions and to strengthen existing ones, but also because the social acceptance of the energy transition at least requires - and its pace depends on the fact - that citizens, and users more broadly, know the reasons of the changes to which they are subjected or requested to adopt. Then, thinking about the issue of trust in the energy field (Büscher et al., 2018), it is worth going back to the fact that technical solutions, such as those based on Artificial Intelligence, Machine Learning and the like, are immaterial, impalpable and invisible. However, they present characteristics considered to be typical of humans, such as perception and decision. The more so for the 'most intelligent' among these solutions. Given the importance of energy for the economy and daily life, the abilities of these tools to perceive and take decisions should be properly pondered. It is also desirable that they are given the possibility to gain the trust of those whom such decisions have effects on.

Energy Feedback and Energy Literacy

One of the reasons of the difficulties in reducing/optimising energy consumption at the final consumption stage largely lies in the 'physical' characteristics of energy: it is not tangible and it is not visible (Hargreaves et al., 2010 and 2013). Bridging the cognitive and spatial distance between places of production and places of consumption can contribute to making energy visible. At least partly, this can be done through more information. Research and literature on energy feedback is mainly focused on electrical energy. This can be explained by its pervasiveness and importance in quantitative terms but also for other reasons that will be made clear in the following.

The best-known example of energy feedback reaching users is energy bills. Energy bills are an example of "indirect" energy feedback (Darby, 2006; Donnelly, 2010), where data reach users only after being processed. Energy bills have an informative content that would not be considered acceptable in consumption contexts other than those related to the energy used in the domestic environment (Kempton & Layne, 1994). The main limitations are that energy bills come to users only months after energy has been used, refer to a period that is too long (generally one/two months), contain information that is somewhat difficult to read and interpret, and contain aggregate data. All these features make it difficult to end-users to identify the causes of energy consumption, as well as the causes that would explain the levels of consumption achieved. Data, when reported in a not too much user-friendly way, make it necessary to use time for their understanding. In turn, in interpreting and understanding data, errors, arbitrariness and incompleteness can manifest. Numerous variables can influence changes in energy consumption levels, as well as in their cost: the price of raw materials, energy taxes, changes in the energy mix, occupancy ratios, outdoor temperatures, and so on. These variables are not, or only partially, considered in conventional feedback systems. Self-metering can be included in conventional feedback systems as well, as an example of "direct" energy feedback. Reading energy counters at regular intervals may allow, with an effort quantifiable in a few minutes per week to: calculate the average daily consumption; make comparisons between different years or months; detect the energy consumption happening even when there is nobody at home and when, theoretically, devices should be off, the so-called "ghost consumption". Finally, other information related to energy consumption can be obtained through the labels glued or printed on most electrical devices. However, although this type of information is available to users, errors relating to what the specific consumption of different devices are is anything but insignificant (Attari et al., 2010; Wynes et al., 2020).

It is already possible from this short list and its description, to get useful advice for the evaluation of conventional feedback systems. Improvements in conventional feedback systems could be numerous. These may concern: swifter and more frequent energy feedback; shortening the feedback reference period; improving accessibility and readability of data; developing solutions that allow data disaggregation. Yet, though largely unsatisfactory, they have been part of the lives of hundreds of millions of families for decades. Moreover, although some information could be already obtained, users do not exceed in efforts to obtain and use them. The time required to get, and subsequently to reflect on, energy feedback, combined with the uncertainty relating to the correct methods of interpretation, can be considered a relatively important barrier.

Illiteracy in the energy field (energy illiteracy) is noteworthy when one considers the number of 'energy slaves' an average westerner controls (Court, 2018; Kümmel, 2011), and that help him/her in almost all practices, in almost any moment of the day and of the year, as well as in maintaining the so-called essential services that every westerner enjoys, more or less directly, more or less consciously, willing or not. But also for the fact that the energy sector is the one that contributes the most to greenhouse gases emissions. Energy illiteracy manifests in the lack of even basic notions. Some of the possible examples refer to the lack of knowledge about how much energy one consumes on average, what is the difference between primary and final energy; what is the order of magnitude of the specific consumption of the main electrical household appliances. At a higher level, it manifests for example in the lack of knowledge about how energy is produced, transported, and distributed; about how energy systems (national or within buildings/apartments) function; how much 'our' energy system is unsustainable, and how much energy is produced from renewable sources at the national and/or the global level. Pursuing energy literacy (DeWaters & Powers, 2011) means giving non-energy-sector actors the competences and notions that will put them in the position to (better) understand whether energy systems performances are in line with the energy transition objectives or not and how consumers can be more than just consumers, and in case they do not feel the need to be more than that, what they can expect the energy transition objectives to be likely to ask them to adapt to.

CASE STUDY

This section will describe the activities some of the authors were responsible for, and that were undertaken within the DIMMER project. DIMMER (District Information Modelling and Management for Energy Reduction) was a three-year EU-funded project (FP7 research programme) that ended in September 2016. It was aimed at realizing a system for the visualization, simulation, and computation of data on energy systems at the urban district level. The expected result of DIMMER was a consistent reduction in energy consumption, by enabling optimised energy distribution policies.

DIMMER activities took place in two pilots situated in Manchester (United Kingdom) and Turin (Italy). The activities that will be described in this chapter refers to the Turin pilot, which was focused on thermal energy, i.e. district heating network. More in details, they refer to an urban district situated in the proximity of, and comprising, the Polytechnic University of Turin's main campus, which was one of the seven buildings selected for experimentation. Besides technical and structural criteria, the seven buildings were selected to include both public and private buildings with different intended uses (the other six buildings being: a private university residence; two public schools; one public office building; and two private apartment buildings). Buildings were also representing different dimensions. In terms of 'population', the two condos had around 70 and 200 inhabitants respectively, while the population of public buildings ranged from the around 20 workers and 150 pupils of the kindergarten to the many hundreds of workers and tens of thousands of students of the Polytechnic University.

The engagement activities that were carried out can be enlisted under two main categories: mapping of the heating socio-technical system, and co-design activities. The 'mapping' activity assumed that energy systems are socio-technical. Put differently and simply, they are assemblages of human and technical elements. Assemblages change over time due to changes (both quantitative and qualitative) of the elements they are composed of. For instance, a new visualization tool can be considered a new element that has to be integrated into an already existing assemblage where, for example, an older visualisation tool or no visualisation tool was present. Understanding which consequences its introduction will bring to the existing assemblage (i.e. to its composition and performance) would have necessarily required an understanding of which was the state of the assemblage before its introduction.

43 interviews and 3 focus groups were carried out during the first half of the project. Interviews were typically led by a single interviewer and were audio-recorded. They were semi-structured and lasted approximately 90 minutes each. Depending on interviewees' role (summary in Table 1), the interview outline was structured in such a way to allow a description of users based on their energy-related activities: inside the buildings, in everyday practices; in managing buildings; in managing the energy production and distribution network; at the district level, for what refers to representatives of institutions having regulatory and/or planning roles. The aim was to understand the features of the actors playing a role in the wider energy network or in the energy systems of the buildings they were users or managers of (Abi Ghanem & Mander, 2014): energy-related roles; energy-related skills and competencies; information and data they already get; desired information and data; used technical devices; main relationships with other actors; motivations and barriers to energy transition enabling activities.

The three focus groups of the first year were aimed at: testing the presence of conflicts and agreements about energy/thermal-related issues within some buildings; understanding the ideas that concepts such as smartness and smart grid move; and at confirming and sharing the first results from interviews. Focus groups involved 18 people among the users and managers of three buildings (university residence, primary school, and public office). Particular attention was dedicated to discussing on project's main aim, and on which energy data could be useful for the specific building and for the wider context the building was situated into (i.e., the urban district). Based on the outcomes of these first activities, the consortium decided for which categories of actors the tool would have been developed.

Co-design activities, which included a proper co-design phase and the users' validation of the first version of the DIMMER tool, involved around 50 participants representing the three categories of selected target users (i.e., private and public building managers; public administrators, which included representatives of the Piedmont Region; and the DH operator). They were aimed at translating users' needs into requirements for the DIMMER visualisation tool, then at testing its first version.

The District Heating Network

District Heating (DH) in Turin started in 1980s and currently almost two-thirds of the population are served by it (i.e., more than 550,000 inhabitants) (Gaudio, 2012). Heated volumes, the majority of which from residential buildings, are at around 56 million cubic meters, having almost reached the maximum capacity given the current infrastructure.

Due to the beneficial effects of DH on air quality, public authorities may encourage the expansion of the DH network to yet unserved neighbourhoods. Due to the minor efforts and costs DH brings to building managers and users, compared to individual centralised heating systems, the expansion is likely to continue insofar there is spare generation capacity to do so. In general, satisfaction with respect to DH is quite high, which is demonstrated by the fact that the total number of 'defections' (i.e., moving away from DH after having experienced it) is close to zero.

Table 1. Interviews. Sample description

	Interviews (43)	Interviewees (49) and roles
District Heating operator	9	Directors/managers: 4 Technical employees: 3 Top management: 2
Public buildings	15, of which: Kindergarten: 5 Primary school: 6 Public office: 2 University: 2	Directors: 3 Employees: 9 Energy manager: 1 Facility managers: 2
Private buildings	13, of which: Two apartment buildings: 8 University residence: 5	Building managers: 3 Facility managers: 3 Residents: 7
Public administration	3	City councillor: 1 City district councillors: 3 Energy manager: 1
Building managers' association	1	Directorate: 2 Technical consultant: 1
Company in charge of heat cost allocators and thermostatic valves	1	Management: 1 Technicians: 2

The DH operator must supply heating during specific hours, and in a way sufficient to reach the desired setpoint temperature. Both the heating schedule and the setpoint temperature are defined by building managers within some limits established by local and/or national regulations.

The DH operator in Turin faces some of the same obstacles and challenges that electrical energy providers face. This is partly because it too produces electrical energy (through CHP plants) that is delivered to the national grid. On the other side, as it happens for electrical energy and energy grids, the DH operator takes advantage from reducing thermal seasonal and daily peak-to-average ratios, to postpone or avoid costly infrastructural integrations, implementations, and revamping.

Peak shaving strategies have been implemented in two different, albeit complementary, ways: trying to change demand through a two-tier TOU (time-of-use) tariff and through technical improvements. Acting on demand by asking (incentivising) consumer participation can lead to more substantial improvements (lowering the peak and / or reducing the peak-to-average ratio) which, however, cannot be as 'sure' as in the case of merely technical improvements. On the one hand, this is because users may not accept the commercial proposal which, if followed carefully, should lead to shifting part of heating demand at night; on the other hand, because while adapting to a new tariff system, they may at some point (perhaps the one in which it would be most useful) decide not to adapt to it anymore.

The DH operator expresses a position of relative closure regarding the sharing of information with other parties (e.g., public administrators and building managers), due mainly to privacy issues. In this field, the technological innovations should be accompanied by legislative and organizational innovations that would make it easier, or possible, to effectively share energy transition enabling information.

Many interviewed professionals of the DH operator reported having had problems in making residential building managers accept the peak-shaver TOU tariff. In recent years, this problem has been accrued by the increased presence of radiator thermostatic valves and heat cost allocators. Thermostatic valves allow end-users to "reduce waste". Jointly with heat costs allocators, they encourage end-users to close the valves when they leave last from home. This looks like a textbook perverse effect (Boudon, 1977; Merton, 1936). Thermostatic valves, an instrument that allows a more efficient management of comfort and thermal consumption in homes, reduces waste at the flat and condominium levels, but this does not necessarily increase the efficiency of the system, both intended as peak-to-average ratio and as the ratio between primary energy consumption and heat consumption. The hours when people are not at home is mostly during afternoons, and therefore not during consumption peak hours (i.e., early morning and late afternoon/early evening). The awareness of having made economies during the rest of the day may thus allow (in an economic and / or psychological sense) to increase the demand for energy services during the hours of presence at home. For a sort of compensation effect, the enactment of peak energy-intensive practices is reinforced by the reduction of energy consumption in off-peak moments. The result would be an increase in the distance between consumption peaks and valleys which, at best, is due to a lowering of consumption valleys (a situation that reduces primary energy consumption, but at the same time reduces the revenues of the DH operator), and in the worst case is due to a lowering of consumption valleys and a simultaneous rise of consumption peaks (a situation that does not reduce, indeed likely increases, primary energy consumption, and which reduces the revenues and increases the costs for the DH operator). Besides, similar reasoning could also apply to annual consumption peaks and valleys.

In this case, the possibilities of acting against the needs of the DH system is hampered by the low dynamism of the variation in the heating supply times. In most cases, buildings remain anchored to a flat rate. That was not because the commercial proposal is not good enough, but – as it will be explained later - because private buildings' administrators usually try to reduce the chances of receiving complaints and, above all, of receiving them during house owners' meetings. It is highly likely that changes, while possibly improving the average situation, will worsen the situation of one or more of the flats, therefore by a number that may already prove sufficient to make the mediation work of building administrators more difficult, especially during house owners' meetings.

Understandably, the DH operator already has many data, some of which in near real-time, collected from sensors that are present all along the distribution network, from power plants to heat exchangers.

Other data the DH operator is interested to have refers to:

- Updated public cadastre of buildings, with a special focus on volumes to be heated. This would allow better plans of network extensions to be developed.
- Users' comfort feedback, specially from the buildings situated in the farthest parts of the network.
- Tool allowing the examination of both a wide-area scale and micro-level buildings' behaviours, within a model related to how the micro-level influences the macro-level, and vice versa.
- Dynamic real-time model of the network, to replace static ones. The dynamic model will help to identify more promptly the problems along the network, by integrating existing information and enabling better diagnostic activities. This will help at more rapidly identifying solutions and make simulations of the entire network (or parts of it).

Table 2 contains a summary description of the functionalities, input data, and visualisation output of the final version of the visualisation tool for the DH operator personnel.

Residential Buildings

Table 2. DH operator. Functionalities, inputs, and outputs of the visualisation tool

Functionalities	Inputs	Outputs
• Modify features and/or add or remove components to the DH system • Include the costs of installation of new components or repairing of existing components • Define the ranges of variation of the parameters for the simulations (e.g., aging speed of components, changes in natural gas price, changes in energy taxes, speed of change of "invariable" features of the buildings, diffusion rate of thermostatic valves, external temperatures) • Choice and simulation of different tariff schemes	• Invariable and variable data (current and historical, obtained from sensors and feedback) about the components of the DH system • Physical-constructive "invariable" features of served buildings (thermal behaviour, volumes, etc.) • Variable features (current and historical consumption, setpoint temperatures, heating schedules) of served buildings • Physical-constructive "invariable" features of buildings that are not served but are located within the range of the network or might be reached through network extensions • Conversion factors: primary energy / thermal energy / emissions / costs • Outdoor temperatures last n years • Current energy taxes	• Display (on table/map) the features of the components of the DH system, and of buildings (both served and not) • Results of the simulations (on table/map): expected results, variations, identification of critical variations • Warnings / visualization / reporting anomalies on components: compared to other components of the same kind; compared to previous performances of the same component

Private residential buildings are managed by building administrators. They are professional, meaning that they normally manage several buildings. They can be part of a building administrators' association. They are responsible for the management of a varied set of aspects, such as maintenance, refurbishment, cleaning, mediation of conflicts, insurances, water seepages, payment of building's bills. Also, building administrators keep contacts with suppliers, pay invoices, prepare annual budgets, urge households to pay their due share of costs, etcetera. They must meet the requests of house owners, in order their appointment to be renewed by them annually. Decisions on extraordinary management must be taken by house owners during their meetings. In normal conditions, house owners' meetings happen once per year, or twice per year in case the building has a centralised heating system or is connected to district heating (one of the two meetings being about heating issues). House owners' meetings take decisions on things like, e.g., installing a new boiler, planning refurbishments, changing provider, changing the rules of apportionment of heating costs by deciding which part will be based on apartment cubic meters and which part on real consumption.

Building administrators communicate householders' requests and complaints to the DH operator. Building administrators are the contact point between the DH operator and final users. End-users are free to contact the DH operator support centre, but the DH operator will not undertake any change to heating settings without a prior request from the building administrator.

Yearly, the DH operator sends the heating invoice to the building administrator. Reportedly, this is not very transparent. More than yearly, building administrators also receive data from the company in charge of reading heat costs allocators. However, this data is almost useless (actually, it is not used) until the DH operator sends the annual invoice (as it contains the costs to be apportioned). Noteworthy, the readings of heat cost allocators are just sent to building administrators. For transparency, they will then forward it to house owners. In any case, there is no data exchange between companies in charge of reading heat costs allocators and the DH operator.

Major decisions are difficult to take in residential buildings. Due to infrequent house owners' meetings and to quite frequent disputes among them, or between them and the building administrators, priority is usually just given to urgent and unpostponable actions.

Building administrators (and their association's representatives as well) reported what happened after the installation of thermostatic valves and heat cost allocators. They reported of flats whose heating expenses increased by up to 100% to the previous year. This could be due to a combination of two effects. One is that the new cost apportionment rules disadvantage some flats (usually those situated at the first and the last floors, those more intensely occupied, and those with a bigger external surface) (Morgenstern et al., 2014); the other is that users may have been pushed to overestimate the expected energy and monetary savings coming from thermostatic valves, thus leading them to demand more of heating services. Therefore, inhabitants from those flats are embittered, upset, and distrustful of building administrators, even if the latter may not be those who contributed to creating expectations of savings. This may lead to open obstructionism, which may explain why building administrators are reluctant to propose house owners to adopt the TOU tariff scheme proposed by the DH operator. This regardless of whether they find the commercial proposal good or not (some buildings administrators think it is not advantageous enough). Building administrators reported that, based on their experience, up to two years may be necessary for the interactions with the upset inhabitants to become 'normal' again (provided no new conflicts arise).

Building administrators are also in charge of requesting the DH operator to change the heating setting for the buildings they manage. Within a building, there are different and often conflicting subjective perceptions of thermal comfort and different times of presence in the building. This makes difficult for building administrators to set the building heating system in a way that satisfies everyone. One of the building administrators reported having bought a stock of thermometers that then gives for free to those inhabitants that more frequently (and, reportedly, unfoundedly) complain about the indoor temperature, to reduce time wastages.

Building administrators would like to have an improved possibility of accessing heating meters (they are out of reach, usually in a locked room in the basement). They complain about a lack of transparency, for example in the annual bill, and a lack of data about the heating systems of the buildings they manage. Finally, they think that improved communication and information exchange from the DH operator may partially help them in reducing, or preventing, conflicts with house owners and tenants.

Table 3 contains a summary description of the functionalities, input data, and visualisation output of the final version of the visualisation tool for building administrators (residential buildings) and building managers (public/non-residential buildings).

People living in residential buildings (as well as non-resident house owners) have a very limited role in the management of thermal issues. Their goal is to assure that thermal comfort in the apartments is reached and, secondarily, to keep the energy bill as low as possible. This limited role should not come off as a surprise, given the limited information and data they have and given the practical problems in taking common decisions. Data they receive is just limited to the total expenses for their building and to the amount they must pay after costs apportionment. Those who have a thermometer can verify if their sensation of thermal discomfort is founded; those whose heat cost allocators have a display can do some self-metering (reportedly, a not-so-common practice). Participation in house owners' meetings does not happen enthusiastically, for the reasons reported above. They can report problems related to heating/

Table 3. Building managers (residential and public/non-residential). Functionalities, inputs, and outputs of the visualisation tool

Functionalities	Inputs	Outputs
• Display data on consumption (tables, graphs) adjusted to outdoor temperatures • Verify and locate consumption anomalies • Perform simulations related to refurbishments: expected consumption, emissions, costs, and payback periods • Visualize the heating settings • Display preferences about heating schedule expressed by householders • Perform simulations related to the change of heating setting and calculate expected consumption, costs and emissions • Compare buildings based on thermal energy consumption: 3D views of buildings with different colours • Receive notifications of change of the heating setting • Have benchmark for energy consumption (e.g., with similar buildings) • Identification of the measures needed to meet carbon targets (e.g., achieve a better energy certificate rating) • Discuss energy info with building users • Possibility to publish consumption data of the managed private buildings	• Historical and real-time energy consumption / expenditure / emissions • Room use features (e.g., number of workers, surface, volume, number of work hours performed, activities, presence of energy intensive equipment) • Physical characteristics of each room • Allocation of rooms/areas/devices to internal administrative units (e.g., Departments) • Historical data of outdoor temperatures last n years • Anomalies (it requires system to detect anomalies) • Model of the energy networks (generation, consumption, and losses) the buildings is linked to • Data (current and historical) on buildings energy consumption and generation • Give priority to either carbon or economic benefits • Requirements from energy certificate rating scheme • Conversion factors: energy / carbon emissions / costs • Heating settings (current and historical) • Thresholds for alerts related to changes in heating settings • District heating and natural gas tariffs (current and historical) • Standard costs for refurbishments (roof insulation, thermal coat, heat accumulators, fixtures, other) • Current fiscal incentives • Standard effectiveness of refurbishments • Buildings' physical features (e.g., thermal behaviour, energy class, volume, surface area, heating system) • Chronology of refurbishments • Building use features (intended use, number of residents / workers etc., number of hours work performed, etc.) • Historical and real-time data on energy consumption / expenditure / emissions for heating • Benchmark(s) for heat consumption	• Graphical ranking of building sections based on energy consumption • Cumulative energy consumption for sections (both adjusted and not) • Colouring of buildings and parts of them (3D) based on energy consumption • Anomalies visualization • Spare electricity and heat generation capacity in each building • Available network capacity required to transport energy to a given building • Suggested operational regime for electricity and heat infrastructure • Automated management of real-time operations • Projections (graphs) of energy consumption / expenditure / emissions • Conversion of heat to environmental (CO_2, primary energy, NO_x, etc.) and monetary units • Expected reduction in energy consumption / emissions from refurbishments • Estimated payback period for refurbishment with different outdoor average temperatures • Curves of energy consumption / expenditure / emissions (historical and current), both adjusted and not • Report (graphics) on consumption • Chronology of the simulations requested • Comparisons between past projections and actual results • Colouring of buildings (3D) based on thermal features

energy to their building administrator and the support centre of the DH operator, but as said before, the intermediation of their building administrator is required.

Table 4 contains a summary description of the functionalities, input data, and visualisation output of the visualisation tool for residents in residential buildings (not developed within DIMMER).

Table 4. Residential buildings – Residents. Functionalities, inputs, and outputs of the visualisation tool

Functionalities	Inputs	Outputs
• Perform simulations related to refurbishments, to the installation of heat accumulators and to the changes to heating schedules • Calculate expected consumption, emissions, costs, and payback period • Express preferences regarding heating schedule to modify it • Visualize heating schedule • Receive notifications about changes of heating schedule	• Standard costs of refurbishment (roof insulation, thermal coat, heat accumulators, others) • Current fiscal incentives • Standard effectiveness of refurbishments • Historical and real-time data on consumption / expenditure / emissions for heating • Buildings' physical features (year of construction, thermal behaviour, energy class, volume, surface area, heating system used, others) • Conversion factors energy consumption / costs / emissions • Heating schedule (current and historical) • District heating tariffs (current and historical) • Preferences about heating schedule (input coming from user)	• Expected reduction of energy consumption / emissions as a result of refurbishments • Estimated payback period for refurbishment with different outdoor average temperatures • New heating schedule based on expressed preferences • Differences in energy consumption / costs / emissions with new heating schedule • Differences in energy consumption / costs / emissions if all households opt for the same heating schedule

Other Buildings

In non-residential buildings (which may include buildings offering residential services, such as university residences and hospitals), three management roles can be identified, which in some institutions may be all present: energy manager, facility manager, and (lacking a better definition) representative manager.

Energy managers usually work in the biggest institutions (in this case, City's Council and University) and oversee the energy management of the many buildings that institution has, meaning they cannot be 'linked' to just one building (e.g., the one in which their office is situated). They propose actions to the facility manager(s) to save energy, analyse environmental and consumption data, and draft plans for ensuring comfort and keeping energy waste and energy consumption low. They usually use professional ICT systems/tools to analyse the energy performance of buildings. They lead on the development and implementation of their institution's environmental and/or sustainability plan.

While the energy manager's role is more that of an analyst, devising the plans to save and supply energy, the facility manager is responsible for their execution. They receive advice from the energy manager (in the cases where there is) on how to save energy and oversee the building management system, e.g., setpoint temperatures, heating schedules. Usually, they carry out mainly organizational and operational activities. They are in contact with suppliers (such as the DH operator), they deal with energy procurement, the maintenance and procurement of energy plants, and collect complaints and grievances from users regarding thermal comfort. They measure, monitor and report on the sustainability performance of the buildings they manage.

These two professionals daily work with energy data. Anyway, they suggested several data as useful to improve their performances:

- Having a benchmark to compare the building energy behaviour with other similar buildings.
- Having a service allowing them to evaluate the consumption trend considering weather conditions.

- Receiving more frequent information about energy consumption from the provider (i.e., not just the annual invoice).

An example of what has been defined here as representative manager is the dean of a school. Of course, energy/thermal issues have a role in allowing the educational service to be properly delivered, but these aspects are out of his/her main field of competence (albeit he/she may have some sensitivity and competence) and could be a distraction from his/her main job. For the most technical aspects related to energy/thermal management, he/she has to ask for the intervention of a facility manager, who may work in the same building or may be situated in a central office dedicated to the facility management of all institution's buildings. Situations are quite varied. In some cases, he/she may not even receive energy bills, which are thus managed by a central facility management office. At best, the energy data he/she receives is very limited.

The three kinds of managers described above reported having many difficulties in common, some of which are also in common with residential buildings' managers. For example: there may be relatively high temperature differences between different areas of the buildings, which lead to users' discomfort and related complaints; they cannot identify which offices or wings of the buildings consume more energy (heat costs allocators, which may also serve as radiator-level heating meters, are not mandatory in public buildings); there is scarce staff spare time to implement measures; those acknowledged with DH tariffs agree with some residential building managers in saying that the peak-shaver tariff is not advantageous enough; they cannot verify to what extent changes in energy consumption are due to the measures they may have been enacted or to external factors (e.g. energy price, weather conditions); they too have limited or no possibility at all to over-ride sensors and computerized energy systems, when there are; they too reported that thermometers may be very useful in founding users' comfort perceptions. Some of these professionals do not have any direct role in the thermal management of the building but just control on maintenance activities. Most of these managers do not receive any information about consumption, mostly in the cases where the control on this aspect is centralized. Finally, representative managers reported having to manage too much information in their working practices, related to their many and main tasks, so that in the end they fear more information could be ineffective.

Depending on the service provided in each specific building, users may be employees, students, service clients or citizens. They do not pay for the energy consumed and they do not have any role in energy consumption control and upon thermal devices, especially where automatised systems are in place.

Table 5 and Table 6 summarise and enhance the descriptions done thus far, making possible a comparison of different roles in public and private buildings.

Public Administrations

Public administrators are professionals having among their many tasks that of reducing energy consumption, energy expenses, emissions due to energy consumption and production, regarding the territorial level the institution they work for operates. In general, they must ensure the compliance of local, national and EU policies and targets and to define the local targets through local action plans for energy and the environment, aimed for example at reducing energy consumption and emissions, and promoting the use of renewables. They are responsible for energy aspects at various levels, from ensuring the energy production and distribution can satisfy local demand, to air quality policies which effects also depend on how and where energy is generated and consumed. Through their energy and facility managers, they

Table 5. Summary of different 'energy roles' in public buildings

	Energy managers	Building managers	Building users
Required to reduce costs	Yes, it is among their tasks	Occasionally	Not directly
Possibility to reduce costs	Depending on available financial resources	Generic	Limited
Required to reduce energy and environmental impacts	Yes, also for image purposes	No	Occasionally
Possibility to reduce energy and environmental impacts	Depending on financial resources	Generic possibility of minor magnitudes	Generic possibility of minor magnitudes
Possibility to verify the effectiveness of ecological practices	Yes, although with big margins of error	No	No
Interested in contributing to the optimisation of the DH distribution network	Yes, upon incentives	No awareness	No awareness
Knowledge of the energy/thermal system of their building	Yes	Vague or wrong	Vague or wrong
Knowledge of the level of consumption of their building	Yes	No, or yearly and only in monetary terms	No
Knowledge of buildings' consumption compared to other buildings	Yes, although with big margins of error	No	No
Interested in knowing it	Yes	Yes, for educational/image purposes	Generic interest
Occasions to come to know it	Yes	No	No
Occasions to participate in the energy/thermal building management	Yes	No	No
Control on other building users' behaviours	No	Depending on personal inclinations	Informal roles

have the role to oversee the energy management of their buildings stock. Regarding the private sector, public administrations, such as municipalities, are not able to intervene directly on energy-saving actions. Nonetheless, they play a role in creating a favourable environment.

Public administrations have a twofold role: i) planning, regulatory and 'community' role: they can decide about incentives, energy and environmental policies and have a cultural and educational role as well; ii) as building owners, they are facility and energy managers. For example, the buildings stock of the Turin Municipality consists of more than 800 buildings, over the half of which host educational services. Noteworthy, energy/thermal issues do not reach the district level of public authorities, besides those related to the buildings in which urban district authorities carry out their activities.

Their requests of data related to the buildings they directly manage are comparable to the requests expressed by the other building managers described above. They suggested energy-saving targets should be defined for all their sectors, and representative managers should be made accountable also for that.

Their needs are bigger in scope given their many fields of competence related to the entire energy life cycle. This can be summarised in a much bigger need of more data (most of which already exists but is not available to them) and in the request of tools performing a much bigger and diversified set

Table 6. Summary of different 'energy roles' in private residential buildings

	Buildings managers	Residents (and house owners)
Required to reduce costs	Moderate	Low
Possibility to reduce costs	Moderate	Low/Moderate
Required to reduce energy and environmental impacts	No	No
Possibility to reduce energy and environmental impacts	Moderate	Low/Moderate
Possibility to verify the effectiveness of ecological practices	Partially and with big efforts	Partially and with big efforts
Interested in contributing to the optimisation of DH distribution network	Yes, upon incentives	No awareness
Knowledge of the energy/thermal system of their building	Yes	In most cases, vague or wrong
Knowledge of the level of consumption of their building	Yes, monthly	Yes, but yearly and only in monetary terms
Knowledge of buildings' consumption compared to other buildings	No, or only for other buildings they directly manage	No
Interested in knowing it	Generic interest	Moderate
Occasions to come to know it	No	No
Occasions to participate to the energy/thermal building management	Yes	Rarely and difficultly
Control on other building users' behaviours	No	No

of functionalities, including the evaluation of environmental performances (e.g., air quality) and social performances (e.g., energy poverty).

Table 7 contains a summary description of the functionalities, input data, and visualisation output of the final version of the visualisation tool for public administrators.

SOLUTIONS AND RECOMMENDATIONS

As it happens in any other (almost any) field and process where ICTs (a very broad category which includes many different technologies, artefacts, and functionalities) are implicated, ICTs have a great potential also for supporting the energy transition. Not only because of the implied computational power which may serve, for example, to optimise energy flows and the real-time balancing of offer and demand but also because of the big amount of data they can contribute to produce and make visible, thus giving the possibility for the decisions taken within energy systems – whether by professionals or by other users, including citizens - to be better informed. Used as a communication tool, they can give knowledge, competencies, as well as increasing motivations. They can help, for instance, at putting together individuals, groups, and organisations to pursue common objectives.

The previous section showed (see Tables 2-4) that the potential of ICT solutions (some of which could be enlisted under the categories of Artificial Intelligence, Machine Learning and Digital Twins) are recognised and somehow desired by certain kinds of users. This seems to be mostly true for 'non-

Table 7. Public administrators. Functionalities, inputs, and outputs of the visualisation tool

Functionalities	Inputs	Outputs
• 2D/3D city/district view, with buildings coloured depending on "static" characteristics (e.g., thermal behaviour index, construction period, roof surface, heating system) • Select and regroup areas • Display data on consumption, with the possibility to choose the period • Adjust data on consumption to outdoor temperatures • Make projections/simulations • Compare past projections with actual data • Visualize socio-economic and demographic data • Correlate socio-economic and demographic data with energy consumption data • Communicate with the population	• Buildings' physical features (thermal behaviour, energy class, volume, surface area, heating system, year of construction, type of building – e.g., Social housing) • Building use features (e.g., intended use, number of residents/workers, number of performed work hours) • Chronology of refurbishments • Outdoor temperatures last n years • Historical and real-time energy consumption / expenditure / emissions for thermal energy • Conversion factors: thermal energy / primary energy / emissions depending on the type of heating system • Demographic, social, economic data (e.g., number of inhabitants, demographic and social variables, level of income) • 'Standard' effectiveness of some refurbishment actions	• Descriptive indicators of different areas (already existing or created by users themselves) of the town (and 2D/3D representation) • Colouring of buildings (2D/3D) based on thermal features • Simulation of energy consumption / emissions for buildings for which real data are not available • Energy consumption / expenditure / emissions curves (historical, current), both adjusted and not • Comparisons between expected results obtained from past projections/simulations and actual results • Ranking of city areas for different variables (both taken individually and not)

residential' users, i.e. those who work on more complex socio-technical systems (e.g., DH network and plants; a local public administration).

The previous section also described a DH system - intended as socio-technical system - and identified some of the barriers preventing it to be optimised and participated (O'Dwyer et al., 2019). Some of the barriers we identified are about data availability and production, others are about the different roles, competences and objectives of the different actors composing the system (see Table 5 and 6), as well as to the real and imagined (Skjølsvold & Lindkvist, 2015; Throndsen, 2017) relationships among them. These may prevent actors to cooperate and this may further prevent data to circulate and be effectively used to get some mutually desired results. All this may lead to prefer technical improvements, requiring no or minor engagement of users, to the DH system. However, hat clashes with the social acceptance and the vision of users' engagement as requisites for a smooth and more effective energy transition. The identification and resolution, in situated socio-technical systems, of these non-strictly-technical problems may be, at least in some cases, a pre-requisite for any ICT-based solution to deploy its full potential.

Among the non-strictly-technical problems, and besides those reported in the previous sections, we would like to highlight the following solutions at reach.

With a given infrastructure, the DH operator aims at reducing the peak-to-average ratio and the height of the thermal season's peaks of consumption. Different tariff schemes and contractual agreements may already prove useful in these cases. Contracts may foresee, possibly under rebates, the possibility for higher flexibility of heating supply provided some pre-established critical peak conditions manifest. TOU tariffs may also be adopted and proposed more incisively. Flat rates, proposed by many energy companies, may well serve the aim of attracting customers, but not that of making them aware of the 'needs' of an energy system. Another solution may be related to diffused heat micro-storage systems, to be installed in buildings (Jebamalai et al., 2020). Of course, all these aspects may face some technical and normative feasibility obstacles that cannot be solved by the DH operator alone and that may require more research and the intervention of regional or national authorities.

For what refers to private residential buildings, certainly the main problem is the 'bottleneck' of building administration, i.e. building administrators' role and house owners' meetings. Apartment buildings are a special field where collective actions could potentially blossom (there are people living in quite a proximity and sharing some common resources), so that it is of the utmost importance this problem is solved. Households receive very few data; they are subjected to (at times) exaggerated expectations that create problems when unmet; they rarely meet with each other so that they do not form a community (let alone a strong community); the time dedicated to homeowners' meetings (approximately 6 hours per year on average) is largely insufficient, and participation is low. At the same time, building administrators reported difficulties in doing more than what they already do. Because of a persistent lack of time, when not of training (e.g., conflict mediation). In general, the difficulties they face are well-documented, yet still unsolved and not taken into due account by national authorities and, at local level, by the DH operator.

At present, the implicit collective requests of householders are quite basic: spending possibly less than the year before to get an equal or improved level of thermal comfort. While waiting for the bottleneck identified above to be solved, and given the very limited possibility of action households have, what can be done to increase users' awareness and commitment towards the energy transition? First, energy literacy should be addressed. Or, better, what can be defined as 'energy system literacy'. It means making clear that the building (and the flat) in which people live is not an isolated entity surrounded by other isolated entities. They all belong to a wider system in which what is done on one point has also repercussions on the others. That means that energy responsible actions are not just a matter of reducing one own's consumption. It is also a matter of reducing it in a specific part of the day (and of the year) whose boundaries varies depending on the moments where others' consumption takes place. That is why it was proposed in DIMMER to use an application (Table 4) aimed at simultaneously targeting a multiplicity of aspects, by taking into account of the current barriers in house owners' meeting: i) translation of individuals' ecological motivations into heating setting changes; ii) increase of the flexibility of heating setting changes; iii) receive more data, and/or more frequently, with particular attention paid to data on the 'ecological performance' of the building and its district.

In public buildings, and more in general in public administrations, difficulties are fortunately not as severe, and seemingly unsurmountable, as those of private residential buildings. In short, data recipients are professionals prone to 'absorb' new data (and forms of data visualisation) and embrace new tools in their work practices.

The request coming from public administration as well as from other professional users is to have at their disposal a system that helps them in identifying anomalies and critical situations. In short, a trained and autonomous 'workmate'. Although it will be possible for a user to 'play' with the applications, it is preferred that the applications themselves 'play' and then forward to user the most significant results of the game. Anomalies and critical situations may concern the variation over time of the operating parameters of components, but also the difference with respect to external components and systems. In short, a building can have similar behaviours over the years, while being anomalous when compared to other buildings. This is where the demand for benchmarking tools comes into play. To evaluate the performance of a system over time and to compare it to other external systems, it is necessary to have tools performing the wider as possible functions of normalisation and correction of data. These operations should refer to, first, outdoor temperatures, then to some physical characteristics of the buildings (e.g., surface, volume, transmittance) as well as 'social' characteristics (e.g., intended use, number of inhabitants, number of man-hours worked).

Finally, from the methodological point of view, it is important to strongly stress once more the need for a better SSH-STEM integration (Foulds & Robison, 2018; Sonetti et al., 2020), above all when a project has the ambition to create a 'definitive' tool that is there to stay. Technical solutions for data processing and visualisation may perfectly work on an isolated technical component or in a controlled experimental setting, but not in a situated energy system whose 'social' barriers may be hard to break and be very specific to that energy system (Xu et al., 2021). This brings to the somehow inescapable flaw that is inherent in investigations like this one. What it provided was a description of a specific socio-technical system at a specific point in time. The changing interrelations among external factors, actors' needs and technical potentialities (real and/or imagined), bring to the need of establishing continuous and participated mapping activities.

CONCLUSION

This chapter describes the results of a sociological investigation carried out within an EU-funded project that was aimed at creating a tool to visualise and compute energy data at an urban district level in order to optimise the local DH network's distribution policies. It suggests two main avenues of interest for the integration of ICT-based solutions (be these 'common' or 'frontier' solutions) in the energy transition endeavour: the more central role to be attributed to non-energy-sector actors (including citizens), and the need to address what has been defined as 'energy system literacy'. It provides a detailed description of a specific DH socio-technical system, of its main frictions, and of the main solutions identified within the project but which may also apply to other and different cases. Finally, it provides a set of recommendations related to diverse fields such as tariff schemes, conflict mediation, and continuous and more robust SSH-STEM integration, that may help enhancing the effectiveness of any ICT-based solution.

ACKNOWLEDGMENT

This research was supported by the European Union's FP7 [grant number 609084]. We want to thank all participants in our research activities, and we want to acknowledge the important contribution from our colleagues: Chiara Colombo, Mauro Bonaiuti and Beatrice Surano.

REFERENCES

Abi Ghanem, D., & Mander, S. (2014). Designing consumer engagement with the smart grids of the future: Bringing active demand technology to everyday life. *Technology Analysis and Strategic Management, 26*(10), 1163–1175. doi:10.1080/09537325.2014.974531

Alcott, B. (2008). The sufficiency strategy: Would rich-world frugality lower environmental impact? *Ecological Economics, 64*(4), 770–786. doi:10.1016/j.ecolecon.2007.04.015

Attari, S. Z., DeKay, M. L., Davidson, C. I., & De Bruin, W. B. (2010). Public perceptions of energy consumption and savings. *Proceedings of the National Academy of Sciences of the United States of America, 107*(37), 16054–16059. doi:10.1073/pnas.1001509107 PMID:20713724

Blumsack, S., & Fernandez, A. (2012). Ready or not, here comes the smart grid! *Energy*, *37*(1), 61–68. doi:10.1016/j.energy.2011.07.054

Boudon, R. (1977). *Effets pervers et ordre social*. PUF.

Boulanger, P. M. (2010). Three strategies for sustainable consumption. *SAPI EN. S. Surveys and Perspectives Integrating Environment and Society*, *3*, 2.

Broto, V. C. (2017). Urban governance and the politics of climate change. *World Development*, *93*, 1–15. doi:10.1016/j.worlddev.2016.12.031

Büscher, C., Schippl, J., & Sumpf, P. (2018). *Energy as a Sociotechnical Problem: An Interdisciplinary Perspective on Control, Change, and Action in Energy Transitions*. Routledge. doi:10.4324/9781315186313

Callon, M. (1987). Society in the making: the study of technology as a tool for sociological analysis. In W. E. Bijker, T. P. Hughes, & T. Pinch (Eds.), *The social construction of technological systems. New directions in the sociology and history of technology* (pp. 83–106). MIT Press.

Clastres, C. (2011). Smart grids: Another step towards competition, energy security and climate change objectives. *Energy Policy*, *39*(9), 5399–5408. doi:10.1016/j.enpol.2011.05.024

Court, V. (2018). Energy capture, technological change, and economic growth: an evolutionary perspective. *BioPhysical Economics and Resource Quality*, *3*, Article 12.

Darby, S. (2006). *The effectiveness of feedback on energy consumption*. Environmental Change Institute - University of Oxford.

DeWaters, J. E., & Powers, S. E. (2011). Energy literacy of secondary students in New York State (USA): A measure of knowledge, affect, and behavior. *Energy Policy*, *39*(3), 1699–1710. doi:10.1016/j.enpol.2010.12.049

Donnelly, K. (2010). The technological and human dimensions of residential feedback. An introduction to the broad range of today's feedback strategies. In K. Ehrhardt-Martinez & J. A. Laitner (Eds.), *People-centered initiatives for increasing energy savings* (pp. 13–24). ACEEE.

Ellegård, K., & Vilhelmson, B. (2004). Home as a pocket of local order: Everyday activities and the friction of distance. *Geografiska Annaler. Series B, Human Geography*, *86*(4), 281–296. doi:10.1111/j.0435-3684.2004.00168.x

Foulds, C., & Robison, R. (2018). *Advancing Energy Policy: Lessons on the integration of Social Sciences and Humanities*. Springer Nature. doi:10.1007/978-3-319-99097-2

Frantzeskaki, N., Broto, V. C., Coenen, L., & Loorbach, D. (2016). *Urban sustainability transitions*. Routledge.

Gangale, F., Mengolini, A., & Onyeji, I. (2013). Consumer engagement: An insight from smart grid projects in Europe. *Energy Policy*, *60*, 621–628. doi:10.1016/j.enpol.2013.05.031

Gaudio, F. (2012). *1982-2012. 30 anni di teleriscaldamento a Torino*. IREN Energia.

Geelenn, D., Reinders, A., & Keyson, D. (2013). Empowering the end-user in smart grids: Recommendations for the design of products and services. *Energy Policy*, *61*, 151–161. doi:10.1016/j.enpol.2013.05.107

Hargreaves, T., Nye, M., & Burgess, J. (2010). Making energy visible. A qualitative field study of how householders interact with feedback from smart energy monitors. *Energy Policy*, *38*(10), 6111–6119. doi:10.1016/j.enpol.2010.05.068

Hargreaves, T., Nye, M., & Burgess, J. (2013). Keeping energy visible? Exploring how householders interact with feedback from smart energy monitors in the longer term. *Energy Policy*, *52*, 126–134. doi:10.1016/j.enpol.2012.03.027

Jebamalai, J. M., Marlein, K., & Laverge, J. (2020). Influence of centralized and distributed thermal energy storage on district heating network design. *Energy*, *202*, 117689. doi:10.1016/j.energy.2020.117689

Kempton, W., & Layne, L. L. (1994). The consumer's energy analysis environment. *Energy Policy*, *22*(10), 857–866. doi:10.1016/0301-4215(94)90145-7

Kümmel, R. (2011). *The second law of economics: energy, entropy, and the origins of wealth*. Springer Science & Business Media. doi:10.1007/978-1-4419-9365-6

Latour, B. (2005). *Reassembling the social. An introduction to Actor-Network-Theory*. Oxford University Press.

Lorek, S., & Spangenberg, J. H. (2019). Energy sufficiency through social innovation in housing. *Energy Policy*, *126*, 287–294. doi:10.1016/j.enpol.2018.11.026

Lutzenhiser, L. (2014). Through the energy efficiency looking glass. *Energy Research & Social Science*, *1*, 141–151. doi:10.1016/j.erss.2014.03.011

Markard, J. (2018). The next phase of the energy transition and its implications for research and policy. *Nature Energy*, *3*(8), 628–633. doi:10.103841560-018-0171-7

Mengolini, A., & Vasiljevska, J. (2013). *The social dimension of Smart Grids. Consumer, community, society*. EU Commission, JRC Scientific and Policy Reports, 31.

Merton, R. K. (1936). The unanticipated consequences of purposive social action. *American Sociological Review*, *1*(6), 894–904. doi:10.2307/2084615

Miller, C. A., Iles, A., & Jones, C. F. (2013). The social dimensions of energy transitions. *Science as Culture*, *22*(2), 135–148. doi:10.1080/09505431.2013.786989

Morgenstern, P., Lowe, R., & Chiu, L. F. (2015). Heat metering: Socio-technical challenges in district-heated social housing. *Building Research and Information*, *43*(2), 197–209. doi:10.1080/09613218.2014.932639

O'Dwyer, E., Pan, I., Acha, S., & Shah, N. (2019). Smart energy systems for sustainable smart cities: Current developments, trends and future directions. *Applied Energy*, *237*, 581–597. doi:10.1016/j.apenergy.2019.01.024

Powells, G., Bulkeley, H., Bell, S., & Judson, E. (2014). Peak electricity demand and the flexibility of everyday life. *Geoforum*, *55*, 43–52. doi:10.1016/j.geoforum.2014.04.014

Princen, T. (2005). *The logic of sufficiency*. MIT Press.

Ramchurn, S. D., Vytelingum, P., Rogers, A., & Jennings, N. R. (2012). Putting the 'smarts' into the smart grid: A grand challenge for artificial intelligence. *Communications of the ACM*, *55*(4), 86–97. doi:10.1145/2133806.2133825

Saunders, H. D. (2015). Recent evidence for large rebound: Elucidating the drivers and their implications for climate change models. *Energy Journal*, *36*(1), 23–48. doi:10.5547/01956574.36.1.2

Schick, L., & Winthereik, B. R. (2013). Innovating relations–or why smart grid is not too complex for the public. *Science & Technology Studies*, *26*(3), 82–102. doi:10.23987ts.55289

Shove, E., Pantzar, M., & Watson, M. (2012). The dynamics of social practice. Everyday life and how it changes. *Sage (Atlanta, Ga.)*. Advance online publication. doi:10.4135/9781446250655

Shove, E., & Warde, A. (2002). Inconspicuous consumption: the sociology of consumption, lifestyles, and the environment. In T. Benton (Ed.), *Sociological theory and the environment: classical foundations, contemporary insights* (pp. 230–251). Rowman & Littlefield.

Silvast, A. (2013). *Anticipating interruptions. Security and risk in a liberalized electricity infrastructure*. University of Helsinki – Department of Social Research.

Skjølsvold, T. M., & Lindkvist, C. (2015). Ambivalence, designing users and user imaginaries in the European smart grid: Insights from an interdisciplinary demonstration project. *Energy Research & Social Science*, *9*, 43–50. doi:10.1016/j.erss.2015.08.026

Smil, V. (2010). *Energy transitions: history, requirements, prospects*. ABC-CLIO.

Sonetti, G., Arrobbio, O., Lombardi, P., Lami, I. M., & Monaci, S. (2020). "Only social scientists laughed": Reflections on Social Sciences and Humanities integration in European energy projects. *Energy Research & Social Science*, *61*, 101342. doi:10.1016/j.erss.2019.101342

Sovacool, B. K. (2016). How long will it take? Conceptualizing the temporal dynamics of energy transitions. *Energy Research & Social Science*, *13*, 202–215. doi:10.1016/j.erss.2015.12.020

Stern, D. I. (2017). How accurate are energy intensity projections? *Climatic Change*, *143*(3–4), 537–545. doi:10.100710584-017-2003-3

Throndsen, W. (2017). What do experts talk about when they talk about users? Expectations and imagined users in the smart grid. *Energy Efficiency*, *10*(2), 283–297. doi:10.100712053-016-9456-5

Viitanen, J., & Kingston, R. (2014). Smart cities and green growth: Outsourcing democratic and environmental resilience to the global technology sector. *Environment & Planning A*, *46*(4), 803–819. doi:10.1068/a46242

Wei, T., Zhou, J., & Zhang, H. (2019). Rebound effect of energy intensity reduction on energy consumption. *Resources, Conservation and Recycling*, *144*, 233–239. doi:10.1016/j.resconrec.2019.01.012

Wolsink, M. (2018). Social acceptance revisited: Gaps, questionable trends, and an auspicious perspective. *Energy Research & Social Science*, *46*, 287–295. doi:10.1016/j.erss.2018.07.034

Wüstenhagen, R., Wolsink, M., & Bürer, M. J. (2007). Social acceptance of renewable energy innovation: An introduction to the concept. *Energy Policy*, *35*(5), 2683–2691. doi:10.1016/j.enpol.2006.12.001

Wynes, S., Zhao, J., & Donner, S. D. (2020). How well do people understand the climate impact of individual actions? *Climatic Change*, 1–14.

Xu, L., Francisco, A., Taylor, J. E., & Mohammadi, N. (2021). Urban Energy Data Visualization and Management: Evaluating Community-Scale Eco-Feedback Approaches. *Journal of Management Engineering*, *37*(2), 04020111. doi:10.1061/(ASCE)ME.1943-5479.0000879

Chapter 5
Planheat Tool:
A Bottom-Up Approach at District Level to Plan Low Carbon Future Scenarios

Matteo Porta
Rina Consulting, Italy

ABSTRACT

More than two thirds of the European population live in urban areas. Cities are places where both problems emerge and solutions are found. They are fertile ground for science and technology, for culture and innovation, for individual and collective creativity, and for mitigating the impact of climate change. Cities are communities where to study environmental, social, economic impact of new energy technologies. A tool to unlock the planning potential of EU cities has been developed thanks to the collaboration of different EU research centers, SMEs and consultancy firms led by Rina Consulting involved in Planheat EU Project (funded by the European Union's H2020 Programme under grant agreement 723757). PLANHEAT Consortium developed and validated an integrated and easy-to-use, GIS-based, and open source tool to support local authorities in selecting, simulating and comparing alternative low carbon and economically sustainable scenarios for heating and cooling.

INTRODUCTION

The EU aims to be climate-neutral by 2050, an economy with net-zero greenhouse gas emissions. This objective is at the heart of the European Green Deal and in line with the EU's commitment to global climate action under the Paris Agreement (EU Commission - Climate Action, 2020). EU Member States are required to develop national long-term strategies on how they plan to achieve the greenhouse gas emissions reductions needed to meet their commitments under the Paris Agreement and EU objectives. These plans need to be updated every five years and they could be made up of an ensemble of regional or local plans and strategies. Member States have the commitment to encourage municipalities and other public bodies to adopt integrated and sustainable energy efficiency plans with clear objectives, to involve citizens in their development and implementation and to adequately inform them about their

DOI: 10.4018/978-1-7998-7091-3.ch005

content and progress in achieving objectives. On the other side, Smart Energy City (SEC) is an emerging urban development strategy in Europe and collaborative planning (an holistic approach that includes different assets of cities' sustainable development) is one of the main elements (Mosannenzadeh, et al., 2017). More and more cities across EU are participating to initiatives and EU funded projects related to matching smartness concepts and solutions and urban planning (Smart Cities Information System, 2020) engaging local stakeholders in their local energy measures (Krog, 2019). The Guidebook 'How to develop a Sustainable Energy and Climate Action Plan (SECAP)' published in 2018 by JRC (Bertoldi, 2018), provides detailed, step-by-step guidance to local authorities to develop effective plans including smart, sustainable and inclusive solutions thus providing elements for the inclusion of smartness concept in the energy planning.

An extensive literary review identify with the term smart city, an umbrella concept that contains a number of subthemes such as smart urbanism, smart economy, sustainable and smart environment, smart technology, smart energy, smart mobility, smart health, and so on (Cocchia, 2014), (Trindade, et al., 2017), (Gudes, Kendall, Yigitcanlar, Pathak, & Baum, 2010), (Lara, Costa, Furlani, & Yigitcanlar, 2016). Smart planning, a new dimension of urban planning linked to procedural innovation in the management of territorial transformations and technological innovation of the generation, processing and distribution of data the creation of new "digital environments" such as GIS, BIM, models of augmented and mixed reality, useful for describing changes in human settlement in real time, represent the on-top element that can include all the instances of a smart vision for a sustainable development.

According to art. 14 of Directive 2012/27/EU (eur-lex.europa.eu, 2012), Member States should carry out a comprehensive assessment of the potential for high-efficiency cogeneration and district heating and cooling in order to define potentials and strategies for the decarbonization of H&C systems. Indeed, the local dimension (cities and regions) of heating and cooling strategy is crucial due to the fact that Europe's level of urbanization is expected to increase to approximately 83.7% in 2050 (ONU, n.d.). Thus, the solutions of the future should be tailored to the local conditions such as: specific territorial characteristics and climate conditions, current and forecasted end-user's demands in locations with sufficiently high population/activity densities and in the proximity of heating or cooling sources, opportunities for further development of district heating and cooling infrastructure as well as of integration of high efficient cogeneration plants, available Renewable Energy Sources, available waste heat sources at urban and industrial level. However, public bodies face the lack of appropriate easy-to-use tools to support them in the definition, simulation and evaluation of suitable strategies for sustainable heating and cooling tailored to local conditions for achieving the ambitious targets set-up in their local plans (Maya-Drysdale, Krog Jensen, & Vad Mathiesen, 2020). Indeed state of the art tools for energy planning are often designed for energy experts rather than for public authorities, are not user friendly, don't allow the visualization of maps for better integration into a comprehensive integrated urban plan and are often expensive.

In this context, PLANHEAT EU funded Project, has developed and validated through real use-cases an integrated georeferenced open source tool for the mapping, planning and simulation of H&C systems at local level. The tool objective is to support local authorities in selecting, simulating and comparing alternative low carbon and economically sustainable scenarios for heating and cooling (from a panel of advanced technologies) that could balance the forecasted demand. The PLANHEAT tool with its bottom-up approach aims to overcome the weaknesses of the currently developed tool with an easy to access online database to retrieve strategic information needed for planning future scenarios.

Background

Separate tools with specific scopes (mapping or planning or simulation) exist but there is no evidence of existing and integrated tools able to combine together mapping planning and simulation capabilities. Moreover, PLANHEAT tool is an open source tool and no license is required. These aspects reaffirm the innovative content of the PLANHEAT concept and its potential impact.

PLANHEAT concept

The PLANHEAT tool is divided in three modules: the mapping, the planning and the simulation modules.

The tool can assess future heating and cooling scenarios taking into consideration both the energy systems used to convert energy from sources to end-user plus the availability of local renewable sources and the current situation represented by the reference baseline scenario. This structure leads to the creation of future scenarios according to different planning criteria and to the calculation of KPIs for the evaluation of the compliancy to the target set at the beginning of the process. Regarding the supply mapping potential of the mapping module different local sources can be mapped in a geo-referenced environment: from "standard" renewables (such as biomass, geothermal and solar thermal) to unconventional sources of waste heat (from sewage systems, aquifers, data centers and supermarkets) till waste heat from industries.

For what concerns the energy conversion systems, both single building equipment and connection to district heating or cooling networks are options that can be taken into account. In particular for the latter, an optimization algorithm has been implemented in the District Heating and Cooling Route Optimizer (DHCRO) able to select considering different operational cost options, the best routing for the design of a completely new network or an extension of a already existing one. A graphical user interface is provided to guide the end user in the different steps of the process.

Keeping in mind the data confidentiality protection and the economic viability requirements, the PLANHEAT software is delivered as a desktop application. A desktop application comes with several advantages. First, the data confidentiality is ensured as the user can use the software directly on its own desktop. All the data is therefore secured on a personal computer unless the end-user decides to share it (internally or publicly). Secondly, a desktop application is easily deployed. The software is downloadable on both the PLANHEAT website and the GitHub repository.

Gathering data is a tedious and difficult task. For that reason, the PLANHEAT users have access to a web database. This database contains a wide range of European data that each user is able to access and use for their studies. This European data are available in addition to the one provided by the end-user to either substitute missing data or simply complete an existing set of data.

Last but not least, the web database is hosted online and the data are downloadable directly from the desktop application. In Table 2 the list of the available databases included in the PLANHEAT web database is provided; for each database the related type of data requested is reported as well as a possible repository of information at Municipality level.

PLANHEAT Tool Architecture

The application is a QGIS based desktop application. The three PLANHEAT modules are implemented as QGIS plugins embedded in the QGIS client. a consequence, the PLANHEAT platform profits from

Planheat Tool

Table 1. List of currently developed tools for energy mapping, planning and simulation

Functionalities	Delivery model	Open-source	Stregth(s)	Weakness(es)
Dynamic Energy Atlas (Mapping)				
Mapping and monitoring through time the current energy demand and renewable energy production in the region; Assess the unexploited potential for renewable energy; Assess the potential for heat exchange and heat networks; Design and implement policies aimed at the development and exploitation of energy landscapes with due consideration of spatial planning and environmental constraints.	Desktop tool	NO	Very flexible system: can be used in data poor and data rich regions, contains a lot of allocation algorithms and algorithms searching for optimal locations of facilities; Spatially-explicit information system equipped with GIS functionalities; Currently 10 applications installed at policy makers in Belgium; 7 years of development, ongoing development of new functionalities.	License required
Neighborhood Evaluation for Sustainable Territories – NEST (Mapping)				
Environmental assessment with life cycle approach at district scale; Mapping of the energetic performance at building scale: use of energy demand, energy consumption and energetic impact of each building; In conjunction with the building sector, NEST takes into account other aspects of a district: water treatment, waste treatment, transport and public lighting.	Desktop tool	NO	3D visualization Its data base is adapted to the current Spanish and French building sector legislation Automatization of the building energy demand assessment The environmental impact is assessed by life cycle approach	data are inserted manually for each building
Heat Roadmap Europe (HRE)/Pan-European Thermal Atlas 4 (Mapping)				
It is now available in two versions, Peta 4.1 and Peta 4.2. First release contained a 100m resolution grid of the modelled heat demand in 2015, a layer showing city areas where district heating systems exist in these countries, as well as a database of modelled excess heat supply. Second release, launched in September 2017, contains 10 layers in total, including the marginal cost of distribution capital.	Web based tool	NO	European coverage	Mapped data are not downloadable
CityPlanner (Planning)				
Sketch, analyse, and export from 3D cities Import CAD, GIS, & other files and web services Share & publish projects and crowdsource ideas	Web based tool	NO	Very interactive tool which helps to better communicate and engage citizens Big number of clients in all Europe	License required
FASUDIR IDST tool (Planning)				
Analysis of KPIs Assessing the Energetic Weak points of buildings Assessing the Feasibility of Heat Networks Identification of Synergies and Interactions Solar Potentials of surface areas Consideration of Financial Mechanisms	Web based tool	NO	Based on CityGML data model Based on a powerful energy simulation software (IES-VE)	The tool is the result of a project which is not currently supported by any partner Applicability of the FASUDIR IDST tool is mainly for the definition of strategies and technical solutions at district level while establishing the retrofitting framework
ACCENT tool (Planning)				
provides maps and data to design actions reducing the carbon intensity of the city. It's designed for local authorities, building sector specialists and energy providers.	Web tool	No	Combine mapping and planning functionalities	Licensed
LEAP (Stockholm Envir. Institute) (Planning)				
LEAP, the Long-range Energy Alternatives Planning System, is a widely-used software tool for energy policy analysis and climate change mitigation assessment developed at the Stockholm Environment Institute.	Desktop application	No	Multi-year calculation. Free for government sector organizations based in developing countries.	Licensed for government sector organizations based in not-developing countries (appr. 3200€);
EnergyProforma (MIT) (Simulation)				
Assess energy use and carbon emissions at the neighborhood scale; Assess energy use and carbon emissions in four categories: operational, transportation, embodied, and renewable; Assess the energy and emissions impact of decisions, such as solar panel installation, or compare community design options.	Web tool	No	Provides neighborhood scale data on energy use efficiency, energy consumption, and carbon emissions	The tool is still in its infancy and not yet ready for widespread use. Currently, the tool is calibrated for Jinan, China, so use in other areas provide inconsistency in the results
EnergyPLAN (Aalborg University) (Simulation)				

continued on following page

Table 1. Continued

Functionalities	Delivery model	Open-source	Stregth(s)	Weakness(es)
EnergyPLAN simulates the operation of national energy systems on an hourly basis, including the electricity, heating, cooling, industry, and transport sectors..	Desktop application	No	Freeware; Hourly simulations of energy systems at national scale	Mapping functionalities not included (Not GIS based /no results visualization on maps); Limited to national framework (not suitable for local analyses at regional and city scale)
EnergyPRO (EMD International A/S) (Simulation)				
In energyPRO user can model virtually any type of technologies from well-known, fossil fuel based production units to state-of-the-art renewables.	Desktop application	No	User-friendly	Licensed; Mainly related to district heating projects
HomerPro (HOMER Energy) (Simulation)				
The HOMER Pro® microgrid software by HOMER Energy is the global standard for optimizing microgrid design in all sectors, from village power and island utilities to grid-connected campuses and military bases.	Desktop application	No	Customizable with modules according to the needs of end-users	Licensed
TIMES/MARKAL (ETSAP) (Simulation)				
The TIMES model generator combines two different, and complementary, approaches to modelling energy: a technical engineering approach and an economic approach. In a nutshell, TIMES is used for, "the exploration of possible energy futures based on contrasted scenarios"	Desktop application	No	Models to represent the time evolution of global, national, regional, state or province, or community energy systems in equilibrium among economic, engineering and environment objectives (4E models); Detailed multi-year analysis; Powerful data management systems.	Free but it requires a GAMS system (licensed) is to run the model, along with a data base management system (from to 3000€ to 15000€ depending on type of institution); Mapping functionalities not included (Not GIS based /no results visualization on maps); No hourly simulations; Limited to 9 specific time intervals.
Mesap PlaNet (Seven2one) (Simulation)				
Mesap (Modular Energy System Analysis and Planning Environment) is an energy-system analysis toolbox, and PlaNet (Planning Network) is a linear network module for Mesap that is designed to analyse and simulate energy demand, supply, costs and environmental impacts for local, regional and global energy-systems.	Desktop application	No	The model has simulated a 100% renewable energy-system The simulation is carried out in a user-specified time-step which ranges from one minute to multiple years, and the total time-period is unlimited.	Licensed (6800 €); Mapping functionalities not included (Not GIS based /no results visualization on maps).

Figure 1. Modules of PLANHEAT tool

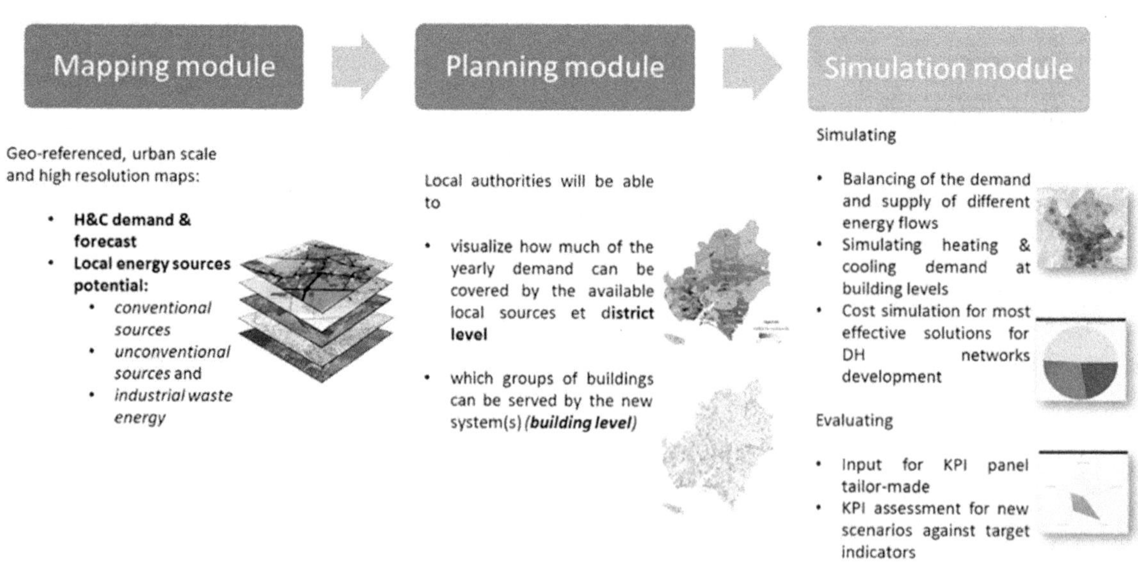

Table 2. List of databases accessible form PLANHEAT web server and related modules functionalities

Type of Data	PLANHEAT Module	Note on the databases (1) or the availability of data at city level (2)
Administrative boundaries of the municipality and districts	Mapping CMM	1) GISCO-NUTS 2) Information available from the city
Aggregated data about heating demand (final energy consumption of each energy vector for the city or the selected area).	Mapping CMM	1) EUROSTAT – DG Energy databases from EU 28 at National Level 2) Data available from local DSOs or Gas Grid managers
Share of Renewable Heating at city level	Mapping CMM Planning	1) SHARES and EUROSTAT database, STRATEGO Data, data from EC,2014[1] 2) Data available from local producers
Aggregated Data about cooling demand	Mapping CMM	1) From STRATEGO project 2) Difficult to be estimated (SEAP)
Share between residential and tertiary services heating/cooling consumption	Mapping CMM	1) EUROSTAT – DG Energy Databases and EU Building Stock Inventory cross reference 2) Difficult to be estimated (SEAP)
Share of DHN end users and of single building heating systems	Mapping CMM / Planning	1)From EuroHeat and Power Statistics 2) From documentation related to heating systems installed at city level
Data About CHP installation and how much thermal energy is recovered (operative factor)	Mapping CMM	1) Data from EEA 2)From local energy utilities data (i.e. compulsory declaration of CHP producer for feed-in tariffs)
Assessment of the quality/efficiency/age of heating system installed at city level and their share	Mapping CMM / Planning	1) DG ENERGY Database (2012) 2) From documentation related to heating systems installed at city level
Assessment of the quality/efficiency of single building envelope and EPC	Mapping DMM	1) From Tabula/Episcope database and from ODYSEE database 2) From local register of EPC
Share of DHW and Space Heating of Final Heating Demand	Mapping the demand	From EU Building Stock Inventory and Literature
Corrective factors for the evaluation of future demand related to the increase of energy efficiency in the heating system	Planning	1) From EED targets at National Level 2) From local SEAP
Energy renovation measures to be applied / required at city/building level for the calculation of future demand	Mapping the demand Planning	1) From Tabula/Episcope database and from ODYSEE database 2) From local SEAP/local project authorized at city level
Increase of the inhabitants or density in specific areas	Mapping the demand Planning	1) From EUROSTAT (national Level and also city Level – GEOSTAT population shapefile 2) From local census
Cadaster data	Mapping DMM	1) From EUROREGIONALMAP - EUROEGEOINFO (preliminary information) and from some National GIS cadaster. OPENSTREET MAP data for the footprint 2) From local cadaster and archive
Land Cover Data	Mapping Demand and Supply	1) From EU Databases and Atlas such as GEOSTAT, LUCAS, DEGURBA, CORINE 2) From local cadaster and archive
Use of the buildings	Mapping DMM	1) EU Builidng Stock Inventory 2) From local archive and permitting procedures
Age of the building	Mapping Demand	1) EU Builidng Stock Inventory 2) From local Cadaster and archive
Presence of Cultural Constraints for building renovation	Mapping demand	1) Tabula databases 2) From local archive and local Cultural Heritage specifications
Municipal wastes produced per inhabitant (incineration process potential)	Mapping Supply	1) Geostat Databases 2) From local waste utility data
Geothermal Potential	Mapping Supply	1) Heat Roadmap Europe, GEODH databases 2) Geological analyses
Sewage Network mass flowrate and number of sewage treatment plants	Mapping Supply	1) Inhabitants density and specific algorithms/correlations from literature 2) Measured data from the sewage local utility
Emission from local industrial plant	Mapping Supply	1) EU ETS databases 2)Emission measured by local industries
Ventilation rate from Underground shaft or car parking	Mapping Supply	1) Local and EU technical normative [Vol/h] 2) Measured data if possible
Biomass Potential in urban and peri-urban area	Mapping Supply	1) BIOMASS FUTURES and STRATEGO database 2) Data related to local environment and forestry agency related to wood crops permitting
Solar Potential	Mapping Supply	1) PV GIS database 2) Irradiation measure at local level
Water Basin Data . Temperature of surface water	Mapping Supply	1) WATERBASE Database 2) Local measures from environmental agency (water quality check)
Local Emission Factor (CO_2) per each energy vector	Simulation	Covenant of Mayor data (country by country)
Primary Energy Factor for DH	Simulation	Reference from Literature – values country by country (E. Latosov, 2016)
Primary Energy Factor for electricity	Simulation	Covenant of Mayor data
Data about electric vehicles and electrified transport sector	Simulation	Data from SHARES Database, EUROSTAT at National level, McKinsey assessment of EU EVs market
Fuel and electricity costs	Simulation	Data from EUROSTAT (updated year by year)
Technology costs (CAPEX, O&M)	Simulation	DG ENERGY Database (2012)

the numerous QGIS functionalities in terms of data management (conversion, correction, visualization, export, etc.) visualization and report generation.

This choice has been consistent with outcomes of both interviews conducted and global desktop usage statistics, according to which the most common OS is Microsoft Windows (various versions). As far as GIS software is concerned, both ArcGIS (various versions) and QGIS (various versions) seem to be in common use. One respondent stated the explicit wish of his municipal government to migrate from ArcGIS to QGIS as part of a general move to open source software. Those two outputs consolidate the choice of creating a cross-platform and open source integrated tool in the context of the PLANHEAT project.

As aforementioned, PLANHEAT tool consists of three interoperable modules namely mapping, planning and simulation module (Figure 2). By using those modules, it is be possible to perform analyses applying the bottom-up approach. The bottom-up approach workflow starts from the information at the bottom of the information chain i.e. the heating, cooling and domestic hot water demand of the buildings (residential or tertiary) under investigation. Once this information is known (acquired from the end-user or calculated by the tool) the workflow proceeds upwards, towards the calculation of the final and primary energy consumption of the whole groups of buildings of the area/district/city considered. On the contrary, the top-down approach (as well present among the tool functionalities but still in development phase) starts from the information collected from available datasets of final energy consumption for heating, cooling and domestic hot water at aggregated level (area/district/city) and moves downwards towards the calculation of the energy demand of the group of buildings included in that territory. In the middle of both approaches, the information about the conversion technologies used to covert final energy into

Figure 2. The platform architecture

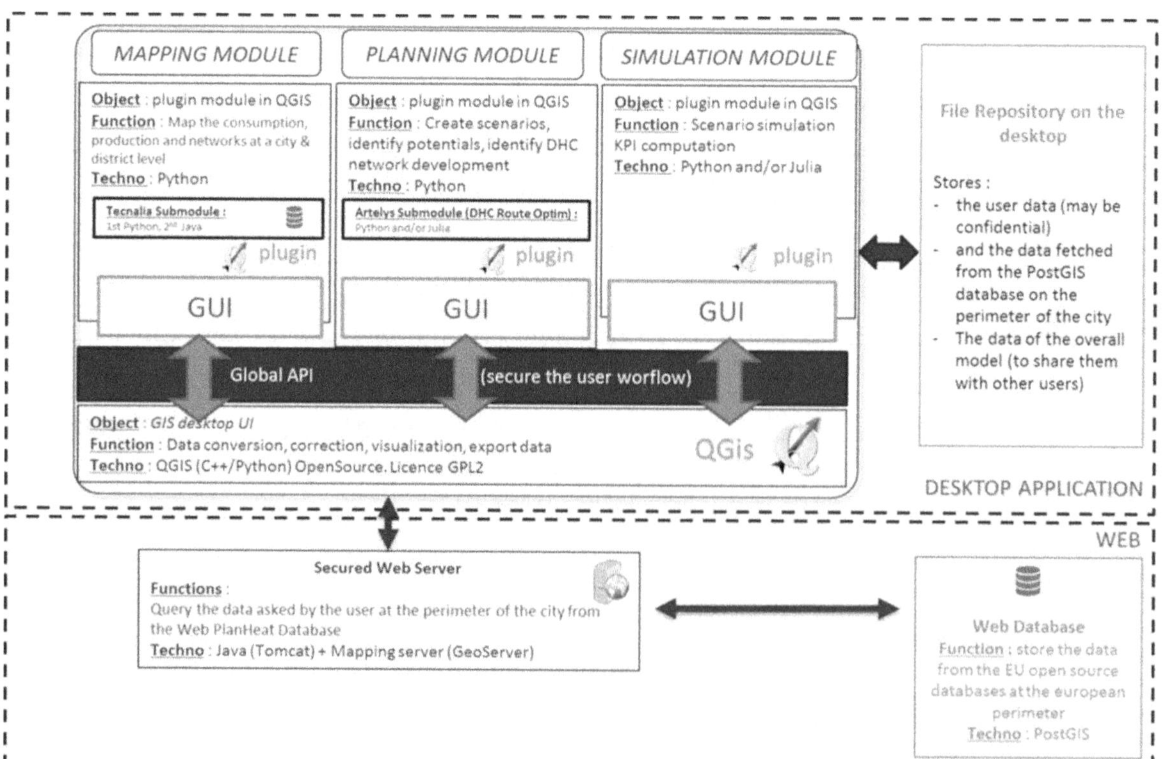

Figure 3. Bottom up approach schematic representation

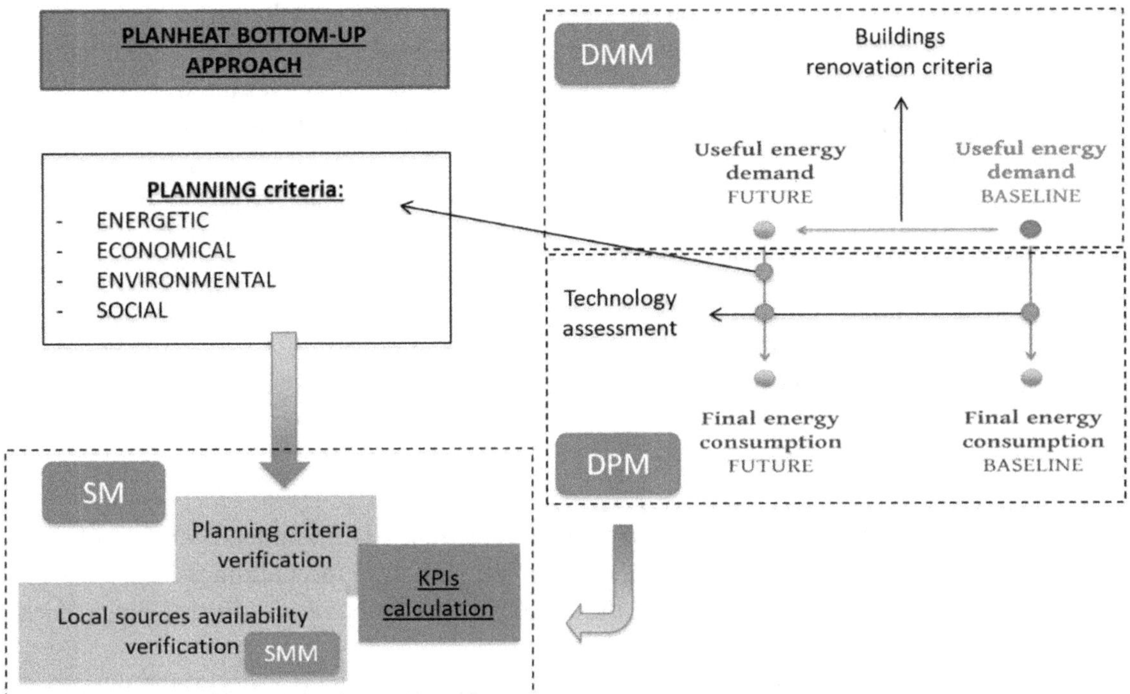

useful energy is needed. Between the two approaches, the bottom-up is the only one that can consider in the analysis the information of single buildings that are characterized at hourly level thus representing a very detailed instruments when precise information are available.

The bottom-up workflow starts from the baseline scenario assessment and specifically from detailed information at building level estimating hourly profile of Heating and Cooling (H&C) demand. Then, the end-user is provided with information about availability of local sources and finally enters the planning module for the definition of future H&C scenarios. Each future scenario is thus processed by the simulation module running numerical simulation on hourly basis to assess the energetic, economic and environmental feasibility.

Figure 3 represents the workflow of the bottom-up approach of the PLANHEAT tool that can be summarized as follow:

- In the District Mapping Module (DMM), every building of the district under investigation is assessed in terms of hourly profile for heating, cooling and domestic hot water demands both for current baseline scenario and for future new scenarios by the introduction of renovation criteria
- In the District Planning Module (DPM), starting from the information coming from DMM, both for the baseline and for the future scenarios the conversion technologies are considered and the final and primary energy consumption are calculated. Moreover a section dedicated to district heating and cooling networks is present with the possibility to consider completely new grids or already existing network to be renovated or extended

- Between the baseline and the future assessments, planning criteria are selected and targets are set for specific KPIs in the energetic, economic, environmental and social domains
- Then, the Simulation Module (SM) is launched and the KPIs selected are calculated. In particular the SM module allows to optimize the mix of technologies used to cover the three different demands by minimizing the operative cost related to maintenance operations and to fuel acquisition
- After KPIs calculation, the targets set with the planning criteria are verified as well as the availability of the local sources coming from the Supply Mapping Module (SMM)

The results obtained from the simulation of future scenarios can be visualized directly in the main GUI of the tool but the hourly profiles resulting from the calculation algorithms can be also exported as .csv files as well as the final list of KPIs. Similarly, the user can directly edit in the GUI the input required while the collection of data from the PLANHEAT database is managed internally by the software.

District Mapping Module (DMM)

The DMM represents the bottom line of the bottom-up approach. Via the implementation of a set of equation and the acquisition a minimum set of information related to the buildings in georeferenced files (raster or shape files) DMM comes up with the complete characterization of the buildings' heating, cooling and DHW demand.

In order to improve the thermal requirements of the façade, roof and openings, the DMM-PLANHEAT plug-in consider 3 types of refurbishment strategies:

- Façade → ventilated façade.
- Roof → add an insulation layer to the current roof.
- Window → replacement of the windows.

For each refurbishment strategy has been defined 3 different efficiency levels: low, medium and high. Due to these levels, DMM-PLANHEAT plug-in considers:

- 3 ventilated façade systems with different insulation thicknesses.
- 3 different insulation thicknesses to add to the roof.
- 3 different combination of windows: double glazing + aluminium frame with thermal bridge break (1); low emissive double glazing + PVC 3 chambers (2), triple glazing + PVC 6 chambers (3).

The main input that the end user should provide is a shape file with the following mandatory information each building of the area under study:

- building ID
- Building height value can be obtained defining directly into the shape file or providing the LIDAR files of the selected area
- Number of floors: In case that the end user does not provide the information about the height value of each building, it could be enough if the end user defines the number of floors of each building.
- Year of construction: This value defines the construction year of each building or the year when this building is registered in the cadastre

Figure 4. Graphical result of the DMM for the city of Antwerp in terms of energy demand for heating

- Use: definition of the main use of the building. Highlight that this tool does not allow to differentiate between ground floor use and the use of the other floors: for example, when the ground floor use is commercial in a Residential building).
- Some other information such as the country of the area under study is directly defined by the GUI of the DMM.

As a result of the process, a csv file is generated with 8760 values of heating, cooling and DHW of the hourly profile for each building under study and the annual demand is than visualized in QGIS environment as shown in Figure 4.

This output can be used to detect the areas with the highest energy demand and to relate these demands to the type of building according to its use, age range, or other parameter associated. In the specific case represented in Figure 4, in the historical district of Antwerp, the treatment of the output shape file shows that the heating and DHW energy demand of more than 23.5% of the buildings and more than 14.56% of the net heated area is greater than 166 kWh/m2,which reflects the average heating and DHW energy demand for existing buildings of Belgium. This value highlights the inefficient energy performance and need to improve all these buildings. In addition to the implementation of passive strategies and considering the heritage of the district, high efficient active systems and renewable energy should also be considered to improve the sustainability of the energy supply system of the historical center of Antwerp.

Supply Mapping Module (SMM)

The Supply Mapping Module of PLANHEAT tool is providing the user with a graphical vision on the heating and cooling demand and the supply in a city or district. In order to facilitate end-user visualization, the PLANHEAT plug-in is embedded in the open-source QGIS software. The Mapping Module interfaces with the Planning Module, which allows the user to define future scenarios via the calculation of different KPIs in 4 main domains: economic, environmental, energetic and social.

The Supply Mapping Module (SMM) in particular enables to map the yearly theoretical-technical potential of sources for low carbon heating and cooling. Both simplified and detailed models are available to overcome the issue of data availability. Potential sources can be classified according to their distribution in space: i) point sources which physical location is known very precisely (such as excess heat supply from industrial plants), ii) line sources (e.g. heat extraction from rivers), iii) surface sources distributed over a full territory (e.g. biomass forestry).

The approach for every source is similar:

- Data on the location and the size of the source are collected (for instance from Eurostat, CORINE land cover map…), pre-processed and compiled in the PLANHEAT webserver;
- Additional spatial constraints can be taken into account on the map;
- The values in these maps (e.g. crop yield, temperature…) are converted to theoretical energy supply source potential by the PLANHEAT methodology;
- A conversion factor can be used to convert the theoretical supply to the technical supply, taking into account an average technological efficiency relevant for the source. In the Planning Module, this conversion will be refined;
- The resulting maps are imported in the PLANHEAT environment of QGIS and can be visualized by the user;

Figure 5. SMM potential supply from biomass (agriculture) [MWh/y/2500m2] for the city of Lecce

Figure 6. SMM potential supply from Urban Waste Water Treatment Plants for the city of Antwerp

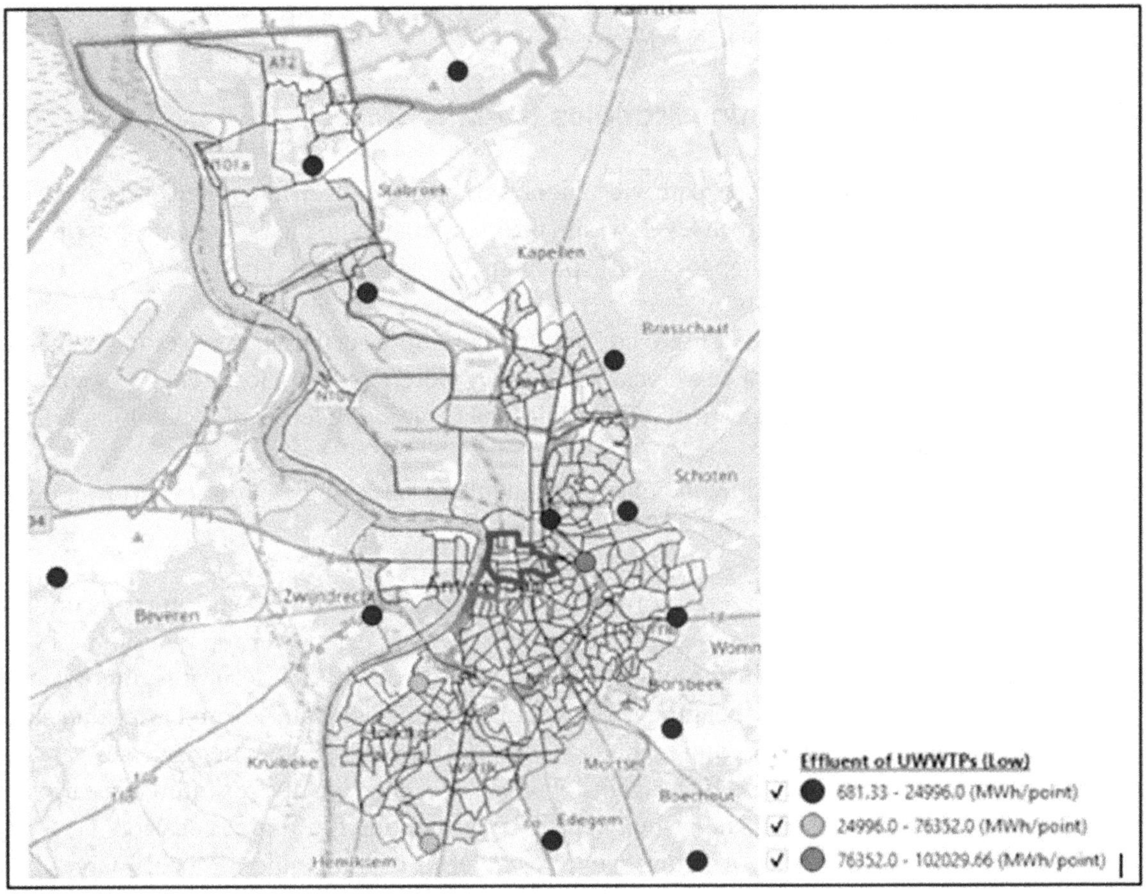

- If the user has better maps at their disposal (e.g. more detailed information on forest or agricultural zones, a map of the sewage system, a map of the subway system, already the supply map) the user can insert these maps as well.
- calculations are done on raster maps. The resolution refers to the level of detail of the mapping and expresses the cell size of the raster maps that will be created. The resolution can vary between 50m and 1km.

In Figure 5 and Figure ***6***, examples of application of SMM to the city of Lecce and Antwerp are provided.

Figure 5 displays the potential of agricultural biomass for energy purposes in the surroundings of the city of Lecce. The SMM is provided with a set of equations that allow the calculation of the potential of different types of biomass by matching information on land coverage, production yields and share of production to be destined to energy purposes. The countryside surrounding the city of Lecce reveals a good potential of exploitation of agricultural biomass.

The presence of several water treatment plants in the surroundings of Antwerp's city center, enable the exploitation of this low enthalpy level source, particularly exploitable in cities projected to the development of 4th and 5th generation DHCNs (Figure ***6***).

District Planning and Simulation Modules (DPM & SM)

According to PLANHEAT overall scope to provide public authorities with an integrated tool able to map, plan and assess energy scenarios at urban level, within the planning module users have the possibility to create future energy scenarios according to specific planning criteria and to get preliminary information about their related technical feasibility.

Within the bottom-up approach, end-users are able to plan future scenarios to cover the demand of heating cooling and domestic hot water at local level using more rationally local energy sources, by analyzing either single buildings or group of similar buildings (similar in terms of technologies used to supply heating, cooling and DHW).

The main functionalities provided by the planning module for planning energy scenarios at district level are:

- Fully detailed baseline scenario starting from information coming from DMM in terms of useful energy demand (hourly profile aggregated on yearly basis and peak power) up to building level;
- Set up of future scenarios importing future useful energy demand (hourly profile aggregated on yearly basis and peak power for heating, cooling and domestic hot water) from DMM and selecting future mix of sources/vectors and related technologies to satisfy future demand;
- Define planning criteria and objectives against which assess the "quality" of future scenarios;
- Preliminary calculation of final and primary energy consumption for future scenario;
- Preliminary check on feasibility of future scenarios in terms of availability of local sources
- Enable end-users to explore the possibility of developing DHC networks (both new district heating and cooling networks or expansion of existing ones) providing optimized design for pipe routes;
- Include information for transport sector both in baseline and future scenarios.
- The starting point for planning future energy scenarios at building/group of building level is the information coming from mapping module about baseline scenario, namely:
- The area of interest (e.g. buildings for which information are available in DMM);
- Calculated useful energy demand by DMM for both in baseline and in future scenarios;
- All energy sources/vectors potentially exploitable in the area of interest.

Following the steps to plan the future scenarios are described.

STEP 0: Import and Visualization of Relevant Information From Mapping Module

Maps and data coming from the DMM (Heating, Cooling and Domestic Hot Water demand of baseline and future scenario) and the SMM (Potential extractable from mapped energy sources) are directly transferred to the DPM or they can be uploaded in case of a subsequent use of the DPM.

Inputs and maps from the DMM/SMM can be charged directly via uploading them as well as drawing them on the Q-GIS environment. In this way the end user is also able to identify/design on a larger map a specific area which becomes the area of analysis of the DPM and automatically all the info related to demand/sources potential on that area are transferred and maps become visible in the Q-GIS visualization layer.

STEP 1: Baseline Scenario Characterization (Residential, Tertiary and Transport Sector)

In the previous step the user is able to charge local GIS Map and find all the info related to the demand of the buildings coming from the mapping module. Knowing the local situation, the end user is now able to define which technologies and sources are used to cover this demand.

To do so the end user has to characterize each building of its scenario via step 1 procedure.

The user can choose building by building (or selecting them on the map or selecting them from building list via its ID) sources and technologies to be allocated and he's also able to select an area of the district and provide information about H&C technologies and sources for that area.

Parameters required for H&C technologies/sources characterization in each building are reported in the following list::

Figure 7. Step 1 "network approach" input technology characterization menu

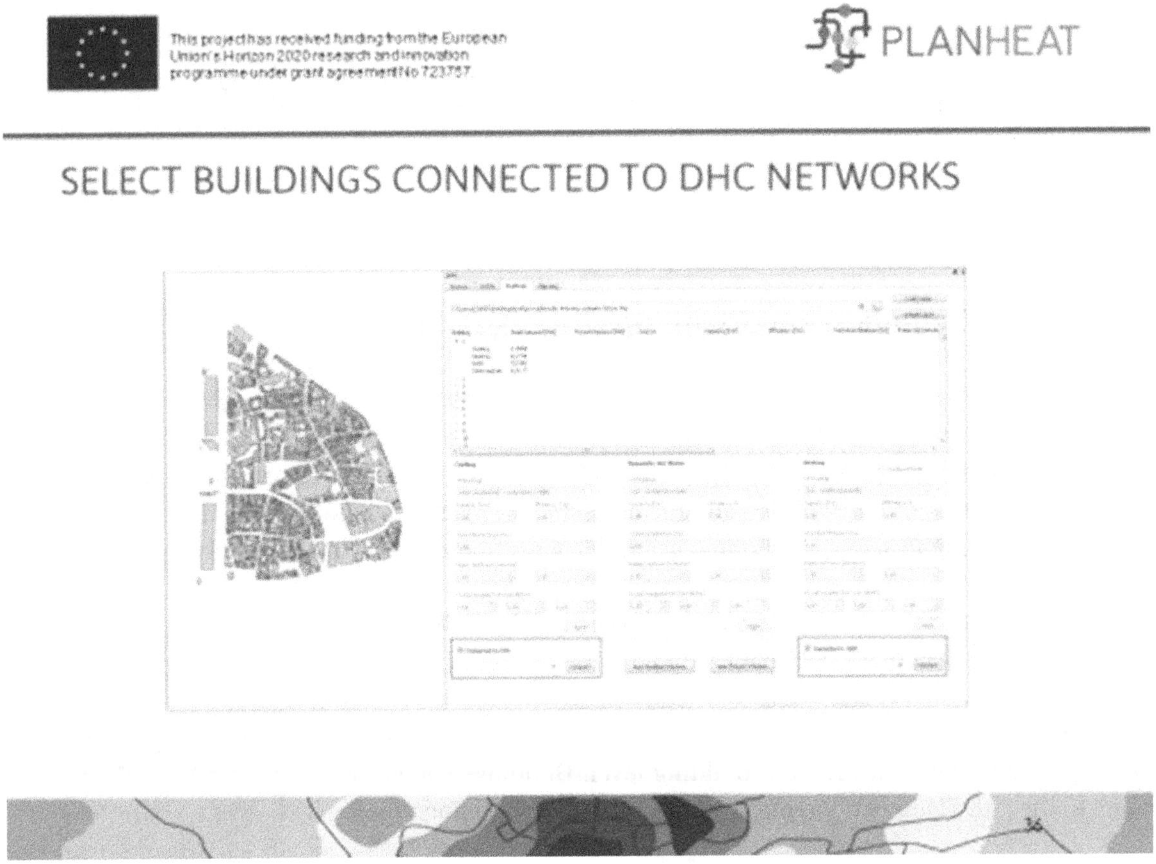

- Capacity [kW]
- Efficiency [%]
- Technical minimum [%]: entire number are suggested
- Ramp up/down [min]: entire number are suggested
- Technology CAPEX [€/kW]
- Fuel costs [€/KWh]
- Other OPEX (maintenance etc.) [€/kWh]

The end user can associate per each "type of demand" (H,C,DHW) one or more generator.

This information could be either input by users or provided by the PLANHEAT database as default values.

In parallel the user is also able to provide information about the presence of local DHCNs (District Heating and Cooling Networks) provide info about networks in the DPM (Figure 7).

Each network has an ID and it has to be defined in DHCN tab. In this tab, the end user is able to locate on the map generation points and to associate per each generation point a source and a technology.

A similar approach would be needed for the characterization of the H&C systems associated to DHN/DCN (which has to be associated to each connected building).

In the last STEP 1 tabs the end user is required to provide relevant data for the calculation of KPIs and simulation performance. Such data are referred to emission factors (CO_2, SO_x, NO_x, Particulate) and primary energy factors per each source and factors useful for the evaluation of the electromobility sector impact on a local H&C scenario. These data can be manually inserted by the end user according to their knowledge or data present in the PLANHEAT database.

STEP 2: Baseline Scenario Assessment (Simulation on Hourly Basis)

When the step 2 button is enabled, the user is able to launch the simulation module and perform a one year simulation on hourly basis of the local scenario calculating relevant KPIs and the related final energy consumption.

STEP 3: KPIs Definition and Planning Criteria Definition

In this step the end user can identify the planning criteria that guide the planning activity, assign a weight factor, select relevant KPIs and related target values for each selected criteria.

When selecting one planning criteria, the end user receives in the dialogue window some "suggestions" for the creation of the scenario and the assessment of H&C technologies that the user identifies in STEP4. According to selected planning criteria, energy sources are prioritized in Sources Tab.

STEP 4: Future Scenario Characterization (Residential, Tertiary and Transport Sector)

Starting from the baseline scenario created in Step 1 the end user is now able to change sources and technologies at building or group of building level. Buildings can be selected directly on the map, according to their ID (Number coming from the DMM) or building typology.

At this stage the end user is also able to optimize the routing of new/existing DHCN via the DHC route optimizer (DHCRO).

At the end of STEP 4 (Future scenario characterization) the end user is able to re-define if needed the sources and transport values of the baseline scenario, before starting the final simulation and benchmarking between baseline and future scenario, as the District Simulation Module button is now active.

STEP 5: KPIs Computation for Future Scenario and Evaluation

As soon as the end user has terminated the creation of the future scenarios, he/she is able to launch the simulation module and perform a one year simulation on hourly basis calculating relevant KPIs: these KPIs are benchmarked with STEP 2-3 ones also to evaluate the scenario effectiveness.

As it could be easily realized the end user has no interaction with the simulation module. Nevertheless, its role is very relevant also to evaluate at a first glance if the created scenario is feasible or not (for example if the solar production can effectively cover a certain peak demand thanks to the integration of a storage or not).

Considering that the DSM provides hourly simulation results per each building, as much complex could be a scenario and as longer it will be in terms of computational time. Nevertheless, the validation procedure on the validation cities during PLANHEAT Project period showed a computation effort up to 10 minutes each 500 buildings considered.

In this sense two different level of simulation can be performed a complete one and a simplified one.

In the first one 8760 hours values are calculated and the management of all H&C generators is optimized hour by hour in order to minimize costs.

In the second one only one "typical/sample" day per month is calculated and the management of all H&C generators is optimized hour by hour to evaluate if the created scenario is technically feasible.

The fact that the end user has no interaction with the DSM is a plus from a certain point of view (hidden simulation engine), but it could become a tricky point in case "loop" simulation that could happen if an end-user doesn't follow, during scenario creation, warning messages sent by DPM wizard related to unavailability of sources or not adequate thermal power capacity.

The results of the simulation are presented via two levels:

- KPI panels and the different related tabs (in the upper part of the result visualization panel)
- Final energy consumption tent menu (in the lower part of the result visualization panel)

PLANHEAT Planning Criteria and KPI

District Mapping Module and Supply Mapping Module provide the background information for the creation of future decarbonized scenarios. In particular SMM allows to verify that the target of energy consumption from local sources set in the planning phase corresponds to the actual availability in the proximity of the area under development.

KPIs are intended both to evaluate their variations between current baseline reference scenario and future scenarios generated in the planning module according to selected planning criteria and to benchmark future scenarios among each other evaluating their performance.

In Figure 8 the whole PLANHEAT KPIs panel is presented and the list of KPIs is provided. KPIs are divided in four main categories (domains).

Figure 8. PLANHEAT planning criteria and KPIs

Economic planning criteria

- Creation of **economically feasible H&C scenarios**
- **Operational cost** reduction
- Reduction of **thermal energy tariffs** for DHC end users
- Achieve acceptable **levelised cost of electricity** (LCOE) **and heat** (LCOH)
- Cost reduction due to **CO2 savings** (Cost per ton CO2 reduced in terms of grants needed and related to other technologies)
- Increase **security of supply**

Environmental planning criteria

- Greenhouse gasses, NO_x, SO_x and particle **emission reduction**
- **Noise pollution** reduction

Energy planning criteria

- Reduction of **primary energy consumption**
- Reduction of **final energy consumption**
- Reduction of **useful energy demand**
- Increased **share of RES**
- Increase the **utilization of local waste heat sources**
- **Reduction of conventional fuels** consumption
- Reduction of **energy import**
- Increase share of DHC in the total final energy supply
- Increase **share of cooling**
- Increase **share of electrical transport**

Social planning criteria

- **Jobs** creation
- Decrease **energy poverty**

Energetic domain takes into consideration parameters related to final, primary and useful energy consumptions; absolute variations with respect the values of the assessed baseline scenario are considered and savings are also calculated.

CEN and ISO norms start the energy performance evaluation of a building from its demands, passing through the distribution systems delivering energy to the different zones, the generation systems providing the energy until the evaluation of the primary energies needed to fulfil the building needs. This is the approach that has been adopted in PLANHEAT Project.

The first step of calculation takes into account the net energy demand at building level to maintain desired thermal comfort levels both for heating and for cooling purposes. This amount depends from different factors such as the internal temperature set point, the external ambient temperature, the building type of use, the inside human activities, the type of transparent and opaque surfaces of the building, the type of insulation of the façade, the age of the building. In particular in the District Mapping Module a detailed analysis at building level is carried out in order to assess buildings performances and consumption in baseline and future scenarios.

The second step of calculation has a bigger impact in district heating and cooling networks. In the assessment of the energy performance of these systems KPIs related to the efficiency of the distribution system are taken into account and calculated.

The third step of calculation considers the technologies that are used to transform the chemical energy of the energy sources (final energy) into energy to be directly used (in form of hot/cold water or air). Different technologies are considered for each energy source together with their specific efficiency.

The last step considers primary energy factors in order to calculate from final energy consumption the primary energy consumption of each source; this is also a method to compare different sources among them and to determine the whole primary energy consumption of the area under investigation.

Environmental domain takes into account parameters that quantify the production of CO_2 and environmental pollutant in the energy generation and transformation processes such as NO_x, SO_x and PM10. These parameters are derived through the application of some conversion factor from the final or primary consumption related KPIs. Absolute variation and savings are calculated from baseline reference scenario to any possible future scenario created by the end-user.

The planning criteria and KPIs considered within the economic domain are intended to evaluate not only the profitability of the intervention proposed in the creation of the future scenario but also to define the cost associated to intervention focusing on CO_2 emission decreasing, to compare costs of production from different local and imported sources, to verify the advantages for the end-user to move from a technology to another. Also within this domain, absolute values and savings are calculated to compare scenarios and define their performance.

Social KPIs and planning criteria focus the attention in the definition of parameters such as energy poverty, job creation and security of supply in all the scenarios assessed in order to quantify variation.

Discussion

PLANHEAT tool wants to address the main issues related to a modern approach to sustainable planning with an easy-to-use instrument, open source, accessible to many type of end-user without the requirements of specific technical skills. Nevertheless, it is undeniable that more and more technical competencies are required to keep up with the innovative solution introduced with the development of the concept of smart cities: not only the cities need to develop technical solutions to be smarter but also methodologies and processes need to adapt to this new frontier towards sustainability. Planning processes, needed for the development of new SECAPs, will more and more rely on the possibilities provided by BIM (Building Information Modelling) methodologies and GIS capabilities, therefore the technical competencies of the end-users (the main target of PLANHEAT tool are Municipalities) need to adequate to the complexity level accompanying these instruments. One of the first criticalities emerged when developing the tool was the variegated level of preparation of different Municipalities across EU: from the ones on the cutting edge to those where digitalization was still to be introduced. This represented a first barrier to overcome since as the tool was intended to be used by the widest range of end-users big, differences among end-users capabilities and resources led to a sort of compromise. The rooms for development are a lot towards a more massive use of digital data.

Nevertheless, there is a barrier that exceeds the level of expertise of the end-user, that is the availability of reliable data. With the colloquial expression "garbage in – garbage out" we refer in general to the facts that reliable results depends not only from the accurateness of the methodology implemented but also from the reliability of the input used. For the specific case of the PLANHEAT tool and in general of the tools having the objective to provide an help to planning processes, a number of preliminary information are requested to run simulations: data needed to calculate buildings' energy demand, data related to the type of conversion equipment currently in use, data related to land use or solar irradiation. Also in this case the more advanced a Municipality is about data availability, the more can rely on the results from every mapping, planning and simulation tool that, at the end of the day, represents a set of transfer functions from some inputs to some outputs.

The aim of every tool currently available to support Municipalities into the development of their SECAP is to develop and constantly upgrade the functionalities in order to provide an holistic planning instrument able to embrace all the domains related to the energetic aspects. The PLANHEAT tool for instance already includes a module (still under development) considering the transport sector of a territory in relationships with the heating and cooling sectors. PLANHEAT is actually focused in heating and cooling sectors but already considers the relationship with the electricity sector including technologies that produces (CHPs) or consume (Heat Pumps) electricity.

A big advantage of PLANHEAT is that is an open source tool that can be not only used but also enhanced by anyone only by accessing the source code publicly available. The idea was to develop in a democratic process a comprehensive instruments to support public entities to achieve the targets set for 2050 and beyond with a limited initial expenditure.

CONCLUSION

Preliminary results on validation cities of the PLANHEAT Project funded by the EC, demonstrated that the tool has big potentialities and can act as an incentive for the less advanced municipalities in terms of data availability, experience, expertise and informatization level, to adequate their systems to the most advanced forerunners. Nevertheless the tool has still big margins of development that are fully exploitable by the energy community being an open source tool.

Other similar tools currently present and representing the SoA have some limitation mainly due to the fact that are licensed or not covering the three aspects covered by PLANHEAT in a unique solutions: mapping, planning and simulation.

The need to develop advanced tools for sustainable planning is requested at EC level and the importance of including those solutions in advanced and smarter environments such as BIM and GIS is recognized as pivotal in the development of the smart cities of the next future.

PLANHEAT tool is currently under development by different actors towards this direction.

ACKNOWLEDGMENT

The tool has been developed in the framework of the Horizon 2020 EC financing program, thanks to the collaboration of all the partners of the consortium playing different complementary roles:

- RINA Consulting (ITA) the Project Coordinator, developer of the City Mapping module wizard, of the District Planning Module, responsible of the validation activities and technical responsible of the validation city of Lecce
- Technical University Delft (NL), responsible for the mapping of the unconventional energy sources module
- University of Zagreb (HR), developer of the Simulation Module, leader of training activities and technical responsible of the validation city of Velika Gorica
- National Observatory of Athens (GR) developer of satellite based heating degree hours and cooling degree hours database
- VITO (BE), developer of the city and the supply mapping modules, and technical responsible for the validation city of Antwerp

- TECNALIA (ESP) developer of the District Mapping Module
- Euroheat & Power (BE), responsible for the dissemination activities
- GEONARDO (HU) responsible for e-learning platform setup and update
- Artelys (FR) developer of District Heating and Cooling Route Optimizer and responsible for the software integration activities as well as developer of the main GUI of the integrated tool.
- The validation city of Antwerp (BE)
- The validation city of Velika Gorica (HR)
- The validation city of Lecce (IT)

References

Bertoldi, P. (2018). *Guidebook 'How to develop a Sustainable Energy and Climate Action Plan (SECAP).* European Commission, Joint Research Centre.

Cocchia, A. (2014). Smart and digital city: A systematic literature review. *Smart City*, 13–43.

EU Commission - Climate Action. (2020). Retrieved from https://ec.europa.eu/clima/policies/strategies/2050_en

eur-lex.europa.eu. (2012). Retrieved from https://eur-lex.europa.eu/legal-content/EN/TXT/?uri=celex%3A32012L0027

Eurostat. (2017). *NUTS database.* http://appsso.eurostat.ec.europa.eu/nui/show.do?dataset=apro_cpnh1&lang=en

Gudes, O., Kendall, E., Yigitcanlar, T., Pathak, V., & Baum, S. (2010). Rethinking health planning: A framework for organising information to underpin collaborative health planning. *The HIM Journal*, •••, 18–29.

Krog, L. (2019). How Municipalities Act under the New Paradigm for Energy Planning. *Sustainable Cities and Society, 47.*

Lara, A., Costa, E., Furlani, T., & Yigitcanlar, T. (2016). Smartness that matters: Comprehensive and human-centred characterisation of smart cities. *Journal of Open Innovation*, 1-13.

Latosov, E., Kurnitski, J., Thalfeldt, M., & Volkova, A. (2016). Primary Energy Factors for Different District Heating Networks: An Estonian Example. *Energy Procedia, 96*, 674–684. doi:10.1016/j.egypro.2016.09.126

Limberger, J. (2014). *Assessing the prospective resource base for enhanced geothermal systems in Europe.* https://www.researchgate.net/publication/320622611_Limberger_et_al_2014_Data-supplement

Maya-Drysdale, D., Krog Jensen, L., & Vad Mathiesen, B. (2020). Energy Vision Strategies for the EU Green New Deal: A Case Study of European Cities. *Energies*, •••, 13.

Mosannenzadeh, F., Bisello, A., Vaccaro, R., D'Alonzo, V., Hunter, G., & Vettorato, D. (2017). Smart Energy City. *Cities (London, England), 64*, 54–65.

ONU. (n.d.). *Department of Economic and Social Affair - Population dynamic*. Retrieved from https://population.un.org/wup/Download/

Smart Cities Information System. (2020). Retrieved from EC-Funded Projects Tracked by the Smart Cities Information System: https://smartcities-infosystem.eu/sites-projects/projects

The Engineering Toolbox. (n.d.). https://www.engineeringtoolbox.com/density-solids-d_1265.html

The PVGIS Database. (2017). https://re.jrc.ec.europa.eu/PVGIS5-release.html

Trindade, E., Hinnig, M. P., Costa, E. M., Marques, J. S., Bastos, R. C., & Yigitcanlar, T. (2017). Sustainable development of smart cities: a systematic review of the literature. *Journal of Open Innovation: Technology, Market and Complexity*.

Vis, D., & van den Berg. (2010). *Best practices and methods Handbook*. http://www.uvm.edu/~cdl/2012workshop/BEE%20Biomass%20Methods%20Handbook%202010.pdf

KEY TERMS AND DEFINITIONS

BIM: Building information modelling.
CAPEX: Capital expenditure.
DHCN: District heating and cooling network.
DHCRO: District heating and cooling route optimizer.
DHW: Domestic hot water.
DMM: District mapping module.
DPM: District planning module.
GIS: Geographic information system.
GUI: Graphical user interface.
H&C: Heating and cooling.
JRC: Joint research centre.
KPIs: Key performance indicators.
OPEX: Operative expenditure.
OS: Operating system.
SEC: Smart energy city.
SECAP: Sustainable energy and climate action plan.
SM: Simulation module.
SMM: Supply mapping module.

ENDNOTE

1 EC, 2014. Communication from the Commission to the European Parliament and the Council on the short term resilience of the European gas system. Preparedness for a possible disruption of supplies from the East during the fall and winter of 2014/2015, COM(2014) 654 final. European Commission. Availabe at (2015-02-19): (https://ec.europa.eu/energy/sites/ener/files/documents/2014_stresstests_com_en.pdf).

Chapter 6
BIM Simulation Lab:
Fostering Digital Transformation in Local Small-Medium Enterprises and Public Administrations

Gabriele Pasetti Monizza
https://orcid.org/0000-0003-2197-6267
Fraunhofer Italia Research, Italy

Christoph Paul Schimanski
Fraunhofer Italia Research, Italy & Free University of Bolzano, Italy

Giada Malacarne
Fraunhofer Italia Research, Italy

Dominik T. Matt
Fraunhofer Italia Research, Italy & Free University of Bolzano, Italy

ABSTRACT

The architecture, engineering, and construction (AEC) sector is facing the digital transformation by promoting the building information modelling (BIM) as a standard methodology for the digital managing of information along the whole lifecycle of a construction work. Although small-medium enterprises (SMEs) and public administrations (PAs) are aware of the BIM benefit, they ask for pilot actions and tools for testing BIM applications in their daily activities in order to measure benefits and difficulties in detail. This chapter discusses the BIM Simulation Lab initiative which aims at establishing a laboratory for an effective and efficient BIM implementation, by promoting a physical space and specific services for supporting the territory in the DT. The authors describe the concept of the lab and they introduce an assessment method that adopts an indirect approach of the BIM benefit assessment leveraging principles from construction cost estimation and probabilistic risk management.

DOI: 10.4018/978-1-7998-7091-3.ch006

INTRODUCTION

The McKinsey Global Institute (MGI, 2016) highlights that only big enterprises have a higher level of digital maturity and, even among the same enterprises' size, some industrial sectors are affected by a low level of digitalization (such as basic good manufacturing or construction). According to this picture, the weakest realities such as small-medium enterprises (SMEs) and Architecture, Engineering, & Construction (AEC) sector are not benefiting from the Digital Transformation (DT).

The AEC sector may be defined an Engineer-to-Order (EtO) industry, according to Wortmann classification (Wortmann, 1983) This implies that every final product (building) is almost unique, a kind of prototype. Every new product has to be designed and engineered time by time, according to specific customers' needs. This strong customization is one of the main reasons for a lack in productivity. Besides that, the AEC identifies different task leaders (architects, engineers, general contractors, facility managers, etc) for each process usually structured in a serial workflow. Serial workflow reduces the efficiency of the information management system by enhancing the propagation of errors (small details neglected at the beginning could become huge problems in latest processes, such as installation and maintenance) because of a fragmented information flow among the task leaders (BSI, 2003 – PMI, 2013). The AEC is looking at the DT as a new opportunity to overcome the lack in productivity by pushing the collaboration in an interdisciplinary environment.

The Building Information Modelling (BIM), also known as Building Information Management, is a methodology for the digital managing of information along the whole lifecycle of a construction work. Handling 3D-geometry (such as technical models, blueprints, etc.) and alphanumeric data (costs, assembly, maintenance, etc.), it aims at implementing an integrated platform that spans project planning, design, execution, operations, and maintenance (Sacks et al., 2018). Thus, the BIM is pursuing the objective of driving the information management system towards an efficient DT. It is globally recognized as a standard methodology trough the ISO 19650 standard - Organization and digitization of information about buildings and civil engineering works, including building information modelling (BIM)-Information management using building information modelling - and the EU countries are adopting the BIM for public building commitments.

Although the benefits identified by the scientific community and the compulsory need of the BIM, referring to a local statistic sample (D.T. Matt et al., 2018), entrepreneurs know the BIM methodology and they are interested in BIM implementations. Less of the half of them are applying the BIM methodology in their daily activities, but most of them do not have a clear idea about the BIM methodology and its benefits. They ask for pilot actions and tools for testing BIM applications in their daily activities in order to measure benefits and difficulties.

Enlarging the statistic sample to the national territory, ASSOBIM measures that the 50% of entrepreneurs already uses the BIM methodology in daily activities, the 11% is planning to introduce it within a year and the 20% is planning to introduce it within 2 years (ASSOBIM, 2019). Among the sample that already uses the BIM methodology in daily activities, entrepreneurs define further aspects:

- The 66% ca. of them is highlighting a low awareness by clients about the added value of the BIM methodology. In other words, the BIM methodology cannot be considered a competitive advantage yet.
- The 50% of them is highlighting that the demand of the BIM methodology is going to raise within the next years and they are conscious of impacts on their enterprises and their processes.

- Most of them declares that introducing the BIM methodology will force deep changes in their procedures, practices and workflows.

Tezel, discussing applications of the BIM and Lean Construction methods, highlights the need of research activities aiming at exploring dissemination strategies of the BIM methodology among the Small-Medium Enterprises (SMEs) focusing on implementation methods (Tezel et al. 2017). These research activities should measure impacts of BIM implementations on processes, technologies and dissemination strategies aiming at informing operators and Public Administrations (PAs) within the AEC sector.

The BIM Simulation Lab is an initiative aiming at establishing a laboratory for supporting local SMEs and PAs towards an effective and efficient BIM implementation. The initiative promotes a physical space and specific services for supporting the territory in the DT. Flexibility and adaptive solutions are applied in order to support external users not only through customized services but also for highly customized products (buildings). These services include a specific assessment tool for measuring impacts of the BIM implementation in daily activities.

Referring to the Italian context, this essay would discuss the BIM Simulation Lab initiative in order to provide a reference case study for similar initiatives in different contexts. In a first section, the authors would introduce a benchmark analysis of exiting initiatives and their services across the Italian territory and a state-of-the-art analysis in assessment methods. In a second section, the concept of the lab and the assessment method would be introduced. Finally, in a third section, the authors would discuss the early results providing an overview of future research activities.

BACKGROUND

Benchmark Analysis

The DT in the AEC sector is related with the DT or the information management system that leads the complex and fragmented workflow of an industry (the Building Industry - BI) producing highly customized products (buildings). As discussed in the previous section, the BIM methodology is the standard approach for leading this transformation of the information management system towards an efficient digital information management system.

Phang identifies specific Critical Success Factors (Rockart, 1979) to be analyzed implementing the BIM methodology among SMEs. These factors define possible barriers and specific aspects that can be used as a reference for an initiative aiming at supporting local SMEs and PAs towards an effective and efficient BIM implementation. Table 1 summarizes the most relevant factors according to a specific literature review (Phang et al., 2019).

Phang highlights that management factors are the most requested ones, far beyond than investment and training factors. This picture offers similar results of ASSOBIM's study (discussed in the previous section): actions aiming at supporting the BIM implementation among SMEs must include management aspects in order to offer a clear understanding of impacts on procedures, practices and workflows. Thus, the success of the DT in the AEC sector through the BIM methodology relies on minimizing implementation impacts. The DT appears as a tailored suit: the less are the changes in consolidated daily procedures, practices and workflow; the higher are the benefits and the adoption rate. Thus, the main challenge is to provide methods and tools for a better understanding of implementation impacts.

Table 1. Critical Success Factors proposed by Phang et al.

Category	Critical Success Factors	Source
Management	Willingness to share information among project participants	Won et al., 2013
Management	Maser BIM model team/manager	Won et al., 2013
Management	Effective collaboration among project participants	Won et al., 2013
Management	Organizational structure to support BIM	Won et al., 2013
Investment	Continuous investment	Won et al., 2013
Leadership	Leadership of senior management	Won et al., 2013
Technology	Information-sharing protocols	Won et al., 2013
Training	BIM training programs	Won et al., 2013
Technology	Technical support for interoperability issues	Won et al., 2013
Management	Standardized work procedures for BIM	Won et al., 2013
Management	Top management support	Son et al., 2015 Tsai et al., 2014
Management	Coordination and integration among the professions	Tsai et al., 2014
Technology	Design validation of BIM tools	Tsai et al., 2014
Management	Definition of project goals before forming teams	Tsai et al., 2014
Technology	Producing models and drawings for construction and fabrication	Liao and Teo, 2017
Technology	Integrating model management tools with enterprise systems to share data	Liao and Teo, 2017
Technology	High accuracy of model-based documentation	Liao and Teo, 2017
Technology	Data sharing and access on BIM platforms	Liao and Teo, 2017
Strategy	BIM vision and leadership from the management	Liao and Teo, 2017
Management	Changes in organizational structure and culture	Liao and Teo, 2017
Training	Training on new skillsets and new ways of working	Liao and Teo, 2017
Strategy	Stakeholders seeing the value of adopting their own part of BIM	Liao and Teo, 2017
Leadership	Owner's requirement and leadership to BIM adoption	Liao and Teo, 2017
Technology	3D visualization enabling design communication	Liao and Teo, 2017
Management	Regulatory agencies' early participation to BIM use	Liao and Teo, 2017
Management	Design coordination between disciplines through clash detection and resolution	Liao and Teo, 2017
Strategy	Gaining competitive advantages from full BIM use	Liao and Teo, 2017
Technology	Standard platforms for integration and communication	Amuda-Yusuf, 2018
Investment	Cost of development	Amuda-Yusuf, 2018
Training	Education and training	Amuda-Yusuf, 2018
Management	Standardization (product and process)	Amuda-Yusuf, 2018
Strategy	Clear definition and understanding of users' requirement	Amuda-Yusuf, 2018
Management	Increased number of subcontractors experienced with BIM projects	Olawumi and Chan, 2019
Management	Greater awareness and experience level within the firm	Olawumi and Chan, 2019
Management	Increased involvement of project stakeholders in green projects	Olawumi and Chan, 2019

The AEC sectors produces highly customized products within a complex and fragmented workflow. This implies that procedures, practices and workflow are highly customized too. Methods and tools aiming at measuring implementation impacts should be a tailored service. Referring to the Italian context, the scientific community is promoting initiatives to support SMEs and PAs in the BIM implementation according to the strategy of a tailored service. According to an early analysis, the Table 2 shows the identified initiatives.

Table 2. Italian initiatives to support the BIM implementation as digital information management system

Name	Promoter	Business focus
DrawingTOthefuture	Politecnico di Torino	Research/University
Laboratorio BIM	Consulting and Management SRL	Private/Consulting
ABCLab	Politecnico di Milano	Research/University
Gicarus Lab	Politecnico di Milano	Research/University
LIE	Università degli Studi di Padova	Private/Consulting
BIS-LAB	Gruppo contec	Research/University
LBIM	Università degli Studi di Firenze	Research/University

These initiatives have been analyzed collecting the information freely available for promotional or commercial purposes. They can be classified according to two main business focuses:

- Private/consulting - initiatives promoted by private entities for consulting or supporting purposes.
- Research/University - initiatives promoted by Universities or no-profit research entities for education and research (fundamental or applied) purposes.

The initiatives are exploiting their services according to these business focuses. Private entities are mainly focusing on supporting or executing specific tasks according to actual standards and state-of-the-art practices, such as:

- Development of Employer Information Requirements (EIR)
- Compliance check with international standards
- 3D modeling for public procurement
- Development of BIM templates and libraries
- BIM design
- BIM commissioning
- Property Management
- Facility Management
- Project Management
- Surveys and mapping through drones
- Training programs for professionals

Universities or no-profit research entities are promoting activities aiming at establishing new methods and standards or at developing prototypes and proofs of concept, focusing on the following topics:

- Guidelines for the BIM implementation in PAs
- Interoperability and data exchange
- Advanced data representation (using Virtual Realty or Augmented Reality techniques)
- Digital fabrication (both off-site and on-site)
- Digital survey of building heritage (mostly focusing on historical heritage)
- Object recognition and reconstruction (ORR)
- Spatial Data Infrastructure (SDI)
- Didactic activities for students

The services are exploited applying specific equipment. Considering that it is not possible to collect homogeneous information about the adopted equipment for each initiative, the analysis cannot define minimum or average requirements. Each initiative is applying different equipment according to its own services. As a general observation, it is possible to highlight the following most applied equipment: drones, laser scanners, devices for Virtual Reality or Augmented Reality and 3D printers. In some initiatives the equipment is officially including only 3D workstations. All the initiatives are adopting standard software solutions available on the market for BIM authoring and for BIM analysis (such as BIM checkers, structural design tools, environmental design tools, etc.). Some of them are developing specific software components using visual scripting or scripting environments.

State of the Art in the Assessment of the BIM Benefits and Impact

As discussed in the previous section, initiatives to support SMEs and PAs in the BIM implementation are facing the challenge to provide methods and tools for a better understanding of implementation impacts. In the last decade, different metrics and approaches have been established in order to assess the benefit of the BIM approach against the traditional ways of project delivery. In this context, the BIM benefit should be viewed from various perspectives: According to Ghaffarianhoseini et al. (2017), BIM benefits can be grouped into a total of 9 clusters:

1. *Technical benefits* which facilitate the interoperability, enabling users to transfer design data and specifications between different software applications.
2. *Knowledge management benefits* which enable comprehensive information about the project, ranging from geometry, spatial information, quantities, etc.
3. *Standardization benefits* which facilitate the provision of information using a neutral and open standard data format (IFC).
4. *Diversity management benefits* related to specific BIM uses and applied to the built environment, such as: as-built renders, retrofit planning, energy management.
5. *Integration benefits* which lead the AEC sector to an integrated working approach across the whole project lifecycle, facilitating the complexity of construction projects.
6. *Economic benefits* which prevent schedule delays, documentation errors and contractual claims.
7. *Planning and scheduling benefits* refer to the use of 4D simulations where materials and components are automatically calculated, ordered and delivered.

8. *Environmental benefits* in a Lifecycle-Assessment (LCA) perspective refer to the advantages throughout the project lifecycle from the early conceptual design to demolition.
9. *Decision support benefits* refer to the opportunity of data sharing among different stakeholders that enhance constant evaluation and information control.

Even if the scientific community is paying higher attention on the investigation of the benefits of BIM one of the largest barriers to BIM adoption is a lack of consistent awareness on the balance between the cost of its adoption and its related benefits. To fill this gap, previous researchers, such as (Barlish and Sullivan 2012), Guerriero (2016) and (Zhou et al. 2017), investigated how to measure and evaluate these benefits.

Depending on the nature of the benefits to measure, both quantitative and qualitative metrics can be used. Examples of qualitative metrics are: sharing ideas across project´s stakeholders; communication among stakeholders; quality of results; information management across the project lifecycle. These metrics are usually evaluated using surveys and interviews with professionals involved in BIM projects. Examples of quantitative metrics are: productivity evaluated through the "gains vs. costs" formula; the number of requests for information, the number of requests for modification, the amount of savings regarding materials and building products, the earlier or later delivery. These metrics are usually evaluated through case study analysis comparing the results collected from BIM projects vs. non-BIM projects.

Even if a case study analysis is the most appropriate method to investigate the benefits of information technologies (Bakis et al. 2006), studies based on this method admit their limits due to the unique nature of a construction project (Barlish and Sullivan 2012; Li et al. 2014). This situation results in a growing skeptical attitude by SMEs towards BIM, since they should take a decision on whether to utilize BIM based on speculated benefits. For this reason, other researchers such as Li (2014) and Guerriero (2016) evaluate quantitative metrics in case studies, as a test in the development of a more general approach. The idea behind the development of a general approach lies with the fact that it could be replicated by all the stakeholders involved in a construction project (including SMEs) and in all type of projects independently from their uniqueness. An example is the approach proposed by Guerriero (2016) which based on return-on-investment (ROI) calculations.

The ROI is defined as follow:

$$ROI = \frac{Gain - Cost}{Cost}$$

The ROI approach proposed by Guerriero (Figure 1) is a combination of quantitative metrics (based on the financial valorization of the gain/losses of productivity and the indirect gains) and another set of metrics which considers the qualitative value of BIM. Moreover, since this approach is adaptive to heterogeneous project environment, it also considers the influence of specific BIM uses and BIM users.

However, many limitations emerge when investigating the relationship between BIM benefits and how these effects ROI. First, the relationship between BIM benefits and ROI is deeply influenced by the maturity level of BIM users and of their ability to perform advanced BIM uses (such as BIM 4D). For examples: tasks performed with a low BIM maturity level takes more time and gains a negative performance value if compared with the same task performed without BIM. Moreover, gathering the data needed to perform both the qualitative and quantitative analysis for ROI calculation is very difficult due

Figure 1. Metrics for BIM ROI assessment (Guerriero et al. 2016)

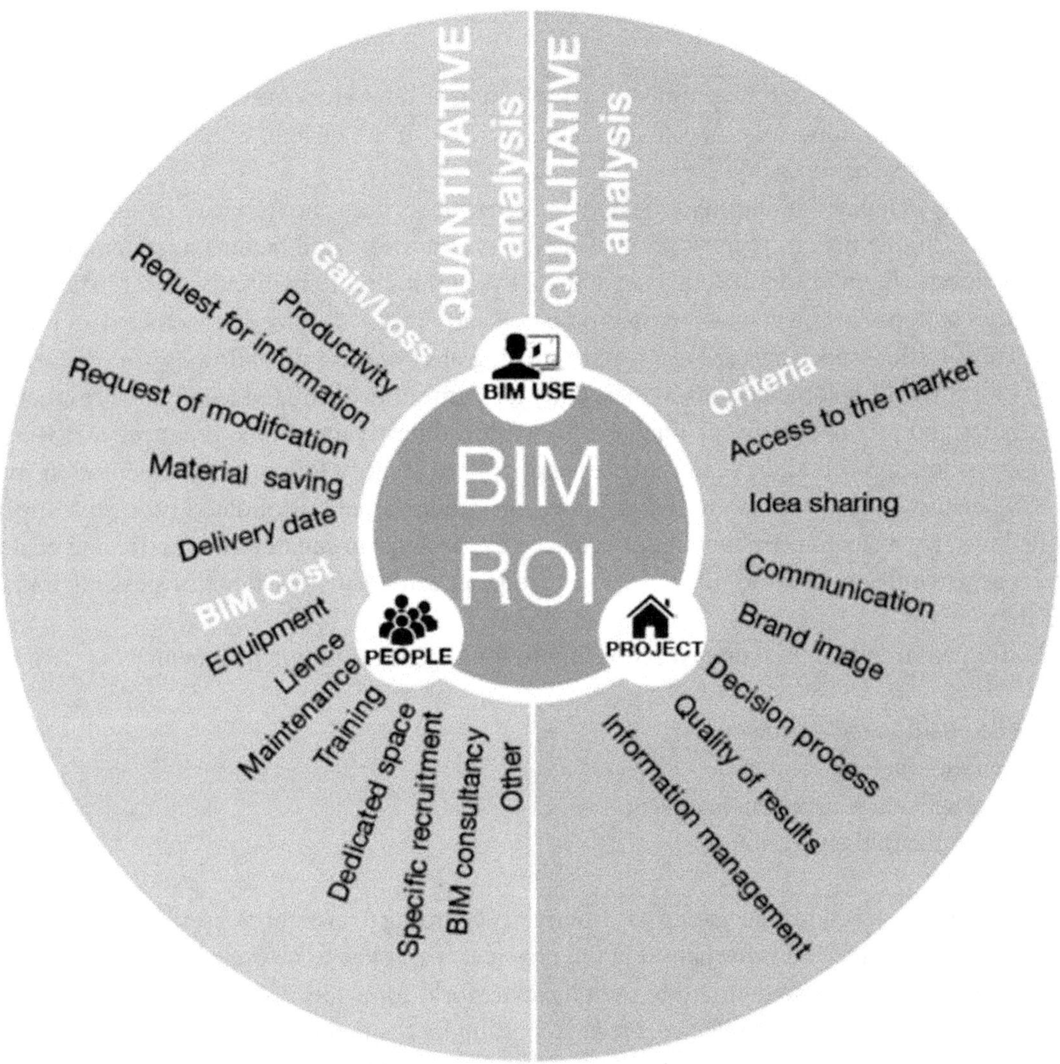

to unclear recording data principles and the unavailability of practitioners in sharing such productivity-related information. For example: some projects adopt BIM from the beginning, some other not; projects use BIM to perform some tasks, while other projects use BIM for some other tasks; some practitioners are not disposed to make sensible information public.

This interplay might affect ROI considerations and lead to an increased difficulty of direct benefit measurement when thinking of BIM implementations. For this reason, and in addition to a BIM-based experimental environment for local SMEs in the AEC sector in South Tyrol as well as supporting software prototype for holistic BIM workflow, the BIM Simulation Lab initiative also develops a framework for project-related evaluation of the BIM benefits following an indirect way by considering different BIM functionalities and their impact on risk management in construction projects. This framework is described later in this essay.

THE BIM SIMULATION LAB

Physical Space and Services

The BIM Simulation Lab initiative aims at establishing a laboratory for managing and visualizing digital information to support the decision-making process in the design, execution, operation and/or maintenance of construction works.

As previously discussed in introduction and background sections, an initiative to support SMEs and PAs in the BIM implementation must provide methods and tools for a better understanding of implementation impacts. In order to fulfill this need, the BIM Simulation Lab initiative establishes a physical space for pilot actions aiming at developing prototypes and proofs of concept, together with local SMEs and PAs. This physical space becomes an isolated testing environment that allows external users (SMEs and PAs) to test BIM implementations in their daily activities. Such isolated testing environment develops prototypes and proofs of concept concurrently with ordinary procedures, practices and workflows. This side-by-side approach allows external users not only to identify tailored implementation strategies of the BIM methodology, but also to directly measure impacts on consolidated procedures, practices and workflows. By a clear understanding of impacts, external users can define specific and customized management strategies in order to fulfill the most frequent category of Critical Success Factors (Phang et al., 2019).

The testing environment is arranged according to four specific sub-environments:

1. Data acquisition environment
2. Data processing environment
3. Data storage and sharing environment
4. Data visualization environment

This subdivision structures the testing environment according to critical steps of a digital information management system, since an information in the physical world has to be transformed in a digital data, until the digital data has to be returned to the physical world for a specific task (such as production of elements, on-site assembly, maintenance, etc.). Accordingly, the equipment of the testing environment follows the same structure in order to identify the proper technology for each critical step. Prototypes and proofs of concept are developed interacting at different levels with each of the identified sub-environments and applying the equipment needed for fulfilling their specific goals and objectives. Thus, the testing environment can develop tailored prototypes and proofs of concept within a common structure of the testing environment in order to enhance interoperability and data exchange.

According to the main business focus of Fraunhofer Italia (the promoter of the BIM Simulation Lab initiative), services to external users exploits activities aiming at establishing new methods and standards or at developing prototypes and proofs of concept, focusing on topics according to the national best practices of the same business focus, such as:

- Guidelines for the BIM implementation in PAs
- Interoperability and data exchange
- Advanced data representation (using Virtual Realty or Augmented Reality techniques)
- Digital fabrication (both off-site and on-site)

- Construction process management
- Digital survey of building heritage (mostly focusing on historical heritage)
- Object recognition and reconstruction (ORR)

The Data Acquisition Environment

The first sub-environment focuses its activities on data acquisition and survey, testing technologies and equipment for acquiring information from the physical world and transferring them in a digital information management system. A digital information management system that applies the BIM methodology arranges data according to two data types: geometries and alphanumeric proprieties (also known as metadata). Each of these data types applies different technologies and equipment for acquisition purposes.

Geometry can be acquired through direct or indirect approaches. Direct approaches imply that the operator commits a specific action between a measurement and the other. These approaches apply traditional equipment such as: rulers, measuring tapes, laser distance meters, ultrasonic distance meters, etc. The traditional equipment does not transform the measure into a digital data automatically, but it commonly asks for a manual procedure by the operator. The lack of an automatic transformation into a digital data can be identified as the main difference with indirect approaches. Indirect approaches apply equipment that measure an object through a semi-automatic or automatic sequence of measurements, such as: total stations, laser scanners, photogrammetry, etc. This equipment transforms measurements into digital data that must be post-processed for filtering and extracting purposes.

Alphanumeric proprieties include all the information more than geometry. This data type is wide, and a universal classification does not exist because it has to be adapted to specific (even case-by-case) needs. This data can represent or identify material type, material proprieties, assembly instruction, safety notices, maintenance notices, deterioration state, etc. The alphanumeric proprieties can be surveyed though different and heterogeneous equipment: cameras, infra-red cameras, climate data sensors, proximity sensors, infra-red sensors, ultrasound/infrasound sensors, penetrometers, etc. According to this picture, it is not possible to define a classification following the genesis and the evolution of the digital data as previously suggested for geometry data. Each equipment should be analyzed one-by-one in order to identify potentials and criticisms within a digital information management system that applies the BIM methodology.

Concurrently testing technologies and equipment, the data acquisition environment develops methods and automation strategies for data acquisition. Starting form facilitating the digital transition and the automation of measurements though direct approaches, this sub-environment also pursues prototypes and proofs of concept that apply advanced data processing and post-production techniques such as: Visual Recognition, Machine Learning, Deep Learning, Artificial Intelligence, etc.

The Data Processing Environment

This sub-environment focuses its activities on data processing and data analysis within a digital information management system that applies the BIM methodology. A BIM model, as a data container, holds the core of the data structure within the BIM methodology framework.

Nowadays, data processing and data analysis are the most developed topic within the BIM methodology. The market offers robust and validated tools though specific software solutions for BIM authoring and BIM analysis purposes, using both open and closed formats. These solutions are often well known by external users, who already adopted them in daily activities. According to this picture, the data processing environment pursues the development of prototypes and proofs of concept for defining interoperability methods and automation strategies for data analysis, beyond focusing on data processing and data analysis within BIM models though software solution available on the market.

Automation strategies for data analysis ensure the control and the update of the data structure within the BIM methodology framework. Once a first BIM model has been defined, it should be maintained and updated in order to provide proper information contents that evolve during the lifecycle of a construction work. Updated and precise information contents become inputs data for automated analysis that aim at facilitating performance analysis and error detections, not only during design and execution but also during operation and maintenance. Thus, automation strategies can be considered as a trigger for establishing a digital twin of the real object (building).

The data processing environment develops methods for the interoperability, offering the opportunity to test software solutions already available on the market focusing on interoperability issues and barriers. External users can test open formats in order to be conscious of available tools and barriers against the open exchange of information. Furthermore, this environment develops methods and strategies for data automation, starting from solutions that facilitate the update and the analysis of data, until the development of prototypes and proofs of concept that apply advanced approaches such as: Machine Learning, Deep Learning, Artificial Intelligence, etc.

The Data Storage and Sharing Environment

The third sub-environment focuses its activities on data storage and data sharing. The BIM methodology promotes collaboration; thus, data storage and data sharing can be considered the success factor for efficient collaborative strategies.

Solutions for data storage and data sharing (also known as CDE – Common Data Environment) are recently introduced in the market. They usually rely on on-premises solutions or cloud solutions, managing closed-format or open-format data. These solutions are known by external actors because usually they are offered side-by-side with software solutions for BIM authoring and BIM analysis. Solutions for data storage and data sharing act as hub for data exchange. Through cloud solutions, data are available everywhere and as soon as they are requested, as they ask for an internet connection as the only pre-requisite.

The interoperability, within solutions for data storage and data sharing, aims at the interconnection among different solutions. Although the interoperability within software solutions for BIM authoring and BIM analysis aims at defining standards for open-format data, the interconnection focuses on the reliability and security of data access. According to these focuses, the interconnection defines the capacity of data to be transferred from a solution to another, without losing information contents and ensuring data tracing.

The data storage and sharing environment develops methods for the interconnection, offering the opportunity to test solutions for data storage and data sharing. External users can test these solutions in order to be conscious of available tools and barriers against the reliability and security of data access. Furthermore, this environment develops methods and strategies for data sharing robustness, data tracing and information reliability through the development of prototypes and proofs of concept that implement specific technologies and approaches such as the blockchain.

The Data Visualization Environment

The last sub-environment focuses its activities on data visualization. According to the specific task (such as survey, building design, production of elements, on-site assembly, maintenance, etc.) that has to be performed, information have to be visualized using the proper technique, during the whole lifecycle of a building.

Independently from the data type (geometries or alphanumeric propriety), information has to be returned and queried at any time. Each information needs different visualization techniques, not only accordingly to its type (geometry, technical drawing, picture, text, technical proprieties, etc.), but also accordingly to its complexity. Complex information must be grouped or deconstructed, focusing on their dependencies and simplifying the control of the dataset (as a BIM model).

Data visualization techniques should allow the operator to reach the needed information as soon as it has requested, through specific technologies that do not impact the task in the physical world. Considering that the information detail increases during the lifecycle of a building, the interconnection between the task in the physical world and the information detail require solutions (methods and techniques) for facilitating the interaction with the datasets. Nowadays, several techniques are available for data visualization purposes which aim at enhancing interactivity and immersivity of operators with digital contents. The main objective of these techniques is to overcome barriers: passivity barriers (as the lack of immersivity due to traditional interaction methods, such as keyboards, mouse, monitors, etc.) and accessibility barriers (as the lack of portability and interactivity). The most known techniques are the Virtual Reality (VR) and the Augmented Reality (AR).

Virtual Reality establishes a virtual environment where an operator can explore and query digital information through different levels of immersivity. A VR environment can be established applying personal devices (such as VR headsets) or applying devices for groups of operators (such as VR caves). Both these devices can generate a 3D visualization of an object through stereography techniques, in 1:1 scale. The 3D visualization is computed through specific graphic engines. Considering the success of VR in videogames industry, the most advanced graphic engines for VR environments come from the videogames industry, such as Unity®, Unreal Engine®, etc.

Augmented Reality aims at enhancing the experience of the physical world though the five senses (sight, smell, hearing, taste and touch) by applying digital information though a brokered fusion. The increasing computational power of portable devices (smartphones and tablets) and wearable devices (smart glasses) allows on-site AR visualisations as soon as internet connection is available. Like for VR environments, the most advanced computational engines for the AR visualisation come from the videogames industry.

Concurrently to advanced visualization techniques such as VR and AR, the success of portable devices based on touch interfaces triggers a new approach to alphanumeric propriety interaction. Beyond advanced visualization of geometry data types, touch interfaces enhance the interactivity with alphanumeric proprieties that can be interactively edited on-site.

The data visualization environment develops methods for enhancing the accessibility, the interactivity and the immersivity of digital information. External users can test these methods in order to be conscious of available tools and developing tailored visualization strategies according to the tasks committed by local SMEs and PAs. This sub-environment promotes advanced visualization techniques, such as VR and AR applying an infrastructure that implements also portable devices. By doing so, external users can develop tailored prototypes and proofs of concept that can be also tested on-site in order to be conscious of possible limitations, barriers and impacts on tasks in the physical world.

The BIM Benefit Evaluation Framework

As briefly mentioned above, the establishment of the testing environment as well as the provision of BIM-based services for external users is accompanied by the development of a framework for evaluating the BIM benefits from the project level perspective. The foundations for this framework have been proposed by the authors in previous research activities (Schimanski et al., in press). It comprises an indirect approach of the BIM benefit assessment that leverages principles from construction cost estimation and probabilistic risk management. This BIM evaluation framework is intended to contribute to quantitative BIM benefit assessment, which goes beyond qualitative BIM benefit, such as the better communication and coordination between project participants mentioned above. The underlying assumption is, that BIM can have a positive impact on certain construction risk factors which are the main drivers for contingency allocation in cost estimation. In this regard, the main construction project risks were worked out from analyses of the literature. The focus was placed on risks whose likelihood of occurrence can apparently be reduced using BIM functionalities. Construction project risks are often assigned a financial impact using stochastic risk analysis, which is reflected in an increased risk contingency in monetary units. Both the likelihood of occurrence and the financial impact of these risks have been sufficiently tested in practice and research. However, it is not known what influence BIM may have on these risk factors. With the proposed evaluation framework, the authors hypothesize that this influence is positive and thus leads to lower project budgets in BIM-supported project deliveries. In detail, a risk register with project risks has been created for the evaluation framework, which only contains risks that can presumably be influenced by BIM. The risk register also contains parameters on the likelihood of occurrence, the financial impact in case of risk occurrence and parameters on the probability distribution of the financial impact (mean and standard deviation). The risk register itself serves as a stochastic model for Monte Carlo simulations, in which a necessary contingency reserve is calculated as many times as needed on the basis of the model parameters and random variables influenced by them.

The total amount of contingency is then calculated as follows:

$$Total\ amount\ of\ contingency\left[\ \right] = \sum_{i}^{j} FI\left(RF_i\right) \quad (1)$$

With:

$$FI = \text{Financial Impact of Risk factor (RF) i} \quad (2)$$

$$RF = \text{Risk factor} \quad (3)$$

$$j = \text{Total amount of considered Risk factors} \quad (4)$$

The assumption of this framework is that especially the likelihood of occurrence of project risks in the BIM-based case is different from a traditional project delivery.

Hence, this framework hypothesizes that overall construction costs can be reduced due to the reduction of the needed contingency caused by the BIM benefits. Figure 1 shows the principal steps for applying the proposed BIM evaluation framework. For the configuration of the single parameters of the probabilistic risk model, the authors used as a first validation step the results of a dedicated questionnaire with construction experts. In this questionnaire the construction experts where asked to which extent BIM might influence certain risk factors in construction. The answers were abstract into configuration parameters for the risk model. The outcomes of this test are briefly described in the "Early Results" section.

Figure 2. Proposed step of the BIM evaluation framework

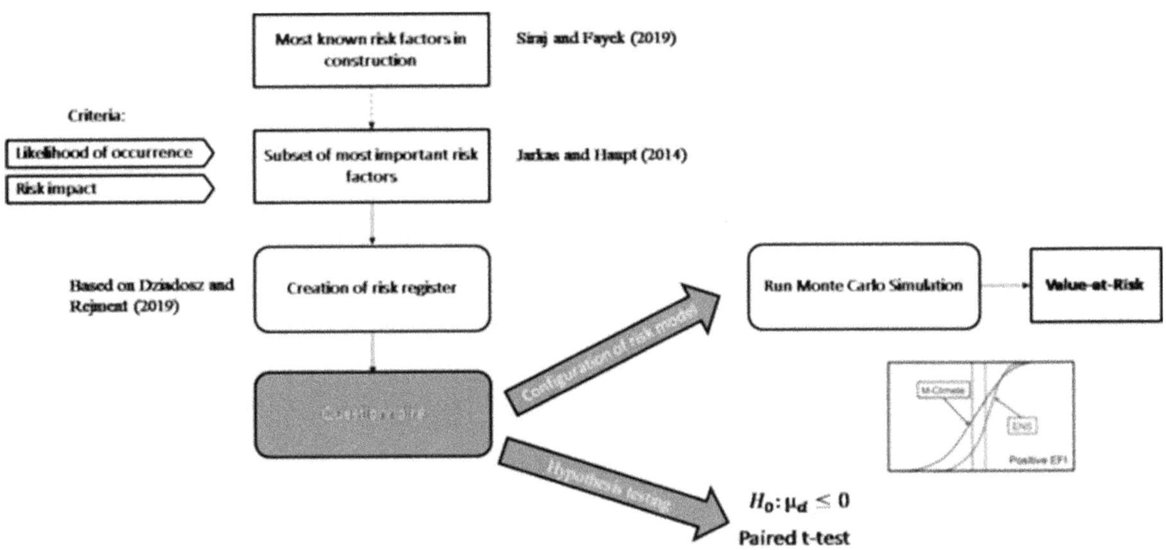

EARLY RESULTS

The Infrastructure of the BIM Simulation Lab

At the actual stage, the BIM Simulation Lab initiative is establishing the infrastructure of the physical laboratory according the arrangement of the testing environment, and it is planning to start spreading services to external users by the end of 2021. The testing environment follows the structure of the four sub-environments as discussed in the previous section. Referring to the activity that each sub-environment plans to develop, the infrastructure implements software and hardware equipment, merging equipment already available at the Fraunhofer Italia institute with new equipment that has to be implemented and purchased within the initiative.

The physical laboratory is temporarily located at the Virtual Reality Lab of the Fraunhofer Italia institute. The Virtual Reality Lab offers enough space for locating all the four sub-environments and it is already equipped with a VR cave system for supporting data visualization. Besides hardware equipment, a specific software architecture has been defined. The software architecture supports the activities in each sub-environment, simulating an information management system that applies the BIM methodology during the whole lifecycle of a construction work.

The core of the software architecture is a system for data storage and data sharing that will act as an interconnection node for the equipment (both hardware and software) of the other sub-environments. The system is implementing an on-premises opensource solution and a commercial cloud solution, both compatible with open-format data. Further equipment, as well as prototypes and proofs of concept, will communicate with each other through these data storage and data sharing solutions in order to simulate a centralized interconnection node. Communication relies both on geometry and alphanumeric data types, adopting a bi-directional communication strategy in order to request, modify and update a specific information content.

Figure 3. Software architecture of the BIM Simulation Lab[1]

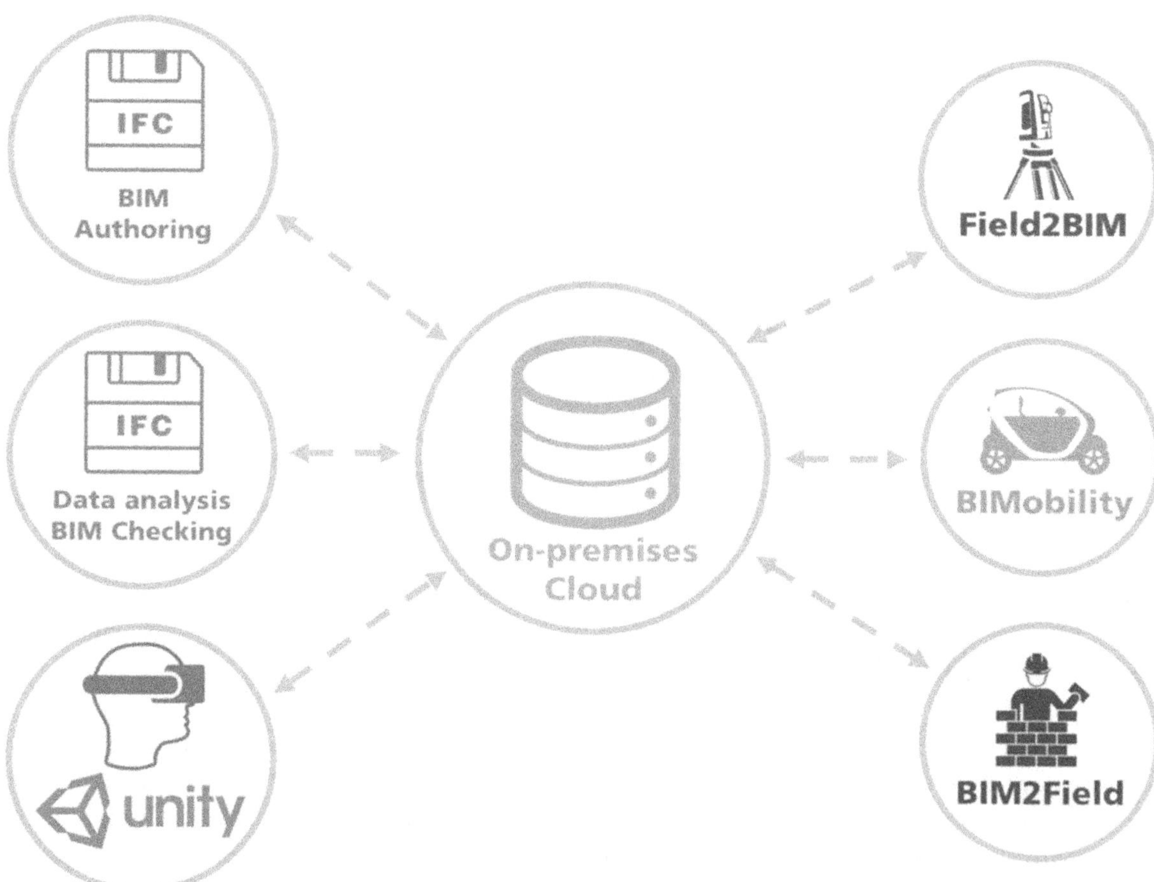

The equipment for each sub-environment is summarized in the Table 3. Some of this equipment was already available at the Fraunhofer Italia institute, some has been implemented and purchased within the initiative.

Table 3. The equipment within the BIM Simulation Lab according to the structure of sub-environments

Data acquisition environment		Data processing environment		Data storage and sharing environment		Data visualization environment	
Equipment	**Type**	**Equipment**	**Type**	**Equipment**	**Type**	**Equipment**	**Type**
Laser scanner	Commercial Hardware	Workstation	Commercial Hardware	Server	Commercial Hardware	VR headset	Commercial Hardware
Handheld laser scanner	Commercial Hardware	BIM authoring	Commercial Software	On-premises CDE	Open-source Software	VR cave	Commercial Hardware
Workstation	Commercial Hardware	BIM analysis	Commercial Software	Cloud CDE	Commercial Software	AR headset	Commercial Hardware
Software for cloud points	Commercial Software	Libraries for data analysis	Commercial/ Open-source Software	-	-	Touch table	Commercial Hardware
Libraries for point cloud analysis	Open-source Software	-	-	-	-	Workstation	Commercial Hardware
-	-	-	-	-	-	Tablet	Commercial Hardware
-	-	-	-	-	-	Gaming engine	Open-source Software
-	-	-	-	-	-	Libraries for 3D visualization	Commercial/ Open-source Software

In order to facilitate external users accessing and understanding capabilities of the testing environment, the initiative is developing three prototypes. These prototypes aim at providing applied examples to external users. These examples develop software tools for specific task along the lifecycle of a construction work, within a digital information management system that applies the BIM methodology. Each software tool refers to a different task and involves different sub-environments within the testing environment of the BIM Simulation Lab initiative. Table 4 shows a description of each software tool and its relationships with the task in the physical world and the sub-environments.

Early Validation of the BIM Benefit Evaluation Framework

In addition to the three mentioned software tools, the presented BIM evaluation framework will be tested and validated. As a first step in this validation procedure, demonstrations, expert interview sessions as well as a questionnaire with local construction experts have been conducted. In this context, a questionnaire-based configuration of the evaluation framework was carried out in a first validation round. For this purpose, a total of 22 actors from the AEC sector were questioned about the likelihood of occurrence of certain construction project risks in the two scenarios "BIM-based" and "traditional" project delivery. The analyzed risk factors were:

- Slow decision-making process (Risk-ID: M1)
- Frequent change orders by client (Risk-ID: T1)
- Errors and omissions in design drawings (Risk.ID: T2)

Table 4. The software tools under development within the BIM Simulation Lab initiative

Tool	Description	Task	Sub-environment			
			Data acquisition environment	Data processing environment	Data storage and sharing environment	Data visualization environment
Field2BIM	A software library for the (semi-)automatic transfer of data from the real world to the digital world through Visual Recognition and Machine Learning techniques	Survey	x	x	x	
BIMobility	An application to support the design of individual buildings and/or entire neighborhoods implementing electrical mobility systems, facilitating the sizing and positioning of specific equipment	Design		x	x	x
BIM2Field	An application for digital construction-site management through the application of Lean Construction Management techniques, such as the Last Planner System (LPS)	Execution	x	x	x	x

- Unavailability or shortage in specified material (Risk-ID: R2)
- Delay in response to requests for information (Risk-ID: M2)

The different opinions in the two scenarios were used to configure the model for stochastic risk assessment according to the proposed BIM evaluation framework. The exact procedure for the configuration of the risk model is described in (Schimanski et al., in press). In brief, however, it can be said that the experts' opinions have been mathematically converted into new likelihoods of occurrence for the risks examined. These and the experts' estimates of the financial impact of the risk events were used to calibrate the risk model for the BIM-based and traditional case. The resulting percentages of the corresponding likelihoods are given in Table 5.

Table 6. Summary of the expert questionnaire results

Ordinal Value	Risk factors									
	M1		T1		T2		R1		M2	
	T**	B**	T**	B**	T**	B**	T**	B**	T**	B**
4	45,5	22,7	55,5	22,7	45,5	18,2	31,8	9,1	45,5	13,6
3	55,5	50	45,5	54,5	55,5	36,4	55,5	36,4	55,5	36,4
2	0	22,7	0	18,2	0	40,9	13,6	45,5	0	50
1	0	4,5	0	4,5	0	4,5	0	9,1	0	0
Mean	3,49	2,91	3,59	2,95	3,49	2,68	3,21	2,46	3,49	2,64
Resulting %*	56,63	47,24	58,26	47,97	56,63	43,60	52,15	39,93	56,63	42,84

* Resulting percentage of likelihood corresponding
** T = Traditional; B = BIM-based

These results are considered for hypotheses testing, making use of a paired t-test:

Question: Does BIM impact the required project budget positively?

Hypothesis: $H_0 : \mu_d < 0$ → The implementation of BIM has no positive impact on the required contingency in cost estimation.

Alternative-Hypothesis: $H_A : \mu_d > 0$ →The implementation of BIM has a positive impact on the required contingency in cost estimation.

Test distribution: t-distribution, 4 degrees of freedom (df) = n - 1

Test level: $\alpha = 0{,}1 \Rightarrow t_T = 1{,}383$ (with t_T taken from t-distribution table)

Test statistics: $$t = \frac{\bar{d}}{\sqrt{s^{*2} / n}} \tag{5}$$

with:

$$d° = \text{pair differences (pd) average} \tag{6}$$

$$s^{*2} = \text{variance of pair differences} = (\text{pd} - \bar{\text{d}})^2 \tag{7}$$

$$\text{n} = \text{number of pairs} \tag{8}$$

Null-Hypothesis H_0 will be rejected and Alternative-Hypothesis H_A will be accepted,

if: $t \geq t_T$ (with t_T taken from t-distribution table)

Conducting the test:

$$df = 4 \quad (9)$$

$$n = 5 \quad (10)$$

Table 7. Paired t-test values

Risk factor	Pair values (resulting percentage according to Table 5)		Pair differences (pd)	$(pd - \bar{d})^2$
	Traditional	BIM-based		
M1	0,5663	0,4724	0,0939	0,0005
T1	0,5826	0,4797	0,1029	0,0106
T2	0,5663	0,4360	0,1303	0,0170
R1	0,5215	0,3993	0,1222	0,0149
M2	0,5663	0,4284	0,1380	0,0190
Σ			0,1162	0,0620

$$s^{*2} = 0{,}062/df \quad (11)$$

$$= 0{,}062 / 4 \quad (12)$$

$$= 0{,}0155 \quad (13)$$

$$\bar{d} = \sum_i PD_i \quad (14)$$

$$\bar{d} = 0{,}1162 \quad (15)$$

$$t = \frac{\bar{d}}{\sqrt{s^{*2} / n}} \tag{16}$$

$$= \frac{0,1162}{\sqrt{0,0155 / 5}} \tag{17}$$

$$= 2,087 \tag{18}$$

$$t_T = 1,383 \tag{19}$$

(taken from one-sided t-distribution table for: df= 4; α= 0,1)

$\Rightarrow t \geq t_T$

⇒ Null-Hypothesis H_0 will be rejected and Alternative-Hypothesis H_A will be accepted!

The t-test shows that the null hypothesis is to be rejected and the alternative hypothesis is to be accepted, which means that BIM has a statistically significant (positive) effect on the risk occurrence according to the questionnaire findings given the considered risk factors. This assessment of the construction experts was taken as an incentive to carry out the Monte Carlo simulation for contingency estimation for the two cases of a traditional and a BIM-based project management in order to evaluate the proven positive BIM influence also in monetary terms for a hypothetical project scenario.

In summary, the results showed that in a BIM-based project delivery fewer contingency reserves have to be kept available at a constant confidence level compared to a traditional project delivery. However, since the results refer only to a hypothetical reference project, which does not yet allow a generalization of the findings.

FUTURE RESEARCH ACTIVITIES

The proposed BIM evaluation framework represents an indirect evaluation methodology and should ideally be linked to the other mentioned prototypes to be developed within the project on a data processing level in order to guarantee a continuous tool chain throughout the BIM Simulation Lab software components. In detail, the other tools will collect data during their pilot phase to improve the configuration of the BIM evaluation approach.

In the area of the BIM2Field prototype, which is based on the conceptual level mainly on approaches of Lean Construction Management enriched with BIM data, the stability of the flow of works and process reliability during construction is supported by software. Parameters for these supported objectives are derived from monitoring data collected on site using the BIM2Field tool. In the course of the planned continuous tool chain, parameters, such as "reasons for non-completion", should be mapped to the configuration parameters of the BIM evaluation framework if certain tasks cannot be implemented as planned, in order to better record the BIM influence on specific construction project risks. The simulation results will then be compared with benchmark data from comparable projects for a traditional project delivery scenario.

The collection of the data is to take place on various pilot construction sites mainly in South Tyrol, Italy, where the BIM2Field is ideally used from start to finish during the execution phase. For this purpose, some physical components of the BIM Simulation Lab (see also Table 3) will be brought to the construction sites and used by the industrial partners that are supporting the BIM Simulation Lab research project. In the area of BIM2Field tools, the touch table will be of importance here. The BIM models previously loaded onto the cloud model server will be made available in IFC format for collaborative and digital process planning in the format of web application. This process planning is defined in particular by the fact that so-called digital sticky-notes, which represent the individual process steps, can be collaboratively positioned on the touch table in the best possible sequences. Moreover, the digital sticky notes can be linked to the corresponding objects of the BIM model. In this way, the metadata intrinsically present in the objects, such as the quantities to be installed, can be digitally linked to the construction site activities, and be made available for a more transparent and automated process planning on-site.

CONCLUSION

This essay discusses an ongoing initiative which aims at establishing a laboratory for an effective and efficient BIM implementation, by promoting a physical space and specific services for supporting the territory in the DT. Besides trying to establish effective services for local SMEs and PAs, the initiative introduces a BIM benefit evaluation framework that leverages principles from construction cost estimation and probabilistic risk management.

Although the initiative has not yet fulfilled its main goals and objectives, according to the early results, it could represent a case-study for similar initiative that are facing the same challenges by facilitating the BIM implementation in SMEs and PAs, towards an effective and integrated Digital Transformation of the AEC sector.

ACKNOWLEDGMENT

The ERDF initiative 1086 BIM Simulation Lab [CUP: B51G17000270001] is promoted by Fraunhofer Italia and financed by the Autonomous Province of Bolzano, within the framework of the European Regional Development Funds (ERDF 2014 – 2020).

REFERENCES

Assobim. (2019). *BIM Report 2019.* Accessed 01/04/2020, available at: https://www.assobim.it/bim-report-2019/

Azhar, S. (2011). Building information modeling (BIM): Trends, benefits, risks, and challenges for the AEC industry. *Leadership and Management in Engineering, 11*(3), 241–252. doi:10.1061/(ASCE)LM.1943-5630.0000127

Bakis, N., Kagioglou, M., & Aouad, G. (2006). Evaluating the business benefits of information systems. In *3rd International SCRI Symposium*. Salford Centre for Research and Innovation, University of Salford.

Barlish, K., & Sullivan, K. (2012). How to measure the benefits of BIM - A case study approach. *Automation in Construction, 24*, 149–159. doi:10.1016/j.autcon.2012.02.008

Bryde, D., Broquetas, M., & Volm, J. M. (2013). The project benefits of building information modelling (BIM). *International Journal of Project Management, 31*(7), 971–980. doi:10.1016/j.ijproman.2012.12.001

BSI - British Standards Institute. (2003). *Poor Communication Costing UK Construction Industry*. http://www.bsigroup.com/en-GB/aboutbsi/media-centre/press-releases/2003/9/Poor- Communication-Costing-UK-Construction-Industry/

Bughin, J., Hazan, E., Labaye, E., Manyika, J., Dahlström, P., Ramaswamy, S., & Cochin de Billy, C. (2016). *Digital Europe: Realizing the continent's potential*. McKinsey Global Institute. https://www.mckinsey.com/business-functions/mckinsey-digital/our-insights/digital-europe-realizing-the-continents-potential

Guerriero, A., Kubicki, S., & Reiter, S. (2016) Building information modeling in use: How to evaluate the return on investment? *eWork and eBusiness in Architecture, Engineering and Construction - Proceedings of the 11th European Conference on Product and Process Modelling, ECPPM 2016*, 537–544.

Li, J., Hou, L., Wang, X., Wang, J., Guo, J., Zhang, S., & Jiao, Y. (2014). A Project-Based Quantification of BIM Benefits. *International Journal of Advanced Robotic Systems, 11*(8), 123. doi:10.5772/58448

Matt, D. T., Marcher, C., Malacarne, G., & Brozzi, R. (2018). *BIM Report Alto Adige 2018. Sondaggio sulla conoscenza e sull'uso del BIM tra gli operatori del settore delle costruzioni*. Accessed 01/04/2020, available at: https://www.fraunhofer.it/content/dam/italia/it/documents/per_il_sito_BIM%20report%20Alto%20Adige_GT_Versione%20per%20PDF.pdf

NRC. (2012). *NRC Publications Archive (NPArC) Archives des publications du CNRC (NPArC)*. Author.

Phang, T., Chen, C., & Tiong, R. (2019). New Model for Identifying Critical Success Factors Influencing BIM Adoption from Precast Concrete Manufacturers' View. *Journal of Construction Engineering and Management, 146*(4). Advance online publication. doi:10.1061/(ASCE)CO.1943-7862.0001773

PMI - Project Management Institute. (2013). *The high cost of low performance: The essential roles of communications*. https://www.pmi.org/-/media/pmi/documents/public/pdf/learning/thought-leadership/pulse/the-essential-role-of-communications.pdf

Rockart, J. F. (1979). Chief executives define their own data needs. *Harvard Business Review, 57*(2), 81–93. PMID:10297607

Sacks, R., Eastman, C., Lee, G., & Teicholz, P. (2018). BIM Handbook: A Guide to Building Information Modeling for Owners, Designers, Engineers, Contractors, and Facility Managers. Wiley.

Tezel, A., & Aziz, Z. (2017). From conventional to IT based visual management: a conceptual discussion for lean construction. *ITcon, 22*, 220-246. https://www.itcon.org/2017/12

Wortmann, J. C. (1983). *A Classification Scheme for Master Production Schedule. Efficiency of Manufacturing Systems*. Plenum Press., doi:10.1007/978-1-4684-4475-9_10

Zhou, Y., Ding, L., Rao, Y., Luo, H., Medjdoub, B., & Zhong, H. (2017). Formulating project-level building information modeling evaluation framework from the perspectives of organizations: A review. *Automation in Construction*, *81*, 44–55. doi:10.1016/j.autcon.2017.05.004

ENDNOTE

1 This content is not sponsored by or affiliated with Unity Technologies or its affiliates. Unity Trademark is trademark or registered trademark of Unity Technologies or its affiliates in the U.S. and elsewhere.

Chapter 7
Smart Cities and Smart Societies:
The Shock, or the New Paradigm for a Smart Society

Roberto Pagani
Politecnico di Torino, Italy

Gian Vincenzo Fracastoro
Politecnico di Torino, Italy

ABSTRACT

The post-shock scenario is outlined: an uncertain future with a "new normality." The embryos of the new paradigm are alongside the powerful discontinuity generated by COVID-19. With examples and anecdotes from Shanghai and China, a transformation already underway is portrayed. No more perfect shock could be thought to reconsider the role of humans on this planet, on our cities. There is a crucial need for resilience of local systems, for short chains, for autonomous energy and food self-sufficiency, for decentralizing essential products. Security and contingency plans are needed and must operate on a global scale, but at the same time at the country and the city level. The future must be reinvented, acting in depth, for shifting from "exploitation" to "cooperation" with natural systems. Topics are education, work, services, transport, food safety.

INTRODUCTION

This paper is the authors' reflection on how the smart city will be influenced by the pandemic, outlining from a privileged point of view the way the most active Chinese city, Shanghai, has reacted to the Covid-19 pandemics. It is not the report of a traditional research work, but the synthesis of the experience of the authors, been involved in many years of dedicated work on smart cities, energy efficient communities, and related topics.

DOI: 10.4018/978-1-7998-7091-3.ch007

China is recurrent in this text, both for the centrality of this country in the development of the pandemic, and for the way it has managed, with firmness and technology, its containment in the national territory and citics.

This article is, then, a sort of live description of what is going on in China, while the Western world is still fighting for the lives of its citizens and the disaster of its economies.

In the first chapter, "Investing in Smart City and Digital Infrastructure in China", the huge effort to build digital infrastructures and smart cities in the Shanghai region is described. Investments of local governments and private parties are addressed to make this region one of the worldwide most attractive for the level of digitalization and IT density. Nevertheless, business originality is still lacking, since top-down protocols prevail on a bottom-up creativity and widely spread innovation.

In the second chapter, "Smart and Healthy Cities in China", the criticalities of the "healthy cities" movement are introduced. Launched in Shanghai in 2016, on the occasion of the WHO Global Conference on Health Promotion, the network hosted city mayors from over 100 countries of the world - including 46 Chinese city mayors. In the following years, they should have combined the smart city with the healthy city principles. On the contrary, all have shown how unprepared they were to combat a pandemic.

The great challenges of humanity bring with them as many challenges for the scientific and technological community to understand the critical issues, and to overcome problems. Often these challenges become innovation factors, sometimes disruptive innovations. In the third chapter, "Smart Technologies for the Healthy post-Pandemic Cities in China", the authors highlight the technological aspects of city living that have strengthened during one year of pandemic, and from which there will probably be no turning back, because they have established themselves as being more efficient than previously. After the first devastating uncertainties, and widespread COVID-19 diffusion, China launched an articulated and extraordinary health response in Wuhan and in Hubei Province, setting up strategies and solutions that were adopted, step by step, throughout the country. Scientific applications and technological innovations, that normally take years to scale-up, have been tested and applied for the first time on a large scale. The shortening of time response has been facilitated by a strong public-private collaboration that has brought the most important and innovative Chinese high-tech giants into play. The technological catalogue put in place to combat the epidemic and limit its spread is documented. These technologies will continue to be part of the city of tomorrow.

The fourth chapter, "The Shock, or the New Paradigm", provides an outlook to the future, not limited to the Chinese landscape. The pandemic shock has produced such a scale-up of so many technologies, processes, ways of working, learning, teaching, traveling and not traveling, that will make the "new normal" to become a "new diversity". Are we only at the beginning of a change in lifestyle and consumer trends? Maybe nothing will be the same as before. It will simply be "different", but it will be...

INVESTING IN SMART CITY AND DIGITAL INFRASTRUCTURE IN CHINA

"How Many Things are Here...! "

An element contributing to mark a part of contemporary European culture from the Chinese culture is the number of things that one can meet in restricted, bounded spaces. In the *western* perception, harmony, beauty, quality, are often associated to essential, almost empty spaces: few things of high aesthetic value that come into harmony with spaces and qualify them. So far, "minimalism" has emerged in the western world to suggest the quality of such essence. In China, it seems to be different, there is a search for excess, for a high number of things that are brought into spaces and determine a diverse harmony. When a Chinese enters a hotel' room he must be able to say: "nice! How many things are here..." Objects, furnishings, knick-knacks, mirrors, stuccos, large and small containers, decorations. "How many things are here" is not only the standard, it is a quality indicator. Perhaps, this different perception of the "number of things" is not limited to the use of spaces, but expands to the whole Chinese imagination, through which economy, development and quality of life are measured.

A big City Investing in Digital Infrastructure

Starting from 2020, the city of Shanghai invests a total 35 billion Euro in 48 megaprojects on "new infrastructure" in the next three years. According to this three-year action plan (2020-22), the city is going to install 100,000 charging stations for electric vehicles. In addition, fully automated production lines are built for 100 automated factories and workshops, bringing 150,000 new businesses to the cloud platform of the city. At the same time, Shanghai accelerates the construction of scientific and research facilities across the whole metropolitan area (Xia, 2020a). With the title of "new infrastructure" Shanghai refers to a digital infrastructure network, charging stations, data centers, artificial intelligence networks and the industrial internet, or internet of things (IoT). Compared to traditional infrastructures, private companies are financing most of the digital infrastructure projects, while a portion is covered by local governments, according to the principle of a Public Private Partnership. Shanghai, which has China's main Information Technology (IT) infrastructure, branched on smart city programs, national science parks, data centers, new generation networks, is also building the first Chinese port with intelligent cross-border logistics in a new free trade area. Some 34,000 new 5G stations to be installed are part of this digital infrastructure. Shanghai Telecom, one of China's leading telecommunications operators, in 2020 has launched a pilot project to convert a dozen public telephone booths in the center of Shanghai into micro stations to support the city's 5G network. In fact, installing small and micro 5G stations in public telephone boxes can help to cover the blind spots of external 5G networks. The appearance of these cabins remains unchanged, thus keeping the perception of the urban landscape with the old-fashioned booths. Shanghai still has over 5,000 public telephone booths, disseminated throughout the streets of the city, and China Telecom plans to equip more and more of them as 5G stations. (Xia, 2020b)

The Great Plan G60 of the Yangtze River Delta

So far, Shanghai's plans for the 5G digital challenge. These plans, however, are not limited to the city, they are part of the larger G60 Sci-Tech plan, which takes its name from the huge infrastructure (car and rail) that connects nine cities belonging to the Yangtze River Delta region. G60 S&T Innovation Valley will serve as a model for an integrated development across the region. In the central government goals, it will also become a test field for implementing five development concepts - innovation, coordination, sustainability, openness, sharing – contributing to the transition from "Made in China" to "Created in China". The G60 Yangtze River Delta Research Institute and the Shanghai Low-Carbon Technology Research Institute have both settled in the Sci-tech G60 corridor. Various industrial parks will draw on the innovations originating from these centers to update their products. The G60 Plan has a unitary strategy, apparently organic, with various vocations for each area of this region: biomedicine, precision equipment, internet plus, integrated circuits, automotive, intelligent manufacturing, household appliances, robots, electronic and electrical appliances, new and advanced materials, artificial intelligence, new energies, appliances, intelligent security (Yinfen, 2019). Nevertheless, the strategic sectors of the nine cities, affected by the G60 plan, are almost identical. So are the implementation principles, based on same incentives provided to international and national investors. In the political expectations of central and local governments, the investment in new digital infrastructures would create new demand and new dynamism for the eastern China. It would also accelerate the construction of a modern industrial system and provide a new engine for high-quality economic development, with very favorable conditions for investment from abroad. Here is the current "storyline" of this big technological and digital infrastructure is described.

Does Gigantism Change the Model?

However, this gigantism of structures and infrastructures does not change the model followed so far by the central and local planning of this country. The guiding principles of all industrial areas are replicated, in open competition to each other, with similar vocations. Robotics, big data, bio-medicine, data centers, artificial intelligence, are the common ingredients of these productive areas, where the specialization of territories is non-existent. The audacity in the infrastructural, scientific, and technological ventures of this region, as well as other areas of China, is not accompanied by the courage to give freedom to business creativity. Still everything is planned according to top-down protocols, printed in their Five-Year Plans. In addition, this self-proclaimed context of innovation keeps some pre-technological residues of this country. Repeated peaks of air pollution reemerged in Shanghai, while had disappeared in the early months of 2020, due to a slower industrial production. The progressive reopening of all factories, with continuous cycles, led to reaching worrying levels of air pollution, typical of the pre-COVID period. Over 30% of days, in recent months, have exceeded the threshold of 100 AQI index (i.e. unhealthy air for sensitive groups); 15% of days exceeded the unhealthy threshold. Between 18 March and 21 March 2020, a sequence of unbreathable air was faced, with particulate PM 2,5 reaching levels above 300 μg/m^3 (micrograms per cubic meter) up to 369 μg/m^3, that is about seven times higher than acceptable limits for health (Shanghai Air Pollution, 2020).

Basically, there is a careful planning for digital innovation, but the opportunity to improve conventional production processes with eco-friendly production has been missing. There was no advantage of three months of total lockdown to select clean productions, at the expense of polluting ones.

SMART AND HEALTHY CITIES IN CHINA

"How Healthy our Cities are? ..."

A city is "healthy" based on a set of parameters, indicators, criteria, levels, to be calculated by standardised procedures. Each *healthy city* adopts these parameters and *feels okay*. No matter if its parameters reveal that acceptable levels of air quality are not reached, there will be room for improvement in the future. It does not matter if the water quality is such as to actually threaten health in citizens, there is room for adding purifiers, and improve one's position in the ranking. It doesn't matter if citizens do not sort their wastes and everything ends up in an indistinct burner or in an indistinct landfill, there is room to build a new excellent plant, or to introduce an archaic differentiation system, to improve one's status as "healthy city". Thus the "Healthy Cities" network was born: cities adopting the same criteria to measure how healthy they are. Shanghai being elected as the network's leading city.

Shanghai Consensus on Healthy Cities 2016

In 2016, on the occasion of the 9th Global Conference on Health Promotion, the World Health Organization launched in Shanghai a new plan to "Promote health in the objectives of sustainable development". In this initiative, the WHO was joined by many city mayors from over 100 countries of the world - including 46 Chinese city mayors. It is well known that mayors play a fundamental role in promoting health, in an increasingly urbanized population.

The declaration signed was formulated as follows: *"We - more than 100 mayors from around the world - have come together on 21 November 2016 in Shanghai, China - united in the knowledge that health and sustainable urban development are inextricably linked, and steadfastly committed to advancing both. We also recognize that health and wellbeing are at the core of the United Nations Development Agenda 2030 and its Sustainable Development Goals. Cities are at the front line of sustainable development and we are convinced that mayors have the power to make a real difference. We must and will be ambitious in localizing the 2030 agenda and we will set health targets to hold ourselves accountable. We recognize that everyone in the city will need to do their part to work towards these ambitious priorities."* Among the ten priorities of Healthy City action areas, the 7th mentions: *"7. make our cities safe from infectious disease through ensuring immunization, clean water, sanitation, waste management and vector control"* (World Health Organisation, 2016a).

The Debate in the Healthy Cities Network

The foundation of the "Healthy Cities" network was at the center of the 2016 WHO conference. Several cities from five continents and a large number of Chinese cities joined this network. Among the best known: SaoPaolo (Brazil), Santiago (Chile), Cienfuegos (Cuba), Mocorito (Mexico), Lima (Peru), Quebec (Canada), Houston (USA), Brazzaville (Congo), Cairo (Egypt), Cork (Ireland), Vologda (Russia), Moscow (Russia), Lods (Poland), Sofia (Bulgaria), Istanbul (Turkey), Hue (Vietnam), Balanga (Philippines), Dunedin (New Zeeland), Queensland (Australia), Wonju (Republic of Korea), Ulanbaatar (Mongolia), Yokohama (Japan). Finally, a number of 46 Chinese cities, led by Shanghai. The debate on Smart and Healthy Cities have characterized the plenary session dedicated to the Mayors Forum, with the city mayors participating to the memorandum signatory. The *healthy cities* features have been listed in detail: a) including *health* among the urban quality measurement indicators, in connection with environmental parameters; b) managing urban improvements in a transdisciplinary way; c) monitoring project data and measurements. Above all, overseeing communication policies on health on an urban scale, a power that only Mayors have. Improving the quality of food, water, physical activity, and adopting stricter regulations were the phrasings used in the conference plenary sessions. A large number of reports highlighted China's policies and actions to introduce urban-scale changes and extend them to the Belt & Road Initiative. In particular, the emphasis on the evaluation aspects of Healthy Cities led to the adoption of homogeneous parameters for all cities, with common criteria for global comparison. The action plan of Shanghai to become "*Healthy City*" was presented. It responds to improvements that can be parameterized in a series of target values, such as: 2.7 years increase in life expectancy; 25%

increase in educational rate, 22% increase in physical activity. It also plans to improving urban fruition, with parks reachable in 10 minutes on foot from any part of the city (World Health Organization, 2016b).

The political nature of the Shanghai event overcame the strategic aspect and darkened out the methodological one.

Where Have the "Healthy Cities" Been During the Pandemic?

The Healthy Cities network met in 2016. Now most of those cities have collapsed on health. They should have combined the smart city with the healthy city principles; they should have been ready to respond with high quality services to the containment of any epidemic, to the wise administration of hospital resources, to the intelligent management of people flows and crowds. On the contrary, all have shown how unprepared they were: not very smart and not healthy, indeed. Our complex systems, as cities are, hardly are prepared to deal with sudden threats. Although there is preparation, adherence to principles, underwriting of agreements, there are no skills to predict when threats are coming, adopting already codified plans. The paradox of being part of a Healthy Cities Network and breaking out the pandemic in one of these cities is evident. It is not enough to subscribe to reputed networks, it is necessary to work day-by-day in deserving it, activating controls, neutralizing incorrect behaviors.

But despite China has been caught off-guard by the epidemic, it has then supported and introduced many digital innovations to fight against it.

SMART TECHNOLOGIES FOR THE HEALTHY POST-PANDEMIC CITY IN CHINA

"It's Today That Matters..."

"*How many things are here!*" and "*how healthy our cities are?*". Both statements determine the new progressing trend. In fact, it is today that matters, *hic et nunc* in the Latin expression. China was the first country to be severely touched by COVID-19. After the initial major uncertainties, that provoked a widespread diffusion of the virus, China launched an articulated and expensive health response in Wuhan, setting up strategies and tactics adopted, from time to time, throughout the whole country. Scientific applications and technological innovations that normally take years to scale-up have been tested for the first time. The great challenges of humanity correspond to equal challenges for the scientific community, to understand the critical issues and to develop technological solutions to comply with criticalities. Often these challenges stimulate disruptive innovations. In China, the shortening of time *from-challenge-to-innovation* has been determined by a strong public-private collaboration, that has brought the most important Chinese high-tech companies into play.

The main areas for these innovations are identification, medical testing, ICT tracing and treating of the population.

Identifying

The first task of new technologies was making the Chinese Government able to identify people.

Two out of three Covid-19 positive people in China in April 2020 had no typical signs of Covid-19, highlighting one of the main challenges of this disease: identifying the infection and stopping its spread.

A wider spectrum of disease involves the need to significantly expand testing to identify cases with atypical symptoms. To prevent further spread, the World Health Organization recommends isolating infected people, and tracing their contacts to identify other potential cases. These identification activities require scientific and technological applications which have been tested in China with use of expanded technical capabilities and solutions. They consist of:

- Localization technologies
- Satellite monitoring
- Drones
- Mass surveillance
- Social distancing measures

Localization technologies

Localization technologies (Figure 1) are known to play a crucial role during times of crisis and catastrophes. A Global Navigation Satellite System (GNSS) helped to "identify" outbreaks and individuals infected, helping to contain the virus, as well as analyzing the pattern of the epidemic. With the help of detailed data, maps and images, China was able to build two hospitals in Wuhan in just a few days, and thousands can be built across the country. Identification systems are commonly used in transport planning: logistics companies use GNSS terminals to be faster in delivering essential goods. In addition, Radio Determination Satellite Service system (RDSS) can transmit in real time, as an alternative to GNSS (BeiDou, 2020).

Satellite monitoring

A Chinese satellite that uses second generation Artificial Intelligence is capable of performing powerful data selection processes (Figure 2).

By combining its processing capacity with geocoding, a health mapping of COVID-19 was created, through which individuals can locate virus carriers and discover the distance between them and the active infection (Watson, 2020).

Drones

In some areas, severely affected by the virus, drones helped health personnel in delivering medical equipment or patient samples, saving time and avoiding contamination of medical samples (Figure 3). In addition, drones have visible QR codes, that can be scanned to find out more information about health directives. Agricultural drones have sprayed disinfectant in the countryside, while other drones, equipped with facial recognition, have been used to warn citizens not to leave houses and to wear masks.

Mass surveillance

China has adopted a rigorous (draconian) surveillance system. Smartphone locations, body temperatures, travel histories and other information about people are collected in a centralized database, whose data are analyzed using Big Data and Machine Learning techniques. Alipay, the mobile payment platform,

Figure 1. localization technologies (Universmartphone, 2020. "Phones Could Track the Spread of Covid-19. Is It a Good Idea?". Appeared elsewhere on the internet. No copyright infringement is intended).

and WeChat, the popular instant messaging application that also has a digital wallet, are used to collect positioning data at the time of transactions. Radio stations also offer a geolocation service for devices that do not have a GPS receiver. Using and crossing all these data, local authorities can find out the number of people with whom an infected person has come into contact and prescribe their self-isolation.

Technologies to support social distancing

The rapid implementation of social distancing is necessary to flatten the coronavirus curve and prevent the epidemic from worsening. But just as the relapse of the coronavirus threatens to cause an economic recession, according to many experts it can also cause a "social recession", that is a collapse of social contact, particularly hard for the populations most vulnerable to loneliness: the elderly, people with disabilities or in precarious health conditions. Many technologies have come into common use to reduce isolation and respond to socialization needs: video connections, games, online activities, psychological assistance. These are part of an extensive catalogue.

Testing

In March 2020, the Suzhou Biomedical Institute of the Chinese Academy of Sciences has developed a nucleic acid test kit that can produce results in 45 minutes. The kit, which also allows for on-site tests, can

Figure 2. Satellite monitoring (dailymail.co.uk, 2020. "Satellite images of Wuhan coronavirus hospital, built in 10 days"2020. Appeared elsewhere on the internet. No copyright infringement is intended).

be stored and transported at room temperature. The research team started working on the kit in January and subsequently moved to Hubei Province to test its effectiveness. The kit requires only 20 microliters of nasal swab or mucus samples to produce a useful result. A single test equipment is capable of analyzing 60 samples per hour. So far patients have been tested with accuracy greater than 99% (Flanders, 2020).

Quick online test

A company in Hong Kong has developed a monitoring platform with Artificial Intelligence for individuals at greater risk of COVID-19 (Figure 4).

It provides self-registration for high-risk populations, who were in close contact with infected individuals, isolated at home, or with fever. The monitoring platform is interfaced with more than 400 hospitals to share information and monitor the daily health status of high-risk groups. Another team has developed a free online screening toolkit called "COVID-19 rapid test". The team drafted clinical guidelines and epidemiological investigation strategies and used big data and artificial intelligence technologies for this toolkit, which helps users to quickly understand their health status, infection risks and offers recommendations based on the results of screening (China Academy of Information and Communications Technology, 2020).

Figure 3. Drones (Geospatial World Net, 2020. "The China way: Use of technology to combat Covid-19". Appeared elsewhere on the internet. No copyright infringement is intended).

IoT to track clinical samples

A Beijing technology company focused on bio-products has integrated artificial intelligence, IoT, and blockchain technologies for the identification of clinical samples and reagents used to fight the COVID-19 epidemic. The cloud system for medical waste, from a leading Chinese high-tech company, helps improving the hospital management of medical wastes. An innovation provided by Tencent Cloud improves the efficiency and safety of body temperature measurement with online registration by residents: an automatic collection of temperatures, coupled with personal information, provides a high functionality to support the offline temperature sensing of an entire community.

Tracing People Through ICT

The availability of a reliable multi-parametric system for monitoring individual mobility has been resolutely supported by the Chinese government. High-tech companies were able to develop, in a very short time, cloud computing applications to provide geolocation of areas with the highest concentration of contagion. The same APPs are able to block access to public establishments, based on the personal movements profile. The access to public transportation, such as buses, trains, or airplanes is also subject to a tracing check, a determining factor for the success of the lockdown and recovery phases. In addition, the well-established e-commerce in China allowed to receive food and basic goods provisions at the entrance of housing compounds, minimizing the psychological and organizational impact of the lockdown.

Figure 4. Quick online test (Sanwa Bio-tech, 2020. "Fight Covid-19: How innovative Hong Kong biotech companies aim to help people battle coronavirus disease". Photo: Norm Yip. Appeared elsewhere on the internet. No copyright infringement is intended).

QR code

The QR code system, launched through Ant Financial's Alipay in Hangzhou on February 2020, assigns users one of three colored QR codes: green, yellow and red (Figure 5). It is possible to access a mini-program both through the ubiquitous Alipay application and with Tencent's Wechat. Users obtain their codes by entering their name, identity number and registering with facial recognition. The green code indicates that the user is not in quarantine and can move freely around the city. Those with yellow and red codes must respectively self-quarantine at home, or undergo a controlled quarantine. The color code status is updated at midnight each day and follows millions of people daily.

Contact tracing app

Closely connected to the QR code, a mobile phone app processes the track of contacts and notifies the user, upon confirmation of any case of infection (Figure 6).

By keeping a temporary record of the proximity conditions among individuals, the application can immediately notify recent contacts of diagnosed cases and request to isolate themselves. This app in China is not strictly mandatory, but it is required to move between neighborhoods, in public spaces and premises, in public transport, to protect the people sharing same spaces. The application, a plug- in for

Figure 5. QR code (photo by R. Pagani, 2020)

Wechat and Alipay, allows a central database to collect data on user movements, to loosen or adopt restrictions on their movements. The database is analyzed by an artificial intelligence algorithm that releases color codes.

Big Data and Artificial Intelligence

Big Data and Artificial Intelligence, combined with the speed of application, made possible to block hundreds of millions of Chinese in their homes during the lockdown, and allowed to trace their movements and contacts during the subsequent recovery phase. Access to information has led to the creation of dashboards continuously monitoring the progression of the virus. A leading company in artificial intelligence has developed a solution to recognize faces even when wearing masks, with a relatively high degree of precision.

Predictive algorithms

Baidu, the Chinese Internet giant, has made its data analysis algorithm and predictive models available to teams fighting the epidemic (Figure 7).

Figure 6. Contact tracing (Baidu, 2020. Appeared elsewhere on the internet. No copyright infringement is intended).

Unlike Ebola, HIV, and influenza viruses, COVID-19 has a single stranded RNA, so it is able to mutate rapidly. The algorithm is much faster than other algorithms that help predict the structure of viruses. It allows reducing the waiting times of virus detectors by two orders of magnitude: from 55 minutes to just 27 seconds (Baidu, 2020).

Screening of large populations

Relevant Chinese companies have developed tools for screening large populations and infrared systems, powered by artificial intelligence, capable of detecting changes in people's body temperature. Over 200 Chinese cities have steadily adopted these cameras, equipped with facial recognition with multisensory technology. In particular, they are installed in airports, hospitals, nursing homes and in the main public gathering places. This camera technology automatically detects people with fever and tracks their movements, recognizes their faces and detects whether or not persons are wearing masks. These systems cross data and identify potentially infected individuals. They can scan up to 200 persons per minute, without interrupting the flow of people.

Figure 7. Predictive algorithms (Baidu, 2020. Appeared elsewhere on the internet. No copyright infringement is intended).

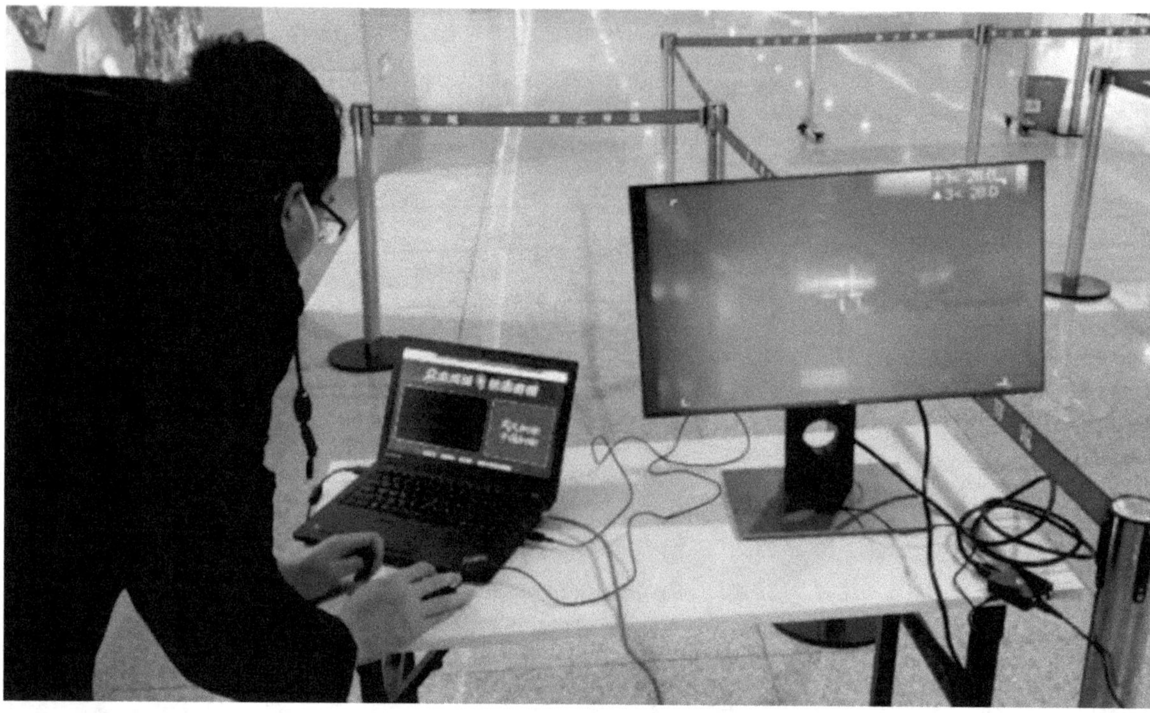

Smart helmet

Some augmented reality companies have developed a smart helmet with video camera to detect the body temperature of people up to 5 meters away. Advanced versions, capable of detecting several people at the same time, have been developed. These can be used in department stores, stations, public places, all equipped with facial recognition (Figure 8).

Treating

Some innovations introduced in China in the treatment of COVID-19 have already been on the market for some years, but had an extraordinary scale-up during the epidemic.

Free online clinics

Online clinic consultations have become a new normal (Figure 9). According to Ali Health data, visitors to the free online clinics had exceeded 2.8 million before the end of January 2020. In the meantime, real online medical service platforms have started online diagnosis services for the symptoms of Coronavirus, to reduce the flow of false positives to hospitals. The Alibaba research institute manages to test coronavirus infection through artificial intelligence applications with an accuracy rate of 96%. This new algorithm relieves pressure on hospitals, completing an identification process in 20 seconds, much faster than the usual 5-15 minutes for a doctor diagnosis (Sun Henan, 2020).

Figure 8. Smart helmet (Abacus, 2020, Photo Kuang-Chi. Appeared elsewhere on the internet. No copyright infringement is intended).

Robotics for health

From preparing meals in hospitals, serving as waiters in restaurants, spraying disinfectants and distributing hand sanitizers, robots were at the forefront in preventing the spread of Coronavirus. A hospital in Wuhan, the epicenter of the epidemic, was run entirely by robots. The hospital in Wuchang, and a manufacturer of cloud-based robotics, have joined together for this project aimed at making the hospital structure completely intelligent and digital. Here most hospital devices are IoT-enabled, while services are performed by robots. Initial patient screening is performed by 5G-enabled thermometers that send immediate updates. In addition, there are rings and bracelets connected to a cloud platform so that it can monitor all alterations in the patient body. The cloud company alone has deployed 100 robots to hospitals across China. Smart transportation robots deliver food and medicine to healthcare workers without human contact (Figure 10.)

Autonomous vehicles

Autonomous vehicles are proving to be of great use in the supply of essential goods, such as medicines and food products. A Baidu's autonomous vehicle platform has joined a startup to deliver supplies and food to a large hospital in Beijing. Baidu has also made available its microcar kits and autonomous driving cloud services to companies that fight the virus (Figure 11).

Figure 9. Free online clinics (Martin, Agence France-Presse, 2020. Photo by STR / AFP) / China OUT (AFP/STR). Appeared elsewhere on the internet. No copyright infringement is intended).

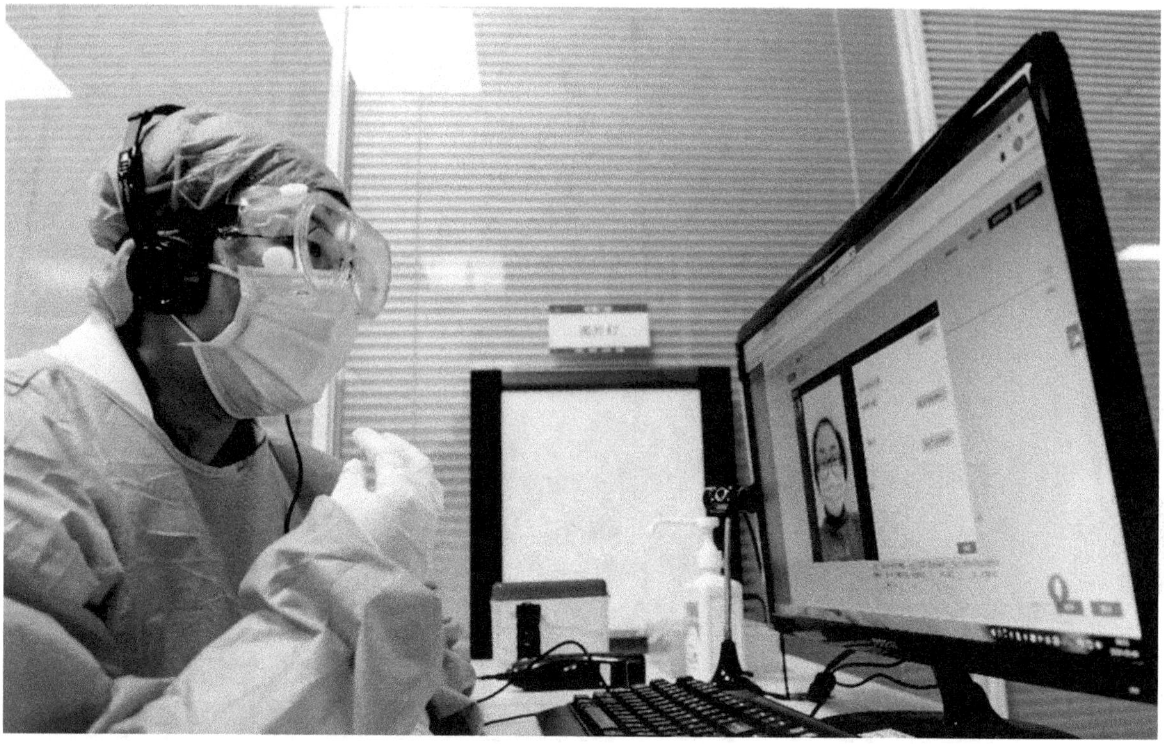

What are the Effects of This Digital Breakthrough?

A question arising in the daily life of post-COVID is how much these technologies, tested with an enormous deployment of resources by both the central and local governments in China, can be sustained over time. The present is uncertain. There is a new normality made up of circumspection, shops having few customers, insecure visitors in shopping centers, offices having their staff connected online from home. There is much more work flexibility, much less urban mobility, while the socialization is far from being regained. Social distance has become a new rule all are getting used to. QR codes testifying the citizens' health seem to persist. Will they last after the pandemic or will they be abandoned? It is difficult to predict, but the guarantee that a QR code offers to the administration of health is a convenient justification for maintaining a safeguard code over time. Masks would gradually be abandoned, although they have a persistence as well, they are becoming a dress-code, with shapes, colors, design that make them one of the most requested and democratic garments, from grandparents to children. The online clinics are nothing new, they were already existing before the pandemic. They were viable alternatives for quick medical consultation, in a country where family doctors do not exist. The tracking and self-isolation systems had an impressive scale-up and produced an unimaginable health justification for governments intending to control flows, mass gathering risks and health. Drones and satellite monitoring have found new fields of application, and these will be enhanced. Drones will increasingly serve the isolated communities and a capillary control of territories will take place by the combination of these two technologies.

Figure 10. Robotics for health (Fannin, 2020. Appeared elsewhere on the internet. No copyright infringement is intended).

Beside and apart from such disruptive technological innovations, there may be unwanted changes induced by Covid-19. One of these is the loss of the "sense of community", which is at the roots of our societies. The fear of contagion has induced an increase of "physical distance" between individuals, but all media have usually called this "social distance", revealing what was actually about to happen. If society becomes just a plurality of individuals without full physical relationships, our society, our cities as we know them, may rapidly vanish. And we should not forget that a plurality of separated individuals is what any autocratic state would most desire. On the same line, the enormous amount of data that new IT technologies may generate and handle about our lifestyles could make permanent the state-driven tracing of individuals justified by the pandemic.

However, trends are already outlined and perhaps nothing will be as before, but it will be.

THE SHOCK OR THE NEW PARADIGM

"We Need a Shock..."

In occasion of a late 2019 conference in Amsterdam on "Energy and the New Vision" a premonitory of what has happened with the pandemic was disclosed. While arguments were suggested by speakers,

Figure 11. Autonomous vehicles (Baidu 2020, Appeared elsewhere on the internet. No copyright infringement is intended)

such as: "people's understanding must increase, decision-makers must be well-informed", a disagreement rose: "it is no longer the time for information and rising awareness of decision makers, it has been done for too long. We need a shock…!".

But now, we all are suffering an existential shock, much bigger than predictable in that statement. Are we only at the beginning of a change in lifestyle and trends? The roots of what is happening with this coronavirus are not disconnected from "sustainability". Indeed, the roots are fully linked with sustainability.

Humans have the weakness of dividing things. We divide agriculture from health, energy from education, nutrition from environment. Each discipline is self-related. We lack the ability to associate and see deep bonds. How many times, in our experts' meetings, have we talked about the need to bring together different budget chapters, the use of space and time, to rationalize our energy and environmental system!

The term "unbundling" was a sort of *code word* about twenty years ago, when liberalization in the energy market started to spread all over Europe after being imported into the UK from the US. The idea was that breaking up large conglomerate utilities into many smaller companies (for instance, power production, transportation, distribution, sales) and optimizing each of them separately would have improved the overall performance of the production chain, minimizing the price to final customers and maximizing the profit of the companies. Unbundling was not only a matter of size. Inevitably, in a market economy, privatization also separated the economical dimension from all the others. A lower price to final user, under this approach, could be immediately perceived by the final user, while the local and global impacts on the environment would not be easily traced back to their specific cause. The search for local optima increased the tension in each ring of the chain at the expense of the least protected party in question, that is, the environment. Pulling too much the rope which was holding the human society together with natural systems, finally made the rope to break down. Only a truly holistic approach can re-establish this link, and the young generations now seem to have clearly understood it, providing a strong thrust towards this new approach.

The crisis of this virus and of the next one (we must already think about the next ...) directs us on how natural resources, food production and the food product cycle are exploited in an unsustainable way, in an unruly, chaotic, monstrously irrational mechanism and aggressive behavior.

Far from letting millennium-style fears be reborn, we must nevertheless humbly admit that we cannot control everything that happens around us. The environment is so complex that we are not able to predict in a deterministic way all the consequences of our actions. In any case, we have to realize that, whatever the human kind will do, nature will find a new equilibrium and those species which will best adapt to the new environment will survive. The issue is whether we will be among them.

"A shock is needed." But who knows if a shock will be sufficient? Once again, sectoral decisions and policies are considered, such as: "a vaccine is vital", "containment measures are needed". All true, in an emergency situation. Actually, the sustainability of all human activities must become the main driver for change. Acting deeply, in the conversion from *exploitation* of, to *cooperation* with nature, if we want to have a future.

In 1982, John Naisbitt (1982) wrote his book Megatrends, by cutting newspaper articles and exploring the future through the news on disruptive innovations.

Emulating the Naisbitt's exercise, we may find the discontinuity created by this shock, and how this will change the paradigm, or better, how it has already changed it. Let's see the different sectors.

Higher Education Regenerates Online

All universities in China have been closed since January, some months later in Europe and in the rest of the world. Universities closed their accesses, but they did not close their activities. They have provided teaching, remotely, they organised degree sessions in video conferencing. Remote teaching has been experimented in various ways already since decades, but only sporadically has entered into regular use. In our universities, we are still inspired by the educational principle of the peripatetic school of Socrates, where the "teacher" taught, by walking, a restricted circle of students. We have increased by two orders of magnitude the students of Socrates classes, and we carry out lessons with classrooms of 300-400 students, without asking ourselves whether this teaching is out of date or not. Was a shock needed to make us understand this? Now we realize that having classes of 100 students connected on teleconferencing platforms is effective; that doing graduation sessions remotely is profitable and fun. This teaching model has now been tested by almost all teachers, with different degrees of success. If this becomes common practice, and if it replaces at least 50% of the conventional teaching, huge savings in terms of resources (time, money, energy, quality of the environment) will be generated, and structural savings will be made. An assessment is possible. In a city with 100.000 university students, taking into account capacity factors and frequencies, roughly 10.000 students would avoid daily commuting to attend lessons, helping to reduce traffic jams and the crowding of public transport during peak hours. Of course, in case of unmitigated full online teaching, the social dimension is missing, the interaction between young people that generates innovation is affectedly reduced. But this interaction is certainly not going to be canceled, it is simply taken out of the isolated and often monotonous conventional teaching. It is going to be concentrated in interactive, creative teaching, with interpersonal sharing, workshops and focus groups. Then much more fruitful. Here is the discontinuity of the shock: can this model be dismissed by universities, after having implemented it on such a large scale? After having developed entirely new training modules, and after having well-thought-out this model and making it more effective for the new education? If it will not be dismissed, universities may explore new teaching approaches made of three components: a) the *disciplinary teaching*, recurrent lecture by lecture, year by year, that could mainly go online; b) the *interactive teaching*, based of laboratory activities, workshops, focus groups, study days, with strong interaction with teachers and among students, therefore certainly performed in presence; c) the *extra-ordinary teaching*, made of special webinars with qualified experts, facilitated by the online mode, an out-of-ordinary experience, somehow rebalancing the discrepancies among the more and less privileged universities.

The Work is Remotely Spread

In Shanghai and in China, companies have closed their doors well before and far more extensively than in other countries. Of course, the economy cannot sustain if companies close, but some companies have almost regularly worked. Those are highly automated companies, in which management and services have been led from home. Virtual meetings were organized, data exchange, discussions on machineries, remote productions were managed. It is reasonable that these working conditions - once more flexible, better organized and out of the emergency - will enforce in the short period. Remote working is nothing new, but it has never been experienced in such a universal way. The scale-up has always been missing and this shock has helped to ascertain that expansion is possible, having behind it a good organization. The factories were born in the second half of the 1700s, they were outside the urban circle because of a massive use of coal, with unavoidable pollution spread in the surroundings. Today, with cleaner industrial production, urban planning has rather changed. New industries can be mixed with urban fabrics, and the remote management can lessen the need for local equipment control, reducing workforce at the production site. By experiencing the shock, entrepreneurs have come up with new ideas. When the epidemic started in China, for the first time people were working from home to avoid the spread of the new coronavirus. Many large companies, including Tencent, JD.com, Suning, Huawei, organized homeworking for all their employees for many weeks. Many cities and provinces took similar measures, imposing mandatory quarantine periods with remote working. Video conferencing companies, such as Tencent's WeChat Work, Alibaba's DingTalk and the American Zoom, were swept up in web traffic, with many users complaining about delays and interruptions. DingTalk had recorded a historic spike in web traffic, and assigned 12,000 additional servers to increase its capacity. Zoom's share price rose 15% in just one day (Davis, 2020). Many companies developed new verification protocols, deadlines and objectives to be achieved during their video conferences. This is a well known history, since all countries have experienced the same conditions, with time lags of few months one after the other. Homes have been transformed into "webinar-factories" (the modern factories), in which professional parents and children are simultaneously involved in their daily businesses or classes. The discontinuity of the shock: are we sure that this model will be dismissed by companies, after having deemed it effective for themselves and efficient for their workers?

The Boom in Decentralized Services

Services are inherently vast and multifaceted, each one peculiar. Here is outlined the service that is becoming foremost important, during the current shock. In Shanghai, food delivery has been routine for many years. People hardly buy in supermarkets, but order on Taobao. Quantities do not matter. Taobao brings disposable packages of minimal economic value and size to everyone' homes. Through Sherpa applications, portions of meals are ordered and delivered in the middle of city parks if a family is having a picnic. Sherpas track with the Wechat localization APP, and deliver the single-portion package in the middle of the park. If the meal is complex, a number of Sherpas will deliver. Delivery services, online purchases, the logistics to reach the final customer when is needed, originate from the "just in time" formula, invented by Toyota to innovate its industrial production in the 1970s. These services, already widespread in China, have had a "scale-up" throughout the world. Amazon had a sort of temporary blackout due to shortages of home delivery slots. In Italy, after the announcement of the lockdown of the entire country. Tens of thousands of connections on the Amazon providers to book door-to-door groceries suddenly shown-up. Of course, the site sold out the delivery slots in a few minutes, as did most of the other food distribution websites. This is the discontinuity made by the shock: a purchase method hitherto reserved to a minimal part of the population extends to a large number of families, with advantages for environment, energy consumption, and use of time. Are we sure that this widespread online food shopping will be dismissed by families, after such a massive use, once tested being more efficient than buying in supermarkets on Saturday afternoons?

Another trend is competing and we do not know which will prevail. The 5G has introduced a disruptive innovation in supermarkets. Customers are simply bringing products into their shelves, put them in their bags and go out. Everything is automated, the recognition of the customer with his biometric coordinates and his contact details, the recognition of the products that are taken from the shelves, the electronic payment without having to show any card or telephone app. There are no cashiers or controls. Only staff to replenish the shelves as they are emptied. What difference is there between this technological innovation allowed by 5G and home delivery after selecting from virtual shelves? Almost none.

Again, this would be a distressing solution if widespread in practice, but another model may prevail. A model where some exceeding shops will perish, while more sustainable shops would flourish, just because the less qualified or less attractive are vanishing. Shops that contribute to ameliorate the urban scenes, the fruition of streets, the living conditions of the local communities would become essential. While the too many shops in malls and supermarkets could naturally disappear.

This new trend is already visible in 2020, in a city like Shanghai, where big malls are less and less attractive and visited by people, while more and more shops on streets are.

Overcoming High Intensity Transport

Airlines around the world have reduced capacity and suspended routes to several countries affected by the Covid-19 virus, starting with the focus of the epidemic, China. Now, as the virus has spread around the world, the infection has affected airlines in every region, particularly some of the busiest transit connections. This devastating situation induces a change of model, which is also a change of paradigm. Are we actually sure that many who have experienced the effectiveness of webinars, not once, but a dozen times in a month, will adapt again to book airplanes and hotels, to move from one city to another, to get in taxis and manage inconvenient logistics for almost the same result? Are we sure that art is accessible only by forming long queues to enter a museum, and going around the halls to look at artworks, together with many others that often preclude the view? Or, on the contrary, will the decentralized model of conferences, the art and cultural fruition, the virtual and augmented reality, cut forever a significant share of air travel around the globe? Isn't the current swirling pattern irrational and unsustainable?

From another perspective, the epidemic has stopped the industrial production, reduced transport and has massively interrupted commercial activity throughout this country and all these measures have contributed to ameliorate the environment. With fewer factories in operation, fewer cars and heavy vehicles on the roads, airplanes in flight, the air quality has significantly improved, and the skies have become considerably cleaner. Are we really persuaded that, after having experienced blue skies, and a more breathable air, the reasons of production will again increase the pollution levels of pre-Covid19? At the moment, the need to restart industrial production made these conditions accepted. But there could be another way: that of trying to maintain the lowest pollution levels, gradually increasing a clean production. Like when, having made a weight loss treatment, we start eating regularly again, but trying to maintain our healthy weight.

Nutrition Respectful of the Natural and Animal World

So far, the paradigm doesn't look very different from the previous one. More sustainable? Not so much. Perhaps more efficient: less travel, less crowding, lower energy consumptions, better use of urban space and time. To understand how the shock will create a "new paradigm" in the area of nutrition it is compulsory to move on, and tackle the serious problem of the relationship between man and nature, in particular, man and the animal world. Until we manage to revolutionize our relationships with animals beyond individual behavior alone, there will only be slight adjustments to the current model.

What did the shock made us understand? Coronavirus is an effect. It is not the root of a pandemic disease. Pandemic disease is the exploitation of nature by man, in particular of that part of the natural world that we deem in our "exclusive service": animals used as food.

China has understood that the consumption of wild animals is at the origin of the virus and has already taken steps to ban its trade. Still, this trade is the basis of a thriving industry in China, calculated for a value of 7 billion US$. It is possible that it will remain under track, but the government's signal was immediate: it must be closed. This shock may not be enough to moderate the enormous consumption of "animals for food". The 12 million pigs that are "converted" into food each year in Shanghai are not eradicable, unless an enormous afterthought takes place. Yet, it is man's exploitation of nature that has generated this virus, and that will generate others, more and more devastating. Our pursuit to find remedies for catastrophic events must be reconsidered in the opposite way. It is necessary to switch from a re-active to a pro-active attitude. It is estimated that only 50 million people (around 3.6%) of the Chinese population are vegetarian. By contrast, 40% of Indians claim to practice vegetarian diets (Vandenberg, 2020).

Despite this, some signals of change are felt. How much has China's pork consumption decreased in the first months of 2020? Pork is a good indicator: the Chinese pork production in 2019 decreased by 5.46 million tons (14%), reaching about 40.5 million tons. In 2020 it will reduce by a further 9.43 million tons (23%). The government has tried to stabilize prices by releasing hundreds of thousands of tons of frozen pork, but in the meantime, consumption has decreased by an equal amount. This is not only due to the 2019 pig disease, that has decimated the pigs population in this country. It is more complex and widespread (Patton, 2020).

Nevertheless, the eating habits of the Chinese are changing. The Chinese government has planned to cut its 1.4 billion inhabitants' meat consumption by 50% by 2030. The dietary guidelines developed by the Chinese Ministry of Health in 2016 recommend a daily consumption of 40g to 75g of meat per person. The measures are released every 10 years, they are designed to improve public health, but also to significantly reduce greenhouse gas emissions (Chinese Resident Balanced Diet, 2016).

Chinese people do not eat less meat because of the epidemic. However, the shock created a discontinuity that simply could accelerate a different nutrition paradigm. If this trend is kept, and becomes structural, it will be the most important mega-trend of the new paradigm.

CONCLUSION

"Toward global equilibrium" was the title of a famous book from the 1970s, subsequent to the well-known "Limits to Growth" (Forrester, Meadows and Meadows, 1973).

Will this shock be enough to modify life models, to make people feel the unsustainability of humans on the planet? The dilemmas of globalization have been disclosed by this shock. There is a crucial need for resilience of local systems, for short chains, for autonomous energy and food self-sufficiency, for decentralizing essential products. Security and contingency plans are needed, and must operate on a global scale, but at the same time at the country and the local level. The list of roles to be reviewed would be very long.

At the time of writing, the perception of the future is distorted by the events, and conformed to them, so nothing is like last year. Today it is simply different.

REFERENCES

Abacus. (2020). *Chinese police now have AI helmets for temperature screening*. https://www.scmp.com/abacus/news-bites/article/3052879/chinese-police-now-have-ai-helmets-temperature-screening

Baidu. (2020). *How Baidu is bringing AI to the fight against coronavirus*. https://www.technologyreview.com/2020/03/11/905366/how-baidu-is-bringing-ai-to-the-fight-against-coronavirus/

BeiDou. (2020). *COVID-19: China's BeiDou Satellites Help Fight Against Epidemic*. POB Point of Beginning. https://www.pobonline.com/articles/102110-covid-19-chinas-beidou-satellites-help-fight-against-epidemic)

China Academy of Information and Communications Technology. (2020). *Data and Smart Applications for Epidemic Prevention and Control*. Source: CAICT, Research Report. http://www.caict.ac.cn/english/research/rs/202004/P020200430479193356866.pdf

Chinese Resident Balanced Diet. (2016). Release time: 2016-05-18 声明:关于《中国居民平衡膳食宝塔(2016)》官方版本发布时间: http://dg.cnsoc.org/article/04/8a2389fd54b964c80154c1d781d90197.html

Davis, K. (2020). *Remote Control: Coronavirus Has Millions in China Working From Home*. Sixth Tone. https://www.sixthtone.com/news/1005160/remote-control-coronavirus-has-millions-in-china-working-from-home

Fannin, R. (2020). *The rush to deploy robots in China amid the coronavirus outbreak*.

Flanders-China Chamber of Commerce. (2020). Three clinical Covid-19 vaccine trials approved in China. http://news.flanders-china.be/three-clinical-covid-19-vaccine-trials-approved-in-china

Forrester, J. W., Meadows, D. L., & Meadows, D. H. (1973). *Toward Global Equilibrium*. Wright-Allen Press.

Hua, X. (2020b). *Intelligent recovery of the old telephone booths*. http://www.xinhuanet.com/english/2020-05/26/c_139089181.htm

Martin, D. (2020). *China retreats online to weather coronavirus storm*. Agence France-Presse.

Naisbitt, J. (1982). *Megatrends: Ten New Directions Transforming Our Lives*. Warner Books Inc.

Patton, D. (2020). *China's 2019 pork output plunges to 16-year low as disease culls herd*. Reuters. https://www.reuters.com/article/us-china-economy-output-pork/chinas-2019-pork-output-plunges-to-16-year-low-as-disease-culls-herd-idUSKBN1ZG08H

Shanghai Air Pollution. Real-time Air Quality Index. (2020). [Data set], https://aqicn.org/city/shanghai/

Sun, H. (2020). *Alibaba says AI can identify coronavirus infections with 96% accuracy*. Nikkei Asian Review. https://asia.nikkei.com/Spotlight/Coronavirus/Alibaba-says-AI-can-identify-coronavirus-infections-with-96-accuracy

Vandenberg, L. (2020). *How Can China Fulfill Its Rising Meat Demand?* The Diplomat. https://thediplomat.com/2020/04/how-can-china-fulfill-its-rising-meat-demand/

Watson, J. (Ed.). (2020). *Biosurveillance Market. 07 mai 2020*. Reports and Data. https://www.globenewswire.com/fr/news-release/2020/05/07/2030018/0/en/Biosurveillance-Market-To-Reach-USD-5-62-Billion-By-2027-Reports-and-Data.html

World Health Organisation. (2016a). *Shanghai Consensus on Healthy Cities*. https://www.who.int/healthpromotion/conferences/9gchp/9gchp-mayors-consensus-healthy-cities.pdf?ua=1

World Health Organisation. (2016b). *9th Global Conference on Health Promotion, Shanghai 2016*. https://www.who.int/healthpromotion/conferences/9gchp/programme/en/

Xia, H. (2020a). *Shanghai to invest 270 bln yuan in new infrastructure in next 3 years*. http://www.xinhuanet.com/english/2020-05/07/c_139038827.htm

Yinfen, Z. (2019). *Innovation Drives Development, Technology Leads the Future: G60 Science & Technology Innovation Valley*. Shanghai International Studies University, SESTA. http://en.shisu.edu.cn/resources/features/G60-Science-Technology-Innovation-Valley

Chapter 8
The DIM Approach for Digital Twin

Matteo Del Giudice
Politecnico di Torino, Italy

ABSTRACT

In the era of connections and information and communication technologies, the building industry is facing the challenge of digitization at the building and urban scale. Several researches have been carried out to generate virtual city models to manage and represent a variety of data to reach the smart city concept. Therefore, the development of building/urban digital twins is directly linked to the definition of innovative methods and tools that are able to collect, organize, query heterogeneous data to make it available for the various involved actors. This chapter aims at presenting the district information modelling methodology that is strictly related to the digital twin concept, starting with data domains, arriving at the various tools developed to reach the users' needs.

INTRODUCTION

Nowadays, Architecture Engineering Construction, and Operations (AECO) industry is crossing a transition period driven by digitalization. The building industry community is investing a considerable effort together with researchers who are investigating in depth the issue of using digital technologies and digitized data to impact how to work and its products get improved. The driving impulse of the construction sector digitization is due to several factors linked to climate change and current economic conditions. The growing demand for energy from existing buildings and the consequent increase of pollution in cities have led the various states and the European Commission to call for new standards to improve the liveability of urban spaces by optimizing the management of available resources. In these terms, the fourth industrial revolution (Industry 4.0) has introduced digital technologies, sensor systems, intelligent machines, and smart materials to the construction industry to increase knowledge of the existing architectural heritage and to improve it with strategies oriented towards structural reliability and energy efficiency. Recently, an amendment of the directive on the Energy Performance of Buildings (EPBD) encourages each EU Member State at accelerating the cost-effective renovation of existing buildings, both

DOI: 10.4018/978-1-7998-7091-3.ch008

public and private, into a highly energy-efficient and decarbonized building stock by 2050, facilitating the cost-effective transformation of existing buildings into nearly zero-energy buildings (2018/844/EU). In this area, several researchers are studying the best processes and tools to transform existing cities into smart cities that European Commission defines as places where traditional networks and services are made more efficient with the use of digital and telecommunication technologies for the benefit of its inhabitants and business. Besides, the achievement of a smart city implies the active involvement of private and public administrations and citizens with their habits. What said before highlights the need to develop systems capable of linking data stored in various databases and providing it to various users. So, it is clear that data management is gaining an essential role in the smart city field to optimize energy saving at building and urban level.

Several projects have been recently developed in order to face the lack of systems where all-district and building-related data is collected in a common digital repository able to be enriched and queried to define new policies towards smart cities framework. Within this context, the European 7th Framework Programme (EU FP7) District Information Modelling and Management for Energy Reduction (DIMMER) project aims at managing all district-related data through the DIMMER platform, available to all the stakeholders involved at the district level. The interoperable process makes the data exchange possible.

BACKGROUND

The city has always been considered as a meeting place where a series of complex dynamics linked to human beings' habits take place. From this point of view, it can be considered a living organism that transforms itself into the function of the society that inhabits it (Pagani & Chiesa, 2017). Vice versa, the population adapts itself to the city in which it lives, assuming behaviors that could change in another city according to its characteristics. The urban transformations that have taken place over the centuries have been driven both by the growing needs of citizens and by the technological innovation that has developed throughout history.

Over the centuries, the management of the city has been subjected to the needs of the population, which in different forms modified the urban ecosystem to improve the usability of space and optimize human activities, finding answers to the need for transformation and urban renewal according to the lifestyle of citizens and the adopted technologies.

These transformations have been often influenced by the demographic, industrial, and economic growth of countries, ignoring the aspects related to the sustainability of interventions, thus developing not very resilient cities to climate change and exceptional events caused by land degradation.

However, in recent years, more attention has been paid to the capacity of cities and buildings to adapt to climate and socio-economic change by setting several constraints to be respected on environmental sustainability.

In this sense, the EPBD directive refers to Smart Ready Technologies (SRT) which should be disseminated in the existing building stock to achieve significant energy savings with a consequent economic gain, while helping to improve indoor thermal comfort so that the building can adapt its energy behavior considering the user's needs.

Currently, the EPBD requires the development of a voluntary European rating scheme to assess the intelligent readiness of buildings, called Smart Readiness Indicator(SRI). It aims at making the added value of building smartness more tangible for building users, owners, tenants, and smart service providers (Verbeke et al., 2018).

The assignment of a value to this indicator implies introducing innovative methodologies and tools in the building industry to monitor the real condition of a certain building, to collect information on its energy consumption, and then to process it to produce refurbishment scenarios.

Thus, in the digital age, Information and Communication Technologies (ICTs) have considerable importance for the development of innovative city models based on intelligent control systems. These systems offer the ability to collect and manage large quantity of data for monitoring the use of energy resources to improve citizens' lifestyles, avoiding energy waste.

These assumptions require that buildings, districts, and cities of tomorrow have to be equipped with a system capable of connecting the territory, urban space, and people's lives to ensure sustainable development and a better quality of life through more efficient services, with a dynamic information offer while respecting data ownership. Through this system, the planning and programming of interventions aimed at transforming the urban environment to optimize it from time to time and orient it according to the needs of the population should be easier. (Testoni, 2016). In this context, the scientific community has been analyzing different ideas of the smart city based on digital platforms able to handle large amounts of data for different uses (i.e. energy efficiency) over the last years. Many definitions are available on the smart city concept (Testoni, 2016), and the majority agrees that ICTs will play a key role in influencing people's behavior in the near future in the field of sustainable development and improvement of the quality of life. This is because the idea of a smart city is based on the need to manage urban processes with innovative methods and tools that can provide answers to the current challenges on climate change. This approach involves a wide range of aspects related to the field of technologies, data and the people who have to use and interpret it through common platforms, providing solutions to today's society.

These platforms are based on the use of graphical interfaces that find innovative ways to visualize the urban environment. Usually, these tools optimize the way to provide various information both graphical and alphanumerical. In particular, developing virtual 3D city models able to collect, monitor, and extract information from multiple datasets is one of the most interesting topics studied by researchers in the last decades, concerning the data management optimization of the city.

These datasets can be exploited to simulate building/district energy behavior to carry out new policies for potential energy refurbishments. In this contest, one of the principal key points consists of the adoption of Building Information Modelling (BIM) methodology, as a set of processes applied to create, manage, derive and communicate information among stakeholders at various levels. It is based on a digital representation of the physical and functional characteristics of a facility that can be created by all participants to the building process, at different times and for different purposes, to ensure quality and efficiency throughout the entire building lifecycle (Osello, 2012). In parallel, at urban level, Geographic Information System (GIS) is used to generate a digital model of the city that collects large amounts of data related both to the buildings and energy heating network (i.e district heating network). The GIS dataset consists of a digital representation of a geographical area in which the graphic elements describe the elements of the city by associating graphic information with alphanumeric information (i.e. construction period) that can be organized according to the study to be carried out.

Many cities are moving in this direction, developing digital platforms capable of displaying information that can be used by various users. For instance, some examples of 3D city models that are under development are listed below:

- Virtual Singapore is a dynamic three-dimensional (3D) city model and collaborative data platform, including the 3D maps of Singapore. It will enable users from different sectors to develop sophisticated tools and applications for test-bedding concepts and services, planning and decision-making, and research on technologies to solve emerging and complex challenges for Singapore (Foundation, 2018);
- Helsinki's 3D city model consists of two next generations of 3D city models: a semantic city information model and a visually high-quality reality mesh model (Helsinki, 2019). The models serve as a platform for designing, testing, applying and servicing the entire lifecycle of the built environment, as well as smart city development;
- Digital 3D modeling of Rennes Metropole is an innovative tool that enables collective thinking on the city of tomorrow together with experts and residents (Eurocities, 2018). This online model is used in various ways, such as for urban mediation with citizens, and for urban development purposes such as sunshine simulation, noise modeling, tree shadow impact on buildings, and so on;
- Urban Data Project is a platform developed by Telensa and powered by the Microsoft Azure cloud platform that creates a trust infrastructure for urban data, one that enables cities to collect, protect and use their data for the benefit of all citizens. The first deployment will be in Cambridge in the UK (Telensa, 2020);
- The New York City Department of Information Technology & Telecommunications (NYC DoITT) provided a 3-D NYC Building Model for the sustained, efficient, and effective delivery of IT services, infrastructure, and telecommunications to enhance service delivery to the City's residents, businesses, employees, and visitors (NYC, 2020).

This information is represented at urban scale with a different level of geometric and alphanumeric detail from that adopted for BIM models. Although the creation of these information platforms has given a strong boost in the direction of smart cities, BIM and GIS datasets represent reality through models and cannot collect information in near real-time, against the Digital Twin (DT) concept (Grieves, 2014). DT requires the identification of a physical environment and its components, the development of a virtual twin, and its connection with the real one based on data. Therefore, the need to manage heterogeneous sets of data, from building to urban scale, has highlighted the importance of Information and Communication Technologies (ICTs). Through interoperability, communication and information sharing are improved by raising the collaboration between all the parties involved in the building process.

The growing availability of 3D city models for displaying energy data, particularly in the Open Geospatial Consortium (OGC) standardized and open data format of City Geography Markup Language (CityGML), have provided a significant boost in the development and visualization of urban energy data (Würstle et al., 2020). CityGML is an open data model and XML-based format for the representation and exchange of virtual 3D city models. It provides a basis for 3D geospatial visualization, analysing, simulation and exploration tools. Thus, it offers the possibilities for numerous and varied spatial analyses related to the building performance simulation (Nouvel et al., 2013). Although the need of collecting and storing specific information on buildings increased, the limit of this process is the fact that the CityGML language does not provide explicit attributes for modelling energy consumption of buildings (Krüger & Kolbe, 2012).

Recently, many researchers investigated the integration of GIS, neighborhood-scale 3D (CityGML format) and BIM (IFC format). Extensive efforts were put into building model extensions, to realize meaningful integration. The key of the integration is merely data conversion. (Jusuf et al., 2017). For this reason, several attempts have been made to integrate data coming from the geomatic standard to the building scale. One of the first examples is an approach for data integration based on a Unified Building model (UBM) which encapsulates both the CityGML and IFC models, thus avoiding translations between the models and loss of information (El-Mekawy et al., 2012). Another approach aims to convert already existing architectural models, stored in the IFC format, into CityGML models; the result of this study was a conversion algorithm that accurately applies the correct semantics from IFC models and that constructs valid CityGML Level of Detail (LOD) 3 buildings by performing a series of geometric operations in 3D (Donkers, 2015). Although Virtual 3D city models have been used mainly for the visualisation or graphical exploration of cityscapes (Kolbe, 2009) they continue to be static models of reality, without a connection network capable of gathering information that can be connected to the urban 3D model.

The introduction of the Internet of Things (IoT) domain facilitates the connection between the virtual model with the physical one, by installing several devices into the physical model. In this way, different types of data related to energy assessment can be collected and then compared with the simulated ones. Thus, the outputs of specific simulations developed by the virtual models have to be subsequently tuned with measured data. At the end of this procedure, the calibrated virtual model can be used to run scenarios for the improvement of energy efficiency and internal comfort. In this context, the extension of this idea at urban scale belongs to the field of an Urban Digital Twin (UDT) that is investigated by several researchers to find a right solution to describe reality with a certain level of abstraction at city level. As it is not the exact copy of reality, but a sophisticated abstraction of ibidem, it can be best characterized as a container for models, data, and simulations. Beyond these challenges, it has great potential to support scenario development processes and testing of these at all scales. (Dembski et al., 2020)

RELATED WORKS

The modeling phase therefore becomes a major theme that need to be faced in a transversal and interdisciplinary way from the urban to the building scale. Many international projects have tackled this issue and have come up with various definitions of urban digital twin:

- Under the EU FP7 "Indicator-based Interactive Decision Support and Information Exchange Platform for Smart Cities" INDICATE smart cities project, a decision support masterplanning tool will be developed, which will; plan, integrate and optimize the sub-systems of the overall city system: The INDICATE tool will achieve this through the integration of real data from a city with the dynamic simulation software, the IES <Virtual Environment>, 3D Urban CAD modelling tools, the ESRI CityEngine package, and Sustainable Urban Indicators, to create a Virtual City Model. This will provide a 3D model of the city, its networks and its buildings, ready for urban simulation (Melia et al., 2015);
- The "Semantic Tools for Carbon Reduction in Urban Planning" (SEMANCO) project has developed a 3D web-based platform which allows the user to implement a methodology proposed for simulating different energy improvements in a city or an area by using semantic data modelling. It enables to look-up the stored information in different sources and facilitates the analysis of results within a determined number of embedded tools to calculate the energy performance of different scenarios. (Corrado et al., 2015);

- The "Optimised Energy Efficient Design Platform for Refurbishment at District Level" (OptEEmAL) project is funded under the European Union's Horizon 2020 research and innovation programme. The platform delivers an optimized, integrated, and systematic design based on an Integrated Project Delivery (IPD) approach and supported by the utilization of BIM models for an integrated, optimized and systemic energy-oriented refurbishment at the district level. This is achieved through the development of a holistic and effective services platform that involves stakeholders at various stages of the design while assuring interoperability through an integrated ontology-based DDM (District Data Model) (Romero et al., 2016);
- The SPHERE Horizon 2020 European Project aims to provide a BIM-based Digital Twin Platform to optimize the building lifecycle, reduce costs, and improve energy efficiency in residential buildings. It mainly focuses on developing an ICT platform able to manage and update both static and dynamic building information across the design, construction/renovation and operations phases from multiple sources, and utilize this information both in operational energy control and for feedbacks from operations to design (Alonso et al., 2019);
- Digital Urban European Twins (DUET) is a H2020 funded project designed to advance policy development in the age of big data and cloud to deliver a trusted, scalable and transferable solution for accelerating the adoption of data-driven, collaborative decision making, and policymaking. It aims to leverage the advanced capabilities of cloud and high-performance computing (HPC) to help public sector decision-making become more democratic and effective, through the development of Digital Twins for policy exploration and experimentation across entire cities and regions. (McAleer et al., 2020);
- The SINFONIA project is a five-year initiative to deploy large-scale, integrated, and scalable energy solutions in mid-sized European cities. The heart of the initiative consists of cooperation between the cities of Bolzano and Innsbruck, working hand in hand to achieve 40 to 50% primary energy savings and increase the share of renewables by 20% in two pioneer districts. The project is mainly given to the transferability and scalability of the solutions deployed in the two pilot districts. In this way, it will define a limited set of district typologies and corresponding refurbishment models, enabling cities to easily assess their needs and efficiently define their long-term refurbishment strategies (Bisello, 2020);
- The aim of the "Optimising Energy Use in Cities through Smart Decision Support Systems2 OPTIMUS project, funded under the EU FP7 between 2013-2016, is to support the decision making process, improving the energy efficiency of buildings, by optimizing the energy use in their premises, and reducing CO2 emissions. The result of the OPTIMUS project is a package of web-based consulting tools for energy managers and energy consultancies, to make more energy efficient and sustainable cities. One of these technological outputs of the project is a semantic framework that enables cities to share and integrate their data from different domains, using Semantic Web technologies. (Sicilia et al., 2015);
- The aim of "Decision support Advisor for innovative business models and user engagement for Smart Energy Efficient Districts" (DAREED) project is to deliver an ICT service platform (and some specific tools) to foster energy efficiency and low carbon activities at neighborhood, city and district level. In this way, the project focuses on reducing energy consumption in buildings by using a system capable of receiving information from various sources, analyze the energy consumption taking place, and provide information and advice to both citizens as utilities and public institutions, involving all stakeholders in the process of improving energy efficiency. (EC, 2016).

The projects mentioned above have as a common objective to define innovative strategies to reduce energy consumption and CO^2 emissions. Some of these projects are more oriented to the development of tools able to support the strategies to be chosen for energy efficiency. Instead, others are focused on the virtual cities models development aimed at visualization and integration, some are aimed at collecting data from IoT devices installed in buildings to monitor the state of art, defining semantically the rules to link together data belonging to different disciplines.

Comparing DIMMER with the projects described above, the real innovation is in the use of different entities (e.g. BIM, SIM, GIS) to develop a District Information Modelling (DIM) domain that is considered one of the starting point of the project. It extends the Building Information Modelling (BIM) at the district scale. The novelty of the DIMMER framework consists of enhancing the potential of smart technologies, starting with the development of a district parametric model integrated by an IoT platform based on open standards of the Web. Thanks to the use of web services this paper shows how the DIM approach can be applied at urban scale to face the energy efficiency challenges, starting from the definition of a DIM standard based on information technology, through several DIM tools for data visualization and energy simulation.

METHODOLOGY

Case Studies

For the DIMMER project two pilot cities have been selected: Manchester and Turin. For each city, a specific number of case studies have been chosen to enhance the replicability of the DIM methodology.

The Turin Pilot consists of the Polito's District close to Politecnico di Torino. The Turin Pilot District is a selection of different kinds of buildings situated in the "Crocetta" District. It is composed of several buildings of different typologies. The Manchester Pilot District is a selection of the University of Manchester buildings situated on 'the Corridor' and covers 38 hectares of 'the Corridor'. The Manchester Pilot District comprises of a diverse mix of building types. (DIMMER consortium, 2016).

Each selected building is part of a different building energy typology, not only for characterizing the selected district and the district's buildings but also to implement and validate DIMMER approach so that there are heterogeneous case studies to define a standardization in the DIM process. For the description of the pilots, a DIMMER matrix was defined as a right tool to underline the heterogeneity of the Turin and the Manchester case studies, concerning the building typology (DIMMERcollaboration, 2016). This idea is inspired by the IEE Project TABULA and EPISCOPE project (TABULA website, 2012), where residential building typologies have been developed for 13 European countries. In this perspective, Figure 1 illustrates the process followed by the DIMMER project, starting from the collection of input data (e.g. archival documentation, maps, CAD files) to the development of BIM models, identifying the building typologies of the selected buildings, through the DIMMER matrix. For characterizing the selected case studies, the DIMMER matrix, visible in the center of Figure 1, has been developed to visualize each building component of the building envelope, related to the building typology. Each building component is described with the name, the U-value, etc. This information is usually stored in different ways depending on each country and for this reason, this matrix can be considered a synthesis of all information

Moreover, users' awareness has been considered one of the essential topics of the DIMMER project, to disseminate the issues and the results of the project engaging common users' in the process of developing smart cities.

Figure 1. Schema for districts and building typologies of the two pilot cities for the DIMMER matrix

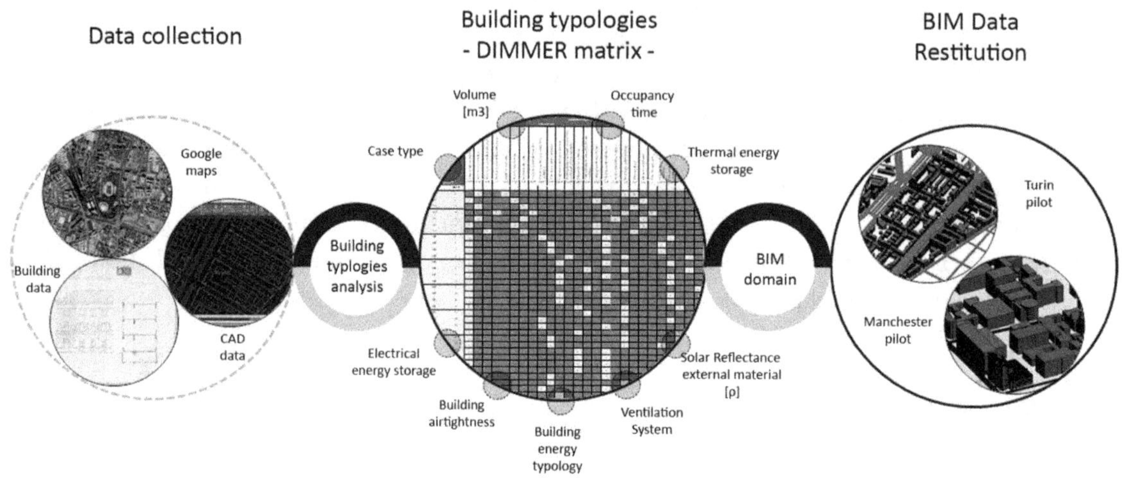

The term Digital Twin is commonly used to describe a virtual replica of physical things that composes reality. In the last years, it is mostly used as an example of revolution and is considered fundamental to transformation, but the broad scope of the concept makes a common definition difficult. (Evans et al., 2019). The construction industry often idealizes this concept as a beacon to be followed in order to give life to the digital transformation of the existing architectural heritage. Although the stakeholders involved consider DT as a virtual replica of the real world, one of the main features consists of the connections that have to be established between the physical and the digital world. Starting from the users' needs, it is possible to create digital twins with different skills that allow them to develop actions with a different degree of maturity. Based on the digital twin maturity spectrum (Evans et al., 2019), Figure 2 shows the various skills that are possessed by a digital twin, starting from simple monitoring and reporting to integration from the various datasets, arriving at prediction and autonomous decision.

The creation of the digital twin is essentially based on three components: i) the physical components, ii) the virtual models; iii) data that connects them. For this reason, the DIM approach involves the development of various building and district data domains that describe reality with graphical databases. Subsequently, a network of sensors inside the buildings must be installed to guarantee the data collection that must be processed to enrich the virtual space making it dynamic. This data can be used later to run simulations aimed at creating possible scenarios to improve the physical world. This "Virtual-Physical" duality is provided by data that is subject to different transformations from the physical world to the real world and vice versa due to the adopted technologies. (Boje et al., 2020). For this reason, the development of a digital twin requires a level of interoperability that enhances the uniqueness of the data and avoids possible data loss. This process requires necessarily a common web platform able to gather, organize, query, and make information accessible for various tools for desktop and mobile, enabling

Figure 2. Graphical representation of Digital Twin idea

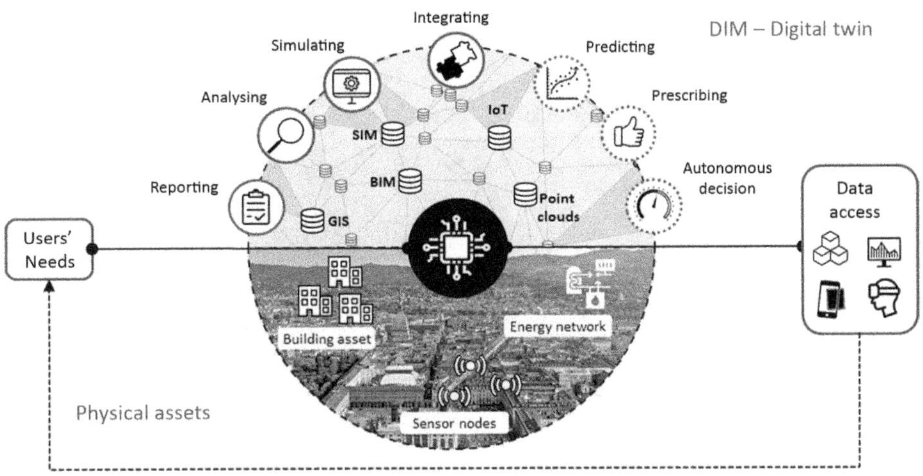

also immersive experiences. Through these interfaces, a virtuous process can be triggered to meet the user's requirements. This process can be used for the active transformation of cities into smart cities.

As this paper aims at presenting the DIM approach based on the extension of the BIM methodology at the district scale, the DIMMER project focused on improving data sharing by using a DIMMER platform, based on an OpenSource Middleware (Linksmart, 2020). It provides an innovative decision support platform for several end users (e.g. public administration, building manager, energy professional) which is one of the major challenges to achieving smart city objectives. The DIMMER Platform provides an easy way to access different kinds of data, fostering different scenarios and demonstrators. It integrates mainly two different types of data sources related to (i) Building Information Models, Geographic Information Systems, and System Information Models; (ii) Sensors and Energy Measurements (Device Connectors). (Osello et al, 2016). This means that interoperability, consists of creating a dataflow that involves heterogeneous data coming from different domains describe below:

- BIM Domain: represents all information about the building scale and it can be used to perform all the simulations needed for its design and operation like lighting and thermal analysis;
- GIS/SIM Domain: is used to manage urban and district data for the collection of a large amount of information useful to permit the integration and the elaboration of various information related to "energy", that are crucial for decision making on energy issues. It is essential to collect data about the building typologies, district heating network and energy demand, useful to analyze at district scale the energy consumption;
- Building Energy Models (BEMs) Domain: is linked with the energy simulation needs at building and district scale. In particular, it is generated from a 3D parametric model (BIM domain) through interoperability that can "filter" the energy model;
- Building Automation System (BAS) Domain: is represented by static and dynamic data of commercial Building Management Systems (BMS), sensor platforms and monitoring systems deployed in the demonstrator districts, and ICT hardware developed as a part of the project.

Figure 3. Methodological workflow from data domains to the DIM tools, using DIMMER Middleware

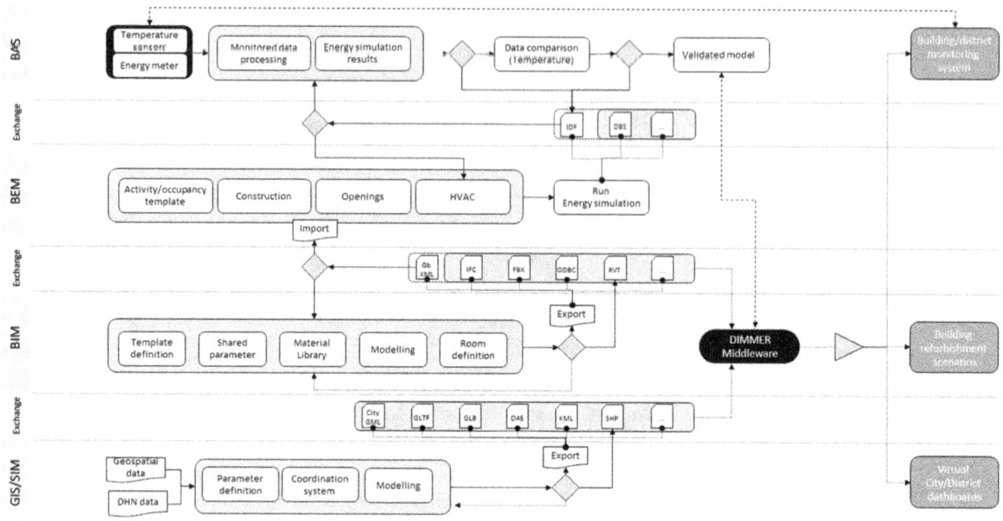

Figure 3 describes the methodological workflow that starts from the creation of each data domain until its usage for specific tools.

Each data domain is characterized by a series of settings (e.g. template definition, shared parameter, material library, modeling, room definition for the BIM domain) concerning the creation of the database. Once the modeling step is concluded, each dataset needs to be exported in various standard exchange formats useful for later data integration with the other ones and its upload to the DIMMER platform. These operations are aimed at producing reality models with a certain level of graphic and/or alphanumeric detail to be then put into play with the other data sets and finally uploaded to the DIMMER middleware platform. This software platform was developed following a "microservice" approach for the management of information flows in a decentralized way. Each data source within the DIMMER middleware is managed in the form of web services to manage data interoperability between the different tools offered.

Afterward, the software platform, suitably structured, will select from time to time the data necessary for a specific use. Then, it exploits selected data to the specific tools that will have to produce graphical interfaces to optimize its usage for the building/district monitor system, the building refurbishment scenario, and the virtual city/district dashboard. In this way, interoperability, fostered by DIMMER middleware, promotes the flow of data from one domain to another one, arriving at the specific tool. The data domains interact with each other by exchanging data flowing from one domain to another, enriching the urban model of information.

Through this infrastructure, the DIM domain is created and standardization of the process has been defined. The framework of the methodology followed is visible in Figure 4.

The BIM model provides different information for several uses such as the validation of the DIMMER strategy that is mainly based on energy saving at the district level through the adoption of innovative technologies that optimize data management. Adding to this, at the building level, BIM model allows professionals to simulate different scenarios of building refurbishment that can be evaluated considering the Return of Investment (ROI). For instance, a real estate manager could use BIM model and the DIMMER platform to simulate a certain building renovation checking the energy usage, the thermal comfort,

Figure 4. BIM approach in the DIMMER project

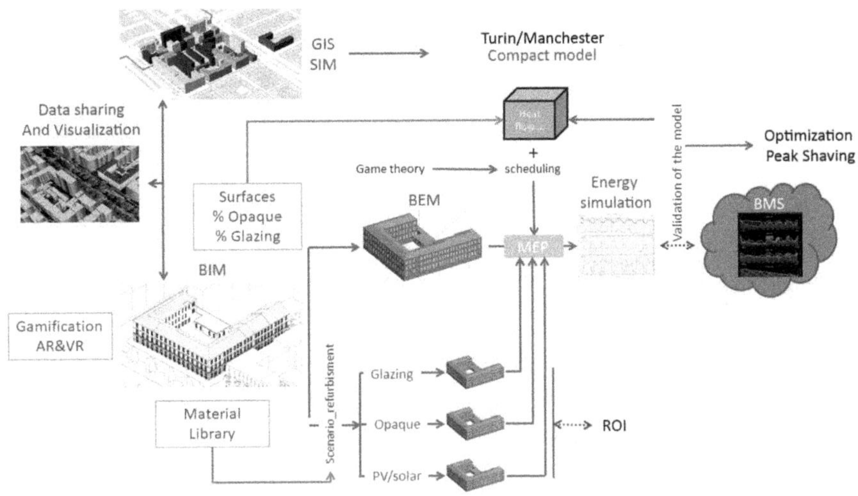

and the ROI, according to his needs. To provide as wide as possible access to the essential information related to the case studies, it is fundamental to correctly input the parameters that are subsequently shared in the BIM. Since the beginning of the entire process, besides the geometrical data, it is necessary to input those parameters useful to perform energy simulations (e.g. the thermo-physical features of the materials or the dimensions of the radiators). To perform the energy consumption assessment of an urban district, firstly reliable energy models of each building have been built. Through the interoperable process, the models have been calibrated comparing the results of simulations with the real data of the consumptions and the indoor temperatures, monitored by sensors and meters installed in the buildings selected as case studies. The validation of the energy models is achieved through an iterative process aimed to calibrate the parameters of the envelope and the systems. In particular, the thermo-physical properties of the material are inferred on the basis of the construction period of the building while the technical data of the systems is collected from the project or on-site by reading labels of the various components of the system (e.g. boilers, heat exchanger, fan coils, etc.).

Figure 5 shows how the development of a BIM model aimed at DIM for energy analysis entails a graphical simplification of the objects to be modeled, by entering the specific information to the related purpose.

Starting from the BIM authoring platform (e.g. Autodesk Revit), the architectural model has been simplified to be transformed into an Energy Analysis Model (EAM) by the gbXML export process. The latter was then imported into the specific calculation software (e.g. Design Builder) for the building energy performance assessment. Through this export/import process some criticalities related to the integrity of the starting model typical of the BIMtoBEM process have been analyzed (Kamel & Memari, 2019). The interoperability process ends using Energy Plus directly as a simulation engine, as DesignBuilder is only a user interface of the EnergyPlus calculation engine and does not allow for deep and complete control of the parameters. Within EnergyPlus it is possible to set detailed information of the components of the system checking each field manually. Energy Plus does not include a graphic interface, but the data visualization is possible through a specific viewer (i.e. Open Studio Results Viewer). This allows

Figure 5. BIM use for energy analysis – characteristics to be considered

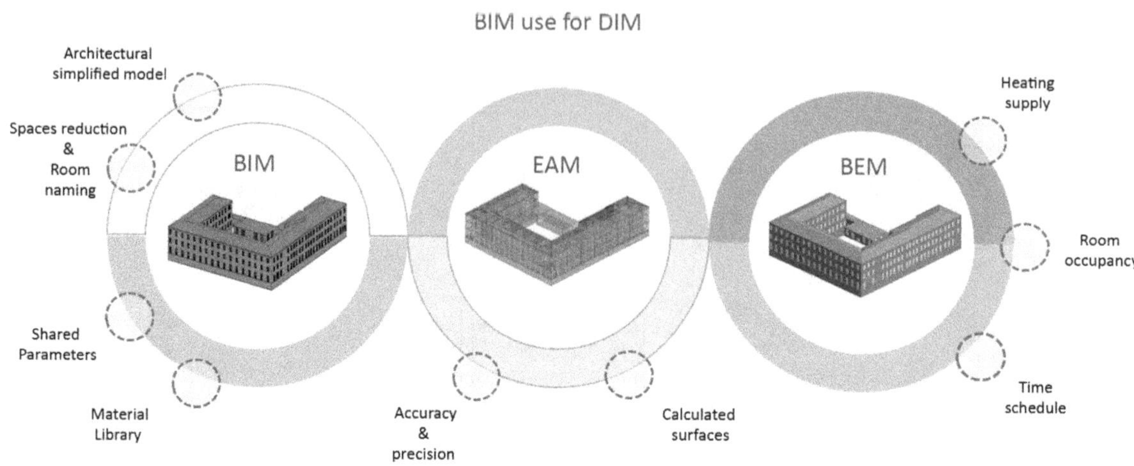

the .sql output to be visualized, managing a large amount of output data (from the water inlet temperature of every single radiator to the boiler thermal energy demand) (Osello, 2016).

Certainly, the simulations developed by the BEM model have been compared with data from the BAS domain (i.e. temperatures, energy consumption) to calibrate and validate the model and use it as a real digital twin. This means that this step consists of an iterative process of comparing the model to actual system behavior in order to refine the model in a step by step manner.

Then, energy refurbishment scenarios (potential refurbishments aiming to increase efficiency) have been hypothesized, such as the addition of a thermostat and the substitution of windows, the refurbishment of opaque surfaces (external walls and roof) and the thermostat, and a better Heating Ventilation and Air Conditioning (HVAC) management (i.e. peak shaving).

Finally, it was possible to obtain as a result the total annual simulated energy consumption for each scenario and the relative percentage of saved energy, calculated through the comparison of the energy consumption before and after the application of the strategy.

What has been said so far confirms the need to define specific levels of detail for the various data domains in order to avoid data loss that would have caused the same information to be loaded several times. Once the structure of the information levels and domains has been defined, it is clear that the amount of information available to the DIMMER project is multiple and heterogeneous. Then filters were placed that depending on the scale of display and detail, to query and identify the correct data at a glance. For this reason, starting from the American Institute of Architects (AIA) and Publicly Available Specification (PAS) international standards on Levels of Development and Detail (LODs), DIM-LODs (Pagani & Chiesa, 2017) have been proposed. This sort of scale contains the semantics to filter information from different domains. Unlike the various LODs definitions available, the DIMMER middleware allows you to obtain various types of information from the architectural to the geomatic sector to query the DIM database each time. Certainly, this capability is due to the flexibility of the DIMMER middleware that among its features allows a possible implementation for specific uses.

The digital maturity of the DIM domain is higher than other kinds of information management, reaching the requirements of the international standard UNI EN ISO 19650 about the definition of Level of Information Need for specific model purposes.

For instance, as the BIM domain contains information on the building, data needs to be transferred correctly to obtain a BEM domain through interoperability. The starting model needs to be simplified in relation to architectural elements, avoiding loss of data. The model is then enriched by moving it from one domain to another one for specific analysis, avoiding waste of data. As urban analysis is not the only objective of the DIMMER project to be achieved through the use of BIM (DIM), the models and information was shared among the partners of the project and exported using several file formats. Through the DIMMER middleware, which is a digital data repository where models are stored or retrieved using different formats the BIM models have been exported and uploaded in the middleware adopting the following formats: .gbXML, .ifc, .fbx, .odbc, .rvt Figure 3. The reason why several formats were uploaded is related to the different uses of BIM for DIM, as far as geometrical and alphanumerical data is concerned. The DIM approach culminates with the development of tools capable of retrieving data from the DIMMER middleware for various purposes to reach the users 'needs. From this perspective, the project partners have developed mainly four tools to optimize data visualization:

- DIMMER dashboard, developed by CSI Piemonte. It integrates GIS data at the district level such as buildings, networks, topology, and the "live" and simulated data shared among the DIMMER middleware components. The BIM models developed for the project have been made available and interactively accessible in the 3D visualization. The Dashboard GUI presents all the possible options and functions at a single click and highlights the most important ones. GUI has a user-friendly interface and provides a data visualization solution which is as flexible as possible;
- DIMMER benchmarking tool, developed by D'Appolonia S.p.A. It is designed to follow a top-down approach, starting from a district level and going into further detail at the building level. Indeed, from the initial page of the tool, the user, in the context of a selected district (as in the case of Manchester and Turin), can select a building to perform energy comparisons and simulation as a benchmark for investigating the energy behavior of the building itself;
- Community portal, a web portal developed by CNet for the DIMMER users (i.e. Public administrators, building estate manager, energy professionals) to engage the building residents/visitors by displaying DIMMER data with them. The Community tool uses the DIM Context layer for the Benchmark Service Provider and the Dashboard to retrieve data to visualize in a private group page;
- District visualizer, developed by Clicks and Links. It consists of a 3D visualization engine able to display data using multiple canvases. It is composed of some layers that present DIM data on a map or model, plus three exemplar apps showing the use of the visualizer, for solar panels and district heating applications. User experience is via screens or virtual reality headsets, allowing collaboration with users in other physical locations.

RESULTS

The investigated DIM methodology has to be considered a good approach based on an intelligent integrated simulation of architecture and systems using parametric software able to manage urban and architectural data.

The standard defined within the District Information Modelling methodology includes several domains, which are integrated into the DIMMER platform and visualized through the DIMMER tools. Starting from GIS and SIM, then integrating BIM, EAM, and BAS domains, data coming from several sources are shared at different levels of detail or development. In this way, the data domains integration is one of the main outputs achieved by the DIMMER project that highlighted the potentialities of developing an open platform that provides alternative scenarios based on user needs. Of course, the creation of a 3D digital and measurable urban model able to collect and query geometric and alphanumeric information can be considered one of the results of this project by following the DIM methodology.

At the end of the process, the target users, chosen during the first phase of the project, will be able to use the DIMMER platform for the data management and visualize the results of the project based on specific use cases. The graph visible in Figure 6 shows the interaction of the different fields to achieve the DIM standard definition (Osello, 2016).

Figure 6. Diagram showing the interactions within the DIM following the DIMMER standard

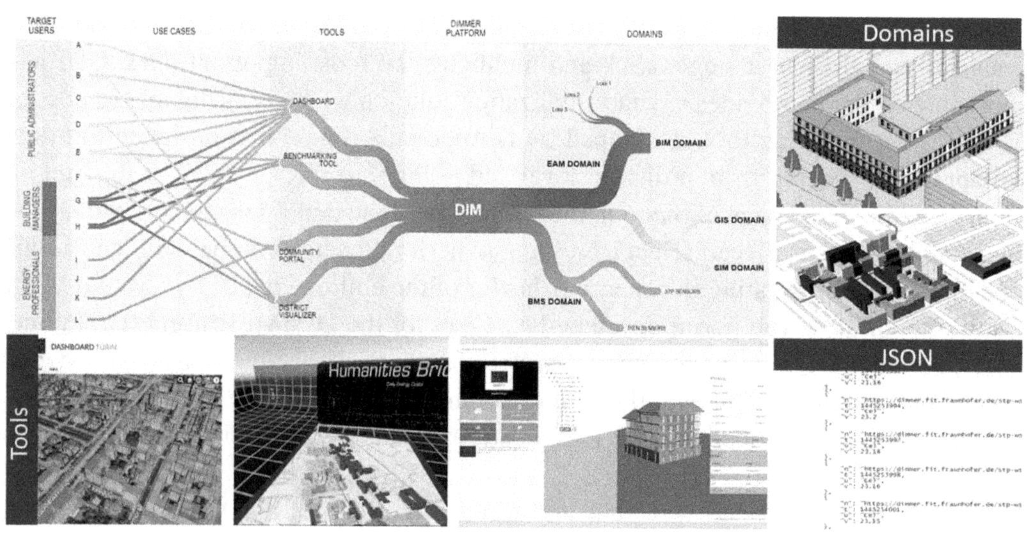

Thanks to multidisciplinary work, several use cases have been identified and classified by target users. Then, the tools developed within the project satisfied them by exposing the heterogeneous information of the DIM domain, connecting it, creating rules to automate the entire information process. The JavaScript Object Notation (JSON) language was adopted to achieve this objective, against the other projects described in the related works where CityGML and IFC have mainly been used.

Thanks to ICTs, the development of an urban digital twin has been optimized, by exploiting the software infrastructure to collect information and distribute it in different ways. The integration of data, coming from the different domains analyzed within the various project research units, resulted in the realization of the DIM, the standard defined by the DIMMER project.

The results obtained by the DIMMER project were then diffused among the citizens through various activities with a gamification approach. To increase the awareness of the participants in these activities in the field of energy saving, in a fun and interactive way, a playful approach has been studied by mixing traditional types of games, adopting innovative technologies such as augmented and virtual reality.

Figure 7. Dissemination activities to enhance the awareness of young generation concerning smart cities topics

Various activities have been conducted from participation in events such as the researchers' night to numerous fairs such as the International Book Fair in Turin, as visible in Figure 7. In these events, the results obtained by the DIMMER project were shown through the development of games such as the DIMMER District Board Game, the ECO card game, the DIMMER's wheel, and involving high school students with the ToBeSmart project for the development of a theatre show.

CONCLUSION

The advent of connected, smart technologies for the built environment promises a significant value that has to be reached to develop digital city models. As at international level, the role of the digital twin is strictly related to a massive amount of data that need to be processed this argument proposes several challenges in terms of digital technologies capability, computing, interoperability, simulation, calibration, and representation. Thus, the building industry is crossing a digital transition from traditional design processes to information modeling, following the idea of optimizing data sharing and integration.

The integration of several data domains can lead to the creation of an extensive repository able to provide a series of useful pointers towards energy saving and reduction of CO_2 emissions, enhancing the liveability of the city. Through the analysis of standard exchange formats that facilitate data transfer the investigated DIMMER platform can be considered one of the examples of urban DT to be adopted by private and public administrators. This interdisciplinary research provided an opportunity to highlight the strengths and weaknesses of the digitization of the city by developing 3D city models that increase the users 'awareness of energy usage.

Through these innovative methods and tools, the obtained results of the DIM approach must encourage the development of smart cities through the adoption of increasingly high-performance DT as a decision support tool.

ACKNOWLEDGMENTS

The DIMMER project received funding from the European Union's Seventh Framework Programme for research, technological development and demonstration under grant agreement n° 609084.

The author is pleased to thank the PhD students Daniela De Luca and Arianna Fonsati for their significant contribution provided within the project.

REFERENCES

Alonso, R., Borras, M., Koppelaar, R. H., & Lodigiani, A. (2019). *SPHERE: BIM Digital Twin Platform*. Proceedings - MDPI. Retrieved from https://sphere-project.eu/download/sphere-digital-twin-definitions-for-buildings/

Bisello, A. (2020). Assessing Multiple Benefits of Housing Regeneration and Smart City Development: The European Project SINFONIA. *Sustainability 2020, 12*(19), 1-28.

Boje, C., Guerriero, A., Kubicki, S., & Rezgui, Y. (2020). Towards a semantic Construction Digital Twin: Directions for future. *Automation in Construction*, •••, 1–16.

Consortium, D. (2016). *Deliverable 1.1 - Districts and buildings selection*. Author.

Corrado, V., Ballarini, I., Madrazo, L., & Nemirovskij, G. (2015). Data structuring for the ontological modelling of urban energysystems: The experience of the SEMANCO project. *Sustainable Cities and Society*, *14*, 223–235. doi:10.1016/j.scs.2014.09.006

Dembski, F., Wössner, U., Letzgus, M., Ruddat, M., & Yamu, C. (2020). *Urban Digital Twins for Smart Cities and Citizens: The Case Study of Herrenberg*. Sustainability - MDPI.

Directive (EU) 2018/844 of the European Parliament and of the Council.

Donkers, S., Ledoux, H., Zhao, J., & Stoter, J. E. (2015). Automatic conversion of IFC datasets to geometrically and semantically correct CityGML LOD3 buildings. *Transactions in GIS*, *20*(4), 547–569. doi:10.1111/tgis.12162

EC. (2016). *DAREED – Decision support Advisor for innovative business models and useR engagement for smart Energy Efficient Districts*. Retrieved from Cordis: https://cordis.europa.eu/project/id/609082/it

El-Mekawy, M., Östman, A., & Hijazi, I. (2012). A Unified Building Model for 3D Urban GIS. *ISPRS International Journal of Geo-Information*, *1*(2), 120–145. doi:10.3390/ijgi1020120

Eurocities. (2018). *Digital 3D modelling of Rennes Metropole*. Retrieved from eurocities: http://wsdomino.eurocities.eu/eurocities/documents/Digital-3D-modelling-of-Rennes-Metropole-WSPO-A8JDUM#:~:text=Rennes%20Metropole%20has%20developed%20a,on%20buildings%2C%20and%20so%20on

Evans, S., Savian, C., Burns, A., & Cooper, C. (2019). Digital twins for the built environment - An introduction to the opportunities, benefits, challenges and risks. The Institution of Engineering and Technology (IET).

Foundation, N. R. (2018). *Virtual Singapore*. Retrieved from National Research Foundation: https://www.nrf.gov.sg/programmes/virtual-singapore

Grieves, M. (2014). *Digital Twin: Manufacturing Excellence through Virtual Factory Replication*. Whitepaper.

Helsinki, C. o. (2019). *Helsinki's 3D city models*. Retrieved from City of Helsinki: https://www.hel.fi/helsinki/en/administration/information/general/3d/

IEE Project TABULA and EPISCOPE web site. (n.d.). https://episcope.eu/welcome/

Jusuf, S. K., Mousseau, B., Godfroid, G., & Soh, J. H. (2017). Path to an Integrated Modelling between IFC and CityGML for Neighborhood Scale Modelling. Urban Science, MDPI, 1-20.

Kamel, A., & Memari, A. M. (2019). Review of BIM's application in energy simulation: Tools, is-sues, and solutions. Automation in Construction, 164-180.

Kolbe, T. H. (2009). Representing and Exchanging 3D City Models with CityGML. In *Proceedings of the 3rd International Workshop on 3D Geo-Information, Lecture Notes in Geoinformation & Cartography*. Springer Verlag.

Krüger, A., & Kolbe, T. H. (2012). Building analysis for urban energy planning using key indicators on virtual 3D city models—The energy atlas of Berlin. *The International Archives of the Photogrammetry, Remote Sensing and Spatial Information Sciences*, *39*(B2), 145–150.

Linksmart. (2020). *Linksmart® - Free, open source IoT platform*. Retrieved from http://linksmart.eu/

McAleer, R. S., Kogut, P., & Raes, L. (2020). Digital Urban European Twins. Change the way you see the city. Project white Paper.

Melia, A., Nolan, E., & Kerrigan, R. (2015). INDICATE: towards the development of a virtual city model, using a 3D model of Dundalk city. In *CISBAT 2015 International Conference "Future Buildings and Districts - Sustainability from Nano to Urban Scale"* (pp. 925-930). Lausanne, Switzerland: Lausanne, EPFL Solar Energy and Building Physics Laboratory.

Nouvel, R., Schulte, C., Eicker, U., Pietruschka, D., & Coors, V. (2013). CITYGML-based 3D city model for energy diagnostics and urban energy policy support. *Proceedings of BS2013: 13th Conference of International Building Performance Simulation Association*, 218-225.

NYC. (2020). *NYC 3-D Building Model*. Retrieved from The Official Website of the City of New York: https://www1.nyc.gov/site/doitt/initiatives/3d-building.page

Osello, A. (2012). *The Future of Drawing with BIM for Engineers and Architects*. Dario Flaccovio Editore.

Osello, A. (2016). Deliverable 1.5 - Districts and Buildings energy typologies. Turin.

Osello, A. (2016). Deliverable 3.1.3 - District Information Modeling: implementation and standard definition. Turin.

Osello, A., Acquaviva, A., Del Giudice, M., Patti, E., & Rapetti, N. (2016). District Information Models. The DIMMER project: BIM tools for the urban scale. In R. Pagani & G. Chiesa (Eds.), *Urban data. Tecnologie e metodi per la città algoritmica* (pp. 244–275). FRANCOANGELI.

Pagani, R., & Chiesa, G. (2017). *Urban Data. Tecnologie e Metodi per La Città algoritmica*. FRANCOANGELI.

Parliament, E. (2018). Directive (EU) 2018/844. European Parliament and of the Council of 30 May.

Romero, A., Izkara, J., Mediavilla, A., Prieto, I., & Pérez, J. (2016). Multiscale building modelling and energy simulation support tools. *ECPPM 2016 - 11th European Conference on Product & Process Modelling*, 315-322.

Sicilia, Á., Madrazo, L., & Costa, G. (2015). *Building a semantic-based decision support system to optimize the energy use in public buildings: the OPTIMUS project. Sustainable Places 2015 event*. Savona.

Telensa. (2020). *Urban Data Project*. Retrieved from Urban Data Project: https://urbandataproject.org/

Testoni, C. (2016). *Towards Smart City*. FRANCOANGELI.

Verbeke, S., Ma, Y., Van Tichelen, P., Bogaert, S., & Gómez Oñate, V. (2018). Support for setting up a Smart Readiness Indicator for buildings and related impact assessment - Final report. Brussels: Flemish Institute for Technological Research NV ("VITO").

Würstle, P., Santhanavanich, T., Padsala, R., & Coors, V. (2020). The Conception of an Urban Energy Dashboard using 3D City Models. In *e-Energy '20: Proceedings of the Eleventh ACM International Conference on Future Energy Systems* (pp. 523–527). Virtual Event Australia: Association for Computing Machinery. 10.1145/3396851.3402650

KEY TERMS AND DEFINITIONS

Building Information Model: It is data rich, object oriented, intelligent and parametric digital representation of the facility, from which views and appropriate data for various users' needs con be extracted and analyzed to generate information that can be used to make decisions and improve the process of delivering the facility.

Building Information Modelling: It is a method based on a building model containing any information about the construction. In addition to 3D object-based models, it contains information about specifications, building elements specifications, economy, and programs.

Data Visualization: Is the graphical representation of information necessary for the understanding of technical details. In this way, by means of visual elements such as dashboards, web sites, Head Mounted Display, tables, data visualization maps help understanding and allow to hypothesize improvement scenarios.

Digital Twin: It is a virtual replica of reality and contains three main parts: physical products in real space, virtual products in virtual space, and the connections of data and information that ties the virtual and real products together.

District Information Modelling: It extends the Building Information Modelling at the district scale, by creating a new kind of domain. The main challenge to face when creating a DIM concerns the ability to manage a big amount of data, coming from different domains. In this field, interoperability plays a key role for data exchange among different domains. It can be considered a first step to optimize data management and the development of a smart city.

Interoperability: It is defined as the ability of two or more systems or components to exchange information and to use the information that has been exchanged.

Middleware: It represents a set of computer programs whose objective is to enable communication and management of data in distributed application. For instance, sensors for energy monitoring communicate with each other using their own computer languages, often created or implemented by different manufacturers. Consequently, the middleware works as a translator, thus allowing the sensors to communicate with each other.

Smart City: It is a place where traditional networks and services are made more efficient with the use of digital and telecommunication technologies for the benefit of its inhabitants and business. A smart city goes beyond the use of information and communication technologies (ICTs) for better resource use and less emissions. It means smarter urban transport networks, upgraded water supply and waste disposal facilities and more efficient ways to light and heat buildings. It also means a more interactive and responsive city administration, safer public spaces and meeting the needs of an ageing population.

Chapter 9
KPIs to Drive Smart City Assessment

Arianna Fonsati
https://orcid.org/0000-0001-8566-3425
Politecnico di Torino, Italy

ABSTRACT

Developing evaluation methods using key performance indicators (KPIs) to drive smart cities assessment represents a very interesting topic that have been deeply analysed over time. The aim of the chapter is to present several different approaches towards the assessment of cities' performance, discussing advantages and disadvantages of such methods that could be integrated in order to get more farsighted results in terms of decision-making strategies. For this reason, the chapter firstly analyses the different context in which KPIs are applied, defining the fields of application and describing the characteristics of such indicators. Then, two examples of European projects using indicators for evaluation purposes are studied in detail, highlighting the different level of application of indicators, used for assessments at city level in the first project (INDICATE) and at a district/building level in the second one (DIMMER). At the end, the use of KPIs as metrics of evaluation to develop multi-criteria decision approaches (MCDA) is discussed by introducing several examples.

INTRODUCTION

The concept of smart city is assuming more and more importance in the policy makers agenda during the last years; "smart" represents a term adopted and popularized by multinationals like Cisco and IBM with their typically American and hi-tech imprint, which over time have become almost a synonym for "technological" (Mattei, 2013). But there is more to it than that; a lot of cities, mainly those with a long history and culture, cannot be drawn in a unique picture, considering only some aspects. For instance, in Italy each town is different from the others, not only in architectural terms, but also in terms of life quality and lifestyle; almost all towns have a complex and challenging history, not homologable in a unique, albeit super-technological design. For this reason, specific indicators have been established in

DOI: 10.4018/978-1-7998-7091-3.ch009

order to define a "smart city": green urban areas, water, air, waste, built heritage, life quality, energy, healthcare, mobility and logistics (*ibidem*).

Within this context, it is undeniable that problems related to cities are becoming even more crucial over time. In this sense, it is possible to highlight that in 2010, 50% of the world's population lived in urban areas and, according to a report from the Worldwatch Institute, this figure is forecast to rise to 70% by 2050 (Bencardino & Greco, 2014). Due to this increase in urban population, governments are required to figure out how to create future spaces for the citizens, taking into account and reasoning on the future quality of life in cities. For this reason, cities have started to think about how to assess their performance in terms of "smartness", mainly from an "energy performance" point of view. In order to change cities behaviors two main activities can be performed: (i) the activity of "sensing", which means collecting information by monitoring specific processes or situations; (ii) the activity of "actuating" which means reacting and finding solutions to the collected information (Ratti, Smart city, smart citizen, 2013). Furthermore, the forces that are actually guiding cities, such as energy, construction, transports etc., allow citizens to take a closer part in the management of the city itself. Indeed, in order to transform urban spaces, bottom-up actions are required, because "there is no smart city, without smart citizens" (Ratti & Claudel, 2017).

The objective of the chapter is analyzing the use of different evaluation methods developed over the years that use Key Performance Indicators (KPIs) to assess the performance of cities. Starting from the meaning and definition of indicators and their first application within the context of European or International projects, going through the analysis of the characteristics that KPIs must have in order to be relevant and meaningful, until their use as metrics of evaluation to develop Multi-Criteria Decision Approaches (MCDA). In particular, two different European projects funded by the 7th Framework Program of the European Union have been deeply analyzed within this chapter. From one side, the INDICATE project, which analysed the use of sustainable indicators at city level in order to define a Common City Index (CCI) to benchmark several cities on the basis of specific common criteria. On the other side, the DIMMER project developed indicators at district and building level in order to benchmark different buildings in terms of energy performance and refurbishment strategies. At the end of the chapter, several experiences highlighting the use of indicators as metrics within the application of multi-criteria decision methods are presented, in order to show the great adaptability of these assessment approaches, which are really helpful in supporting decision making processes towards sustainability.

Background

Sustainability assessment of cities has been very successful in the past years, also thank to the development of multicriteria analysis methods of evaluation which have made great use of KPIs, applied, for instance, to measure performance and efficiency of smart cities.

Using KPIs in order to give an evaluation towards sustainability has its origins in article 40 of Agenda 21: "Countries at national level and international governmental and non-governmental organizations at the international level should develop the concept of indicators of sustainable development in order to identify such indicators. In order to promote the increasing use of some of those indicators in satellite accounts, and eventually in national accounts, the development of indicators needs to be pursued by the Statistical Office of the United Nations Secretariat, as it draws upon evolving experience in this regard" (United Nations, 1992).

Figure 1. Examples of quantitative and qualitative indicators

QUANTITATIVE INDICATORS	QUALITATIVE INDICATORS
Energy consumption	Knowledge about rational use of thermal energy/electricity
	Change of behaviour towards energy efficiency
Renewable energy consumption	Public participation in green policy
	Clean and efficient energy policies
CO_2 emissions	CO_2 reduction strategy
Use of non car transport	Green transport promotion
Municipal waste production	Waste reduction and policies

Several different definitions related to KPIs are available in the literature: “An indicator provides information on the state or conditions of a phenomenon; it is a simplified way to explain a complex concept. In terms of sustainability, it is possible to narrow this definition to a policy relevant variable defined in such a way as to be measurable over time and space” (Astleithner, Hamedinger, Holman, & Rydin, 2004). Another definition, which is particularly set into the context of sustainability assessment is the following: “Indicators provide data to guide policy-making and allow for comparisons to be made across municipalities and regions. The impacts and challenges of sustainability policies and plans on the urban environment can also be shown through indicators” (Munier, 2005). Indicators can be divided into quantitative and qualitative ones. Quantitative indicators are mainly related to monetary analysis and give numeric information on a specific topic and can be calculated from metered data; they can be used in the areas where costs and benefits are expressed in monetary terms and they are strictly related to cost-benefit analysis and new business models development. On the other hand, qualitative indicators enable the measurement of aspects related to user behavior and social aspects, for example they can refer to the quality level of services inside the city from the point of view of people. These indicators cannot be estimated in quantitative terms because they relate to impact that are “non-quantifiable” in monetary terms but are important to assess the benefits related to the project development they are used for. **Figure 1** shows some examples of quantitative indicators with their linked qualitative ones.

Furthermore, the nature of indicators depends on their belonging to a specific domain; indicators such as energy consumption or CO_2 emissions are obviously used to measure the performance of the environmental domain, while GDP and employment rates are grouped under the economic domain. In general, the following three main domains are considered: social, environmental and economic. As far as the selection of indicators is concerned, there are some advice in order to help in choosing the most suitable sustainable indicators to perform a specific assessment. Firstly, it is important to keep in mind that the simpler the indicator the easier its use as a tool of communication; this is the reason why indicators need to be defined with reference to their understanding and usability in order to avoid wrong interpretations. Therefore, indicators must be meaningful in order to support the decision-making by identifying potential ways of improvement and inclusive of covering the most important aspects and data available. Furthermore, it is paramount to select indicators taking into account the available measured data, relying on identified and rated measures. The KPIs should support project activities and achieving project objectives, but at the same time, they should allow for reevaluation and further development. For instance, in terms of project sustainability KPIs are calculated before and after the deployment of a specific solution in order to quantify the savings introduced by the intervention (Banu, 2018). The phase related to the collection of KPIs often involves different critical points, starting from data availability because of privacy issues up to technical issues in collecting the necessary data. Indeed, this problem is greatly impactful at all the different scales. While at building/district level the privacy of sensible data could make some databases impossible to be shared and used as metrics of evaluation, at city level data could be too generic to deeply represent specific situations, because such information is mainly based on historical database rather than real-time data.

Since the Agenda 21, a considerable number of research projects have been developed with the aim of selecting efficient sets of indicators to be used in the evaluation of cities capability to perform activities towards sustainability. In general, the aim of those projects was to help local authorities in the evaluation and the promotion of sustainable developments within their context. Therefore, through the use of indicators as "measure of sustainability", local governments managed to assess their environmental, economic and social performance and were often able to compare their results with those from other cities, thanks to databases where all data was stored during the development of the project.

Several International and European projects had a great impact and most significant results in evaluating environmental performance by using sustainable indicators in the past twenty years, among them it is possible to introduce the following: ECI - European Common Indicators (1993-2003); LASALA – Local Authorities' Self-Assessment of Local Agenda 21 (1999-2002); PASTILLE – Indicators into Action: Local Sustainability Indicator Sets in Their Context (2000-2002); TISSUE – Trends and Indicators for Monitoring the EU Strategy on SD of the Urban Environment (2004-2005); STATUS – Sustainability Tools and Targets for the Urban Thematic Strategy (2005-2006); SIEMENS GREEN CITY INDEX (2009); Informed Cities (2009-2012); CDP Cities – Carbon Disclosure Project (2011); URBAN-NEXUS (2011-2014). From the analysis of the previously cited projects some generic considerations can be drawn: (i) the approach towards indicators can be top-down or bottom-up, the first method is highly technical and require expert policy makers for the definition of goals and indicators, the second one involves extensive consultation with stakeholders to select appropriate indicators; (ii) although there is not a suitable method of assessment for all cities, international renowned frameworks represent a reference point for all initiatives; (iii) the attention to environmental issues brings all experiences together, however the majority of performance evaluation initiatives focused on environmental more than social or economic issues related to them, wrongly separating different aspects of the same topic, because considering social aspects at the same level as environmental and economic one means recognizing that civil society plays an important role in reaching the community sustainable goals; (iv) the collaboration of all stakeholders is paramount, because public participation in the development of sustainable processes is necessary to create awareness among users and a win-win situation for everyone.

In recent years, thanks to the considerable interest towards KPIs management, researches also dedicated to enhance the understanding of KPIs management, defining a customized related taxonomy (Domìnguez, Pérez, Rubio, & Zapata, 2019). This study aimed at providing a taxonomy able to group information related to different issues of interest distributed among several key dimensions, enhancing the understanding of the field and enabling to select the most suitable solution among different perspectives of KPIs management on the basis of the requirements. The resulted taxonomy established five question-based dimensions:

- *What is measured by a KPI?* This dimension aims at identifying the aspects that can be measured through KPIs, presenting different connotations: (i) the difference perspectives under which performance measures are proposed; (ii) the reasons why an indicator has to be used, such as policies, mission statements etc.; (iii) the scope that implies the use of specific metrics.
- *What features are considered in the specification of KPIs?* In this case features are intended as properties, such as calculation aspects, related human resources or relationships among KPIs.
- *What is a KPI measured for?* This dimension is directed to the identification of possible different purposes for which a KPI may be used, that have been grouped on the basis of their aim to evaluate the present or past performance of a system or to predict the future behavior of a system.
- *What artifacts are used for KPI design and specification?* The process of defining, using and manage KPIs is carried out starting from a KPI framework, developing a KPI model and providing KPI specification on the format provided for the KPI, for instance text-based or graphical.
- *What are the characteristics of a KPI management approach?* This dimension addresses the different particularities to manage KPIs used by existing proposals: (i) clear and accurate syntax and semantics; (ii) expressiveness in terms of embracing a broad spectrum of KPIs; (iii) understandability; (iv) standards support within the field; (v) acceptability by user community; (vi) empirically evaluated to validate the applicability of the proposed approach; (vii) customization of KPI definitions to adapt to specific situations; (viii) evolution management, in terms of evolving accordingly to system requirements; (ix) technological support with automatic mechanisms that facilitate the management of KPIs.

One of the main results of such study is to gather and consolidate the existing knowledge ad terminology in the KPI field, developing a taxonomy that needs to be updated on the basis of new possible requirements occurring over time (*ibid*).

On the other side, indicators as metrics of value are widely used within the application of Multi-Criteria Decision Analysis methods, both in the evaluation of quantitative and qualitative parameters. MCDA have origins within complex contexts of decision-making characterized by high level of uncertainty (Watróbski, Jankowski, Ziemba, Karczmarczyk, & Zioło, 2019). **Figure 2** shows the workflow of the decision-making process including the use of evaluation tools to assess different transformation scenarios, highlighting the three main phases: (i) problem framing; (ii) scenarios and alternative transformation projects building; (iii) problem solving. The acronym MCDA is referred to a wide range of techniques, introduced by the end of 60s, handling with multiple quantitative and qualitative criteria, used to make

Figure 2. Decision-making process and the role of evaluation tools. Source: adapted from Faggiani, 2006

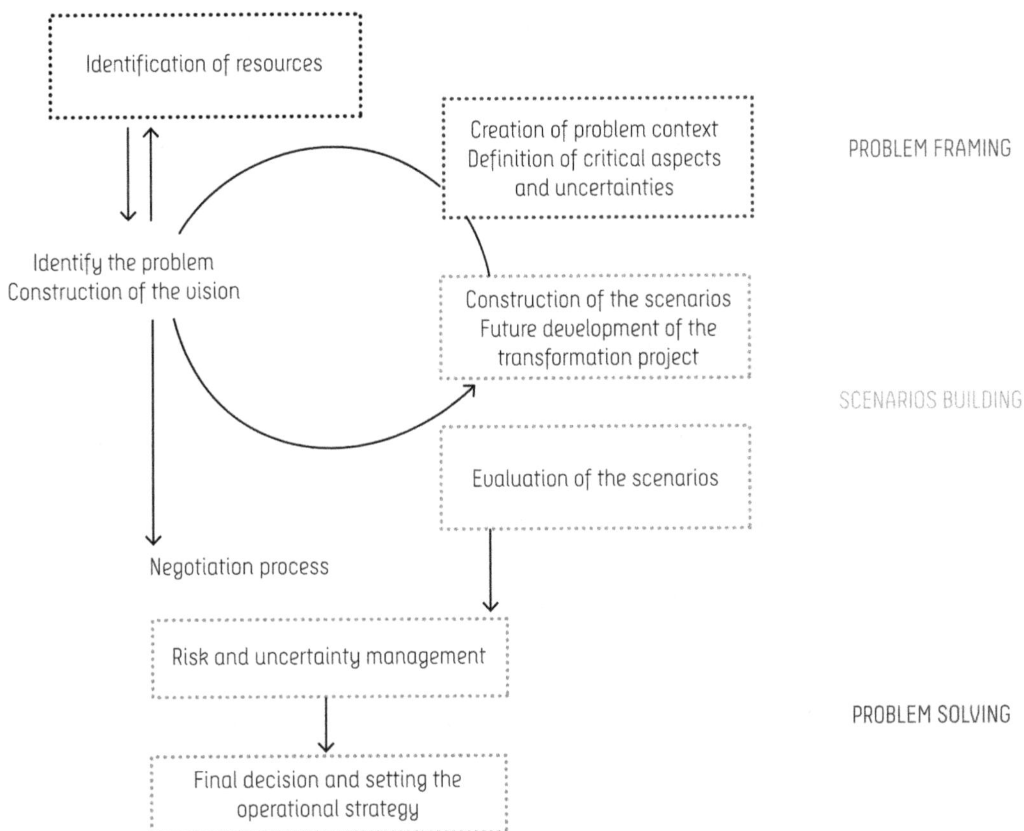

a comparative assessment of alternative projects or heterogeneous measures. These techniques have been designed to support decision-makers towards the integration of different alternatives, reflecting on the options of the actors concerned, into a prospective or retrospective framework (Figueira, Greco, & Ehrogott, 2005). As result, such assessment methods usually produce operational advice or recommendations for future activities, so they lend themselves well to their application within smart cities context.

Such methods, applied to a "macro" level of interest, highlight the necessity of a proper construction and aggregation of sustainability assessment indicators and indexes to measure, for instance, sustainability, because results obtained depend on several factors: (i) the quality of the information available; (ii) the indicators chosen, as representation of the reality used and the interests taken into account; (iii) the direction of each indicator, in terms of deciding if maximizing o minimizing the indicator itself on the basis of its positive or negative effect; (iv) the relative importance of the chosen indicators, indicated by the weighting factor attached; (v) the ranking method used (Munda, 2005).

THE SELECTION OF KPIS BASED ON THE INDICATE PROJECT REQUIREMENTS

Within this context, the Indicator-based Interactive Decision Support and Information Exchanged Platform for Smart Cities (INDICATE) founded by the 7th Framework of the European Union was based on the use of sustainable indicators for assessment purposes related to the performance of cities in terms of "smartness". The main goal of the project was the realization of a supporting tool able to guide potential stakeholders, Public Administration and private investors, towards the right choices to make, in order to improve the energy efficiency of cities. One of the fundamental steps in the INDICATE project involved the definition of a set of sustainable indicators used to aid decision support within the urban environment, through integration with the INDICATE Virtual City Model (VCM) as shown in **Figure 3**. Therefore, these indicators, measuring a number of parameters having a hold on the overall energy efficiency of the urban environment, have been used in the construction of the VCM in order to identify and highlight the most suitable amendments to enhance the energy efficiency of the urban area concerned. The effectiveness of the tool provided by the INDICATE project has been tested on two pilot cities: Dundalk, in Ireland, and Genoa, in Italy (Grey, 2016).

The work package 3 (WP3) of the project concerned the selection of the indicators to be used with the purpose previously enounced. The function of such indicators was to normalize differences in climate, culture, lifestyles, governance and building typology of cities across Europe, in order to generate a Com-

Figure 3. VCM Flow Diagram. (Source: Annex I - "Description of Work" - INDICATE)

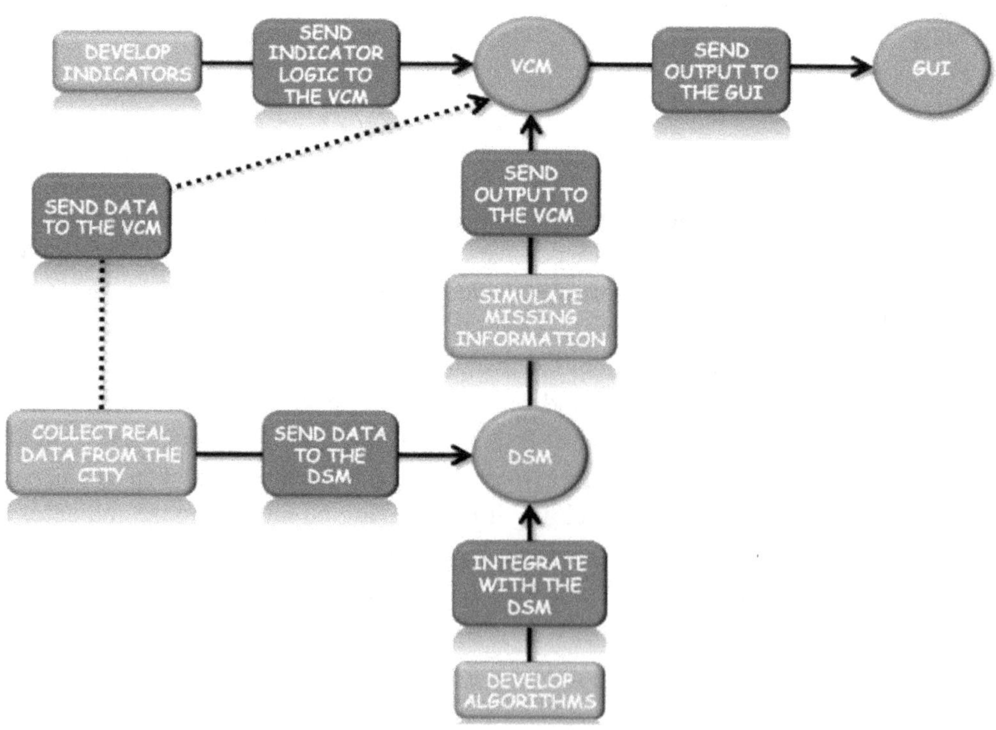

mon City Index (CCI) to facilitate transferability and adaptability of the VCM between urban areas within different member states. In the specific case of the INDICATE project the formulation of indicators was mainly directed towards supporting decision making with the aim to: reduce the cost for urban areas of energy provision, reduce CO_2 emissions and enhance the overall urban energy efficiency. Consequently, the indicators have been selected on the basis of their ability to contribute towards the identification of shortcoming and possible solutions towards those requirements, conducting the analysis at urban scale.

In the selection of criteria, the very first decision to take is the definition of the important domains that have to be measured at a city-scale context.

Figure 4. Selected indicators

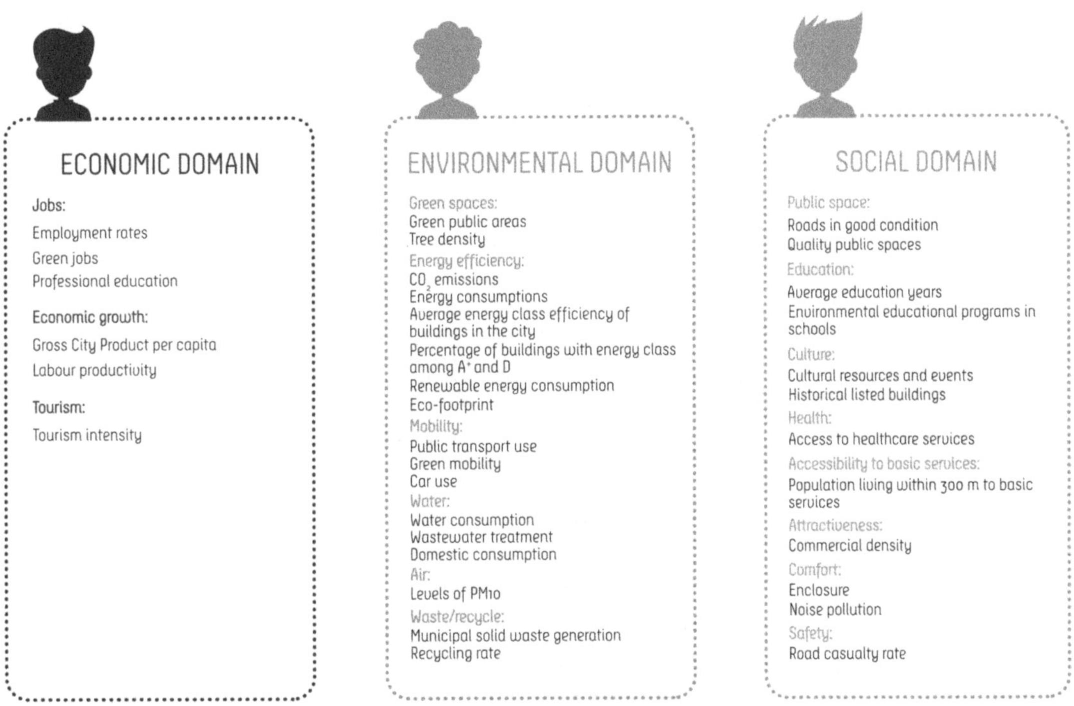

Within this context, a possible procedure of selecting KPIs is presented below. Starting from the project requirements and considering what kind of phenomena a generic city would have the necessity to track, three domains have been selected: economic, social and environmental. Then, each domain has been in turn subdivided in different categories; 3 categories for the economic domain (jobs, economic growth and tourism), 6 categories for the environmental domain (green spaces, energy efficiency, mobility, water, air and waste/recycle) and 8 categories for the social domain (public space, education, culture, health, accessibility to basic services, attractiveness, comfort and safety). **Figure 4** presents the complete selection of indicators for all domains.

Table 1. Indicators for the Environmental domain

CATEGORY	INDICATOR	DESCRIPTION	UNITS OF MEASURE	SOURCE
Green spaces	**Green public areas**	Green urban areas per inhabitant. It defines the area of vegetation as a carbon sink.	[%]	The Urban Ecosystem Europe set of indicators
	Tree density	this indicator quantifies the tree canopy cover within the city area, which contributes to carbon sequestration. It measures the number of trees out of the total municipal area.	[n/km2]	The Urban Ecosystem Europe set of indicators.
Energy Efficiency	**CO_2 emissions**	this indicator measures the total quantity of the city CO_2 equivalent emissions released by residential, tertiary and transport sectors per capita	[tons CO_2/year]	TAPE (Turin); SEAP (Genoa)
	Energy consumption	this indicator measures the total energy consumption (electricity and fossil fuels) of the city	[MWh/year]	TAPE (Turin); SEAP (Genoa)
	Average energy class efficiency of buildings in the city	this indicator measures the total energy consumption (electricity and fossil fuels) of the city	[MWh/year]	Discovery/sensation
	Percentage of buildings with energy class among A+ and D	this indicator measures the total energy consumption (electricity and fossil fuels) of the city	[MWh/year]	Discovery/sensation
	Renewable energy consumption	this indicator reflects the renewable energy consumption, as a % of total energy consumption of the urban area	[%]	TISSUE indicators
	Eco-Footprint (EF)	this represents an aggregated indicator, because it integrates two indicators: (i) the biologically productive area needed to maintain resource consumption and waste absorption; (ii) biologically productive area that a region can provide. This indicator judges whether the regional productive consumption activity is within the carrying capacity of the local or defined ecosystem by comparing the two indicators	[m^2]	Global Footprint Network
Mobility	**Public transport use**	this indicator measures the use of public transport. Units of measure	[passengers/inhabitants]	ISTAT
	Green mobility	this indicator defines the percentage of green mobility within the city (bicycles, electric cars etc.)	[%]	Discovery/sensation
	Car use	this indicator defines the percentage of all journeys under 5 km by private car use. Private car use refers to the use of a vehicle for private journeys for social, domestic or work purposes	[%]	Discovery/sensation
Water	**Water consumption domestic**	this indicator measures the total annual water consumption	[m^3 per head]	European Green City Index Indicators
	Wastewater treatment	this indicator measures the percentage of dwellings connected to the sewage treatment	[%]	European Green City Index Indicators
	Domestic consumption	this indicator measures the amount of water consumed per person for the purposes of drinking, hygiene, cooking, washing of utensils and other household purposes including gardening use	[Liters/capita per day]	STATUE indicators
Air	**Levels of PM10**	this indicator monitors the levels of PM10	[number of days]	ISTAT
Waste/Recycle	**Municipal solid waste generation**	this indicator measures the total municipal solid waste produced in the urban area	[annual kg per capita]	TISSUE indicators
	Recycling rate	this indicator defines the percentage of waste recycled out of total waste disposed at municipal landfill sites	[%]	Strategic Environmental Assessment Indicators (Dublin)

To better understand the characteristics of such indicators, the proposal for the selection of KPIs within the example of the environmental domain is presented below.

The proposal for the environmental domain is reported in **Table 1.**

The indicators proposed refer to an urban level context, highlighting the necessity of collecting data through specific surveys and monitoring activities at the city-level. This explains the generic nature of such indicators, also considering that cities involved within the assessment needed to present homogeneous data in order to proceed with the comparison. In fact, within the selection of indicators it is of paramount importance to consider indicators for which it is possible to find and collect data, otherwise the comparison through that specific KPI cannot be performed. Furthermore, the structure of the previously presented selection of indicators, including details on the description and the reference to the source of the KPI, is very important within the selection phase because, in order to proceed with the normalization technique it is paramount to have in mind the different units of measure and the description to define if the indicator must be maximized or minimized.

However, the overall application of such processes is related to the city scale, to manage the assessment of the city overall performance as "smart city". But what happen at a smaller scale?

Such indicators do not catch the different shadings of performance within the city itself, because the values of indicators are the result of the city performance towards specific domains; for instance one specific district could perform better in terms of renewable energy consumption than other, but this difference is not seen while considering the city as a whole. For this reason, the selected indicators previously described are relevant to their application for the definition of a comparison among several cities, to define for instance, as happened within the INDICATE project, and index of performance at city level, but not to a smaller scale, at which such indicators would not be meaningful.

KPI AT BUILDING AND DISTRICT LEVELS. THE EXAMPLE OF THE DIMMER PROJECT

KPIs are not only relevant at a higher level, such as city level, but might be useful at district and building levels too. KPIs at building level serve as means of making the environmental impacts - and possibly benefits – of buildings visible and understandable to all relevant actors (residents, building managers etc.) and facilitate the consideration and management of an array of environmental issues within relevant decision-making situations. While sustainable indicators at city level are mainly used to realize city-to-city comparisons, indicators at district or building level are used in the development of specific local plans or projects.

The broad acceptance of indicators across different groups of decision-makers in different phases of buildings life cycle is especially important when indicators are used in voluntary bottom-up initiatives to investigate if, and to what extent, a consensus on environmental indicators for buildings as "a common language for green buildings" can be reached among core actors in the building sector and how such indicators could be selected. By choosing the indicators to be taken into account, it is fundamental to check what kind of data can be measured and collected. Furthermore, to be sure that the indicators are then used in practice, they also have to meet demands related to the social environment. Indeed, KPIs allow for a comparison among buildings, that can be new, renovated or in a bad state of preservations; such indicators can be grouped and used to build reference indexes.

In this case also, the approach towards the selection and use of indicators can be bottom-up or top-down. When choosing the bottom-up approach an iterative approach is followed, which consists of:

- **Clustering information** at building level and then at district and city levels;
- **Calculating** the **relevant KPIs** (related to buildings and districts) to assess the impact of the energy efficiency measured first at building level, starting from metered data;
- **Aggregating** the **data** of more buildings to form a district and then calculating the appropriate district related metrics.

On the other side, the top-down approach starts taking into account sustainable indicators at city level and then develop a focus at a smaller scale; in general this kind of approach is used for city-to-city comparison, so it consists of less precise information, because it takes into account the whole city and it would not be possible to collect metered data and/or simulations for all the buildings. Furthermore, in this case most indicators are defined on the basis of data converted into indicators to create a common unit of analysis to show comparability across indicators.

Some examples of indicators useful at building level are included in **Table 2.**

These indicators are more specific and targeted to a smaller level of interest, related to the building and district scale. This scale has been deeply analyzed and developed within another European funded project, and namely the DIMMER project. The District Information Modeling and Management for Energy

Figure 5. DIMMER project concept. Source: DIMMER project website

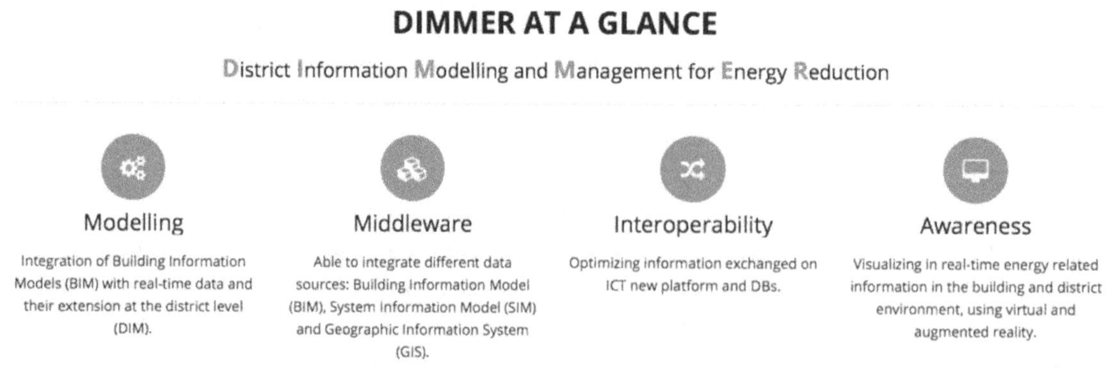

Table 2. Indicators at building level

CATEGORY	INDICATOR	UNITS OF MEASURE
Building energy	Building energy	[kWh/m2/year]
	Tree density	[kWh/m2/year]
	Ratio of building electricity peak and average load	[ratio of kWh or MWh]
Building carbon	Building carbon emission square meter per year	[kgCO2/m2/year]
	Building carbon emission per year	[kgCO2/year]
	Building carbon emissions per kWh delivered	[tonnesCO2 equivalent/capita]

Reduction (DIMMER) has been part of the FP7, as well as INDICATE, it started in 2013 and finished in 2016. The project has been carried out by a consortium of 11 partners from 4 European countries. DIMMER has been defined as a "web-service oriented open platform with capabilities of real-time district level data processing and visualization (Del Giudice et al., 2017). Thanks to the web-service interface, applications were developed to monitor and control energy consumption and production from renewable

Figure 6. Benchmarking Tool functionalities. Source: adapted from Osello, A. (2016). Deliverable 3.1.3 District Information Modeling: implementation and standard definition. Turin.

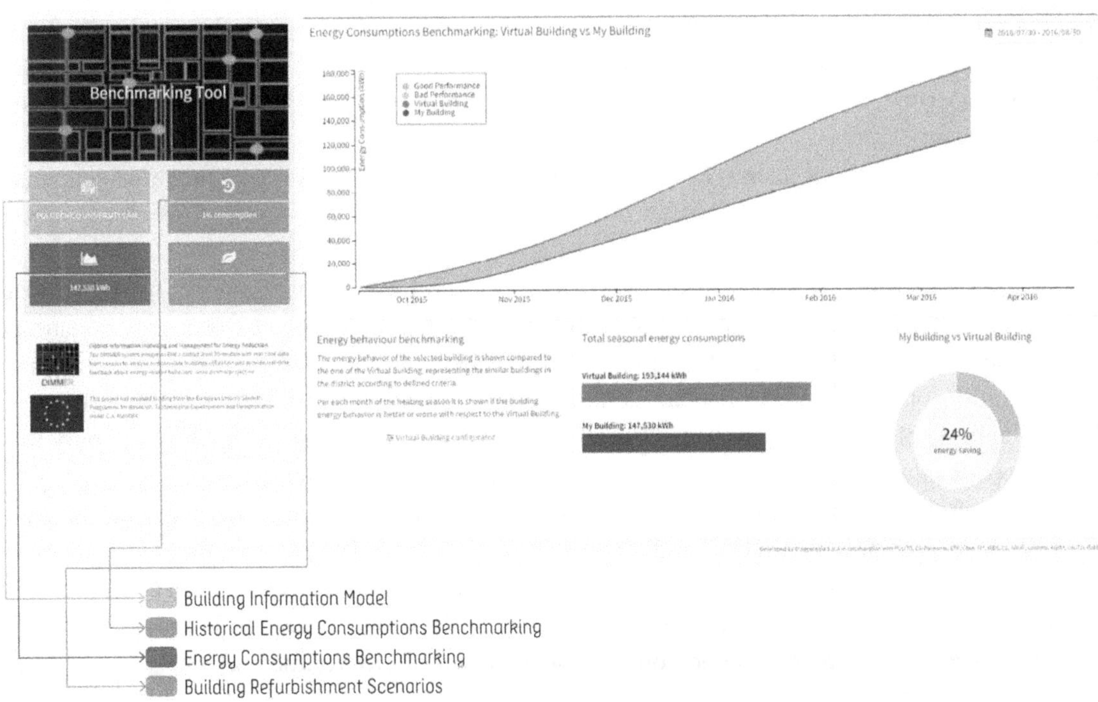

sources. Another result of the project concerned the development of an application to visualize real-time energy utilization, in order to make users aware of their impact on buildings and district performance.

DIMMER managed to combine the use of Building Information Modelling (BIM) and district level 3D models with real-time data from pervasive sensors and user feedback to analyze and correlate buildings utilization and provide real-time feedback about energy-related behaviors (**Figure 5**). Each building was equipped with a smart gateway box that retrieved data from the heat meter and the heat controller transmitted it in real-time to the middleware. Real-time sensors were crucial to characterize the buildings behavior and enable a more effective distribution of the district heating/cooling energy. Real-time information allowed to suggest possible personalized best practices related to the installation/use of renewable energy or devices such as thermal storage systems. In order to validate the DIMMER system, two mixed-up urban districts were selected in two different cities: Turin (Italy) and Manchester (UK). These districts were chosen because they were characterized by heterogenous buildings and users: public (schools, university campus, office etc.) and private (mainly residentials).

Within this project, an example of application of indicators has been applied to one of the tools produced as projects results, the "Benchmarking Tool", which allowed users to assess their building performance through the benchmark to one of the building typologies defined within the DIMMER project. The benchmarking tool represented one of the components of the Web Portal Energy Platform of the DIMMER project and was developed in order to provide the following functionalities (**Figure 6**):

- **Visualization** and deep navigation of the Building Information Model of the pilot buildings, used for the benchmark;
- **Benchmarking of the energy consumption at building level** using properly selected KPIs. In this case the indicators were used to compare the energy consumption of a selected building with the energy consumption of a Virtual Building of reference, representing similar buildings within the district pilot;
- **Historical energy consumption benchmarking** to monitor the energy consumption of a selected building over time, in order to evaluate and measures eventual interventions on the building energy efficiency;
- **Building refurbishment scenarios**, simulating on the virtual model several refurbishment options related to the building envelope.

The last three functionalities have been developed through the use of a specific set of indicators; these metrics enabled the performance of simulation to determine data such as the energy consumption of a selected building, the metrics used are presented in **Table 3**.

The calculation of metrics has been implemented through algorithms useful to directly perform analysis on the basis of the collected data. This application showed a valuable method to apply indicators as metrics of evaluation at a smaller scale (building/district) rather than generalizing it at city level, which of course includes building performances considerably different from each other on the basis of parameters such as year of construction, function etc.

To conclude, it is undeniable that integrating indicators suitable at different scales (building/district and city level) represents a great challenge. First of all, the lack of a common language between the small and big sphere has a great influence on the management of environmental issues within relevant decision-making situations. Then, the fragmentation of information results in data coming from heterogeneous sources, almost impossible to aggregate and use to build a complete and extensive database. For

Table 3. Metrics used in the Benchmarking Tool

CATEGORY	INDICATOR	UNITS OF MEASURE
Actual energy consumption	Energy consumption	[kWh per each month], eventually normalized to KWh/m2 or KWh/m3
	Energy consumption	[kWh] (total consumption), eventually normalized to KWh/m2 or KWh/m3
Historical energy consumption	Energy consumption	[kWh per each month] of the last three heating seasons
	Energy consumption	[kWh] (total consumption per heating season)
	Energy consumption	[kWh/DD] (climate corrected seasonal energy consumptions)
Building refurbishment scenarios	Actual consumption vs. simulated consumption after refurbishment	[kWh of actual consumptions associated to window/roof/façade vs. kWh of simulated consumptions associated to window/roof/façade refurbishment]
	Refurbishment cost	[€/m2] associated to the refurbishment simulation
	CO_2 savings	[%]
	Energy savings	[%]
	Cost savings	[%]

this reason, the application of KPIs-based assessment methods is still limited and not fully integrated among the different levels of consideration.

SMART CITIES ASSESSMENT THROUGH KPIS AND MCDA

The use of indicators is applied to different sectors in relation to the assessment of different aspects of performance. Multi-criteria methods of assessment, for instance, apply the indicators as metrics of reference in the process of evaluation and comparison of alternatives. Moreover, some of the current researches propose algorithms for the control and organization of KPIs for smart cities, in order to obtain a global view of the quality of life conditions associated internally with each KPI, which allows the evaluation of a city from the perspective of different models (Silva, et al., 2020). Within this research, the main concept is to create a tool able to manage different levels of layers, allowing to relate each KPI to a set of data in order to obtain both individual and general measurements. In order to organize KPIs in such a way, a complete design involving the development of algorithms has been implemented. As previously seen, the organization of KPIs is fundamental for assessing cities performance and needs to be properly set to have a better and more comprehensive knowledge of the state of the city. The method from Silva et al., 2020 allows for the creation of a model, customized on the selected KPIs, to be compared with existing models or other created models.

Other researches aim at ranking innovation initiatives, giving the opportunity to countries that want to compare themselves globally regarding innovation to do that using a multicriteria method that adheres to the quantitative analysis of the innovation indicators, the Technique of Ordination by Similarity with the Ideal Solution (TOPSIS) multicriteria method (Silva, Gaviao, Gomes, & Lima, 2017). The objective of this study was to implement a multicriteria method able to analyze the innovation indicators within

22 countries in South America and the Caribbean, using a qualitative approach to understand the dimensions that define these indicators and a quantitative approach related to the analysis of the World Intellectual Property Organization (WIPO) methodology through a multicriteria TOPSIS analysis. The WIPO institution has the role to identify the countries characterized by the highest level of innovation, using each year different methodologies for the assessment, changing for instance the number and type of indicators, developed as the result of studies and researches identifying countries with active innovative perspectives. The application of TOPSIS method showed small changes in ranking positions when qualitative principles and methodology of the innovation indicators are observed in the countries, highlighting the necessity to always consider different domains interacting among each other (social, economic et.) when performing any kind of assessment (*ibidem*).

Another research proposes the use of the Analytic Hierarchy Process (AHP) multi-criteria decision making approach for its application in the smart metropolitan city context, aiming at analyzing the sustainable development of energy, water and environmental systems, through a set of objective performance indicators (Carli, Dotoli, & Pellegrino, 2018). The approach has been applied to the real case of the city of Bari, showing that the AHP approach could support local governments towards the benchmarking of metropolitan areas with respect to sustainable activities, providing indications on how to lead the decision making process towards the promotion of an innovation process of the organizational and management structure of the metropolitan city. The AHP method was developed by Prof. Thomas Saaty in 1977 to deal with complex systems related to making a choice among several alternatives, providing a comparison of the considered options (Saaty, 1980). The output of the analysis involved the final priority vector of the alternatives, where the decision alternatives were the metropolitan areas of the city. The priority vector expresses how much that specific alternative contributes to achieving the goal of the problem statement, and so, in this case the sustainability of energy, water and environment systems of the metropolitan city (*ibidem*). By applying this method to the actual conditions of cities it is possible to benchmark the performance of the city towards aspects related to sustainability through a set of indicators. Such indicators not only reflect the characteristics of the observed real systems, but also the aspects that the decision maker considers as priorities in the sustainable development of metropolitan cities. For this reason, these analyses enable the understanding of the propensity and susceptibility of the individual metropolitan areas towards the sustainable development of energy, water and environment systems.

The application and comparison of several multi-criteria decision approaches together with the use of KPIs towards smart cities assessment presents a lot of examples. Within the analysis presented by Ogrodnik K., 2020 the PROMETHEE MCDA method was used to rank a set of selected Polish cities in terms of "smartness" with respect to six main dimensions: smart economy, smart people, smart governance, smart mobility, smart environment, and smart living, taking into account publicly available indicators of the smart city concept (Ogrodnik, 2020). This method resulted to be an effective and relatively easy tool to evaluate the performance of cities and mainly useful at the stage of development strategies. Furthermore, this research highlighted the need for national, publicly accessible statistical database as repository of metrics to use within such analyses (*ibidem*).

Another example of application of a MCDA method for the sustainability assessment of, in this case, 14 smart "megacities" is reported by Shmelev et al., 2019. Within this research, the authors applied the ELECTRE III multi-criteria decision aid tool to a panel of twenty indicators, considering performance under four different policy priorities: environmental, economic, social and smart (Shmelev & Shmeleva, 2019).

The Elimination and Choice Expressing Reality (ELECTRE III) method was applied to develop a pairwise comparison among the cities, using different sets of weights in order to adopt various assessment perspectives. For this reason, the results of the ranking differ on the basis of the dimension, social, environment etc., considered. The sensitivity analysis, which provides a means for examining the extent to which vagueness about the inputs or disagreements between people makes any difference to the final overall results, was modeled to the particular stakeholder worldview and incorporated by means of the economic, social, environmental and smart priorities.

Within this context, for the purpose of the chapter, an attempt to assess a range of cities towards sustainability, applying two different multi-criteria decision approaches, has been done in order to compare the two evaluation methods and test the application of KPIs as metrics of evaluation. The chosen MCDA methods have been the AHP, previously mentioned, and the Preference Ranking Organization Method for Enrichment Evaluations) PROMETHEE method, which is one of the most recent MCDA method that was developed by Brans (1982) and further extended by Vinckle and Brans (1985). This approach is an outranking method for a finite set of alternative actions to be ranked and selected among criteria, which are often conflicting. Both methods have been applied to the same case study, in order to develop an evaluation model able to define the most sustainable city, among a range of 12, using a set of quantitative indicators as metrics for the evaluation.

Starting from the goal previously enounced, the following five criteria have been selected to perform this evaluation: Social aspects, Environmental conditions, Built environment, Economic development and Resource consumption. Each criterion included several sub-criteria, which in turn are assessed through specific quantitative indicators (**Table 4**). The cities considered for the evaluation are the following: Tokyo, Stockholm, Shanghai, Seoul, Prague, Paris, New York, London, Hong Kong, Copenhagen, Berlin, Beijing. Furthermore, a set of indicators has been taken into consideration in order to perform the comparison among the set of cities; some of them are reported in **Table 5**.

The following section of the paragraph describes the application of the two methods, starting from the AHP and then presenting the PROMETHEE method.

As mentioned before, the AHP is based on a hierarchical structure of criteria and sub-criteria to consider within the evaluation; for this reason, it is mainly used when the evaluation needs to consider a lot of hierarchical levels (sub-criteria). This approach is structured around three main principles: breaking down the problem, perform a pairwise comparison of the various alternatives and define a synthesis of the preferences.

In order to develop the evaluation, four main steps have been followed:

- Define the **hierarchy structure**: goal, criteria, subcriteria and alternatives using Expert Choice 2000;
- **Normalize values** for the different indicators used to measure sub-criteria, deciding to maximize or minimize the indicator;
- **Pairwise comparison** among alternatives for sub-criteria using normalized values. In this case two cities at a time have been compared in terms of their contribution to their specific upper-level criteria. The performance is determined on a 9 points scale, entitled "Saaty's Fundamental Scale" (Saaty, 1980);
- **Weighted evaluation** of **criteria** from the point of view of the Local Government; in this case it is assumed that the decision maker is more interested towards the economic and social aspects than the environmental ones.

Table 4. Criteria, sub-criteria and relative indicators chosen for the ranking of the cities

CRITERIA	SUB-CRITERIA	INDICATORS
Social aspects	Employment	Employment share
	Doctor resource	Number of doctors per capita
	Education	Middle school students share
	Healthcare	Healthcare security coverage
	Air pollution	Concentration of NO2, SO_2, PM10
Environmental aspects	Air pollution	Concentration of NO2, SO_2, PM10
	Industrial pollution	Industrial air pollution SO2
	Air qualified days	Air qualified days per year
	Wastewater management	Wastewater treatment rate
	Household waste management	Domestic waste treated
	Urban density	Urban population density
Built environment	Mass transit usage	Passengers using public transit
	Public green space	Coverage of public green space
	Public water supply	Public water supply coverage
	Internet access	Household access to Internet
Economic development	Income level	Disposable income per urban capita
	Reliance on heavy industry	Service share in GDP
	Capacity investment	Government investment in R&D
Resource consumption	Energy consumption	Energy consumption per unit GDP
	Power efficiency	Residential power consumption
	Water efficiency	Total water consumption

After having developed the process described above, a series of graphs have been produced using the software Expert Choice. Considering the overall result in terms of cities ranking with respect to the goal, from the perspective of the Local Government, mainly interested in economic aspects, the results are the following: 1.Beijing (11,7%); 2. Tokyo (10,7%); 3. London (10,4%); 4. Stockholm (9,6%); 5. Copenhagen (9,4%); 6. Shanghai (8,1%); 7. Paris (7,6%); 8. Prague (7,6%); 9. Hong Kong (7,3%); 10. Seoul (7,2); 11. New York (5,5%); 12. Berlin (5,0%).

Furthermore, the consistency of the model has been calculated; the consistency ratio represents a measure of how a given matrix compares to a purely random matrix in terms of the consistency index. A value of consistency ratio CR $\leq$ 0,1, as in the evaluation model developed, is considered reasonable. The last step of the process is represented by the "sensitivity analysis" on the basis of the final outcome. This analysis involves a "what if" question in order to see if the final model is robust or not; for instance, if changing inputs, judgements or priorities of the evaluation the model does not change drastically, it means that the final answer is stable. In this case, the sensitivity analysis has been performed by assigning the same weight to each criterion; the outcome shows that the ranking does not change in a relevant way, so the final model is robust.

On the other side, the PROMETHEE has been applied following the seven steps described below:

Table 5. Values of some of the indicators used for the assessment

CITY	Social aspects	Environmental aspects	Built environment	Economic development	Resource consumption
	Indicators				
	Employment share [%]	Air qualified days [% of air qualified1 days equal or above level II in a year]	Coverage of public green space [% of public green space in city built area]	Government investment in R&D [US dollars per capita]	Residential power consumption [kWh per capita]
Beijing	0,53	0,113	0,52	932,99	0,699
Berlin	0,5	0,024	0,31	181,9	8,9
Copenhagen	0,52	0,035	0,4	672	5,95
Hong Kong	0,502	0,062	0,4	10,79	1,594
London	0,514	0,027	0,6	405	3,988
New York	0,54	0,021	0,25	21,92	2,6
Paris	0,678	0,025	0,23	0	11,2
Prague	0,515	0,028	0,14	538,35	1,153
Seoul	0,623	0,041	0,28	0	1,283
Shanghai	0,47	0,08	0,43	873,14	0,757
Stockholm	0,52	0,025	0,4	480	6,75
Tokyo	0,501	0,02	0,22	0	2,376

- **Define** number of **actions** (alternatives), **criteria** (sub-criteria that have been grouped afterwards) and **scenarios** using Visual PROMETHEE software;
- Set the **unit of measure** for each criterion;
- Write down the **values** of each action for all criteria, being aware to input values with the right unit of measure in the *Evaluations* section;
- Within the ***Preference*** **section** set:
  - Min/Max, deciding to maximize or minimize the criterion
  - Weights (later using SRF)
  - **Preference function** (function used: linear or usual)
  - Q Treshold (Indifference)
  - P Treshold (Preference)

Considering NOT normalized values;

- Assign appropriate **values to the weights of criteria** using SRF method from 2 points of view: Economic interests and Environmental interests;
- Write down **normalized weights** in the **two** different **scenarios:**
  - _Scenario 1_Environmental
  - _Scenario 2_Economic;
- Compare results among scenarios using PROMETHEE *Rankings and Scenarios Comparisons,*

The SRF method has been applied to give an appropriate value to the weights of criteria on the basis of the priorities considered, using a set of cards. Within this test two scenarios have been selected: the Environmental one, in which decision makers gave priority to environmental aspects by maximizing or minimizing selected indicators, and the Economic one, where decision makers enhanced the economic performance of criteria. From the point of view of the environmental decision maker the most important indicators to consider have been: Air qualified days (Maximized), Energy consumption (Minimized), Household waste management (Maximized). On the other side, from the of point of view of the economic decision maker the most important indicators to consider have been: Employment (Maximized), Reliance on heavy industry (Maximized), Income level (Maximized). The results, in terms of ranking, are synthetized in **Figure 7** for the Economic scenario and in **Figure 8** for the Environmental one.

Figure 7. PROMETHEE Complete Ranking for Economic scenario

PROMETHEE Flow Table

Rank	action	Phi	Phi+	Phi-
1	Tokyo	0,1857	0,3401	0,1544
2	Seoul	0,1357	0,2700	0,1342
3	London	0,0423	0,2247	0,1824
4	Stockholm	0,0113	0,2334	0,2222
5	Prague	-0,0029	0,2702	0,2731
6	Copenhagen	-0,0055	0,1893	0,1948
7	New York	-0,0061	0,1675	0,1736
8	Paris	-0,0342	0,2321	0,2663
9	Beijing	-0,0343	0,2136	0,2479
10	Hong Kong	-0,0397	0,2036	0,2433
11	Berlin	-0,0452	0,2228	0,2680
12	Shanghai	-0,2072	0,1850	0,3921

Therefore, the application of both AHP and PROMETHEE methods to the same case study produced different results in terms of overall ranking among cities. In both cases the same indicators have been used, anyway the outcomes showed different trends of performance.

Figure 8. PROMETHEE Complete Ranking for Environmental scenario

PROMETHEE Flow Table

Rank	action	Phi	Phi+	Phi-
1	Tokyo	0,1338	0,2747	0,1409
2	Seoul	0,1271	0,2322	0,1051
3	Stockholm	0,0882	0,2326	0,1444
4	London	0,0668	0,2451	0,1783
5	Copenhagen	0,0421	0,1981	0,1561
6	New York	0,0325	0,1730	0,1404
7	Prague	-0,0270	0,2215	0,2485
8	Berlin	-0,0527	0,1754	0,2282
9	Beijing	-0,0528	0,2345	0,2873
10	Hong Kong	-0,0669	0,2190	0,2860
11	Paris	-0,0765	0,1935	0,2700
12	Shanghai	-0,2146	0,1867	0,4012

Figure 9 shows the comparison between the results obtained for each method; the most relevant differences are linked with green lines. For instance, Beijing and Seoul are the most "ambiguous" cities, because Beijing in the AHP method is in the first range of cities while using the PROMETHEE method it ends up among the last four cities, while Seoul is in the opposite situation. This aspect highlights how much the definition of the goal and data availability are important in the selection of the proper evaluation method.

To conclude, the experiences previously described showed the great effort that has been done in order to define, organize and use Key Performance Indicators for smart cities assessment. However, in all cases it resulted very difficult to define a universal set of indicators for any kind of development strategy and for any city in the world; for this reason multi-criteria analyses, such as PROMETHEE or AHP, proved to be optimal in taking into account local conditions, mainly due to the possibility to define the weight of criteria and, in the case of PROMETHEE, a preference function too.

Figure 9. Comparison among ranking outcomes from the two multicriteria analysis methods

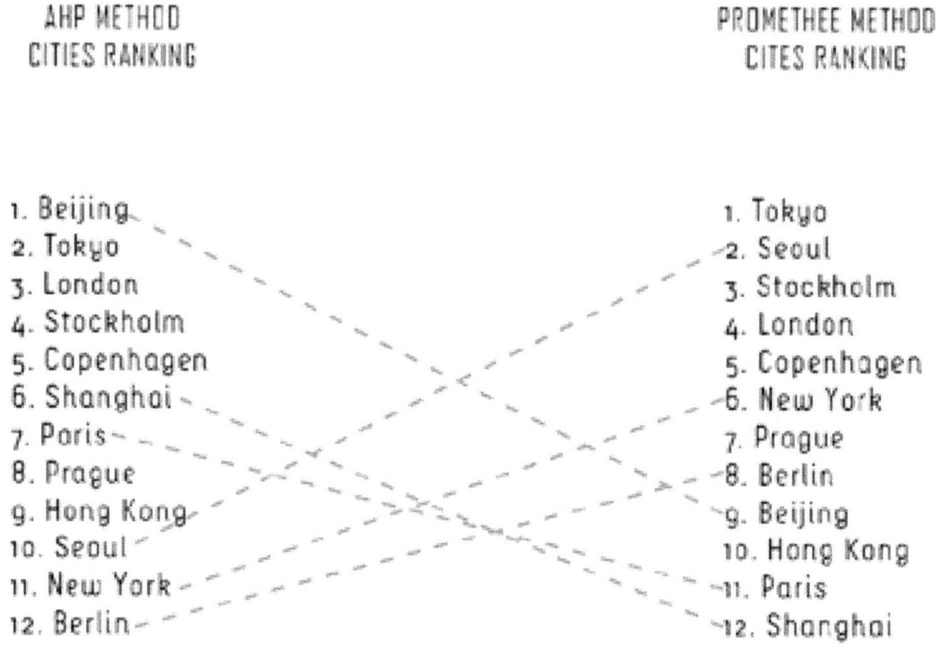

CONCLUSION

This chapter discusses the different definitions and uses of Key Performance Indicators at different levels and for different purposes towards the assessment of sustainability performance of urban areas, highlighting their wide field of application. First of all, when approaching assessment through KPIs, it is of paramount importance to properly select indicators on the basis of the "rules" enounced previously, considering that KPIs must be easy to understand, meaningful and represent available measured data, to perform the evaluation. Since Agenda 21, different projects developed methods and approaches to deal with the assessment of the extent of sustainability reached by urban areas mainly. Such experiences have laid the foundations to the extensive application of KPIs as metrics of evaluation to develop assessment methods towards the smart cities performance. In general, studies related to European or International projects made the effort to benchmark cities in a holistic way, such as the INDICATE project, by developing sustainable indicators and composite indices at city level, without providing pragmatic indications on smaller areas, except in some rare cases such as the DIMMER project in which the focus was at district level, developing performance strategies at building level too, by considering more specific sustainable indicators.

To conclude, during the last years, the interest towards the use of KPIs has grown exponentially, giving birth to a multitude of heterogeneous approaches addressing different aspects in the application of such indicators and creating a flexible and replicable framework of assessment within several fields of application. Potential directions for future research include the evaluation of cities using a combination of selected multi-criteria methods, leveraging the great adaptability of such assessment approaches. Indeed, these approaches are really helpful in supporting decision making processes towards sustainability and easily applicable within smart cities thanks to the activities of "sensing" carried on, which means collecting information by monitoring specific processes or situations; collected data could enrich the population of KPIs thanks to the great number of information coming from a structured monitoring system of different aspects (social, environment and economic) of a smart city. In turn, the results of assessments using MCDA could suggest specific and targeted "actuating" activities, preferring a solution rather than another that may be more significant on the basis of the monitored data.

ACKNOWLEDGMENT

The INDICATE and DIMMER project received funding from the European Union's Seventh Framework Programme for research, technological development and demonstration under grant agreement n° 609084.

REFERENCES

Astleithner, F., Hamedinger, A., Holman, N., & Rydin, Y. (2004). Institutions and indicators – The discourse about indicators in the context of sustainability. *Journal of Housing and the Built Environment*, •••, 7–24.

Banu, G. (2018). Masuring innovation using key performance indicators. Procedia Manufacturing, 906-911.

Bencardino, M., & Greco, I. (2014). The Paradigm of the Modern City: SMART and SENSEable Cities for Smart, Inclusive and Sustainable Growth. In Computational Science and Its Applications – ICCSA 2014: 14th International Conference, Guimarães, Portugal, June 30 – July 3, 2014, *Proceedings, Part II,* (pp. 579-597). Guimarães, Portugal: Academic Press.

Carli, R., Dotoli, M., & Pellegrino, R. (2018). Multi-criteria decision-making for sustainable metropolitan cities assessment. *Journal of Environmental Management*, •••, 46–61.

Del Giudice, M., Osello, A., Fonsati, A., De Luca, D., Patti, E., Brundu, F. G., & Acquaviva, A. (2017). District data management, modelling and visualization via interoperability. In *Proceedings of Building Simulations (BS).* International Building Performance Simulation Association.

Dimmer, Project Description. (2015). Retrieved from DIMMER: http://dimmer,polito,it/project

Domìnguez, E., Pérez, B., Rubio, A. L., & Zapata, M. A. (2019). A taxonomy for key performance indicators management. *Computer Standards & Interfaces*, •••, 24–40.

Figueira, J., Greco, S., & Ehrogott, M. (2005). *Multiple Criteria Decision Analysis: State of the Art Surveys.* Springer.

Grey, T. (2016), INDICATE Indicator-based Interactive Decision Support and Information Exchange Platform for Smart Cities. *Planning, Climate Change and Energy in Ireland - Irish Planning Institute Guide*, 65-72.

Mattei, M. (2013). Quando la città diventa smart. In C. Ratti (Ed.), *Smart city, smart citizen* (pp. 9–17). Egea.

Munda, G. (2005). Multiple criteria decision analysis an sustainable development. In *Multi Criteria Decision Analysis, State of the Art Surveys* (pp. 954–986). Springer.

Munier, N. (2005). *Introduction to Sustainability: road to a better future*. Springer.

Ogrodnik, K. (2020). Multi-criteria analysis of smart cities in Poland. *Geographia Polonica*, •••, 163–181.

Ratti, C. (2013). *Smart city, smart citizen*. Egea.

Ratti, C., & Claudel, M. (2017). *La città di domani, Come le reti stanno cambiando il futuro urbano*. Einaudi.

Saaty, T. (1980). *The analytic hierarchy process*. McGraw Hill.

Shmelev, S., & Shmeleva, I. (2019). Multidimensional sustainability benchmarking for smart megacities. *Cities (London, England)*, •••, 134–163.

Silva, J., Mojica, J., Pineres, A., Rojas, A., Acosta, S., Giuliany, J. G., & Sanabria, E. S. (2020). Algorithms for the Control of Key Performance Indicators for Smart Cities. *Procedia Computer Science*, •••, 971–976.

Silva, M., Gaviao, L. O., Gomes, C., & Lima, G. (2017). A proposal for the application of multicriteria analysis to rank countries according to innovation using the indicators provided by the World World Intellectual Property Organization. *RAI Revista de Administração e Inovação*, 188-198.

United Nations. (1992, June 14). *Sustainable Development*. Retrieved from Agenda 21: https://sustainabledevelopment,un,org/content/documents/Agenda21,pdf

Watróbski, J., Jankowski, J., Ziemba, P., Karczmarczyk, A., & Zioło, M. (2019). Generalised framework for multi-criteria method selection. *Omega*, *86*, 107–124. doi:10.1016/j.omega.2018.07.004 PMID:30671511

ADDITIONAL READING

Mascarenhas, A., Coelho, P., Subtil, E., & Ramos, T. B. (2010). The role of common local indicators in regional sustainability assessment. *Ecological Indicators*, *10*(3), 646–656. doi:10.1016/j.ecolind.2009.11.003

Mascarenhas, A., Nunes, L. M., & Ramos, T. B. (2015). Selection of sustainability indicators for planning: Combining stakeholders' participation and data reduction techniques. *Journal of Cleaner Production*, *92*, 295–307. doi:10.1016/j.jclepro.2015.01.005

Mcalpine, P., & Birnie, A. (2005). Is there a correct way of establishing sustainability indicator development on the Island of Guernsey. *Local Environment: The International Journal of Justice and Sustainability*, *10*(3), 243–257. doi:10.1080/13549830500075537

Milan, L., Kin, B., Verlinde, S., & Macharis, C. (2015). Multi-actor multi-criteria analysis for sustainable city distribution: A new assessment framework. *International Journal of Multicriteria Decision Making*, *5*(4), 334–354. doi:10.1504/IJMCDM.2015.074088

Pires, S. M., Fidélis, T., & Ramos, T. B. (2014). Measuring and comparing local sustainable development through common indicators: Constraints and achievements in practice. *Cities (London, England)*, *39*, 1–9. doi:10.1016/j.cities.2014.02.003

KEY TERMS AND DEFINITIONS

City or Urban Area: Built-up area with a high population density governed by one or more political/administrative bodies.

Domain: An area of interest for monitoring and assessment consisting of several related topics. A group of indicators of such topics belong to a domain. In sustainable development assessment, the following three dimensions are often considered as domains: social, environmental, and economic.

Index: A special type of indicator that consists of two or more indicators that each measure distinct system characteristics in separate units that are normalized and aggregated.

Indicator: An indicator provides information on the state or conditions of a phenomenon; it is a simplified way to explain a complex concept, Indicators may be used directly or as elements in other measurement tools, for instance MCDA.

MCDA: Multi-criteria decision approach or MCA (multi-criteria analysis) offers a number of ways of aggregating the data on individual criteria to provide indicators of the overall performance of options.

Metrics: The specific operational design of an indicator in regard to how the values are measured, parameterized, calculated, normalized, combined, etc. (e.g., as tons per capita, or per unit of GDP).

Smart City: urban area that uses different types of electronic Internet of things (IoT) sensors to collect data collected from citizens, devices, buildings and assets. Insights gained from that data are then processed and analyzed to monitor and manage assets, resources and services efficiently; in return, that data is used improve the operations across the city. The smart city concept integrates information and communication technology (ICT), and various physical devices connected to the IoT network to optimize the efficiency of city operations and services and connect to citizens.

Chapter 10
EEB Project System Integration and Technology Sperimentation Matrix

Francesca Maria Ugliotti
Politecnico di Torino, Italy

ABSTRACT

Today an increasing number of cities are equipping themselves with three-dimensional urban modelling and simulation platforms for energy management to integrate both spatial and semantic data for enabling better decision-making. The work presented in this chapter is the result of the study carried out by Politecnico di Torino within the Energy Efficient Buildings (EEB) project. Collected data on urban and building scale are managed in specialized, independent, and heterogeneous domains such as GIS, BIM, and IoT devices for energy and electrical monitoring. Possible relationships among these datasets in the perspective of system integration have been carried out according to a rich matrix of experimentations. Specific tools, including innovative visualization technologies and web services, are put in place to allow final users to benefit from this data. The infrastructure is intended to establish a common interoperable ground among heterogeneous networks to achieve the goal of smart cities digital twins.

INTRODUCTION

The great challenges of this Millennium such as urbanization, resource efficiency, climate change, responsible consumption and production, globalization, circular economy, and connectivity frame cities as a reference point for future development. Tackling these priorities requires a systemic approach to innovation that aims for a system-wide transformation by affecting the economic, social and environmental dimensions as well as their interconnections. This implies a trans-disciplinary solutions-oriented perspective that integrates technology, advanced models, organization, governance and regulation and involves co-creation of knowledge and co-delivery of outcomes with economic, industrial and research actors, public authorities and civil society (European Commission, 2020). As is known, a Smart City provides "the effective integration of physical, digital and human systems in the built-up environment to

DOI: 10.4018/978-1-7998-7091-3.ch010

deliver a sustainable, prosperous and inclusive future for its citizens" (The British Standards Institution, 2014). In that framework, it is not a single system that defines an intelligent city, but an interconnected network of innovations, "a system of systems" (Mitchell, 2001), focusing on the interaction between network and network and between the city and final users to make sure that the city can be more suitable to the citizen's needs, and the citizen is increasingly active in the creation of new sustainable cities. Although often associated with energy efficiency and sustainability, smart cities are much more than that as outlined in Figure 1, concerning the efficiency of urban operation and services. Over the past years, Information and Communication Technology (ICT) allowed the correlation of data, processes and methods often very different from each other to make them available. At the same time, it links and strengthens networks of people, businesses, infrastructures, energy and spaces, besides providing intelligent organizational and governance tools (European Parliament, 2014). Cities and building managers need to have systems to analyze data for better decisions, anticipate problems to resolve them proactively and coordinate resources to operate effectively (IBM Corporation, 2012). Even more urgent today is to dwell on the concept of resilience exploring "the reactive, recovery and adaptive capacities and also the transformability of urban systems" (Chelleri & Olazabal, 2012). The resilient city is continuously facing an ever-widening range of pressures that focus on time in its various expressions: degradation, human neglect, natural hazard events, as well as transformation, innovation, and the constant evolution of human needs. To govern this evident complexity is imperative to refer to a knowledge system of the built heritage that from closed and static must become increasingly dynamic and updated with real-time information (Shah et al., 2019).

Figure 1. Smart City scenario

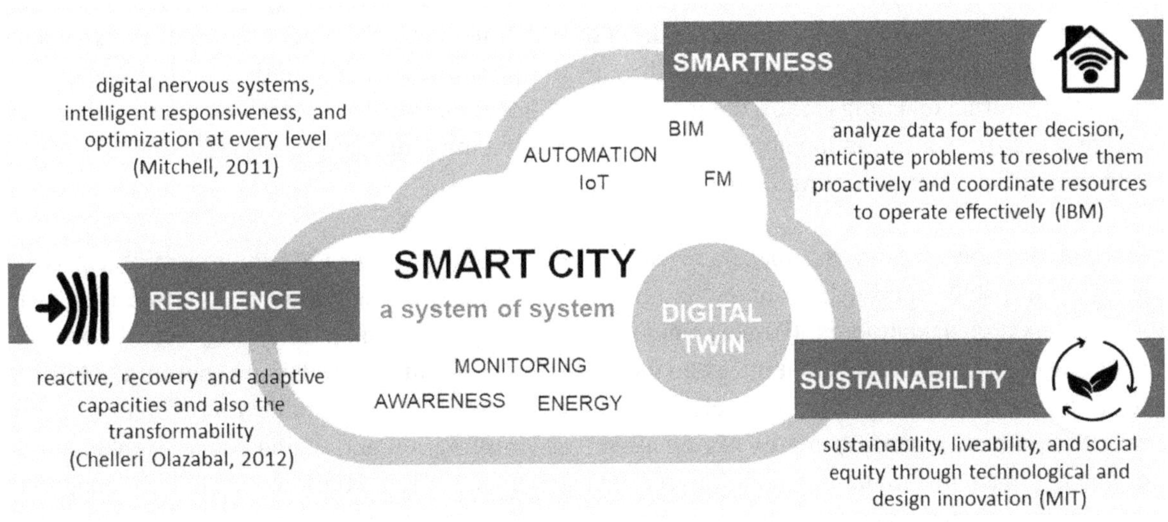

According to this vision, a more future-oriented definition of smart cities already comes from the Chinese National Smart City Standardization. It talks about "a new concept and a new model, which applies to the new generation of information technologies such as the internet of things, cloud computing, big data and space/geographical information integration, to facilitate the planning, construction, management and smart services of cities" (ISO/IEC JTC 1, 2014). Therefore, to works better, a smart city needs to be structured around a digital twin (Wright & Davidson, 2020; Riddhi et al., 2020) representative both of tridimensional representation with information relating to its operation and of how people interact with the built environment. Already successful in manufacturing with predictive and fault prevention functions, the potential of digital twins for buildings, infrastructure and entire cities is yet to be fully discovered. In this scenario, digital models can be seen as the foundation of this state-of-the-art knowledge and management systems. In fact, it is possible to trace a clear connection between Building Information Modelling (BIM), Geographic Information System (GIS) and the Smart City concept as they imported in the world of architectural and territorial design powerful tools able to give excellent data richness to the projects. As buildings have a significant impact on cities in terms of economic, efficiency and quality of life, the strengths of system integration are summarized below.

- The BIM environment is powerful for planning and simulations, enabling best design solutions with a reduction of errors, disputes, risks, cost, and time whether for new construction or renovation due to the computations and visualizations capability.
- Collaboration and data sharing enabled by the BIM methodology are the keywords for the entire construction value chain. Moreover, BIM data optimization during the building lifecycle promotes smarter facilities management services and improve the relationship between the building industry and the other sectors.
- BIM supports all levels of the communication process, exploiting new ways for data visualization and the presentation of projects. Thanks to the three-dimensionality of space and the interaction with new technologies, designers, public administrations as well as citizens can better understand design solutions and implement knowledge about the heritage.
- BIM and GIS can provide a broader perspective to analyze and manage the existing heritage, focusing on costs and energy saving, and, on the other to check the potential use of the assets and control new urban development. According to the possibility of integration with other datasets and the Internet of Things (IoT), the building manager can have a cross-scale dynamics vision and access real-time information about the service installed, making accurate assessments of the asset operating condition, enabling its better usage and utilization.

The construction industry can make a significant contribution to meet the long term objectives by adopting a multi-scale approach, from the building to the city scale, that includes (i) knowledge sharing among players and stakeholders, (ii) performance monitoring and management (iii) use of collaborative building management tools that integrate the whole lifecycle information from sourcing to building construction, facility management and refurbishing and end-of-life, (iv) a behavioural change of end-users and consumers regarding energy saving as well as conservation and maintenance. Hence, only an integrated approach that includes technology, collaborative tools, digital buildings, facility management and user awareness can have a tangible impact on the cities of tomorrow.

The chapter illustrates the results of a preparatory study of the Politecnico di Torino in the digital twin perspective. From the digitalization of the built environment in the urban field, the possibilities of the interdisciplinary use of the deriving information between the different data domains are investigated. The background is the Zero energy buildings in smart urban districts project (Energy Efficient Buildings - EEB), which aims to pursue the objective of increasing energy efficiency of existing buildings and urban districts, through the pervasive use of non-invasive technologies for the real-time monitoring and control of environmental parameters and of the energy production/consumption using smart devices.

Methodology

Over the years, a lot of time and resources are allocated to the "knowledge site" to survey, map, represent, and restore the memory and the current state of a building. For this purpose, technologies are considered fundamental in setting up an integrated knowledge framework from which it is possible to retrieve information for ordinary management conditions. In these terms, the research is directed towards innovative tools of data survey, collection, use and visualization to assess, maintain and prevent the built environment. An in-depth process of data acquisition and interconnection focused on the intrinsic features as well as the immaterial values of a place can raise awareness public administration, technical figures and citizens in terms of historical knowledge and identity. This approach can facilitate regeneration, urban resilience processes, participatory and sustainable planning. In this scenario, it is central to consider which information must be represented and in what way, so that they become descriptors as targets to be found in previously during on-site digital survey and in post-processing as information modelling strategies. Tools as GIS, BIM, IoT, Virtual (VR) and Augmented Reality (AR) can contribute to achieving the vision of digital twin of the built environment to manage the knowledge site more accurately in a digital way sharing and providing information from different perspectives.

Figure 2. Knowledge site methodological framework

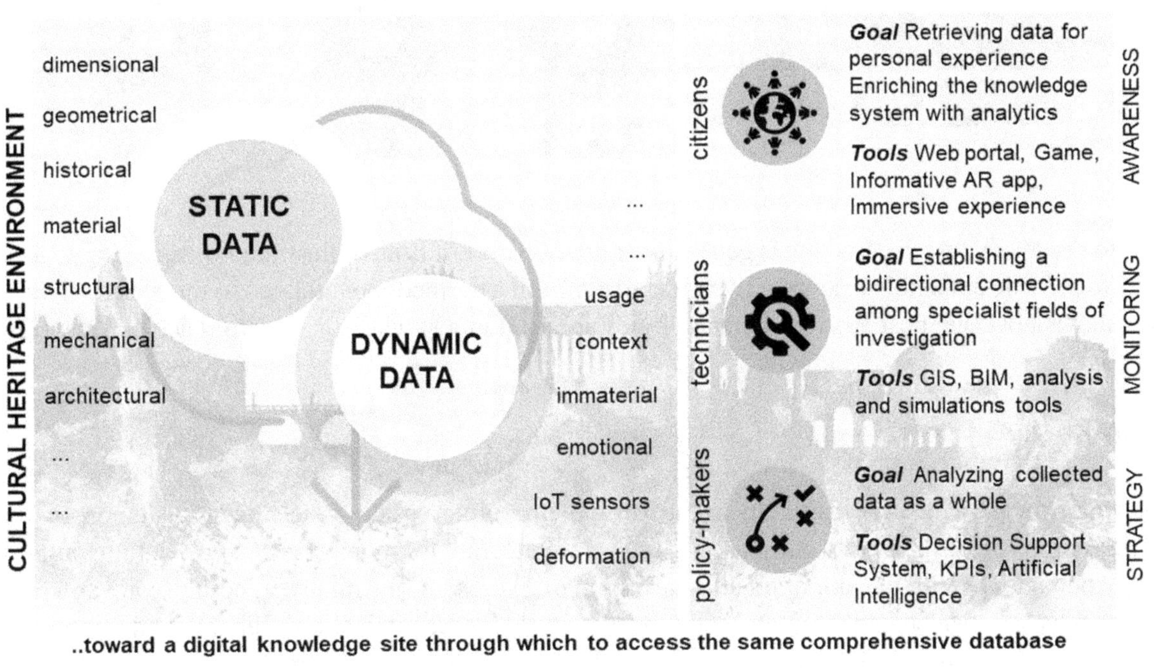

As shown in Figure 2, the mission pursued is to establish a comprehensive database network to which different actors can access according to differentiated levels of information. Within this digital environment, Policy-makers are envisioned to be able to analyze the collected data as a whole, supported by Decision Support Systems (DSS), Key Performance Indicators (KPIs), and Artificial Intelligence (AI). Technicians could establish a bidirectional connection among specialist fields of investigation, taking full advantage of software interoperability for analysis and simulation. Finally, the citizens could use the data to enrich their personal experience, using virtual and augmented reality game-based informative applications, and provide feedback (White et al., 2021). Specifically, as stated in Figure 3, the proposed method seeks to integrate different systems, domains and technologies according to a rich matrix of experimentations to achieve the following goals: (i) modelling; (ii) simulation and monitoring; (iii) data visualization; (iv) user awareness.

Figure 3. System integration and technology experimentation matrix

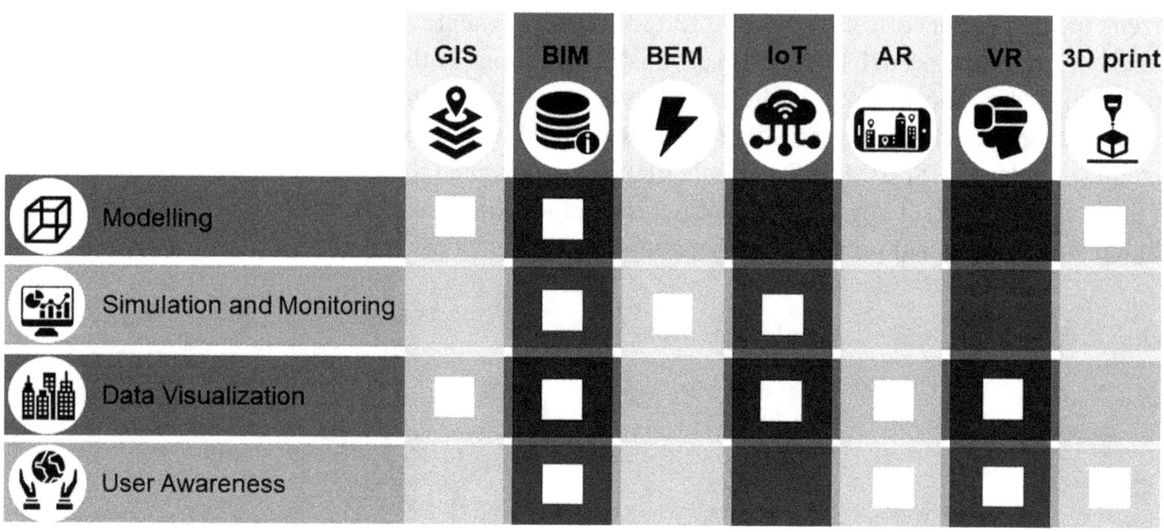

The approach described below is generally valid; however, it is also illustrated by applying to a case study to provide a tangible example. Even if starting from a defined topic related to the energy field, the pilot shows how the most advanced technologies are used in a synergistic way to reach the mentioned objectives.

Case Study

Settimo Torinese, a medium-sized city of the Turin metropolitan area, was selected as a demonstrator of the EEB project to establish a cloud-based software architecture for monitoring and modelling buildings energy behaviour. With the aim of creating a system of interest to several users, both private and public buildings were considered to set information models useful for the realization of energy and management optimization assessments. The city consists of 300 census sections and approximately 3,600 residential buildings with 47,831 inhabitants. In this area, three representative public buildings, including a recently constructed public library and offices, a primary and a secondary school were mapped. The middleware is in charge of enabling the interoperability between the heterogeneous data-sources by three-layers. The bottom is the Data-sources integration layer that integrates heterogeneous hardware and software technologies by abstracting their features into Web Services. Specific IoT gateways were developed for different standards and protocols, such as ZigBee and Spirit. BIM, GIS and weather stations are software data-sources. The core of the middleware is the Services Layer that offers the components for developing generic applications and distributed services (Bottaccioli et al., 2017). Finally, the Application Layer provides a set of API and tools to develop a distributed application for building management and to post-process collected information. From this application context, the integration processes achieved are detailed in the following.

GIS-BIM Modelling

Since the city is the heart of the experiments on digital twins for the construction sector, there is immediately a definite problem in finding and managing information which goes from general to particular and vice versa. Through a GIS system, as anticipated, it is possible to obtain a general and transversal view of territory according to different perspectives and points of view. By superimposing the multiple information layers available, a large-scale representation can be managed, and cross-type queries can be performed. On the other hand, BIM environments provide a complete and detailed description of the characteristics of a single building or complex to promote better maintenance and retrofitting actions. The collection and visualization of data concerning an urban district are still considered a crucial point in the process and a challenging field of research.

Geographic Information System

A geospatial statistical model of the building stock can be developed in the GIS environment to provide a representative picture of the city's distribution of selected factors. As an example, the case of current energy consumption performance is given to critical areas on maps. In particular, a mixed approach methodology was adopted to joints Urban Energy Consumption Mapping (UECM), with Multi-Criteria Decision Analysis (MCDA) (operation research approach) and Stakeholders Mapping (SM) (social research approach) due to their complementarity in fulfilling different tasks in the urban energy planning process (Torabi Moghadam et al., 2017). The main phases developed regards the building stock data collection, geo-referencing territorial data, and the building stock characterization. The thermal energy consumption data for space heating was derived by the energy-use registered by the district heating system considering a period of 4 years. The geometrical features were defined from an *ArcGIS* map. A large building stock technical and performance characteristics were taken from the National Census database (ISTAT, 2011), while the EPC database (APE, 2015) was used for new building construction. The collected data were analyzed and elaborated to create a supporting dataset. Starting from this mapping, a Multicriteria Spatial Decision Support System (MC-SDSS) was built through the *CommunityViz* platform (Communityviz, 2020) and using Suitability Analysis and Interactive Impact Assessment models. The integrated platform allows the user to consider socio-economic, technical and environmental dimensions. *Scenario 360* Advanced Formula Editor was used to structure, edit, display and syncs all the components of the model by using dynamic attribute and indicators. As a result, the baseline and the possible future scenarios can be displayed by maps and charts in real-time (Torabi Moghdam & Lombardi, 2019) as shown in Figure 4. The tool is configured in a participatory and collaborative way to define and analyze the different energy saving-scenarios at urban scale considering various aspects of the problem.

Figure 4. CommunityViz interface (Torabi & Lombardi, 2019)
Source: (Torabi & Lombardi, 2019)

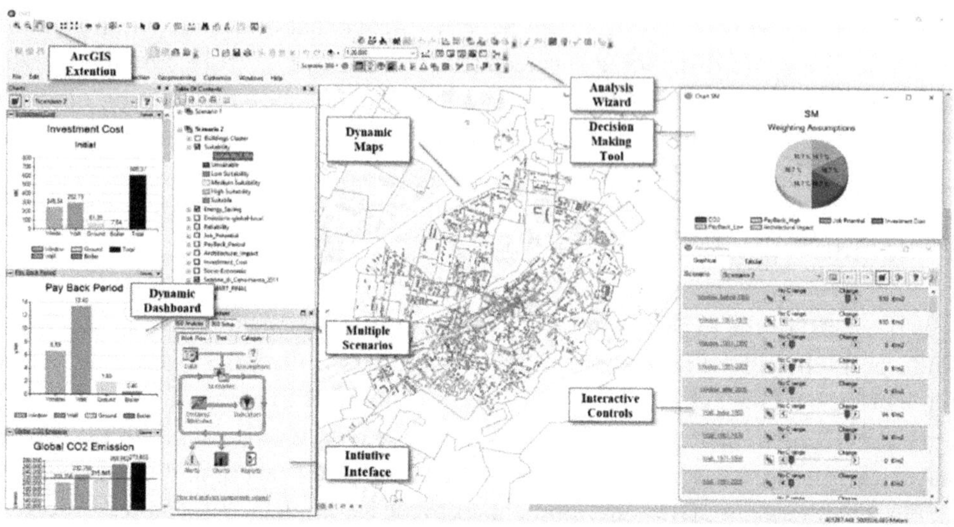

According to previous experience (Vinco et al., 2018), a simulator of solar radiation and photovoltaic production was realized on GIS basis to analyze the potential and the photovoltaic production within intelligent urban districts. Moreover, an innovative methodology for short-term (e.g. 15 minutes) forecasting of Global Horizontal Solar Irradiance (GHI) was tested. The neural network has been trained and validated with a four years solar radiation dataset collected by a real weather station. The GHI forecast output is used as input to the previously mentioned photovoltaic simulator to predict energy production in short-term time period (Aliberti et al., 2018).

Building Information Model

On the other side, single buildings can be analyzed more deeply by creating a BIM as-is model. Starting from the collection of technical documents, the current state of the public buildings selected as case studies was verified by an expeditious on-site survey focusing on possible optimization actions concerning space and energy management. In this sense, a qualitative rather than quantitative digitalization activity was considered. Realizing an effective BIM model does not just mean to implement all the objects and their properties, but it is necessary to organize information so that it can be functional in retrieving data for facility management activities (Teicholz, 2013). Considering the resources currently available for this type of activity, both in terms of data acquisition and updating, a conceptual and simplified representation was set up with a basic but consistent dataset. Accordingly, the models realized include the characterization of the building envelope, the inventory of spaces and intended use, and the mapping of system terminals. These objects are reliable in terms of number, size, and qualitative position in the room. To establish a link with IoT dataset, fictitious elements representative of the sensors installed in the field were introduced. The focus was on establishing a modelling methodology that could be valid and replicable to analyze the assets belonging to the same real estate portfolio. Although the parametric model is a database in itself, consulting it may not be immediate. For this reason, a unified template was created to obtain homogeneous and comparable outputs among buildings promoting immediate use of data by the Facility Manager/Department. The value of this study is traceable in the approach used aimed at making sense of BIM data to use it for lifecycle asset management effectively. Schedules and themed plan views were exploited for optimizing the architectural modelling phase and controlling the digital model in the perspective of interoperability and data integration. By appropriate use of shared parameters and equations within schedules, it is possible to make the most of their format for each different purpose, from analysis to managing or reporting. The accurate and structured inventory of the building elements allows relating information about spaces and relative intended use with the facilities costs and the energy consumption. In this way, it is possible to obtain several Key Performance Indicators (KPIs) for evaluating the utilization rate of the space and occupancy as well as maintenance costs chargeback (Ugliotti, 2017). The relationship between the model and facility management services can take place (i) through direct use of data and functionality setting in the modelling environment (e.g. building registry, space accessibility checks, fire-fighting requirements, minimum dimension for windows to guarantee the amount of sunlight), (ii) by exporting data to other formats through interoperability to perform simulations with specialized software (e.g. energy, computational fluid dynamics, structural, seismic) (iii) through the integration of data with other data sources (e.g. CAFM, CMMS, GIS, IoT).

Figure 5. Public buildings BIM models

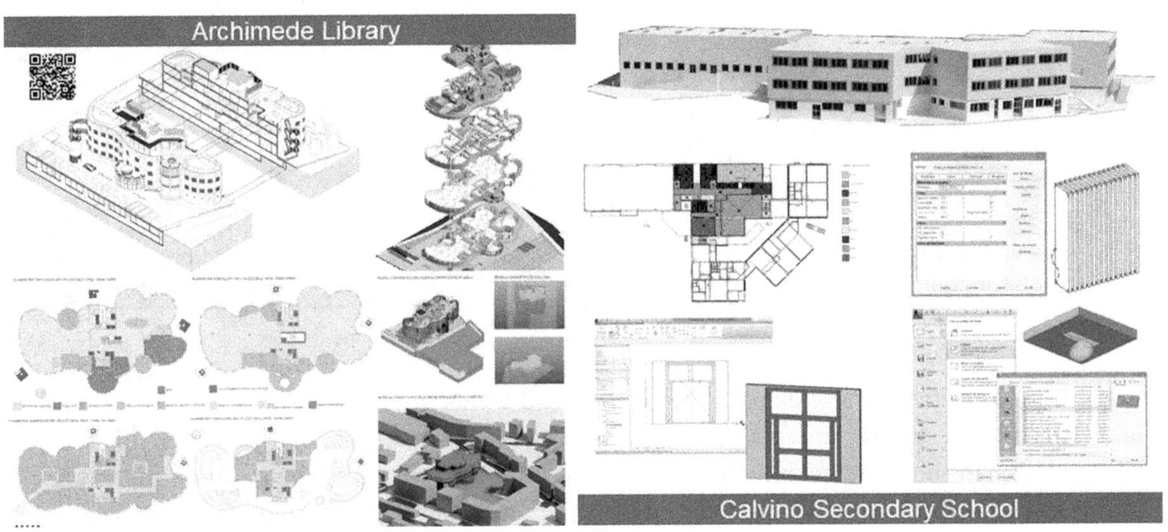

BIM-GIS Integration

In the scenario outlined, urban models as digital geometric twins can give value to the city services dealing with multiple aspects according to specific needs. As far as the district is concerned, a balanced mix of GIS and BIM data is required in order to improve decision-making and information processes. Within the energy assessment purpose, for example, general information such as location, orientation, size, use, energy consumption, as well as the characterization of building envelopes and installations, are some of the crucial subjects. As shown in Figure 6, five LODs (Torabi Moghadam et al., 2016) was identified to define a balance regarding the granularity of the information of the urban scale coming from the two different domains. From the regional level LOD0, more general and contextual, the LOD1 provide the three-dimensional view of the public cadastre, where it is possible to identify the building volumes that make up the city, combined with general data on location and use. Such information is mostly related to the GIS domain. LOD2 is the synergistic representation of the urban fabric, where the buildings are inserted in a network of services and distribution. This is the first stage of connection with BIM. LOD3 is focused on the building system, and LOD 4 introduces details specialist information on equipment and components.

Figure 6. Calibration of the information level according to the different scales of representation (Torabi Moghadam et al., 2016)
Source: (Torabi Moghadam et al., 2016)

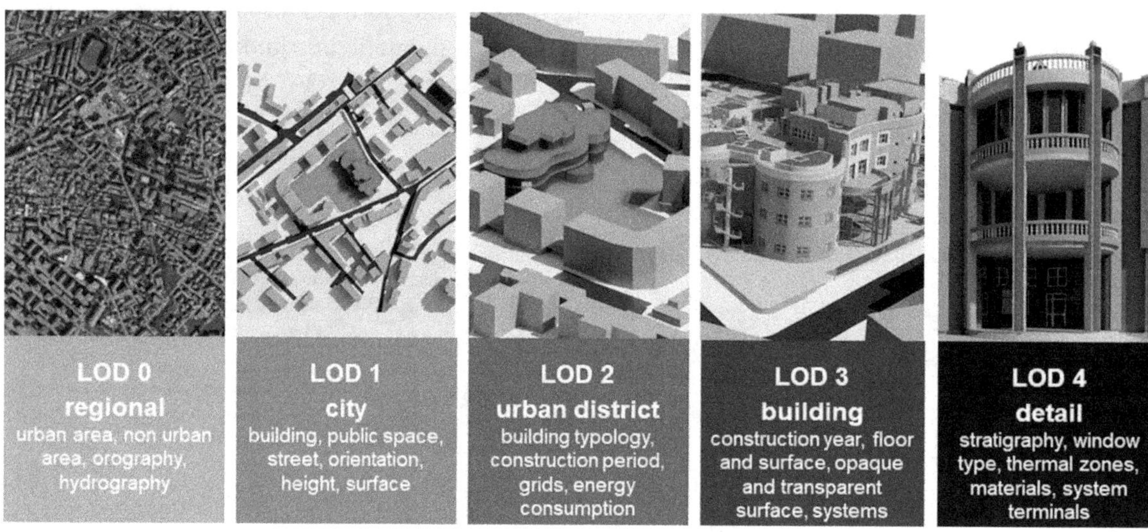

In addition to the different scales of investigation and levels of contents, the primary issue in integrating BIM and GIS systems reflects their diversity concerning modelling environments and reference systems, two-dimensional and georeferenced and three-dimensional objects in local coordinates respectively. Several integration methods were tested to export and convert both geometries and information into BIM-GIS-compatible formats. A preliminary attempt to display the BIM model and the shapefile together refers to the use of the proprietary software, such as *Autodesk Infraworks*, as a common platform for Industry Foundation Classes (IFC) (buildingSMART International, 2020) and CityGML (OGC, 2020) formats, which are considered the two critical standard exchange in the building industry. A further opportunity is to implement the information part of the two systems, for example exporting the *Revit* ODBC database. However, to achieve full integration, there is a need to provide interoperability at the semantic level (Karan et al., 2015) through a web-service interface, which preserves both geometric and alphanumeric information. This approach enables the connection of heterogeneous data sources, enriching the data visualization modes and the query possibilities for a more detailed overview of the built heritage. In these terms, it is possible to investigate buildings/districts/cities according to several knowledge purposes, calibrating the information of interest from time to time.

To pursue this objective, an interoperability methodology (Provera, 2017) aimed at creating a platform for the free interactive visualization of parametric models on an urban scale was tested by using virtual globes and exploiting the 3D Tiles technology (OGC, 2020). The digital globes are scale-bound structured models of celestial bodies in virtual space. They enable distortion-free visualization of earth imagery and digital images on a three-dimensional structure at any scale and from any angle (Aurambout et al., 2008). As presently used extensively in areas such as education, research/collaboration, and disaster response, these systems are revolutionizing display and integration of earth observation data in terms of both to the access and content contribution democratization. For the technical users, virtual globes have vastly reduced the overhead associated with accessing global archives of satellite imagery by eliminating purchase costs and effort required to stage and manage extensive image holdings. Regarding the modern virtual globes, the most outstanding examples in open source domain are *NASA Web World Wind* and *Cesium* as they leverage the latest technologies for Web: JavaScript, WebGL and HTML5. According to this, non-experts users can make links to their earth observation data through Application Programming Interface mapping (API) web services. The globes can be conveniently customized with specific content and diverse sources to meet the needs of case increasing productivity for individual projects and studies. One application is the realistic 3D building models implementation by exploiting different modelling and photogrammetry systems. In this context, 3D Tiles are an open specification for streaming and rendering massive heterogeneous 3D geospatial datasets such as Photogrammetry, 3D Buildings, BIM, Instanced Features, and Point Clouds. The foundation of 3D Tiles is a spatial data structure that enables Hierarchical Level of Detail (HLOD) so only visible tiles are streamed and only those that are most important for a given 3D view (Cozzi, 2015). The BIM-GIS integration project developed pursues the aim to bring to the same platform data and geometries from the two different IT environments. On the BIM side, the main obstacle is the non-capability of direct use of the native format file, such as .RVT for *Autodesk Revit* software, as an exchange code language. Previous researches have studied the interpretation of the gbXML format (Massara, 2016; CSI Piemonte, 2019). In this study, the IFC features capabilities were also interpreted and compared. The interoperability process involves a sequence of conversion and transferring procedures intended to maintain and display information and scenes from sectoral and non-communicating BIM and GIS software. The overall procedure involves three main steps summarized in Figure 7: (i) BIM model optimization for open formats exportation; (ii) 3D tiles generation; (iii) dynamic visualization dashboard. As general considerations, the gbXML standard is characterized by energy-specific data export capability. At the same time, IFC adopts a comprehensive and generic approach to represent the entire building project, covering domains from building construction to building operation.

Moreover, the IFC uses a "top-down" and relational approach, which yields in a highly organized and complex data representation schema. Ideally, it can maintain semantic integrity automatically as it can trace back all the changes when a value of the element in the schema changed. The "bottom-up" gbXML schema is flexible, simpler and easier to understand, facilitating the quicker implementation of extension for different design purposes (Dong et at. 2007). Another fundamental difference concerns the way geometry is generated. gbXML only exports basic rectangular plane surfaces, while the IFC classes correspond entirely to an object including different 2D/3D representation, from simple extrusion to detailed volumes. Another critical remark involves the possibility to fully control the transfer mode by the IFC in terms of objects filtering as well as mapping file customization and additional information. Based on these arguments, a model optimization mode was experimented to obtain a standardized export procedure.

Figure 7. Sequence of conversion and transferring procedures for BIM-GIS district model

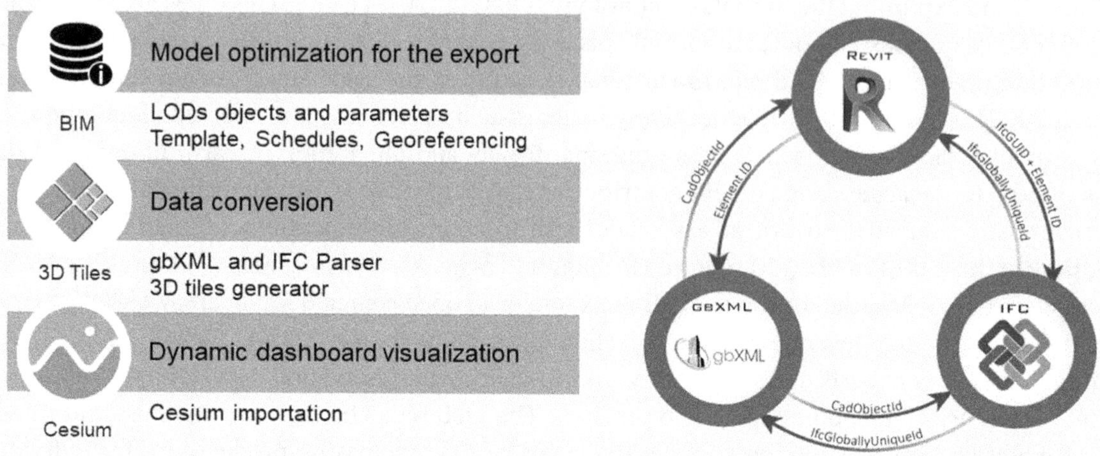

Step 1: BIM Model Optimization

Primarily, it is indispensable to define the objects and the information which have an essential role in the final user experience. The LOD definition represents the capability of the 3D model to increase or decrease the complexity of its geometries and the amount of information to display, increasing the subsequent rendering efficiency on the virtual globes. LOD2, LOD3, LOD4 previously mentioned were considered. Given this, the first step to accomplish in the preliminary modelling phase is the creation of a customized project template. This strategy is intended to limit the proliferation of random data and elements which can cause misunderstandings and redundancy errors. Especially in the IFC exports, the pre-defined environment includes coding and naming systems, object styles and libraries, additional parameters as well as 3D view settings and specific schedules. The latter are used to extract and manage only the elements of interest contained in a specific 3D view and to select the attributes for each level of detail by setting dedicated schedules for each object category. Besides, the model must be positioned correctly in terms of georeferencing and north orientation. The quality of the export must be verified before taking the next steps. As mentioned before, in the case of gbXML format, it is necessary to check the correct export of surfaces from a geometrical point of view to avoid voids, room overlaps, and air-surface generation. Modelling guidelines in this area are widely investigated as they are used for energy simulations, as illustrated in the next section. The *Revit gbXML 3D viewer* is used to manage the energy setting, visualize the level of the rooms by level and the analytical surface examining warnings one by one. In the case of IFC, the customization of the IFC class name parameter through IFC Mapping file and of the IFC project template allows us to make the given setting replicable on different models and projects. The IFC standard selected is the IFC 2x3 coordination View 2.0. “Export only elements in view”, “Export schedules as property sets”, “Export only schedules containing IFC, PSet or Common in the title” are some of the property sets to include. The results of the IFC exportation is checked using *BIMVision*, a freeware IFC model viewer. To establish a connection between the Revit model entities and its exportations in the different exchange formats, it was essential to trace the encoding system for the identification of the elements. In the gbXLM tree structure, every surface is identified by the CADObjectId, which contains a text string reporting the name of the object type and a sequence of eight decimal characters representing the Revit ID. The IFC specification uses a Globally Unique Identifier (GUID) resulting in 22 character length string. Figure 7 shows a summary of the different identifiers to track single building elements and data along the interoperability process. In this project, only two exchange models were analyzed for communication between models, but this method gives the researcher the possibility to relate them in order to find the more satisfying combination of geometries and data visualization.

Step 2: 3D Tiles Generation

The second step involves the conversion and standardization of data giving the Batched 3D Models (B3DM) tileset as output. For the 3DTiles generation, a custom open-source converter software developed in the beta version was used as Middleware (Massara et al., 2019). This includes IFC and gbXML Parser, which converts the model in a 3D mesh surfaces with associate attributes, and the 3DTiles generator, which creates a tileset.json file and a B3DM asset for each model. These parts are implemented with Oracle Java Platform, Enterprise Edition 8 technology (Oracle, 2020). The software is implemented with a decoding tool that through specific query s able to detect and analyze the IfcPropertySet contained in the model and recognize in which visualization level the attributes contained must be shown. The software can suddenly parse the IfcPropertySingleValue extrapolating the "Name" and "Value" of the parameters. In this way, the same object of the model can show different attributes according to the LoD represented on *Cesium*. Each category and type of object requires a dedicated parsing procedure for the extraction of geometry and attributes. It is a long time-consuming road for mapping the full IFC with API.

Step 3: Dynamic Visualization Dashboard

This 3DTiles asset can now be loaded through the glTF format the on the *CesiumJs 1.60* framework, thanks to a HTML calls which load all the files to a specific web server. In addition to the BIM models, the shapefile of the surrounding urban context was imported through the same tile format. In *Cesium*, the server-side sends the B3DM data with attributes based on which the styling can be accomplished in client-side without further data requests from the server, which improves the user experience and reduces the computation load in the server. Figure 8 shows the dashboard interface. In the top-right corner, it is possible to see the view's navigation tools. Pointing the mouse over an object, this will be highlighted and a cloud panel showing the properties attached to the object will appear. The zoom of the view controls the level of detail. A first notable difference in the use of the two formats is that gbXML is not able to handle the original object modelled in Revit, but only the portions with which adjacent rooms come into contact. For example, for LOD 2, which refers to a conceptual representation of the building, the geometry from the district shapefile must be used. Conversely, different entities depending on the LOD can be used from an IFC export. Besides, the colours used for the rendering derive from IfcColor class of the Revit model as RGB coordinates. As in the gbXML case, some geometries can be displayed as transparent surfaces in specific points of view. This issue is probably due to the manageability of the normal attribute object of the glTF model of the building. With the IFC model, the problem is accentuated by the fact that the objects are more complex and composed of a lot of planar surfaces. Concerning the way geometries are generated, discontinuities among analytical surfaces may be resulting in a disconnected representation. Based of its light and ease geometrical structure, gbXML is an interesting way to represent BIM models on a GIS platform. However, its structure based on planar surfaces is inadequate for the visualization and interrogation of detailed building element, and it is limited to a low detail display. In the case of IFC, some difficulties or errors are given by the return of complex objects. Taking windows as an example, the dashboard is currently unable to represent this kind of object, since they are composed by several elements such as glass, frame, handle, each with a particular representation and its reference system. Nevertheless, the higher potential is recognized in the use of IFC for the quality of geometries and attributes managed.

Figure 8. Cesium Dashboard interface

BIM-IoT Energy Simulation and Monitoring

As the existing heritage is one of the focal points of this study, BIM was also exploited to perform energy simulations and certifications in order to promote retrofit actions. As they represent a significant repository of graphical and alphanumeric information, parametric models can include useful data for energy analysis such as: (i) accurate building envelope characterizations in terms of correct stratigraphy, thermal and physical properties; (ii) space typology and occupation profile; (iii) materials; (iv) system schemes and terminal typology; (v) attached documentation relating to interventions and operation. Traditionally the energy modelling activity is carried out according to a traditional approach by manually imputing all the needed information to describe a building and by redesigning its geometry for every plant view. This procedure implies a considerable amount of time, mainly when applied to an extensive heritage and increase the probability of making errors. These difficulties can be overcome thanks to the BIM methodology, optimizing the process and making it integrated. In the ideal case, the geometry and envelope characteristics of the building can be extracted directly from the architectural model to evaluate the energy performance and execute thermal calculations, minimizing misinterpretations and incorrect approximations encountered in practice. Several interoperability tests towards energy-dedicated software have been carried out with the main aim of evaluating the quality of the data exchange. *Edilclima* and

DesignBuilder were selected in order to consider two different types of simulation engines within the stationary and dynamic model calculations method, respectively. It is necessary to admit that a total exchange can not be reached with the current level of software development. For this reason, a range of solutions was developed to optimize the exportation and to speed up the data input phase. That includes modelling guidelines to simplify the architectural model concerning non-energy significant elements and to represent spaces, joints between elements, and complex geometries. Most of the interoperability problems affect any building model, while some specific ones are related to historical or modern buildings due to their particular construction typology. For instance, thermal simulations are based on one-dimensional heat transfers, so they cannot consider curved surfaces – like walls, slabs or ceilings –, which must be approximated with a finite number of flat surfaces (Ugliotti et al., 2016). Since many information contained in the model are not transferred, such as transparent component or shading elements, shared parameters and customized families (Barone, 2017) are used to make the data explicit and more easily accessible to support the filling in the energy software. As mentioned above, the gbXML export must be carefully checked to avoid the analytical surface. The level of adoption of the BIM in the energy field is conditioned by many variables, including the complexity of the building, the analysis to be carried out and the software used, the level of experience of the modeller, if the process is moving towards progressive automation.

Moreover, the building energy modelling and monitoring approach is one of the most challenging as it combines the knowledge of the buildings established by BIM with real data from the field, collected by the IoT devices. This aspect represents the core of the EEB project that pursues building management mainly devoted to energy-managers providing near-real-time and historical information on environmental parameters of buildings and spaces. This integration enables to validate the Building Energy Model (BEM) with real data from monitoring activities beside energy consumption bills. An iterative energy optimization process (Bottaccioli et al., 2017) has studied to manage the refurbishment scenarios starting from the correlation of the *EnergyPlus* simulation engine output and the indoor air temperature and humidity collected and sent in the Store Manager every 15 minutes. As a strong point of these simulations, third-party weather data source from the nearest weather station (i.e. solar radiation, outdoor air temperature and humidity) was used instead of the default Typical Meteorological Year (TMY), which is not representative of the real weather conditions. According to this approach, validation is much more reliable, given the ongoing climate change. Comparing the air temperature charts, measured data and simulation results with real-weather conditions have the most similar trend compared to the TMY simulation. Analyzing the temperature and consumption trends, factors that may affect the energy model can be identified, such as user behaviours, malfunctions and anomalies in the system. By comparing measured and simulated data, it is possible for example, to discover non-regular trends of real indoor temperatures due to faults in on/off schedules of the heating system or efficiency losses of the building-system.

In this context, BIM models can be used to evaluate different refurbishment scenarios (e.g. external/internal coat application, fixtures replacement and power peaks regulation), becoming a new input immediately available for the simulation. This process is iterative and can help building and energy-managers in evaluating the best solution for both energy performances and Return of Investment. As Information and Communication Technologies and Machine Learning techniques look as crucial players in the development of new control policies based on systematic knowledge and prediction of energy behaviour, such as Demand/Response and Demand Side Management applications, an indoor air temperature forecasting method has been studied exploiting a non-linear autoregressive neural network

Figure 9. Proposed energy modeling optimization process (Bottaccioli et al., 2017)
Source: (Bottaccioli et al., 2017)

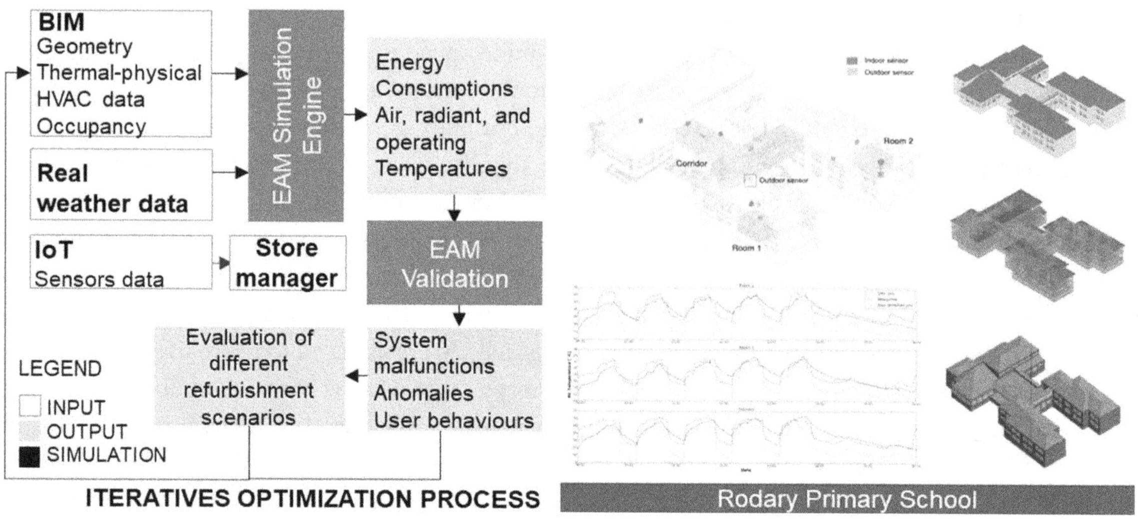

(NAR). It bases its predictions on past values of the series and current and past values of the exogenous driving series, producing the error of prediction. The neural network has been trained and validated with a dataset consisting of six years (four training-set and two validation-set) of indoor air-temperature values (real and synthetic, one sample every 15 minutes) of a building demonstrator (Aliberti et al., 2018). As the monitoring period foreseen by the project was too short for this type of activity, the BIM model and *EnergyPlus* were used again this time to generate a consistent artificial database. The goal is to predict the values of indoor air-temperature (that corresponds to the ahead k-step prediction of the system) in the most extended period with the best accuracy compared to the original output. The same environments considered for the validation of the energy model mentioned above were studied. The analysis of performance indicators highlighted an overall good performance in predicting temperature values up to two hours with 13 regressors. This kind of approach allows energy profiles prediction, for example, to exploit the flexibility of electro-thermal devices or to reshape thermal energy to reduce the peaks in district heating.

BIM-IoT Energy Data Visualization

As a further opportunity to interconnect data from different databases, a functionality to recall and query data from IoT sensors within the BIM environment was implemented. In particular, *Dynamo* software in plug-in mode from *Autodesk Revit* was used to bring dynamic data recorded in *InfluxBD* into dialogue with static data from the digital model. *Dynamo* allows users to operate through Visual Programming processes through which it is possible to link elements to define relationships and sequences of actions that make up custom algorithms. *Dynamo* hosts a rich ecosystem for development and allows the user to create web requests for data access to other servers/clouds. The aim is to retrieve the temperature and humidity data from the installed sensors of the different case study and map them as parameters within the BIM models.

Using this mode, the values measured in the field can be interrogated from a graphical database. To achieve this, first of all, the sensors have been modelled in a simplified way and placed in virtual environments, in the same position as the real ones. Each sensor object is identified within the model by a unique *Revit* identifier and a code that identifies it with respect to *InfluxDB*. The integration methodology, as illustrated in Figure 10, foresees the following procedural steps: (i) connection to the external resource and web request, (ii) correlation of data from an external resource and BIM data, (iii) populating data in the BIM model from external resource; (iv) historical and real-time Temperature and Humidity data query.

Figure 10. BIM and IoT data correlation methodological scheme

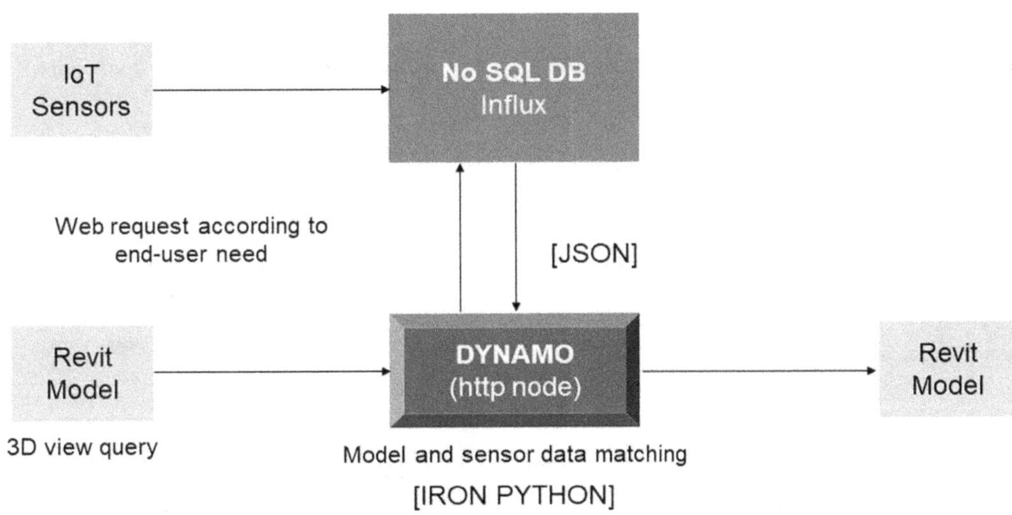

Dynamo, thanks to its ability to interface directly with the modeling software used, makes it possible to manipulate the BIM database in an integrated way, extending its potential. The connection to new external resources, instead, is managed through REST communication interface. The script implemented, reproduced in Figure 11, performs the following actions through the use of nodes and functions.

- Connection and authentication to the *InfluxBD* non-relational database.
- Creation and execution of an HTTP information request to the Web Server.
- Control of the HTTP response of the query and the execution time of the request.
- Extraction and deserialization of the response content through JSON files.
- Selection of the values of interest (Temperature, Humidity and Timestamp) according to real time and by manual query with date and time for the historical series.
- Population of the corresponding attributes associated to the sensor objects with the extracted values.

In the specific case, each web request recalls the measured value range of an individual sensor. Therefore, two http requests have been set for each sensor, one for Temperature values and the other for Humidity. In the case of Rodari Primary School, the data corresponding to the 15 sensors installed were extracted. The values are updated every time the script is executed. As far as the historical series values are concerned, it is possible to specify the date and time by means of classic input from script or *Dynamo Player* interface from *Revit* model. Once the data is mapped inside the BIM model, it can be displayed in different modes. In addition to being visible among the properties of the sensor objects and present of the schedules, it is possible to visualize them graphically through labels of the sensor objects. Through an additional *Dynamo* script the Temperature and Humidity data mapped on the sensor objects have been transferred to the rooms where they are contained. In this way, it has been possible to create thematic maps, as visible in Figure 12, with the measured values at a given time and to obtain an overall view.

BIM-AR-3D Printing Energy User Awareness

Figure 11. Dynamo script for populating temperature and humidity data on sensor objects

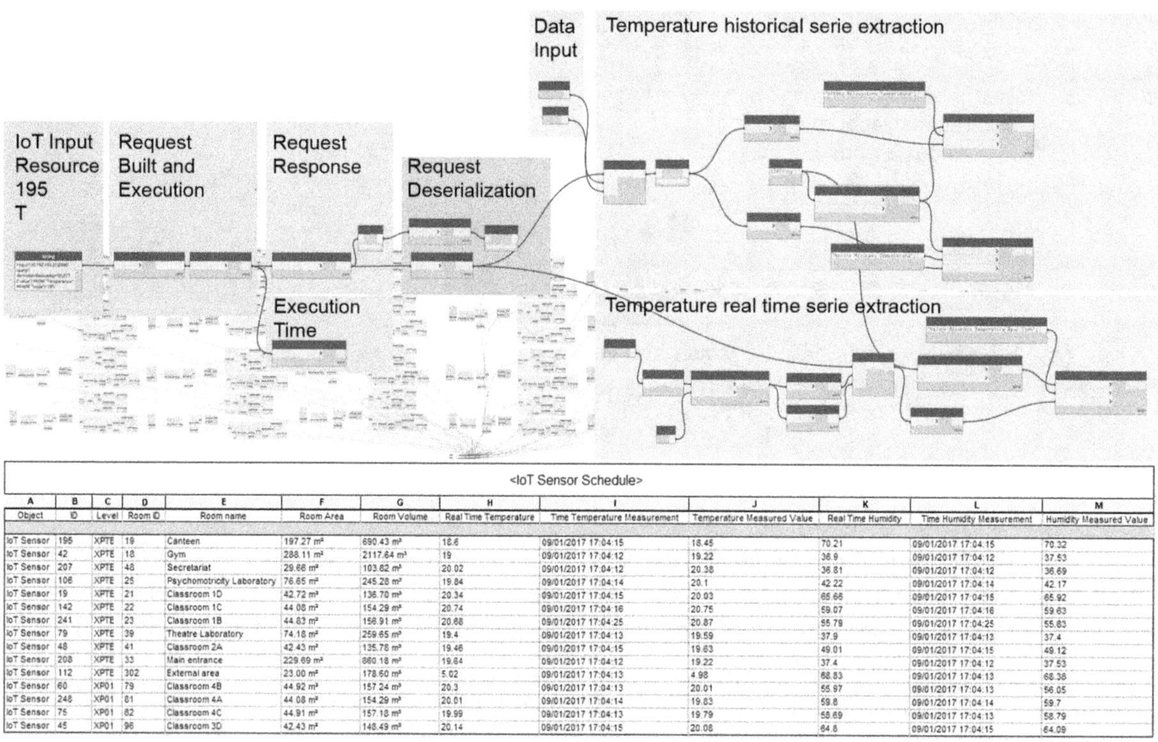

<IoT Sensor Schedule>

A	B	C	D	E	F	G	H	I	J	K	L	M
Object	ID	Level	Room ID	Room name	Room Area	Room Volume	Real Time Temperature	Time Temperature Measurement	Temperature Measured Value	Real Time Humidity	Time Humidity Measurement	Humidity Measured Value
IoT Sensor	195	XPTE	19	Canteen	197.27 m²	690.43 m³	18.6	09/01/2017 17:04:15	18.45	70.21	09/01/2017 17:04:15	70.32
IoT Sensor	42	XPTE	18	Gym	288.11 m²	2117.64 m³	19	09/01/2017 17:04:12	19.22	36.9	09/01/2017 17:04:12	37.53
IoT Sensor	207	XPTE	48	Secretariat	29.66 m²	103.82 m³	20.02	09/01/2017 17:04:12	20.38	36.81	09/01/2017 17:04:12	36.69
IoT Sensor	106	XPTE	25	Psychomotricity Laboratory	76.65 m²	245.28 m³	19.84	09/01/2017 17:04:14	20.1	42.22	09/01/2017 17:04:14	42.17
IoT Sensor	19	XPTE	21	Classroom 1D	42.72 m²	136.70 m³	20.34	09/01/2017 17:04:15	20.03	65.66	09/01/2017 17:04:15	65.92
IoT Sensor	142	XPTE	22	Classroom 1C	44.08 m²	154.29 m³	20.74	09/01/2017 17:04:16	20.75	59.07	09/01/2017 17:04:16	59.63
IoT Sensor	241	XPTE	23	Classroom 1B	44.83 m²	156.91 m³	20.68	09/01/2017 17:04:25	20.87	55.79	09/01/2017 17:04:25	55.83
IoT Sensor	79	XPTE	39	Theatre Laboratory	74.18 m²	259.65 m³	19.4	09/01/2017 17:04:13	19.59	37.9	09/01/2017 17:04:13	37.4
IoT Sensor	48	XPTE	41	Classroom 2A	42.43 m²	135.78 m³	19.46	09/01/2017 17:04:15	19.63	49.01	09/01/2017 17:04:15	49.12
IoT Sensor	208	XPTE	33	Main entrance	229.69 m²	860.18 m³	19.64	09/01/2017 17:04:12	19.22	37.4	09/01/2017 17:04:12	37.53
IoT Sensor	112	XPTE	302	External area	23.00 m²	178.60 m³	5.02	09/01/2017 17:04:13	4.98	68.83	09/01/2017 17:04:13	68.38
IoT Sensor	60	XP01	79	Classroom 4B	44.92 m²	157.24 m³	20.3	09/01/2017 17:04:13	20.01	55.97	09/01/2017 17:04:13	56.05
IoT Sensor	248	XP01	81	Classroom 4A	44.08 m²	154.29 m³	20.01	09/01/2017 17:04:14	19.83	59.8	09/01/2017 17:04:14	59.7
IoT Sensor	75	XP01	82	Classroom 4C	44.91 m²	157.18 m³	19.99	09/01/2017 17:04:13	19.79	58.69	09/01/2017 17:04:13	58.79
IoT Sensor	45	XP01	96	Classroom 3D	42.43 m²	148.49 m³	20.14	09/01/2017 17:04:15	20.08	64.8	09/01/2017 17:04:15	64.09

The engagement of ordinary users towards energy-oriented behaviours is one of the objectives of the dissemination activities of research projects (De Luca et.al, 2017). Since young generation will be the society of the future and one of the case studies is a school, the gamification approach was considered to stimulate children to take an active role in sustainability issues. Specifically, an interactive BIM-based map was created to activate a guided learning path on smart cities strategies exploiting AR and 3D printing. In the first task children can experience VR through the use of customized cardboard to visualize the three dimensional models of the EEB case studies and their surroundings, compare them with 3D printed version. By framing the correspondent QR Code and putting the smartphone in the viewer, their immersive vision is enabled through a stereoscopic rendered view generated by Autodesk

Figure 12. Thematic map of measured temperatures

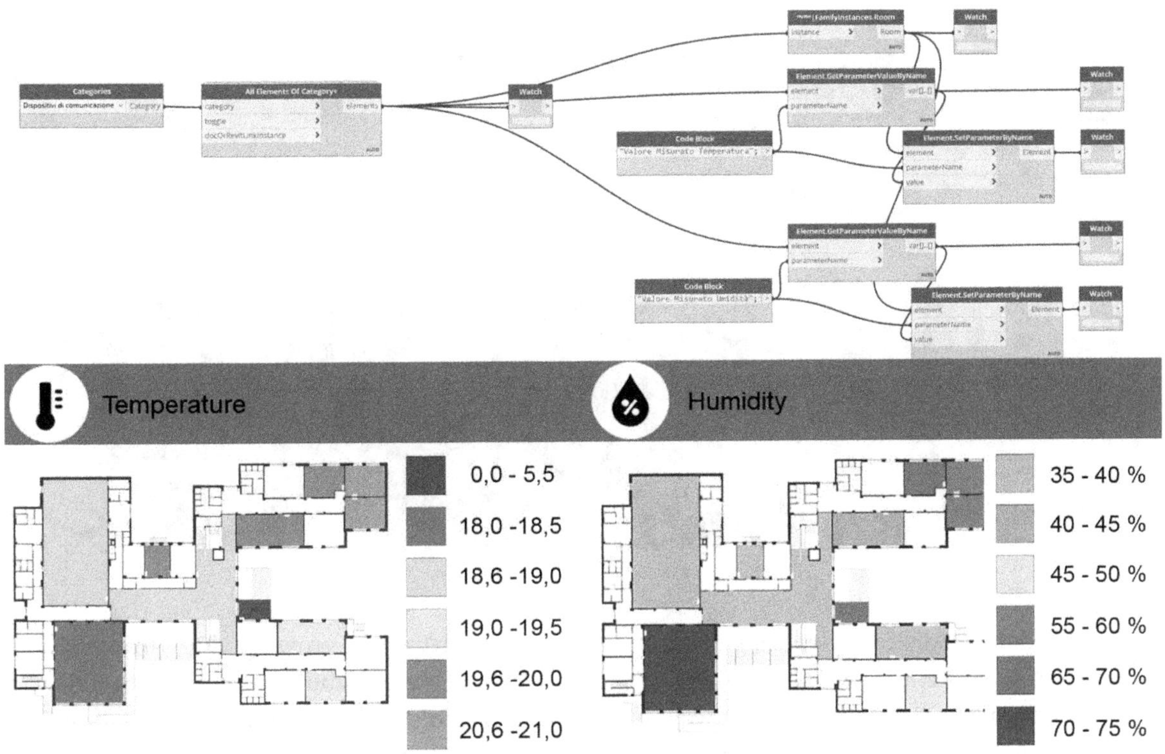

360 cloud platform. As output, players learn to observe the urban context, understanding that there are several fields of application besides buildings, considering for example green areas, squares and streets, up to consider the entire city. In the second task, smart city strategies are investigated through a quiz. Questions are displayed in AR on tablet or smartphone with the use of Aurasma application by framing the training points on the map. Clicking on the device screen at the correct answer, the solution and a brief suggestion for sustainability are displayed. Finally, the last task allows children to experience Augmented Reality by recomposing a jig saw puzzle of the school district. The model was exported through .FBX in Autodesk 3DS Max to allow the AR visualization through the AR-media plug-in. This activity encourage people to work and cooperate together to create something new for a more efficient city and making the system the positive experiences. From an educational point of view, the map helps children both to emphasize the importance and effectiveness of the strategies traditionally used and to encourage the application of new technologies to achieve the targets of a smart city.

FUTURE RESEARCH DIRECTIONS

As many aspects of smart environments are already in use through internet-controlled building management systems (e.g. lighting and temperature regulation within buildings, traffic management systems, real-time information about means of transport), managing the building/city physical assets represent a critical field of technological application to enable better decisions about maintenance and usage what-if

Figure 13 EEB 3D BIM Map (De Luca et.al, 2017)
Source: (De Luca et.al, 2017)

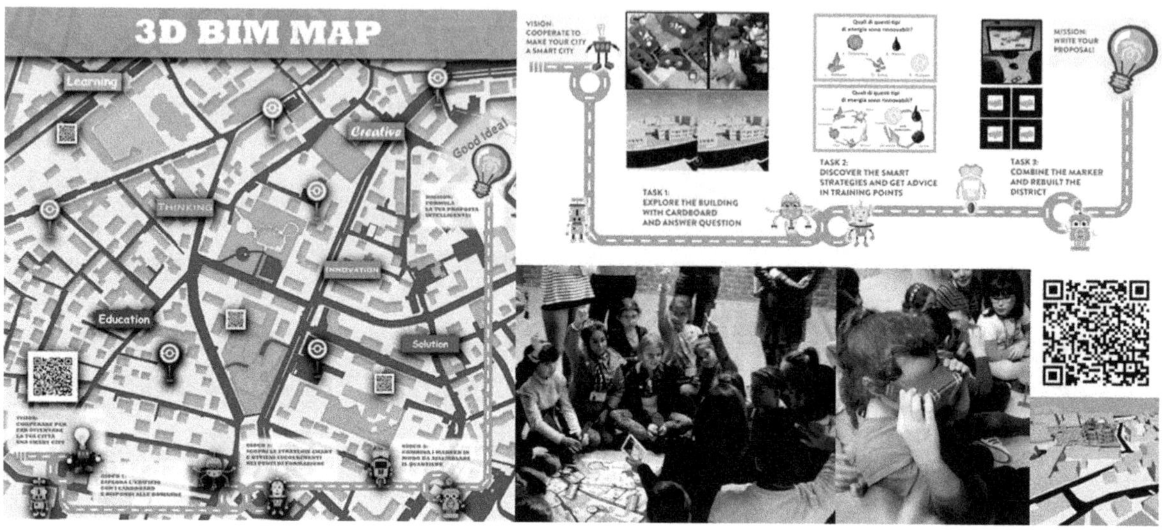

scenarios. Thanks to the experiences gained in other sectors, there is a growing shift from a sustainable static vision of a city towards a resilient dynamic, in which the possibility of human-digital model interaction becomes central. In the same way that the IoT is rapidly spreading through the setting up of increasingly innovative services, experiences such as those illustrated in this research should be encouraged to enable increasingly smart city digital twin-oriented framework. The spatial geometric correlation of dynamic sensor data can enrich not only energy analyses but also structural and seismic evaluations. In this context, in addition to ordinary aspects, a new vision opens up for the emergency management such as earthquake, damages in conflict area, pandemics like Covid-19, which can be monitored before and better managed afterwards. The extension of the controlled parameters and their relationships will increase the resilience of the city.

CONCLUSION

This chapter presents some of the most significant results achieved by the EEB project. The study, while considering strictly energy-related aspects, launches a methodology for the collection, digitalization and integration of information in a framework queryable by different levels of users depending on the tools and services used. The most exciting aspect goes beyond the vertical management of specific domains, but rather lie in the possibility to relate them to each other, constituting an enlarged knowledge environment. The main areas covered involve GIS for the representation at territorial level, BIM for the building scale, IoT sensors for energy and electrical monitoring and the use of innovative visualization technologies such as Augmented and Virtual Reality. The focus is still on the sensorization of 3D models, but the experimental matrix is starting to exploit the capabilities of the Internet of Things and Artificial Intelligence. The most complex aspect to address is certainly to establish an environment that makes data sources accessible that are not homogeneous in terms of assets, scale, and content. Owners, managers, technicians, citizen: everyone is looking for data. It all starts with content selection and storage. For cost and processing time savings, standardization of digital models built in BIM and GIS environments should be aimed at making the data immediately accessible and comparable. In this way, building data become descriptors useful to establish a multidisciplinary picture of an asset. Moving towards a cyber-city allows to democratize the access to information extending the area management capabilities. The combination of tools tested creates a structure able to collect under the same batch different versions of the same model, overcoming the static representation of the scene in favour of a customizable smart 3D streaming where significant advantages in terms of rendering speed and fruition can be experienced. This solution thanks to its web-based nature enable maximum effect in communications and cognitive awareness. In this direction, software must also work in order to be more and more open and interoperable in order to save optimize the analysis phases of alternative scenarios more and more performing, keeping all the necessary parameters under control. The context that comes to outline provides an improvement of shared knowledge within profiles awareness.

ACKNOWLEDGMENT

This research titled "Edifici a Zero consumo Energetico in Distretti Urbani Intelligenti" (EEB) was supported by the Italian Ministry of Education, University and Research for the Cluster 2013 – 2016 [grant number CTN01_00034_594053]. The author is pleased to thank the Master student Fabrizio Provera and the PhD Sara Torabi Torabi Moghadam for permission to exhibit their work.

REFERENCES

Aliberti, A., Bottaccioli, L., Cirrincione, G., Macii, E., Acquaviva, A., & Patti, E. (2018). Forecasting short-term solar radiation for photovoltaic energy predictions. *Proceedings of the 7th International Conference on Smart Cities and Green ICT Systems*, Volume 1: SMARTGREENS, 44-53. 10.5220/0006683600440053

Aliberti, A., Ugliotti, F. M., Bottaccioli, L., Osello, A., Macii, E., Patti, E., & Acquaviva, A. (2018). Indoor air-temperature forecast for energy-efficient management in Smart Buildings. *IEEE Proceedings*, 272-277.

APE. (2015). https://www.certificato-energetico.it/

Aurambout, J., Pettit, C. J., & Lewis, H. (2008). *Virtual Globes: the Next GIS?* Landscape Analysis and Visualisation.

Barone, L. (2017). Interoperabilità BIM per l'analisi energetica. In BIM: verso Il catasto del futuro. Conoscere, digitalizzare, condividere. Il caso studio della Città di Torino, (pp. 105-110). Gangemi Editore spa.

Bottaccioli, L., Aliberti, A., Ugliotti, F. M., Patti, E., Osello, A., Macii, E., & Acquaviva, A. (2017). Building energy modelling and monitoring by integration of IoT devices and Building Information Models. *Building Digital Autonomy forna Sustainable World (COMPSAC),* 914-922.

buildingSMART International. (2020). https://www.buildingsmart.org/standards/bsi-standards/industry-foundation-classes/

Chelleri, L., & Olazabal, M. (2012). *Multidisciplinary perspectives on urban resilience. Workshop Report.* Bilbao, Spain: BC3, Basque Centre for Climate Change.

Communityviz, (2020). https://communityviz.city-explained.com/

Cozzi, P. (2015). *Introducing 3D Tiles.* https://cesium.com/blog/2015/08/10/introducing-3d-tiles/

De Luca, D., Del Giudice, M., Dellosta, M., Fonsati, A., Osello, A., & Ugliotti, F. M. (2017). Augmented and virtual reality for smart cities users' awareness. In *INTED 2017, 11th International Technology, Education and Development Conference*, 4333-4342. IATED Academy.

Dong, B., Lam, K. P., Huang, Y. C., & Dobbs, G. M. (2007). A comparative study of the IFC and gbXML informational infrastructures for data exchange in computational design support environments. *Proceedings: Building Simulation, 2007*, 1530–1537.

European Commission. (2020). *Horizon 2020 Work Programme 2018 – 2020* (European Commission Decision C(2020) 1862 of 25 March 2020). https://ec.europa.eu/research/participants/data/ref/h2020/wp/2018-2020/main/h2020-wp1820-fet_en.pdf

European Parliament. (2014). *Mapping Smart City in the EU.* https://www.smartcities.at/assets/Publikationen/Weitere-Publikationenzum-Thema/mappingsmartcities.pdf

IBM Corporation, . (2012). *IBM Smarter Cities. Creating opportunities through leadership and innovation.* Author.

ISO/IEC JTC 1. (2014). *Smart cities. Preliminary Report 2014.*

ISTAT. (2011). https://www.istat.it/it/archivio/104317

Karan, E. P., Irizarry, J., & Haymaker, J. (2015). BIM and GIS Integration and Interoperability Based on Semantic Web Technology. *Journal of Computing in Civil Engineering, 30*(3), 04015043. doi:10.1061/(ASCE)CP.1943-5487.0000519

Massara, F. (2016). Trying to visualize GIS & BIM information on the web: a solution using Leaflet and Cesium. *FOSS4G Bonn 2016.*

Massara, F., Provera, F., & Ramassotto, A. (2019) Streaming and rendering the Turin 3D geospatial content through GIS and BIM integration. *FOSS4G Bucharest 2019.*

Mitchell, W. J. (2001). *Smart Cities.* https://smartcities.media.mit.edu/

OGC. (2020). https://www.ogc.org/standards/3DTiles

Oracle. (2020). https://www.oracle.com/technetwork/java/javaee/tech/index.html

PiemonteC. S. I. (2019). https://github.com/csipiemonte/ogc-citygml-utils

Provera, F. (2017). *Implementation of BIM methodology for district data visualization.* Master Thesis in Building Engineering, Politecnico di Torino.

Riddhi, D., Siddhi, D., & Hemal, T. (2020). Digital Twins: Current problems in Smart City and Recommendations for future technology. *International Research Journal of Engineering and Technology, 7*(5).

Shah, S. A., Seker, D. Z., Rathore, M. M., Hameed, S., Ben Yahia, S., & Draheim, D. (2019). Towards Disaster Resilient Smart Cities: Can Internet of Things and Big Data Analytics Be the Game Changers? *IEEE Access: Practical Innovations, Open Solutions*, *7*, 91885–91903. doi:10.1109/ACCESS.2019.2928233

Teicholz, P. (2013). *BIM for Facility Managers*. John Wiley & Sons. doi:10.1002/9781119572633

The British Standards Institution. (2014). Smart cities – Vocabulary. *PAS*, *180*, 2014.

Torabi Moghadam, S., & Lombardi, P. (2019). An interactive multi-criteria spatial decision support system for energy retrofitting of building stocks using CommuntiyVIZ to support urban energy planning. *Building and Environment*, *163*, 106233. doi:10.1016/j.buildenv.2019.106233

Torabi Moghadam, S., Lombardi, P., & Mutani, G. (2017). A mixed methodology for defining a new spatial decision analysis towards low carbon cities. *Procedia Engineering*, *198*, 375–385. doi:10.1016/j.proeng.2017.07.093

Torabi Moghdam, S., Lombardi, P., Ugliotti, F. M., Osello, A., & Mutani, G. (2016). BIM-GIS Modelling for Sustainable Urban Development. *NEWDIST*, *2016*(Special Issue), 339–350.

Ugliotti, F. M. (2017). Utilizzi del modello BIM per il FM. In BIM: verso Il catasto del futuro. Conoscere, digitalizzare, condividere. Il caso studio della Città di Torino, (pp. 56-73). Gangemi Editore spa.

Ugliotti, F. M., Dellosta, M., & Osello, A. (2016). A BIM-based energy analysis using Edilclima EC770 plug-in. Case study: Archimede Library, EEB Project. *Procedia Engineering*. *WMCAUS*, *6*(161), 3–8. doi:10.1016/j.proeng.2016.08.489

Vinco, S., Bottaccioli, L., Patti, E., Acquaviva, A., Macii, E., & Poncino, M. (2018). *GIS-Based Optimal Photovoltaic Panel Floorplanning for Residential Installations*. Paper presented at 2018 Design, Automation & Test in Europe (DATE). 10.23919/DATE.2018.8342049

White, G., Zink, A., Codecá, L., & Clarke, S. (2021, March). A digital twin smart city for citizen feedback. *Cities (London, England)*, *110*, 103064. doi:10.1016/j.cities.2020.103064

Wright, L., & Davidson, S. (2020). How to tell the difference between a model and a digital twin. *Advanced Modeling and Simulation in Engineering Sciences*, *7*(1), 13. doi:10.118640323-020-00147-4

KEY TERMS AND DEFINITIONS

3D Tile: Technology used to streaming and rendering massive heterogeneous 3D geospatial datasets such as Photogrammetry, 3D Buildings, BIM models, Instanced Features, and Point Clouds according to different Hierarchical Level of Detail.

Digital Knowledge Site: A constantly evolving process of retrieval, organization, preservation and updating of the building knowledge in a digital environment considering both material and immaterial aspects.

Dynamic Data: Data from monitoring systems or information that does not remain unchanged over time, including emotional and intangible aspects.

Energy Model: A virtual or computerized simulation of a building realized with specialized software, focusing on energy performance calculation or assessment.

Smart City Digital Twin: Digital three-dimensional replica of the physical assets that make up the urban environment (buildings, urban infrastructure, utilities, movement of people and vehicles) integrated with real-time information which is structured as full-fledged system of interconnected things capable of interacting with humans not only to monitor but to prototyping and predict.

Urban Model: Computer-based digital environment used for visualizing the distribution of indicators on graphical maps and testing planning policies on the future form of cities.

Visual Programming: Computer language that allows programming activities even for non-expert users through the graphic manipulation of the elements and not through written syntax.

Chapter 11
Connected BIM Models Towards Industry 4.0

Daniela De Luca
Politecnico di Torino, Italy

Monica Dettori
Politecnico di Torino, Italy

Matteo Del Giudice
Politecnico di Torino, Italy

Anna Osello
Politecnico di Torino, Italy

ABSTRACT

In the era of the fourth industrial revolution, the cyber physical systems, intended as enabling tools to generate an autonomous system, able to facilitate the relationships between different and distant objects and subjects, allow to digitalize the production system, in order to better outline what constitutes the smart factory. The benefit of such systems is the ability to associate to physical objects and virtual or digital models useful information related to the analyzed object, such as life cycle, geometry, mechanical properties, and parameters related to management and maintenance. This contribution aims to evaluate building information modeling methodology in the industrial context, as a cyber-physical system, developing flexible 3D parametric models as a data set, where information can be visualized and optimized management, using different visualization tools. The research has underlined the importance to share information between virtual and real worlds through virtual and augmented reality (VAR) systems.

DOI: 10.4018/978-1-7998-7091-3.ch011

INTRODUCTION

The fourth industrial revolution is in progress and is evolving, it is characterized by a series of physical and digital technologies such as artificial intelligence, cloud computing, adaptive robotics, augmented reality, additive manufacturing, and Internet of Things (IoT). The term Industry 4.0 represents one of the most widespread fields of research in recent years. (Ustundag & Cevikcan, 2018)

Industry 4.0 embraces a range of topics from mechanization, automation, digitalization, networking to miniaturization. (Lasi et al. 2014)

New technologies and digital innovations have enabled the industrial sector to improve production processes, plant logistics, and energy efficiency. At the same time, progress in technology has facilitated the management of huge amounts of data and, above all, has allowed developing analytics based on big data sets to optimize production quality, to save energy, and to improve equipment service. Data from different sources, from equipment and production systems to business and customer management, are increasingly being collected, analyzed, and subsequently used to support choices. (Rüßmann et al., 2015)

In this context, data warehousing is a technological trend for business decision support process. It focuses on collecting, cleaning, and storing large volumes of information, and it is possible to fix three main uses of data warehouse. First, it is used for standard reports and charts presentation and allows data from different transaction systems to be consolidated in the warehouse and used in reporting. Second, it supports a type of query and reporting called dimensional analysis, and it can facilitate the comparison of the results between different dimensional values in particular time periods. Third, data warehouse enables a new technology called data mining that can automatically recognize patterns in data that can help end-users to describe existing data and predict future behavior. Data warehousing's success depends on the use of online information retrieval, artificial intelligence, and graphical user interface tools. (Ma et al., 2000)

From the point of view of data management, metadata plays a key role on the digital archive. They describe company data's meaning and structure and how they are created, consulted and used. (Devlin, 1997)

The following functions can be highlighted:

- Directory to help the analyst identify the contents of the data warehouse;
- Guide to mapping data when it is transformed from the operating environment to the data warehouse environment;
- Guide to algorithms used as a summary between current detailed data, slightly summarised, and highly summarised data.

The more data is stored, the more cloud and visualization systems will have to manage this amount of metadata in the best way, finding solutions and technologies suitable for the target purpose. (Dolk & Kirsch, 1987)

Moreover, visualizing data management through digital models allows us to show metadata through simple graphic representations. The interaction of them in real-time and the communication of the changing spaces with the management system allow us to optimize the developed digital models' connection. (Brooks, 1997)

Metadata and its efficient management requires the adoption of innovative methodologies to make its structure clear and connected to different uses and targets. For this reason, Building Information

Modelling (BIM) can be introduced as a technological process capable of optimizing the connection between the real and virtual worlds.

This contribution focuses particularly on using BIM as a repository of different kinds of data and, thus, a digital collection of industrial building data during its life cycle. The BIM methodology has also been used as a Cyber-Physical System (CPS) to develop new scenarios for energy efficiency and internal logistics to optimize an industrial plant. Multiple databases can be inserted and connected to cloud platforms that can also return data from sensors outside the digital model. Recording a huge amount of information needs a virtual model that constantly monitors and synchronizes data from all the platforms connected.

The industrial environment is characterized by an increase in heterogeneous data that requires optimized management by users to improve its use and sharing. Therefore, in the developed connected model, interactions between data and users are made possible by interoperable systems and user-friendly platforms.

Organization and energy efficiency are two major issues towards Industry 4.0, for this reason the information analyzed within this article relates to the energy sector and the plant's production or logistics chain.

BACKGROUND

The rise of new technologies and the increasing use of digital systems have created new challenges in different areas, such as industry and construction. Production operations' digitalization and industrial plants' energy efficiency have implemented new skills and a strong increase in the automation of information processes. In this regard, new technological solutions have been developed to generate modular and efficient processes, to optimize resources and reduce waste, especially in the energy and logistics sectors. Many factors affect social, economical and political transformation, and define new strategies applied in the short term and with strong effects also in the long term. (Lasi et al. 2014)

A virtuous example is the energy sector that, in recent years, has seen an increase in the use of sensors and efficient data collection techniques for the rationalization of energy resources.

Optimizing resources efficiency involves, on one hand, improvement of product management within the building and, on the other hand, social changes that require more attention to the sustainability of energy resources. A second example in which several factors influence products' production and storage is the automation of items within specific areas. In this context, different technologies are used to identify and localize products within different industrial departments, such as QRCode and voice assistant, to connect specific devices, update data, and reduce human errors. The constant progress has highlighted the need to use Cybernetic Systems and Digital Twin Virtual models. It monitors and synchronizes information constantly thanks to its connection to a network that enables any updates and changes to the real model. By leveraging Information and Communications Technology (ICT) systems, these virtual models can capture data from the real world and modify them as needed. These models can be connected to different nature sensors that generate a data flow, which is difficult to manage with traditional systems. The sensor connected to the network generates a multitude of interchangeable data at any time and at any condition. It creates benefit by transforming the data into useful information for the process. (Facchinetti & Della Vedova, 2011)

In this sense, the use of new technologies generates capacitive virtual models that exploit the systems' dynamism to collect and transfer more and more information. In fact, the amount of data, called Big Data (Lee et al., 2013), can be coordinated thanks to ICT platforms that connect data within physical models.

In recent years, technological systems have changed significantly and are increasingly influencing the industry in processing data. The future society is looking to improve the way data is collected and how it is produced and managed from different sources. For these reasons, in literature, it is possible to classify the ten main trends of Technological Innovation (IT), that help users consider data as aggregate forms and tools, that can simplify daily operations. (Gartner, 2020)

Among these trends, the most significant that can best help interconnect users, and digital models are: Internet of Behavior (IoB), Total experience, AI engineering and Machine Learning and Hyperautomation. The behaviors that humans can assume in the face of technological evolution are strongly linked to feedback and scalability of the instrumentation used. The totality of the lived experience will improve the complexity of the Information Management System and the technologies deployed. The development of machine learning techniques combined with artificial intelligence facilitates data transmission reliability within the Digital Twin, making it scalable and more manageable over time. The use of automation systems in traditional processes introduces innovative processes that allow Industry 4.0 to incorporate user needs better. Still, at the same time, it can require great efforts of economic and social human resources.

These trends are also affecting the definition of CPS because of their cross-cutting nature across multiple industries and uses. The most important definition in literature, like the methodology used in this paper, is the use of CPS as the ability of physical elements to communicate and interact with the real world, thus collaborative computational element systems that control physical objects. (Baheti & Gill, 2011)

Through these systems, three sequential scenarios are generated that compose and outline the importance of CPS development:

- Generating and acquiring data from the real world through sensors applied to objects or people;
- Generate the acquisition of data in a single database through specific software;
- Creating tools to support the decision-making process.

With virtual models, simulations can be done to determine the causes of possible problems and manage the system by different users even remotely. (Bloem et al., 2017)

In recent years, simulations have not only been used by computer experts but they have become a daily tool used by engineers as design, test, and validation tools. (Rosen et al., 2015)

In Architecture, Engineering, and Construction (AEC) Industry, Building Information Modeling acts as a container for all building information and can be used for building performance and logistics management. Buildings use up to 40% of global primary energy and contribute to more than 30% of CO_2 emissions (Costa et al., 2013); therefore, several studies have been carried out. These ones aim to reduce the energy consumption of buildings through the installation of sensors and to create a Building Energy Management System (BEMs) to monitor data inside the building. The reduction of consumption can be carried out by evaluating and forecasting energy consumption, and these new technologies allow us to support energy-saving decisions. (Zhang et al., 2020)

BIM modeling enables complex logistics to outline quick and sustainable communication strategies. Adopting this methodology allows users to efficiently manipulate the database and spaces, with maximum control of any gaps. Increasing BIM adoption in the industrial sector enables logistics to achieve high-performance standards by accelerating current information management mechanisms. (Whitlock, 2018)

BIM and CPS as Virtual Simulation System for Industry 4.0

The industrial sector has had to face many transformations in the last decade, changing internal relationships and methods of sharing energy and economic resources but also the way the products are managed inside the building. (Gersch & Goeke, 2007)

This transformation has outlined a new industrial scenario where the convergence of several emerging technologies, enhanced by the IoT, highlights how smart CPS can create value for industrial activities. CPS's advantage is due to their interaction with data coming from the sensors, generating virtual images also related to the entire life cycle. Based on the whole information, the system can react autonomously to various situations. (Frank et al., 2019)

To best meet these aims, cyber-physical models assume a high degree of complexity to take advantage of traditional modeling systems and digital reconstruction of elements such as laser scans and more complex tools with a real time return of information (Vignais et al., 2012)

With an efficient energy simulation system and an optimized product location, this new data acquisition system allows the virtual model to interact with other systems and connected users within a new intelligent industry model. These digitization processes and connectivity increase the value of adopting new technical skills and operations abilities required by Industry 4.0. (De Lange et al., 2017)

In operational activities, mechanical processing, purchasing, warehousing, and logistics, it is necessary to introduce new methods to optimize information through BIM.

Thanks to recent advances in Automated Data Capture (ADC), the BIM methodology facilitates integrating real objects or spaces into virtual information models, enabling a bidirectional data flow in real-time. (Srewil & Scherer, 2017)

There are currently several solutions to connect virtual models to real BIM-based models. The first one allows us to connect physical objects to their digital twins using Radio-frequency identification (RFID) technology, laser scanning, or GPS. The second allows them to acquire data through sensors and adapt to different scenarios through their space position. (Srewil & Scherer, 2016)

Using specific hardware and software, it is possible to control and increase in the best way the relationship between man and machine. Moreover, through immersive or mixed reality simulations, the operator acquires the necessary data for the correct functioning of the entire system and makes any changes to improve the energy efficiency of the building and the production line. (Paelke, 2014)

For this reason, during management and maintenance operations, even remotely from the plant, the use of Mixed Reality (MXR) and Virtual Reality (VR) tools is common, facilitating connections of CPS between users and optimizing the possible scenarios of energy recovery and logistics of the plant. These technologies can become a simulated environment that can be seen with wearable or mobile devices that encapsulate a person's physical presence or object in a real or imaginary world. (Gottschalk, 2009)

Within this technological transformation, digital parametric models can be developed and become real CPSs with a constantly evolving optimization of the process to respond to real inputs. (Vachálek et al., 2017)

Therefore, this BIM model becomes a virtual copy of what happens in the real system. The inclusion of sensors, actuators, and high precision connectivity systems inside the factory, transforms the building into a Cyber-Physical System able to detect any variation or fault of the plant and functional components through sharing and monitoring via network. (Zuehlke, 2010)

Some projects will describe the need to develop virtual models to provide digital information through different visualization tools and energy efficiency scenarios in literature. Three projects are analyzed below:

- IMMORTAL Project: has developed an integrated working methodology for creating cyber-physical systems, adapted to search for innovative solutions about maintenance and management of industrial plants and control of faults in an automated way. The development of CPS outlines the reliable design, fault management, and gap found automatically. Error management is the key objective of the project. Besides, it is possible to manage different real-time needs dynamically, preserving control, system security, privacy, reliability, and energy efficiency. Based on the mapped data, the user, on one hand, checks the reconfiguration of the system (e.g., isolates the error) and, on the other hand, informs the software about the procedures for the correct operation of the system. The project has developed different methodologies, such as holistic CPS modeling, reliability metrics, and new failure models, verification and testing methods, dynamic, semi-formal, and formal analysis tools, and dynamic reliability of failure detection mechanisms on the plant. (Immortal Project, 2018);
- COGNITO Project: the capture, recognition, and automatic representation of human sensory activities are one of the technological challenges faced by the project. The project aims to design a personal assistance system, where Augmented Reality and Mixed Reality systems support users to identify faults and give guidelines for solving problems previously encountered. Operator movement data is stored in the management system and transmitted to the virtual model. The project involves using multiple sensors that provide a large amount of data at high speed and can be visualized through devices connected to the network. The elements displayed can be 3D virtual objects, images, videos, sounds, or an even more detailed description of the activity provided to the user using a text/audio converter. In conclusion, the project wants to monitor the users' activity about the work area, connecting the information to the underlying flow models, and offering new display features through mixed systems. (Cognito Project, 2012);
- REMAIN Project: it develops a correct methodology to increase an efficient use of energy and materials in smart factories. The aim was to optimize production by improving renewable energy systems and the recovery of waste energy. The project proposes innovative solutions concerning digital models able to simulate energy performance during different phases, such as production and energy maintenance of the building. In this way, renewable energy is used through new products and installation technologies, with optimized energy planning tools. Another aim of the project is to seek more innovative and cost-efficient technologies to help factories become more efficient by better exploiting wasted energy, integrating renewable energy, and optimizing operations. The project has made it possible by creating tools for decision-making, allowing plant managers to choose the best strategies to reduce energy use and minimize environmental impacts, increasing information about possible financial investments. (REEMAIN Project, 2017)

METHODOLOGY

Case Studies

The issues related to the digitization and energy efficiency of existing buildings are faced by selecting two case studies examined in two multidisciplinary projects whose objectives are:

- Data integration and visualization for the optimization of production and storage environments;
- Management of building data and evaluation of energy efficiency from building and plant components.

The first objective includes the knowledge of the production process and the correct management of the logistics chain and its warehouse in order to optimize the production cycle's supply and storage. The Dislo-Man project has been selected to respond to this need and to find solutions that bring together different sectors and skills for a new digital twin factory concept. The second one focuses mainly on managing an existing building whose plant complexity and size provide energy resource optimizations. The examined case study is part of a project for digitizing an industrial plant, currently active, which refers to the energy efficiency and innovative maintenance fields of a widespread industrial heritage. These needs require the adoption of a new factory model called Industry 4.0, where the rational use of energy sources should allow both supplier and consumer to be through a new management system at the same time.

The Dislo-Man platform designed in Figure 1 is composed of an innovative infrastructure that combines ICT systems for the integrated, dynamic, and independent management of the production chain, optimizing resources related to employees, materials, production systems, and energy. The platform can improve the industrial process and define new data available to promote the best solutions for the digitization of daily operations. In addition, at the logistic level, flows can be optimized, reducing time and costs. The advanced management of the systems throughout the building's entire life cycle guarantees all aims mentioned. Moreover, the Dislo-Man platform underlines the data transfer capability for visualization, optimized information, and production management. Within this context, virtual models considered as real CPS have been developed through the BIM methodology. In this contest, BIM model is designed to transfer and to visualize the information required for logistic purposes, into the ICT platform. These digital models deliver management and logistic data to different cloud systems to detail and simulate new behaviors and facilitate users' work.

Among the project's general objectives, the communication platform will have to define the best strategies in adopting systems and technologies to enable the operators to use wearable sensors for indoor tracking and facilitate the handling of materials. Moreover, these sensors allow us to communicate via a gateway in the Real-Time field data from different sources.

Another essential element that makes the industry's management more efficient is the automatic identification of operators and their functions, with the possibility of using mobile instruments to facilitate their tasks and data collection. Within the case study, the integration with sensor networks to detect the parameters related to the production and sharing of these is done through cloud or middleware. The platform itself becomes a storage box capable of communicating the collected data through standard formats and sending them to the different devices connected to the network. The methodological and practical innovation brought to the project through the BIM model concerns 3D models 'creation of the

Figure 1. DISLO-MAN Platform

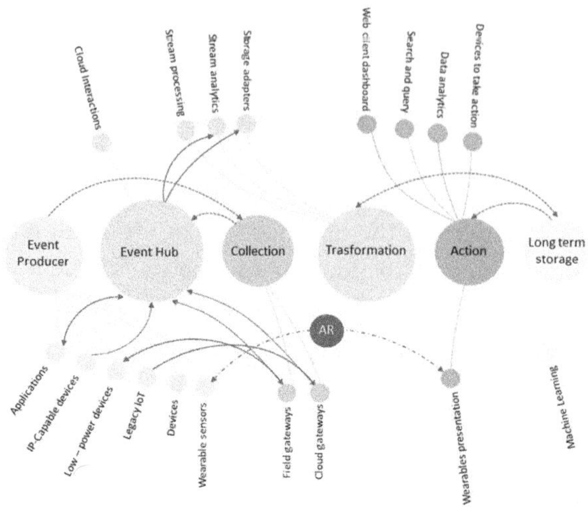

building and individual components as efficient cyber-physical models. Out of different technologies adopted within the platform, AR and VR tools are being tested for information visualization and communication at different development degrees.

All these methods allow the different pilots of the DISLO-MAN project to improve the efficiency of their management and maintenance system with a reduced effort in terms of resources and an increase in operational quality. Proposing more and more innovative solutions for the management of the single component and regulating the industrial plant's correct functioning is the innovation of the DISLO-MAN platform.

The importance of virtual systems activity to reduce energy consumption has been explored within the second case study, as highlighted in Figure 2. In the second one, a digital model is useful for outlining new refurbishment scenarios and for optimizing building management. Therefore, starting from a BIM-based approach, a digital model has been developed to obtain an energy one that can provide specific building simulations. Meanwhile, a sensor network has been installed to monitor energy data (e.g., temperature and humidity) and developed a virtual and learning plant model for energy efficiency. This one should predict indoor temperature and consumption simulations based on data acquisition and Building Energy Model (BEM) simulations. The advantage of this approach is to develop a Model Predictive control for energy optimization, and a Digital Twin Factory prototype based on the connection between BIM/BEM models and IoT devices integrated systems.

Before creating a BIM model, an important activity is data collection to gather all the information associated with the building, from architectural data up to system data and thermal properties. In this way, it has been possible to develop a model characterized by all the data collected and all the information necessary for the following analysis in energy platforms. Certainly, one of the most important steps is the model calibration with the plant's real energy consumption. The simulated consumption has been compared with the company's data that owns the plant, and the BEM model has been calibrated to obtain values with a maximum deviation of 10%.

Visualization, control, and query energy data and system data applications can be developed, both starting from BIM model and simulation results. In this way, specific virtual models allow the plant manager to discover the right decision between different energy refurbishment scenarios. Analyzing the methodology in two case studies allows us to exploit digital models' potentialities for specific needs. In this sense, the BIM models incorporate the requirements of Industry 4.0 that can be fulfilled through interoperability, facilitating the exchange of incoming and outgoing data between platforms.

Figure 2. Digital platform for energy efficiency and maintenance

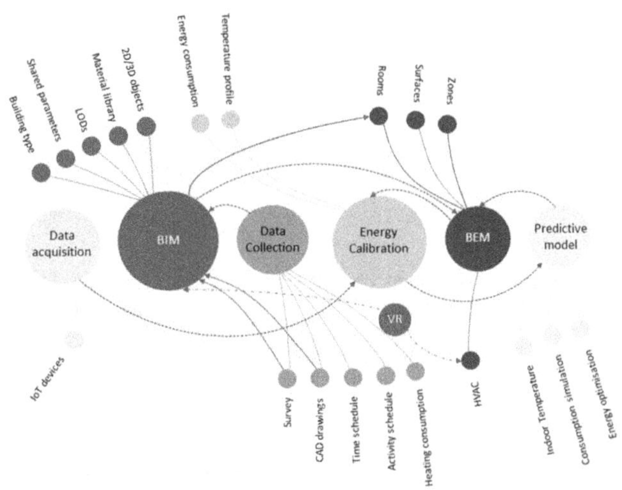

DIGITAL MODEL USES AND PROCEDURES WITHIN THE TWO CASE STUDIES

Growing demand for goods leads to an increase in production levels therefore, it is necessary to generate a sustainable evolution of demand by implementing innovative management systems. Technological innovation can meet this demand by bringing multiple benefits to the industrial world. For example, in the predictive maintenance of individual machines and all systems, including energy systems, the development of virtual environments defines a better space management system, evaluating aspects related to comfort and waste reduction. In the end, these high-efficiency environments, especially from the energy point of view, allow planning and optimization of supply and proper delivery of heating and cooling systems. All this to achieve a transparent, dynamic, and shared management flow of information. (Lee et al. 2014)

For this reason, the industry's sustainable development is based on the creation of smart factories in which products and services communicate with each other with an intelligent network composed of sensors, cloud systems, and emerging technologies like Augmented, Mixed and Virtual Reality, to facilitate data and objects visualization.

It is important to incorporate a smart technology network into buildings to implement strategic processes within the city. This allows industries to improve the connection with various resources (e.g. energy and logistics) concerning decision-making and sustainable development. The Digital model plays a key role in the control and monitoring of a building or a single component because the increase of

Figure 3. Connected BIM model uses into Dislo-man platforms and energy efficiency project

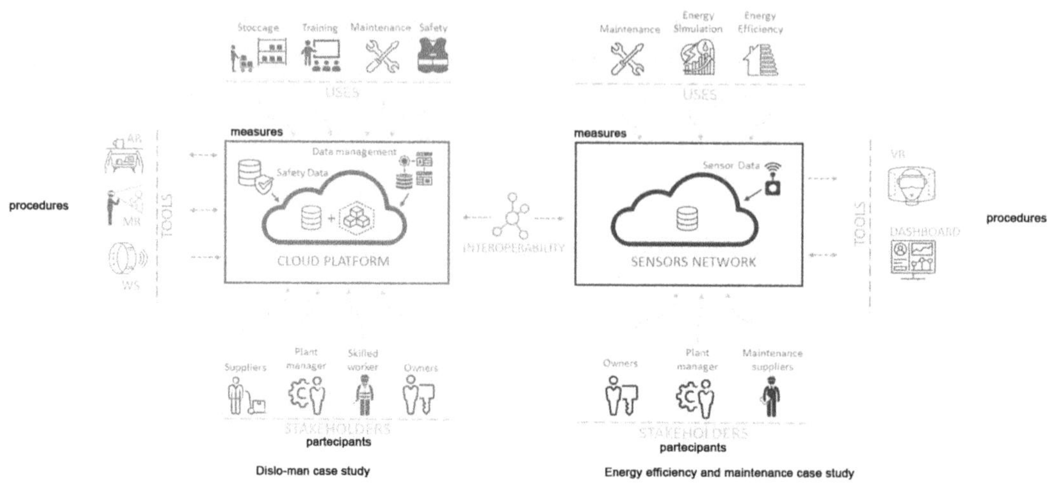

management activities during the operational phase of a building allows BIM methodology adopted to optimize data exchange between different components and users. (Bonci et al., 2019)

This contribution aims to investigate the topics mentioned above and find out some improvement from data management and visualization concerning the industrial building stock. Moreover, digital simulation models and the connection to graphic databases have become an innovative element to knowledge of the built environment. The two projects selected as example want to introduce a new approach to the city vision, which is composed of different layers that become information coming from digital models. In this way, it is possible to generate a Digital Twin aligned to the needs of new smart societies. New strategies have been defined to improve the man-building and man-machine/digital technology relationship. The information management process inside industrial buildings has been improved by analyzing the data flow and hypothesizing new energy efficiency scenarios. CPS can simulate the plant's real operation in an efficient, collaborative, and resilient way. For these reasons, based on the BIM, these systems generate an automated monitoring of buildings and their components during their normal operation, through a customized simulator and it defines their energy efficiency and functional or logistics elements.

Figure 3 shows the data-sharing procedures between different platforms and users within two case studies, highlighting participants, procedures, and useful measures for sharing data and virtual models. It can be extended to many uses and can involve users with several functions, so it is possible to measure differences between them through visualization and monitoring tools.

Both projects are based on the relationships between different stakeholders and active participants in defining their sector's innovative needs. Once the uses have been identified, the procedures and protocols used to transfer real data in the different virtual scenarios through cloud platforms and sensor networks can be outlined. Finally, the validation of the procedures and standards used in the two case studies are measured through AR, VR, and MXR tools by all the users in the sector who evaluate specific parameters such as maximum capacity, rotation index, average product stock, and internal temperature, humidity consumption, energy occupancy.

The process's development shows that data communication between different participants can be very laborious and difficult to apply to large databases if closed and non-standardized exchange formats are used. Also, if a shared ICT platform is not used, communication between applications and digital models can be slow and inefficient. To improve information sharing, it is necessary to apply the standardization of information systems and the definition of compatibility standards to be adopted between systems. Besides, it is of fundamental relevance to establish correct information and geometric content within these digital models to reach the sharing optimization. According to the uses and the degree of detail to be achieved, the BIM model needs an organizational structure such as the identification of the Level of Development (LOD) that defines the quantity and quality of data and the use and the objective to be achieved. (UNI, UNI 11337-4:2017, 2017)

By choosing the right LOD and the necessary information, the communication turns out to be working and with a high-reliability level. Digital models design, using the BIM standardization mentioned above can provide an updated virtual representation of the building and its assets through three-dimensional reconstructions based on real data or future scenarios with the interaction within the simulations. (Del Giudice et al., 2019)

BIM CONNECTION FRAMEWORK

Virtual simulation tools give the BIM model the possibility to connect multiple uses within a virtual environment to analyze the entire building, the properties of the elements, examine the installations' efficiency, and optimize society behavior. (Motamedi et al., 2017)

The development of collaborative methods made it possible to use, for both case studies selected, the BIM models that allow coordination of information adaptable to multiple needs through open-source formats and standards.

Figure 4. Interoperability framework

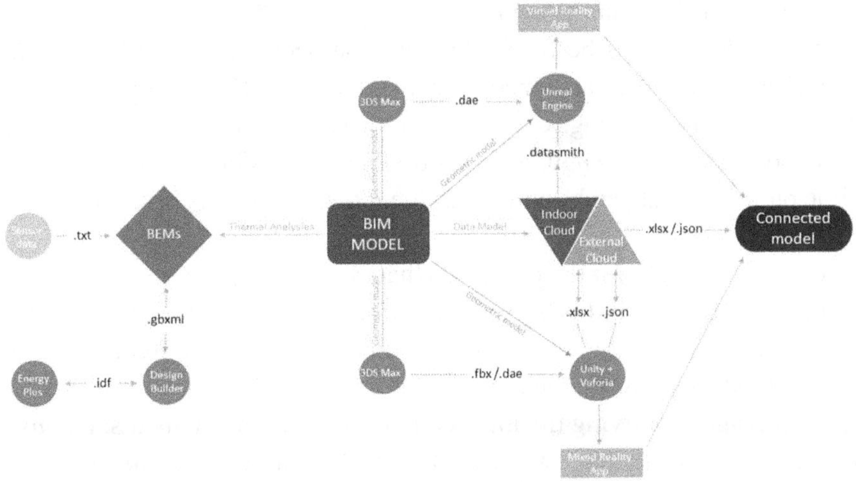

The use of the digital model shown in Figure 4 follows two purposes. The first one concerns the management and visualization of data within a goods storage area, while the second one focuses on the optimized management of an energy plant related to the industry. These two purposes have the possibility to connect the digital model through internal and external cloud platforms to the parametric digital modeling software.

For the first case, the digital model has been imported on the Unity 3D platform to define better interactions with the real environment. To lose the least amount of information and keep the geometries with the digital model's materials, it is useful to use the international exchange format Filmbox (FBX) and rework the surfaces by simplifying triangulations on a bridge program such as 3Ds Max. It is possible to connect directly the modeled object to the different data management and collection platforms, using different exchange formats. To date, the best format in terms of geometric transposition is Collada. While for the transfer of the virtual model's alphanumeric parameters, the most interoperable formats are datasmith and JSON.

Many ICT platforms use the JSON format to connect databases from different domains. The advantage of using this format is related to the correct visualization of the modeled objects' alphanumeric encodings and the possibility of modifying and updating the database within the cloud. Finally, to add information to the real world, it is important to use specific Mixed and Virtual Reality software. In particular, for this phase, the additional software used is Vuforia.

In recent years, a new type of format is emerging, called datasmith. It allows many alphanumeric properties to be transferred between digital and visualization systems that exploit Virtual Reality principles. For example, it is possible to transfer information details between digital modeling software and Unreal Engine for VR application development using the dedicated plugin. Being a new format and little used by different platforms, there are still some interoperability problems. One of these is that the association of the two components does not happen automatically. Still, the user has to invoke the alphanumeric properties through scripts, and these data are not editable. For this reason, the study highlighted not the bidirectionality of the exchange but the uniqueness of the process. On the other hand, for an industrial building's energy data management, the BIM to Building Energy Model (BEM) has been investigated. The goal is to export the building's data to the energy software, designed to run energy simulations.

In recent years, some research has been carried out to analyze the connection between two different platforms. Green Building XML (GbXML) and Industry Foundation Classes (IFC) are the main data exchange format used among BIM and BEM tools. (Dong et al., 2007)

In this contribution, GbXML has been selected, and the necessary steps for the geometric and alphanumeric data exchange have been analyzed, starting from the BIM model. In addition, gbXML has the advantage of facilitating the exchange of building information between the manufacturer's database, BIM models, and energy simulation engines. It is possible to transfer detailed descriptions of a single building or multiple buildings that can be entered as input data for energy analysis and simulation tools. (Jalaei & Jrade, 2014)

An important result to highlight for the process is that all thermal properties have been imported into BEM tools and read correctly. Since it is a text containing the model information, during the GbXML file generation, it is common that the generated analytical surfaces can create air gaps that can lead to problems during data acquisition by the simulation software. These issues can be solved by acting within the BIM platform and modifying the intersections between the elements, paying attention to the connections between walls, roofs, windows, and other architectural elements. Moreover, HVAC data cannot be exported inside by the BIM model. Thus, to obtain simulations, results are required to model this part again.

RESULTS

Today the smart city is composed of buildings that communicate data through ICT platforms to catalog and manage the flow of information in an optimized way. This brings multiple advantages to the relationship between users-city and technologies. Information platforms best capture these advantages by providing flexible and adaptable systems to the urban fabric in which they operate. In smart cities with sustainable buildings, data collection and analysis must occur through specific transmission protocols and system architectures developed with standard formats. Only in this way, the data-building-user connection is independent of the cloud environment used as a connector.

The key element generated by this new relationship is the digitization of cities and their buildings through management servers and rest protocols. For these reasons, we have highlighted the formats and data visualization and transmission platforms tested to minimize the gap between real and virtual models in the results section. Different platforms have been analyzed to define standard formats more interoperable with the digital model at the information and geometric level. Also, two specific applications that take advantage of VR and MXR are presented for different case studies. These applications can connect the previously developed Digital Twin with data collection systems in which users are asked to validate the process and the reliability of the cataloged information.

Another outcome to highlight is how the Dislo-man platform investigated the most effective strategies for integrating software and visualization technologies for sharing interoperable information models. As processes and tools evolve, it can become a true demonstrator applicable to industrial partners to test over time the evolution of data and the need to connect its players in a dynamic and shared process. Although the platform is currently used only in part, some technologies are not fully exploited due to difficulties in managing instrumentation and strict security protocols. However, the second project analyzed is still in the implementation phase, integrating more the network of sensors in a virtual model that can train in the cloud data and simulate scenarios even more efficiently thanks to the introduction of new energy systems. Moreover, the use of Virtual Reality presented in this contribution becomes the starting point to integrate data in Real-Time obtained by the network sensors and the simulations, giving the user the possibility to manage their own behaviors better.

Some critical issues have emerged from both projects that can be overcome and implemented with more stable and user-friendly technologies with the ever-evolving technological progress and long-term monitoring of data in individual pilots. The main problems encountered are listed below:

- A multiplicity of one-way software and formats not easily integrated with BIM-ICT platforms;
- Visualization tools with functionality unrelated to bidirectionality between data and geometric model;
- Need to develop open-source web protocols but with high security for industrial data privacy;
- In BIM to BEM transfer some energy system data are missed;
- An automated simulations data visualization is not easy to be implemented.

VISUALIZATION PLATFORMS FOR BIM CONNECTION

The platforms tested for this contribution, present on the market, allow us to use, in future projects, the BIM and visualization tools in an interoperable way without data interruption and waste of resources.

In this direction, the software that uses AR, VR, and MXR procedures connected to the digital model can validate the measures and procedures adopted.

Depending on the intended use and purpose, it is possible to equip users with custom and connected applications that focus on the digital model's geometric and alphanumeric interoperability. However, some platforms need basic knowledge of programming languages to set up small interactions such as clicking a button and displaying images or data. This complexity is linked to the reason why they become real plugins to be integrated within more complex software to develop video games. Some platforms have development limitations due to demo versions that are released for short periods. From the study carried out, in Figure 5, different platforms have been cataloged from the market, they are open sources and they have close correlations with digital models of simulations such as BIM.

Figure 5. Goals and platforms for BIM Connection

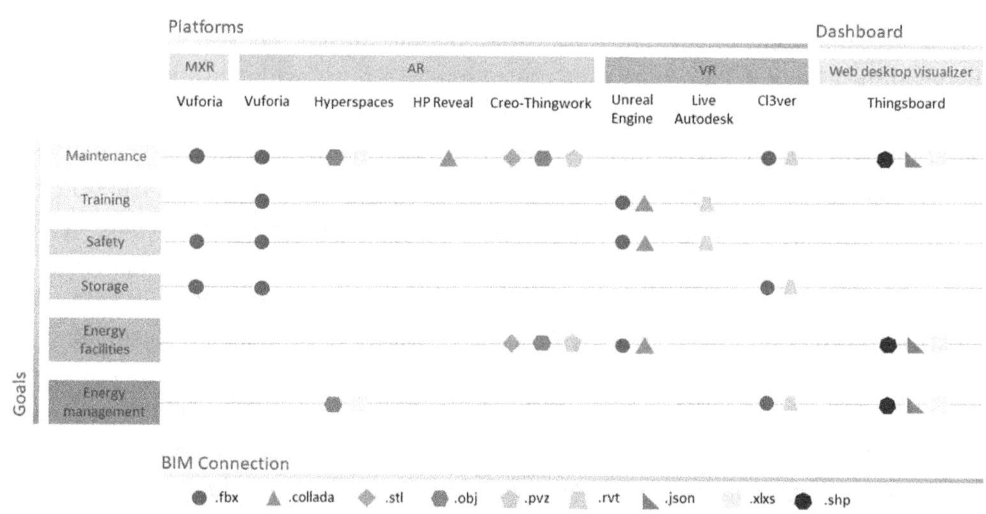

Below there is a list of tested software and achieved results about parametric digital models' interaction, focusing on these models' data acquisition and visualization modes.

- The Vuforia platform: it is connected to Unity3D, it allows us to create specific targets for the geometric model visualization in Augmented and Mixed Reality, and with the addition of virtual buttons for data entry.
- Hyperspaces platform: it is an Augmented Reality platform that allows the user to develop applications for different sectors and purposes without using scripts or programs for visual programming. The platform presents different interactions to associate information through either images or a three-dimensional digital representation of the virtual model.

- HP Reveal platform: this platform facilitates visualization of additional content in the real world with Augmented Reality even for non-developers. The creation of a target allows us to insert additional content, images, videos, or video links on the web.
- Thing work platform is a software of Augmented Reality that offers a wide range of interactions, becoming a true IoT platform. In particular, the additional module CREO allows, through reality increase, to develop a complete application to view videos through direct links to the website.
- Cl3ver: this software allows the user to activate more features and degrees of diving through the virtual reality system. The program provides the insertion of video through external URL calls, technical information, and decomposition of the elements with a description of the individual parts.
- Live Autodesk: is a software for a virtual tour, directly connected to the parametric digital model through a plugin. The user from its desktop or headsets can navigate the model and activate some interactions, such as analyzing the technical properties entered during the modeling phase.
- Unreal Engine: is a game development engine that, in recent years, has been widely used to create dynamic and real-time interactions with a digital model. In fact, the user can enter any data through scripts and allow the visualization and editing and updating directly on the cloud.
- Thingsboard: is a platform that could be developed to collect, monitor, and visualize energy information of the building asset. Several web-based tools allow us to reach these goals, and one of these is offered by ESRI, ArcGIS Online, Infographics. The dashboard can also be customized according to the required information and can help the user answer questions about energy, consumption, or any other information related to the building. The main advantage is the possibility to represent ranges of attribute values with color scales.

It is important to underline how the interoperability between different software becomes an important tool to evaluate the correct connection and multidisciplinarity of BIM modeling in the industrial field. In addition, it is well known how still different platforms can be investigated and used to promote and improve data accessibility and exchange, but also help the user in deciding new strategies and developments.

DATA VISUALIZATION FOR ENERGY AND MANAGEMENT SYSTEM

The improvement of energy efficiency of a building is a relevant and very current issue, especially for industrial buildings, often dated and characterized by very high energy consumption. To face this problem, it is necessary to set up platforms that allow us to check and monitor energy data coming from sensors and, at the same time, to develop applications that support the system maintenance. A large amount of research has been done and is ongoing to develop these monitoring platforms, particularly web platforms, starting from a GIS and BIM interoperable model. (Cecchini et al., 2019)

The second project focuses on collecting, querying and using the BIM database as a management tool and decision support during the performance evaluation of buildings and the definition of efficiency strategies. The aim of this study is the creation of a platform for data visualization related to the system components of an industrial building. The query of information is oriented to support Facility Management (FM), an activity with the objective of guaranteeing comfort, safety, functionality, efficiency of the built environment through the integration of people, networking of locations, combination of processes and use of technology. (International Facility Management Association, 1980)

Therefore, the obtained results is a platform that allows the maintenance operators to enter within the model and consult information during the operational activities. The technology used for interaction with the model is Virtual reality, so the user finds himself completely immersed in a virtual environment that, in this case, is an industrial building. To reach this goal, data smith has been selected as the exchange format to connect platforms. It is a set of tools and plugins in Unreal Engine that allows the user to import entire scenes already created and complicated equipment that can be previously generated in a series of industry-standard design platforms. This new methodology of visualization, through the datasmith format, enables us to connect and make the digital model more interoperable by setting some limits on the editing of information. In fact, by converting this format into standard systems, it is possible in Figure 6 to invoke these data through programming languages within the VR platform and connect them to your management system.

Figure 6. Application for data visualization of energy and management system

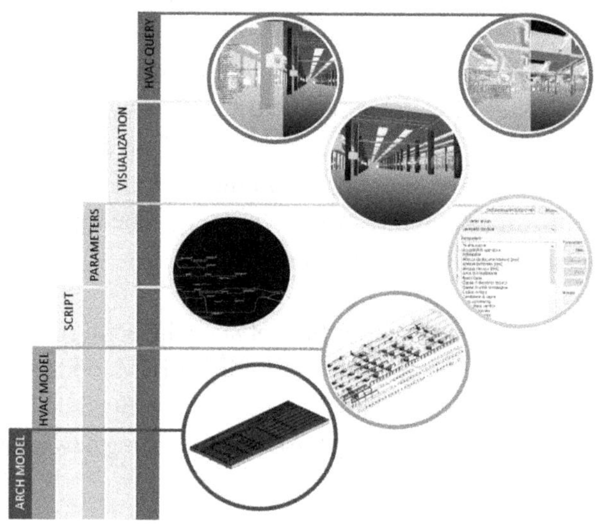

Moreover, the program offers reading the parameters by single objects through visualization systems based on a web browser. Every property became a single cell to write or modify part of codes. Through direct control, the different connected users will verify the information and define the errors on site. To transfer information, it uses a plugin that allows us to share information coming from Autodesk Revit to Unreal Engine, used as the VR platform to create the application. The most important benefit of this format is preserving alphanumerical information previously inserted inside the model, even though they can only be viewed by creating an additional script. This was created using the Blueprints Visual Scripting system in Unreal Engine, a programming language characterized by a node-based interface that lets the user generate application components within the program editor. This language allows us to define object-oriented (OO) classes or objects in the engine. It is a very advanced and powerful tool that gives the user more flexibility to create scenes and manipulate data. (Unreal Engine, n.d.)

Data visualization platform has been developed through the above-mentioned programming language, and it enables us to visualize and query all HVAC data contained inside the BIM model. Moreover, to improve data visualization, a widget has been added, with a white background, to facilitate reading the parameters when the user selects each system component. The analysis of the obtained results has highlighted how a new smart method for data visualization can be exploited in the industrial field and can help in the choice of new energy scenarios and new business strategies.

DATA VISUALIZATION FOR DATA MANAGEMENT PRODUCTS

In the industry sector, the concept of data management is of fundamental relevance to optimize the management system and the activities within the building. From the optimized management, it's possible to use the data warehousing concept to combine two different needs; information accessibility by each user with ease of finding the data needed to control daily operations and manage their database in different places without loss of flow. (Reeves, 1995)

The Dislo-man platform, thanks to its ICT infrastructure, is allowed to manage data through special scripts that recall what is loaded on the platform and transfer them to the BIM model. The resulting database is shared on the cloud through the standardization of collected data. Leveraging interoperable software, this management process integrates real-time information modified by users through innovative visualization tools. For this reason, within the project, a software application has been realized for the digitization of operational best practices and their representation in mixed reality through semi-immersive viewers. This application, connected to the benefits of the whole Dislo-man platform, helps the industrial sector to better evaluate the performance of the production and logistic activities related to human activity. This application makes the Dislo-man platform interconnected, interoperable and allows the Industry 4.0 user to equip themselves with non-invasive wearable technologies without violating current security systems and optimizing their work. The application presented in this contribution combines the connected database with the visualization of digital objects following four main development phases. The first part extracts the database related to the created digital model from the Dislo-Man platform and displays it through a viewer. After this part, identify the user needs for visualization and interactions.

Figure 7. Application for data warehouse management in Mixed Reality

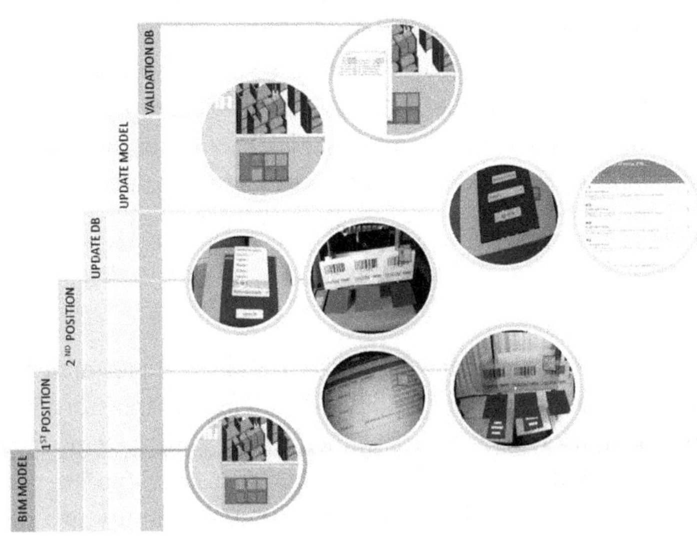

The third phase can interact with the virtual model and modify alphanumeric properties. The last part uses the programming language to develop forms of communication. The Mixed Reality application developed in Figure 7 enables the user to carry out the picking and handling operations of the products, thanks to the digitalization of the models and updating the database inside the real warehouse.

In particular, the department's specialized employee, with his mobile device connected to the internal management system, can identify the products present by selecting the identification code.

Equipped with a viewer of MXR, he can visualize their real position and, thanks to a minimap, keep under control his own position according to the object. Once the product has been identified, within the application, special scripted commands allow the employee to select a new location and send the data to different departments. In this way, the normal administration activities carried out daily are tracked, and the execution time is halved, thanks to intelligent visualization aids. To extrapolate the database from the platform, it is necessary to filter useful information such as the parameters of the single sector with coding Name, Area, Volume at the number of safety outputs, and finally the specific parameters of the single products: Product Code, Object ID, Placement Area, Product Situation. The application carried out on Unity 3d uses the parametric digital model. The warehouse's maximum capacity is simulated, with over 1084 products to be controlled during the working day. In addition to the three-dimensional visualization, using the JSON format, it is possible to connect the previously exported database and make it updatable.

Once the barcode has been read through the app, a virtual plane appears within the viewer to identify the correct location of the product within the department. To facilitate the operations of cataloging and identifying the product, a special QR Code has been created, considering all the data inserted inside the virtual model. When the QRcode is scanned, the product code appears on the device. Through virtual planes of different colors, the application will identify the correct position and validate the employee's work. The new position is emphasized in a different color to be easily detected in the warehouse through a minimap. To make the minimap less static, it is also possible to track the employee's movement and position through a well-designed tracking system, using wearable sensors that communicate with the environment and the virtual model. A virtual minimap has been developed to facilitate the operations of searching for positions within the warehouse that remain active throughout the application's execution. The entire system can leverage the potential of the visor's voice commands to communicate with multiple users within the building and report any problems or failure to take up the product. In this way, the user will always have direct control over the real warehouse. Analyzing the needs that the different industrial partners of the DISLO-man project have expressed, digital model's validation is a complex operation: several factors affect the effectiveness of this technological experimentation; the design time of complex infrastructure and the optimization operations on data collection and dissemination. In the end, the simplification of procedures related to developing a virtual model that best responds to real stresses.

CONCLUSION

In this era, the building industry must leverage digitalization's advantages underlined by research illustrated in this contribution. To overcome the technological challenges of transforming the industrial system into an optimized environment, users need to equip themselves with systems for real-time monitoring of machinery and spaces, mobile devices to visualize data, man-machine integration, and specialized intelligent and competent digital models. To allow information continuity in a digital way of the individual component and the building, virtual models must simulate the real behavior in detail. Interoperability between different tools and software will guarantee the reliability of the digitization, evaluated through specific parameters. Due to the lack of convergence between physical and virtual space, the data generated by the production, management, and maintenance systems within the industrial process are fragmented and stored in individual databases without communication. Therefore, the way data is generated and used in the cyber-physical system is important to merge the physical and virtual product making design, production, service efficient, smart, and sustainable. All this must be supported by a solid ICT infrastructure that uses cloud platforms to store and process inputs in intelligent components or Digital Twin. Technological progress and development of related models also make it possible to improve employees' well-being and health as work increases and knowledge changes. In this way, knowledge is linked to the person and shared by a network of actors able to preserve information for the new generations. Within the decision-making process, the deployment and the use of digital models will be enhanced to test real-time virtual representations of system according to specific parameters: production planning, system configuration, maintenance planning, failure prevention, and energy impact. Certainly, many steps have to be taken in data sharing to achieve Smart cities and Smart factories' goals that play a key role in optimized management in parallel with interoperability.

FUTURE RESEARCH DIRECTIONS

The digitalization and automation of the production environment and the creation of a digital knowledge chain has enabled communication between products, their environment and business partners through innovative tools highlighting the connection between models, spaces and people. New technologies and processes have underlined the need to study new methodologies that create new skills and knowledge within a strongly growing sector. The methodology analyzed in this contribution and the technologies used for the visualization of connected models can help to simplify the way to communicate with the digital model. The correct visualization of information during the construction phases, the life cycle of the building, or simply to analyze new scenarios in the use of energy and logistical resources become key elements that creates the city of the future. The case studies analyzed in this contribution can become a clear example of a new way of connecting different skills and analyzing innovative scenarios to be applied to the industrial sector and to the construction sector, where BIM methodology is widely used. Therefore, innovation and overcoming technological barriers become, within the current digital society, the key to connect the virtual world with the real one. Thanks to the evolution of technological trends, the procedures presented in this contribution can be implemented with even more interconnected systems where the centrality of people, the independence of location, and resilient delivery, make the City System dynamic and interoperable. In the future, even with the continuation of these projects and new demonstration sites, it is possible to apply specific trends, such as the internet of behavior, total experience, machine learning, artificial intelligence, and hyperautomation.

ACKNOWLEDGMENT

All the authors are pleased to thank the students Nicola Rimella, Marco Di Giacomo for the authorization to expose their thesis works. In addition the authors thank the POR 2014-2020 of Piedmont Region

(Italy), FESR (European Fund of Regional Development), and MIUR (Ministry of Education, University and Research, Italy) under the program "Fabbrica Intelligente" (Smart Factory), Action 3, "DISLOMAN" project, Dynamic Integrated Shop fLoor Operation MANagement for industry 4.0.

REFERENCES

ArcGIS Pro Documentation. (n.d.). *ArcGIS Pro* [Software]. https://pro.arcgis.com/en/pro-app/help/mapping/navigation/infographics.htm

Baheti, R., & Gill, H. (2011). Cyber-physical systems. In T. Samad & A. M. Annaswamy (Eds.), *The Impact of Control Technology: Overview, Success Stories, and Research Challenges* (pp. 161–166). IEEE Control Systems Society.

Bonci, A., Carbonari, A., Cucchiarelli, A., Messi, L., Pirani, M., & Vaccarini, M. (2019). A cyber-physical system approach for building efficiency monitoring. *Automation in Construction*, *102*, 68–85. doi:10.1016/j.autcon.2019.02.010

Brooks, P. L. (1997). Visualizing data. *DBMS (Redwood City, Calif.)*, *10*(9), 38.

Cecchini, C., Magrini, A., & Gobbi, L. (2019). A 3d platform for energy data visualization of building assets. In *Proceeding of IOP Conf. Series: Earth and Environmental Science 296*. IOP Publishing. 10.1088/1755-1315/296/1/012035

Cognito Project. (2012). Retrieved from Cognito Project EU: https://cordis.europa.eu/project/id/248290

Costa, A., Keane, M. M., Torrens, J. I., & Corry, E. (2013). Building operation and energy performance: Monitoring, analysis and optimisation toolkit. *Applied Energy*, *101*(C), 310–316. doi:10.1016/j.apenergy.2011.10.037

Devlin, B. (1997). *Data Warehouse: From Architecture to Implementation*. Addison-Wesley.

Del Giudice, M., De Luca, D., & Osello, A. (2020). HBIM Data Management and Visualization for Healthcare Facilities. In C. M. Bolognesi & C. Santagati (Eds.), *Impact of Industry 4.0 on Architecture and Cultural Heritage* (pp. 239–260). IGI Global., doi:10.4018/978-1-7998-1234-0.ch010

De Lange, P., Bähre, B., Finetti-Imhof, C., Klamma, R., Koch, A., & Oppermann, L. (2017). Socio-technical Challenges in the Digital Gap between Building Information Modeling and Industry 4.0. STPIS@ CAiSE, 33-46.

Dolk, D. R., & Kirsch, R. A. II. (1987). A relational information resource dictionary system. *Communications of the ACM*, *30*(1), 48–61. doi:10.1145/7885.7888

Dong, B., Lam, K. P., Huang, Y. C., & Dobbs, G. M. (2007). A comparative study of the IFC and gbXML informational infrastructures for data exchange in computational design support environments. In *Proceedings of Building Simulation*. IBPSA.

Gersch, M., & Goeke, C. (2007). Industry Transformation – Conceptual Considerations from an Evolutionary Perspective. *Journal of Business Market Management*, *1*(2), 151–182. doi:10.100712087-007-0006-5

Gottschalk, P. (2009). Maturity levels for interoperability in digital government. *Government Information Quarterly*, *26*(1), 75–81. doi:10.1016/j.giq.2008.03.003

Gourlis, G., & Kovacic, I. (2017). Building Information Modelling for analysis of energy efficient industrial buildings – A case study. *Journal of Renewable and Sustainable Energy Reviews*, *68*, 953–963. doi:10.1016/j.rser.2016.02.009

Frank, A. G., Mendes, G. H. S., Ayala, N. F., & Ghezzi, A. (2019). Servitization and Industry 4.0 convergence in the digital transformation of product firms: A business model innovation perspective. *Technological Forecasting and Social Change*, *141*, 341–351. doi:10.1016/j.techfore.2019.01.014

Immortal Project. (2018). Retrieved from Immortal Project EU: http://h2020-immortal.eu/

Industry 4.0 in Big Data environment. (2013). *German Harting Magazine*, *26*, 8–10.

International Facility Management Association. (1980). *What is Facility Management?* https://www.ifma.org/about/what-is-facility-management

Jalaei, F., & Jrade, A. (2014). Integrating building information modeling (BIM) and energy analysis tools with green building certification system to conceptually design sustainable buildings. *Journal of Information Technology in Construction*, *19*, 494–519.

Lasi, H., Fettke, P., Kemper, H.-G., Feld, T., & Hoffmann, M. (2014). Industrie 4.0. *Wirtschaftsinformatik*, *56*(4), 261–264. Advance online publication. doi:10.100711576-014-0424-4

Lee, J., Kao, H. A., & Yang, S. (2014). Service innovation and smart analytics for industry 4.0 and big data environment. *Procedia CIRP*, *16*(1), 3–8. doi:10.1016/j.procir.2014.02.001

Ma, C., Chou, D. C., & Yen, D. C. (2000). Data warehousing, technology assessment and management. *Industrial Management & Data Systems*, *100*(3), 125–135. doi:10.1108/02635570010323193

Mantis Project. (2018). Retrieved from Mantis Project EU: http://www.mantis-project.eu/

Motamedi, A., Wang, Z., Yabuki, N., Fukuda, T., & Michikawa, T. (2017). Signage visibility analysis and optimization system using BIM-enabled virtual reality (VR) environments. *Advanced Engineering Informatics*, *32*, 248–262. doi:10.1016/j.aei.2017.03.005

Paelke, V. (2014). Augmented reality in the smart factory: Supporting workers in an industry 4.0. environment. *Proceedings of the 2014 IEEE Emerging Technology and Factory Automation (ETFA),* 1-4. 10.1109/ETFA.2014.7005252

REEIMAN Project. (2017). Retrieved from REEIMAN Project EU: http://www.reemain.eu/

Reeves, L. (1995). Data mining. *Data Management Review*, *5*(7), 12–19.

Rüßmann, M., Lorenz, M., Gerbert, P., & Waldner, M. (2015). Industry 4.0. *The Future of Productivity and Growth in Manufacturing Industries, Boston Consulting Group*, *9*, 1–14.

Srewil, Y., & Scherer, R. J. (2017). Construction objects recognition in framework of CPS. *2017 Winter Simulation Conference (WSC),* 2472-2483. 10.1109/WSC.2017.8247976

Srewil, Y., Ismail, A., & Scherer, R. J. (2016). A method to integrate virtual-physical construction environment in framework of CPS approach. *Smart SysTech 2016; European Conference on Smart Objects, Systems and Technologies, VDE*, 1-8.

UNI. (2017). UNI 11337-4:2017. Edilizia e opere di ingegneria civile - Gestione digitale dei processi informativi delle costruzioni - Parte 4: Evoluzione e sviluppo informativo di modelli, elaborati e oggetti.

Unreal Engine 4 Documentation. (n.d.). *Unreal Datasmith* [Software]. https://docs.unrealengine.com/en-US/Engine/Content/Importing/Datasmith/index.html

Ustundag, A., & Cevikcan, E. (2018). *Industry 4.0: Managing The Digital Transformation*. Springer. doi:10.1007/978-3-319-57870-5

Vachálek, J., Bartalský, L., Rovný, O., Šišmišová, D., Morháč, M., & Lokšík, M. (2017). The digital twin of an industrial production line within the industry 4.0 concept. *2017 21st International Conference on Process Control (PC), Strbske Pleso*, 258-262. 10.1109/PC.2017.7976223

Vignais, N., Miezal, M., Bleser, G., Mura, K., Gorecky, D., & Marin, F. (2012). Innovative system for real-time ergonomic feedback in industrial manufacturing; Ltd and The Ergonomics Society. *Elsevier, 44*(4), 566-574.

Zhang, Y. Y., Kang, K., Lin, J. R., Zhang, J. P., & Zhang, Y. (2020). Building information modeling–based cyber-physical platform for building performance monitoring. *International Journal of Distributed Sensor Networks, 16*(2), 1–21. doi:10.1177/1550147720908170

Zuehlke, D. (2010). SmartFactory—Towards a factory-of-things. *Annual Reviews in Control, 34*(1), 129–138. doi:10.1016/j.arcontrol.2010.02.008

Gartner. (2020). Gartner Special Report "Top 10 Strategic Technology Trends for 2020." *Gartner IT Symposium/Xpo™ 2020.*

Whitlock, K., Abanda, F. H., Manjia, M. B., Pettang, C., & Nkeng, G. E. (2018). BIM for construction site logistics management. *Journal of Engineering, Project, and Production Management, 8*(1), 47.

ADDITIONAL READING

Bahar, Y. N., Pere, C., Landrieu, J., & Nicolle, C. (2013). A thermal simulation tool for building and its interoperability through the Building Information Modeling (BIM) platform. *Buildings, 3*(2), 380–398. doi:10.3390/buildings3020380

Büchi, G., Cugno, M., & Castagnoli, R. (2020). Smart factory performance and Industry 4.0. *Technological Forecasting and Social Change, 150*, 119790. doi:10.1016/j.techfore.2019.119790

Chang, K. M., Dzeng, R. J., & Wu, Y. J. (2018). An Automated IoT Visualization BIM Platform for Decision Support in Facilities Management. *Applied Sciences (Basel, Switzerland), 8*(7), 1–18. doi:10.3390/app8071086

Fonseca, L. M. (2018). Industry 4.0 and the digital society: concepts, dimensions and envisioned benefits. In Proceedings of the international conference on business excellence, Sciendo, 12(1), 386-397.

Jakl, A., Schöffer, L., Husinsky, M., & Wagner, M. (2018). Augmented Reality for Industry 4.0: Architecture and User Experience. In FMT, 38-42.

Kamel, E., & Memari, A. M. (2019). Review of BIM's application in energy simulation: Tools, issues, and solutions. *Automation in Construction*, *97*, 164–180. doi:10.1016/j.autcon.2018.11.008

Negri, E., Fumagalli, L., & Macchi, M. (2017). A review of the roles of digital twin in cps-based production systems. *Procedia Manufacturing*, *11*, 939–948. doi:10.1016/j.promfg.2017.07.198

Quint, F., Sebastian, K., & Gorecky, D. (2015). A mixed-reality learning environment. *Procedia Computer Science*, *75*, 43–48. doi:10.1016/j.procs.2015.12.199

Roblek, V., Meško, M., & Krapež, A. (2016). A complex view of industry 4.0. In: SAGE journal, 6(2), 2-11.

Rodič, B. (2017). Industry 4.0 and the new simulation modelling paradigm. *Organizacija*, *50*(3), 193–207. doi:10.1515/orga-2017-0017

KEY TERMS AND DEFINITIONS

Augmented Reality: Is the addition of three-dimensional or informative elements to human perception through mobile devices that increase knowledge about an object.

Building Energy Management System: Is a building management system in which sensors connected to servers that transmit data to the computer, monitor and control the plant and security equipment of the building.

Building Information Modelling: Is a method based on the realization of a digital model that contains all the information useful for the building life cycle and its components.

Cyber Physical System: Is a computer system that interacts with the physical world in a dynamic and direct way through the ability to process data through a computer.

Data Visualization: Is the graphical representation of information necessary for the understanding of technical details. In this way, by means of visual elements such as tables, data visualization maps help understanding and allow to hypothesize improvement scenarios.

Data Warehouse: Is a system used for the analysis and collection of data from heterogeneous sources. In this way a unique repository containing real time and historical data reports is obtained, to generate data to be transmitted to the whole company.

Digital Model: Is a process in which an object or building with the same geometric and alphanumeric characteristics as the real one is reproduced in a computerized way. Once the 3d data are found they are transferred into the computer for the realization of the three-dimensional shapes and characteristics.

Interoperability: Is defined as the skill of several interrelated systems to communicate information for increased understanding of the element and to use this data for future scenarios.

Mixed Reality: Is a set of technologies and methods used in Augmented and Virtual Reality to produce environments and visualizations of information about real objects that coexist and interact in Real Time.

Virtual Reality: It is a virtual environment reproduced on the computer or inside viewers in which the observer is totally immersed and can interact with the surrounding environment.

Chapter 12
Smart City and Digital Twins:
Definitions, Methodologies, and Applications

Sara Giaveno
Politecnico di Torino, Italy

ABSTRACT

The chapter proposed aims at facing the various implications underlying the smart city concept based on digital twins. The structure of the text is articulated in three main themes: the use of the term "smart city" and the role that technologies had in its definition; the "3D city model" meaning and the integration procedures between BIM (building information modeling) and GIS (geographic information system); the classification of 3D city models by use cases. The chapter can provide researchers with a detailed dissertation aimed at clarifying both the theoretical and technical features belonging to smart city and its related innovative technologies.

INTRODUCTION

The "Smart City" definition was coined in the last decades due to the necessity to incorporate into the urban design discipline the issues related to sustainability, social inclusion and new internet technologies. Smart City is, in fact, a relevant new concept that has been adopted not only in developed countries but also in developing countries. The concept aimed at fostering a sustainable city development through the optimal management of the resources, offering a comprehensive, high-quality life for the citizen (Vassileva et al., 2016). The governance of the contemporary city has to deal with social, economic, political and technical dynamics that are strictly related and profoundly influence each other. Understanding the complexity of this phenomenon requires an indispensable connection between scientific disciplines and humanities (Ferraris, 2018) then today, more than ever, the design should incorporate a wave of relational and technical knowledge, and it should be described in its complexity as a product of negotiation between actors, rules, laws and collective will (Armando & Durbiano, 2017). City governance can not be intended anymore as a deterministic matter based on standards and index for its regulation. However, the city has

DOI: 10.4018/978-1-7998-7091-3.ch012

to be considered as the place where life, relationship and communication develop. Communication plays a crucial role in the city discourse has been improved in the recent years by growing digital products and their power in influencing our daily working habits, teamwork relationships, creative process and a wide range of human products (Floridi, 2011). Information and Communications Technology (ICT) and its power in transmission, reception and processing of data and information are increasingly influencing the method to transform the industry of architecture, engineering and construction (AEC). Therefore, it must be deepened the discourse concerning the methodologies through which is it possible to manage the architectural and the environmental entities belonging to the city. In this field, Building information modelling (BIM) and geographical information systems (GIS) are gaining importance due to the fact that by the integration of the micro-level buildings themselves and the macro-level representation of the external environments of buildings is it possible to establish a comprehensive view of the built environment based on data integrated. This chapter, therefore, presents a broad reflection focused on the description of definitions, methodologies and applications that can be considered to achieve the Smart City goals. In particular, the Smart City issues will be deepened by the following three levels: the social and the environmental dynamics underlying the contemporary city, the technical innovations in design practice that led to the creation of urban digital twins, the applications of digital city models for urban governance.

Background

The chapter seeks to describe the state of the art regarding the definition, the methodologies and the applications to achieve the Smart City goals. To address this challenge it has been collected several critical and technical publications belonging to the sector of Smart City, in particular by following these topics: the contemporary city governance, the Digital Twin concept, the technical integration between BIM and GIS and digital applications for urban transformations. Following this framework, the first objective of the chapter is to clarify the meaning of the Smart City concept (Mohanty et al., 2016), the historical evolution of it (Hénard, 1910; Eremia, 2017; Aurigi, 2005; Brockwell & Horrocks, 2020) and the role of the smarts technologies in it (Mohanty et al., 2016; Kolarevic, 2003). The second objective is to clarify the role of BIM and GIS in terms of city management. The discourse is addressed considering both the single methodology (Bolton, 2018; Hao et al., 2019) and the integration of their data-driven models by facing the issues related to a different scale, datasets, detail of information and visualization (AIA, 2008; (Gröger et al., 2019). The last objective of the chapter is providing a review of the increasing types of applications of 3D city models, focused on visualization and spatial planning (Batty, 2000; Biljecki et al., 2015; Ross, 2000; Saran et al., 2018). Moreover, the chapter provides a taxonomy of virtual models use cases (Biljecki et al., 2015) to show thematic simulations over the city (Bahu et al., 2013; Kurakula, 2007; Chen, 2011; Kaden & Kolbe, 2014).

The Smart City Concept and the Role of the Internet Technologies

The first objective is to clarify the meaning of smart city and to highlight the social dynamics that led to its definition. In a simplistic explanation, a Smart City is a place where traditional networks and services are made more usable, efficient, and sustainable with the use of information, digital and telecommunication technologies, to improve its operations for the benefit of its inhabitants. The different components of a smart city include smart infrastructure, smart transportation, smart energy, smart healthcare, and smart technology, smart governance and smart citizens (Mohanty et al., 2016).

These are the characteristics that make a city “smart”. Technology, and more precisely ICT and related Internet of Things (IoT) and Big Data, represents the core because it is the enabling factors for transforming cities to smart cities efficient and responsive.

Figure 1. The major smart city components and their relation with IoT (Mohanty et al., 2016)

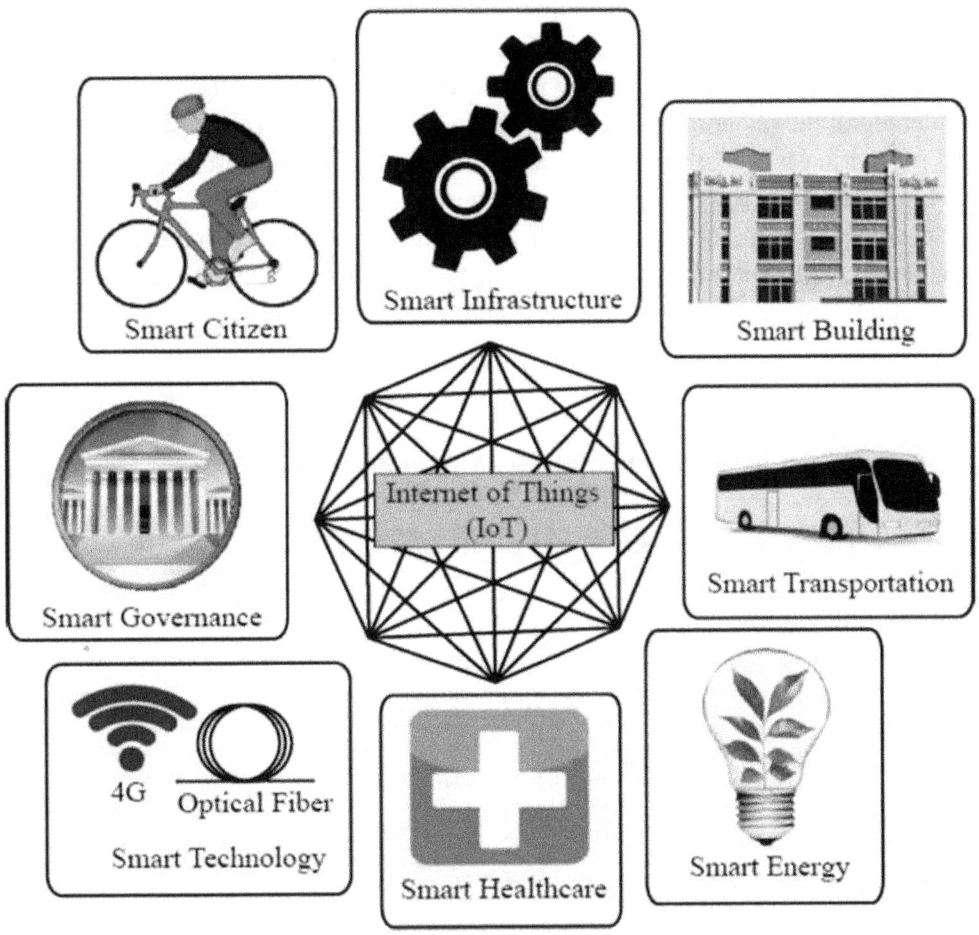

By going deeply into a much more technical explanation of the concept a smart sustainable city is an innovative city that uses ICT and other means to improve quality of life, the efficiency of urban operations and services, and competitiveness while ensuring that it meets the needs of present and future generations concerning economic, social and environmental aspects (Mohanty et al., 2016).

The urban development discipline had to face the issue of sustainable growth since the end of 1900, and the beginning was sanctioned with the publication of “Garden Cities of Tomorrow”(Howard, 1902). He was the first that underlined the necessity to inquire about the influence which modern science may exercise upon the urban planning and particularly on the aspects of the "Cities of the Future" (Hénard, 1910). The definitions that describe the cities of the future progressively improved during the second half of 1900 to include social, economic and governance changing related to community management and growth. "Sustainable city" become after 1950 the most popular term related to the future urban development then in the late 90s the exponential growth of information and telecommunication technology led to the creation of the term "Digital city" (Eremia, 2017). This definition promoted the idea that digitalization, through informatics platforms and digital networks, could efficiently support the management of the complex environment of the cities concerning urban, social and environmental aspects (Aurigi, 2005). However, since 2009, the term was gradually replaced by a new one called "smart city" due to the necessity to incorporate a new element of sustainability and the implications related to internet technologies. This dynamic led to a gradual adoption of many declinations of the term like "smart growth" and "smart development" related to city issues. In fact, the smart city concept rose in strict relation with the demographic phenomenon and the necessity to regulate it. It is predicted that around 70% of the world population will live in urban areas by the year 2050. At present cities consume 75% of the world's resources and energy, which leads to the generation of 80% of greenhouse gases. These dynamics make the concept of smart cities a necessity. It represents a strategy to mitigate the problems emerging by rapid urbanization and urban population growth, such as reduction of energy consumption, water consumption, carbon emissions, transportation requirements, and city waste (Brockwell & Horrocks, 2020).

Figure 2. Components and characteristics of smart cities. (Saraju et al., 2016)

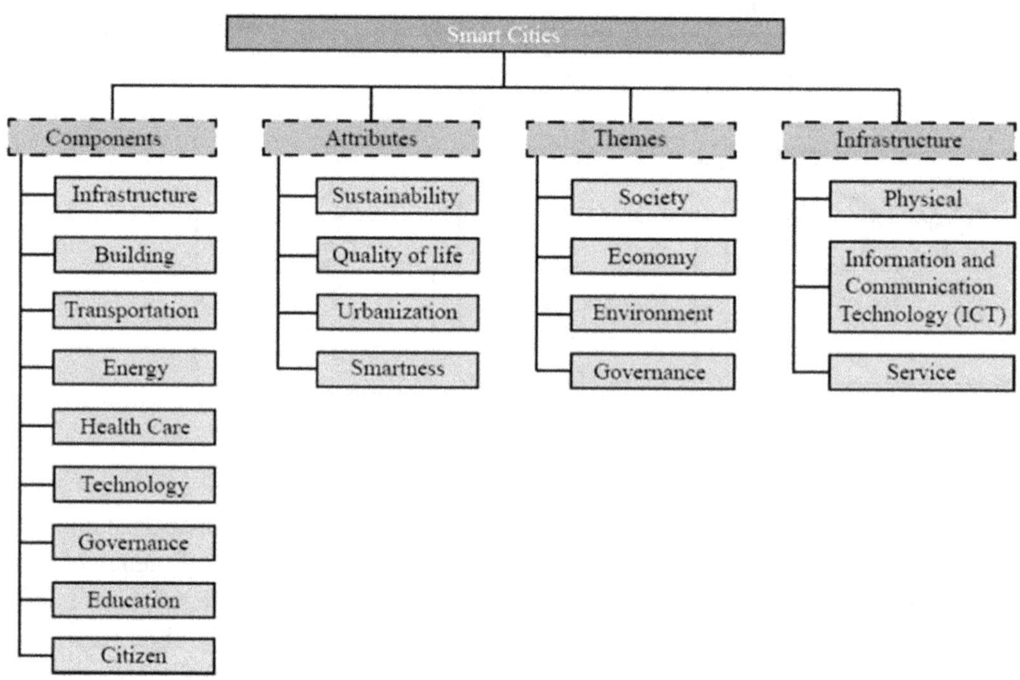

The smart city discourse comprehends, in fact, various attributes defined "smart" (Saraju et al.,2016): The Smart City includes multiple smart components belonging to four themes: Society, Economy, Environment and Governance. More precisely, the characteristics are declined in 9 smart categories: Infrastructure and Transportation, Building, Energy, Technology, Governance, Education, Citizen and Health Care. These components have various attributes that must be ensured, first of all, Sustainability and Quality of life (Saraju et al., 2016)

The critical factor of the scheme is the discourse concerning Smart Infrastructure and more precisely, Internet of Communication Technology Infrastructure, such as fibre optics, Wi-Fi networks, wireless hotspots as well as service-oriented information systems. ICT is the core of the discourse it provides to the city its factors of "smartness", that can be listed as intelligence, interconnection and instrumentation. By going more deeply, the IoT, cyber-physical systems (CPS), and Big Data are key technologies in the context of ICT. The two closely related emerging technology frameworks IoT and Big Data (BD) make smart cities efficient and responsive (Mohanty, 2016). The IoT framework consists of various components including electronics, sensors, networks, firmware, and software and all of the physical objects that are interconnected such as smartphones, sensors, wearable devices, homes, buildings, vehicles, and energy systems. This high net of interconnection generates a great quantity of data, called Big Data that could represent a considerable opportunity in smart management of the city from building to city scale.

The building scale in the smart city includes smart technology and all of the elements listed for smart infrastructures, such as sensors, hardware, software and data network. The most important characteristics are that a building to be considered "smart" must be able to connect and communicate with other technologies and buildings, people and the environment. This last issue introduces a critical point that is the smart city governance. Planning in the smart city has to include the scale both of the building and the urban scale in a way that efficiently manages big data and technologies that belong to the two-scale.

Then, in conclusion, the challenges for building smart cities are various and complex. These design challenges include innovative governance, data management, social issues and public safety, respect of the natural environment. How does it possible to manage these dynamics related to the city? Managing such a complex system of tendencies and their regulation requires a new methodological approach to city transformation in terms of design. Digitally driven design processes, characterized by dynamic, open-ended and unpredictable but consistent transformations of three-dimensional structures, are giving rise to new architectonic possibilities (Kolarevic, 2003). Digital tools are expanding the domain of design to "multidisciplinary" dimensions of social, economic and political matters, and to "multiscale" dimension of design from building to the territory.

Digital Twins in the Context of 3D City Models

The discourse on 3D city models must be introduced by a short definition concerning what Digital Twin represents. The Digital Twin refers to a comprehensive physical and functional description of a component, product or system, which includes all information that could be useful in all lifecycle phases (Boschert & Rosen 2016). In details, the Digital Twin represents the living model of a physical system which continuously adapts to operational changes based on the real-time data from various IoT sensors and predicts behaviours with the help of machine learning and artificial intelligence (Kaur et al., 2020).

The ability to predict and simulate is the major success factor for the use of Digital Twins. In fact, today, modelling and simulation is a standard process in system and service development where simulation-based solutions are realized for optimized operations and failure prediction. In this sense, simulation merges the physical and virtual world in all life cycle phases (Boschert & Rosen 2016). The applications of Digital Twins are, in fact, various in context and scale. It is possible to employ them from the manufacturing of a single element to the representation of building and city scale. Digital Twins are adopted to create a representation of city models that can manage the big data related to the Smart City discourse: current and historical, static and dynamic, geometric and semantic, microscopical and macroscopic (Zhiliang & Yuan, 2019). Once collected, data must be managed by technology, and currently, Building Information Modeling (BIM) and its integration with Geographic Information System (GIS) is the methodology most adopted.

In particular, BIM focuses on the micro-level representation of buildings themselves, then on the dynamic and virtual representation of a physical object or system across its lifecycle, using real-time data to enable understanding, learning and reasoning (Bolton, 2018, pp. 776–808).

GIS provides the macro-level representation of the external environment and can be used to store, to manage and analyze data describing the urban scale (Zhiliang & Yuan, 2019).

Therefore, these applications are crucial for smart city discourse because, through their combination, it is possible to build an integrated a 3D virtual model of all of the characteristics, geometries and information, belonging to the urban environment. Planning a built environment requires, in fact, two levels of models, the building scale that can be supported by BIM, the city scale supported by GIS. The practical applications of BIM-GIS technologies can be summarised in 4 broad categories (Hao et al., 2019, p.45):

1. Life cycle applications in AEC (Architecture Engineering Construction) projects:

it is possible to manage all of the information related to building materials by the use of BIM and GIS by adding into the models the quantities and time planning. This process provides a highly transparent process in managing information, in the visualization of the supply chain and in providing data to suppliers.

2. Building energy management:

it is possible to define energy models for the various design phases. Energy mapping can be developed to support decision-making processes in order to evaluate energy consumption and to improve energy performance programs. Moreover, BIM, GIS and their integration with IoT can be used in real-time monitoring

3. Urban governance:

digital models created by BIM-GIS integration can represent past, present and future conditions, and then they have great power in simulation. This kind of models could be strategic tools for the transformation management of the territory with various applications, for example, for emergency evacuation procedures, infrastructures improvement, climate adaptability of buildings and their environments and moreover.

Then, what is precisely a 3D city model? The 3D city models are not new, however, they appear to have been predominantly used for visualization, but today they are increasingly employed in several domains and for a broad range of tasks and seem the best way by which managing the complex issues related to contemporary cities. A 3D city model is a representation of an urban environment with a three-dimensional geometry of common urban objects and structures, with buildings as the most prominent feature (Biljecki et al., 2015). Studying the applications of BIM and GIS, the first issue to face is how to solve their integration at the technical level.

BIM-GIS technical integration

A typical 3D city model is derived from various acquisition techniques, for instance, photogrammetry and laser scanning, extrusion from 2D footprints, synthetic aperture radar, architectural models and drawings, handheld devices and procedural modelling (Biljecki et al., 2015, p. 2844). 3D city models aim at providing a digital representation of architectural and environmental entities and at simulating performances of design choices throughout the life cycle of an AEC project.

The main issue is that data are created, managed and visualized by BIM and GIS in different ways. BIM and GIS are two types of data-driven models which work at different scales, namely physical building scale and geographic space scale (Hao et al., 2019, p.47) For this reason, must be fixed some rules in the use of the models concerning detailing, information sharing and collaboration procedures:

1. The use of standards: IFC, CityGML

IFC (Industry Foundation Class) and CityGML (City Geography Markup Language) are two major semantic models currently used to represent BIM and GIS models. Respectively, IFC has become the standard model for modelling buildings, then CityGML for modelling urban areas (Hao et al., 2019). However, there are some crucial points to be considered: IFC contains much more detailed information than CityGML, a complete mapping between IFC and CityGML is quite challenging (Liu et al., 2017). The geometry conversion between IFC and CityGML is mainly conducted to lower LODs in CityGML, and it becomes more difficult when a higher LOD is requested (Hao et al., 2019, p.45).

2. Level Of Detail: LOD

The concept of LOD is different between BIM and GIS, then the conversion between IFC and CityGML needs a matching in terms of LOD. The concept of LOD (Level of Development) in BIM is proposed by American Institute of Architects (AIA) to represent the expected completeness of BIM models at each stage of the life cycle of a building and it can be used to standardize the precision of BIM models (Hao et al., 2019). The LOD is defined and classified into five levels from building conceptual design to as-built design as follows (AIA, 2008):

a. LOD 100: symbolic information of the entire building such as building height, area, and volume, corresponding to conceptual design;
b. LOD 200: approximate information of generic systems, objects, or assemblies, corresponding to preliminary design;
c. LOD 300: detailed information of specific systems, objects, or assemblies, corresponding to detailed design;
d. LOD 400: detailed information of specific systems, objects, or assemblies with complete fabrication, assembly and installation information, corresponding to construction design;
e. LOD 500: a field verified representation with information of constructed assemblies, corresponding to as-built design.

The concept of LOD (Level of Detail) for 3D GIS is introduced in CityGML, and the LOD can also be listed into five levels (Gröger et al., 2019):

LOD 0: Digital Elevation Model (DEM)+Digital Orthophoto Map (DOM) at the representation scale of regional landscape;

a. LOD1: block models at the representation scale of city and region;
b. LOD2: basic models at the representation scale of district and street;
c. LOD3: standard models at the representation scale of building exterior;
d. LOD4: premium models at the representation scale of building interior.

This list represents the classification of the elements in BIM and GIS. However, IFC contains much more detailed information that CityGML so the geometry conversion between these two formats is ma is mainly conducted to lower LODs in CityGML, and it becomes more complicated when a higher LOD is necessary.

Once the standards and the formats have been defined, it can be chosen three different procedure of integration between BIM and GIS models (Hao et al., 2019, p.48):

1. BIM leads and GIS supports:

The integrated model is dominated by the BIM model, and the GIS model is used as an auxiliary model to import or integrate data from the GIS model into the BIM model to provide data support.

2. GIS leads and BIM supports:

The integrated model is dominated by the GIS model, and the BIM model is used as an auxiliary model to convert the BIM model into a GIS model to provide data support.

3. BIM and GIS equally involved:

The integrated model emphasizes the interoperability between BIM and GIS models, the integration of data from these two models on third-party platforms, and the practical application of the advantages of these two models.

However, the integrations show some issues. First of all, there is a mismatch of geometric information when IFC is converted to CityGML. Secondly, it is necessary to use a platform of collaboration to match the BIM and GIS models, and thirdly data quality requires a high level of accuracy during the phase of collection, measuring and statistical analysis (Hao et al., 2019, p.48).

Moreover, Hao (2019) reassumes that mode 1 has a better ability to express the building project itself and its internal information. However, it has a lack of ability to integrate with information of other buildings or the surrounding environment. Mode 2 has excellent spatial data processing capabilities but a lack of detailed attribute information for building entities. In conclusion, Mode 3 has a powerful capability in the information processing for building and surrounding but increases the burden of computer processing.

The chart below reassumes the three methodologies and defines accurately all of the characteristics of the models belonging to different processes.

Figure 3. Methodologies used by three modes of BIM-GIS integration. (Hao et al., 2019).

Methodologies used by three modes of BIM-GIS integration.

Mode name	Methodology
Mode 1: BIM leads and GIS supports	1. Use the semantic web standards to declare terms and their relationships in a BIM. 2. Use BIM to build a 3D model and create a simulation of the construction process by associating it with the building plan. 3. Propose a multi-purpose geometric network model (MGNM) based on BIM and explore the strategy of indoor and outdoor network connections. 4. GIS-based vector data and information systems are imported to the parametric 3D-BIM platform.
Mode 2: GIS leads and BIM supports	1. Focuses on GIS-information-based methods, adoption of open standards such as XML, and open application programming interfaces (APIs). 2. Mapping IFC building service systems data into Utilities Network ADE- CityGML. 3. Propose a spatial extract, transform and load (spatial ETL) workflow for the effective integration of IFC, CityGML and Sketchup. 4. GIS mapping has been used to consider the effect of construction projects on the surrounding areas and to delineate the influence areas of construction projects.
Mode 3: BIM and GIS equally involved	1. Integrate IFC and CityGML at different levels of detail, and implement automatic data mapping between IFC and CityGML at different LODs. 2. Use the unified building model (UBM) to encapsulate both the CityGML and IFC models, thus avoiding translations between the models and loss of information. 3. Integrate BIM and GIS into integrated geospatial information model (IGIM), which offers the flexibility to access and process datasets from GIS and BIM through RDF directed graph, and use the GMO mapping algorithm with well-established classification rules. 4. Propose a supplier selection framework by integrating BIM and GIS in a resilient construction supply chain (RCSC).

With the development of BIM and GIS techniques, their integration is in increasing demand. By reassuming, when BIM provides micro-scale information concerning the building, GIS is focused on macro-scale data about the urban environment. By the BIM and GIS combination is then possible to obtain a comprehensive 3D model able to run simulations aimed at achieving public engagement and performing sustainability analysis and other people-oriented applications (Hao et al., 2019).

Classification and Applications of 3D City Models

During the last decades, 3D city models have become valuable for several purposes beyond visualization. The increasing types of applications greatly expanded the possibilities of use of the models. Several studies tried to classify the 3D models, and in a 2000 survey, Batty (2000) provided a conceptual study on the use of 3D city models, focused on visualization and spatial planning. They have segmented the use of 3D city models into 12 categories of industries:

a. Emergency services;
b. Urban planning,
c. Telecommunications,
d. Architecture,
e. Facilities and utility management,
f. Marketing and economic development,
g. Property analysis,
h. Tourism and entertainment,
i. E-commerce,
j. Environment,
k. Education and learning,

1. City portals. Their list serves as an excellent

First of all, the term use case has been defined in software engineering as a sequence of actions that provide an objective or subjective value to a user (Jacobson, 1992). We second this definition: applied to GIS and 3D city modelling, a use case can be seen as a meaningful set of spatial operations that accomplish a goal a user wants to achieve with a spatial data set (Biljecki, 2015). In this context, Ross (2010) listed several applications and provided a general taxonomy of 3D use cases:

1. applications that are based only on geometry (e.g., estimation of the shadow);
2. analyses based on geometry and semantic information (e.g., estimation of the solar potential);
3. analyses based on domain-specific extensions and external data (e.g., noise emission calculation).

The concept of use case is fundamental in understanding the potential of the application of 3D city models and their application. The chart below describes the spatial operations (dark), the most important use cases (visibility analysis, energy applications, estimating population, noise mapping), their applications (dashed), and the relations between spatial operations.

Figure 4. 3D models use cases and applications. (Biljecki, 2015).

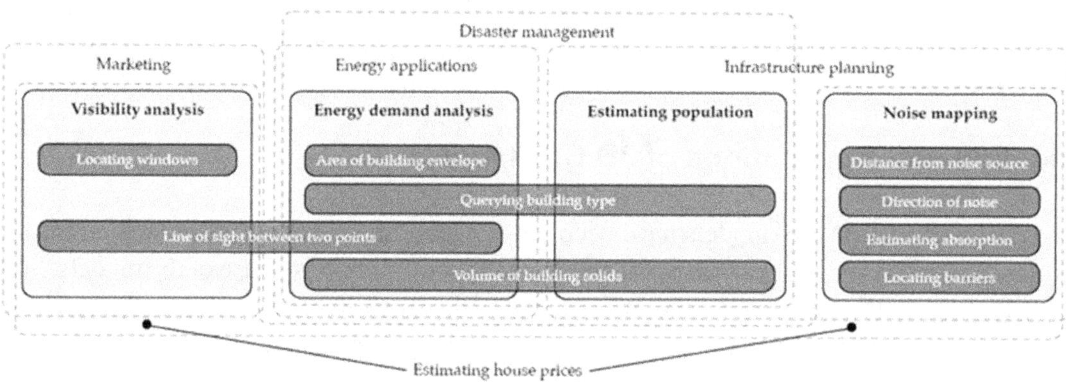

Biljecki (2015) tried to define a specific taxonomy of use cases by collecting them into two groups:

1. *Non-visualization use cases* do not require visualizing the 3D models and the results of the 3D spatial operations (e.g., estimation of solar irradiation, energy demand estimation, classification of building types etc.).
2. *Visualization-based use cases*. This group include:
    a. use cases that require running computation, but where visualization is essential, and the use cases would not make much sense without it (e.g., navigation, serious gaming);,
    b. visualization-only use cases such as communication of urban information and virtual reality, which do not necessarily rely on spatial operations, but where 3D city models have been

found as an essential component (e.g. visibility analysis, 3D cadastre, urban planning, facility management lighting simulations etc.).

In particular, the non-visualization use cases could be listed as follow: Estimation of solar irradiation, energy demand estimation, aiding positioning, determination of the floor space, classifying building types.

The visualization uses cases are: Geo-visualization and visualization enhancement, visibility analysis, estimation of shadows cast by urban features, estimation of the propagation of noise in the urban environment, 3D cadastre, visualization for navigation, urban planning, visualization for communication of urban information, reconstruction of sunlight direction, facility management, automated scaffold assembly, emergency response, lighting simulations, radio-wave propagation, computational fluid dynamics, estimating the population of an area, routing, forecasting seismic damage, flooding, change detection, volumetric density studies, forest management, archaeology. It is necessary to underline the power of the visualization use cases in providing a comfortable and visual reading of the results for non-expert users.

By taking into account these premises, it is interesting to focus on some visualization-base use cases that are considered to show some applications.

Starting from the fact the reduction in terms of energy consumption and Co2 emissions is nowadays the most crucial issue in the building sector (ASVIS, 2018), one of the most valuable application is represented **energy demand estimation**. Recent years have seen the advent of this application where 3D city models are employed to combine the data of the volume of buildings, a number of floors, type of the building, and other characteristics to predict the energy demand for heating and or cooling (Kaden & Kolbe, 2014). In particular, the 3D city model is used to combine the data of the volume of buildings, the floors, the typology of the construction to predict energy demand. This simulation is crucial to prevent emissions and costs, and planning also retrofitting actions in terms of energy efficiency. In the case reported below, it is possible to easily visualize the heat demand and the related estimation of consumption per year.

It is needed to highlight that, according to existing researches, the most critical issue related to city governance is the global energy consumption. Urban energy demand and supply is a far more complex system than a single building, which needs to improve the existing system while putting forward new system in a cooperative way (Yamamura et al., 2017). The energy system is related to the energy generation, the energy consumption by facilities, transportation system, and any other city infrastructures. In fact, it is progressively occurring a shift in energy consumption from industry sector to urban life with two-thirds of CO2 emission generated by people living and working in the cities (Yamamura et al., 2017).

Therefore, Smart City planning has to deal with environmental issues across the different scale. The model has to work such as a multi-layer system that combines BIM and GIS data across the city, from the building scale to the city, and simulate the energy demand and consumption. In this context, it is relevant the case of the EU research project called DIMMER (District Information Modeling Management for Energy Reduction) which integrates the use of BIM and GIS methodologies to perform urban analyses for energy saving. The project aims to develop a web-service oriented open platform with capabilities of real-time district-level data processing and visualization. Thanks to the web service interface, applications can be designed to monitor and control energy consumption and production also from renewable sources (Del Giudice & Patti, 2014). In terms of carbon emission is interesting, also the SEMANCO experience (Semantic Tools for Carbon Reduction in Urban Planning). The SEMANCO integrated platform provides access to widely dispersed energy-related data about cities stored by many different organizations. In this way, the project platform supports improved energy analysis based on the

Figure 5. Estimation of the heat demand of buildings. (Bahu et al., 2013).

assessment of the existing. The platform allows to create a multi-level energy model of an urban area ad can be used to analyze the energy performance of buildings, neighbourhoods, districts and regions.

Strictly linked to energy efficiency purposes is the **estimation of solar irradiation**. This application represents one of the most crucial use cases in 3D City Models. The model is employed in the calculation of the area exposed to the sun, in the calculation of the better orientation, in improving design characteristics of the building in terms of indoor lighting and the estimation of thermal comfort. Another strategic application both for the building scale and the block is the **estimation of noise in the urban environment**. The 3D model is used to mapping the noise pollution with the aim to mitigate them and to define where it is necessary to put noise barriers. The model in this case provides a visual disposition of noise levels and their value in a range of decibel.

Estimation of noise in the urban environment could also be implemented by applications able to simulate the **estimation of the population in an area**. By considering the size of a building, it is possible to estimate the number of residents, which is useful for several application domains.

Concerning the urban scale for the new construction project, it is crucial to consider **the estimation of shadows cast by urban features**. This application consists in fact in positioning the new building into the urban fabric in order to evaluate its impact in terms of shadow cast (Biljecki et al., 2016). This use case is also instrumental in developing other evaluations such as the estimation of solar potential in the building, the thermal comfort and the shaded area of soil for public places design. The simulation of the impact of new construction could also be strategic in preliminary design phases to evaluate, with very low costs of investments, the design alternative that better fits into the urban environment.

Figure 6. A 3D noise simulation. (Kurakula, 2007).

The Municipality is also increasing its interest in developing a property database in 3D by improving the creation of a **3D cadastre** (Osello & Ugliotti, 2015). By this application, it is possible to provide digital methodologies to manage the building architectural patrimony from the public side to the private and fostering an easy way to read it. The model became a digital place where storing, managing and sharing data that are physical, geometrical and legal. In the same field also **classification of building types** is a method through which can be extracted the 3D geometries of the building and classifying them to marketing and real estate utilization.

This application is, in fact, instrumental for another one, that is **urban planning** (Chen, 2011). Simulations are key means by which Municipality and urban planners can organize the urban transformation scenarios. Urban Planning is, therefore, one of the main applications for 3D city models because 3D geoinformation is ubiquitous in urban planning for various tasks. Many Municipalities are increasing their request in the user-friendly platform through which managing urban transformations, here some cases:

The experimentations concerning Urban Planning models are also conducted at universities with relevant and useful simulations concerning various outputs. For example, The Code Lab within the School of Architecture at Carnegie Mellon University investigates relationships between people, spaces and computational ideas and processes.

Moreover, interesting researches have been developed by the MIT Design Lab concerning augmented reality and urban experience, public transportation experience, smart mobility, smart tourism and moreover. In fact, in the field of urban planning, are relevant to the applications aimed at a **visualization for communication of urban information to the citizen**. The experiments are focused on sharing urban information to citizens to promote knowledge sharing concerning the state of the art of the urban fabric and the related possible enhancement (Wu et al., 2010). Public participation in the urban process is, in fact, a crucial point in contemporary city discourse. Many studies affirm that digital tools and ICT could be instrumental in facilitating and promote participatory democracy and many municipalities, such as Amsterdam and Barcelona are moving toward this direction by fostering project pilots for smart governance (Bria et al., 2018). The application for urban governance are then various, and under explo-

Figure 7. Virtual Singapore project by Dessault Systèmes. Urban planning for Shuqun Secondary \school. https://www.youtube.com/watch?v=Dix-8SNxlAo, visited 07/2020.

ration, it can also be listed the communication of cultural heritage and the virtual tours for tourists. In this context are fundamental the experimentations concerning **visualization for navigation**. Navigating urban spaces using 3D city models is instrumental for gaining confidence with the built environment.

3D City model also finds broad utilization in managing and simulating potentiality and vulnerability of the territory, such as **seismic damage**, **flooding**, **forest management**, **emergency response** where is crucial the role of GIS and thematic simulations able to calculate risks and to provide adaptable solutions.

Figure 8. Virtual Rennes project by Dessault Systèmes. Transformation scenario. https://www.youtube.com/watch?v=jeee4OwOPEc, visited 07/2020.

Figure 9. New Urban Analysis Toolkit. Using Computation to Understand How Pedestrians Use Market Square. CODE LAB (CMU). http://code.arc.cmu.edu/news/urban-analysis-framework/, visited 08/2020.

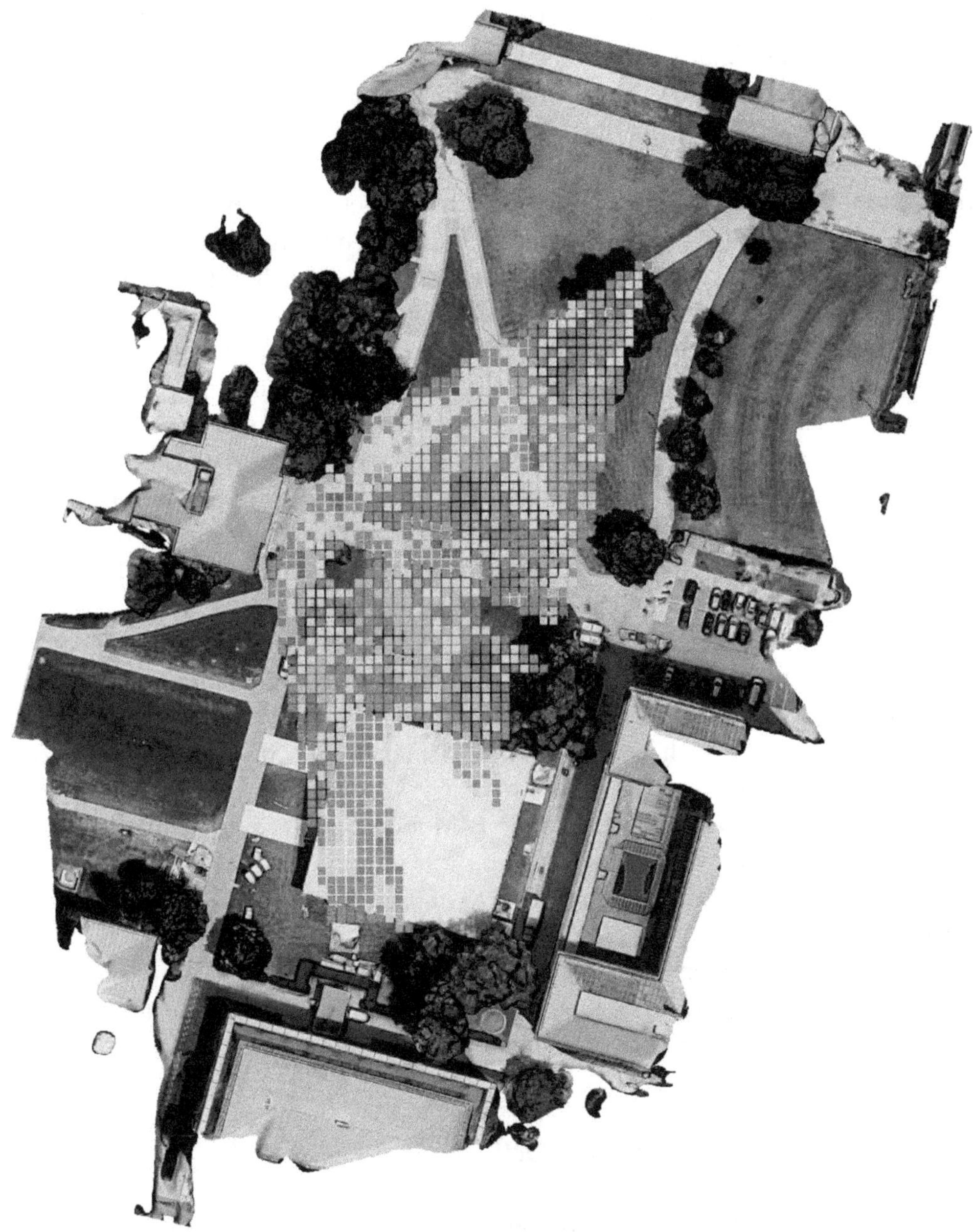

In the last decades then 3D city models are increasingly employed in several domains and for a broad range of tasks beyond visualization. It is crucial to highlight that technology and progress in communications devices is the engine of these simulations. The smart cities need to have some key features such as intelligence, interconnection, and instrumentation which smart technology can provide. Internet of things (IoT), Cyber-Physical Systems (CPS) and Big Data are key technologies in the context of ICT and by their combination, it is possible to make cities efficient and responsive. Big Data is the source of information for thematic simulations in smart cities, and they are generated from a large collection of sensors, databases, emails, web sites, and social media. Big Data has several characteristics, including complexity, volume, variety, variability and veracity (Moharty et al., 2016). Therefore, the discourse concerning smart technologies plays a key role in the Smart City definition. A typical 3D city model is derived from various acquisition techniques, for instance, photogrammetry and laser scanning, extrusion from 2D footprints, synthetic aperture radar, architectural models and drawings, handheld devices, procedural modelling, and geoinformation (Biljecki et al., 2016). Then how to deal with this complexity? The collaborative modelling with GIS and BIM, taking communities as immediate scale, unifies data across the city and building to support the comprehensive analysis. The integration between BIM and GIS can then merge data for construction process and management concerning the urban environment. Geometrical and semantic representations allow generating various simulations focused on improving the decision-making process in the life cycle of the AEC project, in energy management, urban governance and moreover. It is necessary then to highlight the great challenge and advantage that represent the integration between these two technologies. However, from the technical point of view, the accuracy of datasets integration has to be enhanced by improving interoperability between IFC and CityGML.

The crucial points that must be highlighted in conclusion are (Hao et al., 2019):

1. The accuracy of data mapping is required using a third-party digital platform (Esri, ArcGIS);
2. The current challenge is due mainly to the fact that data in BIM and GIS models cannot be fully converted to each other, which leads to difficulties in data collection of building energy consumption in a largescale area;
3. GIS data need to be fully integrated into the BIM model platform to facilitate the application in the life cycle of AEC projects in a larger area;
4. Their integration should mainly be developed to improve energy-saving management at the building level and city level;
5. To expand the application in urban governance, 3D modelling standards of urban buildings and policies supporting BIM and GIS data to be efficiently shared among construction participants;
6. Fostering user-friendly experience of both methodologies.

SOLUTIONS AND RECOMMENDATIONS

It is predicted that around 70% of the world population will live in urban areas by the year 2050. At present cities consume 75% of the world's resources and energy, which leads to the generation of 80% of greenhouse gases. Thus, in the next few decades, there can be a severe negative impact on the environment (Biljecki et al., 2015). This dynamic led to a necessary definition of Smart City.

Given the rate of urbanization that the planet will undergo in the next 15 years in an unprecedented mobilization of financing, creative architecture, new challenges of engineering, and collaborative innovation across disciplines (Brockwell & Horrocks, 2020) the Smart City related innovations seem to be the engine of new challenges. In this context, Brockwell (2020) listed three driving key improvements in the design of cities:

1. Visualization—Unprecedented ability to have insight into the current state, trends, and urgent issues using big data, analytics, machine learning, and artificial intelligence. Augmented and virtual reality visualizations combined with business intelligence, dashboards, and time-based 4D, and time-cost based 5D visualizations composed from live data sources make understanding the context of change and design possible.
2. Generative design—Tools for visual programming like Dynamo combined with highly accurate geospatial data and machine learning can create designs that are beautiful, least-cost, and most efficient. Innovations in materials and construction techniques apply here too.
3. Simulation—The ability to use agent-based simulation to predict the impact of design changes in proactive ways in a timely fashion. Together, these approaches can reduce the cost and duration of transportation infrastructure projects while taking better care of the social, health, and economic consequences of our designs. Without these innovations, it is unlikely we will be able to keep up with the level of urbanization that is coming.

Moreover, related to the Smart City social sphere, Brockwell (2020) highlighted five points:

1. Health—More than freedom from illness, good health is invigorating, inspiring, increases personal, business, and collective productivity. Good health reduces stress, generally improves social cohesion, maximizes mobility, and facilitates learning.
2. Prosperity—Everyone wants their wealth to improve over time. Economic insecurity is an enormous drag on our economic future and the health of citizens. Seems obvious—but the impact is underestimated by many.
3. Family and community—Unless you're a wilderness survivalist full time, you live in a social world inhabited by a network of friends, neighbours, family. The quality of social interactions—the physical ones so much more important than the virtual ones. Evidence suggests communities with cohesive social bonds have substantially higher longevity.
4. Security—This ranges from freedom from imminent harm to the predictability of your daily actions, which is more important. I want to be able to walk to school without fear; I want to know that my income is not precarious; I want to know that financial status is secure.
5. Purpose—Whether it be spiritual, artistic, communal, entrepreneurial, or familial, human beings need to feel a sense of purpose in their lives to ensure long-term wellness. These qualities of life support each other—prosperity correlates with health; community correlates with longevity; security correlates with prosperity, and so on. Underpinning these factors of well-being are other social needs: education, social housing, employment insurance, pensions that help support the criteria for wellness for citizens on a broad scale.

The list by Brockwell (2020) aims at clarifying the most crucial aspects, both technical and social related to city discourse. However, the analysis has to be implemented by explaining two main issues related to Smart City that are the terminology and the features. The city "smartness" includes all of the dynamics that are not only deterministic but social, economic, political and technological. Then the city became the result of the sum of smart components including smart transportation, smart grid, smart components, smart health care and smart governance (Mohanty et al., 2016). The lack of sensibility concerning this complexity led over the years to ambiguous terminology and boundaries of use cases belonging to 3D City Models. Moreover, the scarce clarification of the entity of a 3D City model improved such confusion. Therefore, the chapter intended to explain in clear terms the crucial concept belonging to Smart City by capturing the most relevant features and terminology by a comprehensive bibliometric analysis. The second objective is to highlight the strict dependence that contemporary cities have towards communication technologies and internet networks. These technologies are improving the performances of methodologies, such as BIM and GIS. The increasing possibility of sharing and merging information between datasets, thanks to the improving of standards and formats, is leading to the creation of 3D City Models that are able to integrate construction projects and the surrounding environment. Neither BIM nor GIS can fully facilitate project management and decision-making in the construction industry alone (Hao et al., 2019) but their integration has to be highlighted as the most important way by which city governance is made possible through the use of 3D City Models.

FUTURE RESEARCH DIRECTIONS

The topics addressed in the chapter are crucial for stakeholders belonging to the various field from design to engineering, geomatics, energy, urban planning and moreover. The interest in the topic is not only for academic researchers, but it can be strategic for professionals and especially Municipality. It may be valuable to define clearly the terminology and the features belonging to 3D City Models in order to progressively improve their strategic role in urban governance and thematic simulations over the city. The valuable role and demand of 3D City Models are highlighted by the use cases cited, but new use cases and applications are expected, such as (Biljecki, 2015):

1. Recent advances in augmented reality and virtual reality;
2. Developments in the fusion of computer graphics, GIS and BIM;
3. The majority of use cases rely on buildings, and not many use cases require models of other thematic classes, such as vegetation and bridges. It is expected that, in the future, more use cases will take advantage of thematic features other than buildings;
4. Some application domains that have traditionally relied on 2D and/or 2.5D data are likely to embrace 3D use cases where the third dimension is essential. An example here is the house price models which can be augmented by already available 3D use cases such as estimating the environmental noise at a location;
5. The use of 3D indoor models is increasing, therefore, is envisaged more use cases with integrated indoor-outdoor models, such as existing cases of facility management and navigation.

Moreover, considering BIM and GIS integration, it could be interesting to focus on issues including using a unified model based on the data format of IFC and CityGML for Smart City applications on multiple purposes across the whole lifecycle (Ma & Ren 2017).

CONCLUSION

The chapter proposed contains various contributions focused on the description of methods, processes and tools that can be adopted to achieve the Smart City goals. In particular, the aim is to clarify three crucial points concerning Smart Cities based on Digital Twins: the use of the term "Smart city" and the role that technologies had in its definition; the "3D city model" meaning and the integration procedures between BIM and GIS; the classification of 3D city models by use cases. BIM – GIS integration and its applications represent a strategic research topic for the future development of urban planning, then for society. The discourse is mainly grounded on all of the aspects of sustainability that could be successfully improved through this integration. The applications of BIM and GIS has been discussed from various points of view: definition and terminology clarifications, the role in the contemporary city discourse, the issues related to social discourse, the technical solution for their integration, the practical solution for the city governance. The discourse is framed in a broad speculative view concerning the digital phenomenon and the performances of digital tools in reshaping the features of the design practice: the methodological approach to design, the regulations and the consistency of design experimentations. ICT and digital tools are needed not only for the cause of design but their strategic role is recognized in almost all issues of contemporary society. This means that digitalization is gradually influencing not only the technical side of the architectural practice but also its political, economic and social sphere. This phenomenon requires a wide critical analysis concerning the global issues where digital is imposing its presence. Design practice has to jump from the scale of the single entity to a broad view on the city, and it has to interpret and to solve the political, economic, social and environmental issues linked to the city. Avoiding technical approach to embrace a multidisciplinary view in design is the most important outcome coming from the text proposed. The complexity of the issues related to the city requires not only the inclusion of various disciplines but also of different scales of the project which have been kept divided until now.

REFERENCES

AIA. (2008). *Document E202 Building Information Modeling Protocol Exhibit.* https://www.smacna.org/resources/resource/2008/01/04/aia-e202-building-information-modeling-protocol-exhibit

Armando, A., & Durbiano, G. (2017). *Teoria del progetto architettonico.* Carocci Editore.

ASviS Alleanza Urbana per lo sviluppo sostenibile. (2018), *L'agenda Urbana per lo sviluppo sostenibile.*

Aurigi, A. (2005). *Making the Digital City: The early shaping of urban internet space.* Ashgate.

Bahu, J.M., Koch, A., Kremers, E., Murshed, S.M. (2013). *Towards a 3D spatial urban energy modelling approach.* ISPRS Ann. Photogramm. Remote Sens. Spat. Inf. Sci. 2013, II-2/W1.

Batty, M., Chapman, D., Evans, S., Haklay, M., Kueppers, S., Shiode, N., Smith, A., & Torrens, P. M. (2000). Visualizing the City: Communicating Urban Design to Planners and Decision-Makers. Technical Report Paper 26. Centre for Advanced Spatial Analysis (UCL).

Biljecki, F., Ledoux, H., & Stoter, J. (2016). Does a finer level of detail of a 3D city model bring an improvement for estimating shadows? Springer International Publishing.

Biljecki, F., Stoter, J., Ledoux, H., Zlatanova, S., & Çöltekin, A. (2015). Applications of 3D City Models: State of the Art Review. *ISPRS International Journal of Geo-Information*, *4*(4), 4. doi:10.3390/ijgi4042842

Bolton, R., McColl-Kennedy, J. R., Cheung, L., Gallan, A., Orsingher, C., Witell, L., & Zaki, M. (2018). Customer experience challenges: Bringing together digital, physical and social realms. *Journal of Service Management*, *29*(5), 776–808. doi:10.1108/JOSM-04-2018-0113

Boschert, S., & Rosen, R. (2016). Digital Twin—The Simulation Aspect. In P. Hehenberger & D. Bradley (Eds.), *Mechatronic Futures*. Springer. doi:10.1007/978-3-319-32156-1_5

Bria, F., Sagarra, O., Rodríguez, J., Balcells, P. (2018). *Deployments of pilots in Barcelona (Decentralized Citizens Owned Data Ecosystem)*, V.1.0, 31.

Brockwell, S., & Horrocks, R. (2020). *BIM, GIS, and Smart Cities: Challenges and Opportunities*. Autodesk University.

Chen, R. (2011). The development of 3D city model and its applications in urban planning. *Proceedings of the 19th International Conference on Geoinformatics*. 10.1109/GeoInformatics.2011.5981007

Del Giudice, M., & Patti, E. (2014). *BIM and GIS for district modelling*. Research Gate.

Eremia, M., Toma, L., & Sanduleac, M. (2017). *The Smart City Concept in the 21th Century*. Science Direct. *Procedia Engineering*, 181.

Ferraris, M. (2018). *Scienza Nuova*. Rosenberg & Sellier.

Floridi, L. (2011). *The philosophy of information*. Oxford University Press. doi:10.1093/acprof:oso/9780199232383.001.0001

Gröger, G., Kolbe, T., Nagel, C., & Häfele, K. (2019). *OGC City Geography Markup Language (Citygml) Encoding Standard v2. 0. OGC Doc*. http://www.opengeospatial.org/standards/citygml

Hao, W., Yisha, P., & Xiaochun, L. (2019). Integration of BIM and GIS in sustainable built environment: a review and bibliometric analysis. In Automation in Construction. Elsevier.

Hénard, E. (1910). *The Cities of the future, Royal Institute of British Architects. Town Planning Conference, London, UK.*

Howard, E. (1902). *Garden Cities of To-Morrow*. Faber and Faber.

Jacobson, I. (1992). *Object-Oriented Software Engineering: A Use Case Driven Approach*. Addison-Wesley.

Kaden, R., Kolbe, T.H. (2014). Simulation-based total energy demand estimation of buildings using semantic 3D city models. *Int. J. 3-D Inf. Model, 3.*

Kaur, M. J., Mishra, V. P., & Maheshwari, P. (2020). The Convergence of Digital Twin, IoT, and Machine Learning: Transforming Data into Action. In M. Farsi, A. Daneshkhah, A. Hosseinian-Far, & H. Jahankhani (Eds.), *Digital Twin Technologies and Smart Cities. Internet of Things (Technology, Communications and Computing)*. Springer. doi:10.1007/978-3-030-18732-3_1

Kolarevic, B. (2003). *Architecture in the Digital Age. Design and Manufacturing*. Routledge.

Kurakula, V. (2007). *A GIS-Based Approach for 3D Noise Modelling Using 3D City Models* (Master's Thesis). International Institute for Geo-information Science and Earth Observation, Enschede, The Netherlands.

Liu, X., Wang, X., Wright, G., Cheng, J., Li, X., & Liu, R. (2017). A state-of-the-art review on the integration of building information modeling (BIM) and geographic information system (GIS). *ISPRS International Journal of Geo-Information*, *6*(2), 53. doi:10.3390/ijgi6020053

Ma, Z., & Ren, Y. (2017). *Integrated applications of BIM and GIS: an overview*. Procedia Engineering, Science Direct.

Mohanty, S., Choppali, U., & Kougianos, E. (2016). *Everything You wanted to Know about Smart Cities*. Research Gate.

Osello, A., & Ugliotti, F. M. (2015). *BIM, GIS, AR per il facility management*. Dario Flaccovio Editore.

Ross, L. (2010). *Virtual 3D City Models in Urban Land Management—Technologies and Applications* (Ph.D. Thesis). Technische Universität Berlin, Berlin, Germany.

Saran, S., Oberai, K., & Wate, P. (2018). Utilities of Virtual 3D City Models Based on CityGML: Various Use Cases. *Journal of the Indian Society of Remote Sensing*.

Vassileva, I., Dahlquist, E., & Campillo, J. (2016). The citizens' role in energy smart city development. *Energy Procedia*, *88*, 88. doi:10.1016/j.egypro.2016.06.055

Wu, H., He, Z., & Gong, J. (2010). A virtual globe-based 3D visualization and interactive framework for public participation in urban planning processes. *Computers, Environment and Urban Systems*.

Yamamura, S., Fan, L., & Suzuki, Y. (2017). Assessment of urban energy performance through integration of BIM and GIS for smart city planning. *Energy Procedia*, *180*, 1462–1472. doi:10.1016/j.proeng.2017.04.309

Zhiliang, M., & Yuan, R. (2019). *Integrated application of BIM-GIS: An overview*, Science Direct. *Procedia Engineering*, 196.

ADDITIONAL READING

Abdullah, H., & Kamara, J. M. (2013). *Parametric Design Procedures: a new approach to Generative-Form in the Conceptual Design Phase*. Research Gate.

Batty, M. (2013). Big Data, smart cities and city planning. *Dialogues in Human Geography*, *3*(3), 274–279. doi:10.1177/2043820613513390 PMID:29472982

Bria, F. & Morozov, E. (2018). *Ripensare la Smart City*, Codice editore.

Caputi, M., Odorizzi, P., & Stefani, M. (2015). *Il Building Information Modeling*. Hoepli.

Ciribini, L. C. (2016). *BIM e digitalizzazione nell'ambiente costruito*. Grafill.

Cudzik, J., & Radziszewski, K. (2019). Parametric design in architectural education. World Transactions on Engineering and Technology Education, Research Gate, Vol.17, No.4.

EUBIM Task Group. (2016). *Handbook for the introduction of Building Information Modelling*. European Public Sector.

KEY TERMS AND DEFINITIONS

3D City Model: It is a representation of an urban environment with a three-dimensional geometry of common urban objects and structures, with buildings as the most prominent feature.

BIM Model: The digital representation of physical and functional characteristics of a facility.

Governance: The terms typically refers to the highest decision-making level, includes the definition of objectives, strategy, control systems, organizational structure.

ICT (Internet and Communication Technologies): The set of methods and techniques used in the transmission, reception and processing of data and information.

Information Society: The information society is a context in which new information and telecommunication technologies play a fundamental role in the development of human activities.

Smart City: The concept aimed at fostering a sustainable city development through the optimal management of the resources, offering a comprehensive, high-quality life for the citizen.

Use Case: A meaningful set of spatial operations that accomplish a goal a user wants to achieve with a spatial data set.

Chapter 13
BIM Bin:
Waste Management Through BIM and Digital Twin

Ricardo Codinhoto
University of Bath, UK

Olivia Becher
Oxford University, UK

Jonathan Neil Heron
University of Bath, UK

Vincenzo Donato
https://orcid.org/0000-0002-3947-8402
Università degli Studi di Firenze, Italy

ABSTRACT

A groundswell of opinion exist about the present and future use of smart cities and digital twin technologies and processes and, despite increasing use of information modelling, artificial intelligence, and internet of things, many challenges remain in designing and implement integrated smart systems in large scale contexts. Often, the big picture is shadowed by fragmented processes, and there is a disconnect between the problem and the solution. This chapter aims to address this inverted approach, based on a solution looking for a problem by focusing on the problems of developing integrated solutions for smart cities based on digital twins. The narrative in this chapter is informed by a research project exploring the digitalisation of facilities management processes in Bath, UK. The conclusion is that the development of digital twins goes far beyond linking digital models to sensors.

DOI: 10.4018/978-1-7998-7091-3.ch013

INTRODUCTION

We want to improve our cities, but neglect critical issues that can lead to improvement. We crave efficiency and yet ignore the many areas that enable increased efficiency. We crave health but fail to deliver an integrated care service focused on a preventive approach. We spend a lifetime inside buildings that are poorly designed, built and maintained. Buildings are often too hot, too cold, too damp, with high levels of embedded CO2. They contribute to cities' poor sustainable performance. Cities can be overly crowded, incredibly unsafe, excessively polluted and fail to provide the right conditions for citizens to thrive. We also pay the price for not looking after our cities adequately, for not reacting quickly enough, and for not preventing problems before they occur. The response to COVID19 and the blast in Beirut are contemporary examples of these failures. Although the desired alignment between the different areas involved in the design, construction, operation and maintenance of cities has been continuously evolving as a reflection of enhanced technology, there is still much to do.

Early research, such as Hollands (2008), shows that from time to time, it is essential to revisit the troublesome multifaceted smart cities concept. Albino et al., (2015) and Caragliu et al., (2011) share this view. For both, fast developments in technology expands our understanding of smart cities and what can be done to make them smarter. Here, smart cities are defined as "an urban (data) ecosystem that emphasises the use of digital technology, shared knowledge and cohesive processes to underpin citizen benefits in sectors such as mobility, public safety, health and productivity." Essential to this definition is the concept of a digital twin. This way, we frame smart cities within a digital twin-based solution to address the key challenges that we face now and will face in the future. These include a growing population, increased traffic and pollution, increased waste generation, reduced access to homes and public spaces, and resources such as water and energy. This approach also offers a solution to tackle global warming as a result of uncontrolled carbon emissions in the context of tighter city budgets and ageing infrastructure. The problem and the solution are complex, for realising smart cities requires great efforts and a multidisciplinary integrated approach.

Here the smart cities subject is explored through the lens of the AEC sector. Much like everything else, AEC is undergoing a digital transformation. Technologies that heretofore were restricted to digital solutions are permeating the pre and post construction building industry. Companies can now create building data ecosystem grids to capture and analyse data trails from building users and determine how to enhance building performance. These building data ecosystem grids seed the development of smart cities when connected to one another. In other words, on the one hand BIM becomes a platform for data integration; on the other hand it becomes data ecosystems built around service analytics platforms that tie the ecosystems together, thus creating a digital twin. Such a transformation has been fostered by several disruptive Information and Communication Technologies (ICT), such as Building Information Modelling (BIM), Artificial Intelligence (AI), the Internet of Things (IoT) and most recently the Internet of Behaviour (IoB). However, the transformation is still slow, and while a unified platform does not exist, we will continue working with parts of a bigger reality, that is 'dizygotic' digital twins.

The part of the reality studied in this chapter is related to the use and operation phase of building assets. Previous studies in this area have focused on technologies for data collection, storage and sharing (Fialho et al., 2019; Ye et al., 2018). As explained later in this chapter, these studies promote digital ways to improve building efficiency and their point of departure tends to be the implementation of technology. This is often achieved through BIM-supported simulation tools that predict the performance of newly designed buildings. However, considerably less attention has been given to service performance. It is widely acknowledged in the literature that the volume of resources used during the life cycle of buildings worldwide is the largest of all sectors. Buildings still represent the largest energy-consuming sector in the economy, with over half of global electricity consumption being responsible for approximately one-third of global carbon emissions (Abergel et al., 2017). We also know that the operation phase uses roughly 80% of the cost and resources of a building during its life cycle. However, very little detailed information exists about services inefficiencies that can inform a value-based digital transformation.

Therefore, this chapter explores service analytics as a contributor to the creation of digital twins. Focus is placed upon inefficiencies in the context of waste management by seeking to respond the following research question: can BIM and IoT improve the efficiency of waste management. Two aspects makes this a relevant problem to study. On the one hand, the amount of waste that is generated every year is vast and requires action. On the other hand, governing bodies within cities and businesses are struggling with increased cost of running their services and increased efficiencies are sought everywhere, including from waste management. To start exploring this subject, the University of Bath campus provides a context to exemplify the difficulties related to establishing a twin system with enough information granularity to enable efficient waste management. We argue that the principles adopted in this research can be generalised and used to support waste management within cities.

BIM BIN: SMART WASTE MANAGEMENT

Waste and its management is a universal problem. The existing literature highlights that circa 2.2 billion tonnes of waste are generated per year worldwide. Also that 99% of products released on the market are disposed of within six months. Overall this contributes to large amounts of landfill gas emissions to be produced, where 60% is CO_2 (carbon dioxide), and 40% is CH_4 (methane). Worth reminding readers that the impact of CH_4 on O_3 (ozone) is 25 times higher than CO_2. Waste generated in construction adds to that. Widely acknowledged in the literature, the construction industry contributes significantly to waste in landfills. Circa 30% of the total waste generated in most European countries are disposed of in landfills (Dyer, 2012). In 2014, an estimated 202.8 million tonnes of waste was generated in the UK Of this, 60% was generated by construction, demolition and excavation (DEFRA, 2016). Despite these alarming figures, proposed solutions are failing to resolve the problem, and currently, so much waste is being disposed of that landfill spaces are becoming increasingly scarce.

Reportedly, only a very few robust and scalable initiatives and programs exist to attenuate waste generation. A review of approaches for the management and analysis of Municipal Solid Waste Management Systems (MSWMS) shows that WM practices are best if tailored to local practices (Wilson et al., 2012) and requires the engagement of relevant stakeholders (Guerrero et al., 2013; Abas and Wee, 2014). Also, Life Cycle Analysis (LCA) are extensively used to integrate multiple stages of WM, however, it is argued that they are inappropriate for guiding complex management decisions (Lazarevic et al., 2012). Irrespectively of appropriateness, without being mandatory in many countries, the use of such initiatives has been left to the goodwill of organisations seeking compliance with "Duty of Care Regulations" and sustainability credits. Despite helpful, waste management platforms remains a relatively inaccurate approximation of the reality as critical elements of information (such as materials origin and usage) are not considered within existing methods.

Paradoxically, cities of the world are struggling with achieving rising environmental targets whilst also tackling reduced operational budgets. Despite city infrastructure constant evolution enabled by technology, its various components are not connected efficiently. At a small scale, it is argued that effectively planned facilities and supporting services can create significant business returns (Alexander, 1996). The rationale is based on introducing professional FM methods to support increased efficiency (Amaratunga et al., 2002; and Lennerts et al., 2003), on-going responsibility for the continuous operation of services (Atkin and Brooks, 2009) and a holistic view of the dynamics of the workplace amongst the product design and production of the physical workspace, and also between people and processes, and people and their environment (Alexander, 1996). A great extent of the environmental impacts of the built environment takes place in the use and operation phase (e.g. energy and water consumption, CO2 emission, waste generation); and yet, the management of resources is based on piecemeal information databases.

In this regards, it is widely accepted that digital technologies can enhance our ability to connect data at a systemic level, thereby enhancing efficiency. Effective information management forms a critical aspect of our ability to coordinate processes so to achieve its required output (Alexander, 1996) whilst providing valuable information to planners, designers and corporate decision-makers (Atkin and Brooks, 2009). However, the efficient utilisation of information, its management and its supporting technology have been somewhat problematic (Barrett and Baldry, 2003; Abel et al., 2006; and Bainbridge and Finch, 2009). At building scale, managers can have access to a variety of data sources, but the opportunities to utilise or manipulate data are frequently unexploited (Barrett and Baldry, 2003). Often, the systems that can generate the necessary information required by senior executives to make decisions are lacking (Alexander, 1996), and current solutions rely on the duplication of data that leads to the over-processing of data and information overload (Barrett and Baldry, 2003).

Evidence also shows that BIM, as a platform for data integration, has great potential to support the management of operations of large scale complexes (Kassem et al., 2015). In this respect, the use of BIM has been investigated as a way to support the reduction of unnecessary processes through better integration of building information (Ballesty et al., 2007; Eastman et al, 2018). However, while its use within the design and construction phase is well studied (Eastman et al, 2018), its adoption for the management of large complexes is still embryonic (Cracknell, 2014).

Even most large public owners who have been early adopters of BIM, such as GSA, USACE or Senate Properties, have used BIM more in managing their construction projects than into managing activities after construction. In the UK, one of the reasons for that is that the development of the Government Soft Landing approach was an afterthought and there was, (and there still is), a lack of a strategy to engage public and private organisations to adopt BIM beyond their projects. As highlighted by Codinhoto and Kiviniemi (2014), there is very little that has been implemented extensively in building use and operation and even less have been measured in terms of improvements made due to BIM. For them, articles reporting on case studies about the application of BIM to FM focuses mostly on Hard FM and in particular on the accuracy of 3D as-built models and the link to digital statutory/ maintenance/ supplier information available through online services. In this respect, whilst research about BIM-FM for building specific is increasing, large cases (such as cities) are almost non-existent (Fialho et al. 2020). Large complexes data connection can increase information transparency and impact on significant gains of efficiency through enabling the sharing of resources and the systematic management of service provision. However, very little is known about how it can be achieved.

With regards to waste, several gaps exist in the literature relating to BIM's applications to MSWMS. Research studying the applications of BIM for MSWM is almost non-existent. Such research as there is in this area focuses on improving MSW collection on a strategic level, without considering the implications at an operational level. Also, no research has been carried out on the applications of BIM to waste generation through modelling an individual's experiences of, or navigation through, a building. To fill the above knowledge gap, the BIM Bin research presented here was proposed to explore how BIM, as a platform for digital and sensor data integration, enables gains of efficiency related to waste management.

Research Design and Method

In this project, the University of Bath campus, seen as a mini-city, provided an opportunity to investigate waste management. As a typical Higher Education Institution (HEI), the University is composed of buildings with various purposes, including accommodation, education, and catering. It has a population of almost 20,000 people (students and staff), and it processes circa 280 tonnes of waste (general and recyclable). Due to the scale of work involved in modelling the campus waste management operations, the library building was selected as a pilot, and three other accommodation buildings (not included in this chapter) were used for further investigation. The library is a five-storey building at the centre of the campus, and it can receive circa five thousand students per week at peak times. The research approach adopted was design science research (Kuechler and Vaishnavi, 2008); as such, the research design and data collection techniques were based on problem clarification, suggestion, development, evaluation and conclusion as described in the following.

Problem Clarification, Suggestion and Development

Problem Clarification

Waste collection optimisation is a multifaceted problem. So, to identify the intricacies of waste collection, a series of interviews were conducted throughout the research with five members of staff involved with waste collection from a managerial and operational perspective. The interviews revealed that the library building has a waste collection regime based on two full collections per day during weekdays irrespective of whether bins are full or not. Each level has a different number of bins. Also that there are three categories of waste (general, recyclable and paper). That each waste category is gathered and transported in a trolley to an area near the service lift in each floor before being taken to the main waste bin outside the building. It is worthy noting that none of the waste generated within the campus (or in the city) ends in a landfill. That is because recycling waste is sent directly to local recycling centres and general waste is sent to local energy from waste facilities.

It was also found that from the one hand, waste collection is constrained by sanitary measures that prevent the development of smells and pests; from the other, it is constrained to specific times of the day, usually when building users are not on site. Thus, measures to increase efficiency must incorporate a range of considerations than only volume and time taken to complete waste collection. However, volume and time are a good departing point for increasing efficiency.

With regards to the library population, data obtained from interviewees (Figure 1) shows that at peak time, the building receives circa five thousand students per week. In this respect, evidence shows that a direct correlation exists between waste generation rates and population (Karak et al., 2012; Jambeck et al., 2015). Thus, term times will experience more significant accumulation of waste than holiday periods and reaches its peak during exam weeks when occupation levels are much higher.

Suggestion (for improvement): Kanban-inspired alternate collection regimes to alter the number of bins per collection

Figure 1. Library building occupation variation

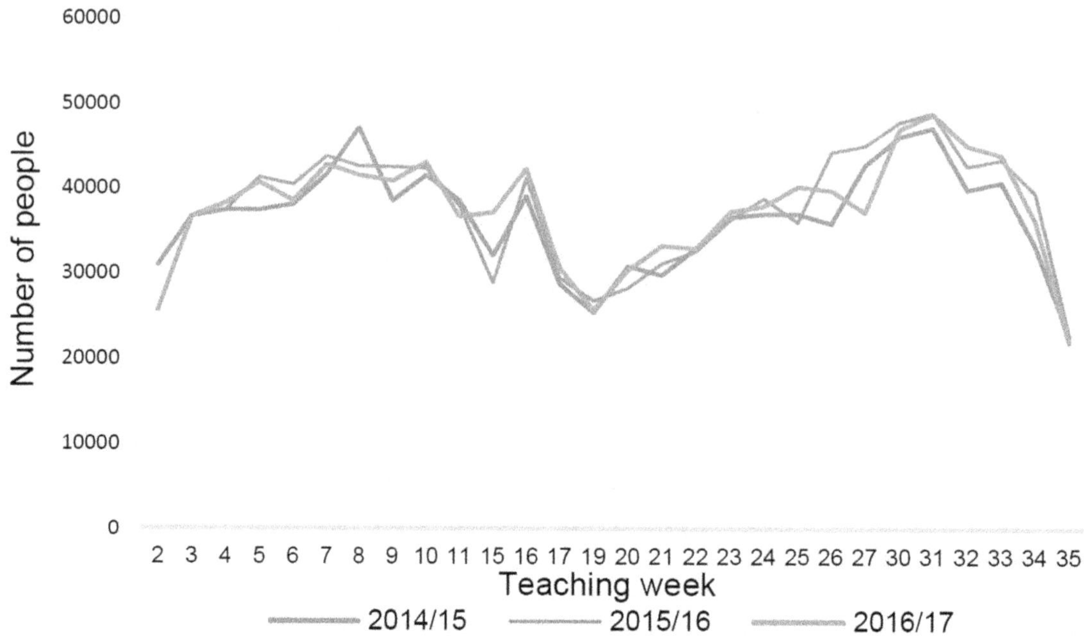

To minimise the number of bins per collection, bin collection was conceptualised as the inverse of a supermarket inventory problem; instead of re-stocking produce when shelves are empty, porters empty bins when they are full. Kanban systems optimise re-stocking by only supplementing stock once it reached critically low levels. As such, the Kanban concept was applied to waste (bin) collection by changing the collection regime to only incorporate full bins.

The Kanban-inspired bin periods method is based on 'bin collection periods'. As shown in Figure 2, this concept is based on the fact that specific bins will reach maximum capacity more rapidly than others. According to the frequency with which bins become full, and the maximum time between collections, 'bin collection periods' can be assigned to a bin.

Currently, two bin collection events occur per day at 05:30 and 13:00. Therefore, a simplified model of bi-daily collection regimes was proposed: a morning collection regime (C1) and an afternoon regime (C2). The height of waste in each bin was measured by standing a ruler against the outer surface of the bin, at approximately 05:00 am on four weekdays during term time. Dimensions of the bins were taken to allow mean waste height measurements to be converted to volumes, which were subsequently converted to % maximum capacity (% utilisation). Within the model proposed, bins that reached a % utilisation greater than, or equal to, 50% just before a collection, require emptying at each collection. These were denoted T1, whereas, those below 50% utilisation just before collection were assigned a period of T2. Bin periods were assigned for all bins of the library according to the below flowchart.

The survey of bin height revealed that at peak periods almost all general waste bins reached a % utilisation greater than 50% just before a collection, whereas most of the paper and recycling bins did not. Therefore, T1 bin periods were allocated to all general waste bins and T2 for paper and recycling

Figure 2. Collection regimes assigned using bin periods

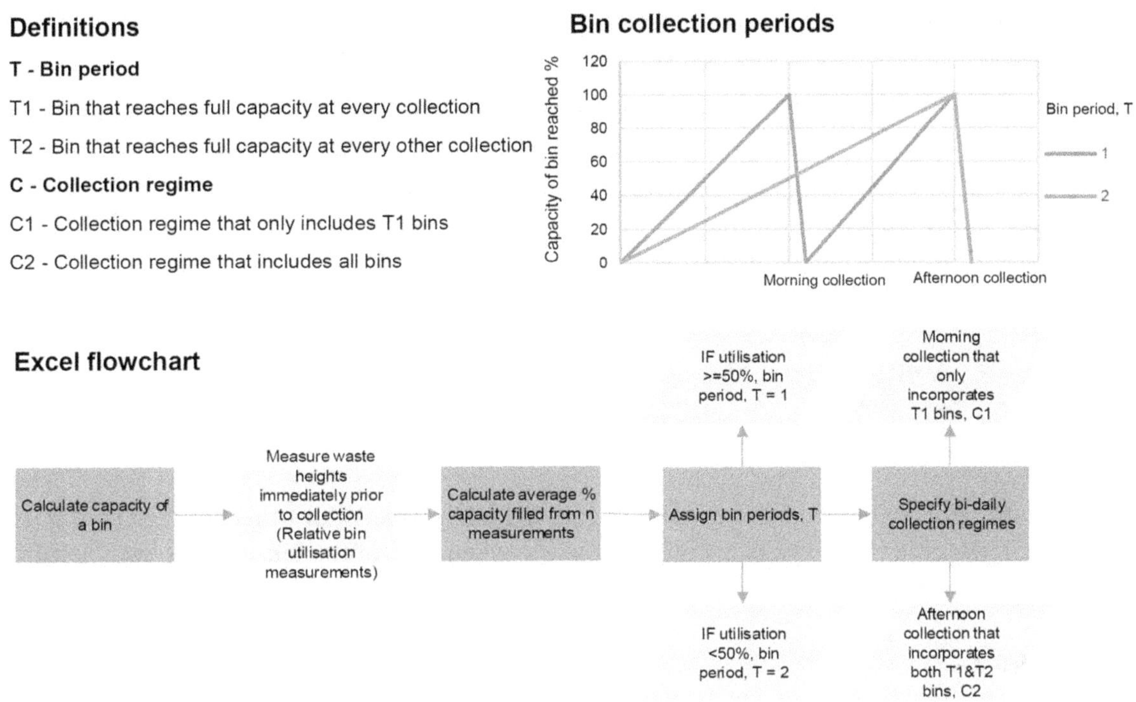

bins. Level 1 was not assigned a C1 regime because it contains only five bins, making the difference between C1 and C2 collection regimes insignificant in comparison to the other levels.

Combined with occupation information, it is possible to forecast which bins will be over 50% full on a given day and which will not, allowing bin collection regimes to be coordinated with occupancy trends. To test this hypothesis, the weight of different bins on level 3 were analysed concerning how many people entered level 3, over holiday and semester periods. To do that, library entrance data was gathered by accessing mobile phone logs within the Wi-Fi system on an hourly basis, for each level of the library. This data provided a good indication of the number of library occupants per floor (excluding occupants not connected to the Wi-Fi and those transitioning between floors within the hourly time-step). Also, the weight of five bins was monitored using an 'Ohuas ranger 3000 compact bench scales' of 30kg capacity connected to a laptop (via an RSVP output to USB converter), positioned below bins to output weight measurements into an Excel spreadsheet at one-minute time intervals. This way, all assumptions related to waste increase rates and occupation were confirmed as shown in Figure 3.

Below in Tables 1 to 4, the data from measuring the mean waste height in 'cm' of each bin in floors 1, 2, 4, and 5 of the library building is presented. In addition to the heights, the volume of waste (capacity in cm^3) in each bin corresponding to waste height is presented using the width and length of the bins. Each number associated with the bin denotation was used for identification purposes. The level measured is in centimetres and the capacity as a dimensionless ratio of the volume of waste compared to the total volume of the bin. The total number of bins on the four floors combined is 59, and the number of bins that were given a T1 status and then a C1-2 collection regime was 19. Out of the 19 bins over 50% capacity, 13 were general waste bins, three were for cans and plastic (recycling), and the other 3

Figure 3. Level 3 population versus the weight of waste in bins; term and holiday time

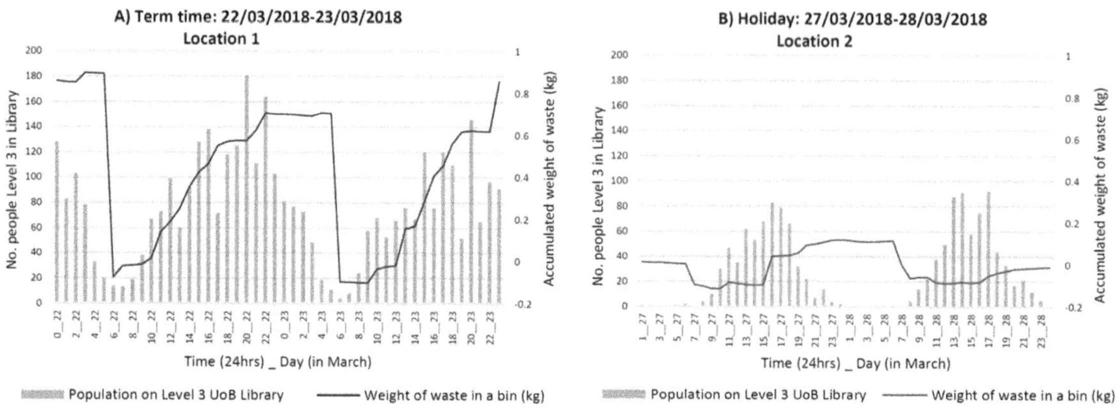

were paper bins with the reduced size compared with the other bins. Whilst this research considers the optimisation in the collection of waste; collecting waste when 68% of the bins are not even half full, is not an efficient collection method.

Development

The Shortest Path

Table 1. Level 1 bin heights and corresponding capacity

Floor 1		Tuesday		Wednesday		Thursday		Friday				
		Height	Capacity	Height	Capacity	Height	Capacity	Height	Volume	Average Capacity	Bin Period	Collection Regime
GW	1	60	0.6	70	0.7	70	0.7	50	0.5	0.625	T1	C1-2
	2	45	0.45	45	0.45	40	0.4	50	0.5	0.45	T2	C2
	3	55	0.55	30	0.3	50	0.5	30	0.3	0.4125	T2	C2
CP	1	55	0.55	55	0.55	70	0.7	55	0.55	0.5875	T1	C1-2
	2	35	0.35	30	0.3	45	0.45	45	0.45	0.3875	T2	C2
	3	30	0.3	35	0.35	40	0.4	45	0.45	0.375	T2	C2
P	1	35	0.35	45	0.45	50	0.5	60	0.6	0.475	T2	C2

Taking into consideration that enough information about the waste collection regime in place, the occupation of the library building and the volume of waste generated, the next step of the research was to increase efficiency through digital means.

The first task was defining the shortest path for the collection of waste (efficiency of time and effort in the waste collection). For that, the logic of the Travelling Salesman Problem (TSP) algorithm in Python script (adapted from Anon., 2015) was inverted and applied on the bin collection regime proposed. The TSP is a well-known Non-deterministic Polynomial acceptable (or NP-hard) problem in mathematics. This essentially means that finding an exact (lower bound Held-Karp) solution for any

Table 2. Level 2 bin heights and corresponding capacity

		Floor 2										
		Tuesday		Wednesday		Thursday		Friday				
		Height	Capacity	Height	Capacity	Height	Capacity	Height	Volume	Average Capacity	Bin Period	Collection Regime
GW	1	50	0.50	50	0.50	35	0.35	50	0.50	0.46	T2	C2
	2	45	0.45	50	0.50	55	0.55	40	0.40	0.48	T2	C2
	3	70	0.70	65	0.65	30	0.30	35	0.35	0.50	T1	C1-2
	4	80	0.80	70	0.70	50	0.50	50	0.50	0.63	T1	C1-2
	5	60	0.60	50	0.50	40	0.40	40	0.40	0.48	T2	C2
	6	30	0.30	45	0.45	30	0.30	30	0.30	0.34	T2	C2
	7	80	0.80	100	1.00	65	0.65	55	0.55	0.75	T1	C1-2
	8	60	0.60	100	1.00	80	0.80	35	0.35	0.69	T1	C1-2
	9	60	0.60	55	0.55	35	0.35	40	0.40	0.48	T2	C2
CP	1	40	0.40	55	0.55	50	0.50	40	0.40	0.46	T2	C2
	2	45	0.45	50	0.50	40	0.40	30	0.30	0.41	T2	C2
	3	40	0.40	30	0.30	35	0.35	25	0.25	0.33	T2	C2
	4	25	0.25	25	0.25	30	0.30	15	0.15	0.24	T2	C2
	5	50	0.50	50	0.50	45	0.45	45	0.45	0.48	T2	C2
	6	15	0.15	20	0.20	30	0.30	20	0.20	0.21	T2	C2
P	1	25	0.25	30	0.30	50	0.50	60	0.60	0.41	T2	C2
	2	30	0.30	45	0.45	25	0.25	75	0.75	0.44	T2	C2
	3	45	0.45	50	0.50	25	0.25	25	0.25	0.36	T2	C2
Pa	1	40	0.67	50	0.83	35	0.58	45	0.75	0.71	T1	C1-2
	2	30	0.50	35	0.58	35	0.58	35	0.58	0.56	T1	C1-2
	3	45	0.75	60	1.00	25	0.42	25	0.42	0.65	T1	C1-2

more than a trivial number of visited locations requires an excessive amount of computational power (Yu, 2014). The TSP Python script implemented incorporates composite heuristics: one that constructs the tour by incrementally adding new unvisited locations that are nearest to the last visited, and another which switches the paths and adopts improved path combinations (Anon., 2015).

The TSP Python script was inserted into Dynamo BIM 0.9.1.4062 (adapted from Venkov, 2017) and connected to a LOD 200 Revit model of the library building to assess the kanban bi-daily collection

Table 3. Level 4 bin heights and corresponding capacity

		Floor 4										
		Tuesday		Wednesday		Thursday		Friday				
		Height	Capacity	Height	Capacity	Height	Capacity	Height	Volume	Average Capacity	Bin Period	Collection Regime
GW	1	45	0.45	65	0.65	65	0.65	65	0.65	0.60	T1	C1-2
	2	70	0.70	60	0.60	55	0.55	55	0.55	0.60	T1	C1-2
	3	65	0.65	70	0.70	60	0.60	60	0.60	0.64	T1	C1-2
	4	80	0.80	70	0.70	85	0.85	60	0.60	0.74	T1	C1-2
	5	80	0.80	50	0.50	30	0.30	45	0.45	0.51	T1	C1-2
	6	40	0.40	45	0.45	25	0.25	45	0.45	0.39	T2	C2
	7	35	0.35	60	0.60	60	0.60	35	0.35	0.48	T2	C2
	8	50	0.50	40	0.40	50	0.50	40	0.40	0.45	T2	C2
CP	1	70	0.70	85	0.85	40	0.40	60	0.60	0.64	T1	C1-2
	2	75	0.75	60	0.60	80	0.80	45	0.45	0.65	T1	C1-2
	3	25	0.25	30	0.30	30	0.30	25	0.25	0.28	T2	C2
	4	25	0.25	35	0.35	30	0.30	35	0.35	0.31	T2	C2
P	1	55	0.55	40	0.40	30	0.30	25	0.25	0.38	T2	C2
	2	50	0.50	45	0.45	50	0.50	30	0.30	0.44	T2	C2
	3	30	0.30	50	0.50	20	0.20	15	0.15	0.29	T2	C2

Table 4. Level 5 bin heights and corresponding capacity

Floor 5		Tuesday		Wednesday		Thursday		Friday				
		Height	Capacity	Height	Capacity	Height	Capacity	Height	Volume	Average Capacity	Bin Period	Collection Regime
GW	1	55	0.55	50	0.50	45	0.45	60	0.60	0.53	T1	C1-2
	2	60	0.60	40	0.40	25	0.25	50	0.50	0.44	T2	C2
	3	60	0.60	55	0.55	30	0.30	60	0.60	0.51	T1	C1-2
	4	60	0.60	75	0.75	60	0.60	55	0.55	0.63	T1	C1-2
	5	55	0.55	50	0.50	30	0.30	45	0.45	0.45	T2	C2
	6	45	0.45	25	0.25	35	0.35	25	0.25	0.33	T2	C2
CP	1	30	0.30	20	0.20	40	0.40	30	0.30	0.30	T2	C2
	2	35	0.35	40	0.40	30	0.30	30	0.30	0.34	T2	C2
	3	50	0.50	45	0.45	30	0.30	35	0.35	0.40	T2	C2
	4	30	0.30	30	0.30	40	0.40	20	0.20	0.30	T2	C2
	5	5	0.05	15	0.15	10	0.10	20	0.20	0.13	T2	C2
P	1	45	0.45	30	0.30	10	0.10	15	0.15	0.25	T2	C2
	2	40	0.40	40	0.40	15	0.15	35	0.35	0.33	T2	C2
	3	45	0.45	20	0.20	50	0.50	20	0.20	0.34	T2	C2
Pa	1	25	0.42	25	0.42	25	0.42	15	0.25	0.38	T2	C2
	2	30	0.50	15	0.25	10	0.17	15	0.25	0.29	T2	C2

regimes through scenario-testing. Note: the Dynamo workflow (Steps 1 to 6 below) could not circumvent wall boundaries or other obstacles, limiting its practical application to bin collection regimes. 'Corridor nodes; and 'Lift nodes' were added to manipulate the collection path to minimise the extent to which paths intersected walls and to ensure it went via the staff lift, respectively. Other elements of the waste collection path were also incorporated into the model, as well as the necessary model assumptions:

Figure 4. **Step 1:** *assess the kanban bi-daily collection regimes through scenario-testing*

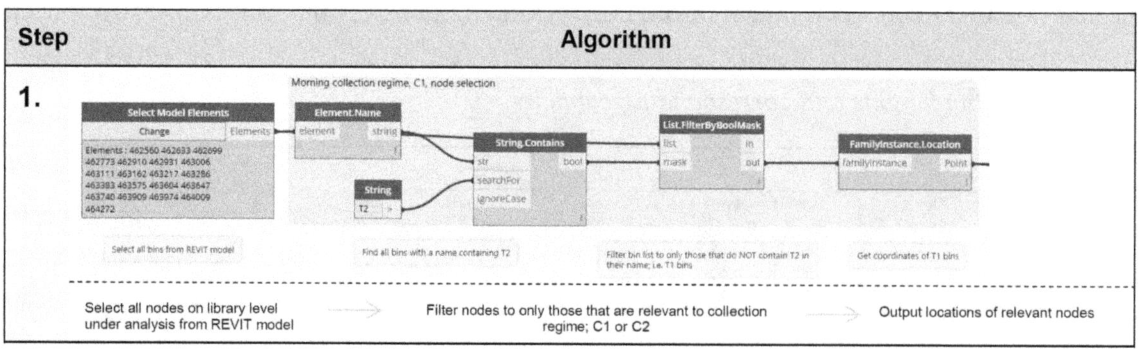

As learnt from the portering team, the lift usually gets full once the bins on level 3 have been loaded. This was accounted for as part of the analysis of C2 on level 3 (under the assumption that the lift does not reach its capacity before the end of C1).

The TSP script was run on each level of the Revit model. The optimum paths found shown in Figure 10 (blue representing C1 and red C2). The larger green circles represent T1 bins emptied at every collection, and the smaller ones represent T2 bins (emptied every other collection). 'Corridor nodes' were

Figure 5. **Step 2:** *add variable - time for walking between bins with trolley – assumed walking speed with trolley: 0.5 m/s (adapted from Venkov, 2017)*

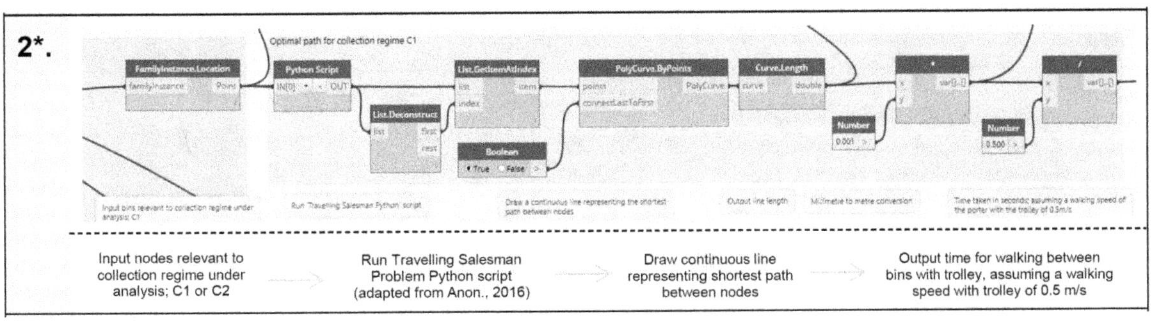

Figure 6. **Step 3:** *add variable - time taken to empty and replace each bin liner – assumed value: 15 secs/bin*

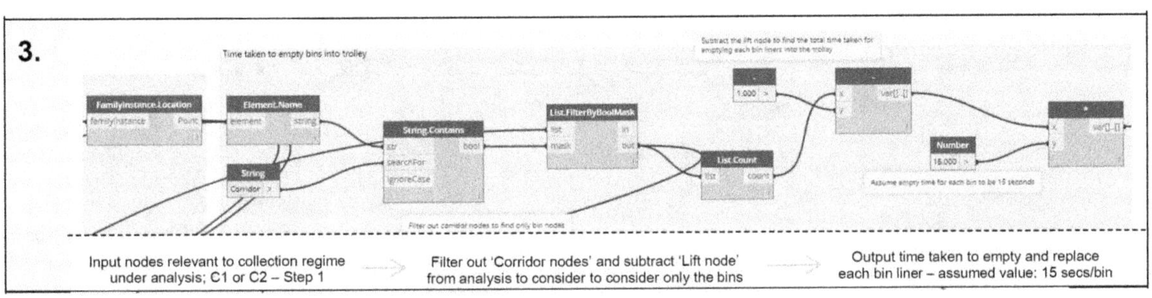

Figure 7. **Step 4:** *add variable - travelling to the lift and back when trolley reaches capacity - approximately six bins*

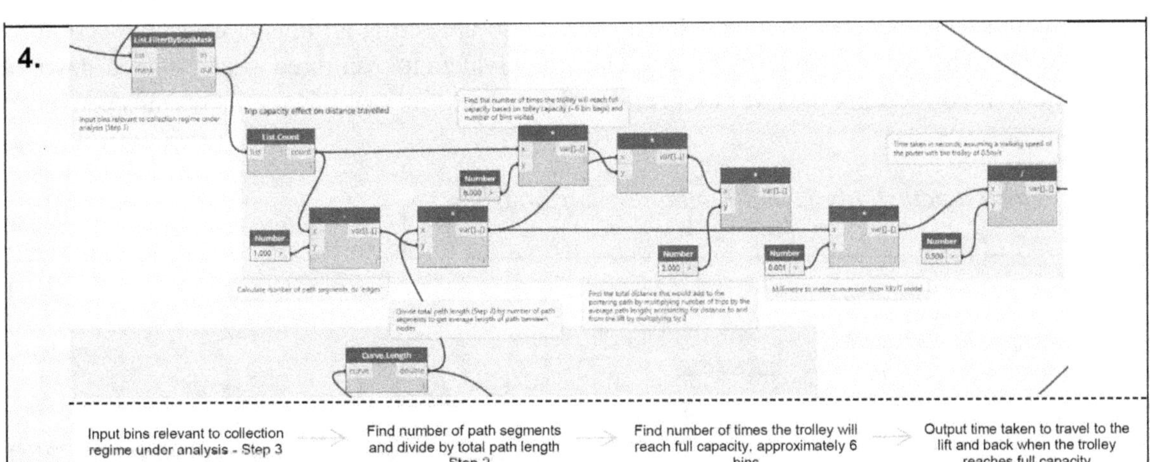

Figure 8. **Step 5:** *add variable - unloading each trolley bag into the lift as well as from the lift on Level 1 – assumed time: 10secs/bin*

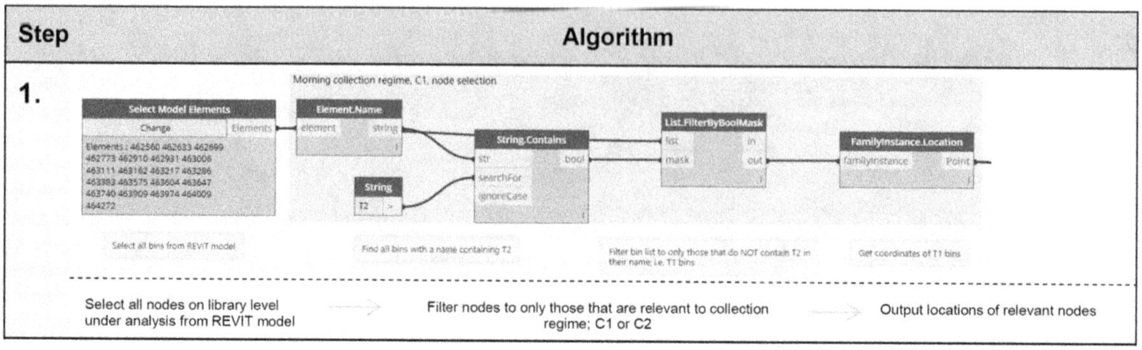

Figure 9. **Step 6:** *add variable - transportation to the level below via the lift – assumed lift speed: 2m/s*

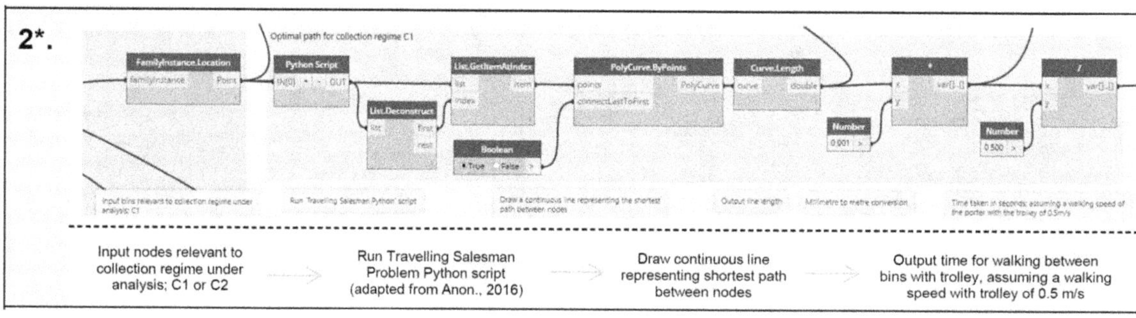

added to minimise the extent to which the shortest paths intersected walls (blue and red rectangles for C1 and C2, respectively). 'Lift nodes' were also added to manipulate the path to go via the staff lift (light blue rectangle).

Results of scenario-testing between C1 and C2 are presented in Figure 11 above. The united benefit of assigning alternate collection regimes, in comparison to collecting all bins at every collection, was found to be 128 hours per year (in level 3 only). This is equivalent to over three weeks' standard working hours per floor. If considering the five floors of the library building as uniform, the savings per year per

Figure 10. Shortest path analysis and Kanban scenario-testing

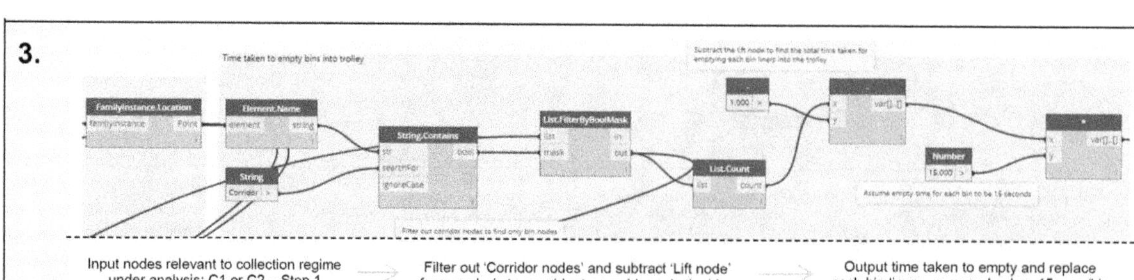

Figure 11. Time measured in seconds for each stage in bin collection for each collection regime

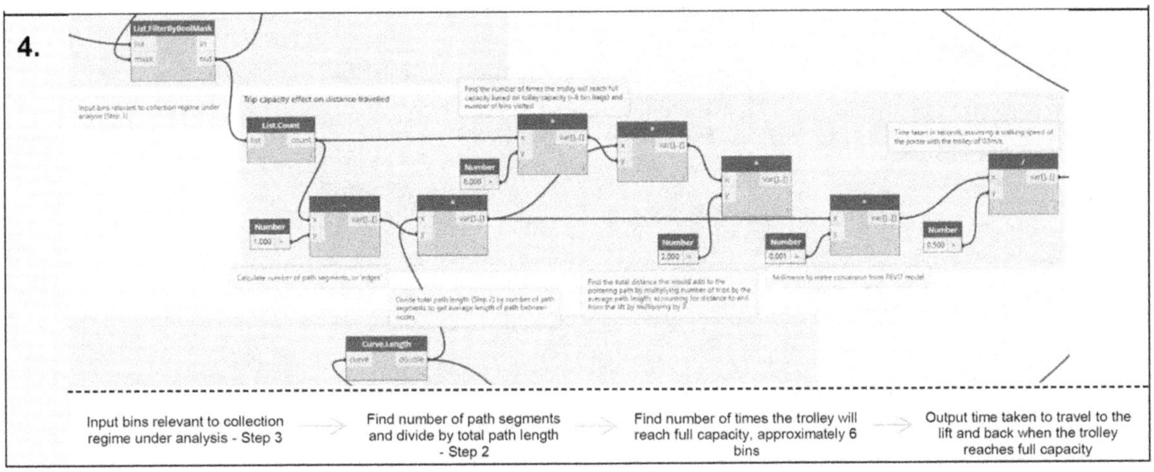

building is 80 days. That represents nine thousands and five hundred days of savings if extended to the 118 buildings within the university campus. It should be noted that this benefit contains a large margin of error as it is limited by the assumptions made in the model and does not consider other conditions of the collection path, such as picking up litter. It also, does not consider different building uses and sizes. However, the results suggest that, if this approach were implemented across multiple buildings around the campus, a significant reduction in time taken for bin collection, and hence annual FM expenses, could be achieved.

Bin Layout Optimisation Using Network Visualisation

In this development, bin layouts were optimised to reduce bin collection time. For that, Graph Theory was used following previous research by Biagini et al., (2015) and Donato (2017) demonstrating how graphs can be applied to optimise building attributes. The existing Revit model was prepared for input into Gephi 0.9.2. This was done through conceptualising bins as the nodes of a network, in which edges represent the collection path during bin collection. The information was entered through a Revit plug-in (Donato, 2014) for parameterising rooms and circulation spaces into nodes and edges, and exporting as an Excel file. To ensure the nodes extracted by this plug-in symbolised the bins, a room was allocated per bin in Revit using Voronoi diagrams. Voronoi cells divide space according to a set of points, or centres; such that locations within each cell, are closest to its centre than any other centre. Therefore, large open spaces were divided according to bin locations, by creating Voronoi diagrams in MatLAB R2016b. The coordinates of bins were extracted from AutoCAD 2017 drawings generated from the Revit model.

Figure 12 shows how the Revit floorplan of Level 3 was prepared for modelling the morning collection regime, **C1**, using level 3 as a network. Black and grey objects represent the existing walls, doors and furniture. Large and small circles represent **T1** and **T2** bins, respectively. Blue lines represent the Voronoi cell boundaries used to divide open spaces into a room per bin. Doors within these walls were added to represent connectivity between adjacent rooms. The edges of the network (red lines) were undirected and weighted by the straight distances between bins, accounting for doors and walls.

Figure 12. File preparation for bin network visualisation

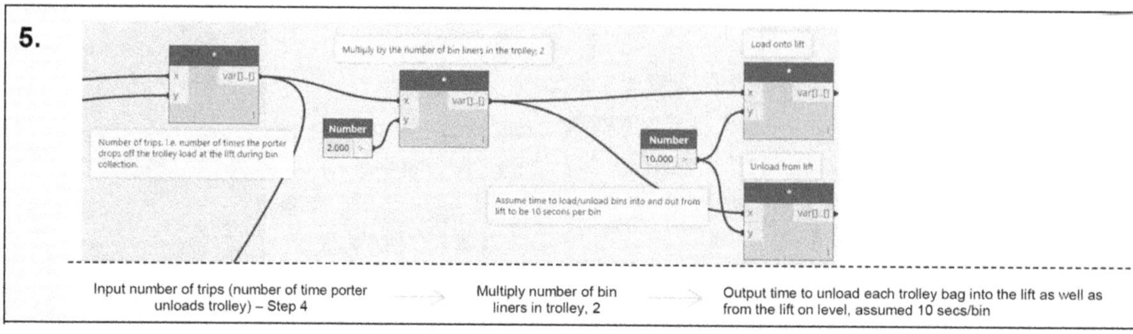

The Excel spreadsheets, populated by the Revit plug-in (Donato, 2014), were subsequently inputted into in the open-source network visualisation software Gephi 0.9.2. The resulting analysis involved parameterising complex bin layouts, concerning the distances between bins, for easy comparison between multiple networks (Table 5). Five different bin networks were visualised in Gephi 0.9.2. Preliminary regression analysis was subsequently completed to compare which statistics were strongly correlated to collection time, found from the shortest 'collection' path algorithm.

Table 5. Network parameters available from the analysis in Gephi 0.9.2

Nodes	Points of interest in the network.
Edges	Connections between nodes.
Degree	An average number of edges per node; the connectivity of the network.
Average Weighted Degree	Average of the weights of all the edges per node (Totet, 2013).
Network Diameter	Size of the network; distance between farthest two nodes in the network (GitHub, 2015).
Graph Density	Graph density is equal to 1 if all nodes are connected to all other nodes in the network; completeness of the graph (GitHub, 2015).
Clustering Coefficient	An average of how complete each nodes' cluster is in a network (GitHub, 2015).
Modularity	The strength with which a network can be divided into communities or clusters (GitHub, 2015).
Average Path Length	The average distance between all node-pairs; compactness of the network (GitHub, 2015).

Having seen how bin collection can be optimised by implementing a Kanban-inspired alternate collection regime system, the next step was to investigate how the overall layout of bins can be optimised to reduce collection time. Five bin layouts were analysed in a 'Circular Layout' in Gephi 0.9.2 (Figure 13). Darker and thicker edges represent longer distances between nodes. This can be seen for the 'Level 3 C1 bin network' and 'Level 3 C2 bin network. In these graphs, the weakest line in 'Level 3 C1 bin network' is between '01_GW_T1' and '02_GW_T1', which are the two closest bins. Preliminary linear regression analysis was then completed to investigate which Gephi 0.9.2 statistics correlate most strongly with a bin collection time.

Figure 13. Bin network visualisation in Gephi 0.9.2

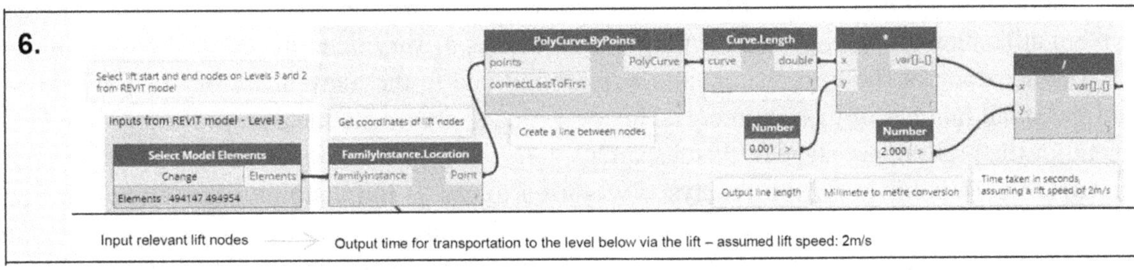

Results suggested that the statistics that provide the best correlation with overall bin collection time were 'Network Diameter' and 'Av. Path Length' (highlighted in grey in Table 6) and that 'Graph Density' provides the best correlation with bin collection time per bin Length' (highlighted in blue in Table 6). These findings could be used to optimise bin layouts concerning 'Graph Density' to reduce bin collection time for a given number of bins. However, only the C1 networks for levels 2, 4 and 5 were completed due to time constraints of the research project; a more extensive dataset would be required to provide conclusive evidence to support bin layout design decisions.

Table 6. C1 and C2 bin network visualisation and statistic comparison

Network statistic	Level 3		Level 2	Level 4	Level 5	Bin collection time	
	C1	C2	C1	C1	C1	Seconds	Seconds per bin
Nodes	5	31	8	8	6		
Edges	7	40	9	10	7		
Av. Degree	2.3	2.6	2.3	2.5	2.3	62%	-77%
Av. Weighted Degree	33	4	36	28	41	-89%	83%
Network Diameter	3	10	4	4	3	97%	-88%
Graph Density	0.47	0.09	0.32	0.36	0.47	-91%	94%
Av. Clustering Coefficient	0.33	0.25	0.14	0.14	0.36	91%	-80%
Modularity	0.24	0.46	0.32	0.24	0.17	4%	47%
Av. Path Length	1.67	4.25	1.96	2.00	1.67	97%	-87%

Exploring a Theory to Explain Behaviour Using Bins

The data available enabled researchers to explore potential theories explaining waste disposal. Having analysed the bins as a network and bin layout optimisation using network visualisation,' it was hypothesised that the bin layout not only impacts bin collection time but also which bins are more utilised than others. Findings from the literature suggest that convenience plays an essential role in recycling behaviour (Wan et al., 2012; Miafodzyeva and Brandt, 2013). For this research, the convenience of using a bin was defined by a bin's visibility to the user and the footfall rates experienced at its location. The following 'waste management metrics' were assigned to simplify the problem:

Y variable - Relative bin utilisation: the height of waste in each bin on level 3 was measured at approximately 05:00 on four weekdays during term time. These measurements were used to generate a relative bin utilisation graphic, by representing bins as circles of varying radii according to mean waste height. The accuracy of these measurements was limited due to the unevenness of waste inside each bin and the small number of measurements. However, this approximate measure was considered to be sufficient for the preliminary stage of this research.

X1 variable - Visibility Graph Analysis: was used to assess bin visibility. The branch of Space Syntax analysis that is most applicable for measuring bin visibility is Visibility Graph Analysis (VGA); For that, DepthmapX 3.0 (Li and Klippel, 2012) was chosen for modelling visibility because it is an open-source software that performs variations of Space Syntax analysis on plan representations of spaces, on a city or building scale. As with network visualisation in Gephi 0.9.2, the power of DepthmapX 3.0 analysis lies in its ability to parameterise specific attributes of a layout for ease of analysis through Graph Theory concepts. Thus, a VGA was carried out by inserting a simplified DXF AutoCAD drawing into DepthmapX 3.0 software. Figure 14_bookmark39 shows the VGA of different levels of the UoB library to illustrate how different plans affect which regions have the highest levels of visibility.

Figure 14. VGA of floor levels 2, 3 and 4 of the UoB Library Building

X2 variable - Footfall analysis and movement: footfall analysis was done with SmartMove software and used to assess movement within the building. SmartMove is a Rhinoceros plug-in that also uses Graph Theory concepts to simulate complex movements and interactions of people through various spatial configurations, such as multi-directional people flow (Sharma and Tabak, 2008). Data input included DXF AutoCAD 2017 drawings of the level 3 that were imported into Rhinoceros 5 (64-bit). The capacity of all workspaces in the library was surveyed and inserted into the SmartMove simulations of each level. Sequences of movement through space were then assigned by assuming that all workspaces were full and half of the population entered and exited from each 'Entrance/Exit' space. This provides an approximate simulation of the overall pattern of movement on the level by assuming people separate

Figure 15. Footfall analysis of levels 2, 3 and 4 of the UoB library building

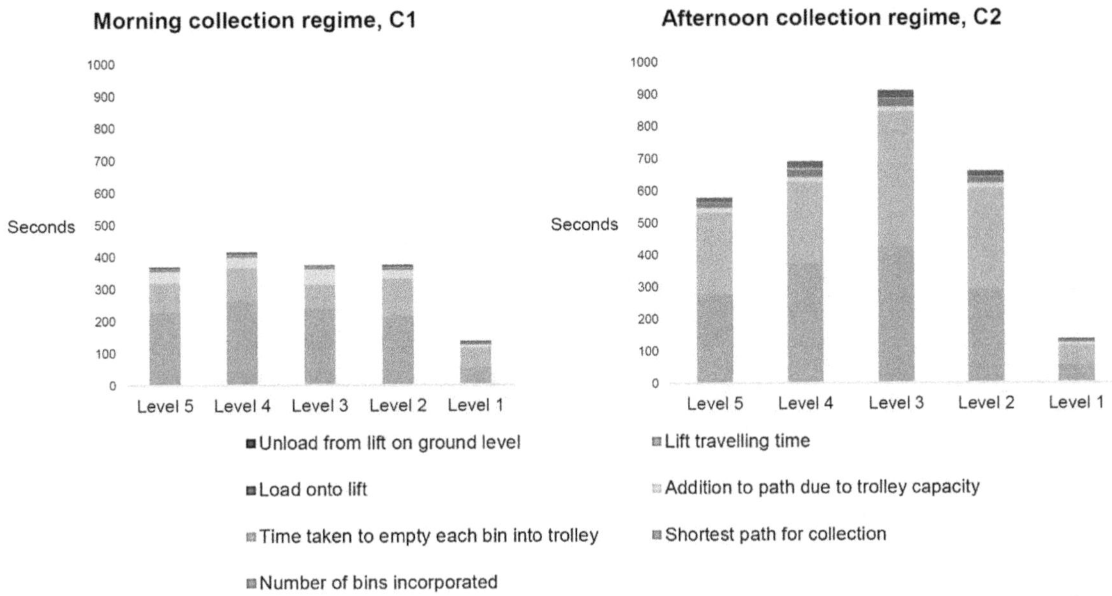

in proportion to the capacity of library workspaces. The footfall analysis was conducted, and a heat map of the most visited areas was generated (Figure 15).

Results of the comparison between bin utilisation, visibility and footfall analysis on Level 3 of the UoB Library are shown in Figure 16. In 'X1. VGA versus Y. relative bin utilisation' it is not possible to distinguish different visibility levels between bins that are highly utilised and those that are not. However, in 'X2 - Footfall analysis versus Y. relative bin utilisation' there is a much clearer variation in footfall rates with bin utilisation. The most utilised bins are adjacent to the 'Exit/Entrance' spaces, which experience the most significant rates of footfall.

Both visibility and footfall likely influence bin utilisation. However, the results of this high-level analysis indicate that footfall provides a better indication of bin utilisation than visibility. A possible explanation for this is that occupants are likely to be, or become, familiar with where bins are, without them having to be within their fields of vision.

Finding that bin utilisation correlates with the regions of most significant footfall aligns with the hypothesis that convenience plays a significant dominant influence over recycling, based on findings from (Wan et al., 2012; Miafodzyeva and Brandt, 2013). The quantification of individual patterns of behaviour, in terms of rates of footfall, can be used to justify and guide management decisions (Browne et al. 2014). Recent research has shown promising results in how 'choice architecture' interventions could increase recycling rates (McCoy et al., 2018). However, this research was severely limited for replicability across other HEIs due to the omission of a systematic procedure for bin rearrangement (McCoy et al., 2018, pp. 613). Choice architecture interventions for recycling are rooted in the concept that placing bins in different areas will increase their utilisation. This research shows that SmartMove could be used to inform bin placement in such interventions based on footfall analysis. However, the only high-level, visual analysis was undertaken, which is limited to providing exact directions for bin placement.

Figure 16. Visual correlation between footfall, visibility and bin utilisation

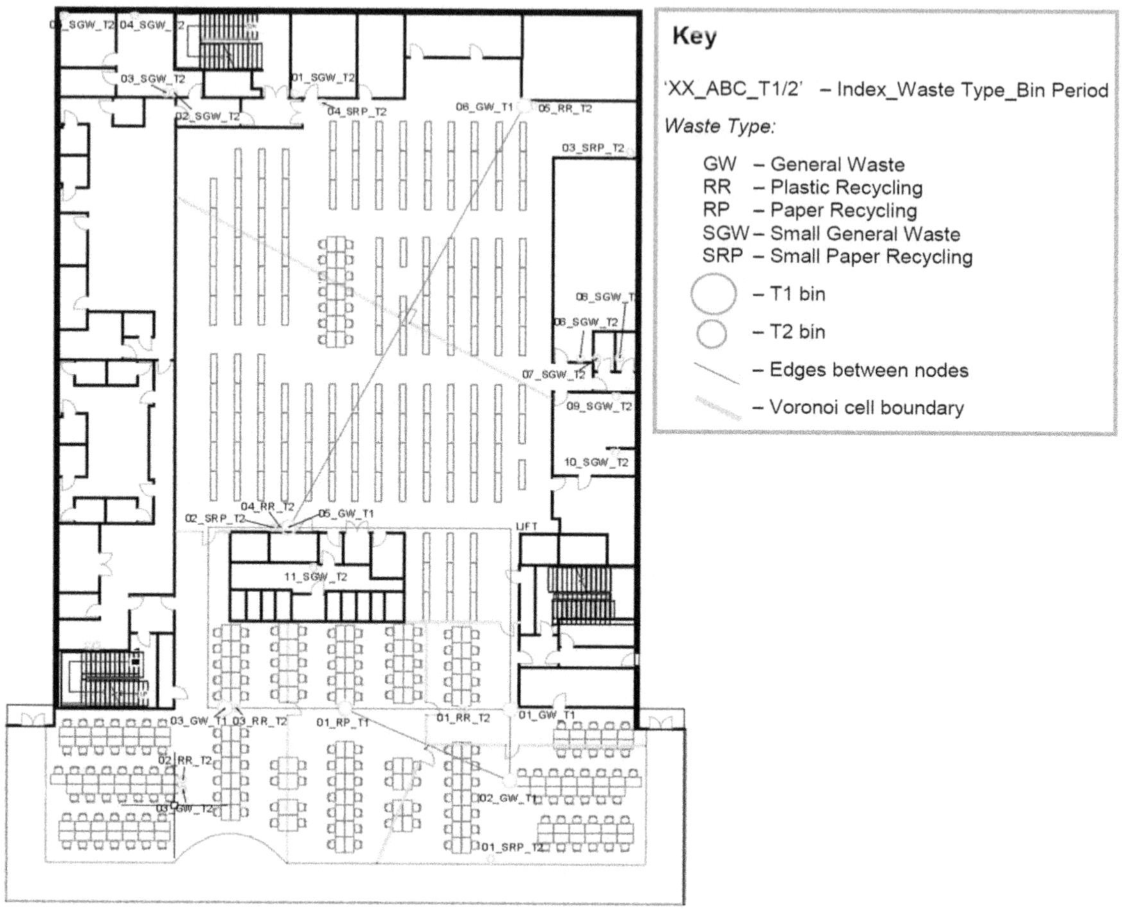

IoT and Digital Twinning

To this point, all data analysed produced an intermediate understanding of how waste is and can be managed more efficiently that is based on analogue surveying methods. This section expands on that; finding real-time accounts of the waste produced on the 3rd and 4th floor of the library building through the addition of sensors for measuring waste generation through time connected to the digital model. This way, an enhanced understanding of which bins must be collected and which would generate a collection regime more precise. For that, an ultrasonic sensor was assembled to gather real-time data of waste height. Sensor components and their roles are presented in the following. The assembled sensor can be seen in Figure 17.

Ultrasonic Sensor HC-SR04: The HC-SR04 was chosen due to the availability of the device and the time and budget restraints within this research. HC-SR04 uses sonar to identify objects within a range of 2 to 400. It uses ultrasonic transmitters (i.e. a trig pin, and receivers, and echo pin) and distances are calculated by measuring the time between the transmission and return of a sonar wave (Anon, 2019).

NodeMCU: was used to record sense data from the HC-SR04 remotely. It was chosen because it is based on a single-board microcontroller, with an XTOS operating system running on the ESP8266 which allows a connection to a chosen Wi-Fi. A programming code within NodeMCU and continuous power supply are required for continuous data collection

Arduino: was used to generating the code mentioned above, written in Arduino in the C+ language (available by contacting the corresponding author). The code enabled the interaction with the HC-SR04 and the Wi-Fi, finding the time when each distance is calculated, and then sending the information to an Excel sheet in an online database.

Firebase online database: was used to increase the NodeMCU storage space to a period of seven working days.

Figure 17. The assembled sensor used in the research

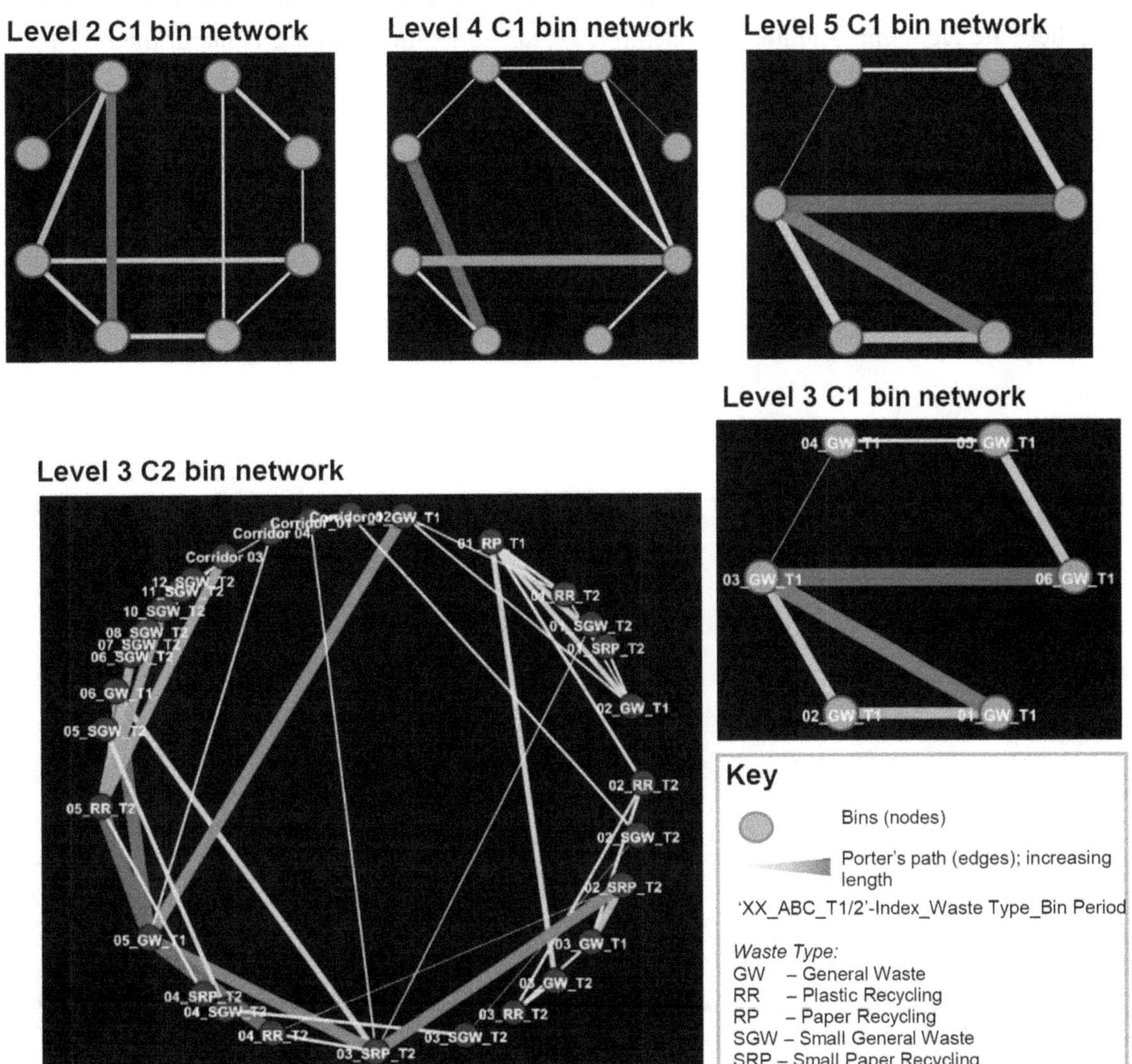

Battery pack: the HC-SR04 requires a constant supply of 5 volts; thus, a rechargeable battery pack was utilised with the capacity of seven days of energy supply.

Connections: four female to female DuPont jumper wire cables were used for connecting the NodeMCU and the HC-SR04 (one for the trig pin, another for the echo pin, and the other two for the ground and voltage). One micro USB cable joined the NodeMCU with the battery pack.

The test procedure was defined in a series of incremental stages to assure that the sensor worked correctly. The following tests were run: 7-day energy consumption test combined with a static test (no distance variation); 2-day in locus single bin test (connection to waste bins, tampering prevention, data capture and reading intervals of 2 and 3 minutes); 4-day in locus multiple bin test (multiple data sources, data capture and reading intervals of 5 minutes) this was done in levels 3 and 4 of the library also to investigate data collection from bins with different capacities. The results of the last two tests are presented in the following.

With regards to multiple sensors at level 3, real-time data from five of the general waste bins are presented across a span of 4 days in this segment. Below, Figure 18 shows the real-time waste levels of general waste bin sensors 1 to 5, respectively. In the four days of testing, due to the bi-daily collection regime, there are eight opportunities for each bin to reach maximum height. None of the bins reached maximum height, as shown in the graphics where no peak reaches 70cm in height. These bins are col-

Figure 18. Level 3 general waste heights bin sensors 1 to 5

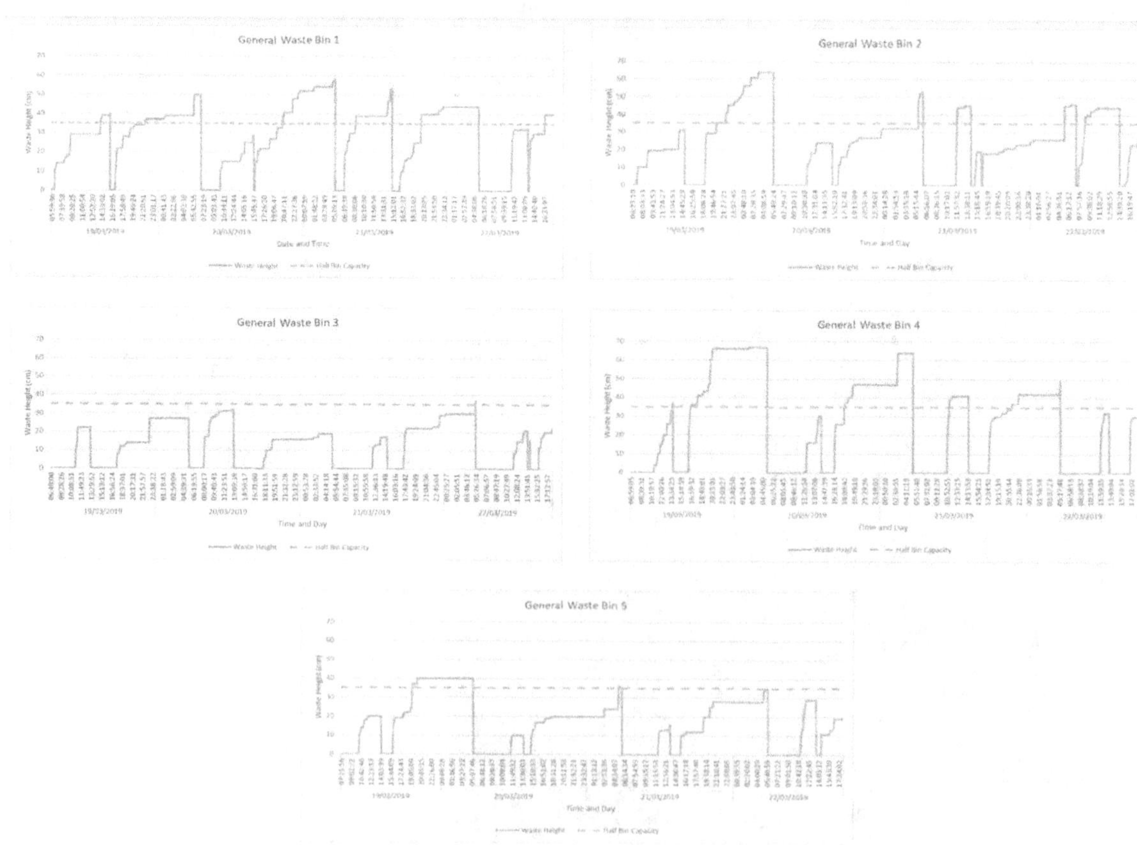

lected twice a day, yet none reach the level where they need to be collected, demonstrating some level of inefficiency in the waste collection process.

A secondary finding is the number of times before collection the waste reaches half height. From the 05:00 collection time to the 13:30 collection time, collectively, a half capacity threshold was reached 6 times, despite having 20 opportunities to do so. The following waste generation period between 13:30 and 05:00 collection time increases from this to 11 times out of a possible 20. This further reinforces the notion that the current collection regime, or potentially the number of bins should be altered for improved efficiency and reduced waste production. Also, by observing the number of times the total waste produced in 24 hours, it was found that all bins produced equal to or more than the bin height ten times out of 20 opportunities; with this number coming from only general waste bins 1, 2, and 4. General waste bins 3 and 5 only reach maximum capacity if the time threshold is extended to 32 hours.

With regards to multiple sensors at level 4, real-time data from two waste bins (general) is presented across a span of 4 days. The outcome of the tests is presented in Figure 19. The gap in time denotes the difference between the sensor unattached and attached to the lid due to external tampering. Similarly to the bins in level 3, the results show that of the bins that accumulated the most waste neither reach maximum height. Half-height is reached three times by GW3, and twice by GW6, just. Within 24 hours maximum height would be accumulated twice by GW3 alone.

Figure 19. Level 4 general waste heights bin sensors 3 and 6

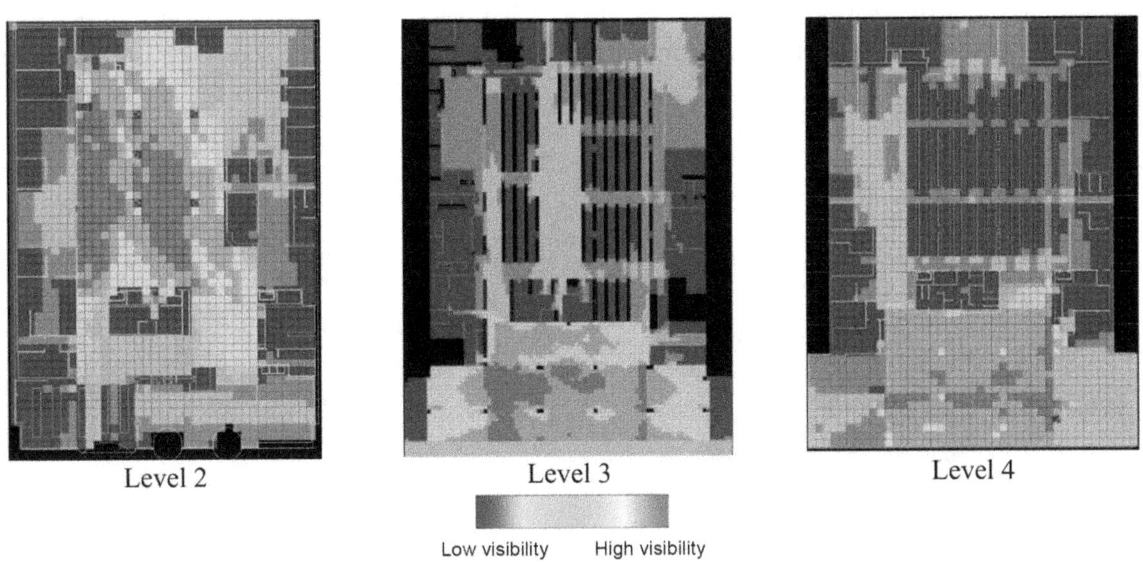

Finally, with regards to the link of sensor data with the digital model aimed at creating a visual device showing a live bin loading ratio visualisation graphic based on a traffic light system (from green representing empty to red representing full). For this purpose, a code was created in Dynamo that linked the recordings from the online Excel database to the bin models. Figures 14 to 16 present the code used to import the day, time, and waste height into Dynamo, thus changing the bin (model) colour based on

Figure 20. Code: Excel to Dynamo import - Time

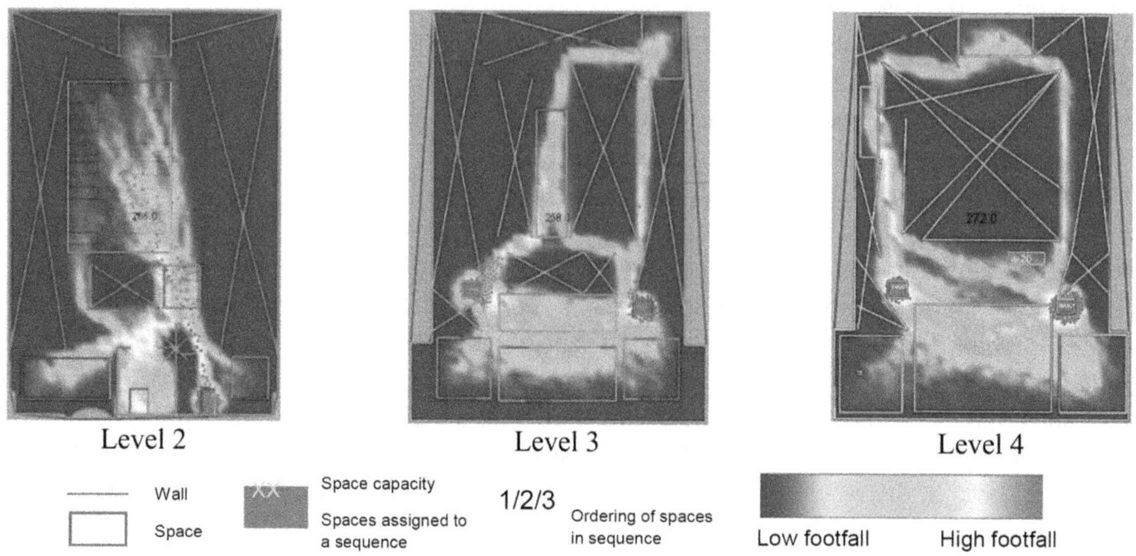

the waste height of the bin. A slider in the code changed the time and day, allowing for a variation in colour to be seen in each bin as the heights increased.

An image of the five general waste bins on floor 3 that were tested can be seen in Figure 23. The colours of the bins represent the loading ratio just before the collection took place. The results show that

Figure 21. Excel to Dynamo Import - Day

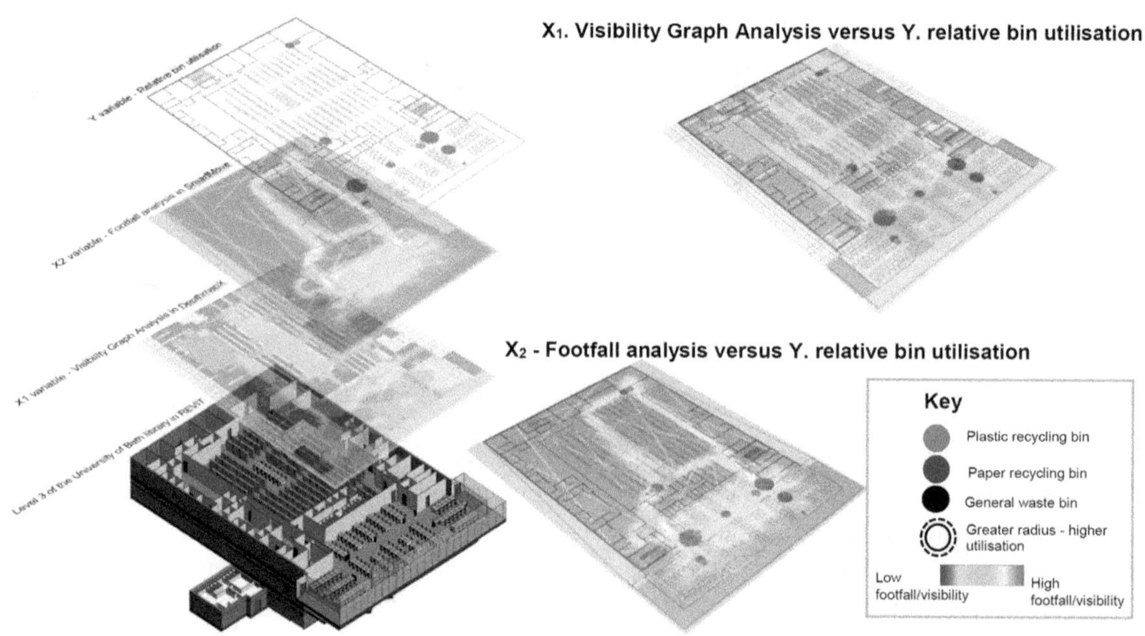

Figure 22. - Excel to Dynamo Import - Waste Height

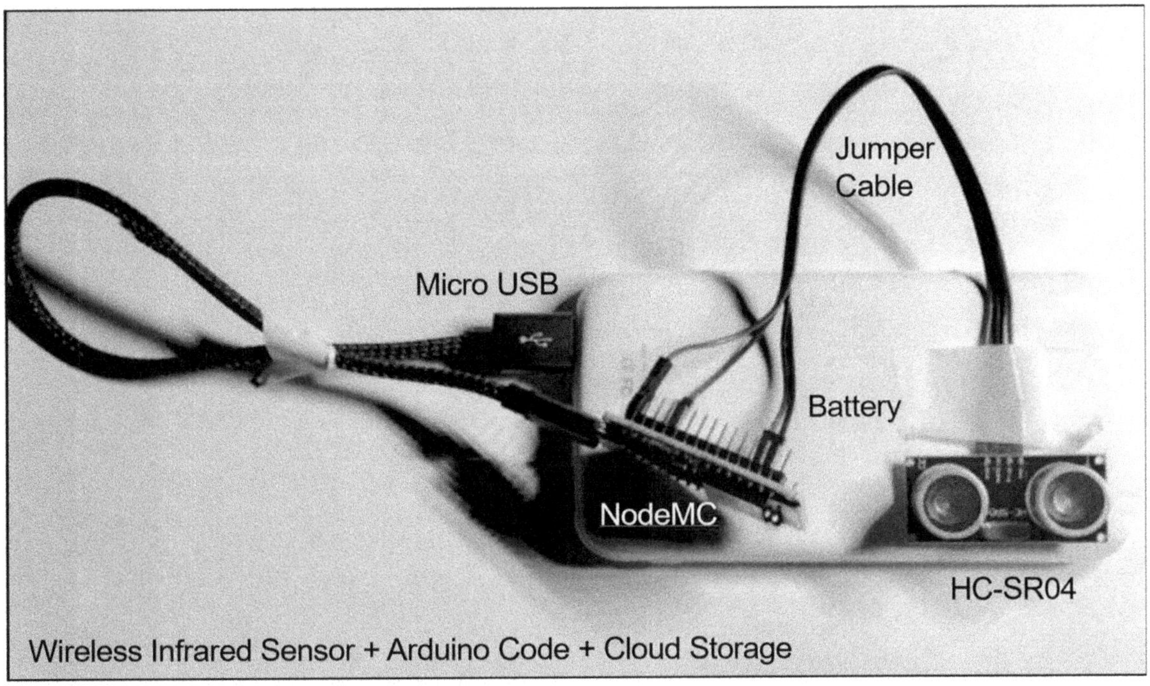

four bins out of the 30 shown have a distinct red/orange colour signifying a relatively full bin - all at the 05:00 collection time. Another 11 present yellow implying the waste height is about half; and the rest is a green or blue, which means less than half full. This graphic shows that 89% of the bins collected in almost an entire week are not full, and 57% are not even half full. Referring back to the premise of this research, this result corroborates with previous ones providing further evidence of a need to optimise the waste collection in an attempt to increase service efficiency.

EVALUATION AND DISCUSSION

The 'Implementation of BIM for waste management' elements addressed the research gaps found in BIM's applications, found from the literature review. BIM was implemented to scenario-test (Hamidi et al., 2014) a Kanban-inspired bi-daily bin collection system (Thürer et al., 2016) which indicated significant potential savings in bin collection time during the operational stages of a building's life (Codinhoto and Kiviniemi, 2014) through. The advantages of this method include its ease of application and flexibility to the unique characteristics of different contexts through the input of parameters such as bin locations, walking speed and trolley capacity (Wilson et al., 2012; Laurent et al., 2014a). Also, the collaboration between stakeholders (Guerrero, Maas and Hogland, 2013), including facility managers and members of the portering team, can be facilitated in the use of one BIM model (Cheng and Ma, 2013). Initial findings suggest that bin collection regimes could be forecast based on annual occupancy trends. Thus, providing a BIM model that responds to the operations of a building throughout its life to reduce costs (Wang et al., 2013).

Figure 23. Level 3 bin loading ratio visualisation graphic 'just' before collection 20/03 - 22/03

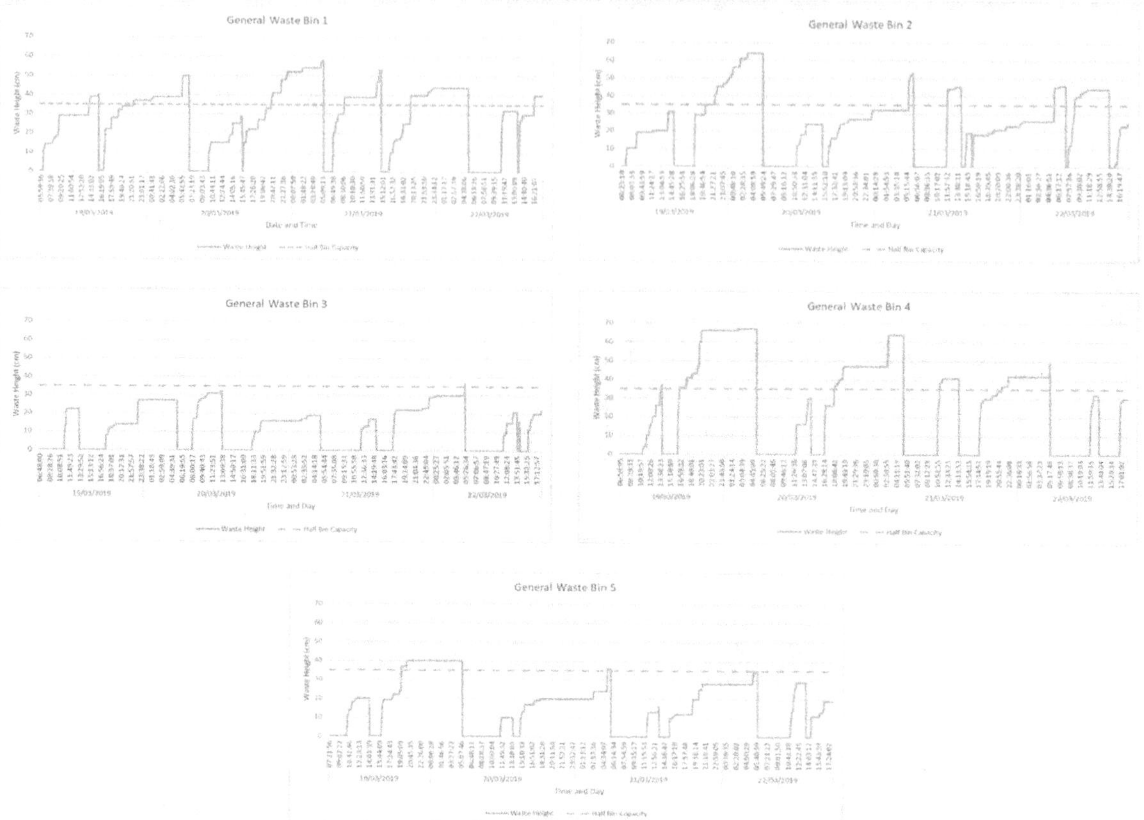

Potential Further Optimisation Solutions

With regards to the results obtained through this research, here plausible improvements that can be made to the current system are introduced, none of which are based on digital twin technology.

Removal of Bins + Changes in Collection Time

This research has found through live monitoring waste collection that the amount of general waste generated per week is consistently less than half of the bin capacity available. Therefore, as a suggestion for improvement, the proposed collection regime can be maintained while half the bins can be removed. The bi-daily collection regime remains the same because calculating the capacity of the bins required, considering the number of times they were emptied. Also, the time of collection has to change, from 13:00 to 11:00, and from 05:00 to 19:00. This time change allows for both an 8 hour and 16-hour collection window to remain constant, permitting the same amount of generation in bins, but also means that the collection of bins is visible to the students.

New Possible Collection Times

The second suggestion for further improvement is to implement a new collection regime. Results show that maximum loading capacity has never been reached. In addition to this, as the waste accumulated between the 05:00 to 13:00 window is consistently smaller than the 13:00 to 05:00, over 24 hours, on average, the maximum general waste height still does not exceed the capacity of the bin. Thus, by generalising the general waste height results to the other bins, the new collection regime for all bins is suggested to change to once a day. The new suggested collection time for the single-day collection regime should be between 21:00 and 01:00. It appears that maximum waste generation is usually between this time when considering all the waste produced in a day.

Estimation of Optimisation

Analogue data collected at the early stages of this research has revealed a potential savings of 128 hours per year per floor resulting from an alternate collection regime over the collection of each bin at every collection time. 128 hours per year accounts to approximately one-third of the overall time spent each year by porters collecting waste. This figure is limited by assumptions made in the absence of live data, and it does not include other factors, such as general clean-up and litter picking occurring as part of bin waste removal.

Through either of the suggested further optimisation regimes, it is conservatively estimated that approximately 50% of the collection time can be saved. This reduction means that approximately 192 hours per year per floor within the UoB library could be saved. A 192 hours reduction in time spent corresponds to approximately five working weeks. If extrapolated to the whole five storeys, that can lead to 120 days being saved per building and 14,000 days for the university campus. This figure is admittedly imprecise as not all buildings are the same, or have the same use and users, and waste collection regime. Nonetheless, it shows that efficiency gains can be reached.

Reflection on the Scalability of the Method

The focal point of this research was the library building; therefore, the proposed method omitted any consideration of WM operations at a municipal-level. Having said that, BIM models are increasingly widespread in the construction industry, and this trend is likely to continue in future years. This suggests that the applications of BIM for waste collection could be readily replicated across other buildings within other institutions. In addition, WM metrics found in the existing literature were different at institutional and municipal levels; suggesting that research should be completed on how to integrate MSWM at multiple scales. BIM also plays a crucial role in the Smart Cities initiative. The Smart Cities initiative aims to improve coordination between municipal services, such as public transport and utilities, to achieve greater efficiency and sustainability on a city scale. Given BIM's applications in monitoring and managing building assets in a holistic way, it follows that BIM would be a natural facilitator of the Smart Cities initiative. In the context of this research, innovations, such as Kanban-inspired bi-daily collection regimes, could be implemented across other institutions through BIM. Thus, bin collection services both within a building and between buildings could be integrated digitally.

FUTURE RESEARCH

The use of sensors for monitoring of waste bin loads enables a more accurate estimation of potential landfill CO_2 and CH_4 generation per organisation. To do that, research to characterise waste bin content concerning CO_2 and CH_4 generation is needed, as well as the creation of camera sensors that synchronously identify and classify waste being disposed of in bins.

Also, the use of sensors for monitoring of waste bin loads associated to the shortest path algorithm and the BIM model can be studied along with intelligent robot wheels coupled with waste bins. This way, robot wheels movement can be triggered when bins are full, and the coordinates generated through the shortest path analysis can inform the direction the robot wheels should follow.

CONCLUSION

This chapter explores the concept of smart cities enabled by digital twins by presenting a study focused on the digitalisation of waste management processes. Much of what was presented in this research was based on trial and error, in the absence of existing supporting literature focused on the same subject. The concept of smart is of great importance in this research. For a smart(er) waste collection system enabled by information and communication technologies can, as demonstrated, increase the operational efficiency of organisations of all scales and types. It can also inform all stakeholders (users and service providers) and improve the quality of services while inducing positive behaviour that promotes ecological and societal welfare.

This research started with the premise that in general increasing the efficiency of waste management is beneficial. However, the relationship between waste management and long-term benefits in terms of resources use in the form of an equation does not exist. After this experiment, we feel safe believing that seeking waste management efficiencies is sensible.

Also unknown is the relationship between building design and waste generation and management. It's a phenomenon culturally dependent too. So, by using an exploratory case of a library building, the impact of any measures might well be different if students use the building, than if staff or any other group does. A well-established science of waste management that can be used for creating digital twins does not exist, and it may 20-40 years for it to be developed. It is not sensible to wait such a long time considering the current environmental climate. Thus, starting from a system that might approximately order services from slightly inefficient to probably better is sensible and can indicate the right direction to governing and managerial bodies.

The direct answer to the research question is yes, BIM and IoT can enhance the efficiency of waste management. The desired science for waste management based on equations has not been developed as a result of this research. However, more is known about the relationship between waste generation, displacement and collection and policy can be now developed. Here, there is a lesson for those willing to develop digital twins in areas where science does not exist. Developing the science is part of the problem and makes the contribution of the research more relevant for it also covers a knowledge gap.

The question as to whether the scalability to the city level of the presented twinning process is feasible remains unanswered. We believe it is feasible, although not without a great deal of effort to advance the knowledge about waste management in urban contexts.

Nonetheless, it is clear that the potential efficiency gains that can be obtained through this process are large. More importantly, this process can also trigger changes in policy and in behaviour that would lead to better sustainable outcomes.

REFERENCES

Abas, M. A., & Wee, S. T. (2014). Sustainable Solid Waste Management in Malaysia: The Concept of Multi-Stakeholder Governance in Solid Waste Policy Implementation. *Public Policy and Administration Research*, *4*(10), 26–35.

Abel, J., Diez, K., & Lennerts, K. (2006). A Standard Database Model for Computer-Aided Facility Management in Hospitals. *Proceedings of the Joint International Conference on Computing and Decision Making in Civil and Building Engineering*, 535–542.

Abergel, T., Dean, B., & Dulac, J. (2017). *Towards a zero-emission, efficient, and resilient buildings and construction sector: Global Status Report 2017*. UN Environment and International Energy Agency.

Albino, V., Berardi, U., & Dangelico, R. M. (2015). Smart Cities: Definitions, Dimensions, Performance, and Initiatives. *Journal of Urban Technology*, *22*(1), 3–21. doi:10.1080/10630732.2014.942092

Alexander, K. (1996). *Facilities Management : Theory and Practice*. E & F N Spon. doi:10.4324/9780203475966_chapter_3

Amaratunga, D., Haigh, R., Sarshar, M., & Baldry, D. (2002). Assessment of facilities management process capability: A NHS facilities case study. *International Journal of Health Care Quality Assurance*, *15*(6), 277–288. doi:10.1108/09526860210442047 PMID:12500653

Anonymous. (2015). *Python Traveling Salesman Greedy Algorithm (Algorithm)*. Stack Overflow. Available at: https://stackoverflow.com/a/30552710

Anonymous. (2019). *Complete Guide for Ultrasonic Sensor HC-SR04 with Arduino*. Random Nerd Tutorials. Available from: https://randomnerdtutorials.com/complete-guide-for-ultrasonic- sensor-hc-sr04/

Atkin, B., & Brooks, A. (2009). *Total Facilities Management, Facilities* (3rd ed., Vol. 12). Wiley-Blackwell.

Bainbridge, M., & Finch, E. F. (2009). Getting the attention the facilities manager deserves. *Facilities*, *27*(7/8), 277–290. doi:10.1108/02632770910956139

Ballesty, S., Mitchell, J., Drogemuller, R., Schevers, H., Linning, C., Singh, G., & Marchant, D. (2007). Adopting BIM for facilities management: Solutions for managing the Sydney Opera House. Cooperative Research Centre for Construction Innovation. Report, pp. 1–32.

Barrett, P. S., & Baldry, D. (2003). *Facilities Management: Towards Best Practice, Facilities Management* (2nd ed., Vol. 2). Wiley-Blackwell.

Biagini, C., Donato, V., & Pellis, D. (2015). Preliminary design through graphs: A tool for automatic layout distribution. *Iconarp International Journal of Architecture and Planning*, *2*(2), 1–13.

Browne, A. L., Pullinger, M., Medd, W., & Anderson, B. (2014). Patterns of practice: A reflection on the development of quantitative/mixed methodologies capturing everyday life related to water consumption in the UK. *International Journal of Social Research Methodology*, *17*(1), 27–43. doi:10.1080/13645579.2014.854012

Caragliu, A., Del Bo, C., & Nijkamp, P. (2011). Smart cities in Europe. *Journal of Urban Technology*, *18*(2), 65–82. doi:10.1080/10630732.2011.601117

Cheng, J. C. & Ma, L. Y. (2013). A BIM-based system for demolition and renovation waste estimation and planning. *Waste Management, 33*(6), 1539-1551.

Codinhoto, R., & Kiviniemi, A. (2014). Bim for FM: A case support for business life cycle. In S. Fukuda, A. Bernard, B. Gurumoorthy, & A. Bouras (Eds.), *Product Lifecycle Management for a Global Market* (pp. 63–74). Springer-Verlag Berlin Heidelberg. doi:10.1007/978-3-662-45937-9_7

Cracknell, D. (2014). *Sector BIM Strategy: Improving BIM Education & Training*. The Higher Education Academy. Available at http://www.heacademy.ac.uk/assets/documents/disciplines/built _environment/BIM/BIM_Skills_CIC_David_Cracknell.pdf

DEFRA – Department for Environment, Food and Rural Affairs. (2016). *UK statistics on Waste*. Governmental Report. Available at: https://www.gov.uk/government/uploads/system/uploads/attachment_data/file/593040/UK_statsonwaste_statsnotice_Dec2016_FINALv2_2.pdf

Donato, V. (2014). *Transformability of Existing Building: An approach based on BIM Technologies* (PhD thesis). University of Florence and TU Braunschweig.

Donato, V. (2017). Towards design process validation integrating graph theory into BIM. *Architectural Engineering and Design Management*, *13*(1), 22–38. doi:10.1080/17452007.2016.1208602

Dyer, T. D. (2012). *Construction and demolition waste*. Springer. doi:10.1007/978-1-4419-0851-3_118

Eastman, C., Teicholz, P., Sacks, R., & Liston, K. (2018). *BIM Handbook: A Guide to Building Information Modeling for Owners, Managers, Designers, Engineers, and Contractors*. John Wiley & Sons, Inc.

Fialho, B. C., Codinhoto, R., & Fabricio, M. M. (2019). Trends in BIM and IoT for Reactive Maintenance. *Proceeding of the 36th CIB W78 2019 Conference*.

Fialho, B. C., Codinhoto, R., & Fabricio, M. M. (2020). *BIM and IoT for the AEC Industry: A systematic literature review*. XXIV SIGraDi Congress "Transformative Design" School of Architecture and Design of the Pontific University.

Guerrero, L. A., Maas, G. & Hogland, W. (2013). Solid waste management challenges for cities in developing countries. *Waste Management, 33*(1), 220-232.

Hamidi, B., Bulbul, T., Pearce, A., & Thabet, W. (2014). Potential application of BIM in cost- benefit analysis of demolition waste management. In *Construction Research Congress 2014: Construction in a Global Network*. Construction Institute of ASCE.

Hollands, R. G. (2008). Will the real smart city please stand up? Intelligent, progressive or entrepreneurial? *City*, *12*(3), 303–320. doi:10.1080/13604810802479126

Jambeck, J. R., Roland Geyer, R., Wilcox, C., Siegler, T. R., Perryman, M., Andrady, A., Narayan, R., & Law, K. L. (2015). Plastic waste inputs from land into the ocean. *Science*, *347*(6223), 768–771. doi:10.1126cience.1260352 PMID:25678662

Karak, T., Bhagat, R. M., & Bhattacharyya, P. (2012). Municipal solid waste generation, composition, and management: The world scenario. *Critical Reviews in Environmental Science and Technology*, *42*(15), 1509–1630. doi:10.1080/10643389.2011.569871

Kassem, M., Kelly, G., Dawood, N., Serginson, M., & Lockley, S. (2015). BIM in facilities management applications: A case study of a large university complex. *Built Environment Project and Asset Management*, *5*(3), 261–277. doi:10.1108/BEPAM-02-2014-0011

Kuechler, B., & Vaishnavi, V. (2008). On theory development in design science research: Anatomy of a research project. *European Journal of Information Systems*, *17*(5), 489–504. doi:10.1057/ejis.2008.40

Laurent, A., Bakas, I., Clavreul, J., Bernstad, A., Niero, M., Gentil, E., Hauschilda, M. Z. & Christensen, T. H. (2014a). Review of LCA studies of solid waste management systems–Part I: Lessons learned and perspectives. *Waste Management, 34*(3), 573-588.

Lazarevic, D., Buclet, N., & Brandt, N. (2012). The application of life cycle thinking in the context of European waste policy. *Journal of Cleaner Production*, *29*, 199–207. doi:10.1016/j.jclepro.2012.01.030

Lennerts, K., Abel, J., Pfründer, U., & Sharma, V. (2003). Reducing health care costs through optimised facility management-related processes. *Journal of Facilities Management*, *2*(2), 192–206. doi:10.1108/14725960410808203

Li, R., & Klippel, A. (2012). Wayfinding in libraries: Can problems be predicted? *Journal of Map & Geography Libraries*, *8*(1), 21–38. doi:10.1080/15420353.2011.622456

McCoy, K., Oliver, J. J., Borden, D. S., & Cohn, S. I. (2018). Nudging waste diversion at Western State Colorado University: Application of behavioural insights. *International Journal of Sustainability in Higher Education*, *19*(3, I3), 608–621. doi:10.1108/IJSHE-05-2017-0063

Miafodzyeva, S., & Brandt, N. (2013). Recycling behaviour among householders: Synthesising determinants via a meta-analysis. *Waste and Biomass Valorization*, *4*(2), 221–235. doi:10.100712649-012-9144-4

Sharma, S. B., & Tabak, V. (2008). Rapid Agent Based Simulation of People Flow for Design of Spaces. In H. J. P. Timmermans & B. de Vries (Eds.), *Design & Decision Support Systems in Architecture and Urban Planning*. University of Technology Eindhoven.

Thürer, M., Pan, Y. H., Qu, T., Luo, H., Li, C. D., & Huang, G. Q. (2016). Internet of Things (IoT) driven system for reverse logistics: Solid waste collection. *Journal of Intelligent Manufacturing*, *2016*, 1–10.

Venkov, D. (2017). *Shortest Path through Points (connect the dots)*. Available from: https://forum.dynamobim.com/t/shortest-path-through-points-connect-the-dots/17610/4

Wan, C., Cheung, R., & Qiping Shen, G. (2012). Recycling attitude and behaviour in university campus: A case study in Hong Kong. *Facilities*, *30*(13/14), 630–646. doi:10.1108/02632771211270595

Wang, Y., Wang, X., Wang, J., Yung, P., & Jun, G. (2013). Engagement of facilities management in design stage through BIM: Framework and a case study. *Advances in Civil Engineering*, *2013*, 1–8. doi:10.1155/2013/189105

Wilson, D. C., Rodic, L., Scheinberg, A., Velis, C. A., & Alabaster, G. (2012). Comparative analysis of solid waste management in 20 cities. *Waste Management & Research*, *30*(3), 237–254. doi:10.1177/0734242X12437569 PMID:22407700

Ye, Z., Yin, M., Tang, L., & Jiang, H. (2018). Cup-of-Water theory: A review on the interaction of BIM, IoT and blockchain during the whole building lifecycle. In *ISARC. Proceedings of the International Symposium on Automation and Robotics in Construction* (*Vol. 35*, pp. 1-9). IAARC Publications.

Chapter 14
Representing a Digital Twin City Model Using Open Source Tools and Integrating It With Dynamic Sensor Data

Fabrizio Massara
CSI Piemonte, Italy

Tatsiana Hubina
CSI Piemonte, Italy

Sara Mannoni
CSI Piemonte, Italy

Adelaide Ramassotto
CSI Piemonte, Italy

Fabrizio Barbero
CSI Piemonte, Italy

ABSTRACT

This work presents the developments of representing a part of the city districts of Manchester, UK and Turin, IT initiated within the FP7 DIMMER project completed in 2016 and continued in the last years by the Center of Excellence GIS of CSI Piemonte. The DIMMER system integrates BIM (building information modelling) and district level 3D models with real-time data from sensors and user feedback to analyze and correlate buildings utilization and provide real-time feedback about energy-related behaviors. The emerging standard 3D Tiles endorsed by the OGC was applied to represent CityGML data of two districts of Turin, Italy and Manchester, UK. 3D Tiles allows for a high level of detail (LOD) visualization that displays increasing detail of geometric features and their alphanumeric properties. Currently, the OGC 3D Tiles technology is mature, and the OGC CityGML specification will be soon updated to version three, with the adoption of exciting innovations like the support of time-dependent properties defined Dynamizers.

DOI: 10.4018/978-1-7998-7091-3.ch014

INTRODUCTION

The results achieved in 2016 thanks to the European 7th Framework Programme (EU FP7) District Information Modelling and management for Energy Reduction (DIMMER) project (Osello, 2016) showed that was possible and effectively useful the integration of building data at city scale with energy-related data from sensors at building and district level: with a wise choice of representative use cases the results could be extended in the near districts or to similar pilot cities to compute and simulate energy consumptions / savings and effects on emissions, so that appropriate decision and strategies could be adopted at administrative level by competent authorities.

The relation between sensors and buildings in the case of the *Turin pilot*, and overall electricity consumption in case of the *Manchester University Campus* was defined at level of the Building Information Modeling (BIM) model: every representative model had one or multiple readings of energy or electricity consumption that could be compared with the volume of the construction and its materials.

The objective of the chapter is to show that the adoption of City Geography Markup Language (CityGML) as descriptive format of the district can improve the capabilities of relating the buildings with energy production, distribution and consumption networks: the new *Dynamizers* concept in the forthcoming definition of CityGML version3 can also make possible to simplify the computations at district level by reducing them at the scale of every single building. The previously stated relation between sensors and buildings can be explicited in a more detailed CityGML format of the building in which geometry, materials and sensors placement are included in the same model.

Background

The authors started to represent city districts in tridimensional format in 2010 as prototypes often published to the end users or by partecipating to open data contests and challenges. All figures presented in this chapter come from work related to the authors and / or to the participation to projects and international contests or challenges.

METHODOLOGY

The task of representing cities in a 3D environment is challenging. Indeed, the game industry achieved excellent results by adopting dedicated and costly rendering engines (Andrade, 2015). Public authorities have access and own a large variety of datasets such as cadastral databases, aerial and satellite imagery, local and small to medium-scale laser and drone scans. While the former are usually displayed in a lot of bidimensional Geographic Information System (GIS) applications and frameworks, the latter generate complex points clouds and tridimesional structures that requires dedicated and specialized viewers and editors. All these data differ from extension to form and resolution, and one of the complex tasks is to integrate them in a simple and usable way for the end-user. The tridimensional representation of the city and of the environment allows for effective thematization and classification of various phenomenona in dense urban environments.

Here below a list of data sources starting from the simplest to the more detailed, not necessarily exhaustive, is provided. They must be harmonized and displayed at different scales defined by the user according to the concept of progressive Level of Detail (LOD) (Biljecki, 2016). The OGC CityGML's classification (Gröger, 2006) defines the level of detail of data considered from the simplest, bidimensional LOD0 to the complete LOD4.It is worth considering that the city, as *CityGML* specification suggests, is made not only of buildings but also from roads, street signs, trees, rivers and all sorts of anthropic and natural elements. Thus, the same principles of twin models apply not only to the city but to the open country and to landscapes, too.

After an evaluation of the visualization tools available, the steps needed to convert and adapt the various kind of datasets to the streaming and rendering of possibly huge geospatial content are considered.

In the results section, then, examples of conversion of *BIM* models into *3D Tiles* format are presented, and a final example section shows different representation of sensors' definitions as *CityGML Dinamizers*.

Source Datasets, an (un-) Exhaustive List

- Data from cadastral sources. Usually, this is vector data derived from manual or automatic digitization of raster images or conversion from local datasets, data can be considered in LOD0 (Figure 1a). The height information, if present, is available in attribute tables. Further, the 3D volume is not available and must be derived by extrusion to recreate an estimation of the structures. Typically, footprint polygon data is not optimized. Distance between consecutive points can vary a lot and so the number of points and the winding order. That is why the polygons should be optimized and harmonized in the same dataset: the winding order is necessary for the correct representation of the 3D structures, data is now represented by average volume and can be classified as LOD1 (Figure 1b).

The data is usually limited in size, and attributes are often very rich and allow for useful classification based on properties and semantic of objects: the model allows to use classification based by color of buildings or color and size of bidimensional or tridimensional polygons representing administrative or natural borders of territory.

Some examples are total population by census sections as colored bidimensional polygons (Figure 2b), population by census sections classified as tridimensional extruded polygons (Figure 2c), residential buildings represented as extruded colored polygons, highlight of industrial sheds in a city (attributes available are among others use destination, year of construction, height, area, volume, etc.), visualization of color ramps of heights of buildings (shorter buildings in blue, taller in yellow, and so on) (Figure 3a to 3c).

- Data from elaboration from previous sources, and/or integrated with manual or automatic recognition. This kind of datasets contains details of the properties and geometry of the corresponding real-world objects. This data includes, for example, building parts, walls, closures or roof geometry. These datasets come from manual digitization and recognition of elements, and parametric modelling of tridimensional structures such as *OpenStreetMap* data: in this case building parts and rooftops classification is present and data can be considered as LOD2. Also, they can be *CityGML* datasets manually or automatically edited from terrestrial or airborne scans, in LOD2 as well- Some examples are *OpenStreetMap* generated buildings, view of *CityGML* simple and textured models of *"Manchester Piccadilly"*and *"Heald Avenue"*, *Manchester, UK* (Figures 2a, 4a, 4b, 5b).

Figure 1. From left to right, 1a) example of footprint data from the cadastral source (screenshot thanks to QGIS 3.14 Pi), 1b) the same element from the previous figure extruded by its height (screenshot thanks to Cesium ion), 1c) example of visualization of city elements such as buildings, fencing, gates, sidewalks, flowerbeds, trees and other furniture

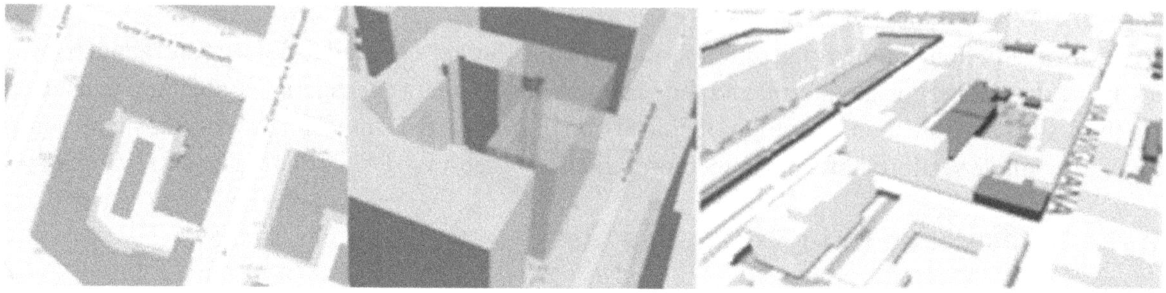

Figure 2. From left to right, 2a) parametric representation of buildings from OSM data (screenshot thanks to OSM Buildings1), 2b) example of classification based on municipality's administrative borders, 2c) another example of classification based on municipality's administrative borders

- Data from airborne or terrestrial scans, such as photogrammetric or laser (or similar) acquisition and elaboration: typically, these are point clouds from which polygon meshes and applied textures can be derived by simplification and triangulation procedures.

Figure 3. From left to right, 3a) example of classification of buildings by use destination, 3b) another example of classification of buildings by use destination 3c) example of classification of buildings by height

Figure 4. from left to right, 4a) manually edited, 4b) automatically edited, 4c) 3D Tiles format datasets of districts of Manchester, UK, the visualization of this last dataset was ready to be displayed in Cesium ion and the quality of the model and the textures is impressive

This kind of data is the biggest in size and demands a lot of resources for its elaboration and visualization. LOD mechanisms are used for displaying the dataset at different scales, and the optimizations are to be decided case by case. From one hand, geometric separation from points can reach the order of the millimeter for close acquisitions. On the other hand, it can be challenging to assign semantic properties to the dataset that is visualized as a continuous big polygon mesh. Examples are simple or textured models of "Manchester Piccadilly" in Manchester, UK and of monuments "Porta Napoli", "Colonna di Sant'Oronzo" and "Il Sedile (Palazzo del Seggio)" in Lecce, IT (Figures 4c, 5a, 6a to 6c). In these examples the model (highlighted in red) is integrated in the city model but is not displayed when not needed due to the LODs configuration, freeing resources on the client. As the user triggers the LOD limit the model is now displayed in the client in full detail in LOD3: other models, even if present, are not shown so that the web browser displays only a subset of the available data and is freeing unnecessary far resources.

Figure 5. from left to right, 5a) view of a 3D Tiles dataset, 5b) automatically edited dataset in CityGML version 3

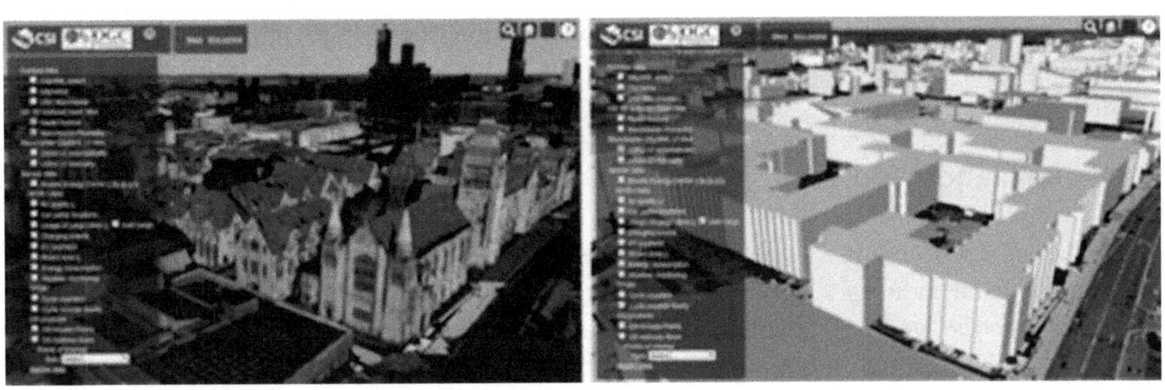

- Data from parametric 3D models like BIM or other 3D sources such as Keyhole Markup Language (KML), Extensible 3D Graphics (X3D) or GL Transmission Format (glTF) models (including 3D Tiles converted format) and in general every 3D model generated in parametric form. Such models

Figure 6. From left to right, 6a) model of "Porta Napoli" monument of the Municipality of Lecce, IT, 6b) the previous model is now displayed in the client in full detail in LOD 3, 6c) two additional models near from each other in the center of the city of Lecce

are usually smaller in size than the corresponding photogrammetric equivalent and typically have a significant number of attributes and semantic references inside them (Figure 12b, 12c).

- Data from *Geographic Information Systems* (*GIS*) sources. This category includes *DEM (Digital Elevation Model)* (and /or *DTM – Digital Terrain Model*) elevation data, orthophoto background imagery, *OGC Web Map Services* (*WMS*) and *OGC Web Features Services* (*WFS*) data from standard *OGC* services of the most different sources and purposes: from land use to administrative borders, weather properties, and so on.
- Dynamic data from sensors, commonly time dependent. This data is the most challenging kind to be displayed and is related to energy consumption, traffic data, air quality, weather measurements and so on. It can be related to physical sensors, for which coordinates specify the position of the sensor, or to average measurements on areal surfaces like city districts or regions, so that they can be represented in conjunction with *GIS* services like the one cited in the previous paragraph.

All these different and heterogeneous sources of data should be harmonized, organized and displayed in a convenient form to the end-user. The right choice should always face the adoption of common standards for the publishing of services, and delegate to services and visualization applications their use or further conversion.

Before the chapter on data elaboration and harmonization, the authors propose a brief survey of services and of end-user applications. Although, in this case, the list is not exhaustive but restricted to the authors' limited experience.

Data Visualization Tools

As stated before, representing a model of landscapes and cities is not a trivial task. Attractive solutions come from the game industry, from flight simulators and other specialized applications that from time to time privilege the landscape and terrain continuum, local structures or dedicated tasks. All these visualization tools share the almost impossibility of personalizing data without using proprietary software or studying specialized data structures. However, it keeps development simple and is not to be considered a defect.

Various frameworks and tools were developed over the years to represent complex structures in specific formats like *KML* (from *Keyhole, Inc*. and then *Google*), *X3D* (successor of *VRML* from *Web3D*), and *OGC CityGML*.

At a lower level, game engines always represented 3D content in *OpenGL* or *DirectX*, and some visualization tools still adopt these technologies.

The authors conducted first attempts in the field in 2010. The three million of extruded buildings of *Piedmont region* (*IT*) in *NASA World Wind* (*Java* version) were visualized: the experiment required to generate tiled layers of *OpenGL* geometries that were added to the application at runtime (source data: cadastral datasets like at previous point 1, Figure 7a).

Then in 2011, the city buildings of a *Turin* district were visualized in *Google Earth* plugin in internet navigators within the *Turin Smart City Dashboard* (*Italian city*) using the styling based on attributes: extruded instances of polygons derived from cadastral building footprints were used (Figure 8a).

The gained experience allowed the authors to participate in the *Web3D City Modeling Competition of Rotterdam* (*NL*) that required to display about three hundred thousand of textured buildings for the city. For the task, *CesiumJs* framework and *Google Earth* application were chosen (Figures 7b and 7c).

The results of these representations were excellent in terms of quality of data displayed. However, they lacked in performance instead.

For small datasets, the results were outstanding. However, the caching and resource releasing mechanism was not tuned by the authors as needed. For what concerns *Google Earth* and *KML*, the "*Regions*" feature could not be understood in its full power so that in case of different depth of detail, the solutions became very slow. For *CesiumJs* instead, the custom implemented *LODs* loading algorithm was too slow for the complexity of data. The results of the attempt were promising. Nevertheless, only more time to fine-tune the solutions could allow acceptable navigation of the whole dataset, huge indeed. Inevitably, a new *LOD* mechanism should be used and adapted to the different characteristics of source data.

Figure 7. From left to right, 7a) visualization of geometry and attributes for 3 million of extruded buildings in NASA World Wind (Java version), 7b and 7c) Web3D City Modeling Competition, display of about 300k textured buildings of Rotterdam (NL) in Google Earth and CesiumJs.

In 2016, with the participation to the *FP7 DIMMER* project, the need of displaying two different pilot city districts, for *Manchester, UK* and *Turin, IT* together with: i) *GIS* data at *district* level: buildings, networks, topology;ii) *BIM* data at *building* level: twenty representative "*case buildings*" analyzed in detail, whose properties were extended to the whole district; iii) *SIM* and *DIM* data at the network level: district heating/power networks, energetic and environmental readings from the sensors (Brundu,

2017); addressed the choice to use the more mature *CesiumJs* open-source visualization tool from *AGI*, that was adopting an early technology to solve the problem of streaming massive amounts of geospatial data: *3D Tiles* (Cozzi, 2018).

The adoption of *3D Tiles*, even in an unsupported form, allowed to satisfy the concept of level of detail considered in the project and permitted to visualize *GIS*, *BIM*, *SIM*, and *DIM* data according to the predefined scales and levels of visualization at design time (or as needed by the applications) (Figures 8b, 8c, 9b).

Figure 8. From left to right, 8a) Turin Smart City Dashboard (IT), classification of city buildings based on attributes in Google Earth plugin, 8b) DIMMER Dashboard, Turin pilot, visualization and classification of GIS data at district level integrated with BIM data at a building level, 8c) DIMMER Dashboard, Manchester pilot, visualization and classification of GIS data at district level combined with SIM data at a network level.

Nowadays, *3D Tiles* is an *OGC* standard. The technology has evolved to support Photogrammetry, 3D Buildings, BIM/CAD, Instanced Features, and Point Clouds. As early users, the authors summarize in the present document their experience in continuing to adopt this technology in the last few years. A parallel technology is recently developing in terms of 2D maps, sharing base concepts with *3D Tiles*: *Vector Tiles*. It will be fascinating to see a meeting point between them in some near future. Recently,

Figure 9. From left to right, 9a) source model of primary school "Coppino" in BIM editor, 9b) DIMMER Dashboard, Turin pilot, level of details applied to geometry and attributes of the simplified exported BIM model, from LOD1 to LOD4: note that LOD concept applies not only to geometry but to attributes and sensor readings, too.

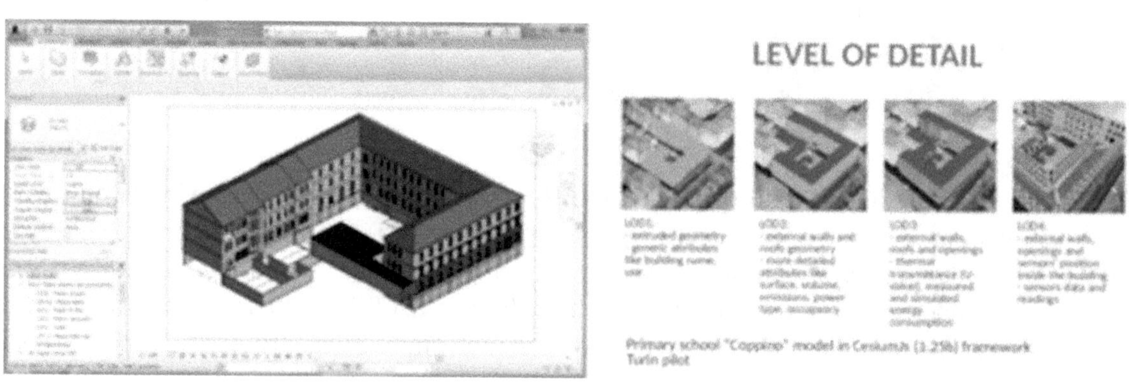

good news has come: *NASA* renewed its support to *NASA World Wind*, so hopefully, in the future, there will be an alternative and a good competitor in the *open-source globes* panorama.

Results

Some assumptions must be made to create and maintain a framework for displaying geospatial data. First, required (or available) datasets must be listed, then a target visualization tool must be chosen. Further, the optimal datasets should be created, converted or processed in real-time, depending on the static or dynamic characteristics of such data. If *Google Earth* is chosen as target visualization tool, probably the best option is to organize data as

Keyhole Markup Language (KML), adopting the "*Regions*" feature to stream the needed and often massive amounts of data available. *Google Earth* correctly supports the standard *OGC* services like *WMS* and *WFS*. Thus, it is possible to design diversified access to different sources of data.

The *X3D* file format is a royalty-free, platform-independent open standard file format and run-time architecture to represent and communicate 3D scenes and objects that can run on many platforms.

The combination of *X3D* format and the *HTML5* can render 3D models in most web browsers, and there is no need for additional or proprietary applications. *X3D* is widely used as an import/export format from various editing tools, open-source or proprietary, and supports real-time networking and geospatial applications as extensions. Further, various visualization tools are available as *HTML/javascript* libraries like *x3dom* for datasets organized as *X3D* sources and services. A *run-time X3D* application has natural access to online resources, so it is easy to consume and display standard *OGC* services according to the requirements.

Data Formats and Processing in 3D Tiles for Cesium ion Visualization

The most important and exhaustive format for the city (and landscape) storage and representation is *CityGML*: an *OGC* standard extension of *Geographic Markup Language* that can represent attributes and geometric properties of a great variety of elements of cities and related territories, not only anthropic but natural, too, semantically ready.

The current stable standard version is number *two*. However, the *beta* version of the next third release is already available. Version *three* includes several new exciting features.

CityGML viewers, editors, processing libraries, and management servers exist, and they give excellent results: the authors propose their experience in datasets management and integration to use visualization tools like *CesiumJs* (now *Cesium Ion*) based on *3D Tiles* (Figure 10).

3D Tiles is an *OGC* standard proposed and developed within the *Cesium* project by *AGI Inc*. This standard is now mature and has consistent support in terms of tools for its generation, validation, and publishing.

In extreme simplification, the concept that stands at the base of *3D Tiles* is the same that serves the *2D raster* tiling mechanism, extended in 3D and generalized: every *tile* can have *zero* to *n children* not necessarily, as in 2D, in geometric progression, and allowing the tiles to adapt to the density of data. In other terms, the tile is not divided by four equal children but has children of convenient size if and where the data is present: if no data is present, the child is omitted. The structure is described in an index file named *tileset.json* that is used by the end-user application to perform optimizations based on the resources available. In the case of *CesiumJs*, the visualization is delegated to *WebGL APIs*, and so

Figure 10. Schema of the architecture adopted in the examples: a flexible J2EE service-oriented architecture caches and adapts data in suitable OGC's compliant services

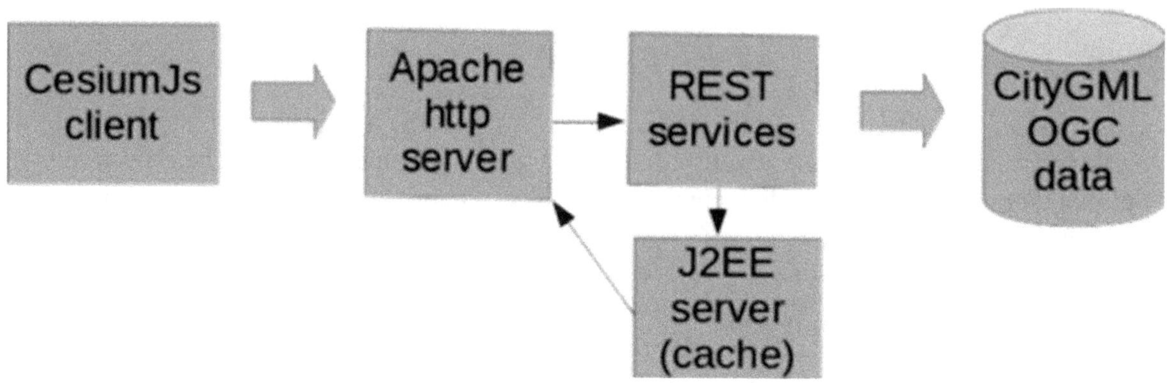

the performance is good even in smartphones or outdated systems. Data and models are organized in *glTF* binary models, specially designed to be consumed without further processing by client *GPUs*. The *LOD* concept is naturally integrated into 3D tiles, so that with *ADD* or *REPLACE* directives placed in the *index* file the models can represent different levels of detail according to the definition of *geometric error*. The visualization of tiles in a 3D environment is not a simple task. When the user tilts the view possibly thousands of tiles appear in the user's field of view: the optimized algorithms in the supported viewers free the designer to consider these visualization problems.

As previously noted, various tools and applications are available to generate or export data in *3D Tiles* format. The following paragraphs show the use of such data as file systems in *zip* or *directory* format. The option to create such archives on the fly is always available in properly designed services and offers the most flexible mechanism to adapt to dynamic data.

Figure 11. From left to right, 11a) model in LOD1 generated by extrusion from the cadastral footprint and known height: the tooltip shows information from the attribute tables, 11b) approximated parametric model in LOD1/LOD2 from OSM data, 11c) interpolated model in LOD1 from LIDAR data: buildings and other elements are clearly distinguishable from natural ones like trees.

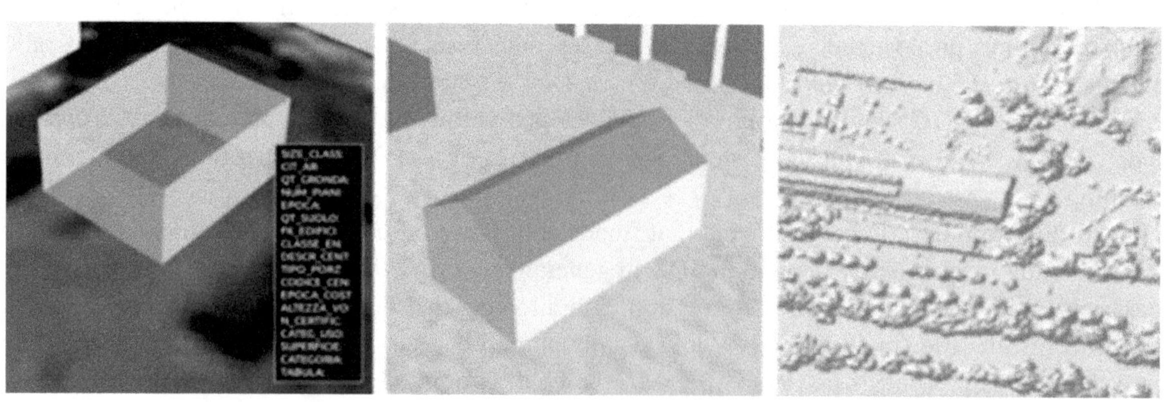

Figure 12. From left to right, 12a) model automatically generated from satellite and /or aerial and/ or UAV imagery, 12b) simplified model from BIM source models and/or photogrammetric close-range sources in LOD1 integrated into the city district, 12c) simplified model from BIM source models and/or photogrammetric close-range sources in LOD3: the tooltip shows data derived from families of source BIM elements, an ifcwall in this case.

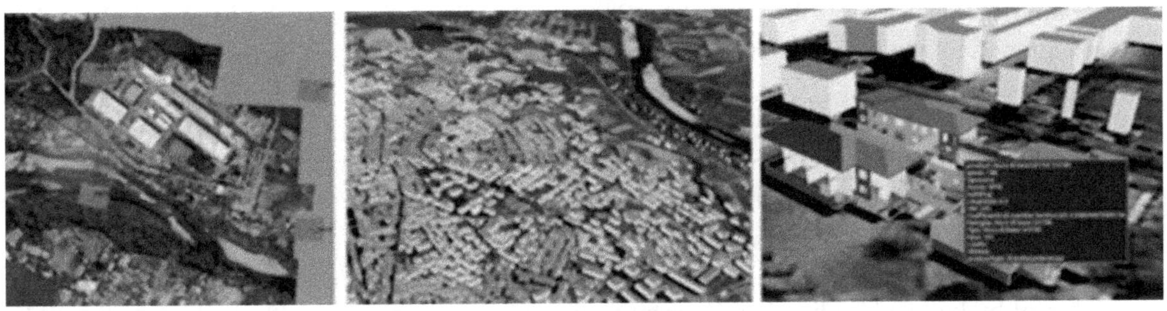

Following the (un-)exhaustive datasets list provided above, hereunder, the authors describe the steps needed to generate *3D Tiles* structures from the different sources of data:

- Data from cadastral sources (see Figure 11a): such data can create tiles according to *LOD0* (planar) and *LOD1* (extruded) specification; usually data is continuous in latitude and longitude, so the conversion, optimization and generation tools should:
    - Subdivide the data in tiles with longitude and latitude steps to fit the dataset extension: special attention must be drawn to avoid duplicated (or missing) elements intersecting the borders of each tile;
    - Generate the 3d structures (oriented and indexed triangles in space) by extrusion from footprint by known building height;
    - Apply convenient indexing and reference to parametric and/or simplified models in case of city elements like street signs, benches, trees and so on (or create in their place parametric models);
    - each tile is published as a *Batched 3D Model* where every element has its properties indexed and retrievable as Batch Table entries: the group by of properties is delegated to the parsing and generating application per part, element, or group of elements according to the required output;
    - Generate the corresponding *tileset.json* index files and recursive structure for every required level of detail (at least *LOD0* and *LOD1*).

The resulting output in *zip* or *directory* format can be saved and included in a proper main *tileset.json* index file of an upper level, and the geometric errors (the main level of detail directives) can be harmonized. This process can be executed occasionally in case of static data, and repeated when needed, or delegated to a service with a proper caching mechanism.

- Data coming from the elaboration of previous sources, and/or integrated with manual or automatic recognition (see Figure 11b): this kind of data contains information from which *LOD1* (extruded) and *LOD2* (details with walls, closures, roofs and other building elements) models can be generated; in case of *OSM* models, they are parametric and require dedicated procedures for the generation of the 3D models. If the sources are in *CityGML*, it is necessary to map the corresponding geometries, attributes, and *LODs* to the corresponding *3D Tiles* format, and to take in consideration all the extension notes listed previously. The conversion, optimization, and generation tools should:

    - subdivide the data in tiles with longitude and latitude steps to fit the dataset extension: special attention must be given to avoid duplicated (or missing) elements intersecting the borders of each tile;
    - generate the 3D structures (oriented and indexed triangles in space) by extrusion from footprint by known building height or by dedicated procedures that take into accounts building parts, referenced models and/or roof structures:
    - apply convenient indexing and reference to parametric and/or simplified models in case of city elements, like street signs, benches, trees (or create in their place parametric models);
    - each tile is published as a *Batched 3D Model* where every element has its properties indexed and retrievable as *Batch Table* entries: the group by of properties is delegated to the parsing and generating application per part, element, or group of elements according to the required output;
    - generate the corresponding *tileset.json* index files and recursive structure for every required level of detail (at least *LOD0* to *LOD2*).

The resulting output and execution notes are shared with the previous case.

- Data from airborne or terrestrial scans, such as photogrammetric or laser (or similar) acquisition and elaboration (see Figures 11c, 12a): this kind of data is often in 3D form as *point clouds* (usually *LAS* format, or *XYZ* coordinates of points with color and attributes data) or *mesh* format (triangle list of optimized *meshes* derived from previous *point clouds*). Attributes usually are not available and the datasets, even if very detailed in resolution (and so not easily displayable by end-users systems), present themselves as a continuum of points or triangles huge in size.
- If the datasets are not yet available in 3D Tiles format, the conversion tools should optimize them and generate at least the levels of detail necessary to harmonize them with the display application.

In this case, the *3D Tiles* tiling mechanism is beneficial and, with proper adjustments, even very detailed local datasets can be loaded without interfering with the end-user usage of the application including low-end devices like *smartphones*.

- Data from parametric 3D models like *BIM* or other 3D sources like *KML*, *X3D* or *glTF* (including *3D Tiles* converted format) models and in general every 3D model generated in a geometric and parametric form (see Figures 12b, 12c): in this case the conversion and optimization procedures should take into account the size of the source model and to adapt *LODs* and complexity of structures and attributes to the destination format. This way every *LOD* in the destination model con-

Figure 13. From left to right, differences between test dataset derived from aerial photogrammetry of "Piazza Vittorio" square, Turin, IT (13a) and corresponding extruded model (13b): in this case, the model can be annotated with tooltips that reflect the properties of the city parts, and such attributes can be assigned to the distinct parts of the whole city mesh

siders geometry and attributes at different levels of detail, mapping 1 to 1 source data if available in this form, or summarizing data if needed and possible (see Figure 9b).

- Data from *GIS* sources, this category includes *DEM* (and/or *DTM*) elevation data, orthophoto background imagery, *WMS* and *WFS* data from standard *OGC* services of the most different sources and purposes: usually, these data doesn't require any conversion in case of *OGC's* standard formats. In the case of *DEM* and *DTM*, some optimization is suggested at a lower level by adopting dedicated publishing servers. Some additional effort might be requested for related properties classification (as shown for example in Figures 2b and 2c).
- In every case, info boxes and queries on layer data require some adjustments to interface with the application's standard query mechanism.
- Dynamic data from sensors, commonly dependent from time: for this kind of data, *CityGML* couldn't offer help in the past (see Figure 9b: in *DIMMER* project sensors were integrated into the building only by position, not by relationships with other elements, and the connection should be created logically by adding custom placeholders; the sensors were represented in the correct building only in the *LOD4* level). Now, with *CityGML* vthree proposal of *Dynamizers*, this behavior can be improved (with *Dynamizers* sensors and other features can directly be connected

Figure 14. From left to right, examples of annotated photogrammetric models of the City of Turin, 14a) "Conte Verde" monument, Piazza Palazzo di Città, 1847, detail of inscription and annotation tooltip, 14b) "Frejus Tunnel Memorial", Piazza Statuto, 1879, detail of the stone subject, 14c) "Vittorio Emanuele II Plaque", Real Palace, Piazza Castello, 1884, detail of the inscription and the relative annotation.

to the corresponding dynamic properties like temperatures or dimensions, so that they effectively change according to their corresponding readings).

To summarize the different input formats, libraries, and applications needed to convert them to the desired output, the following schema illustrates some of the various actors involved:

- possible input formats: *GeoJson* data, *ESRI ShapeFiles*, *PostGIS* databases, *KML*, *X3D*, *glTF* and *IFC* models, images (as *JPG*, *PNG*, *RAW*, *TIFF*), *LAS* or *XYZ point clouds*, terrain *meshes*;
- visualization and conversion tools: *QGIS*, *VisualSFM*, *MeshLab*, *CloudCompare*, *Blender*, *GIMP*;
- utility and conversion libraries used in the following examples are written in *Java Enterprise*, use *Spring framework*, *JSDAI* libraries for reading *IFC*, *JTS* library for geometry-related parsing and conversions, *Gson* and *Jackson* for *JSON* management, *Poly2tri-core* for extrusion, *JCSG* for *Solid Constructive Geometry*, *Vecmath* for data processing, *CesiumJs* libraries for conversions, *JQuery* for page interaction, and the list is not exhaustive, other utilities in other languages exists in open source form;
- Output *OGC* standard formats: *WMS*, *WFS*, *3D Tiles*, *CityGML* and others.

At the end of the workflow, as stated in the assumptions made, and with the help of the standardization of formats, all these open-source tools and libraries make it possible to consume standard services in visualization tools. Nowadays, the best results are achieved in *Cesium ion* application. However, other solutions should follow shortly. In recent browsers, the mobile support is good thanks to the *WebGL API* bindings.

The following sections illustrate practical examples of BIM models conversion in 3D Tiles and sensors' dynamic properties visualization with further detail.

Examples

Case study: the representation of BIM models in 3D Tiles

As an excellent example of data conversion, it can be useful to summarize the steps needed to convert a complex structure like a *BIM* model to the *OGC 3D Tiles* format.

Some commercial tools can do the task, and some libraries exist too. However, to have under control the whole process and to achieve maximum flexibility, a custom conversion library was developed. As the reader could imagine, this was not an easy task.

The library requires to have as input a *BIM* model in *IFC* exported format, and performs the following steps:

- Reads and parses the *IFC* model (in *STEP* format) as java objects, made by geometry and attributes;
- Converts the geometry in a format suitable for *opengl* visualization, essentially indexed triangles;
- Generates *Batched 3D models* in *3D Tiles* format and corresponding indexes and attributes (essentially *Batch Tables*) for every *LOD* required in the output.

Each of these steps is not trivial and requires some work: the custom library is completed for the first and last step and still lacks several geometric conversions in the second step. Probably, it will be replaced in the future with a complete open source library like *IfcPlusPlus* that seems promising, at the cost of linking existing *java* code to *c++ native bindings*.

The library was implemented after the completion of the *DIMMER* project in 2016, within the Master's thesis "*Implementation of BIM methodology for district data visualization*"2 and was tested using various models and completed with *GIS/BIM* interoperability (basic writing of *IFC* features capabilities) (see Figures 12b and 12c, *LOD1* to *LOD3* representation of *BIM* model of *Primary School "Rodari"*, *Settimo Torinese* in *CesiumJs*).

To show the complexity and the number of the objects involved, the count of the *IFC* elements for two different models is reported in the following table.

The authors propose the following algorithm in pseudo-code for the generation of the first *LOD* levels:

- Input parameters to the method: *IFC* file name, coordinates of the model (longitude, latitude, height above sea level), scale of the model (*IFC* models have usually implicit units, from millimeters to meters);

Table 1. Count of elements for two example IFC models: every element has its geometric definition and properties; some structures are easily buildable, some others are more complex, especially the ones generated by addition or subtraction from base types: constructive subtractive geometry (CSG) libraries can deal with them anyway

Primary School "Rodari", Settimo Torinese	Kindergarten "Puffi", Torino
Ifcslab: 66	ifcbooleanresult: 78
Ifcwall: 260	ifcfooting: 19
Ifcbuildingelementproxy: 5	ifcbooleanclippingresult: 49
Ifcwindow:780	ifctrimmedcurve: 16
Ifcdoor: 684	ifcslab: 301
Ifcflowterminal: 5	ifcbeam: 423
Ifcwallstandardcase: 1050	ifcwindow: 521
ifcbooleanclippingresult: 148	ifcdoor: 242
Ifcroof: 27	ifcwallstandardcase: 675
	ifccolumn: 135
	ifcstair: 15
	ifcrailing: 12
	ifcbuildingelementproxy: 210
	ifcwall: 7
	ifcflowterminal: 460
	ifcramp: 3
	ifcfooting: 19
	ifccovering: 1
	ifcroof: 9

- Open the *IFC* model and parse the *STEP* file structure;
- Generate corresponding *java* objects filled with every value retrieved;
- Generate *LOD0* model with geometry and attributes derived from corresponding simplified model in *IFC*, if present, or by generating simplified footprint from available data (see Figures 9b, 12a);
- Generate *LOD1* model with geometry and attributes derived from corresponding simplified model in *IFC*, if present, or by generating simplified bounding box from available data (see Figures 9b, 12a);
- Generate remaining models (*LOD2* to *LOD4*) with geometry and attributes derived from corresponding models in *IFC*, if present, or by showing appropriates elements at different levels of detail (see Figures 9b, 12a, 12b), for example:
  - *Ifcbeam*, *ifcslab*, *ifcwall* and *ifcroof* at *LOD2* (without *ifcdoor* and *ifcwindow*);
  - The same elements of the previous case with *ifcdoor* and *ifcwindow* at *LOD3*;
  - All the available elements at *LOD4* (allowing to show/hide *ifcroof*, *ifcwall* and outer elements in the representation in order to see the interior of the buildings);
- Organize geometry and attributes in the lods as different 3D Tiles models, with different scales of visualization corresponding to the proper geometric errors (usually every LOD model generates a different 3D Tiles level, but further subdivisions in children tiles can be needed for very large models);
- Convert coordinates of the models in a format suitable to be displayed in the correct geographic coordinates and scale (the model should be placed in the correct position of longitude, latitude, height above the sea level and in the correct scale in meters);
- Generate as output the final model in zip or directory format.

Case study: the application proposed to the OGC CityGML Challenge, Manchester, UK, 2019

In 2019, the authors proposed an application based on *cesiumjs* framework to the *ogc citygml challenge of manchester, UK*. The challenge required the visualisation of different sources of data, such as *geojson*, *esri's shapefiles*, orthophoto imagery, *3d tiles* datasets and *citygml* v.two and v.three datasets, and integration of various dynamic properties from sensors with the city model to study the concept of the new *dynamizers* feature present in the third version of the specification.

A brief technical description of the application is available subsequently. As for an architectural diagram stands what is schematized in figure 10.

The application

- The web application is based on the open-source *CesiumJs* framework, adapted to consume and visualize external *REST* services by using the following *javascript* frameworks:
  - *Jquery* framework, basic interaction with *dom* elements on the page;
  - *Bootstrap* framework, composition and layout of the page;
  - *Chart.js* framework, visualisation of *chart* data;
- The web application is hosted on an *Apache HTTP server* and consumes external services provided by the project;
- The custom terrain is provided as a service by:

- *Cesium Terrain Server* open source server.

CityGML v.two datasets are commonly parsed by *open source* libraries, even if the task is difficult. The model is usually very verbose (a *GeoJson* specification is coming to make things a bit simpler): one of the best *Java* libraries comes from the *OGC Schemas* project, providing JAXB bindings for XML Schemas defined by the Open Geospatial Consortium (OGC), but in this case, the *CityGML* v.three specification was not published yet, so a custom converter was necessary.

CityGML version three

- The conversion from *CityGML* v.three datasets to *3D Tiles* was performed by applications based on *Java 2 Platform, Enterprise Edition (J2EE)*
- The parsing of *CityGML* v.three data was done with *XStream* library
- Other tools used: *QGIS*, *Blender* and some other *open-source* library

The conversion gave good results, all kinds of source datasets could be easily converted (and those yet in *3D Tiles* format only required a simple configuration in the visualization tool), results can be seen in Figures 4a to 4c, 5a, 5b and in the next paragraphs.

Perhaps the most interesting task in designing the application was the study of the *Dynamizers* problem, and their relations to *sensors'* concept, in other terms how to effectively integrate available sensor readings in the *CityGML* model of the city.

In the next paragraphs will be listed some results from this proposal, not necessarily definitive but hopefully useful for further developments in the field.

The original *CesiumJs* application was integrated with two panels: the first one, as usual, presented on the left, to act as a *Table Of Contents* to organize and display the different layers of visualisation, the second (shown on the right) to control the sensors' time of the application to perform the queries and obtain the various readings (Figure 15a).

Dynamic data from sensor feeds was exposed as *REST Web APIs*.

A brief schematization and some examples of the sensors implemented as *Dynamizers* can be found in the next paragraphs.

CityGML version three *dynamizers*

- Sensor data life in the application:
  - load sensor attributes and position;
  - initialize sensor;
  - update sensor;
  - destroy sensor data and visualisation;

Sensor: *air quality*

- type: one value per property per station, geographic;
- values: last value, streams;

- algorithm: query value for every property for every station as single or multiple values and visualize according to the desired display type;
- in case averages are required compute classification for colors or heights of objects;
- display:
    - billboards with labels;
    - boxes with classification based on current sensor value and overall sensors maximum and minimum values, or fixed values and thresholds;
    - information panels showing station name, current and/or average and/or historical data, timestamp of values, charts for stream data.

Figure 15. Air quality sensor, 15a) sensor positions, 15b) billboard with labels, 15c) extruded boxes with classification, 15d) information panel

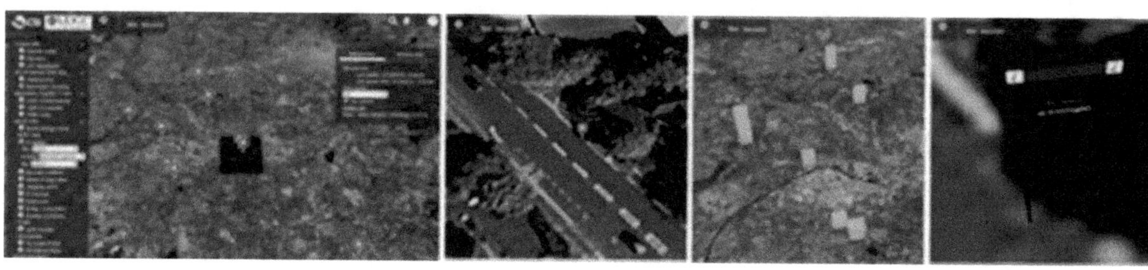

Sensor: *car parks occupation*

- type: one value per station over total value available, geographic;
- values: last value, streams;
- algorithm: query value for occupied and total places for every station as single or multiple values, compute free places (or equivalent operation if free places are offered instead of total) and visualize according to the desired display type;
- in case averages are required compute classification for colors or heights of objects;
- display:
    - billboards with labels;
    - boxes with classification based on current and total sensor value and overall sensors maximum and minimum values, or fixed values and thresholds;
    - information panels showing station name, current and/or average and/or historical data, timestamp of values, charts for stream data;
- live simulation:
    - initialize sensor and view on park entrance; at every change in occupation value simulate the entry or exit of cars to and from the park;
    - initialize sensor and view on the park extents; fill with vehicles the number of occupied spaces; at every change in occupation value drive one car from the entrance to a free place, or from a park place to the exit.

Figure 16. Car parks occupation sensor, 16a) sensor positions, 16b) extruded boxes with classification, 16c) information panel and live simulation with car models on park entrance, 16d) live simulation of car park traffic with car models

Sensor: *usage of cargo bikes*

- type: one value per property per station, geographic varying during time;
- values: last value, streams;
- algorithm: query value for every property for every station as single or multiple values and visualize according to the desired display type;
- in case averages are required compute classification for colors or heights of objects;
- display:
    - billboards with labels where location follows the position value over time;
    - model of a cargo bike in place of the billboard.

Sensor: *rivers level*

- type: one value per property per station;
- values: last value, streams;
- algorithm: query value for every property for every station as single or multiple values and visualize according to the desired display type;
- in case averages are required compute classification for colors or heights of objects.
- display:
    - billboards with labels;
    - boxes with classification based on current sensor value and overall sensors maximum and minimum values, or fixed values and thresholds.
- live simulation:
    - height of extrusion of a geometry representing the level of the river placed on the terrain model

Sensor: *energy center*

- type: one value per property per building
- values: last value, streams

Figure 17. Usage of cargo bikes and rivers level sensor, 17a) usage of cargo bikes positions, 17b) rivers level extruded boxes with classification, 17c) information panel and live simulation with extrusion of geometry representing the level of the river inside the terrain model

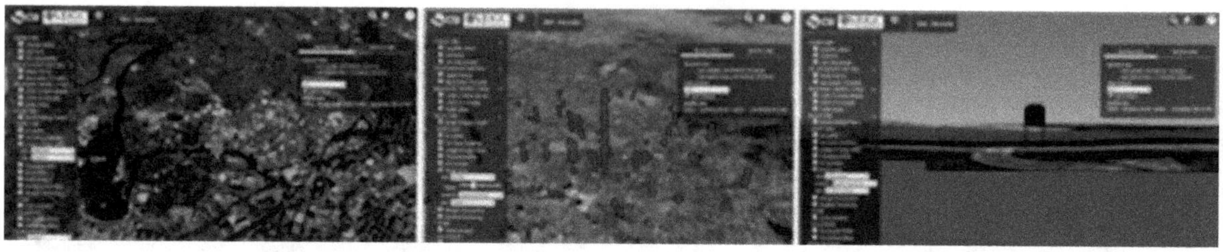

- algorithm: query value for every property for every station as single or multiple values and visualize according to the desired display type
- in case averages are required compute classification for colors or heights of objects
- display:
    - classification by coloring geometries based on current sensor value and overall sensors maximum and minimum values, or fixed values and thresholds

Figure 18. Adapting the terrain model to the rivers level sensor, 18a) preparing the DTM to the depression area of the river, 18b) the river area without water, 18c) the river area filled with water (an extruded polygon of correct height)

CONCLUSION

The work performed by authors after the *DIMMER* project proves the importance of dissemination of the initiative in the geospatial field. Indeed, it has been relatively easy to continue the ideas acquired during the project.

The "*Digital Twin*" concept has proved to be effective in relation with cities, starting from simulations to the effective monitoring of energy, traffic and people dynamics.

The technology is ready to process and visualize in real-time all data related to city life and management. It is always better if services and data are published according to international standards and directives.

Figure 19. Energy center sensor, 19a) classification of mesh portion of the geometry representing the building by physical property (in light yellow, in this case, the energy consumption: the same concept can be applied for internal temperature for example), 19b) classification of mesh portion of geometry representing the energy produced by PV panels (in light yellow, colored panels), 19c) classification of mesh portion of the geometry representing the internal temperature of the buildings (in a light yellow, colored skylights)

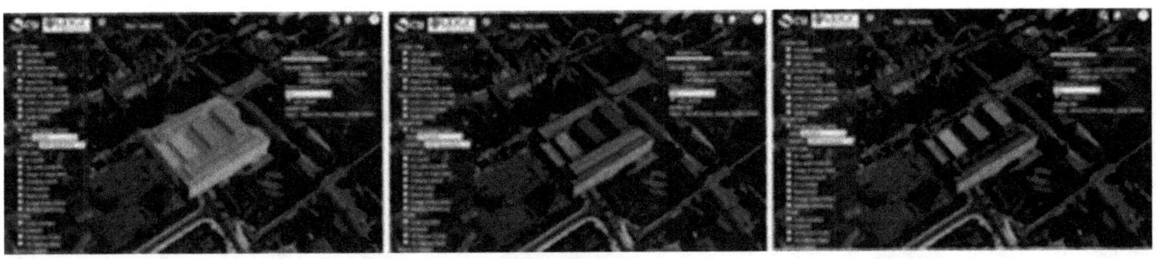

Significant progress was made in the last ten years since the first specifications of *CityGML*. The same data, once monolithic and archived in big files, can be hosted now on databases and dedicated servers, and the same attributes of the city can be varied over time to adapt to reality in real time.

CityGML should be an archiving format for the city and the landscape. However, it is better to delegate the visualisation (the old concept of presentation and content) to lighter and faster formats, such as *3D Tiles* and *REST* services.

The authors would recommend that the outstanding visualization tool from *AGI*, *Cesium ion*, could have other *virtual globes* as competitors in the geospatial field. In the future, the authors hope to support this possibility, along with the implementation of new features of the original specification.

In the previous chapter, dedicated to sensors, the authors described how a sensor could modify properties in the digital twin model (*billboards*, *info panels*, *river heights*, *tree colors*).

Furthermore, the question arises whether it is possible to create a virtual sensor inside the model so that real properties can be read in the virtual model.

The reader will find an answer below. In the next example, the digital model replicates in real-time the live view of a *camera* (in this case applied to *street traffic*) and performs calculations of the passage of vehicles in a volume that exists only in the *CityGML* (*twin*) *model*: various volumes can be defined for different targets (*people*, *cars*, *buses* and so on) (see Figure 20). The adoption (and the developments) in detection systems, augmented and virtual reality can have unexpected and exciting results, the work will continue even in these directions.

ACKNOWLEDGMENT

All figures present in this chapter come from work related to the authors and / or to the participation to the following projects or contests:

- Visualization of geometry and attributes for 3 million of extruded buildings in NASA World Wind (Java version), prototype, about 2010
- *Turin Smart City Dashboard (IT)*, classification of city buildings based on attributes in *Google Earth plugin*, about 2011

Figure 20. CityGML live vehicle counter sensor, 20a) the yellow structures determines the volumes to be intercepted by the car counters, one for direction: the car models are positioned and moved in the twin view reflecting reality with only computation delays, 20b) the feed from a camera is analyzed by YOLO v3 and are recognized vehicle types, positions and speed; such data is published as a streaming service: the CityGML aware middleware performs the actual vehicle count according to the desired position and direction of the sensors.

• *Visualization of the CityGML data of the City of Rotterdam in Google Earth and CesiumJs*, Web3D Consortium - Web3D City Modeling Competition 2015 - https://www.web3d.org/competition/city-modeling-2015

• *DIMMER Project*, dashboard of visualization and classification of GIS data at district level combined with SIM data at network level, and representation of simplified BIM models with level of details applied to geometry and attributes from LOD1 to LOD4, 2013 to 2016

• *2nd Lecce Open Data Contest*, visualization of open data from the city district of *Lecce, IT*, 2016

• *Trying to visualize GIS & BIM information on the web: a solution using Leaflet and Cesium*, Foss4G Bonn 2016 - http://foss4g2016.org/talks.html#140

• Example of visualization of city elements of the *Municipality of Turin*, IT, prototype, 2019

• *Streaming and rendering the Turin 3D geospatial content through GIS and BIM integration*, Foss4G Bucharest 2019 - https://talks.2019.foss4g.org/bucharest/speaker/9DES9G/

• *Dashboard of visualisation to support the capabilities of version 3 of CityGML, including dynamic data feeds from a range of sensors, OGC's CityGml Challenge, Manchester, UK*, 2019

• 3D Urban Content and Sensors' Data in Real Time From CityGML v3 Draft Specification and Dynamizers, Foss4G Italy, Turin 2020

The DIMMER project received funding from the European Union's Seventh Framework Programme for research, technological development and demonstration under grant agreement n° 609084.

REFERENCES

Andrade, A. (2015). Game engines: A survey. *EAI Endorsed Transactions on Game-Based Learning.*, *2*(6), 150615. doi:10.4108/eai.5-11-2015.150615

Biljecki, F., Ledoux, H., & Stoter, J. (2016). *Generation of multi-LOD 3d city models in CityGML with the procedural modelling engine random3dcity.* In ISPRS Annals of the Photogrammetry, Remote Sensing and Spatial Information Sciences, Volume III-4/W1. 1st International Conference on Smart Data and Smart Cities, 30th UDMS, 7–9 September 2016, Split, Croatia.

Brundu, F. G., Patti, E., Del Giudice, M., Osello, A., Ramassotto, M., Massara, F., Marchi, F., Musetti, A., Macii, E., & Acquaviva, A. (2017). A scalable middleware-based infrastructure for energy management and visualization in city districts. *EAI Endorsed Trans. Cloud Syst.*, *17*, 1–10. doi:10.4108/eai.28-6-2017.152753

BuildingsO. S. M. (n.d.). https://osmbuildings.org/?lat=45.05373&lon=7.65550&zoom=19.6&tilt=30

Cozzi, P., Lilley, S., & Getz, G. (2018). *3D Tiles Specification 1.0.* Open Geospatial Consortium (OGC). https://www.ogc.org/standards/3DTiles

Gröger, G., Kolbe, T. H., & Czerwinski, A. (2006). *Candidate OpenGIS® CityGML Implementation Specification (City Geography Markup Language).* Open Geospatial Consortium (OGC). https://portal.ogc.org/files/?artifact_id=16675

Osello, A., Acquaviva, A., Del Giudice, M., Patti, E., & Rapetti, N. (2016). District Information Models. The DIMMER project: BIM tools for the urban scale. In R. Pagani & G. Chiesa (Eds.), *Urban data. Tecnologie e metodi per la città algoritmica* (pp. 244–275). FRANCOANGELI.

Provera, F. (2017). *Implementation of BIM methodology for district data visualization.* https://webthesis.biblio.polito.it/6408/1/tesi.pdf

APPENDIX

Acronyms

3D Tiles: https://www.ogc.org/standards/3DTiles
API: Application Programming Interface
Batched 3D Model: https://github.com/CesiumGS/3d-tiles/tree/master/specification/TileFormats/Batched3DModel
Batch Table: https://github.com/CesiumGS/3d-tiles/blob/master/specification/TileFormats/BatchTable
BIM: Building Information Modeling
DEM: Digital Elevation Model
DIM: District Information Modeling
Direct3D: graphics application programming interface (API) for
Microsoft Windows, part of DirectX
DTM: Digital Terrain Model
EXPRESS: STEP format https://www.iso.org/standard/63141.html
GIS: Geographic Information System
glTF: GL Transmission Format
GML: Geographic Markup Language
GPU: Graphics Processing Unit
HTML5: last version of HTML, Hypertext Transfer Protocol
IFC: Industry Foundation Classes https://www.buildingsmart.org/standards/bsi-standards/industry-foundation-classes/
JAXB: Java Architecture for XML Binding
KML: Keyhole Markup Language
LAS: (LASer) file format designed for the interchange and archiving of lidar point cloud data
LOD: Level Of Detail
OGC Schemas project: https://github.com/highsource/ogc-schemas
OpenGL: Open Graphics Library
SIM: System Information Modeling
Vector Tiles: https://docs.mapbox.com/vector-tiles/reference/
VRML: Virtual Reality Modeling Language
X3D: Extensible 3D Graphics
x3dom: open-source JavaScript framework https://www.x3dom.org/

Names

AGI: Analytical Graphics, Inc. https://www.agi.com/
AIA: American Institute of Architects https://www.aia.org/

CSI Piemonte: Consorzio per il Sistema Informativo https://www.csipiemonte.it/web/it
ESRI: Environmental System Research Institute https://www.esri.com/it-it/home
Google: https://www.google.com
Manchester City Council: https://www.manchester.gov.uk/
NASA: National Aeronautics and Space Administration https://www.nasa.gov/
OGC: Open Geospatial Consortium https://www.ogc.org/OS: Ordnance Survey http://www.ordnance-survey.co.uk/
OSM: Open Street Map https://www.openstreetmap.org/
OSM Buildings: OpenStreetMap Buildings https://osmbuildings.org/
UNI: Ente Italiano di Normazione https://www.uni.com/
Web3D: Web3D Consortium https://www.web3d.org/
YOLO: You only look once is a state-of-the-art, real-time object detection system

Chapter 15
Digital Twin for Smart School Buildings:
State of the Art, Challenges, and Opportunities

Valentina Villa
Politecnico di Torino, Italy

Bernardino Chiaia
Politecnico di Torino, Italy

ABSTRACT

Industry 4.0 is encouraging the introduction of pioneering technologies even in the construction industry. Along with the development of high technology, such as augmented reality, virtual reality, and cloud computing, the development of digital twin has been growing. This contribution aims to present the potential of digital twin in the construction field, suggesting a framework that outlines the many different possible applications in construction, with reference to school buildings. First, it summarizes the current overview of digital twin applications in building construction. Then it shows that significant steps that are being taken beyond the digital model, even if the implementation of the digital twin concept in its full meaning is still a long way off. The research is moving in this direction and the evolution of the current state of the art, combined with the experience gained in the industrial sector, will soon bring a new revolution in the construction industry.

INTRODUCTION

Recently, new paradigms and technologies have been introduced in the construction sector since the term "Smart Building" entered the digital building revolution. First, the Digital Twin that integrates systems such as Artificial Intelligence (AI) and Automatic Learning (ML) to connect data, algorithms and context. Actually, deriving the definition from El Saddik (El Saddik, 2018) who wrote "A digital twin is a digital replica of a living or non-living physical entity. By bridging the physical and the virtual world, data is

DOI: 10.4018/978-1-7998-7091-3.ch015

transmitted seamlessly allowing the virtual entity to exist simultaneously with the physical entity." the Digital Twin in the AEC industry is configured as a virtual replica of the building, which shows in real time, through the virtual model, what happens in the physical environment. This is possible thanks to the use of sensors distributed in the environment that process the signals, elaborate them through analytical techniques and use them to simulate use scenarios or to simulate reactions to the real situation.

A rapidly growing technology is the Internet of Things (IoT) paradigm, which is paving the way for Smart Cities and Smart Buildings. The IoT has allowed to break down the data barrier between vertical domains on which the traditional ICT system was based, allowing the exchange of information between different application domains. The definition of a common model of web-based protocols for data exchange allows to transfer this data between objects, making the various parts of a network interactive.

This is where the Digital Twin concept was born, which is a candidate to revolutionize the building management model through predictive analysis and dynamic simulations based on real-time data. In the world of construction, digital models are now part of common practice, Building Information Modeling (BIM) has pioneered the virtual modeling of buildings and, in the most frontier applications, has touched, in part, the Digital Twin.

Traditionally Digital Twin was referred only to a part of the system (a turbine, an engine, etc.); Currently, thanks mainly to the Internet of Things development, they have become much more complex and are able to connect asset systems. This increasing complexity allows to combine asset data with process data, increasing the ability to solve complex problems. The last frontier is to associate data coming from people, from users, creating so-called Smart Buildings.

By optimizing systems and connecting people, owners and managers can use the Digital Twin to control the way environments are used, reduce costs, accordingly, improve occupancy rates and thus increase the overall value of the public good. In order to maximize the effectiveness of Digital Twin, you need a correct setting of high-performance databases and advanced machine learning algorithms. More generally, four key components are needed: data, algorithms, KPIs and context.

This contribution aims to present the potential of Digital Twin in the field of construction, suggesting a framework that outlines the different possible applications in the building sector, with reference to school buildings.

Background

The application of technologies such as BIM, Digital Twin, Virtual Reality, Augmented Reality, cloud computing, etc. is revolutionizing the world of Real Estate because it allows you to change the level of property management from a building level to building stock level. As a result, their application is valuable for the management of public real estate assets, which is configured as a strategic infrastructure on the territory. We are talking about hospitals, schools, office buildings and all those buildings that create a network of real estate units that need a structured and optimized management.

School buildings are an important heritage for every nation. A study carry out form the UK government, called "Delivering cost effective and sustainable school infrastructure" shows that "the annual budget for routine maintenance is on average between 1.5-2% of capital costs or between 45-60% of the capital costs over the life of the building". In Italy schools are currently at the center of attention for the renewal that is approaching the teaching and learning methods thanks to the introduction of digital.

The Italian property heritage of schools is very large (there are over 40,000 buildings) and very outdated (more than 60% are over 45 years old) (open data school construction from single source school data portal 2018/2019 Italian Ministry of Education, University and Research). The overwhelming majority of buildings were built before the introduction of the technical regulations on school building (DM 18/2/1975) and before the definition of special requirements for seismic zones (Law n. 64/1974). In summary, it was found that only 48.5% of Italian school buildings have a static certification and only 10.1% are built according to seismic safety criteria (open data school safety from single source school data portal 2018/2019 Italian Ministry of Education, University and Research). More than 8% also have structural problems detected, such as impairment of vertical load-bearing structures, floors or roofs. Therefore, more than ever before there are urgent interventions related to the safety of buildings and environments, as well as functional and flexible learning environments for the new educational system. Even if the Italian demographic projections are definitely decreasing, the need for safe and suitable spaces for school activities is more than ever central for the renewal of the country. The key points are security, sustainability and innovation in teaching. To the theme of building safety, at present, is now added the use in safety of spaces according to the directives issued for the Covid-19 pandemic management, which has blocked the entire school system. Even the choices of restarting the school activities are challenging and disclose a basic lack of knowledge of the available spaces and the quality of the performance guaranteed within the school environment. In this sense, Digital Twin are an effective way to respond and can be a decisive policy-making tool for governments.

The IoT is expanding significantly in all areas of every citizen's life. It is defined as "an open and comprehensive network of intelligent objects that have the ability to self-organize, share information, data and resources, reacting and acting in the face of situations and changes in the environment" (Madakam, 2015). Analyzing the evolution of innovative Internet and WSN technologies, the IoT has emerged as an all-pervading global computer network where data collected by the most accessible and available sensors and actuators can be used for analysis-based control of resource data or physical environments (Siavash, 2019).

One of the main elements of a smart city is a smart building. According to the research (Jia, 2018) there are three main features that identify a smart building: components, functions and results. The components consist of several interconnected parts of building parts, plant components, detection and control infrastructure. All these components perform according to their functions, which define the smartness and effectiveness of the building and ultimately translate into specific results, such as health, comfort, productivity and energy efficiency (Mothlag, 2018); applying this process, especially to public buildings of collective interest, would benefit the environment, society and the economy.

From BIM to Digital Twin for Smart School Buildings

Industry 4.0, the ongoing automation of traditional manufacturing and industrial practices, using modern smart technology, is encouraging the introduction of cutting-edge technologies in the production and manufacturing sectors. Together with the development of advanced technologies, such as Augmented Reality, Virtual Reality and Cloud Computing, the development of the Digital Twin is in fact becoming part of the ongoing research of many companies and universities, especially in the manufacturing field. (Xu, 2019)

In the area of construction, the real revolution has already been occurring for some years and has radically changed the way of thinking, designing, planning, constructing, and managing a building project. The introduction of BIM, defined by the National Building Information Model Standard Project Committee of the United States as "a digital representation of physical and functional characteristics of a facility. A BIM is a shared knowledge resource for information about a facility, forming a reliable basis for decisions during its life cycle; defined as existing from earliest conception to demolition" moves very far the design activity based on Computer Aided Design (CAD).

It is not just an object-oriented modeling that supports geometric and relational attributes, but it changes the perspective and vision of the whole process. In addition, BIM provides a collaborative platform to manage not only the project and its components, but also the relationships between stakeholders. (Villa, 2016)

Much has been achieved since the introduction of the BIM concept proposed in a 1974 publication by Charles M. Eastman (Eastman, 1974). In fact, there are subsequent definitions that expand the concept of Digital Model (Estman, 2011) (Wolk, 2014). The main emphasis of the BIM in fact, has shifted from the model to the information contained in it (for example, type of material, date of installation, costs and maintenance schedules).

The evolution of BIM has moved according to the BIM maturity levels (Sack, 2018) also contained in the PAS 1192 standards and the Digital Built Britain protocols. Therefore, starting from the 1990s, during which three-dimensional CAD modeling software was used and the information was mainly contained in paper documents, we move on to level 1. In the 2000s companies started to use a Common Data Environment (CDE) to include all the information related to the digital model, even if the technology did not allow them to share models with each other.

Since almost a decade people have been working on the collaboration concept both on sharing files and models and on implementing the Industry Foundation Class (IFC) and Construction Operations Building Information Exchange (COBie) interchange files.

Currently the goal is towards level 3, when a single shared Digital BIM Model will be connected to a central cloud-based repository that will ensure collaboration between all stakeholders throughout the building's life cycle.

From this perspective, BIM is a platform that provides an interoperable database of building information to improve planning, construction, and maintenance during the life cycle of a structure (Volk, 2014) (Tang, 2010). In particular, the BIM model is not intended as a static database that parallels the design and construction of the building.

Additionally, the model proactively supports the design process because data related to building specifications, time planning, cost estimation and facility management (e.g., 4D, 5D and 6D) (Succar, 2009), (Bryde, 2013), (Wong, 2014) help to reduce costs by reducing design and construction issues (Lee, 2015).

In a current scenario, BIM can also be used to simulate the management of construction site operations during construction, for example, and can therefore support and optimize the development of the building program (Coraglia, 2017).It is also used for the complete visualization of the project by merging models from all disciplines, for the verification of project consistency, for Clash Detection, for the estimation of construction time and costs (Volk, 2014).

Efforts have been underway for several years (Sacks, 2018, p. 19) to ensure that BIM collects data input from IoT sensors and devices in real time (Tang, 2019). In fact, the natural evolution of this lifecycle modeling methodology is the Digital Twin concept, already developed and implemented more extensively in the manufacturing industry.

The visualization of the Digital Twin of a building should be based on a 3D CAD model extracted from the BIM or a custom 3D model of the building. But, according to the Digital Twin definition developed from Digital Built Britain: "What distinguishes a digital twin from any other digital model is its connection to the physical twin". Indeed, BIM has already undergone an intrinsic evolution of its original nature. Following the described path, with the technological evolution, with the BIM model have been progressively associated some information on the FM detected by sensors, going to define the Facility Management Visual Analytics System (FMVAS) that integrated the Computerized Maintenance Management System (CMMS), a sort of precursor of the DT. The automated information exchange process allows data to be input and output from the BIM model, but two-way real-time data exchange still requires specific developments. The paper of ((Kritzinger, 2018) well outlines the classification levels of the Digital Twin implementation (see Figure 1). The first level is the Digital Model, i.e. the digital representation of physical space that does not involve any automatic data exchange between physical and virtual space. At the second level we find an intermediate system, called Digital Shadow, which is based on the Digital Model and provides an automated flow of one-way data between physical and virtual space. The Digital Twin is at the third level and brings everything much further: a dimension in which both the digital and virtual spaces are bi-univocally linked.

Basically, the BIM corresponds to the Digital Model. The studies that link BIM models to sensors correspond to the Digital Shadow; data are collected and linked to the BIM model, but changes made on digital models do not bring changes the physical space.

If an emergency is detected, for example, the system reports the problem early but does not take corrective action on its own, which should be the case with a Digital Twin.

Beginning tests of interaction between BIM model and real-time data have been made on energy models of school buildings with the aim to design a retrofitting of the assets through the optimization of performance and energy efficiency. This was called Cognitive Buildings (Di Giuda, 2018), namely the introduction of IoT systems inside buildings, with the aim of improving comfort and monitoring

Figure 1. Classification levels of the Digital Twin implementation
Source: inspired from (Mahmoud, 2020)

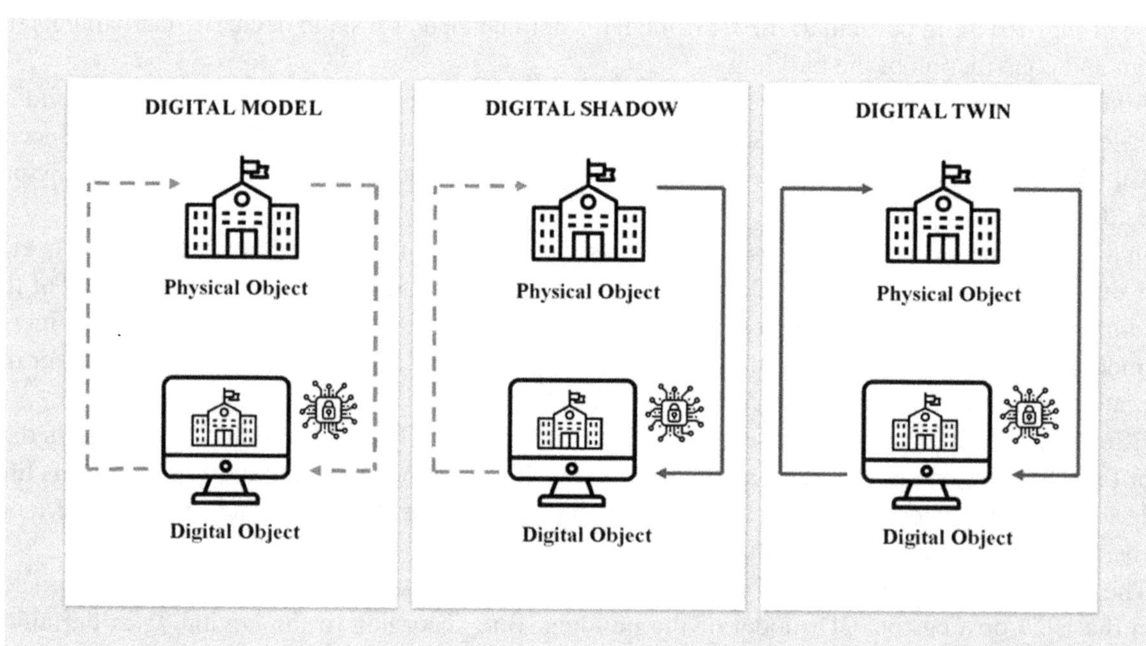

efficiently. The methodology used was based on the principles of circular economy for the built environment, in order to increase energy savings by reducing waste and optimizing the usable space of the building. (Ellen MacArthur Foundation, 2013).

The capability to connect both the data detected and the sensors inside the BIM model allowed to report the real conditions inside the virtual environment (Pasini, 2016) and to compare them with the energy model to refine the setting to accurately simulate the behavior of the building (Ciribini, 2017).

The building Digital Twin use many Wireless Sensors Networks (WSN) to create a real time view of the asset. This dynamic visualization allows real-time analysis, decision-making processes, building efficiency and comfort improvement (Khajavi, 2019).

Digital Twin in Construction

"The Digital Twin is a set of virtual information constructs that fully describes a potential or actual physical manufactured product from the micro atomic level to the macro geometrical level. At its optimum, any information that could be obtained from inspecting a physical manufactured product can be obtained from its Digital Twin." (Grieves, 2016)

This is the definition given by Grieves, considered the father of the Digital Twin concept. In 2002 he presented, at a conference of the Society of Manufacturing Engineers in Troy, Michigan, what he called the Conceptual Ideal for Product Lifecycle Management (PLM).

This intuition contains all the elements that characterize the Digital Twin or rather it considers that every system is composed from two parts: the physical system (real space) that has always existed and a virtual system that contains all the relative information to the physical system. These two systems are linked and a flux of information between the physical system and the virtual one is created (Grieves, 2016).

In accordance with its initial definition, the concept of digital twin is based on the presence of three different parts: the physical product (or building), the virtual model of the same one, and the connections between them.

Information that are available in the virtual model can therefore be used in the physical environment, as well as the inverse. From the literature (Grieves, 2016) we can distinguish: the Digital Twin Prototype (DTP), the Digital Twin Instance (DTI) and the Digital Twin Aggregate (DTA). DTP occurs before there is a physical product and includes all projects, analyses and processes necessary to realize the real product. In the construction industry DTP is equivalent to the BIM model. The DTI is the evolution of the model, namely the digital model which corresponds to every single part of the realized project. The DTA is the aggregation of the several DTI; in this step information are cross-referenced and algorithms can be processed, for example, for performance evaluation or predictive maintenance.

The principles and frameworks containing these three components are therefore in line with the Digital Twin concept, but the literature also distinguishes by the level of data integration. Some virtual representations do not allow automatic bidirectional data exchange, whereas this is the case with fully integrated Digital Twins. To solve the definition of uncertainty, (Kritzinger., 2018) suggests the three subcategories in the Digital Twin classification proposed above.

The first implementation of Digital Twin was in 2010, in the aeronautical world, when the National Aeronautics and Space Administration (NASA) of the Apollo program created two identical spacecraft for mirroring the conditions of the spacecraft throughout the mission (Campos-Ferreira, 2019; Schleich, 2017). In the construction industry we are still far from mirroring the entire building system. The com-

Figure 2. Overview from real building asset to digital twin

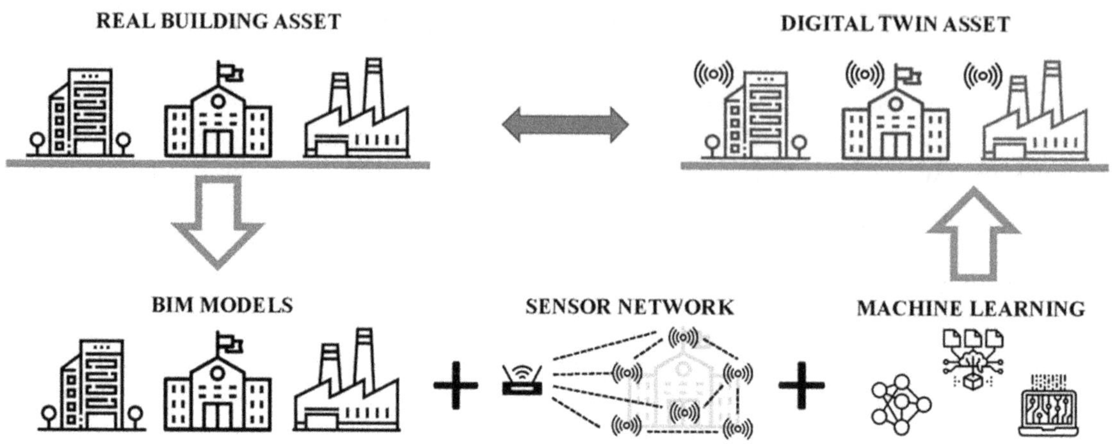

plexity of the structures, components and systems that define the performance of a building creates some critical issues for a correct modeling of the real behavior. Modeling and simulation play a decisive role in supporting design activities and validating system properties (Siavash, 2019). It is indeed the simulation of the behavior of a building that allows to improve its design and to better detail design optimizations.

With this approach, the Digital Twins are set up for detecting breakdowns predictively and for smart facility management. Digital Twin is in fact implemented to monitor the physical asset and improve its operational efficiency by analyzing the parameters in real time (Khajavi et al., 2019). Throughout the life cycle a proper information management in the initial phases is fundamental for the success of the project, but it is the construction and O&M phases that is the real challenge. The concept and design phases are complex but represent only around 30-40 percent of the total project cost (Jiang, 2013). The operation and use phase are the stage that requires the highest costs over the entire life cycle (60-80%) (Jiang, 2013).

For this reason, the applications of the Digital Twin principles of a building, are often oriented to the proper management of the latter aspects (operational and maintenance purposes), allowing facility managers to perform what-if analyses and, ultimately, to improve energy use and user comfort (Khajavi et al., 2019).

The data collected with the Digital Twin during the operation and maintenance of the building can be saved in a database that will be used by architects as a background for future projects (Qi, 2018).

Modeling and simulation also play an essential role in the design phase. We have seen how the BIM model has proven to be revolutionary in terms of project quality and management capabilities. The main difference between simulation in the design phase and the use of a Digital Twin is that the latter requires a physical asset and a Wireless Sensor Network, while the former does not (Tao, 2018). A following study (Saddik, 2018) coined a broader definition: “digital twins will facilitate the means to monitor, understand, and optimize the functions of all physical entities, living as well as nonliving, by enabling the seamless transmission of data between the physical and virtual world.".

Digital Twin is a concept applicable to many areas, from industry to predictive maintenance, from aeronautics to automotive and is completely transforming industries and production lines. Digital Twin was one of the top ten strategic technology trends in 2018 and, based on future research forecasts, the digital twin market will reach $15 billion by 2023 (Market Research Future, 2017).

The main investigations on the application of the Digital Twin concept in the construction sector define the digital twin of a building as "the interaction between the real world internal building environment and a digital but realistic virtual representation model of the building environment, thus offering the opportunity of monitoring and collecting data real-time".

In this definition, an indoor environment indicates, for example, information on air temperature, air flow, humidity and lighting conditions, while a virtual digital environment indicates the computational fluid dynamics and luminance level (Siavash, 2019).

The basic logic of the digital twin is based on the see-think-do rule, i.e.: real-time data are collected for a certain period of time, processed according to machine learning algorithms, and action is taken by optimizing the behavior of the building or part of it (ventilation, heating, access control, etc.). The advantages of creating a digital twin are intrinsically linked to these three aspects: the collection of data and their visualization in real time within the building environments, the analysis of any data gaps and the optimization during use and maintenance.

There are many differences between a BIM model and a Digital Twin. The first important difference is that BIM model is designed to improve the efficiency of design and construction and is mainly used in these phases (Liu, 2017), while Digital Twin is designed to monitor a physical asset and improve its operational efficiency and to allow predictive maintenance (Qi, 2018). The second major difference is that BIM is not set up to display and manage real-time data; it is still used in the industry for design, construction, maintenance and interoperability activities, which require simulations but not real-time data processing (Bruno, 2018); in essence, the Digital Twin is the digital counterpart of a physical asset and works in extension of the current BIM platform.

The digital twin works specifically with real time data provided by sensor systems to record and analyze the structural and environmental parameters of a physical resource for the purpose of performing highly accurate simulations and data analysis (Qi, 2018). The third difference between the two concepts is related to the type of data required for the construction of each model. Although BIM is suitable for the integration of cost estimation and time planning data to improve the efficiency of a construction project (Lee, 2014), Digital Twin is designed to integrate real-time sensor readings to analyze and improve the interaction of the building with the environment and users (Tao,2019).

However, it is clear that the school buildings asset, as illustrated above, was constructed for the most part long before the 1990s. Therefore, most buildings do not have BIM models of the project on which to implement a Wireless Sensor Network as the first step toward Digital Twin. The same situation occurs for more than 80% of the buildings in Europe, whose construction dates back well before 1990 and therefore does not have the BIM model (Volk, 2014; Arayici, 2008; Armesto, 2009; Dickinson, 2009; Attar, 2010).

For existing buildings, the effort to reconstruct a BIM model is relevant: problems are found from the geometric survey as well as the reconstruction of the documentation and information needed to be included in the model (Volk, 2014). There are significant Italian case studies in the literature that have proposed an accurate methodology for the construction of BIM models of existing school buildings (Di

Table 1. Comparison between BIM and DT of building

	BIM	DT
First concept	Charles Eastman (1974)	Michael Grieves (2002)
Focus	Information model and management Clash Detection and Code Checking Time and Cost estimation Collaboration and Interoperability	Predictive Maintenance What-if analysis Resource consumption efficiency
Stages (RIBA)	Design Construction	Use Maintenance
Main Users	Client Architect MEP and Structural Engineer Project Manager	Client Facility Managers Users
Technologies	Industry Foundation Class (IFC) Common Data Environment (CDE) COBie	Wireless Sensor Network (WSN) Data Analytics Machine Learning Algorithm

Giuda, 2015) but have not had an important follow-up such as systematically being extended to a national asset. On the contrary, existing buildings can take an important advantage from the implementation of a network of sensors that can constitute a Digital Twin for the improvement of operation and maintenance (Qi, 2018). The integration of Wireless Sensor Network (WSN) and data analysis are two of the necessary components for the creation of a Digital Twin (Tao, 2018).

The methodology proposed to approach the school heritage refurbishment in the perspective of Smart School Buildings starts from the setting of the correct BIM modeling, which is the beginning of the digital life of a building. BIM is an integrated process particularly valuable for the involvement of users in all phases of design and management. This is why it is particularly significant during user-involvement processes. It is in fact much more than a design tool. Thanks to BIM, smart building data is collected through contextual information in order to improve the integration between building and network (B2G). The proposed plan for sensorizing school buildings creates a powerful infrastructure for Smart Cities. BIM and Digital Twin, if structured in an integrated and synergic way, can give a significant added value to the redevelopment of existing heritage to build the Smart School Buildings of the future.

The use of Building Automation contributes significantly to the efficiency and safety of a building, which becomes an ecosystem, more or less complex, of devices connected to each other with the users who live the spaces. Combining new technologies, such as the Internet of Things, with new methods of collaborative management of the process, such as Building Information Modeling, allows to have Smart buildings able to optimize three main aspects:

- Real-time evaluation of the safety of the environments, in relation to the use and the crowds density. The situation of spaces is constantly monitored to ensure the correct flow of people and the security conditions of the environments.
- Optimization of consumption, through the system control and the data collected, it will be possible monitoring the operation remotely, both by users and by maintenance technicians responsible for the proper operations.
- Continuous control of the use of the spaces with the warning of special alerts in the presence of groupings. As a matter of fact, we are working for the reopening of the schools after the lockdown.

Some of the suggested measures include, i. e. the absence of assemblies, seminars and meetings. It is also likely that students will be able to access 30% of the schools in order to maintain longer distances and less risk of further mass infection. Present your perspective on the issues, controversies, problems, etc., as they relate to theme and arguments supporting your position. Compare and contrast with what has been, or is currently being done as it relates to the chapter's specific topic and the main theme of the book.

Smart School Buildings

The school organization has assumed an important role as a guide for social evolution by promoting new models of learning, class organization, updating of digital tools and introducing personalized educational paths oriented to the centrality of the user for an advancement in knowledge and for the growth of students' skills. The project is user-centered and closely linked to the requirements deriving from ambitions and needs: for example in elementary school the learning space is practically unchanged and pupils develop a sort of familiarity and ownership with the usual space. In addition, it is important to remember that existing buildings used for school purposes were not always born with this task and the rigid distribution that characterizes them is not easily adaptable to the new concept of learning spaces (e.g. the space for educational purposes must be adaptable to comply with the configuration of the classroom to individual, collective or group study) (Tagliabue, 2017).

The suitability of spaces for different activities by gathering people from different and changing environments over time, promotes the integration of the school building with the district and the city. This idea is supported by new technologies that allow the interaction between the building and the user, to implement the services and comfort actively provided for an enriched user experience: this means "cognitive".

This idea leads the building to create complex connections with the main infrastructure of the district facilitating its role as a service provider that manages people flows and behavioral interactions.

Figure 3 diagrams the various areas related to the Smart School Building. The aspect associated to the Energy Management includes the real-time control of the consumptions and the equipments. The Services block includes the tracking of the student and his activities within the school. The evaluation of the use of the spaces includes a scheduling of classrooms and laboratories, an optimization of the use of the asset buildings and a synergic response to emergencies. Additionally, the Safety and security aspect includes sanitization systems, user monitoring and the application of Covid-19 procedures.

The perspective of the circular economy is included in the new model of the school as a civic center, expanding the spaces and the use of the same both in time and in use by the community (Tagliabue, 2016). However, the need to ensure flexibility in employment, usage time, type of users and occupancy schedules creates enormous concern, for example, in light of the Covid-19 pandemic. Management policies are confusing and uncertain, so the use of crowd simulation models, for example, could greatly facilitate national and international strategic choices.

In this context new technology represent a big opportunity in order to improve the safety and comfort of school buildings. With BIM model and, for example, BIM-based gaming and crowd simulation (Simeon, 2013) it is possible to simulate the way of use of buildings and rooms and this is beneficial for reducing people's movement interference and validating security pathways. The connectivity of the school building needs a two-way relationship with the city and the users who are the core. The ability to

Figure 3. Smart School Buildings areas

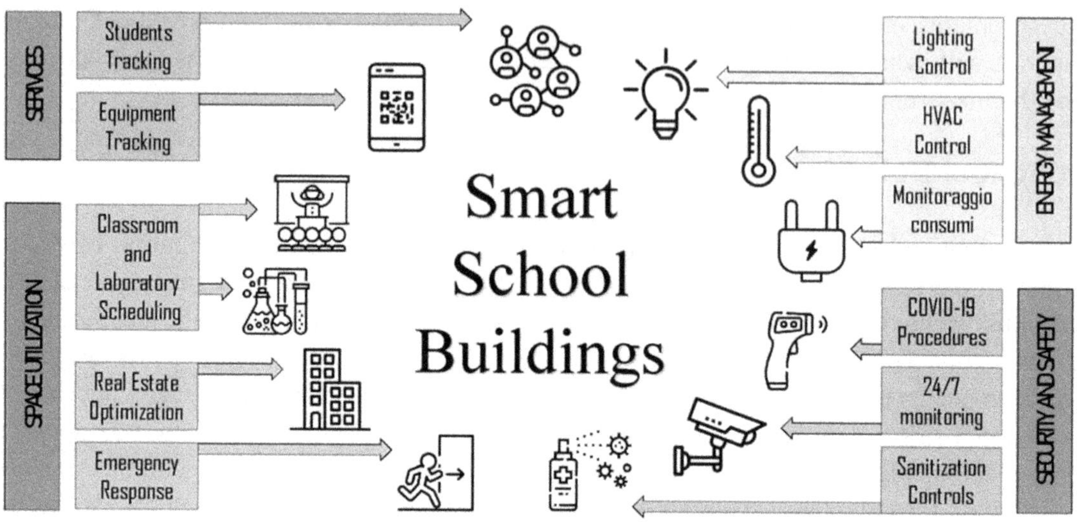

embrace smart technologies and IoT (Internet of Things) for organization, logistics, event coordination, learning growth and social networking management is an upcoming scenario.

Safety and structural monitoring

The first check to be carried out to enable the use of an existing building is related to the seismic safety of the structures and non-structural elements. In this regard, it is worth referring to the experimental research already carried out in the seismic vulnerability assessment with BIM-based approaches and the use of sensors. (Domaneschi, 2019; Cimellaro, 2019; Villa, 2018; Cardoni, 2019). In the reported examples, the positioning of MEMS and force balance triaxle acceleration sensors allowed researchers to detect with absolute precision the behavior of the building and its residual resistance by detecting ambient vibrations. The use of the BIM model has allowed the comparison of the real building behavior with the digital model to optimize the verification of calibrations. Analyzing the construction periods of existing Italian school buildings, it can be noted that the structures are mainly made of reinforced concrete and studies on the behavior of the floors (Kezmane, 2017) are important in view of a redevelopment project.

Safety in Building Use

Many experiments are ongoing in this area, although none of them has yet approached the full vision of the Digital Twin. The use of Virtual Reality is the most immediate tool to verify the safety of use of the building. In this regard we mention the simulations on emergency management and fire evacuations (Cimellaro, 2019) through VR and crowd simulation tools. The latter can improve the management of existing buildings and the design of new ones. The advantages are mainly related to the large amount

Figure 4. Example of a sensorization plan of a structure with the indication of the accelerometers configuration
Source: (Cardoni, 2019)

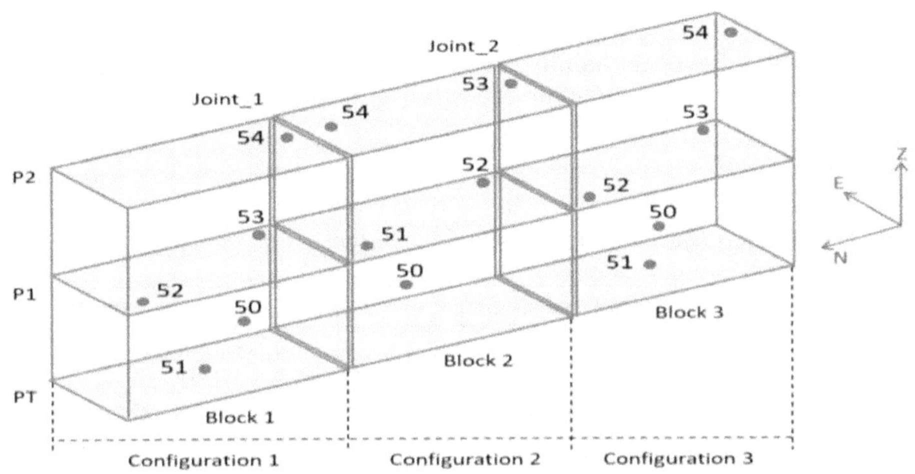

of data that the building generates in the long term, so more information can be processed about how it is used, the type of users and the moments that can create malfunctions or dangerous situations. The analyses can provide predictive information for the operational phase of the building in order to define precise and specific management plans.

Permanent monitoring using Digital Twin helps determining if the building's internal conditions provide the required comfort in the operational phase. The results can be used both to correct weaknesses and to inform subsequent projects. This process helps to define if the performance is positive or negative and, as a result, to correct discomforts and inefficiencies. Benchmarks and standards for energy performance and indoor air quality monitoring are available, making it easier to analyze and evaluate data. (Tagliabue, 2016).

Predictive Maintenance and Facility Management

Deriving from the industrial sector, the focus of Digital Twin research are the predictive maintenance and monitoring of plant systems in order to reduce waste and FM costs.

The Digital Twin in fact provides a new paradigm for fault diagnosis. In the current trials (Xu, 2019) the Digital-Twin-assisted diagnosis method uses Deep Learning and realizes both fault diagnosis and maintenance diagnostics. The model is trained, data from the physical space is analyzed and using Deep Transfer Learning real-time monitoring and predictive maintenance is performed. A first difficulty, especially related to the construction industry, is related to the limits that there are in the collection of complete data on possible faults to properly train the diagnosis model. Complex buildings have simple installations, but they operate with undefined boundary conditions that are difficult to control and standardize. Table 2 shows a comparison between the traditional diagnostic system and the one using Digital Twin.

A methodology proposed for the prompt generation of the data being processed is to add, to the data coming from the physical model, also a series of dynamic simulations of possible critical situations which can be reproduced with the Digital Twin. This method takes up some of the studies already carried out for the retrofitting of existing buildings (Desogus, 2018). In a virtual space in fact it is much easier to obtain the simulation data in different conditions, adjusting the parameters and creating various simulation scenarios.

Table 2. Comparison between Dfdd and traditional fault diagnosis methods. Source: (Xu, 2019)

	Traditional fault diagnosis	**DFDD**
Data Source	Monitoring data of physical entity	Simulation data of virtual space and monitoring data of physical entity
Data Distribution	Same distribution of training and test data	Different distribution of source and target data
Model Training	Different situations need to be modeled separately	Models can migrate in different situations
Diagnosis Period	Operation and Maintenance phase	From design phase to operation and maintenance phase

The efficient use of school buildings in terms of occupancy, heating, cooling, cleaning and space management in occupied areas and during operating hours can optimize consumption and lead to significant cost savings. Digital Twin can trigger savings in terms of resources, costs and time, as well as increased user satisfaction. Referring to the feedback given by the definition of the correct Machine Learning algorithms for data processing, it may take a long time to obtain valid and consistent input.

CONCLUSION

The release of affordable, powerful semiconductor chips (Chui,2010) has allowed the Internet of Things (IoT) to take off. This technology has a major impact on every aspect of our lives and economy (Baranwal, 2016) (Leminen, 2012). Developments such as connected and autonomous cars (Gerla, 2014), drones (Motlagh, 2016) and smart homes (Wang, 2013) are all examples of the integration of IoT into systems, enabling the creation of completely new concepts.

Smart buildings are rising as the next frontier in the development cycle of architectural structures (Minoli, 2017). In fact, collaborations are currently underway between world tech giants such as Amazon, Google and Microsoft to integrate programmable services such as heating, cooling and appliances in new residential buildings. School buildings present a complexity linked both to the mainly functional, dimensional and structural aspects that make them very different from the residential context. The existing school heritage, due to the prevalence outdated and with significant issues in itself, leads to having to consider multiple aspects and the resulting problems are, nowadays, hard to overcome. In order to have a complex systems simulation, such as the school buildings, it is necessary indeed to combine different simulation systems. Often there are problems connected to the interoperability of the systems of collection and visualization data that make impossible to realize integrated simulations of the complex system.

Concepts like the digital factory, already currently applied to some experimental plants, aim to simulate the real-time behavior of entire production systems but are limited to this area of application, very specific, and are bound to specific algorithms of simulations data from the specific production chain.

Summarizing, there are, at present, several modeling and simulation technologies and tools available, but the simulation of complex systems in their entirety is still a great challenge. The component of space use and randomness given by building users is one of the fundamental problems. A new methodology of real building modeling (EDT) is being worked on to create interacting EDT networks that replicate real world scenarios (Schluse, 2018). Although the Internet of Things concept is currently applied to many common objects, the use of enabled sensor networks to build a Digital Twin of a smart building has not yet been studied.

The study presented by (Nasaruddin, 2018) contains the conceptual framework for a transition from a physical room to a Digital Twin, but does not contain any empirical experimentation to support the described theory. The study presented by (Verma, 2018) presents the concept of a Digital Twin of a building but without providing data of a real application. It presents the case of Kone's use of a Digital Twin to reduce elevator maintenance costs and improve service. The Fraunhofer Building Innovation Alliance is studying the Digital Twin of buildings and has highlighted the potential benefits on the life cycle of the building (Fraunhofer, 2019).

This chapter summarizes the current overview of Digital Twin applications in the construction world. The analysis shows that significant steps are being taken beyond the simple digital building model but that the application of the Digital Twin concept in its full meaning is still a long way off. The attention of the research is moving in this direction and the evolution of the current Digital Shadow, combined with the experience that is maturing in the industrial sector, will soon bring new revolutions within the construction sector.

FUTURE RESEARCH DIRECTIONS

Alongside the realization of Smart School Building is taking more and more space a new concept of education as a transfer of skills and knowledge. The Massive Open Online Courses (MOOC), which make university courses available for free or at a minimal cost, are now a disruptive and consolidated technology that challenges traditional classroom teaching. In this paradigm shift in pedagogy, we are experiencing a flood of opportunities to customize education for more effective teaching. Different students based on their natural talent, interest and background have different needs. Although universities and schools are recording more and more data on students, the feedback loop is not complete. (Rasheed, 2020)

A digital twin for students will help counteract this weakness. With the rise of artificial intelligence driving bots, education and personalized counseling, in accordance with the needs of a digital twin for students, will be increasingly available for free.

A custom adaptive learning framework has been built (Yu, 2017) (Peng, 2019) using an intelligent learning environment. Another significant example can be found in the study (Mitrofanova, 2019) where the authors developed a concept of intelligent education incorporating data mining tools. E-learning is undergoing an increasingly rapid transformation process, moving to mobile learning (m-learing), ubiquitous learning (u-learning) and smart learning (s-learning) (Adu, 2014), also thanks to the use of IoT in building an intelligent educational process (Abdel-Basset, 2019).

As illustrated above, however, the interaction between user and building and between the pupils themselves is a fundamental condition for the implementation of the users' cognitive skills, and therefore, for social progress. For this reason, in our opinion, it is important to preserve the social thinking by integrating everything that can be used to improve the effectiveness of teaching in innovative and technologically advanced spaces.

Digital data acquisition and management systems have been tested in order to assist the evaluation process (Sabir, 2018) to map in a more targeted way the educational objectives and the results achieved by students (Osman, 2018). In the study "Smart education in the context of industry 4.0" are indicated the main skills that have become fundamental for the application of the Digital Twin in order to wait for the Industry 4.0 revolution (Assante, 2019). In the wake of the above, analyzing the current gap between universities and the demands of the world of work, combining the digital twin of the education sector with that of other industrial sectors would be more efficient in training targeted talent. (Rasheed, 2020)

REFERENCES

Abdel-Basset, M., Manogaran, G., Mohamed, M., & Rushdy, E. (2019). Internet of Things in smart education environment: Supportive framework in the decision-making process. *Concurrency and Computation: Practice and Experience, 31*(10).

Adu, E. K., & Poo, D. C. (2014). Smart learning: A new paradigm of learning in the smart age. In *Proceedings International Conference Teaching and Learning Higher Education (TLHE).* Singapore: National Univ. Singapore.

Arayici, Y. (2008). Towards building information modelling for existing structures. *Structural Survey, 26*(3), 210-222.

Armesto, J., Lubowiecka, I., Ordóñez, C., & Rial, F.I. (2009). FEM modeling of structures based on close range digital photogrammetry. *Automation in Construction, 18*(5), 559-569.

Assante, D., Caforio, A., Flamini, M., & Romano, E. (2019). Smart education in the context of industry 4.0. *Proceeding IEEE Global Engineering Education Conference (EDUCON),* 1140–1145.

Attar, R., Prabhu, V., Glueck, M., & Khan, A. (2010). 210 King Street: A dataset for integrated performance assessment. *Proceedings Spring Simulation Multiconference.*

Baranwal, T., & Pateriya, P. K. (2016). Development of IoT based smart security and monitoring devices for agriculture. *Proceedings IEEE 6th International Conference Cloud Systems Big Data Engineering,* 597-602.

Bennacer, L., Amirat, Y., Chibani, A, Mellouk, A., & Ciavaglia, L. (2015). Self diagnosis technique for virtual private networks combining Bayesian networks and case-based reasoning. *IEEE Transactions on Automation Science and Engineering, 12*(1), 354-366.

Bruno, S., De Fino, M., & Fatiguso, F. (2018). Historic building information modelling: Performance assessment for diagnosis-aided information modelling and management. *Automation in Construction, 86*, 256-276.

Bryde, D., Broquetas, M., & Volm, J. M. (2013). The project benefits of building information modelling (BIM). *International Journal of Project Management., 31*(7), 971-980.

Cardoni, A., Noori, A. Z., Marasco, S., Domaneschi, M., Villa, V., Ansari, F., Caldera, C., & Cimellaro, G. P. (2019) Dynamic characterization and vulnerability assessment of a school building in Italy. *Structural Health Monitoring 2019: Enabling Intelligent Life-Cycle Health Management for Industry Internet of Things (IIOT) - Proceedings of the 12th International Workshop on Structural Health Monitoring*, 459-466. 10.12783hm2019/32148

Chen, Y. Q., Fink, O., & Sansavini, G. (2018). Combined fault location and classification for power transmission lines fault diagnosis with integrated feature extraction. *IEEE Transactions on Industrial Electronics, 65*(1), 561-569.

Chui, M., Löffier, M., & Roberts, R. (2010). *The Internet of Things* (Vol. 2). McKinsey.

Cimellaro, G. P., Domaneschi, M., De Iuliis, M., Villa, V., Caldera, C., & Cardoni, A. (2019). Fire emergency evacuation in a school building through VR. *COMPDYN Proceedings*, 2046-2055. 10.7712/120119.7057.19951

Cimellaro, G. P., Domaneschi, M., Villa, V., Noori, A. Z., Pietropinto, M. V., Cardoni, A., & Marasco, S. (2019) BIM-based approach for seismic vulnerability analysis: A case study. *Advances in Engineering Materials, Structures and Systems: Innovations, Mechanics and Applications - Proceedings of the 7th International Conference on Structural Engineering, Mechanics and Computation*, 2203-2208. 10.1201/9780429426506-380

Ciribini, A. L. C., Pasini, D., Tagliabue, L. C., Manfren, M., Daniotti, B., Rinaldi, S., & De Angelis, E. (2017). Tracking users' behaviors through real-time information in BIMs: Workflow for interconnection in the Brescia Smart Campus Demonstrator. *Procedia Engineering, 180*, 1484-1494.

Coraglia, U. M., Simeone, D., Cursi, S., Fioravanti, A., Wurzer, G., & D'Alessandro, D. (2017). A simulation model for logical and operative clash detection. *Sharing Computable Knowl., 2*, 519-534.

Darong, H., Lanyan, K., Bo, M., Ling, Z., & Guoxi, S. (2018). A new incipient fault diagnosis method combining improved RLS and LMD algorithm for rolling bearings with strong background noise. *IEEE Access: Practical Innovations, Open Solutions, 6*, 26001–26010. doi:10.1109/ACCESS.2018.2829803

Desogus, G., Di Giuda, G. M., Monni, G., Quaquero, E., Sanna, A., Tagliabue, L. C., & Villa, V. (2018). The building information modeling for the retrofitting of existing buildings. A case study in the University of Cagliari. *International Conference Proceedings of the Seismic and Energy Renovation for Sustainable Cities*, 141-150.

Di Giuda, G. M., Quaquero, E., Villa, V., Tagliabue, L. C., Desogus, G., Sanna, A., & Ciribini, A. L. C. (2018). Towards the cognitive building: information modeling for the energy audit. *TEMA, 4*(2).

Di Giuda, G. M., Villa, V., Paleari, F., & Schievano, M. (2015). *BIM modeling of the existing school heritage for investment planning in Sostenibilita' ambientale e produzione edilizia: La ricerca scientifica sulla Produzione Edilizia nell'era delle nuove sfide ambientali nel Settore delle Costruzioni*. Ingenio.

Dickinson, J., Pardasani, A., Ahamed, S., & Kruithof, S. (2009). A survey of automation technology for realising as-built models of services. *Proceedings 1st International Conference. Improving Construction Use Through Integration Design Solutions*, 365-381.

Domaneschi, M., Villa, V., Cimellaro, G. P., Caldera, C., Zamani Noori, A., Marasco, S., & Ansari, F. (2019). Integrating BIM with on site investigation for seismic vulnerability assessment. *COMPDYN Proceedings*, 544-553. 10.7712/120119.6938.19957

Eastman, C., Fisher, D., Lafue, G., Lividini, J., Stoker, D., & Yessios, C. (1974). *An Outline of the Building Description System*. Institute of Physical Planning Research Report (No. 50).

Eastman, C., Teicholz, P., Sacks, R., & Liston, K. (2011). *BIM Handbook: A Guide to Building Information Modeling for Owners, Managers, Designers, Engineers and Contractors*. Wiley.

Ellen MacArthur Foundation. (2013). *Towards the circular economy: Opportunities for the consumer goods sector*. Ellen Mac Arthur Foundation, Sun, McKinsey Center for Business and Environment.

Fraunhofer. (2019). *Building from Design to Demolition*. Available: https://www.bau.fraunhofer.de/en/_eldsofresearch/smartbuilding/digital-twin.html

Gerla, M., Lee, E. K., Pau, G., & Lee, U. (2014). Internet of vehicles: From intelligent grid to autonomous cars and vehicular clouds. *Proceedings IEEE World Forum Internet Things (WF-IoT)*, 241-246. 10.1109/WF-IoT.2014.6803166

Grieves, M., & Vickers, J. (2016). Digital Twin: Mitigating Unpredictable, Undesirable Emergent Behavior in Complex Systems. In *Trans-Disciplinary Perspectives on System Complexity* (pp. 85–114). Springer.

Jia, R., Jin, B., Jin, M., Zhou, Y., Konstantakopoulos, I. C., Zou, H., Kim, J., Li, D., Gu, W., Nuzzo, P., Schiavon, S., Sangiovanni-Vincentelli, A. L., Spanos, C. J., & Arghandeh, R. (2018). Design automation for smart building systems. *Proceedings IEEE, 106*(9), 1680-1699.

Jiang, H., (2013). *A System Dynamics Model for Manpower and Technology Implementation Trade-off and Cost Estimation*. Electronic Theses and Dissertations 202.

Kezmane, A., Chiaia, B., Kumpyak, O., Maksimov, V., & Placidi, L. (2016). 3D modelling of reinforced concrete slab with yielding supports subject to impact load. *European Journal of Environmental and Civil Engineering*, 1–38. doi:10.1080/19648189.2016.1194330

Khajavi, S. H., Motlagh, N. H., Jaribion, A., Werner, L. C., & Holmström, J. (2019). Digital Twin: Vision, Benefits, Boundaries, and Creation for Buildings. *IEEE Access, 7*, 147406-147419. . doi:10.1109/ACCESS.2019.2946515

Khajavi, S. H., Motlagh, N. H., Jaribion, A., Werner, L. C., & Holmstrom, J. (2019). Digital Twin: Vision, Benefits, Boundaries, and Creation for Buildings. *IEEE Access: Practical Innovations, Open Solutions, 7*, 147406–147419. doi:10.1109/ACCESS.2019.2946515

Kritzinger, W., Karner, M., Traar, G., Henjes, J., & Sihn, W. (2018). Digital Twin in manufacturing: A categorical literature review and classification. *IFAC-Pap., 51*(11), 1016–1022. doi:10.1016/j.ifacol.2018.08.474

Lee, H. W., Oh, H., Kim, Y., & Choi, K. (2015). Quantitative analysis of warnings in building information modeling (BIM). *Automation in Construction, 51*, 23–31. doi:10.1016/j.autcon.2014.12.007

Lee, S. K., Kim, K. R., & Yu, J. H. (2014). BIM and ontology-based approach for building cost estimation. *Automation in Construction, 41*, 96-105.

Leminen, S., Westerlund, M., Rajahonka, M., & Siuruainen, R. (2012). *Towards IoT ecosystems and business models in Internet of Things, Smart Spaces, and Next Generation Networking*. Springer.

Liu, Y., van Nederveen, S., & Hertogh, M. (2017). Understanding effects of BIM on collaborative design and construction: An empirical study in China. *International Journal of Project Management, 35*(4), 686-698.

Lu, W., Li, Y., Cheng, Y., Meng, D., Liang, B., & Zhou, P. (2018). Early fault detection approach with deep architectures. *IEEE Transactions on Instrumentation and Measurement, 67*(7), 1679–1689. doi:10.1109/TIM.2018.2800978

Madakam, S., Ramaswamy, R., & Tripathi, S. (2015). Internet of Things (IoT): A Literature Review. *Journal of Computer and Communications, 3*(05), 164–173. doi:10.4236/jcc.2015.35021

Market Research Future. (2017). *Global Digital Twin Market is Estimated to Grow at a Cagr of 37% From 2017 to 2023*. Available: https://www.marketresearchfuture.com/reports/digital-twin-market-4504

Minoli, D., Sohraby, K., & Occhiogrosso, B. (2017). IoT considerations, requirements, and architectures for smart buildings in Energy optimization and next-generation building management systems. *IEEE Internet of Things Journal, 4*(1), 269–283. doi:10.1109/JIOT.2017.2647881

Mitrofanova, Y. S., Sherstobitova, A. A., & Filippova, O. A. (2019). *Modeling smart learning processes based on educational data mining tools in Smart Education and e-Learning 2019*. Springer.

Motlagh, N. H., Khajavi, S. H., Jaribion, A., & Holmström, J. (2018). An IoT based automation system for older homes: A use case for lighting system. *Proceedings IEEE 11th International Conference Service-Oriented Computer Application (SOCA),* 1-6.

Motlagh, N. H., Taleb, T., & Arouk, O. (2016). Low-altitude unmanned aerial vehicles-based Internet of Things services: Comprehensive survey and future perspectives. *IEEE Internet of Things Journal, 3*(6), 899–922.

Nasaruddin, N., Ito, T., & Tuan, T. B. (2018). Digital twin approach to building information management. *Proceedings Manufacturing Systems Division Conference*, 304.

Osman, A., Yahya, A. A., & Kamal, M. B. (2018). A benchmark collection for mapping program educational objectives to ABET student outcomes: Accreditation. In *Proceedings 5th International Symposium on Data Mining Applications*. Springer.

Pasini, D., Mastrolembo Ventura, S., Rinaldi, S., Bellagente, P., Flammini, A., & Ciribini, A. L. C. (2016). Exploiting Internet of Things and building information modeling framework for management of cognitive buildings. In *Smart Cities Conference (ISC2) in IEEE International* (pp. 1-6). IEEE. 10.1109/ISC2.2016.7580817

Peng, H., Ma, S., & Spector, J. M. (2019). *Personalized adaptive learning: An emerging pedagogical approach enabled by a smart learning environment in Foundations and Trends in Smart Learning*. Springer.

Qi, Q., & Tao, F. (2018). Digital Twin and big data towards smart manufacturing and industry 4.0: 360 degree comparison. *IEEE Access, 6*, 3585-3593.

Qi, Q., & Tao, F. (2018). Digital Twin and Big Data Towards Smart Manufacturing and Industry 4.0: 360 Degree Comparison. *IEEE Access: Practical Innovations, Open Solutions*, *6*, 3585–3593. doi:10.1109/ACCESS.2018.2793265

Qi, Q., Tao, F., Zuo, Y., & Zhao, D. (2018). Digital Twin service towards smart manufacturing. *Proceedings CIRP, 72*, 237-242.

Qin, Q., & Tao, F. (2018). Digital Twin and big data towards smart manufacturing and industry 4.0: 360 degree comparison. *IEEE Access, 6*, 3585-3593.

Rasheed, A., San, O., & Kvamsdal, T. (2020). Digital Twin: Values. *Challenges and Enablers from a Modeling Perspective in IEEE Access*, *8*, 21980–22012.

Rosen, R., von Wichert, G., Lo, G., & Bettenhausen, K. D. (2015). About the importance of autonomy and Digital Twins for the future of manufacturing. *IFAC-PapersOnLine, 48*(3), 567-572.

Sabir, A., Abbasi, N. A., & Islam, N. (2018). *An electronic data management and analysis application for abet accreditation*. arXiv:1901.05845.

Sacks, R., Eastman, C., Lee, G., & Teicholz, P. (2018). *BIM Handbook: A Guide to Building Information Modeling for Owners, Designers, Engineers, Contractors, and Facility Managers*. Wiley. doi:10.1002/9781119287568

Saddik, A. E. (2018). Digital Twins: The convergence of multimedia technologies. *IEEE Multimedia, 25*(2), 87-92.

Schleich, B., Anwer, N., Mathieu, L., & Wartzack, S. (2017). Shaping the Digital Twin for design and production engineering. *CIRP Annual, 66*(1), 141-144.

Schluse, M., Atorf, L., & Rossmann, J. (2017). Experimentable Digital Twins for model-based systems engineering and simulation-based development. *Proceedings Annual IEEE International System Conference*, 628– 635.

Schluse, M., Priggemeyer, M., Atorf, L., & Rossmann, J. (2018). Experimentable Digital Twins—Streamlining Simulation-Based Systems Engineering for Industry 4.0. IEEE Transactions on Industrial Informatics, 14(4), 1722-1731. doi:10.1109/TII.2018.2804917

Schluse, M., Priggemeyer, M., Atorf, L., & Rossmann, J. (2018). Experimentable Digital Twins: Streamlining simulation-based systems engineering for industry 4.0. IEEE Transactions on Industrial Informatics, 14(4), 1722-1731.

Simulation of Systems in Materials Handling. Logistics and Production–Fundamentals, VDI-Standard 3633 Part 1, 2016. Available: www.vdi.eu/3633

Succar, B. (2009). Building information modelling framework: A research and delivery foundation for industry stakeholders. *Automation in Construction, 18*(3), 357-375.

Tagliabue, L. C., Consalez, L., Mastrolembo Ventura, S., & Ciribini, A. L. C. (2016). The Educational Building in the Circular Economy: an Extended Concept of Asset Quality. In Convegno Internazionale ISTeA 2016, BACK TO 4.0: Rethinking the Digital Construction Industry, Naples.

Tagliabue, L. C., & Villa, V. (2017). *Il BIM per le scuole. Analisi del patrimonio scolastico e strategie di intervento*. Hoepli.

Tang, P., Huber, D., Akinci, B., Lipman, R., & Lytle, A. (2010). Automatic reconstruction of as-built building information models from laser-scanned point clouds: A review of related techniques. *Automation in Construction, 19*(7), 829-843.

Tang, S., Shelden, D. R., Eastman, C., Pishdad-Bozorgi, P., & Gao, X. (2019). A review of building information modeling (BIM) and the Internet of Things (IoT) devices integration: Present status and future trends. *Automation in Construction, 101*, 127-139.

Tao, F., Cheng, J., Qi, Q., Zhang, M., Zhang, H., & Sui, F. (2018). Digital twin-driven product design, manufacturing and service with big data. *International Journal Advanced Manufacturing Technology, 94*(9-12), 3563-3576.

Tao, F., Cheng, Y., Xu, L. D., Zhang, L., & Li, B. H. (2014). CCIoT-CMfg: Cloud computing and Internet of Things-based cloud manufacturing service system. *IEEE Transactions on Industrial Informatics, 10*(2), 1435-1442.

Tao, F., Sui, F., Liu, A., Qi, Q., Zhang, M., Song, B., Guo, Z., Lu, S. C. Y., & Nee, A. (2019). Digital twin-driven product design framework. *International Journal of Production Research, 57*(12), 3935-3953.

Verma, U. (2018). *What are Digital Twins in Smart Buildings?* Available: https://inbuildingtech. com/bms/digital-twin-commercial-office-building/

Villa, V., & Di Giuda, G. M. (2016). *Collaborative Design and Collaborative Contracting a Defiant Challenge in Digitizing the analogical thoughts in architecture: a menace or a promise?* Ingenio.

Villa, V., Domaneschi, M., Cimellaro, G., & Caldera, C. (2018). BIM-Based Approach for Seismic Risk Analysis. In Streamlining Information Transfer between Construction and Structural Engineering. AAE-06-1-AAE-06-6. doi:10.14455/ISEC.res.2018.145

Volk, R., Stengel, J., & Schultmann, F. (2014). Building Information Modeling (BIM) for existing buildings: Literature review and future needs. *Automation in Construction, 38*, 109-127.

Wang, M., Zhang, G., Zhang, C., Zhang, J., & Li, C. (2013). An IoT-based appliance control system for smart homes. *Proceedings IEEE 4th International Conference Intelligence Control Information Processing and Cloud Computing Process*, 744-747.

Wong, K., & Fan, Q. (2013). Building information modelling (BIM) for sustainable building design. *Facilities, 31*(3-4), 138-157.

Xu, Y., Sun, Y., Liu, X., & Zheng, Y. (2019). A Digital-Twin-Assisted Fault Diagnosis Using Deep Transfer Learning. IEEE Access, 7, 19990-19999. doi:10.1109/ACCESS.2018.2890566

Yin, Z., & Hou, J. (2016). Recent advances on SVM based fault diagnosis and process monitoring in complicated industrial processes. *Neurocomputing, 174*, 643-650.

Yu, H., Miao, C., Leung, C., & White, T. J. (2017). Towards AI-powered personalization in MOOC learning in npj Sci. *Learn., 2*, 15. PMID:30631461

KEY TERMS AND DEFINITIONS

Digital Model (DM): A Digital Model is a digital representation of an existing or planned physical object that does not use any form of automated data exchange between the physical object and the digital object. The digital representation might include a more or less comprehensive description of the physical object.

Digital Shadow (DS): Based on the definition of a Digital Model, if there further exists an automated one-way data flow between the state of an existing physical object and a digital object, one might refer to such a combination as Digital Shadow. A change in state of the physical object leads to a change of state in the digital object, but not vice versa.

Digital Twin Aggregate (DTA): A DTA is an aggregate of many DTIs. The DTIs may be co-located within one entity (e.g., 100 motors in a single factory) or across entities (e.g., 100 motors across 25 factories). It is well-established a group behavior is not the sum of individual behavior. Likewise, in the future, DTAs might reveal unknown and unexpected insights.

Digital Twin Instance (DTI): A DTI (created from the DTP) is the twin of a physical asset. The DTI stays linked to the physical asset through its lifecycle. The DTI, typically, contains data relating to in-use conditions as captured through the sensors, historical state, predicted state, asset and warranty information, service records, etc. While a DTI starts with the baseline information from its prototype, over the course of the lifecycle, the DTI gets enriched with operational data.

Digital Twin Prototype (DTP): DTP is the prototype of the physical asset. It's like a recipe for creating an asset. The prototypes, dependent on the situation, will contain information pertaining to the physical attributes, properties, operating parameters, bill of materials, part numbers and more.

Chapter 16
Exploiting BIM and Sensor Data Through Web-Based CAFM:
The AR4FM Project

Umberto Di Staso
Territorium Online, Italy

Marco Piovano
Fraunhofer Italia Research, Italy

Ambra Barbini
Fraunhofer Italia Research, Italy

Dominik T. Matt
Fraunhofer Italia Research, Italy & Free University of Bozen-Bolzano, Italy

ABSTRACT

Technological progress and the evolution of the directives concerning the construction sector have led to a significant digitalization of information concerning buildings. One of the challenges that have arisen concerns the use of data in the phases of the life cycle following the construction. How can this information be exploited? Is it possible to use it directly in a facility management tool that is within the reach of expert users and not? The aim of this chapter is reporting the developed framework to support the management and maintenance operations in buildings, defined within the project AR4FM - Augmented Reality for Facility Management. In particular, AR4FM aims to create an innovative software ecosystem that uses the most modern technologies to support facility management (FM) operations through a set of bespoke applications. The first tool is the web-based ICT platform usable through web-browser, while the second tool refers to a mobile App, called AR4FM mobile app, which will enable visualization for both mobile and wearable devices, such as smart-glasses.

DOI: 10.4018/978-1-7998-7091-3.ch016

INTRODUCTION

Context

Ensuring an efficient functionality of a building or infrastructure, Facility Management (FM) can significantly affect the economic and environmental impacts of the life cycle. One of the main problems of FM processes is connected to the management of data and documents, developed from the early stage of design, but fragmented and linked exclusively to a specific discipline or phase of the life cycle. This aspect, together with the possible loss of information during the several steps of the building process, results particularly expensive and time consuming for the facility manager that needs to reconstruct a coherent and structured database.

After the financial-economic crisis of 2008, the attention for the optimization of industrial processes and waste reduction has increasingly grown, leading to relaunch and update the FM software market, to support the digitalization of operational and maintenance tasks, as well as planning and monitoring activities. The newest FM software can have different focuses and specialties, for example, the generation of detailed and comparative analysis, collecting and organizing the considerable amount of data in a systemic way, the support in the management of documents and files, deadlines and facility consumption (i.e. energy, water, goods) or the accounting management for operations of different complexity.

At the same time, the whole construction industry is dealing with a digital transformation, promoted by the European Union and implemented by national governments with different strategies. In Italy, the New Procurement Code (D. Lgs. 50/2016) preannounced the mandatory adoption of digital tools, such as Building Information Modelling (BIM), for public projects above specific thresholds of economic value, which will progressively decrease from 2019 until 2025.

The spread of BIM implementation, together with the evolution FM software, will progressively lead to a digitalized management of real estate assets. As an information management methodology of the construction process, BIM requires an approach oriented to the creation, sharing and management of data. Similarly, FM software are strongly connected to data acquisition, organization, processing and visualisation. Considering the trend of BIM implementation and digitalization of the construction sector, an informative 3D model could support the development of strategic and operational FM activities.

During the life cycle of a construction, a wide amount of data can progressively be collected and stored within a Building Information Model, in order to be exchanged and further developed by different AEC operators (i.e. designers, suppliers, manufacturers, clients, building companies, facility managers), according to a flexible, integrated and multidisciplinary process. At the same time, considering an existing building and the amount of available data and their heterogeneous nature, the management of data results particularly complex and the exchange of data requires special attention. To avoid working with a model overloaded of unnecessary or redundant data and to prevent the loss of essential data for the planned applications, it could be helpful to apply targeted filters.

The integration of data modelled by different discipline has been studied especially focusing on the design phase, while remains to explore data integration potential during the operational phase. There are several overlapping data between the design phase and the operational phase, therefore BIM data modelled during the design phase, as well as BIM data developed from the survey of an existing building can serve as input data for FM software, in order to optimize the process.

Project Objectives

After detecting a strong interest for FM digitalization processes among companies and public bodies in South Tyrol (Italy), Territorium online and Fraunhofer Italia, within a FESR (European Regional Development Found) project, started creating and testing a highly innovative online platform as a support tool for planning and monitoring FM activities, in order to further develop Augmented Reality (AR) applications to support operational and maintenance FM tasks.

Territorium online (TOL) is an Italian GmbH, with an operating office in Bolzano (South Tyrol). TOL has years of experience in Spatial Data Infrastructure set-up and information management for Facility Management. In addition, since 2016, has invested in its R&I department in order to increase its expertise in the BLE (Bluetooth Low Energy) and Computer Vision sensor sector. TOL also features strong development capabilities on mobile applications in a high-performance QT/C++ environment. The adoption of this technology allows not only the creation of high-performance software, but also to enhance the investment made on devices equipped with different operating systems, thanks to the possibility of multi-platform compilation.

Fraunhofer Italia (FHI) is a non-profit, applied research organization, located in Bolzano. FHI develops methodologies and strategies aimed at improving the performance and efficiency of processes, with the aim of increasing productivity and quality, in each phase of a construction life cycle, from design to construction to facility management. The technical-scientific team of the Process Engineering in Construction (PEC) addresses companies to provide activities, research services and innovation support, such as: adoption and implementation of Building Information Modelling (BIM) within private and public entities, building management and maintenance systems through Facility Management (FM), application of digital technologies to construction, such as Virtual Reality and Augmented Reality. The continuous challenge of the PEC team is to generate innovative ideas, which, starting from a theoretical scientific basis, are processed and transformed into concrete solutions, ready to be adopted by companies. In this way, the development of the innovative potential of SMEs is promoted without ever losing sight of the main reference trends such as sustainability, efficiency and digital technologies.

Combining expertise and background knowledge, FHI and TOL are collaborating for the digital integration of FM data at three different levels: from the existing building to the Building Information Model (static data) and to the IoT system (dynamic data), from the static and dynamic data to the developed Computer Aided Facility Management (CAFM) online platform and from the online platform to the mobile and wearable AR devices (Fig.1). For each level has been necessary to consider available data and perform specific selection of information based on the scenario purposes.

Figure 1. Workflow of integration

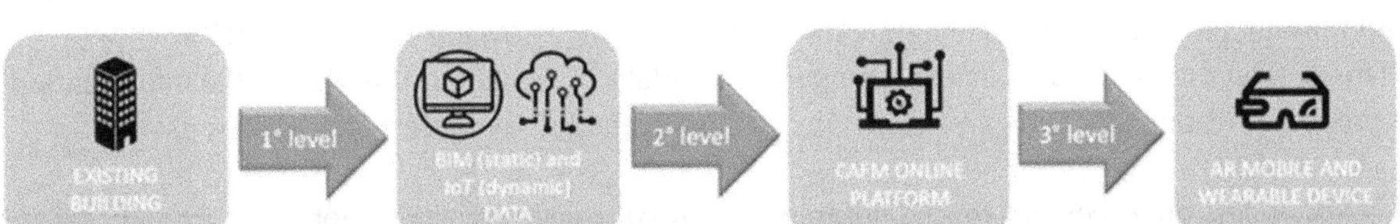

Among the project objectives the authors highlight the followings:

- a decision-making support system and tools for maintenance strategies optimization, supported by building's digital data,
- the development of a next-gen CAFM system, able to integrate, use and update BIM data,
- a smart CAFM system, able to provide real time, geo-referenced and targeted information to web, mobile and wearable devices, improving the quality of maintenance activities,
- the integration of IoT technologies to constantly monitor facility parameters, which support the reduction of waste production and energy consumption.

State of the Art

The progressive digitalization of the building sector is leading not only to the adoption new technologies and tools, but also to the integration of methods and applications. For example, BIM-based and augmented reality (AR) application, which integrates a location-based management system, has been developed and studied to support construction activities (Ratajczak et al., 2019). The integration of BIM and Internet of Things (IoT) has also different possible applications, including an enormous potential to radically change the way construction and maintenance of facilities are currently performed (Shahinmoghadam et al., 2019).

Considering the digitalization of FM, there are considerable expected potential connected to the connection of BIM and FM, in particular for the integration of BIM data with CAFM platform (Lo Turco et al., 2015), however, several open issues have been detected, mainly related to data exchange among AEC operators and interoperability between digital tools and applications (Parsanezhad et al., 2019). BIM has not yet reached an optimal degree of interoperability and data exchange, for the transition between phases up to operation and maintenance, which remains a major challenge for the construction sector (Osello A, et al., 2016).

During project processes the collaboration between expert and professionals need to be improved (Naghshbandi et al., 2017) and would be useful to establish roles and responsibilities of the involved AEC operators, clearly defining who and when must insert different data in the model. For users and clients, it would be also useful to have guidelines for the efficient maintenance of different building and infrastructure typologies (Matarneh et al., 2019). In general, it would be necessary a change in the procedural and cultural mindset of the construction sector, in order to exploit the potential of digital methods and tools (Kelly et al., 2013). The difficulties in integrating Building Information Models within FM platforms are also connected to both in using open formats, such as IFC and common specifications, such as COBie (Yakcinkaya et al.) and the lack of best practices and clear integration requirements from the early stage of project development (Naghshbandi et al., 2017). The quality of information should be better defined in order to use BIM data for FM applications and processes should be standardized and tested through case of studies to integrate different data sources (Matarneh et al., 2019).

This project mainly focuses on exploiting the potentials of digital data, as well as the optimization of FM processes through data integration, trying to answer the following research questions:

- Which data are necessary for the optimization of BIM and FM integration?
- How should be BIM data organized in order to be easily integrated in FM platform?
- How can be performed the association between BIM and FM platform?

- How modern web-based and mobile technologies can exploit them?

EXPLOITING BIM DATA: THE AR4FM APPROACH

Methodology

The aim of the second chapter, **Methodology**, is to illustrate in detail the principles and methods adopted within the AR4FM project. In particular, the adopted choices for defining and structuring of the information that will be visualized in the two tools developed in the project are presented: the Web-based ICT Platform for the visualization of the information in cloud through web browser and AR Mobile Platform for the visualization directly in the field of the useful data for the operator by means of augmented reality through mobile devices and/or smart glasses.

The visible and manipulated information are all offered by the same server infrastructure. The information collected, processed and displayed on the Web-based ICT Platform is divided into two macro-categories: static and dynamic data. The static data are extracted and integrated directly from BIM models, developed in this context for existing buildings for operation and maintenance phase. These data are not updated automatically but are closely linked to any maintenance work that involves updating the building information over time. The dynamic data, instead, refers to real time data acquired via sensors identified according to the needs to be monitored. In order to define an adequate working methodology, particular attention has been paid to the measurement phase, BIM modelling and subsequent integration with the Web-based ICT Platform, which will be discussed in the *Tools* chapter.

Another important aspect concerns the indoor localization, realized through the use of BLTE Beacons. In this context, the system is able to identify the position of the operator at room-level where the operator itself is located, and automatically filter relevant information according to proximity algorithms. The choice to focus our attention on room-based positioning system has been influenced by the small number of BLTE devices needed to develop this functionality: just one for each involved room, thus increasing the possibility of replication of the entire system in third-party buildings and minimizing the overall hardware equipment.

Survey and BIM Modeling Phase

The case of study of the AR4FM project relates to existing buildings, focusing on the fire-fighting system, as FM application. At the present stage of the project, the head offices of the AR4FM partners: Territorium Online and Fraunhofer Italia have been digitized.

By dealing with existing buildings, one of the main issues concerns the lack of reliable and up to date information, as well as the difficulty to acquire archival material or documents, useful to verify the current state and the accuracy of the information. For this reason, the investigation phase must be carefully planned and executed, including document retrieval and/or any geometrical survey, to map all the missing information related to the building or to specific assets, in order to establish a complete and updated picture of the construction. It is extremely important to have clear in mind what will be the final set of use cases: with this understanding, the surveyors can concentrate their attention only on relevant information, avoiding such details that may increase the complexity of the BIM without providing any additional added-value. Therefore, the pre-modelling and acquisition phase plays a key role for the development of the digitalization project.

The Building Information Model collects all the necessary data, that can be used as knowledge base, able to support properly the management of the building as well as the maintenance activities. Such a structured database can be related to CAFM platforms, specialized in accessing to all the information related to the asset through an integrated system of alphanumeric and graphic data.

The interoperable FM-BIM database allow an update and real time check of the information related to the building and its component, assuring the constant monitoring of the construction.

For the project AR4FM the survey phase has been planned analyzing documents and CAD drawings available for each selected building. Since the documents were found to be partial and outdated, the authors decided to carry out an inspection to make an inventory/survey of the spaces and of the fire-fighting devices, in order to standardize the information for the considered buildings. After analyzing and establishing all the necessary data to reach the objectives, the survey has been carried out, mapping all the missing data. In order to avoid acquiring superfluous data, a standardized procedure has been defined, following a hierarchic structure considering both spaces (e.g. buildings, locals) and firefighting assets (fire extinguishers, smoke detectors, speakers, sprinklers, etc.). For example, for each local has been established a code to classify, identify and subsequently integrate with the _pit-FM[1] Facility Management data management platform. Further details about _pit-FM will be provided in the next section. Below are illustrated the standardized sheets used to map the information during the survey:

Figure 2. Data organization

FRAUNHOFER ITALIA office - data organization			
BUILDING		**ROOM**	
General Information	Name	**General Information**	Code
	Address		Intended use
	Municipality		False ceiling
	Postal Code		Fixed furniture
	Province		Hight
	Number of Floors		Exposure
	Construction year		Number of occupants
	Designer		Surface
	Intended use		Volume
	Code	**FIREFIGHTING (es. Fire-extinguisher)**	
Technical characteristics	Total net area	**Classification code**	Code
	Covered area	**General Information**	Position
	Gross volume		Typology
	Net volume		Manufacturer
Cadastral data	Cadastral map		Date of acceptance/inspection
	Land parcel		Weight
	Subalterno		Extinguishing potential
	Municipality urban data		Serial number
Documents	Technical sheet	**Documents**	Technical sheet
	Certifications		Previous maintenance sheet
	Further documents		Certification sheet

Starting from the information collected in the previous phase, the BIM of the selected building has been created using specific modelling software such as Autodesk Revit. During the BIM modelling phase, special attention is paid to the Level of Development (LOD) of the identified elements, under both graphic and alphanumeric point of view. LOD defines the content and reliability of BIM objects at different stages of the project. By "content" means geometric information, structured data and related documentation. By " reliability " means for what uses and to what extent users downstream of the information can trust the accuracy and quality of that content. The objectives underlying the project therefore determined the setting of the modeling and this led us to define a standard to be replicated on different buildings.

These choices are made considering functions and methods integration with the _pit-FM. Below is the BIM model developed for the TOL and Fraunhofer Italia offices:

Figure 3. BIM office Fraunhofer Italia

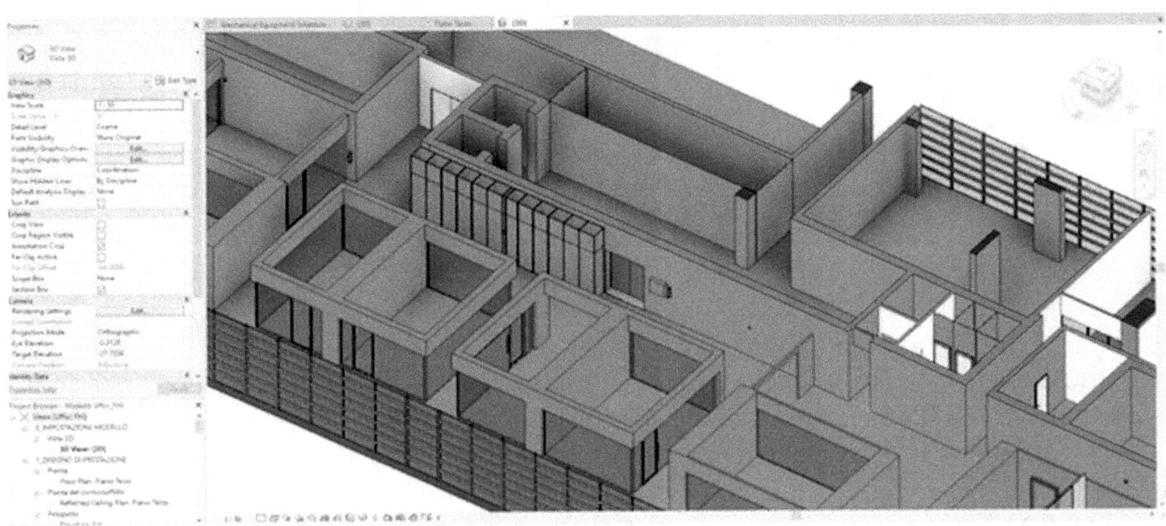

The possibility to access a structured and centralized digital database allows a considerable saving of time and resources, consumed by searching for information, often incomplete and disorganized. In addition, BIM offers Facility Managers the possibility to rely on data, that are always up to date during the decision-making process, allowing to optimize the planning and budgeting of maintenance activities. BIM software don't generally support the planning and management of maintenance activities. However, the structured informative content allows to provide cohesive data that can be properly integrated to FM systems, able to relate data, operators and processes within an efficient logic of continuous improvements and reviews.

BIM - _Pit-FM Integration

The use of the BIM methodology for FM activities favours a 360-degree control of the work, from a technical, economic and energy point of view, guaranteeing a constant awareness of the level of performance

Figure 4. BIM office Territorium Online

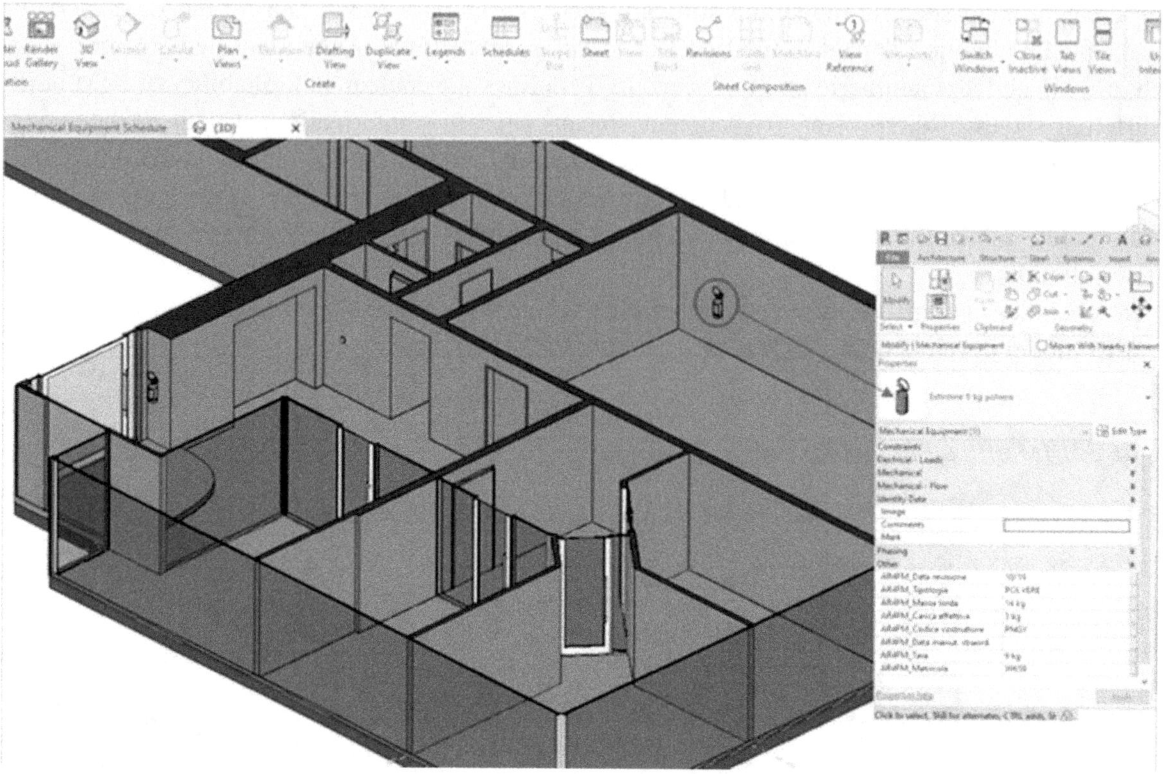

of the structure and/or of any restructuring and necessary interventions. The interaction between BIM models and CAFM/CMMS platforms, as said, plays a fundamental role in the definition of an innovative process for the digitalized management of real estate assets. The _pit-FM information platform plays a central role in implementing specific processes for the management of spaces, assets and maintenance.

The AR4FM infrastructure uses two types of databases for data management: _pit-FM for Facility Management information and an SQL Database for spatial information. _pit-FM is a powerful and flexible CAFM/CMMS process management system for the design, management, administration and maintenance of facilities. The flexible software concept and adaptability ensure targeted visualization of conditions, interactions and work processes. The two databases are communicated via specific functionalities developed in the context of the middleware level to link the FM data with the geographical component. More information about the software infrastructure created will be available in the *Tools* section.

The AR4FM project exploited the potential of interoperability between Autodesk Revit[2] modelling software and the digital building management and maintenance system named _pit-FM. _pit-FM provides a plugin for Revit to import BIM data. As said, the BIM is imported in the CAFM software named _pit by the _pit-FM Ingestion System. The ingestion system is based on the concept of XML-based mapping between the BIM classes and the _pit-FM classes. The aforementioned procedure allows to define a 1:1 correspondence between the data source and target _pit-FM storage. Mapping between the BIM and _pit classes have been previously and manually defined through an intermediate table that allows you to define the correspondence between the BIM fields and the _pit-FM platform. In the aforementioned phase, it is crucial a deep understanding and knowledge of both software.

Figure 5. Mapping between _pit-FM and BIM classes

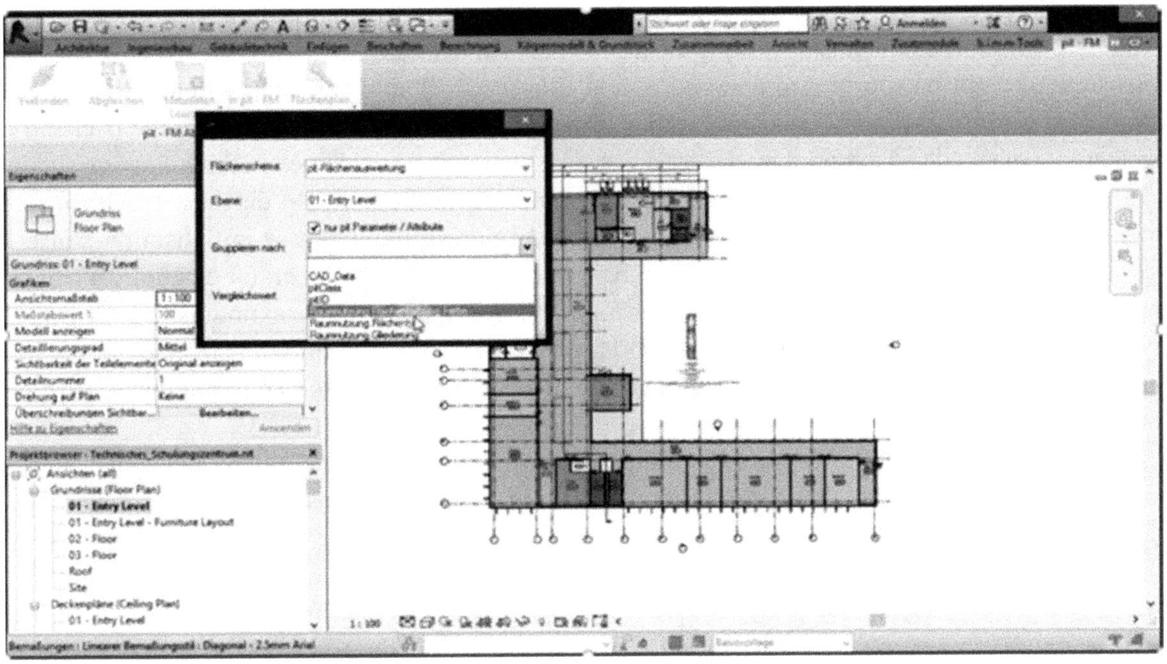

Based on BIM models developed using Autodesk Revit software, you can then import all of the information you believe to be important for your goals into the _pit-FM database, ensuring the correctness of the relationships between objects and rooms in the building in question (for example, the room refers to a floor that is in a particular building).

Optionally, the _pit-FM ingestion system allows to synchronize model updates with the BIM data source, by the mapping defined during the ingestion process. It should be noted that the phase of importing the BIM data into the system infrastructure, based on database reports, is a semi-automatic and, optionally, bidirectional process. Bidirectional process means that it is not only possible to update the imported relational data but, also, to decide to synchronize the BIM model to the updated information.

The interoperability and bi-directional synchronization of the information from the BIM model with the _pit-FM database and vice versa allows to keep the building data constantly updated over time. In this way the main information is always kept up to date, in relation to each other and consistent between the two systems.

For example, an operator can work in the BIM model, directly connected to the _pit-FM system, and make the necessary changes. By updating a data in the BIM, it is possible to synchronize again the relational database; vice versa, by updating a data from the Web or Mobile interface of the FM platform the changes can be later integrated and visible in the digital model.

Ingested data are provided by the AR4FM backend to both the AR4FM Web-Based ICT Platform and the AR4FM Mobile App to be exploited by managers and operators. The data are stored within this platform and can be viewed in the field through mobile devices and/or smart glasses during maintenance operations.

Moreover, in addition to the software infrastructure aimed at data ingestion, specific sensors have been placed into the partners premises to collect dynamic data. The communication between the devices and the AR4FM backend is managed by a secure HTTPS connection.

Sensors are built on top of Arduino and Shield Ethernet technologies and includes measurement tools able to get information every configurable amount of time. Data are then sent to the backend infrastructure, collected and exposed by the set of services that will be explained in the following section.

Talking about how static and dynamic data are organized in the _pit-FM database, each sensor is associated to an object that has been previously ingested during the database-initialization phase. In this way, sensors and related measurements can be related to any stored object; for example, it can be a temperature sensor for a room containing servers or a pression sensor for fire extinguishers. Having the logical connection between the sensor and the monitored object, working rules and related alerts can be easily set, enabling managers and operators to get a notification when sensor's measurements are out of the normal working range.

This approach allows the maximum flexibility and exploitability of the entire dynamic data system, since the main focus is on how the information is acquired, sent and internally managed and not on the measurements typology itself.

Tools

The aim of the present chapter is to illustrate in detail the tools that have been developed within the AR4FM project. The first output concerned the development of the AR4FM Web-based ICT Platform which aims at offering a browser-based visualization of static and dynamic information provided by the central infrastructure, which is in turn obtained by the ingestion of BIM model data and integration of IoT technologies.

Further latest developments concern the AR4FM Mobile App, compatible both with mobile devices and smart glasses. This application will offer functionalities for both on-field visualization in Augmented Reality mode and real-time operator positioning inside the pilot premises.

Many Facility Management tasks require to access to data in real time and on-site, which is currently not possible with traditional software and representation techniques. An innovative Facility Management system, like the one provided in AR4FM, has a tangible impact in the process optimization, quality, and timing.

The following sections describe the background knowledge to build up the AR4FM framework, the AR4FM Web based ICT platform and the AR4FM App for mobile devices and Smart Glasses.

Background Knowledge

The AR4FM ecosystem is based on the concepts of Model Driven Architecture[3]. Model Driven Architecture® (MDA®) is an approach to software design, development and implementation spearheaded by the Object Management Group (OMG). MDA separates business and application logic from the underlying platform technology. Platform-independent models of an application or integrated system's business functionality and behaviour, built using UML and the other associated OMG modelling standards, can be accomplished through the MDA on virtually any platform.

These platform-independent models document the business functionality and behaviour of an application separately from the technology-specific code that implements it, isolating the core of the application from the underlying technology and its relentless churn cycle while enabling interoperability both within and across platform boundaries. The AR4FM framework implements a model-driven runtime (MDR) environment that can execute a platform independent model for a specific purpose instead of transforming it.

Based on MOF-enabled transformations, the MDA unifies every step of the development of an application or integrated suite from its start as a Platform- Independent Model (PIM) of the application's business functionality and behaviour, through one or more Platform Specific Models (PSMs), to generated code and a deployable application. The PIM remains stable as technology evolves, extending and thereby maximizing the software return of investment. Portability and interoperability are built into the architecture. A **model** in the context of MDA is a structured set of information representing some aspect of a system based on a specific set of requirements. The model is related to the system by an explicit or implicit mapping. A model should include the set of information about a system that is within scope, the integrity rules that apply to that system, and the meaning of terms used.

In the AR4FM framework a model defines three main aspects of an application:

- Structural (Classes/Entities): an entity model describes inter-related things of interest in a specific domain of knowledge. An Entity Relationship model is composed of entity types (which classify the things of interest) and specifies relationships that can exist between instances of those entity types. An entity defines a data or information with attributes, relations, and identifiers, that can be implemented in databases. Entities can define behaviours as well related to the entity class (queries, hooks, method...)
- Behavioural (Activities): an activity is a major task that must take place to fulfil an operation contract. An activity can be a simple invocation of an operation, a step in a business process or an entire business process. Activities can be decomposed into sub-activities. From technical aspect an activity is based on a Petri Net (Peterson, J. L., 1981) coordination mechanism. Therefore, the activity represents user-defined behaviour coordinating actions. An activity is the specification of

Figure 6. The Activity representation by a Petri-Net

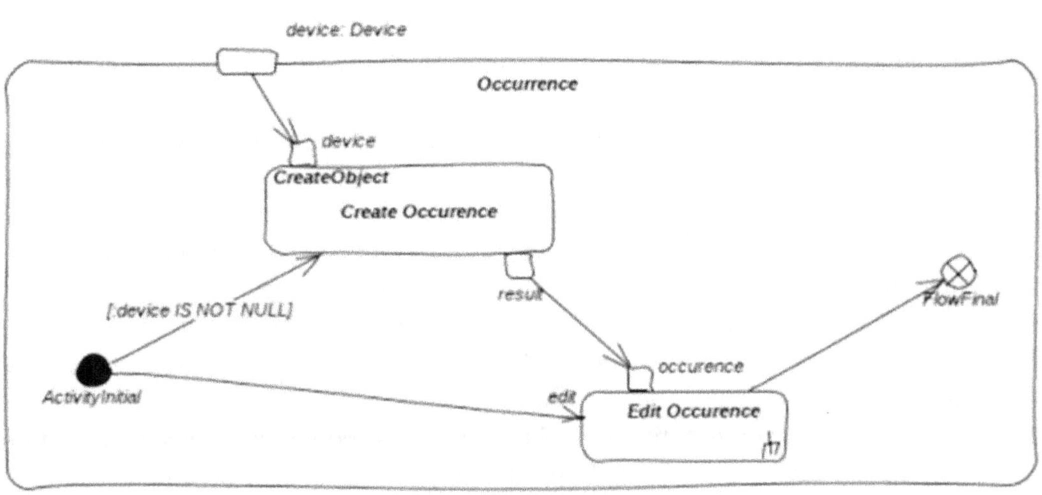

a parameterized sequence of behaviour (workflow). The basic purpose of an activity is to capture the dynamic behaviour of the system. An activity has different nodes and defines the transition between them. The transitions are triggered by events/conditions or by the user. While navigating between the nodes, parameters can be passed between the nodes and the activity.

An activity is always a self-contained unit; The inner workings are not visible from outside and can only be invoked by trigger or events. Even data can only be transferred or read out by means of parameters. The activity has its own context (class) on which it operates. The activity has an exclusive access to the data model (Item Model), which can be reduced by defining a filter. The item model will never be shared with other activities (the data might, but not the model).

Each activity also has a *selection model* which works as follows:

- **No**: No can be selected, only a current is supported.
- **Single** Only a single item of the model can be selected. If one item is selected a previous selection will be reset.
- **Multi** Many items can be selected.

The activity defines nodes and edges; the nodes are special implementations, providing some activity specific logic. The edges define the flow between the nodes. The activity provides two types of edges:

- **Transitions** define the permitted flow between the nodes.
- **Object Flows** define the values that can be transmitted between the nodes.

To the external environment, activity exposes only parts of the available information:

- **Parameter** Are used to provide information to and from the activity. That is the only way to share data with the activity.
- **Trigger** These are *named transitions* that influence the workflow of the activity and state changes in the Petri Net. The transition defines a navigation between different nodes in the Petri Net activity. The transition must be defined between two control nodes or actions. Implicit & explicit transitions are supported.
    - An implicit transition can be invoked automatically by the runtime, without any external event.
    - An explicit transition has a name and must be called by the user or any external trigger event.

Transition can have a guard, a filter expression that must be valid. For example, a transition can invoke the action A if certain conditions are satisfied, otherwise the action B. If multiple transitions are valid at the same time, the system offer instruments to declare transitions priority.

The operations that an activity can perform, each of one represents an instance singe node described in the Petri Net, are called Actions. An Action is the fundamental unit of executable functionality. The execution of an Action represents some transformation or processing in the modelled system. Actions provide an *executable node* within Activities and may also be used within Interactions. Action name's and declarations are inspired by the UML 2.0 specifications (Eriksson, H. E. et al., 2003).

The following list represents a subset of all the already implemented actions, each of them allows data interaction and manipulations.

- **CallBehavior**: The CallBehavior action is the most advanced implementation of an action. The action will invoke another activity and is responsible for the primary application flow. On invocation the calling activity will passivated until the called activity has terminated.
- **CallOperation**: A call operation action invokes a remote operation. For example, it can call a remote script that takes in input specific data and provides a result based on internal rules.
- **CreateObject**: A CreateObjectAction is an Action that creates an instance of the specified Classifier. By default, the classifier is the context of the owning activity. By this action is possible to create new data. For example, it can be mapped by a button used for creating a new damage notification.
- **DestroyObject**: A DestroyObject is an action that destroys objects. For example, it can be mapped by a button used for removing a closed notification.
- **CloneObject**: A CloneObjectAction is an Action that creates a copy of the object provided in input.
- **SetSelection**: A SetSelection action applies a current or selection on the item model. By default, the action chooses the first entry of a dataset.
- **ReadSelf**: A ReadSelf action is used for download, from the central infrastructure, data belonging to a specific entry. For example, the ReadSelf action can be used when its required to display all the data belonging to a specific room to see its details.
- **WriteVariable**: The WriteVariable action changes the attribute values of the current entity.
- **WriteStructuralFeature**: the WriteStructuralFeature commits changes to the remote infrastructure. It is generally mapped on a *save* button.
- **ClearStructuralFeature**: A ClearStructuralFeature action is an action that resets all values of a structural feature to the original ones. For example, it can be mapped by a button to discard changes.
- **UserAction**: The user action is designed to provide interaction with a user or a system. The user action uses the title and the description to interact with a user (Message Dialog) and a template used to trigger related actions.
- **MapAction**: The map actions support different panning and zooming actions on the related map. For example, it can be automatically triggered when the system recognizes a room change by the operator using its internal BLTE indoor positioning algorithm.
- **StartInputAction**: The input event is a generic event that invokes device inputs (QR-Code Scanner, Position from map, Bluetooth, NFC) to retrieve a value and process in the activity.

The implementation of an activity also provides events to inform the outside world about status changes; In addition, the activity provides some special methods to store the current states and to share with the user interface (e.g., sorting, search criteria) the activity is stored. An activity can be defined with an associated *basefilter*: the role of the *basefilter* is to limit the amount of available data under certain conditions. In the next section is presented how this base filter has been used in the firefighting management scenario.

The following example shows how an activity can be declared.

```
<activity name="acDocument" context="Document" filter="cruser eq $USER"
selection="Single">
<node name="Initial" type="Initial"/>
<node name="FlowFinal" type="FlowFinal"/>
<node name="FinalNode" type="FinalNode"/>
<template name="web.ui.xml" target="web"/>
<template name="mobile.ui.xml" target="app"/>
<template name="glasses.ui.xml" target="glass"/>
<action name="act1" type="DestroyObject"/>
<action name="actCreateDocument" type="CreateObject">
<pin name="pnResult" direction="Out" type="None"/>
</action>
<action name="actEditDocument" type="CallBehavior" behavior="EditDocument"/>
<transition source="actEditDocument" target="FlowFinal"/>
<transition source="act1" target="FlowFinal"/>
<transition source="actCreateDocument" target="actEditDocument"/>
<transition name="trCreate" source="Initial" target="actCreateDocument"/>
<transition name="trDelete" source="Initial" target="act1"/>
<transition name="trUpdate" source="Initial" target="actEditDocument"/>
</activity>
```

The activity has a well-defined life cycle.

Activities must define a valid template. A template describes how the activity is represented in the User Interface and the mapping between UI components (buttons, lists, tables, etc.) and triggers. Mapping are usually declared by name. For example, a button named "trSave" triggers a transition named in the same way, linked to a **WriteStructuralFeature** action that pushes current changes to the remote server.

Finally, the system allows the declaration of user interfaces based on the target device: Web, Mobile and Glasses.

Once the Application configuration is completed, the same **model**, represented by the JSON translation of the XML-based configuration, is provided from the central server to client applications, independently by the target device: Web, Mobile and Glasses are developed in order to accept the same **model** and provide the same functionalities. Due to screen-size difference, different templates must be used to organize the same information in different ways, based on target devices.

Having this high-level knowledge of how the system is configured on all its components, the following section will offer a deep description of available platform and related usages.

AR4FM Web-based ICT Platform

The AR4FM Web-based ICT platform has been developed using modern web technologies such as the Google GWT project[4]. GWT is a development toolkit for building and optimizing complex browser-based applications. Its goal is to enable productive development of high-performance web applications without the developer having to be an expert in browser portability, XMLHttpRequest, and JavaScript. The GWT SDK provides a set of core Java APIs and Widgets. These allow to write AJAX applications in Java and then compile the source to highly optimized JavaScript that is browser-independent, including mobile browsers for Android and the iPhone.

The possibility to share the same configuration among different types of devices, with the possibility to vertically define the UI used by each of them, enabled the consortium to think about the possible job profile of the end user. Considering mainly the environment where the end-user will perform its work-operations, the AR4FM Web-based ICT platform is designed mainly for planning and controlling purposes (office scenario with large monitor), while the mobile App is designed for on-field activities performed by operators and, finally, the smart-glasses version will be designed for supporting on-field maintenance processes (no possibility to edit information, just viewing).

Talking about the look and feel of the User Interface, the authors decided to be as much as possible compliant within the Material Design Specifications[5] provided by Google. The decision to be inspired by such guidelines has been taken in order to offer to end-users a framework where the feeling about User Interface is very familiar, due to the huge amount of Google Web and Mobile application using this kind of design available on the market, mainly for free (Gmail, Drive, Android OS). Moreover, the

Figure 7. The web-based ICT Platform user profile selection

available literature offers the good results of deep studies made by Human Computer Interface experts related to the end-user perception among the usage of Material Design (Karana, E. et al., 2013; Bollini, L. 2017; Drygielski, M. et al., 2017).

As mentioned in the previous section, the AR4FM Web-based ICT platform starts displaying all available Use Cases that has been previously configured and are provided by the JSON-based platform-independent structural model. In detail, the organization of use case has been separated by available real estates. Obviously, the available set of real estate available depends by the rights set for each logged user.

Figure 8. The web-based ICT platform scenario selection

Figure 9. The web-based ICT platform firefighting scenario detail

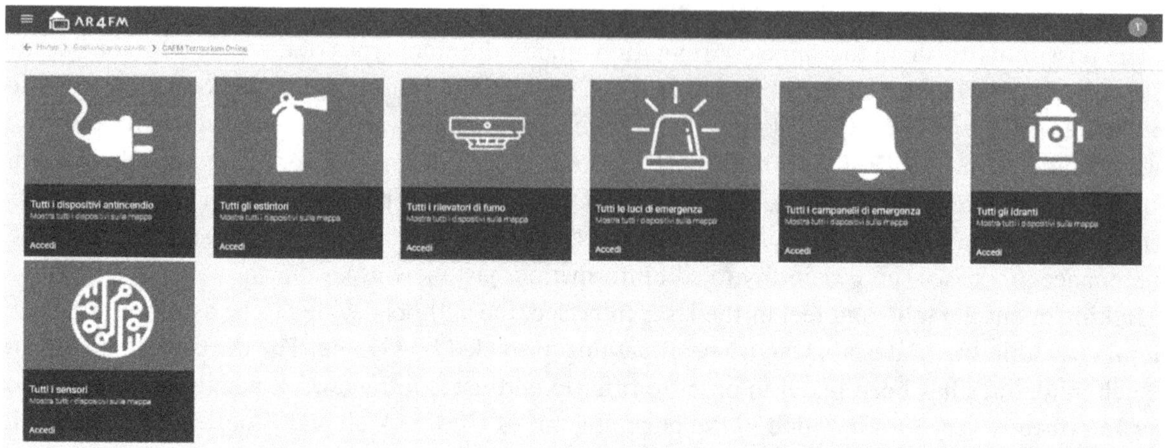

For each real estate root use case, the main menu offers the possibility to select configured sub-scenarios related to buildings managements or third-level available use cases belonging to the firefighting management scenario where relevant items has been acquired during the digitalization of the real offices data into the BIM data format. The navigation into the use cases tree is managed by the availability of a breadcrumb bar that allows to come back to the parent component.

Within the fire management scenario, the typology of each device belonging to the initial offices survey are distinguished, having the possibility to visualize all the available devices or just the ones belonging to a specific category.

In this context, it is important to underline how all the use cases belonging to the firefighting management scenario, shares the definition of the same activity. What is changing between each one is the related configured *basefilter*, that allows to define which data will be provided and visualized by end users.

Figure 10. The web-based ICT platform device selection and corresponding GIS tool

Figure 11. The web-based ICT platform spatial features

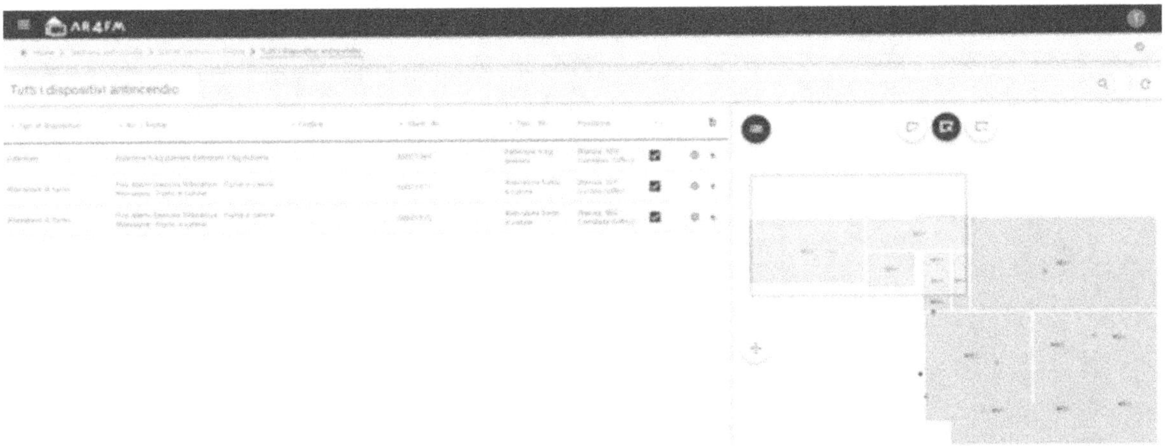

By clicking on an icon, the related activity and the associated Petri Net will be enabled, and the status on the cycle set to the entry point. The following image represents the activity where all available devices are listed.

The added value of the AR4FM framework is to integrate under a single environment both alphanumeric and spatial information, that are connected each other's and extrapolated by the ingested BIM during the import process. This interconnection will enable the availability of advanced functionalities

Figure 12. The web-based ICT platform device detail user interface

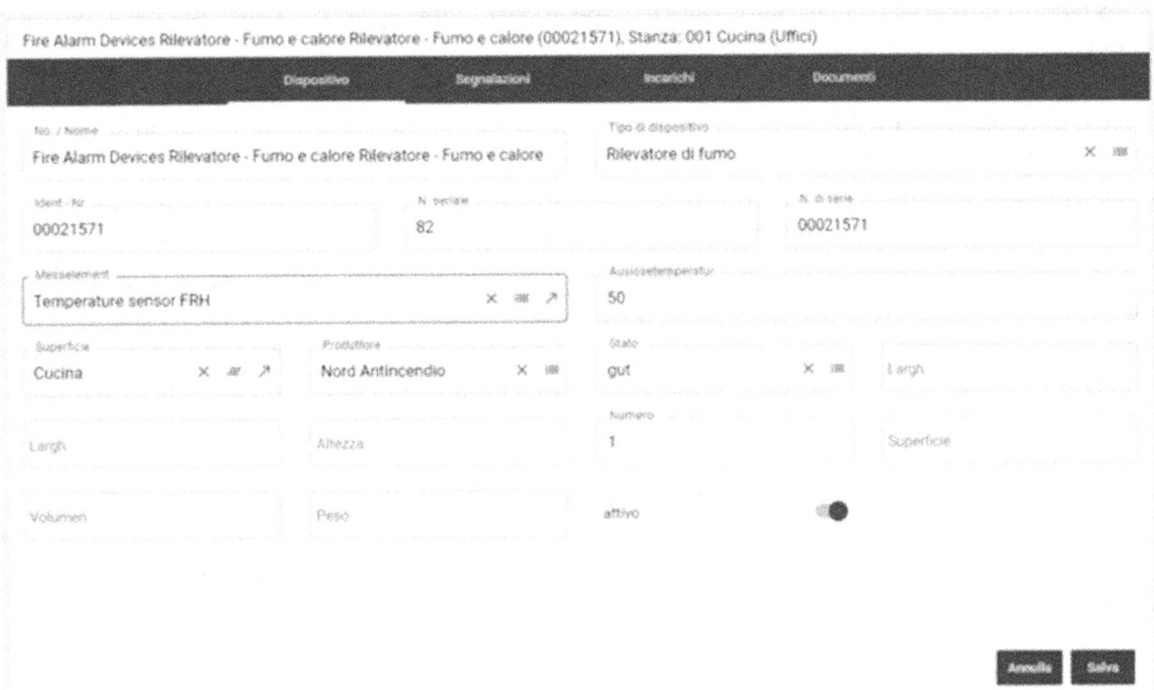

Figure 13. The web-based ICT platform beacons association table

that are usually offered in the domain of Geographic Information Systems (GIS), such as spatial intersection or spatial union. For example, by drawing a rectangle area on the map available in the right portion of the activity UI, the list of devices will be updated accordingly displaying the set of devices retrieved by the *contains* spatial query.

From the firefighting devices list, data can be filtered, sorted, updated, and exported. Each available action is mapped into the corresponding activity by a specific *transition*. As mentioned before, one of the most important concepts of the AR4FM framework is the complete declaration and mapping at configuration level of available transitions, UI and business logic. In the following image is shown the look and feel of the detail UI interface related to a specific firefighting device, which has been previously selected from the list of available items.

In the section related to the AR4FM Mobile App, is described how spatial information will be further exploited through the AR4FM indoor positioning system. Even in the context of indoor positioning system, the AR4FM Web-based ICT platform is used to setup the system. Following the same navigation approach presented for devices visualization and information update, data related to each available room belonging to buildings can be extended by adding, removing, and updating BLTE devices.

From a device detail page, the manager can update the alphanumeric information, upload documents, planning new maintenance tasks that will be processed by operators and insert new notifications (e.g. the device not working). As it is possible to see from the previous picture, the selected device can be referenced with both a sensor element (temperature sensor) and a building structural element, that can be a room, like in this example (Kitchen), a level or even the entire building. References are crucial for querying purposes.

The communication between the AR4FM Web-based ICT Platform and the remote infrastructure is managed HTTP OData REST APIs. OData (Open Data Protocol)[6] is an ISO/IEC approved, OASIS standard that defines a set of best practices for building and consuming RESTful APIs.

AR4FM Mobile App

The AR4FM Mobile App is the mobile-based app version of the AR4FM Web-based ICT platform. The AR4FM Mobile App has been developed using of Qt[7]. Qt is a cross-platform application development framework for desktop, embedded and mobile. Supported Platforms include Linux, OSX, Windows, Android, and iOS. Qt is a framework written in C++. A pre-processor, the MOC (Meta-Object Compiler), is used to extend the C++ language with features like signals and slots. Before the compilation step, the MOC parses the source files written in Qt-extended C++ and generates standard compliant C++ sources from them. Thus the framework itself and applications/libraries using it can be compiled by any standard compliant C++ compiler like Clang, GCC, ICC, MinGW and MSVC. The usage of Qt guarantees the development of performance-based mobile app and the cross-platform functionality allows the development of the same source code for IOS/Android Mobile devices and Android-based smart-glasses.

The AR4FM Mobile App includes all the capabilities of the web-based version plus offline data management and synchronization, QR-code scanning and processing, room-based indoor positioning system based on BLTE beacons devices. Bluetooth Low Energy devices operate in the 2.4 GHz license-free band, and so share the same indoor propagation characteristics as 2.4 GHz Wi-Fi transceivers. The beaconing, or advertising mode, permitted in the BLE standard enables a very short, un-solicited message at very flexible update rates. These messages can be used to allow a device to detect proximity to a specific location, based on the Received Signal Strength (RSS). In this way, location specific triggers can be provided to the user.

In the *background knowledge* section, the most important nodes that may be configured for fulfilling the application requirements have been listed and presented. As for the indoor positioning functionality provided by the AR4FM Mobile App, the involved node is the one so called **StartInputAction.**

The **StartInputAction** accepts, from its definition, multiple types of inputs:

- QR/Bar code: Invokes the QR-Code / Barcode scanner;
- GPS: Invokes the GPS positioning;
- BLUETOOTH: Uses the Bluetooth device;
- NFC: Uses a Near Field Communication.

Formerly, the indoor location service is implemented leveraging on the functionalities provided by the **StartInputAction** node in the application model definition. The **StartInputAction** acts as a Bluetooth scanner that, every configurable amount of time, retrieves the list of BLTE beacons near the target mobile device. As mentioned before, using the RSS approach, the AR4FM Mobile App can identify the nearest beacon. From the beacon's metadata, thanks to the device registration showed before, it is possible to retrieve the room where the relationship with that specific beacon has been previously defined.

The compatibility with QR and Bar Code inputs allows the AR4FM Mobile App to read information from the device camera and trigger specific configurable actions. For example, configuring a **StartInputAction** that can read and interpret a bar code placed on a fire extinguisher and automatically display the activity containing the device detail is trivial.

RESULTS AND FUTURE DEVELOPMENTS

In the present chapter the **Results** of the project will be presented. The whole process of ingestion of static data coming from the BIM model, presented in chapter 2 has been improved after many iterations and refinements, becoming a semi-automatic, well-structured, and rules-based workflow. The ingestion engine used is based on the ingestion system of the _pit-FM Facility Management software. The _pit-FM ingestion system requires the definition of semantic mappings between the BIM classes and those offered by the Facility Management software itself for storing information. In the current phase, the consortium completed the definition of mapping for use cases related to data management in a firefighting scenario. Therefore, it is possible to import both geometric and semantic information related to rooms, floors, buildings, and fire management devices, such as fire extinguishers and hydrants. BIM-based information belonging to Territorium Online and Fraunhofer Italia premises are already ingested by the system. The complete workflow is composed by the following phases:

1. Building survey: the initial phase allows to acquire information regarding the building by an on-field survey or, if existing, using footprints or digital information. In the initial phase it is extremely important to have clear in mind what will be the final set of use cases to cover: with this understanding, the surveyors can concentrate their attention only on relevant information, avoiding such details that may increase the complexity of the BIM without providing any additional added-value.
2. Creation of the Building Information Model: starting from the information collected in the previous phase, the BIM of the selected building will be created using specific modelling software such as Autodesk Revit.
3. The BIM is imported in the CAFM software named _pit by the _pit-FM Ingestion System. The ingestion system is based on the concept of XML-based mapping between the BIM classes and the _pit-FM classes. This procedure allows to define a 1:1 correspondence between the data source and target _pit-FM storage. Mapping between the BIM and _pit classes have been previously and manually defined.
4. Ingested data are provided by the AR4FM backend to both the AR4FM Web-Based ICT Platform and the AR4FM Mobile App to be exploited by managers and operators.
5. Optionally, the _pit-FM ingestion system allows to synchronize model updates with the BIM data source, by the mapping defined during the ingestion process.

Client applications are currently under the validation and refinement phase: the consortium is testing both instruments with the production data ingested and provided by the framework backend. Moreover, particular attention is paid to the indoor location engine based on BLTE beacons devices: as expected, one of the most crucial point to face in order to make the indoor positioning system working properly is to find the best suitable environmental position of each beacon.

By having a tested and stable environment exploitable by the AR4FM Web-based ICT platform and AR4FM Mobile App, the consortium will start to design the set of user interfaces required for the smart-glasses version of the client application. As underlined many times in chapter 3, the strongest added value of the AR4FM framework is the MDA approach, that allows to share core information between available platforms. The design of user interfaces for smart glasses will follow scientific Human Computer Interaction guidelines, based on the available bibliography related to smart-glasses applications not only

Figure 14. AR4FM framework

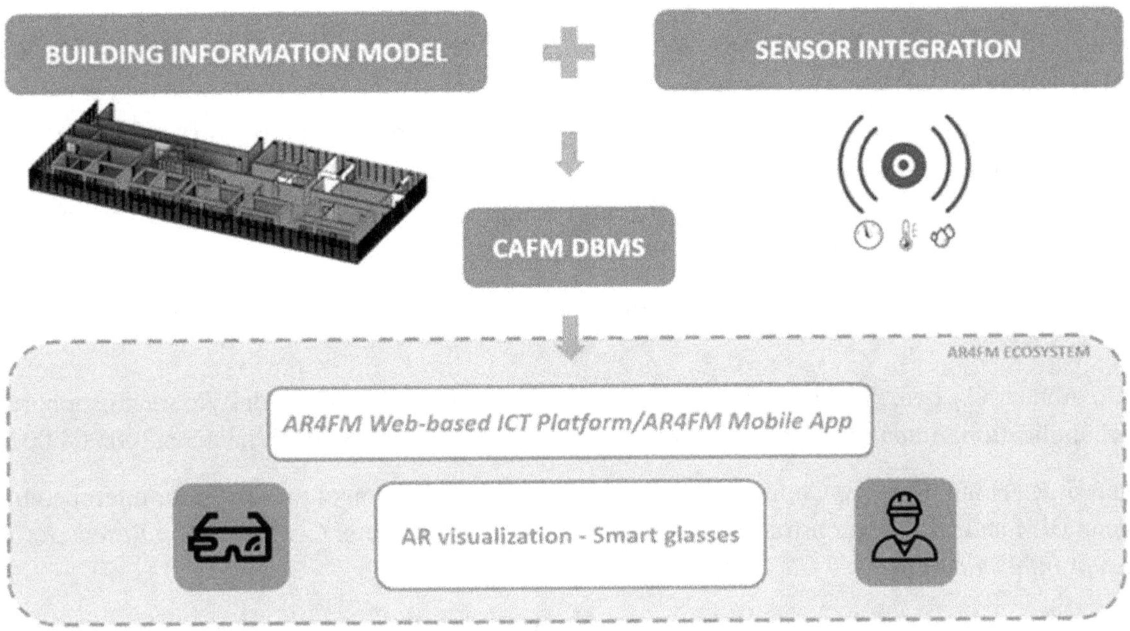

developed for facility management/industrial applications, but following a holistic approach. The final product will be tested by stakeholders and feedbacks will be collected by interviews.

Mid-term developments will be focused on replacing the manual XML-based model definition by a bespoke Rapid Application Development – like environment.

Long-term developments will be focused on the Web and Mobile 3D BIM Viewer: this component, where feasibility testing has been already made by a prototype module for the AR4FM Mobile App, will allow both 3D visualization of the BIM and querying by picking on specific components belonging to the model. This approach has been alrady subjected in several research papers (Sampaio, A. Z., & Simoes, D., 2017; Fu, C. et al., 2006): our expectation is to create a module fully exploitable for mobile devices, especially in terms of performances and interoperability with our Geographic Information System (GIS) and CAFM backend (Hwang, J. R., et al. 2013).

ACKNOWLEDGMENT

AR4FM, ERDF 1064 CUPB51B17000850007, is funded under the European Regional Development Fund of the Autonomous Province of Bolzano (EFRE ERDF 2014-2020) and involves two partners covering trade (Lead Partner) and applied research. The consortium is composed by:

- Territorium Online (TOL, Lead Partner) whose task consists ins the development of the information systems for CAFM/CMMS management and GIS solutions.
- Fraunhofer Italia, Centre for applied research, whose task consists in the digitalization of building data to provide static information in BIM format and research activities on modern indoor location and visualization techniques in AR mode.

REFERENCES

Bollini, L. (2017). Beautiful interfaces. From user experience to user interface design. *The Design Journal, 20*(sup1), S89-S101.

Drygielski, M., Indyka-Piasecka, A., Piwowarczyk, M., Telec, Z., Trawiński, B., & Duong, T. H. (2019, September). Usability testing of data entry patterns implemented according to material design guidelines for the web. In *International Conference on Computational Collective Intelligence* (pp. 697-711). Springer. 10.1007/978-3-030-28377-3_58

Eriksson, H. E., Penker, M., Lyons, B., & Fado, D. (2003). *UML 2 Toolkit, CafeScribe* (Vol. 26). John Wiley & Sons.

Fu, C., Aouad, G., Lee, A., Mashall-Ponting, A., & Wu, S. (2006). IFC model viewer to support nD model application. *Automation in Construction, 15*(2), 178–185. doi:10.1016/j.autcon.2005.04.002

Hwang, J. R., Hong, C. H., & Choi, H. S. (2013, May). Implementation of prototype for interoperability between BIM and GIS: Demonstration paper. In *IEEE 7th International Conference on Research Challenges in Information Science (RCIS)* (pp. 1-2). IEEE.

Karana, E., Pedgley, O., & Rognoli, V. (Eds.). (2013). *Materials Experience: fundamentals of materials and design*. Butterworth-Heinemann.

Kelly, G., Serginson, M., Lockley, S., Dawood, N., & Kassem, M. (2013, October). BIM for facility management: a review and a case study investigating the value and challenges. In *Proceedings of the 13th International Conference on Construction Applications of Virtual Reality* (No. 30-31, p. 5). Academic Press.

Lo Turco, M. (2015). Rappresentare e gestire patrimoni immobiliari: Il BIM per il Facility Management. *Territorio Italia, 2*, 33–38.

Matarneh, S. T., Danso-Amoako, M., Al-Bizri, S., Gaterell, M., & Matarneh, R. (2019). Building information modeling for facilities management: A literature review and future research directions. *Journal of Building Engineering, 24*, 100755. doi:10.1016/j.jobe.2019.100755

Naghshbandi, S. N. (2017). BIM for facility management: Challenges and research gaps. *Civil Engineering Journal, 2*(12), 679–684. doi:10.28991/cej-2016-00000067

Osello, A., Davardoust, S., Semeraro, F., Lucibello, G., Barone, L., Dellosta, M., ... Ugliotti, F. M. (2016). The centrality of Representation with BIM. *DISEGNARECON, 9*(16), 5–1.

Parsanezhad, P. (2019). *Towards a BIM-enabled Facility Management: Promises, Obstacles and Requirements* (Doctoral dissertation). KTH Royal Institute of Technology.

Peterson, J. L. (1981). *Petri net theory and the modeling of systems*. Prentice Hall PTR.

Ratajczak, J., Riedl, M., & Matt, D. T. (2019). BIM-based and AR application combined with location-based management system for the improvement of the construction performance. *Buildings, 9*(5), 118. doi:10.3390/buildings9050118

Sampaio, A. Z., & Simoes, D. (2017). Maintenance of buildings supported on BIM methodology. *Proceedings of XIV DBMC-Durability of Building Materials and Components, Ghent, Belgian*, 83-84.

Shahinmoghadam, M., & Motamedi, A. (2019). Review of BIM-centred IoT deployment: state of the art, opportunities, and challenges. In *ISARC. Proceedings of the International Symposium on Automation and Robotics in Construction* (Vol. 36, pp. 1268-1275). IAARC Publications.

Talamo, C. (2014). *La gestione integrata delle informazioni nei processi manutentivi*. Dall'anagrafica degli edifici ai sistemi BIM.

Yalcinkaya, M., & Singh, V. (2014, July). Building information modeling (BIM) for facilities management–literature review and future needs. In *IFIP International Conference on Product Lifecycle Management* (pp. 1-10). Springer. 10.1007/978-3-662-45937-9_1

ADDITIONAL READING

Eastman, C., Teicholz, P., Sacks, R., & Liston, K. (2011). *BIM Handbook: A Guide to Building Information Modeling for Owners, Managers, Designers, Engineers and Contractors* (2nd ed.). John Wiley & Sons.

Osello, A. (2015). *Building Information Modelling - Geographic Information System - Augmented Reality per il Facility Management*. Dario Flaccovio Editore.

Osello, A., & Ugliotti, F. (2017). *BIM: verso Il catasto del futuro. Conoscere, digitalizzare, condividere. Il caso studio della Città di Torino*. Gangemi Editore spa.

Semeraro, F., Ugliotti, F., Osello, A., Del Giudice, M., & Rapetti, N. (2016). *The DIMMER project user awareness: Innovative Gamification approach through Augmented Reality*. IATED Academy.

KEY TERMS AND DEFINITIONS

Augmented Reality (AR): AR refers to the extension of the physical space with virtual information to be displayed into the reality. This enables to enrich the visualization of objects within specific contexts.

Building Information Modelling (BIM): BIM is a methodology for managing project data in digital format throughout the life cycle of the work: from design to construction, management and maintenance. One of the fundamental results of this process is the Building Information Model, the digital description of all aspects of the built work.

Computer-Aided Facility Management (CAFM): Software platforms for the integrated management of real estate assets.

Geographic Information System (GIS): GIS is a computerized information system that allows the acquisition, recording, analysis, display, return, sharing and presentation of information derived from geographic data.

Indoor Localization: Indoor localization aims to establish the correct position of the maintenance operator inside the building. An indoor positioning system (IPS) is a network of devices used to locate people or objects where GPS and other satellite technologies lack precision.

Internet of Things (IoT): Internet of things is the expression used to describe the extension of the Internet connection to the most varied types of objects. The data collected using special sensors can be exchanged and communicated via the Internet and objects can be monitored and managed remotely.

Interoperability: Interoperability is the possibility to exchange data between different software platforms, not only during the construction phase of the work but also throughout its entire life cycle, from maintenance to decommissioning.

Level of Developments (LOD): LOD defines the content and reliability of BIM objects at different stages of the project. By "content" means geometric information, structured data and related documentation. By "reliability" means for what uses and to what extent users downstream of the information can trust the accuracy and quality of that content.

ENDNOTES

1 https://www.pit.de/
2 www.autodesk.it/products/revit
3 https://www.omg.org/mda/
4 http://www.gwtproject.org/
5 https://material.io/design
6 https://www.oasis-open.org/news/pr/iso-iec-jtc-1-approves-oasis-odata-standard-for-open-data-exchange
7 https://www.qt.io/

Chapter 17
From Sketches and Installations to Bioinspired 5D Printing Models:
Representation Interactions for Smart Cities

Silvia Titotto
Federal University of ABC, Brazil

ABSTRACT

This chapter opens up discussions upon the relevance of interaction of representations and data visualization modes for smart cities design, planning, and development that occur beyond paper and computer drawing. Although many practitioners usually relate smart cities and digital twins design exclusively to CAD/CAM/CAE and BIM methods, processes, and tools, a wider pool of techniques and forms of expression might be the key to a more accurate and comprehensive way of simulating the several kinds of alterations that happen in the planned built environment. The chapter deals with the study of concepts that relate to both physical and virtual prototyping, which underlines an interdisciplinary approach to design and the impact of integrating biologically inspired principles from different backgrounds to the field of smart cities design. In this regard, biomimetics and additive manufacturing may play key roles in smart city's modeling design and the frontier technology of 5D printing reveals real-time decision-making programmable 4D printing process as a potential future development.

INTRODUCTION

This book chapter aims to discuss the relevance of interaction of different representation and data visualization modes for smart cities design, planning and development that occur beyond traditional methods such as paper sketches as well as computer drawings and simulations.

DOI: 10.4018/978-1-7998-7091-3.ch017

In the majority of articles from journals with selective editorial practices, smart cities and digital twins design are often related exclusively to BIM and CAD/CAM/CAE methods, processes and tools (Bernabei et al, 2015; Brown, 2009). Digital twins technologies mainly relate to virtual duplications of real systems and they provide a platform to review interactions, consequences of decisions and activities within the real system (Farsi et al, 2020). Although those ones remain largely important and well diffused among practitioners, a greater pool of forms of expression and techniques that could both virtually and physically improve and optimize the prototyping and the demonstration cities models' alterations experimentally via laboratory research remain underdeveloped yet (Al-Turjman & Imran, 2020). Utilizing the full complexity allowed by additive manufacturing might be the key to unlocking the massive potential of this technology for architectural applications – and biomimetics might be pivotal in this regard (Du Plessis et al, 2019; Miyauchi & Shimomura, 2019).

Biologically inspired processes and techniques (Bar-Cohen, 2019; Noble & Kaminski, 2015) might be the key to simulate more accurately and comprehensively several different kinds of alterations that happen in the planned built environment (Bordegoni & Rizzi, 2011).

This chapter deals with the study of transdisciplinary concepts for the development of physical and virtual modeling design. For this investigation, data from other fields such as material science, mechanical engineering, aerospace design and visual arts are key for putting together information of interest for technological innovation for the representation of twin cities and smart cities. This perspective underlines the significance of interdisciplinary and transdisciplinary approach to research with the aim at bringing to light pieces of knowledge that need a subtle translation from one field to another.

The impact of integrating the principles of different academic backgrounds in the field of smart cities design is promising for contributing to real innovation to modeling. Its main contribution reckons on the thinking upon the extension of the problem and the thematic skeleton of research on architectural design, city planning, contemporary art, energy engineering and their relationships for smart cities development.

Firstly this chapter introduces concepts related to sketches, installation artworks and three-four-five dimensional additive manufacturing for architectural purposes. In sequence, it presents the contextualization of the biologically inspirational factor for the surge of 4D printing and its knowledge to represent design (Dimassi et al, 2021). Then via bioinspired examples including the complex dynamics of termite mounds, it examines some mechanical mechanisms to be used to better predict design shape alteration and it discusses sustainable possibilities of materials for 4D printing (Stoychev et al, 2017) for the representation and optimization of smart cities visualization via physical models. From that, the specific case of bioinspired 4D printed hydrogels is introduced as an example of dual relationship to nature via bio-inspiration, as their form and function merge along with movement and responsiveness. Finally, the chapter mentions the expansion of representation modes from more traditional fields to surprisingly examples merged from science materials. Then the chapter shows future areas to be explored and avenues for new trends in digital cities representation such as 5D printing, a dynamic real-time decision-making programmable 4D printing process. After that, the chapter is finally concluded with the previous remarks.

BACKGROUND

Graphic representation of the built environment used to be traditionally made via sketches and technical drawings but in the last decades, more and more computational approaches were incorporated so that the necessary simulations for the smart cities could be planned and performed according to given parameters.

This chapter discusses technological innovative solutions for smart cities representation based on the merge between traditional and very low technological approaches such as sketches and paper/wood/glue physical modeling as well as high technology solutions such as responsive prototypes (Titotto et al., 2015; Titotto, 2014) that may rely solely on more specific parametric design or combine the specificities of the design together with programmable materials, smart materials, or even metamaterials (Shin et al, 2017). The reason for the use of artificial intelligence and more advanced functional materials in the physical models is enhancing efficiency (Zaheer, 2021) and giving extra support to the ever-changing representation of Architectural and Urban design that is under constant mutation as it is expected for a smart city. And there is a broad understanding that smart city technology will produce great solutions for social, economic and environmental problems (Lai et al, 2021).

From the point of view of this chapter's author, the basic insight for this new kind of "kinetic" representation of the urban space departs from the observations of biological structures and systems – such as the social insects for instance – which respond and react to environmental stimuli, as a way of addressing how nature continuously organizes and plans their living spaces.

Although many social insects living in dense colonies (Bullock et al., 2012) face the problem of maintaining temperature, respiratory gas and humidity levels within tolerable standards, insects solve their problems using easily found structures or by building their own nests or shelters.

A particularly impressive example of insect architecture is found at the taxonomic rank of infraorder Isoptera, which are eusocial insects (Sumpter et al, 2012) popularly known as termites. Although their bodies individually measure only a few millimeters in length, they are recognized together for its ability to build complex large-scale structures without central support. The way they organize their strategies to control and mitigate losses after mound damage after a natural disaster for example can teach a lot to architects and urban planners about the representation of a kinetic and a responsive environment.

In the case of termites, for instance, their continuous re-planning and re-building of their complex city-like environments deeply rely on their relationships with the fungus. Fungi help to keep the perfect thermobalance within the mounds as well as the rebuilding of the mounds after a flood for instance, so termites maintain important relationships with the fungi that cohabit in their constructions.

Although different hypotheses such as constant and floating external wind and internal metabolic heating have been proposed for the ventilation of termite mounds, the absence of direct and effective measurement of internal air flows in fieldwork prevented the development of a definitive mechanism for this physiological function. In other words, the technical difficulties of direct measurements at the location of the airflow in an intact termite mound and its correlation with the internal and external environment are conditions that did not allow the differentiation between any of the hypotheses.

At first sight, it might sound a bit unusual to use nature, engineering and art to build up the foundations of this innovative representation method (Vincent, 2001), but, in fact, for decades or even centuries, interdisciplinarity and transdisciplinarity have been ways of giving continuous support and input to researchers from different fields to work together and to investigate several design possibilities based on people's lifestyles in different contexts.

For making this kind of representation more feasible to the naked eye, the biomimicry lessons may be tested cost-effectively on a smaller scale via installation artworks.

A way of beginning to look at this desired merge is via Installation Art for being a kind of media and spatial creativity that enables visualization of planned environments at a given scale (Reiss, 1999; Bonnemaison & Eisenbach, 2009). It arises from the relationship between art and architecture as the conception of a place of dialogue between them, which pioneers a new field of research in architectural and city design.

As critical spatial practices installations, artworks are formulated in the form of less established proposals for settlements that cross the field of architectural design (Beesley et al, 2015; Kwon, 2004), thoughts concerning new sets of connections between the subject and a spatial order show the promise of the use of the concept of art installations in smart cities design.

In the 1980s, the introduction of NURBS – to define virtual 3D forms by effectively skinning a linear skeleton to be manipulated via control points – was witnessed by architects and engineers, but only two years later that Autodesk released CAD programs for IBM personal computers (Titotto & Dobelis, 2017). It was when early adopters centered on design and engineering started to use that technology for experimental and professional purposes, but the development of software such as SolidWorks ACIS, Solid Edge and Parasolid helped to spread 3D modeling software in the 1990s (Titotto & Dobelis, 2017). Parametric modeling via software such as Rhinoceros, for instance, enables the graphic representation of more complex geometry, especially when it involves possibilities of internal shape alterations (Figure 1).

Figure 1. Parametric modeling of a hollow building based on termite mound internal geometry
Source: Author's design.

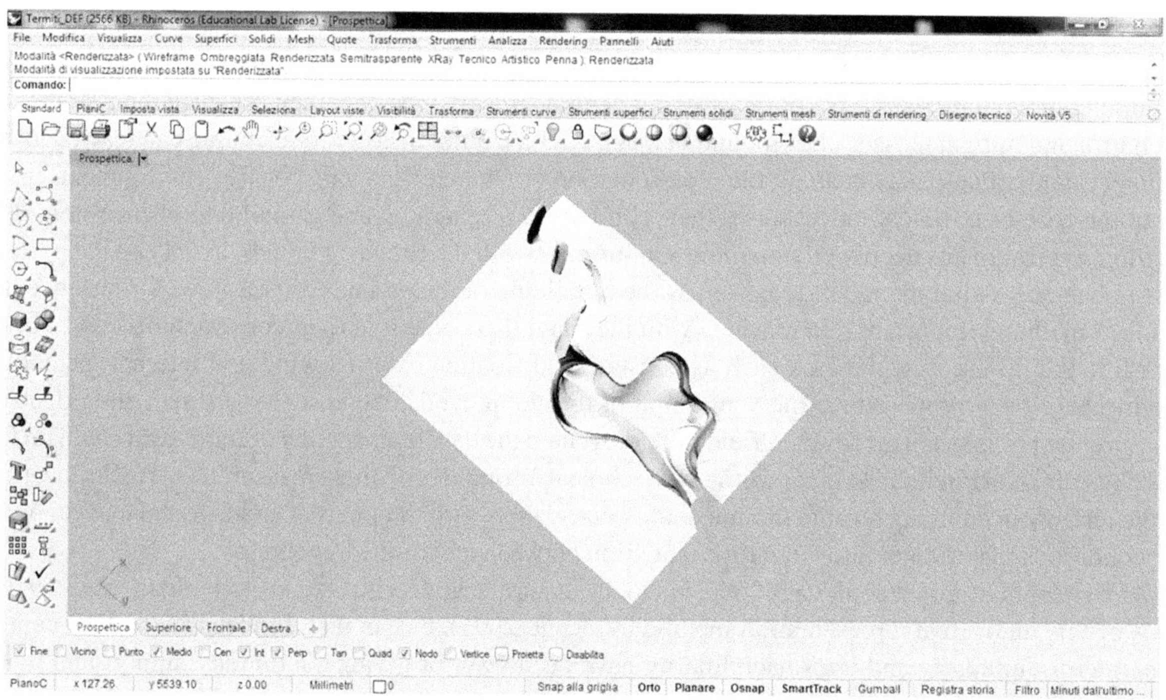

In the last decades, human-centered spatial data is casting the growth of computational social science (Ye & Lin, 2021) and the use of BIM has been playing an important role in the construction field. Whilst it increases the possibilities of design idea representation in most cases, it can also provide feedback on the feasibility of production through geometric analyses that reveal surface errors (Titotto & Dobelis, 2017).

With the emergence and broader use of digital technologies the desire to incorporate CAD/CAM/BIM into urban and architectural design curricula has inevitably accelerated. It is such a strong trend that several developed nations had already integrated CAD /CAM/BIM classes in their National Curriculum for Secondary schools by the end of the last century (Fullwood, 2002).

Computing design and techniques can work as a bridge to smart applications applied to smart cities (Balusamy et al, 2021). The past decade consolidated the trend of representing with more and more accuracy via the use of parametric design and virtual reality on top of the additive manufacturing of physical models printed in resin layers. 3D printing is a kind of additive manufacturing that allows the fabrication of objects using layer-by-layer material deposition (Chee et al, 2017; Choi et al, 2017), leading to accurate programming of some geometrical properties as well as the dimensions of intricate printed objects (Choi et al, 2015; Kirihara & Nakata, 2021).

Every day more and more researchers seem to be interested in investigating a broader application of additive manufacturing for innovative applications and their access to a wider range of materials (Tibbits, 2014). Following this approach, 4D printing encourages the prospect of performing kinetic, adaptive and performative accomplishments via unhackneyed design possibilities (Odling-Smee & Turner, 2011).

This modeling methodology seamlessly integrates materials design (Barthelat, 2015; Barthelat, 2007) with digital fabrication processes within the context of self-assembly manufacturing (Vamvakidis, 2017). Bioinspired self-assembly undertakes to enable more and more advances in a variety of applications in very diverse fields as well as the creation of new categories in the creative industry such as living cities that alter shape and behaviour according to their surrounding environment.

For the time being, 3D printing is much more accessed by the general public (Mendoza, 2015) than the 4D but currently all those technologies are being more or less introduced to the Architectural world via interdisciplinary partnerships. To date, 5D printing does not have peer-reviewed research results to the Architectural field yet as it is still at the frontiers of technology. It is expected to bloom in the next few years after more development is obtained in laboratory research.

In terms of Physics, 4D (Figure 2), meaning the four common dimensions, usually alludes to three-dimensional space (3D) and this builds on the dimension of time to the other three dimensions of length, width, and depth. Whereas 3D printing manufactures motionless objects, the prominent technology of 4D printing permits the fabrication of kinetic structures that alter their form over time while keeping in contact with certain external stimuli. It also agrees with sustainability principles, as it demands very low to no energy expenditure. The difference from 4D printing to 5D printing is that in 5D printing the model movements are continuously controlled via machine learning (Nagar et al, 2020) and different kinds of responsiveness to stimuli may happen at the same time.

Although a number of authors had already referred to correlations between advancements in additive manufacturing and biologically inspired engineering (Fratzl & Burgert, 2009), (Ionov, 2012), it was via predicting connections among several self-evolving and self-assembly systems observed in the natural biological environments and additive manufacturing projects that Skylar Tibbits pioneered the field (Tibbits & Cheung, 2012). Tibbits performed a variety of conceptual prototypes at Massachusetts Institute of Technology (MIT) since 2011 (Tibbits et al, 2017; Tibbits, 2014) and introduced the concept of 4D printing to the broad public in his TED talk in the year of 2013.

The fourth dimension in 4D printing emerges from the capacity of the printed objects to transform their functionality and/or shape overtime during the time they are exposed to certain environmental stimuli as well as during their post-printed existence (Champeau et al, 2020; Ge et al, 2013). It means that, in this case, the fourth dimension can be understood as time because the printed structure has the ability

Figure 2. Transition from 2D to 3D and 4D
Source: Author's design.

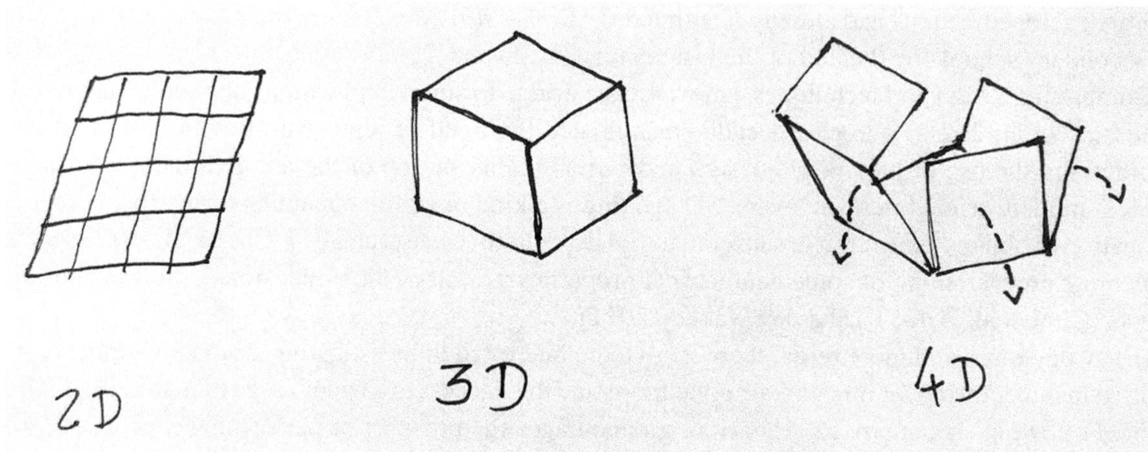

to change its shape, functionality and properties along time under environmental stimuli, for example, temperature, ion concentration, pH, magnetic and electric fields, light or even hydration/dehydration (Champeau et al, 2020).

As per definition 4D printing demands alteration of form and/or function of the printed object over time, improvements in material properties are still demanded for each application to be sustainable and to continue to be efficient over a number of cycles. This way, self-healing, biodegradation, speed response, actuation capability, besides improved toughness (Fratzl & Burgert, 2009) should become mainstream characteristics for self-assembled 4D printed objects in the future.

The potentiality of 4D printing was approved for tissue engineering and soft actuators but in order to kick off printed devices to many if not all the other fields of engineering and creative industries, more technology integration of additive manufacturing, programmable materials and parametric design is highly needed.

The pioneers of 4D printing unfolded the field by associating it to specific stimuli throughout passive energy. It means that the structure was not to be placed after active sources of energy to calibrate its flow to the system (Ge, 2016). Recently some research groups began to consider the probability of using diversified energy sources. It might include active energy to be applied locally (Ge, 2016).

In this case, to resist responsiveness and stimuli variation, one of the most important factors is the capability to return on their own to the primary arranged status. On the other hand, in a few special cases, a partial reversion cycle may really change the reaction and the properties of the material. Anyway, the response and reversion processes are usually expected to not happen in the same amount of time (Pei, 2014) as it can be observed in many examples in botany and also invertebrate sea life.

Figure 3. Preparation for additive manufacture via production of layers at Cura software
Source: Author's design.

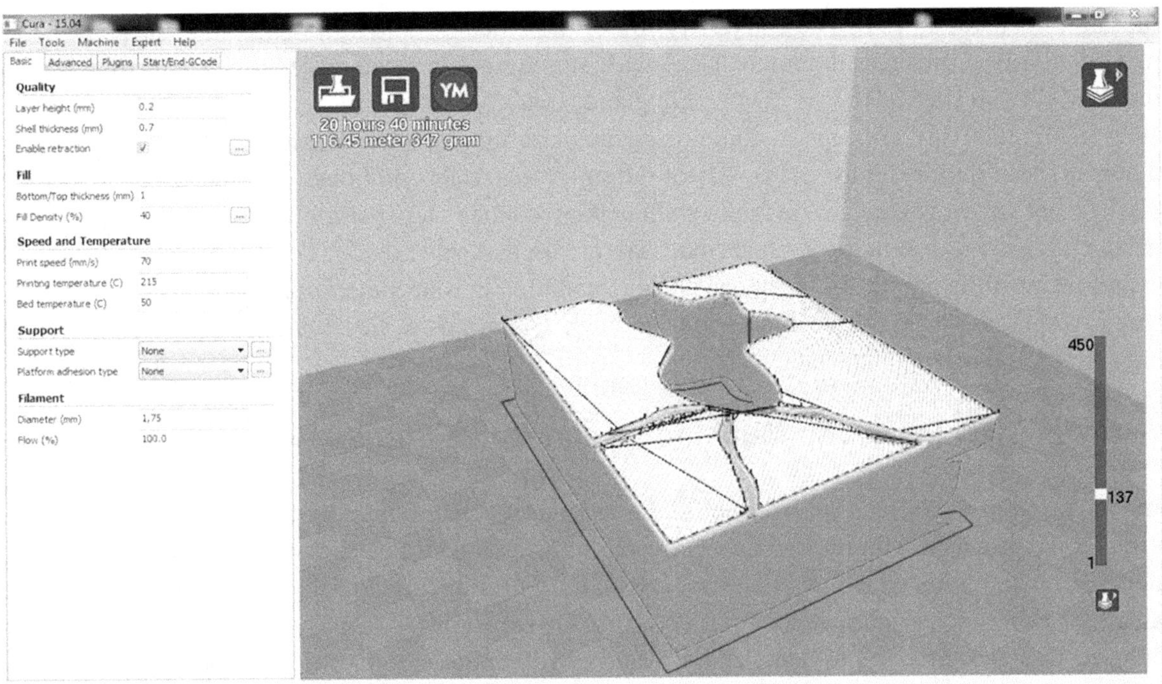

4D PRINTING AS REPRESENTATION AND SIMULATION TOOLS FOR CHANGING CITIES

Merging Various Backgrounds and Approaches for Greater Efficiency

The usage of simulation models throughout the whole work lifecycle has evolved (Tarkhov & Vasilyev, 2020). One of the newest trends in Construction Engineering is the concept of 4D printing – as stated before, a kind of additive manufacturing that enables the printing of design layers that are sensitive to environmental variations and permits shape alterations. It may prove very useful for instance in the simulations of daily morphological changes at the city networks. This kind of additive manufacturing has received so much attention in 2019 that it was even highlighted in a full page of the 12th Annual Edition of the 2019 Tech Trends Report by Amy Webb, who is The Future Today Institute's founder.

In fact, 4D additive manufacturing has been expected to develop broadly in several areas related to Architecture and Civil Engineering such as building and security applications as well as other cities modeling. Besides that, this kind of additive manufacturing changes the shape and its dynamic behaviour makes clear its biomimicry origins as several future applications relate to mimicking biological structures along with their life span main movements.

In architectural applications, 4D printing has been developed for several practices. For example, as a kinetic hydrogel skeletal skin with thermal and light-responsive capacities that also work as an active reciprocal retrofit to optimize the energy efficiency of old buildings and their built environments (Raviv et al, 2014; Tibbits, 2014). And optimum control of winds and other climate conditions via building façade design have also been expanded via a soft reactive "bio-structural" hydrogel skin by the Interdisciplinary Design environments for Engineering and Architecture Laboratory (IDeEA Lab), Deakin University Sculpting Regeneration Lab, The University Illinois at Chicago. The development of these ideas may comprise high-performance computers, large-scale stereographic video projection, 7.1 surround sound, oculus rift VR and state of the art 3D modeling and simulation software.

For the digital twins' concept, it would mean the possibility of tracing movement and shape modifications via the stimuli-response properties of the smart materials combined with the parametric design itself, in order to permit changes on their swelling degree when facing stimuli of temperature, light, pH, hydration/dehydration. When non-smart materials are used, despite results being more limited, they can also present swelling properties in aqueous media, resulting in shape alterations that depend exclusively on passive energy. This way, the simulations of shape alterations at smart cities do not rely on electric energy for example.

Smart Materials (or not so Much) for Smart Cities Simulation

Some 4D printing pioneers described the field as a kind of ''materials science chamber'' where different materials, properties, and micro and macro architectures could be mixed and matched to produce material behaviours and capabilities, in which actuation, sensing, and logic, which would not have been enabled via other industrial processes (Lee et al, 2017; Lin et al, 2018). They proved to limit it as the development of 4D printing has revealed more and more possibilities for the conception of smart printing using materials designed to forecast change processes (Momeni, 2017).

4D printing of metals involves mostly shaped memory alloys, whilst a number of publications have emerged describing alternative stimuli for polymers and composites. On the other hand, despite covalent reactions in hydrogel self-healing mechanisms frequently require the conditions used for polymerization, such as pH or UV light (Kirihara & Nakata, 2021), self-healing hydrogels in particular for soft robotics for physical modeling technologies seem to be much closer to broad architectural use in the next decade.

So far hydrogels are taking a key role in this in accordance with its capability of optimizing its formulation to customize compositions in which it should be possible to create multi-layer hydrogels with different density, swelling and shrinking properties.

The advent of 4D printing has brought to light several possibilities for the development of smart printing by the use of materials that are designed to forecast alteration processes.

Shape memory polymers are already reasonably well known for their application in 4D printing (Ge et al, 2013). A late innovation that is expected to revolutionize the deployable mechanism design is the improvement of programming for shape-memory materials that could be applied to the design of deployment mechanisms (Liu et al, 2014).

Shape memory devices are prone to eliminate the need for traditionally complex mechanical deployment systems as well as systems that independently control deployment. Shape memory mechanisms may also reveal very concise designs that include less non-structural parts. This way they are expected to be considerably lighter in weight.

Elastic Memory Composites (EMCs) belong to the family of shape memory materials (Fetchko et al, 2004). The greatest advantages of EMC materials in comparison to shape memory alloys and shape memory ceramics are that they are a lot less dense, they have higher damping and also higher strain capacities. Hence EMC materials present desirable characteristics for deployable spatial objects and structures and for this reason they have been so much investigated within the additive manufacturing research field (Straubel et al, 2011).

Apart from these polymers with shape memory characteristics, hydrogels are another type of polymers that have been seen as very important to the development and applicability of 4D printing (Chin et al, 2017). In fact, to date, smart hydrogels are the favourite materials for the sensitivity and responsiveness to environmental stimuli, but non-responsive hydrogels are also relevant to researchers as they may expand the range of materials (Yang et al, 2017).

Self-morphing hierarchical structures require sequential shape-changing at both system and component levels (Teoh, 2016).

Another reason for that is obtaining the programmed and complex shape alterations by dehydration and/or rehydration via adjusting internal structures of crosslinking and density or via their combination with other kinds of materials. In reality, for most smart hydrogels, the stimuli produce swelling alterations. This advantage has been used to generate shape transformations in hydrogel sheets built under traditional techniques and it seems to be the reason for being chosen as kind of a favourite for 4D printing (Ding et al, 2017; Thérien-Aubin, 2013).

Although 4D printing via the use of hydrogels for city modeling presents a more straight-forward avenue for stimuli-response properties and the advantage of not demanding programming after printing, there are a few disadvantages that need to be overcome too.

Due to their softness, when hydrogels are used, for example, the stiffness of printed structure tends to be relatively low and the swelling mechanism depends on the diffusion transport (Champeau et al, 2020). It results in slow speed response. And if the environmental condition is changed, the next loss of the solvent in the hydrogels may result in instability of the actuated shape. The reason for that is that the total hydrogel dehydration may result in brittle, porous materials, that represent a challenge for high-temperature systems or interfaces susceptible of evaporating water molecules into the atmosphere but it is not the case for most applications for smart cities modeling (Mandon et al 2017; Seepersad, 2014).

Low Tech First for a More High Tech Approach to Smart Cities Representation

Although most contributions to smart cities representation are related to high-performance software via CAD/CAM/CAE parametric approach (Fuller et al, 2020; Julin et al, 2018), the possibility to have low tech initial sketches is an interesting approach to bring a wider and solid study of project viability.

Smart cities are directly connected to how an urban environment behaves due to its metabolic dynamics. This approach is completely different from a more traditional graphic representation where the solidity and inertia of the sketches are baselines for any project graphical portrayal.

Bringing biomimetics into context (Ehrlich, 2017), there are numerous microstructural compositions from biological organisms whose dynamic morphologies change shape on-demand (Bhushan, 2018; Hwang, 2015; Ivanic et al, 2015; Bandyopadhyay et al, 2015; Tavakoli & Tang, 2017; Gao et al, 2016; Gladman et al, 2016) due to modifications on their environment (Ionov, 2012; D'Arcy Thompson, 2011; Vincent, 2006). Examples in the natural world are retracting leaves (Mimosa pudica L.) and pigmentation loss of the flower skeleton (Diphylleia grayi). In the case of the mimosas (Figure 4), the mechanism of shape alteration happens in real-time, according to the sequence of the touched leaves (Figure 5). The responsiveness does occur immediately after the touch but there is still a slight delay (Figure 6) so that a process of dehydration happens within the leaves until final configuration aftertouch (Figure 7).

Figure 4. Mimosa pudica L. before touch
Source: Author's photograph.

Figure 5. Mimosa pudica L. after the touch happened locally at three leaves partially retracted
Source: Author's photograph.

Figure 6. Mimosa pudica L. immediately after full touch, still partially retracted
Source: Author's photograph.

Figure 7. Mimosa pudica L. after touch at its final spatial configuration
Source: Author's photograph.

Insights for a more accurate representation of smart cities may be perceived via immersive installation artworks that respond and that may even evolve according to the movements of people around them. Some of those kinds of bioinspired responsive installation artworks are powered by protocells as some of the artworks conceived by the architect Philip Beesley. The translation of the concept of "metabolic architectures" can be understood as artificial four-dimensional living structures that may be even capable of breathing, regenerating and growing, instead of inanimate, fixed objects displayed in a "white cube" of the museum architecture. Therefore revision in depth of the concepts of naturalness and artificiality urges more investigation and one way of doing so is via experimental disruptive bioinspired design.

Over the past years, the Institute for Biologically Inspired Engineering at Harvard University has also investigated the mimicking of biological structures in one of its research trends. Via the use of multimaterials printed in 4D technology research has been conducted for developing dynamic alterations as answers to changes in humidity and/or temperature (Gladman, 2016; Lewis & Gladman, 2017).

The Self-Assembly Lab at the International Design Center of the Massachusetts Institute of Technology (MIT) has pioneered in examining the self-assembly phenomenon for 4D printing. The architect Skylar Tibbits and his colleagues departed from the idea that those systems are scale-independent and may be used for self-constructing and manufacturing systems, from a simple set of responsive building blocks, energy and interactions (Fox & Kemp, 2009) that can be designed in a wide range of materials and machining processes. This can happen in the case of additive manufacturing, in which programmable materials get involved in environmental stimuli, and also via the use of vernacular techniques whose results are generated randomly. For instance, it is the case of agitation of particles that Tibbits experimented and exhibited.

Graphical Representation and Modeling of Sustainable Bioinspired Structural Performance

Bioinspiration and computational design have formed a fruitful ground for technological advances. Sometimes that kind of cooperation meets the exhibition hall at a first stage of research that presents possible results at a tiny scale so that the concepts are applied later to broader use in Architecture and Urban Planning. For instance, at the University of Stuttgart, the Institute for Computational Design (ICD) has worked together for many years with the Institute of Building Structures and Structural Design (ITKE) so that a new research pavilion is designed every year. This tradition relates to the famous architect and engineer Frei Otto and his membrane structures lab that put together technological innovation and artistry for several decades. Otto won the RIBA Royal Gold Medal in 2006 and was awarded the Pritzker Architecture Prize in 2015, shortly before his death. Although Frei Otto was not leading the 2016–2017 pavilion, the university team examined two species of leaf miner moths whose larvae spin silk 'hammocks' stretch between connection points on a bent leaf. Along with more bioinspiration yet, the final structure was built via spinning fibres in much the same way as the leaf miners but in this case drones were used instead (Primrose, 2020).

Another example to illustrate bioinspired sustainability is the wind tree for capturing energy. The wind tree is an elegant structure that can be located in urban areas and generate electricity for local use. Its key mechanism is the multitudes of "aeroleafs" that are micro wind turbines mounted vertically (i.e. a VAWT) thereby minimizing space requirements and reducing noise (Primrose, 2020). So the aeroleafs are able to take wind from any direction, as they are remarkably light and almost without inertia. In some urban areas the wind speed may reach extremely variable ranges over very short distances. Therefore, to leverage on each available Watt of energy, an electronic card with a microprocessor keeps each aeroleaf autonomous so that it rotates at the optimum speed for the own exposure (Primrose, 2020).

The lessons given by the termite mounds are especially valuable to smart cities, as they are one of the closest examples found in nature to the challenge of recreation of building strategies soon after natural and provoked disaster as well as daily maintenance.

Over the last 150 years, several professionals from different fields have been interested in termite mounds. In the mid-19th century, the Scottish naturalist James Rennie published the book Insect Architecture (1869), containing research on termite constructions. But it was only in the early 1960s that Martin Lüscher published an article in Scientific American (1961) about what would be the first attempts to understand the air conditioning effect on termite mounds in a thermosyphon model and the possible advantages regarding energy performance for its use in architecture.

From those preliminary models, the search for typological solutions dates back a few decades. An example is the Eastgate Center in Harare, which was built in 1996 under the design of Mick Pearce (Gleich et al., 2010). However, Scott Turner and Rupert Soar (2008; 2009) found that the Zimbabwe building had been modeled under a misconception of how termite mounds really work. Since then, many assumptions about its operation have been revised.

It is important to acknowledge the limitations of translations and transitions between ecosystems (Titotto, 2013) but with greater research maturity in the field of computational parametric models, physical prototypes (Ladley & Bullock, 2004) and their respective checks available via kinetic modeling, technological innovation and optimization of the biologically inspired urban planning are expected to be more feasible, because, in nature, the concepts of efficiency, form and function are inseparable (Thompson & Bonner, 2013).

The opening of possibilities for new paradigms of high-efficiency construction projects concerning energy in recent research results from collaborations on thermoregulation processes of termite mounds between interdisciplinary researchers in the fields of environmental science and technology, architecture and physics, from Harvard University, Massachusetts Institute of Technology (MIT), State University of New York and Nottingham Trent University, which brought about significant changes in the paradigms followed by other relevant researchers of those fields.

Two of the most recent paradigm changes have emerged from the 2015 publications in the Proceedings of the National Academy of Sciences of the United States of America in which King, Ocko and Mahadevan (2015a; 2015b) discuss their studies of about two dozen termite mounds and on the discovery that they resemble the functioning of a lung (Hólló, 2015) that would breathe once a day, driven by changes in temperature between day and night. In other words, the air currents inside the termite mounds expel the carbon dioxide that accumulates in the activity of the subterranean termites. This contrasts with long-held assumptions by biologists that termite mounds existed both to dissipate nest heat or ventilation in response to external airflows.

Based on an understanding of the method by which termites take advantage of external temperature changes to create airflow, the development of analogous models for human architecture can mean real technological innovations from the point of view of environmental performance (Linardou, 2008). In other words, the graphic representation of structures that are not simply inspired by natural biological life seeking energetic and constructive efficiency - but that are, in a sense, as alive as its inhabitants and the environmental systems where they will be inserted.

SOLUTIONS AND RECOMMENDATIONS

Frequently the improvement and development of man's technical skills has been described as a continuation of evolution with other means (Persiani, 2019). Nowadays and in the near future, technology is to be pushed towards a further boundary: boosting the human capacities and augmenting its abilities to control its surrounding space via digital twins and smart cities.

Instead of copying nature directly, it is of interest to exclusively mimic the procedures that nature uses to produce its structures, aiming at how this may inspire the creation of smarter processes that thermoregulate and point out at shape alterations. While the current civil construction generally isolates the buildings and keeps its users within a barrier from the outside, maintaining a constructive paradigm of saving energy leads to the accumulation of damp and humid air in the construction that may contribute to its performance failure. In the natural biological world, the opposite is true - walls are structured and permeable to allow the passage of gases, maintaining the temperature and humidity inside. An analogy can be made to the human breathing process in which the muscles around the lungs are used to push carbon dioxide out and let oxygen in. Thus, a delicate balance of temperature and humidity is preserved at the bottom of the lungs. Likewise, the African termite mounds surveyed by Soar, Turner, King, Ocko and Mahadevan, instead of having muscles to breathe, are built by termites in a way that they use the complex energy of turbulent winds to do the same thing.

Better integration of additive manufacturing and traditional representation techniques are expected to launch potential topics in the fields of smart cities. This will result in a more precise visualization of smart cities via physical models in reduced scale with soft-robotic functions besides the virtual approach of the digital twins.

These attributes, along with the assembly of complex and customized geometries with spatial control, make the 4D printing techniques along with the other representation modes an excellent combination for exploring and developing the smart printing applied to design and to related fields of engineering such as energy and safety.

This viewpoint presents both theoretical and practical-experimental implications and approaches using advanced intelligence to improve the understanding of complex socio-environmental systems (Kamilaris et al, 2021). The benefits to be passed onto future generations are broadly based on technological innovation to the built environment via observation of the biomechanics of living beings and applications of its principles to smart cities design (Titotto, 2014).

A number of innovative solutions have been inspired by the observation of living beings considering both technical and behavioral points of view. Biomimetics plays an important role in that. Stimuli reaction and shape modification are possible features resulting from the adoption of these hybrid technologies and integrating sustainable materials with specific deployment mechanisms in the context of self-assembly.

On top of that, by working together with the fields of art and technology uncover structural and energy principles to the smart cities planning along with poli-dimensional production of installation models and prototypes (Titotto, 2014).

To date, the first advantage of the swelling approach to hydrogels in 4D printing is that they do not require programming after printing in spite of some drawbacks. In the next years, it is expected that a few of the main challenges that 4D printing currently faces – for example from materials properties to bed size and limitations on the scale, print speed and software tools - are not going to necessarily be downsides; rather, those challenges are opening up a picture of the next best possibilities for Architectural modeling.

FUTURE RESEARCH DIRECTIONS

It is given that the observation of the influence of different designs on the 4D transformation of different materials for architectural models will be a highly interesting and demanding concept in the near future and may motivate researchers to create a new field of study that relates to Building Information Modeling.

In spite of the fact that actuators with soft-robotic functions and biomimetic advanced materials with simplistic and programmable fabrication processes continue to be scarce today in the Architectural field (Ding et al, 2017), it is expected that the technology integration of additive manufacturing and smart materials is creating several potential topics in highly interdisciplinary fields such as those related to advanced computational and statistical analysis (Khan et al, 2021) and urban planning in the near future.

As more and more researchers are inspired to think about novel application scenarios for 4D printing in the Architectural scenario, the low cost and accessibility of smart materials might also encourage exploration of new design possibilities with them by implementing kinetic (Khan, 2020), adaptive and performative capabilities to physical city models.

Certainly, the field of 5D printing is a good bet for revolution in physical modeling. So it is expected that the modeling techniques will be very optimized by the development of soft robotics, artificial intelligence and material science within the next decade.

CONCLUSION

More traditional representation modes such as sketches and CAD/CAM/CAE models seem to be on a route to collide and merge with virtual reality and pre-programmed shape transformation enabled by smart materials and artificial intelligence. It shows that even where there is still uncertainty for the field nomination, beyond the currently widely known 3D printing, 4D and 5D printing are also certain to hit and revolutionize smart cities physical modeling within the next decade.

In addition to a faster and more continuous construction process of 3D sketches as well as 4D and 5D physical models along with primal design in parallel with the architectural project are more likely to identify errors still in the earlier stages of digital twins. Consequently, within the parametric design, this approach is more likely to correct them time-wise. It also leads to an initial better perspective of costs, because the further the process goes on the wrong path, the higher the costs will be to adjust the design and planning mistakes.

The bioinspired premises for implementing innovative solutions for the kinetic representation of smart and twin cities can be summed up as follows:

- Adaptation is functional: living systems manage to control flows of matter and energy in natural ecosystems;
- Adaptation emerges from multiples scales: subsidiary scales relate from minor to major shape and flux alterations;
- Adaptation is an interaction: functionality usually involves interactions of diverse forms and their adaptive integration and embodiment; and
- Structure is process: the capability of adaptation includes the continual reshaping of the environmental context where a process occurs and the ideal modeling will incorporate this sort of dynamism into its design.

Aiming at a more accurate interface with digital twins, it is expected that the further development of additive manufacturing may:

- Encourage multi-dimensionalities of smart cities modeling;
- Optimize representation paradigms; and
- Provide further input to researchers to explore a myriad of other design spaces.

ACKNOWLEDGMENT

This research was supported by the Brazilian National Council for Scientific and Technological Development (CNPq) research group "4D Printing & Biomimetics" (4DB) at the Center of Engineering, Modeling and Applied Social Sciences (CECS) of the Federal University of ABC (UFABC). And it was partially funded by the Coordination for the Improvement of Higher Education Personnel (CAPES) (grant #11/2017) and the Alexander von Humboldt Foundation (AvH) (grant #05/2018).

REFERENCES

Al-Turjman, F., & Imran, M. (Eds.). (2020). IoT Technologies in Smart Cities. From sensors to big data, security and trust. The Institution of Engineering and Technology.

Allam, Z. (2021). *The Rise of Autonomous Smart Cities: Technology, Economic Performance and Climate Resilience*. Palgrave MacMillan. doi:10.1007/978-3-030-59448-0

Balusamy, B., Chilamkurti, N., & Kadry, S. (Eds.). (2021). *Green Computing in Smart Cities: Simulation and Techniques*. Springer Nature AG Switzerland. doi:10.1007/978-3-030-48141-4

Bandyopadhyay, A., Bose, S., & Das, S. (2015). 3D Printing of Biomaterials. *MRS Advances*, *2*, 108–115.

Bar-Cohen, Y. (2019). *Advances in Manufacturing and Processing of Materials and Strcutures*. CRC Press. Taylor & Francis Group, LLC.

Barthelat, F. (2007). Biomimetics for next generation materials. *Phil. Trans. R. Soc. A*, 365.

Barthelat, F. (2015) *Architectured materials in engineering and biology: fabrication, structure, mechanics and performance*. Institute of Materials, Minerals and Mining and ASM International.

Beesley, P., Chan, M., Gorbet, R., & Kulić, D. (2015). Evolving Systems within Immersive Architectural Environments: New Research by the Living Architecture Systems Group. Next Generation Building, 2(1), 31-56.

Bernabei, R. (2015). CAD/CAM and Jewellery Design Education. Making Futures 2015.

Bhushan, B. (2018). Biomimetics: Bioinspired hierarchical structured surfaces for green science and technology. *Materials Science*, 292.

Bonnemaison, S., & Eisenbach, R. (2009). *Installations by architects: experiments in Building and Design*. Princeton Architectural Press.

Bordegoni, M., & Rizzi, C. (2011). *Innovation in Product Design – from CAD to Virtual Prototyping*. Springer.

Brown, P. (2009). CAD: Do Computers Aid the Design Process After All? *Intersect: The Stanford Journal of Science. Technology in Society*, *2*(1), 52–66.

Bullock, S., Ladley, D., & Kerby, M. (2012). Wasps, termites and waspmites: Distinguishing competence from performance in collective construction. *Artificial Life*, *18*(3), 267–290. Advance online publication. doi:10.1162/artl_a_00065 PMID:22662914

Champeau, M., Heinze, D., Viana, T., De Souza, E., Chinellatto, A., & Titotto, S. (2020). 4D Printing of Hydrogels: A Review. *Advanced Functional Materials*, *1*(31), 1910606. doi:10.1002/adfm.201910606

Choi, J. (2015). 4D Printing Technology: A Review. *3D Printing and Additive Manufacturing*, *2*(4), 159-167. doi:10.1089/3dp.2015.0039

Chua, C. K., Yeong, W. Y., & An, J. (2017). 3D Printing and Bioprinting in MEMS Technology. *Micromachines*, *8*(12), 229. doi:10.3390/mi8070229 PMID:30400417

D'Arcy Thompson, W. (2011). On Growth and Form. New York: Create Space Independent Publishing Platform.

Dan, R., & Tibbits, S. (2014). Active Printed Materials for Complex Self-Evolving Deformations. Scientific Reports, 4. doi:10.1038rep07422

Dimassi, S., Demoly, F., Cruz, C., Qi, H. J., Kim, K. Y., André, J. C., & Gomes, S. (2021). An ontology-based framework to formalize and represent 4D printing knowledge in design. *Computers in Industry, 126*. doi:10.1016/j.compind.2020.103374

Ding, Z., Yuan, C., Peng, X., Wang, T., Qi, H. J., & Dunn, M. L. (2017). Direct 4D printing via active composite materials. *Science Advances*, *3*(4), e1602890. doi:10.1126ciadv.1602890 PMID:28439560

Du Plessis, A., Broeckhoven, C., Yadroitsava, I., Yadroitsev, I., Hands, C. H., Kunju, R., & Bhate, D. (2019, March). Beautiful and Functional: A Review of Biomimetic Design in Additive Manufacturing. *Additive Manufacturing*, *27*, 408–427. doi:10.1016/j.addma.2019.03.033

Ehrlich, H. (Ed.). (2017). Extreme Biomimetics. Springer International Publishing AG. doi:10.1007/978-3-319-45340-8

Eid, He, X., Bahr, R., Lin, T.-H., Cui, Y., Adeyeye, A., Tehrani, B., & Tentzeris, M. M. (2020, December). Inkjet-/3D-/4D-Printed Perpetual Electronics and Modules: RF and mm-Wave Devices for 5G+, IoT, Smart Agriculture, and Smart Cities Applications. *IEEE Microwave Magazine*, *21*(12), 87–103. doi:10.1109/MMM.2020.3023310

Erb, R. M., Sander, J. S., Grisch, R., & Studart, A. E. R. (2012). Self-shaping composites with programmable bioinspired microstructures. *Nature Communications*, *4*(1), 1712. doi:10.1038/ncomms2666 PMID:23591879

Farsi, M., Daneshkhah, A., Hosseinian-Far, A., & Jahankhani, H. (Eds.). (2020). *Digital Twin Technologies and Smart Cities*. Springer Nature Switzerland. doi:10.1007/978-3-030-18732-3

Fetchko, P. (2004). Deployment Optimization of a Boom for Falcon SAT-3 using Elastic Memory Composite Material. *18th Annual AIAA/ USU Conference on Small Satellites*. SSC04-VIII-4.

Fox, M., & Kemp, M. (2009). *Interactive Architecture*. Princeton Architectural Press.

Fratzl, P. & Burgert, I. (2009). Actuation systems in plants as prototypes for bioinspired devices. *Phil. Trans. R. Soc. A, 6*, 1541-1557.

Fuller, A., Fan, Z., Day, C., & Barlow, C. (2020). Digital Twin: Enabling Technologies Challenges and Open Research. *Access IEEE, 8*, 108952–108971. doi:10.1109/ACCESS.2020.2998358

Gao, B., Yang, Q., Zhao, X., Jin, G., Ma, Y., & Xu, F. (2016). 4D printing for biomedical applications. *Trends in Biotechnology, 9*(9), 746–756. doi:10.1016/j.tibtech.2016.03.004 PMID:27056447

Ge, Q., Qi, H. J., & Dunn, M. L. (2013). Active materials by four-dimension printing. *Applied Physics Letters, 103*(13), 131901. doi:10.1063/1.4819837

Ge, Q., Sakhaei, A. H., Lee, H., Dunn, C. K., Fang, N. X., & Dunn, M. L. (2016). Multimaterial 4D Printing with Tailorable Shape Memory Polymers. *Scientific Reports*. doi:10.1038rep31110

Gladman, A. S., Matsumoto, E. A., Nuzzo, R. G., Mahadevan, L., & Lewis, J. A. (2016a). Biomimetic 4D printing. *Nature Materials, 15*(4), 413–418. doi:10.1038/nmat4544 PMID:26808461

Gladman, A. S., Matsumoto, E. A., Nuzzo, R. G., Mahadevan, L., & Lewis, J. A. (2016b). Biomimetic 4D printing – Supplementary Information. *Nature Materials, 15*(4), 413–418. doi:10.1038/nmat4544 PMID:26808461

Gleich, A., Pade, C., Petschow, U., & Pissarskoi, E. (2010) Potentials and Trends in Biomimetics. Nova York: Springer. doi:10.1007/978-3-642-05246-0

Holló, G. (2015). A new paradigm for animal symmetry. *Interface Focus, 5*(6), 20150032. Advance online publication. doi:10.1098/rsfs.2015.0032 PMID:26640644

Hwang, J., Jeong, Y., Park, J. M., Lee, K. H., Hong, J. W., & Choi, J. (2015). Biomimetics: Forecasting the future of science engineering, and medicine. *International Journal of Nanomedicine, 10*, 5701–5713. PMID:26388692

Ionov, L. (2012). Bioinspired micro origami by self-folding polymer films. *Macromol. Chem. Phys., 214*, 1178-1183.

Ivanic, T., Tadic, Z., & Omazic, M. A. (2015). Biomimicry-an overview. Holistic. Approach. *Environ., 5*(1), 19–36.

Jin, O-Chang, Wonjin, Ju, & Myoung-Woon. (2015). 4D Printing Technology: A Review. 3D Printing and Additive Manufacturing, 2(4), 159-167. doi:10.1089/3dp.2015.0039

Julin, A., Jaalama, K., Virtanen, J.-P., Pouke, M., Ylipulli, J., Vaaja, M., Hyyppä, J., & Hyyppä, H. (2018). Characterizing 3D City Modeling Projects: Towards a Harmonized Interoperable System. *ISPRS International Journal of Geo-Information*, *7*(2), 55. doi:10.3390/ijgi7020055

Kamilaris, A., Wohlgemuth, V., Karatzas, K., & Athanasiadis, I. N. (Eds.). (2021). *Advances and New Trends in Environmental Informatics: Digital Twins for Sustainability*. Springer Nature Switzerland AG. doi:10.1007/978-3-030-61969-5

Khan, I. A., Liew, C. S., Loy, Y. Y., Lu, W., Myint, P. N., & Soh, K. B. A. (2015). *4D printing with smart materials.* https://www.slideshare.net/Funk9 8/4d-printing-with-smart-materials

Khan, M. A., Algarni, F., & Quasim, M. T. (Eds.). (2021). *Smart Cities: A Data Analytics Perspective*. Springer Nature Switzerland AG. doi:10.1007/978-3-030-60922-1

Kirihara, S., & Nakata, K. (Eds.). (2021). *Multi-dimensional Additive Manufacturing. The Japan Welding Engineering Society (JWES)*. Springer Nature Singapore Pte Ltd. doi:10.1007/978-981-15-7910-3

Kwon, M. (2004). *One place after another*. MIT Press.

Ladley, D., & Bullock, S. (2004) Logistic constraints on 3D termite construction. In *Fourth International Workshop on Ant Colony Optimization and Swarm Intelligence* (pp. 178–189). Berlin: Springer-Verlag. 10.1007/978-3-540-28646-2_16

Lai, C. S., Lai, L. L., & Lai, Q. H. (2021). *Smart Grids and Big Data Analytics for Smart Cities*. Springer Nature Switzerland AG. doi:10.1007/978-3-030-52155-4

Lee, A., An, J., & Chua, C. (2017). Two-Way 4D Printing: A Review on the Reversibility of 3D-Printed Shape Memory. *Materials & Design*, *3*(5), 663–674.

Lewis, J. A., & Gladman, A. S. (2017). *Method of 4d printing a hydrogel composite structure*. US Patent App. 14/954,228.

Lin, L., Zhang, L., & Guo, Y. (2018). Mechanical properties and shape memory effect ofthermal-responsive polymer based on PVA. *Materials Research Express, 5*(1).

Liu, K., & Jiang, L. (2011). Multifunctional Integration: From Biological to Bio-Inspired Materials. Perspective ACSNano – American Chemical Society, (5), 6786–6790. doi:10.1021/nn203250y

Liu, Y., Du, H., Liu, L., & Leng, J. (2014). Shape memory polymers and their composites in aerospace applications: A review. *Smart Materials and Structures*, *23*(2), 023001. doi:10.1088/0964-1726/23/2/023001

Mandon, C., Blum, L., & Marquette, C. (2017). 3D–4D Printed Objects: New Bioactive Material Opportunities. *Micromachines*, *8*(12), 102. doi:10.3390/mi8040102

Mendoza, A. J. (2015). *Legal and Social Implications of the 3D Printing Revolution*. CMC Senior Theses. Paper 1032. https://scholarship.claremont.edu/cmc_theses/1032

Miyauchi, A., & Shimomura, S. (2019). *Industrial Biomimetics*. Taylor & Francis Group. doi:10.1201/9780429058837

Mome.ni, Hassani, Liu, & Ni. (2017). A review of 4D printing. *Materials & Design, 122*, 42–79. doi:10.1016/j.matdes.2017.02.068

Nagar, S. V., Chandrashekar, A. C., & Suvarna, M. (2020). Optimized Additive Manufacturing Technology Using Digital Twins and Cyber Physical Systems. In *Cyber-physical Systems and Digital Twins: Proceedings of the 16th International Conference on Remote Engineering and Virtual Instrumentation*. Springer Nature Switzerland AG.

Noble, D., & Kaminski, C. (2015). Introduction for bioinspiration. *Interface Focus*, *5*(4), 20150052. doi:10.1098/rsfs.2015.0052

Odling-Smee, J., & Turner, J. S. (2011). Niche Construction Theory and Human Architecture. *Biological Theory*, *6*(3), 283–289. doi:10.100713752-012-0029-3

Pei, E. (2014). 4D Printing-revolution or fad? *Assembly Automation*, *34*(2), 123–127. doi:10.1108/AA-02-2014-014

Persiani, S. (2019). *Biomimetics of Motion. Nature-Inspired Parameters and Schemes for Kinetic Design*. Springer International Publishing AG.

Primrose, S. B. (2020). *Biomimetics – Nature-inspired design and innovation*. Wiley Blackwell. doi:10.1002/9781119683360

Raviv, D., Zhao, W., McKnelly, C., Papadopoulou, A., Kadambi, A., Shi, B., Hirsch, S., Dikovsky, D., Zyracki, M., Olguin, C., Raskar, R., & Tibbits, S. (2014). Active Printed Materials for Complex Self-Evolving Deformations. *Scientific Reports*, *4*(1), 7422. doi:10.1038rep07422 PMID:25522053

Reiss, J. (1999). *From 'Margin to Center: The Spaces of Installation Art*. MIT Press.

Rennie, J. (1869). Insect Architecture. Bell and Daldy.

Seepersad, C. C. (2014). Challenges and Opportunities in Design for Additive Manufacturing. 3D Printing and Additive Manufacturing, 1(1), 10-13. doi:10.1089/3dp.2013.0006

Shin, D. G., Kim, T. H., & Kim, D. E. (2017). Review of 4D printing materials and their properties. *International Journal of Precision Engineering and Manufacturing-Green Technology*, *4*(3), 349–357. doi:10.100740684-017-0040-z

Stoychev, G., Razavi, M., Wang, X., & Ionov, L. (2017). 4D Origami by Smart Embroidery. *Macromolecular Rapid Communications*, *38*(18), 1700213. doi:10.1002/marc.201700213 PMID:28759164

Straubel, M., Block, J., Sinapius, M., & Hühne, C. (2011). Deployable Composite Booms for Various Gossamer Space Structures. *52nd AIAA/ASME/ASCE/AHS/ASC Structures, Structural Dynamics and Materials Conference*. 10.2514/6.2011-2023

Tarkhov, D., & Vasilyev, A. (2020). *Semi-empirical neural network modeling and digital twins development*. Elsevier.

Tavakoli, J., & Tang, Y. (2017). Hydrogel based sensors for biomedical applications: An updated review. *Polym.*, *8*(12), 364. doi:10.3390/polym9080364 PMID:30971040

Teoh, J. E. M., An, J., Chua, C. K., Lv, M., Krishnasamy, V., & Liu, Y. (2016). Hierarchically self-morphing structure through 4D printing. *Virtual and Physical Prototyping*. Advance online publication. doi:10.1080/17452759.2016.1272174

Tibbits. (2014). 4D printing and universal transformation. *Acadia 2014 - Design Agency*, 539-548.

Tibbits, S. (2014, January/February). Multi-material Shape Change. *Architectural Design.*, *84*(1), 116–121. doi:10.1002/ad.1710

Tibbits, S., & Cheung, K. (2012). Programmable materials for architectural assembly and automation. *Assembly Automation*, *32*(3), 216–225. doi:10.1108/01445151211244348

Titotto, S. (2013). *Branching morphogenesis: from growth patterns of vascular networks to biomimicry devices* (PhD Thesis). University of Sao Paulo & Politecnico di Torino.

Titotto, S. L. M. C. (2014). Models and prototypes of biomimetic devices to architectural purposes. *SCIRES-IT - SCIentific RESearch and Information Technology, 4,* 55-62. http://caspur-ciberpublishing.it/index.php/scires-it/article/view/11117

Titotto, S. L. M. C., & Dobelis, M. (2017). Teaching CAD/CAM: exchanging experiences from Brazilian and Latvian courses In *Engineering Graphics BALTGRAF-14 Proceedings of the Fourteenth International Conference BALTGRAF Tallinn 2017*. Tallinn: TTU - Tallinn University of Technology.

Titotto, S. L. M. C., Oliveira, D. R., & Ferrante, A. (2015). Space and energy: relationships among architects from nature. *SCIRES-IT - SCIentific RESearch and Information Technology*, *5*, 115-120. http://caspur-ciberpublishing.it/index.php/scires-it/article/viewFile/11425/10618

Turner, J. (2009). *The Extended Organism: The Physiology of Animal-Built Structures*. Harvard Univ Press.

Turner, J. (2011). Termites as models of swarm cognition. *Swarm Intelligence*, *5*(1), 19–43. doi:10.100711721-010-0049-1

Turner, J. S. (2000). Architecture and morphogenesis in the mound of Macrotermes michaelseni (Sjöstedt) (Isoptera: Termitidae, Macrotermitinae) in northern Namibia. *Cimbebasia*, *16*, 143–175.

Turner, J. S., & Soar, R. C. (2008) Beyond biomimicry: What termites can tell us about realizing the living building. *Proceedings of the 1st International Conference on Industrialised, Integrated, Intelligent Construction (I3CON)*, 221.

Vamvakidis, G. (2017). Fabricating Gradient Transparency: A Bio-Inspired Digital Design Model and Prototyping Methods. 3D Printing and Additive Manufacturing, 4(1), 41-52. doi:10.1089/3dp.2016.0050

Vincent, J. (2001). Stealing ideas from nature, Deployable Structures, 51-58.

Vincent, J. (2006). Biomimetics: Its practice and theory. *Journal of the Royal Society, Interface*, 3. PMID:16849244

Yang, C., Wang, B., Li, D., & Tian, X. (2017). Modelling and characterization for the responsive performance of CF/PLA and CF/PEEK smart materials fabricated by 4D printing. *Virtual and Physical Prototyping*, *12*(1), 69–76. doi:10.1080/17452759.2016.1265992

Ye, X., & Lin, H. L. (Eds.). (2021). *Spatial Synthesis: Computational Social Science and Humanities*. Springer.

ADDITIONAL READING

Eid, A., He, X., Bahr, R., Lin, T.-H., Cui, Y., Adeyeye, A., Tehrani, B., & Tentzeris, M. M. (2020, December). Inkjet-/3D-/4D-Printed Perpetual Electronics and Modules: RF and mm-Wave Devices for 5G+, IoT, Smart Agriculture, and Smart Cities Applications. *IEEE Microwave Magazine*, *21*(12), 87–103. doi:10.1109/MMM.2020.3023310

Gatenholm, P. (2015)... *Biomacromolecules*, *16*, 1489. doi:10.1021/acs.biomac.5b00188 PMID:25806996

Jefferson, M. (2015). There's nothing much new under the sun: The challenges of exploiting and using energy and other resources through history. [doi:10.1016/j.enpol.2015.01.022]. *Energy Policy*, *86*, 804–811. doi:10.1016/j.enpol.2015.01.022

Lantada, & … . (2013). Comparative Study of CAD-CAE Programs taking Account of the Opinions of Students and Teachers. *Computer Applications in Engineering Education*, *21*(4), 641–656. doi:10.1002/cae.20509

Lyons, N., & Wilker, M. (2012). *Interactive Project Management: Pixels, People, and Process*. New Riders.

MarkstedtK.MantasA.TournierI.Martínez ÁvilaH.HäggD.;

Nadeau, J.-P., & Fischer, X. (2011). *Research in Interactive Design: Virtual, interactive and integrated product design and manufacturing for industrial innovation*. Springer.

Oliveira, D. R., & Titotto, S. L. M. C. (2015). Biomimetic analysis of Hymenoptera wasp nests. *Journal of Civil Engineering and Architecture Research*, *2*, 1173–1180.

Thérien-Aubin, H., Wu, Z. L., Nie, Z., & Kumacheva, E. (2013). Multiple Shape Transformations of Composite Hydrogel Sheets. *Journal of the American Chemical Society*, *135*(12), 4834–4839. doi:10.1021/ja400518c PMID:23464872

Wigdor, D., & Wixon, D. (2011). *Brave NUI world: designing natural user interfaces for touch and gesture*. Elsevier.

KEY TERMS AND DEFINITIONS

4D Printing: Smart materials and sophisticated product designs that are programmed to prompt the printing object to transform itself into another structure over the influence of external passive energy input such as temperature, light, pH, or other environmental stimuli.

Additive Manufacturing: The processes to turn digital blueprints to physical objects by building them layer-by-layer. The term AM encompasses several technologies and it includes subsets such as Rapid Prototyping (RP), 3D Printing, Direct Digital Manufacturing (DDM), additive fabrication, and layered manufacturing.

Biomimetics: An interdisciplinary field in which principles from biology, engineering, and chemistry are applied to the synthesis of machines, synthetic systems, and materials, or those whose functions mimic one or more biological processes.

Biomimicry: A field whose practices mimic the strategies found in nature to develop technological innovation and solve human design challenges.

Graphical Representation: The use of graphs and charts to visually clarify, analyze, display, and interpret functions, numerical data, and qualitative structures in general.

Hydrogel: Cross-linked polymer chains built as a network and that are hydrophilic. They can be sometimes found as a colloidal gel where water is the dispersion medium. A three-dimensional solid object may result from the hydrophilic polymer chains being held together by cross-links.

Installation Art: A three-dimensional or poli-dimensional visual artwork that is generally created for a specific place and it is designed to alter the space perception.

Chapter 18
Quality of Urban Walking Routes:
Interaction of Knowledge Systems for Integrated Representations

Maurizio Marco Bocconcino
Politecnico di Torino, Italy

Mariapaola Vozzola
Politecnico di Torino, Italy

Anna Rabbia
Fondazione Sviluppo e Crescita CRT Torino, Italy

ABSTRACT

The northern area of Turin is involved in complex processes of transformation and regeneration articulated in a plurality of functions that support social revitalization and housing policies alongside temporary residential and commercial interventions. A specific district of Turin, Pietra Alta, is presented as an application case of the indicators and graphic codes that the methodological approach of the research is gradually outlining. The results, guided and supported by effective modes of representation, broaden the understanding of the hierarchy and criteria of the needs to move independently on foot and confirm that, given a safe and comfortable environment, people seek utility, a sense of belonging, and pleasantness as additional and distinct needs to enhance their experience of living in the city. Using survey data and assuming relief given by geo-referenced interviews, the proposed work emphasizes the integration of residents' perceptions and objective measures to understand the impact of environmental features on the behavior of neighborhood residents.

DOI: 10.4018/978-1-7998-7091-3.ch018

INTRODUCTORY AND METHODOLOGICAL ASPECTS

Walkability is part of a broader research area that examines the connections between land use and transport and in particular how the two can be used to create more liveable communities. Much of the literature on land use/transport focuses on urban design features such as sidewalks, landscape enhancement, availability of public transport and the number of roads that can be used to influence our transport choices. But walking, like any other human behavior, is also largely influenced by cultural factors, circumstances, individual preferences and characteristics, and environmental factors. In recent decades, numerous studies in the field of urban planning and public health have identified key features of the built environment associated with active transport, physical activity and related health outcomes, such as lower rates of obesity, cardiovascular disease, diabetes and stroke. Environments built to support walking, cycling, are predictive of a higher proportion of active mode trips and higher rates of walking or physical activity. Researchers in public health, medicine and urban planning have focused on identifying individual and group characteristics and environmental factors that influence walking as physical activity. Many scholars have addressed the issue of urban quality by considering walkability, i.e. the quality of accessibility considered in relation to pedestrian mobility, as an important requirement.

General Purposes

Walkability is considered a key factor in promoting health and environmental objectives. In the field of health, walkability is associated with the goal of the "active community", while in the sustainability literature, walking is seen as a way to potentially reduce the ecological footprint, minimise car travel, reduce energy consumption and limit the invasion of unused land. Recent studies on pedestrians claim that it is the single most important factor determining healthy and vibrant urban places. While it is important to recognise that there are limits, such as the fact that pedestrian areas may not be fully accessible or compromised by crime and other factors. The ability of urban areas to be walkable, i.e. suitable for pedestrians, is now recognised as a central aspect of sustainable cities. Based on social equity, the definition of pedestrians should be broadened to include those who use different aids to move, as required by common sense and current legislation. Walkability is a concept that is becoming increasingly established in urban quality and environmental well-being assessments as a means of developing more liveable communities. Pedestrivability is measured in qualitative and quantitative terms by surveying a variable set of attributes of the built environment, the indices are generally constructed using existing and easily detectable data. The identification of the network of "passable" paths is an essential prerequisite for the analysis of walkability. The network can be used to deduce access to destinations, walking comfort and pedestrian safety, all central aspects of the walking concept. The level of accessibility of an area is largely influenced by the characteristics of the built environment, such as pedestrian services, the mix of urban and territorial functions and proximity to destinations, and is also influenced by individual characteristics such as the level of familiarity with the area and demographic as well as gender factors. With the help of geographic and building information systems and other advanced technologies, researchers can now parametrically model, measure and understand the physical and environmental characteristics that affect walkability. The walk score is generally considered by academics to be a reliable measure for the walkability of the neighbourhood, but it has also been criticized in some peer-reviewed studies for being too simplistic and for neglecting to incorporate many elements considered associated with walkability. The quality of the pedestrian routes is assessed based on various attributes relevant to the walkability, linked to the characteristics of the roads and the surrounding environment that contribute to making the route pleasant, safe and attractive. What the concept of walkability can grasp is, in fact, the quality of accessibility: how and to what extent the urban environment is able to facilitate walking and to offer itself as a platform for daily life centred on pedestrian mobility. Walkability should be understood not only as an objective to be pursued to improve the quality of life of the inhabitants, but also as a tool that designers and architects can use to reorient methods of assessing the quality of urban life in smart cities. The concept of walkability is proposed here as one of the primary factors of urban skills and qualities.

Urban planners are concerned about the environmental qualities that make the best places to walk, not only as healthy physical activity but also for the sensory and experiential pleasure that can derive from such environments. While environmental planning and design have to address various levels of walking needs, aspects of the environment that affect the sensory and social qualities of the environment are particularly significant for the field of urban quality assessment. Survey and representation of walkability in urban spaces is the subject of the contribution we propose, within the complexity of the evaluation processes and the broad landscape of measurement tools; the research is drawn up around the role of indicators in describing architecture, city and territory for the evaluation and monitoring of strategies aimed to assess urban quality starting from planning walkability. Walking research abounds, as said, in qualitative studies that deal with microscale analysis of the environment, even supported by very advanced analysis tools, but often do not consider the spatial component of the environment itself, focusing more on the reticular distribution of paths and interferences.

Methodological Approach

This paper proposes an empirical examination of the relationship between physical characteristics, the use of the city and the social characteristics of the microscale environment and people's behaviour and perception of walking. In addressing the quality of urban space on the microscale of the environment, the proposed chapter will investigate tools for measuring and representing the perception and effects of safety, well-being and the ways and modes of walking in that part of the city between urban and urban environment and what for a long time has represented the "rural countryside" where green areas are prevalent, also of a natural character and connected to watercourses, but the connections between these spaces and inhabited places are still not consolidated. Using survey data and assuming a survey from geo-referenced interviews, the proposed work emphasizes the integration of user perceptions and subjective measures to understand the impact of environmental characteristics on the behaviour of walkers on main and secondary roads at the neighbourhood level. Compared to previous research, this study demonstrates the importance of social qualities in supporting walking and how the interaction of different computing environments for network analysis and the distribution of volumes and functions of the buildings provides significant evaluation elements. The results, guided and supported by effective modes of representation, broaden our understanding of the hierarchy and criteria of needs to walk independently and confirm that, given a safe and comfortable environment, people seek utility, a sense of belonging and pleasantness as additional and distinct needs to improve their experience of living in the city's neighbourhoods.

Our research in progress starts from objectives achieved in previous researches, briefly presented as a reference framework, and proposes a method to investigate and represent what best describes the levels of urban walkability, related to urban quality and environmental comfort.

Figure 1. Pietra Alta - Planimetry of the study area: the Pietra Alta district. Inside the planimetry, elaborated in GIS environment, are represented the different types of walkways: pavements and driveways; the different types of buildings: building and low building; green and trees; bicycle paths; parking and bus stops.

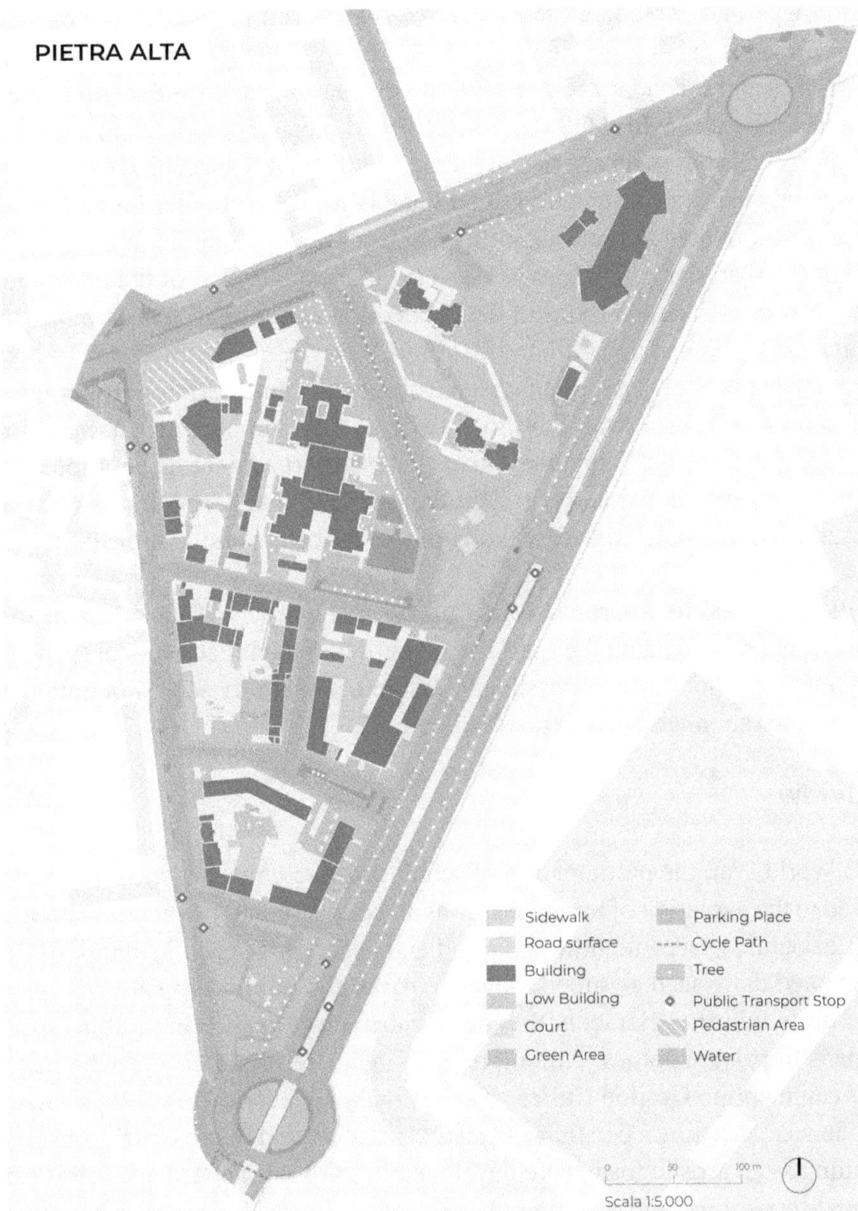

The literature regarding tools to support analysis of walkability is divided into three categories: conceptual frameworks for the study of walkability, non-GIS methodologies and GIS-based approaches. In this proposal we consider the fourth approach, WEB/GIS/BIM strongly oriented to graphic representation as an analysis and design and project tool. We believe the study presented contributes to the literature on the measurement of walkability by proposing a hybrid auditing method that combines the efficiency of a Geographic Information Systems (GIS) and Building Information Modeling (BIM) approach with web-based data logging integrated into OpenStreetMap (OSM), a collaborative project aimed at creating free content world maps, and in Google Street View (GSV), a well-known feature of Google Maps and Google Earth that provides panoramic views along the streets. The document first reviews the various methods of face-to-face and secondary source auditing that have been developed over the past decade. It then verifies the feasibility of a hybrid OSM/GSV/GIS/BIM method compared to conventional methods of analysis. The approach is based on a series of elements objectively measured as quickly available and accessible, subsequently using a combination of GIS and BIM in a particular field of application: that of urban quality assessment related to the walkability of places and its graphic representation as a tool for analysis and design of urban space. The aim is to see to what extent some of the relevant methodologies used could be made more efficient in terms of data collection and storage. The combination of information from OSM and GSV and updated GIS and BIM levels can be an effective way to obtain data on the physical environment that strongly integrates personal observation and traditional photographic detection.

The northern suburbs of Turin (Italy) are today affected by complex urban transformation and regeneration processes articulated in functional mixes that support social regeneration and housing policies alongside temporary residential and commercial interventions. A specific district of Turin, Pietra Alta, will be presented as an applicative case study of the indicators and graphic codes identified in the methodological approach.

The study will be applied to a particular area of Turin (Figure 1). The results of the study can be further analysed by importing data on the spatial distribution of the inhabitants in space to be combined with the viability maps. In this way it is possible to assign a passability score not only to the cells in the urban space, but also to the inhabitants living there.

Literature Review

Since the Second World War, the participation of scholars, researchers and citizens in urban design has been closely linked to the evolution of urban design as an activity mainly concerned both with the design and management of public space and how it is lived and used. This space includes all the places in the city that are frequented daily such as streets, squares, parks and public services.

Urban design deals with the elements that can contribute to qualifying public space, such as the facades and gardens of private homes. Among the exponents of urban design who have dealt with its participation, we can mention Gordon Cullen, Kevin Linch, Giancarlo De Carlo. Cullen addressed the themes of the urban vision, setting a peculiar aspect that involves and impresses the observer. The themes of sight and planning as an artistic form were developed in the book Townscape (1961) and later in The Urban Landscape. Morphology and planning (1976). Lynch focused his research activity on the study of people's perception of the urban landscape. His scientific contributions, as well as representing a turning point for urban theory, range in a wide conceptual field, from environmental psychology to the geography of perception. Lynch's most famous book, The image of the city (1960) is the result of an investigation into the way city visitors perceive urban space and organize spatial information during their experiences. Finally, De Carlo was among the first to experiment and apply the direct participation of users in the design process in architecture. The state of the art of research on the human perception of the built environment is based both on questionnaires and field observations of people interacting with the environment, and on the use of more frontier methods, to investigate the relationship between spatial characteristics and the senses through tools that collect and process relevant data, including various sensors and social media, to investigate what types of places people find most comfortable (Girardin et al. 2009, Resch et al. 2015).

The study of the city is not new to the introduction of science and advanced research fronts. Following the increasing availability of this type of resources in the last decade, social sciences have introduced quantitative parameters and indicators, often labelled as computational social science (Lazer et al. 2009), related to environmental perception, feeling, emotion, social connection, previously limited to qualitative modes of investigation (Moretti 2016). In particular, the emerging network science that studies complex networks has made a significant contribution to field research (Borner, Sanyal and Vespignani, 2007). Network science takes an exclusively spatial perspective on urban data, focusing on the relationships and interactions between people, places, institutions, at different scales of reading. Manuel Castells (1996) introduced in urban studies the concept that by abstracting cities as social-spatial networks of interaction, network science helps to unravel common structural elements shared by most urban systems, allowing to set up cognitive and predictive models of development (Batty 2013). Infrastructuring the urban fabric

Figure 2. Sitte. An excerpt of some representations from architect Camillo Sitte's book, Der Städte-Bau nach seinen Künstlerischen Grundsätzen (1889), in which the quality of urban space and its perception is emphasised (Sitte, 1889)

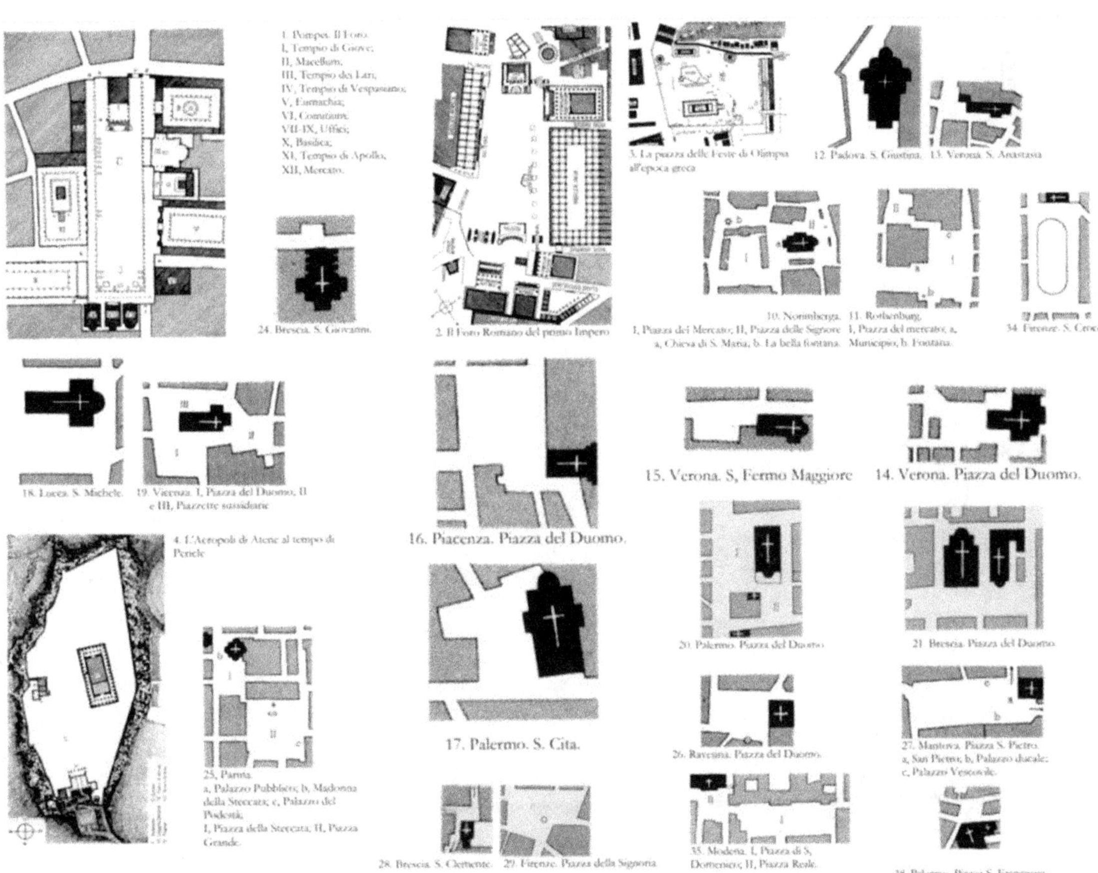

with sensor networks has opened up to real-time representations of the state and condition of places. The growth of social media is leading to new forms of participation and activism, alongside the traditional forms of participation in the city's project; citizens voluntarily play monitoring and reporting roles, a phenomenon that has been defined as the increase of the "expert amateur" (Kuznetsov and Paulos 2010).

The term "walkability" is not new, but has been introduced by many scholars already in earlier times. We can already find the first intuitions in the studies of the architect Camillo Sitte, who already in 1889,

Figure 3. Lynch. An excerpt of some representations from architect Kevin Lynch's book, The view from the road (1964). In his text, Lynch showed that people perceive the urban space they inhabit through common elements and mental schemes, the main ones being routes, i.e. streets, places where people can walk and all the environments people use to move around the city (Lynch, 1964).

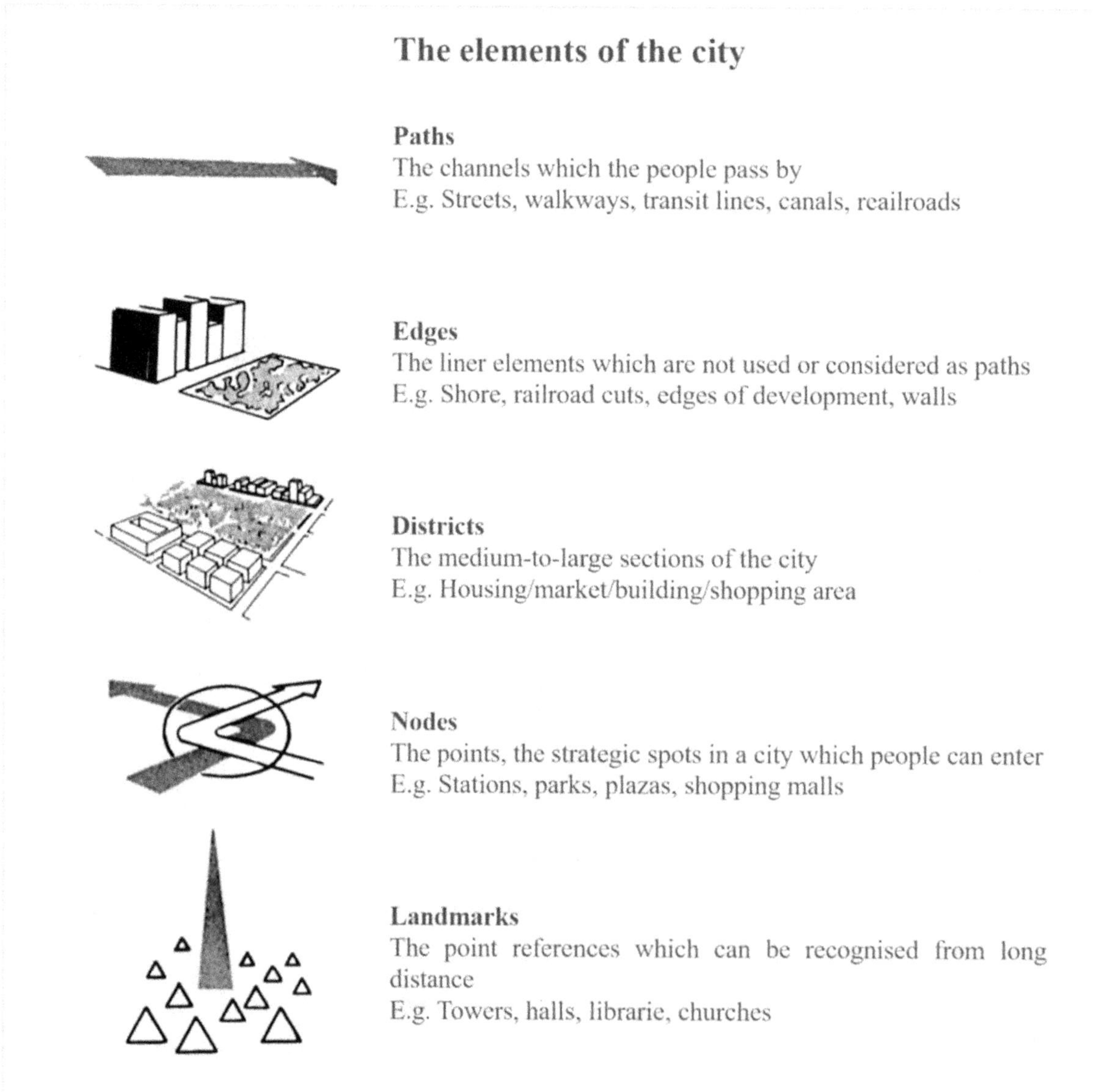

in his treatise Der Städte-Bau nach seinen Künstlerischen Grundsätzen, stresses the need to design and build cities by referring to the quality of urban space and its perception (Sitte, 1889) (Figure 2).

From the second half of the twentieth century, it is possible to find other important urban planners who put the quality of perception of urban space at the centre of urban design, first of all, Kevin Lynch, who in 1960, in his treatise The Image of the City, underlines how the set of urban routes, understood as a network of habitual lines of movement within the urban structure, define the routes that today could be better defined as "prevailing" routes, or "the most potent means" (Lynch, 1960) for the redesign and re-design of the city as a whole (Figure 3). In 1961, Jane Jacobs introduced in the first part of The Death and Life of Great American Cities her critical and innovative vision for urban planning: guiding her choices on the use of streets and paying particular attention to the function played by the pavement (Jacobs, 1961).

Some Issues Related to Case Study of Pietra Alta in Torino

The survey and the urban project see illustrious references in the research; in particular, significant perceptual, visual approaches related to the reading of the urban space have been highlighted through authors, scenes collected apparently in an extemporary way, in spurts, dense with layers of reading, as well as analytical approaches, which break down the image of the city and operate synthesis at different levels of interpretation. The attention of the research group is also directed to the codes of representation, the graphic language that gives access to knowledge and allows the conception and realization of conscious projects.

The urban survey is a very lively field of research, sensitive to the innovation of methods and tools of investigation. Access to an ever-increasing amount of data allows us to answer complex questions about the use of urban space.

Within this context, characterised by such a complex and at the same time fragmented reality, a Social Housing project was implemented in 2011, aimed at creating a place of participation and meeting between different generations and cultures. The project converted a former Italian Post Office building into a structure intended for various forms of social housing. Sharing24, in Via Ivrea n. 24, therefore represents, for the neighbourhood and the city, a starting point for the enhancement of the Pietra Alta neighbourhood, where residents and tourists can experience moments of sharing and reciprocity. Hence the need to understand how to extents this intervention has improved the perception of the area and how the urban fabric has responded to the new redevelopment.

The district to the north of Turin, Pietra Alta from the name of a Roman milestone that indicated the distance from the city, is also physically delimited by large infrastructures, the railway and the large avenues together with the river Stura to the south. The most significant elements are the presence of green areas, in particular the Parco Stura Nord to the south, the worker's village SNIA Viscosa to the north, the towers of Vittorio practically at the entrance of the freeway to and from Milan, numerous public transport lines including metropolitan rank, there is the railway station Stura and an interchange parking lot to the north of the district.

This is a place where several projects of the City of Turin related to the recovery and urban redevelopment have tried to correct the many layers not coordinated in a unitary design. The different phases of growth of the village and also the mixture of public and private buildings are among the main reasons for this heterogeneity. The settlement of new centres for large commercial distribution has had important repercussions on small local commerce. The building fabric is sparse, not very compact, not homogeneous in fact.

This area has been chosen as the starting point of wider work on the social impact of urban regeneration interventions, as it represents the first case of social housing in Turin (2011) and therefore boasts a very significant time horizon of analysis. Not only. It also still represents one of the most important interventions on housing policies in terms of social quality, both in terms of investment and in terms of the peculiarity of the manager and the mix of functions installed.

In the illustrative apparatus, a series of indicators have been reported, and the literature illustrates several of them. At the first level of reading, we have stuck to those that can be considered most recognized. New ones are emerging and derive from new possibilities of access to data, numerous, correlated to the use of public space (for example, real-time recordings on the example of what happens for vehicular traffic, but also asynchronous readings of the places, very useful in periods of closure as the present ones, which are repeated).

The quality of urban paths, with synthetic term walkability in the Anglo-Saxon literature, walkability, its assessment, includes estimation approaches that are defined as objective, others that are more properly subjective and must be weighted. The work presented here tries to incorporate both methods by trying to combine perceptual elements and therefore variable because of the subject, the pedestrian, the person (which will be collected through questionnaires) and objective parameters derived from direct readings (or deferred as we said) and information models of the digital type where different technologies interact for data management.

The study of the accessibility of public spaces and urban walkability is part of a broader research area that examines the connections between land use and mobility and in particular how the two can be used to create more livable communities. Walkability is also considered a key factor in promoting health goals and is associated with the goal of "active community"; in the sustainability literature, it is seen as a way to potentially reduce the ecological footprint, minimize car travel, reduce energy consumption, and limit the occupation of undeveloped land. Its quality can effectively result in healthy and vibrant urban places. It is a concept that is gaining traction and frequency in social and environmental well-being assessments, particularly as a means of developing identity and place recognition.

In the context of the complexity of assessment processes and the wide range of measurement tools, research, which should be significantly completed next year, is developing around the role of indicators in describing urban and social quality for its ex-ante evaluation and monitoring.

Transformation and regeneration processes are complex, and it is complex to articulate them in a plurality of functions that support local revitalization and housing policies alongside social infrastructure interventions.

STATE OF THE ART

The level of walkability in an area is largely influenced by a great number of variables of built environment features like pedestrian amenities, land-use mix and proximity to destinations, and is also affected by individual characteristics like the level of familiarity and known with the area, and by demographic factors such as gender. With the aid of Geographic Information Systems (GIS) and other advanced technologies, researchers can now measure, model and understand the physical and built-environment features that influence walkability better than ever before (Leslie et al., 2007).

But what does 'walkability' really mean? And what does it refer to? When analysing the definitions in the literature, some authors refer to comfort, safety or a visual experience[1] (Southworth 2005 and Tsiompras and Photis 2017). However, the environment in which the citizen walks should not be considered exclusively as a transit area, an area through which one can reach a certain point of interest, but rather create an opportunity to stay and perform different types of activities. Based on this understanding of the term, walkability can be defined as the extent to which the environment considers the various needs of pedestrians from an urban perspective.

The first scientific articles related to walkability appeared in the early 1990s in the field of urban design and development and focused on the assessment of the quality of the urban environment and suburbs (Southworth and Owens 1993, Southworth and Ben-Joseph 1995 and Southworth 1997). The existence of a direct link between the quality of the pedestrian environment and people's willingness to choose walking as their main mode of transport has become increasingly established over the years, finding numerous findings in international research (e.g. Ariffin and Zahari 2013 and Rafiemanzelat et al. 2017).

One of the most important researches in this area was that promoted by Saelens et al. (2003), which examines a relationship between individuals' perception of the environment, physical activity and body weight. Also significant were the studies promoted by urban planner Lawrence D. Frank with psychology researchers. They examined a relationship between urban form and the physical activity of its inhabitants (Frank et al. 2005), neighbourhood mobility, BMI (Body Mass Index), active transport modes and air quality (Frank et al. 2006), as well as neighbourhood walkability, car use and obesity (Frank et al. 2007).

In all the researches, there are presented three main approaches to measuring walkability: objective, subjective and their combination. An objective approach operates with various types of models such as representation 2D and 3D model (Cervero and Kockelman 1997) or as commercial projects such as Walk Score (Kocher and Lerner 2007).

A subjective approach, on the other hand, focuses on the perception and other behavioural variables of individuals, and it is focused on the urban contest quality. The most used approach is a combination of the previous two. Parallel, the literature is divided into different categories of tools and representation for the study of walkability: from conceptual frameworks for the study of the walkability index, to non-GIS methodologies or GIS-based approaches, from which the first investigations at territorial level originated, until arriving in recent years to the new CIM models, understood as City Information Models, where it is possible to analyze and visualize the city through spatial queries represented directly within the Smart City Model. As an example of conceptual frameworks, we analyze the frameworks introduced in McMillan's studies, such frameworks can be used for the reading of the urban form and at the same time for the measurement of the pedestrian accessibility of a neighbourhood according to the users who travel the streets (Duncan et al., 2012). One of the conceptual frameworks proposed by McMillan deals with a daily action of families living in the neighbourhood: through the reading of a child's home-school route, it is possible to relate the urban form of the urban fabric to factors such as the safety of the streets and the city's public transport services available. The safety of a street, through the reading of the illumination index, and therefore the presence of an adequate number of streetlamps, can influence the choices of a family: adequate street lighting can encourage families to walk the home-school and school-home routes; on the other hand, inadequate lighting, which does not give safety to parents, can encourage parents to take their children to school by car. The urban form is therefore not considered to be the direct cause of the child's journey to school (McMillan 2005), but the parents' decision-making process is the real cause, and this process is influenced by elements linked to the assessment of the safety

Figure 4. Alghero. Three representations for the Walkability Scores related to three categories of destinations: commercial, services and green and recreational (leisure) areas (Blečić et al, 2015)

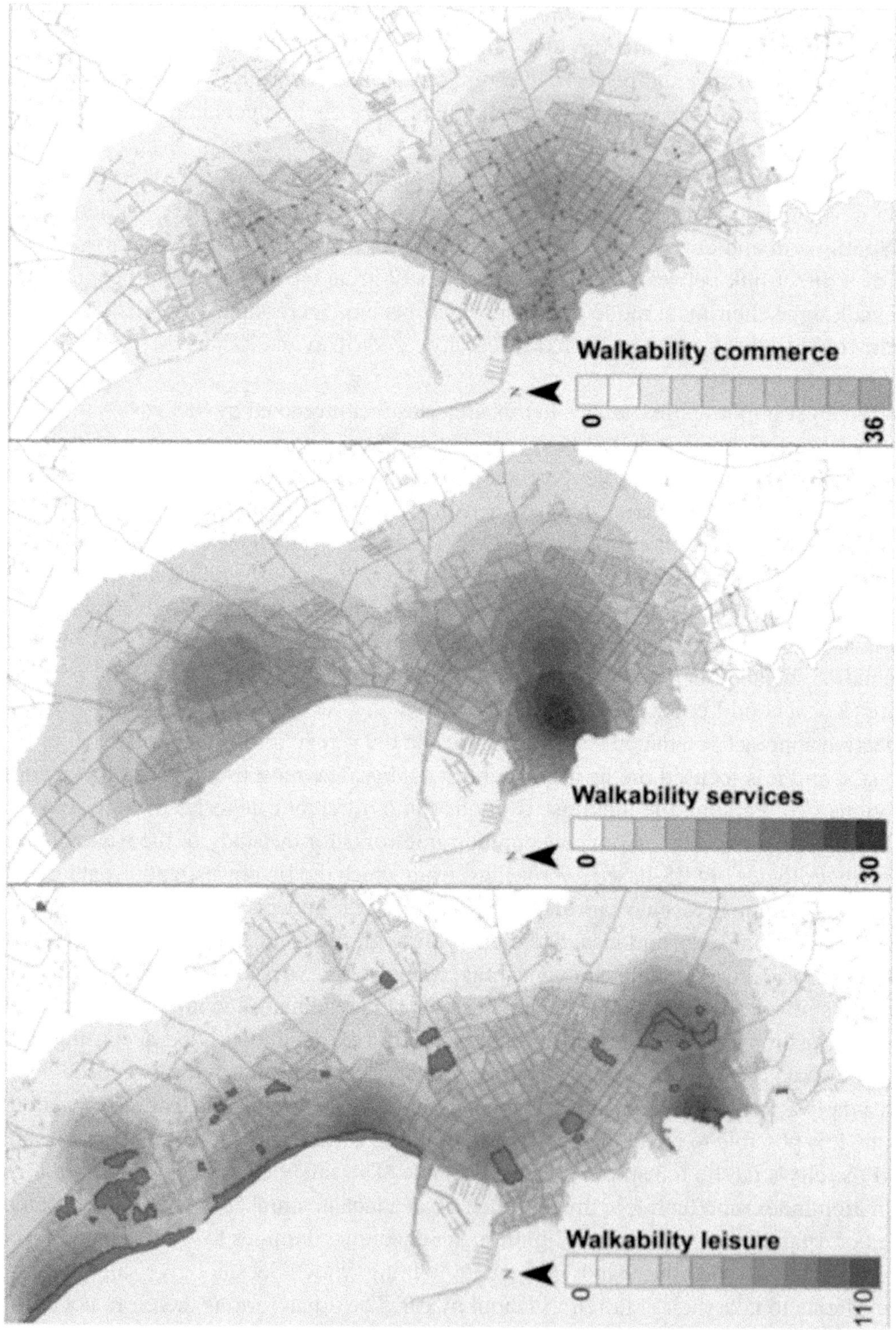

of the neighbourhood, which in turn is assessed according to the number of street lamps along the route and the level of street lighting (McMillan, 2005).

Some scholars, on the other hand, have introduced the issue of walkability as an important requirement for the assessment of urban quality. Taking a look at the national context, one of the spatial and multi-criteria models aimed at measuring the quality of paths is the one presented by Ivan Blečić (Blečić et al, 2015) (Figure 04). The proposed methodology is based on the assessment of urban walkability and the presentation of the results obtained from the computerised assessment tool - Walkability Explorer WE - developed for support during the decision-making phases of urban planning. Within the Walkability Explorer work environment, it will be possible to consider not only the data related to the physical characteristics of the analysed routes but also qualitative indices, such as the sense of welcome, the liveliness of the space and the sense of accommodation, and to elaborate ad hoc maps for different targets: tourists, students, families, etc...

Figure 5. The Prince's Foundation. The 'Walkability' heat maps describe the degree to which newer settlements have succeeded, or not, to emulate the lessons of historical places such as Clifton in Bristol (The Prince's Foundation, 2020)

Recent international studies include the report Walkability and Mixed-Use. Making Valuable and Healthy Communities, developed by The Prince's Foundation. The report aims is to highlight the importance of the score related to the walkability of the places to be designed: to realize environments livable by users only through pedestrian paths, will improve the quality of life of citizens and to realize cities on a human scale, making it possible that the non-residential space is both convenient and flexible, so that it can adapt over time to a wide range of uses (The Prince's Foundation, 2020). The main objective of the research is to create master plans where the use of the car for daily trips - home|work home|school home|shopping - are reduced to zero, and where instead of all services - office, school, bus stop, stores - are accessible through pedestrian paths or bike lanes. Through the elaboration of walkability heat maps, we describe how the new urban centres have managed to achieve the goal of making available to the inhabitants the services of first necessity within 5 to 10 minutes of walking (Figure 5).

Perception of Walkability and Urban Quality

For walkability analysis using a non-GIS methodology, an example is represented from the questionnaire developed and proposed by the Partnership for a Walkable America: the questionnaire consists of five simple questions - Did you have room to walk? Was it easy to cross streets? Did motorists drive well? Was it easy to follow the safety rules? Was your walk pleasant? How does your neighbourhood stack up? - with scores ranging from 5 to 30 points (Pedestrian and Bicycle Information Center, et al., 2005). From these questions are possible to have a photography of the perception of the Citizen about the city and the pedestrian areas and know where it is necessary to improve.

The first studies that used GIS to measure the walkability and accessibility of some streets were carried out in 1997 by Aultman-Hall, Roorda et al.: their research aimed to demonstrate the possibility of creating suburban neighbourhoods with accessible routes for pedestrians, by measuring walking distances between one's home and sensitive places, such as schools and public spaces. Later, in 2001, Randall and Baetz, based on the studies of Aultman-Hall, Roorda et al, deepened the analysis by associating the variable distance with the variable 'directivity' of the pedestrian route, to assess the walkability of the neighbourhood. Pedestrian route directivity (PRD) has been defined as a ratio between the distance of the route and the geodetic distance (straight line) (Randall and Baetz 2001). A ratio of 1.0 would be an excellent path. Using these two measures, Randall and Baetz (2001) developed an ArcView GIS extension called PRD Evaluate that calculated both path distance and directivity for a neighbourhood. Also in the early 2000s, Talen (Talen, 2003) introduced in its evaluations related to the walkability of neighbourhoods, not only directly measurable data, such as distances and linearity, but also data related to the perception of places, the quality of the environment, safety and services that can be reached within the neighbourhood. Based on studies introduced by Handy and Clifton in 2000. Handy and Clifton's study involved a detailed analysis of land use, with particular emphasis on the location and reach of commercial establishments, the types of services available to inhabitants and the individual's choice. This study aimed to determine whether there is an equal distribution between neighbourhood residents and retail establishments (Handy and Clifton 2000), showing that more traditional neighbourhoods are more accessible than conventional ones.

EMPIRICAL ANALYSIS OF WALKABILITY USING INDEXES

Quantitative and Qualitative Element

The introduction of indicators representative of the urban and social evolution of a territory in which new interventions with a mix of functions, including social housing, arise is one of the starting points to highlight changes in the form and identity of the place. The urban quality ought to be assessed through a cognitive matrix, which represents a complete picture of the state of the art and outline possible future

Table 1. Analysis of current status reading indexes for walkability near the Social House Building Sharing24 in Turin.

TYPE	INDEX	INDEX DESCRIPTION
EDUCATION	Level of accessibility of educational services for children	The index describes the incidence: a. number of residents aged 0 - 2 years living within 300mt from a nursery (public or private) out of the total number of residents in that age group; b. number of residents aged 3 - 5 years living within 300mt from a pre-school (public or private) out of the total number of residents in that age group.
	Level of accessibility of basic school services	The index describes the incidence: a. number of residents aged 6 - 10 years living within 500mt from a primary school out of the total number of residents in that age group; b. number of residents aged 3 - 5 years living within 700mt from a secondary school out of the total number of residents in that age group.
	Presence of infrastructure for slow mobility on home-school routes	The index describes the presence of spaces and paths for slow mobility (pedestrian and/or bicycle) within a significant area - 500mt - of the school structures present in the territory (nursery schools, pre-schools, primary schools, secondary schools).
SOCIAL & HEALTHCARE FACILITIES	Level of accessibility of facilities for social and health services	The index describes the incidence of the population residing within 1.000mt and 2.000mt from a structure intended for social and health activities.
SPORTS FACILITIES	Level of accessibility of sports equipment	The index describes the incidence of the population residing within 1,000mt and 2,000mt from a recreational/sporting facility.
CULTURAL EQUIPMENT	Level of accessibility of facilities for cultural activities	Theindex describes the incidence of the resident population within 1.000mt and 2.000mt from a structure intended for cultural activities on the total number of residents.
ROAD NETWORK	Level of functional adequacy of the road network	The index interprets the level of adequacy of the road network from a functional geometric point of view. Specifically, the actual size of the section of each road segment unitarily identified by the graph is compared with the ideal size attributed to the segment itself following the functional classification of the roads operated pursuant to the Highway Code (Legislative Decree 285/1992 and Ministerial Decree 05.11.2011).
	Level of qualitative adequacy of the road network	The index interprets the level of adequacy of the road network from a qualitative point of view. Specifically, each road segment identified by the graph is verified in relation to the presence of ecological and infrastructural equipment (road masts and spaces for slow mobility).
	Street lighting	The index assesses whether or not the road illumination present is sufficient.
	Street cleaning	The index assesses the level of cleanliness of the roads, differentiating pedestrian routes from driveways and cycle paths.
	Level of road maintenance	The index assesses the level of maintenance of the roads, differentiating between pedestrian and bicycle paths.
PUBLIC & LIGHT TRANSPORT	Population served by public transport service stops	The index describes the share of the population, out of the total, living within 300mt from a public transport stop.
	Urban transport quality	The index provides an indicative target that includes an evaluation of the journey time from school to school, from home to work. The index shows the number of public transport available (bus, tram, underground).
PUBLIC & LIGHT TRANSPORT	Cyclopedonality Index	The index describes the linear extension of paths and cycle/pedestrian spaces available in the consolidated city for every 100 residents.
	Continuity of the cycle network	The index describes the degree of interconnection of the cycle network. Specifically, it measures the incidence of the real cycle network in relation to the ideal cycle network.
	Pedestrian Road Index	The index describes the linear extension of routes and pedestrian spaces available in the consolidated city for every 100 residents.
	Reaching the city centre	The index describes the ease with which the city centre can be reached.
	Roads to get out of the neighbourhood	The index represents the number of streets with which it is possible to leave the neighbourhood.
	Population served by car sharing	The index describes the share of the population, out of the total, that can use car sharing in the neighbourhood.
PUBLIC SPACES	Incidence of outdoor public spaces used as squares or meeting places	The index describes the incidence of outdoor public spaces used as squares or social gathering places on the total number of publicly owned spaces.
	Possibility of using social gathering spaces	The index is borrowed from the set of European Common indexes (ECI): its definition refers to the concept of accessibility as "a quarter of an hour walk" by ISTAT and the European Environmental Agency (this corresponds to %00mt on foot for an elderly person, 300mt as the crow flies).
	Adequacy of outdoor public spaces used as squares or meeting places	The index describes, through a survey sheet, the quality status of public outdoor spaces, monitoring some specific aspects also in order to plan requalification interventions.
	Adequacy of parking spaces	The index verifies the presence of an adequate number of parking spaces.
	Permeable car parking spaces	The index verifies the incidence of car parks built with permeable pavements on the total number of new car parks built.
PUBLIC GARDEN	Usable urban green	The index describes the per capita extent of green areas that can be used in urban areas.
	Total green areas	Surface area of the different green areas on the total municipal surface area.
	Green area functionality	The index expresses in percentage the presence of the different functional areas present in the municipal green areas: a. children's area; b. seniors' area; c. barrier-free routes .
	Aesthetic quality green area	Historical value, floristic variety, presence of water, context and background.
	Security	Safe pedestrian access, safe cycling access, fence.
	Services and furniture	Drinking water, bicycle parking, benches and baskets, lighting.

continued on following page

Table 1. Continued

TYPE	INDEX	INDEX DESCRIPTION
PUBLIC GARDEN	Maintenance	Turf, paths, benches and baskets, lighting, cleaning.
BUILDINGS	Building density I Full and empty balance	The indicator describes the density of the buildings inside the lot: i.e. the presence of workshop space between the houses, representing the balance between built areas and voids.
	Building aesthetics I Potentially degraded buildings	The indicator reproduces the territorial distribution of residential buildings that are not used continuously and/or permanently in relation to the presence of aggregation spaces.
	Aesthetics of buildings I Degraded walls, soaked with writing and drawings	The indicator reproduces the territorial distribution of the perimeter walls of the buildings and of the lots that have writing or murals that are not pleasing to the eye.
	Building aesthetics I Building colour	The indicator reproduces the distribution of residential buildings with light coloured facade surface finish in relation to the totality of the buildings.
	Volumetry of buildings I Size of fronts on the street	The indicator reproduces the ratio between the length of the continuous front on the street and the height of the front itself, in relation to the width of the street on which it is located.
	Historical buildings recovered and/or restored	The indicator describes the incidence of historical buildings subject to total recovery and/or restoration of the historical heritage to be subject to recovery.
	Regeneration interventions in built-up areas	The indicator verifies the renewal trend of the building heritage.
COMMERCIAL BUSINESSES	Variety of shops	The indicator shows the number of categories of shops in the neighbourhood.
	Distribution of commercial activities	The indicator represents the distribution of commercial activities within the neighbourhood.
	Businesses in activity	The indicator describes the liveliness of the commercial/production fabric of the neighbourhood.
	Access to shops selling basic necessities	Number of residents living within xxxmt of shops selling basic necessities out of the total number of residents in the neighbourhood.
	Potentially degraded commercial spaces	The indicator describes the state of health of the commercial spaces in the neighbourhood, verifying their state of decay through the census of empty commercial spaces.
SOCIAL HOUSING	Covering the demand for social housing	The indicator describes the incidence of the demand for social housing that has been met compared to the overall demand. The indicator is calculated on residential housing supply only (ERP).
	Impact of the SH offer	The indicator describes the incidence of the number of Social Residential Housing (ERS) on the total housing stock in the neighbourhood.
	Increasing SH assets through interventions on existing assets	The indicator describes the incidence of the number of Social Residential Building (ERS) dwellings resulting from interventions on the existing housing stock, compared to the total number of new ERS dwellings.
WALKABILITY	Pedestrian traffic I Sidewalk	Width of the pedestrian section; Flooring I maintenance quality; Slope; Separation of pedestrian area from driveway area; Crossings with slides.
	Road system	Number of driveways; One-way street; Flooring I maintenance quality; Speed limit; Slope of the route.
	Quality roads	Stop/parking along the road; Presence of activities and services; Presence of sessions along the route; Presence of elements for shadow realization; Night lighting.
	Cycling	Presence of cycle path; Slope of the route; Flooring I maintenance quality.

improvement scenarios. Walkability is quantified area-wise by measuring a varying set of built environment attributes: approaches to studying walkability, indices are generally built using existing and easily accessible data (Hollenstein D. et al., 2016).

In order to be able to read the degree of the walkability of the area, some significant indices (Table 1) have been introduced, to analyse both qualitative and qualitative aspects of the urban fabric surveyed. To read the relationship between the urban fabric and the pedestrian, we have introduced some significant indexes, to analyse both qualitative and qualitative aspects of the urban fabric. To define a set of indicators that can characterise and define in detail the peculiarities and peculiarities of the studied area, some case studies of walkability analysis have been analysed[2], both nationally and internationally, from which some indices have been deduced and from which new ones have been integrated, born from the needs of the area under analysis. The table shows the indicators that have been defined to qualitatively assess the area. Later we will also introduce some representation encodings related to them.

Figure 6. Quality of Walkability - Map of the Pietra Alta area with representation of the Quality of Walkability

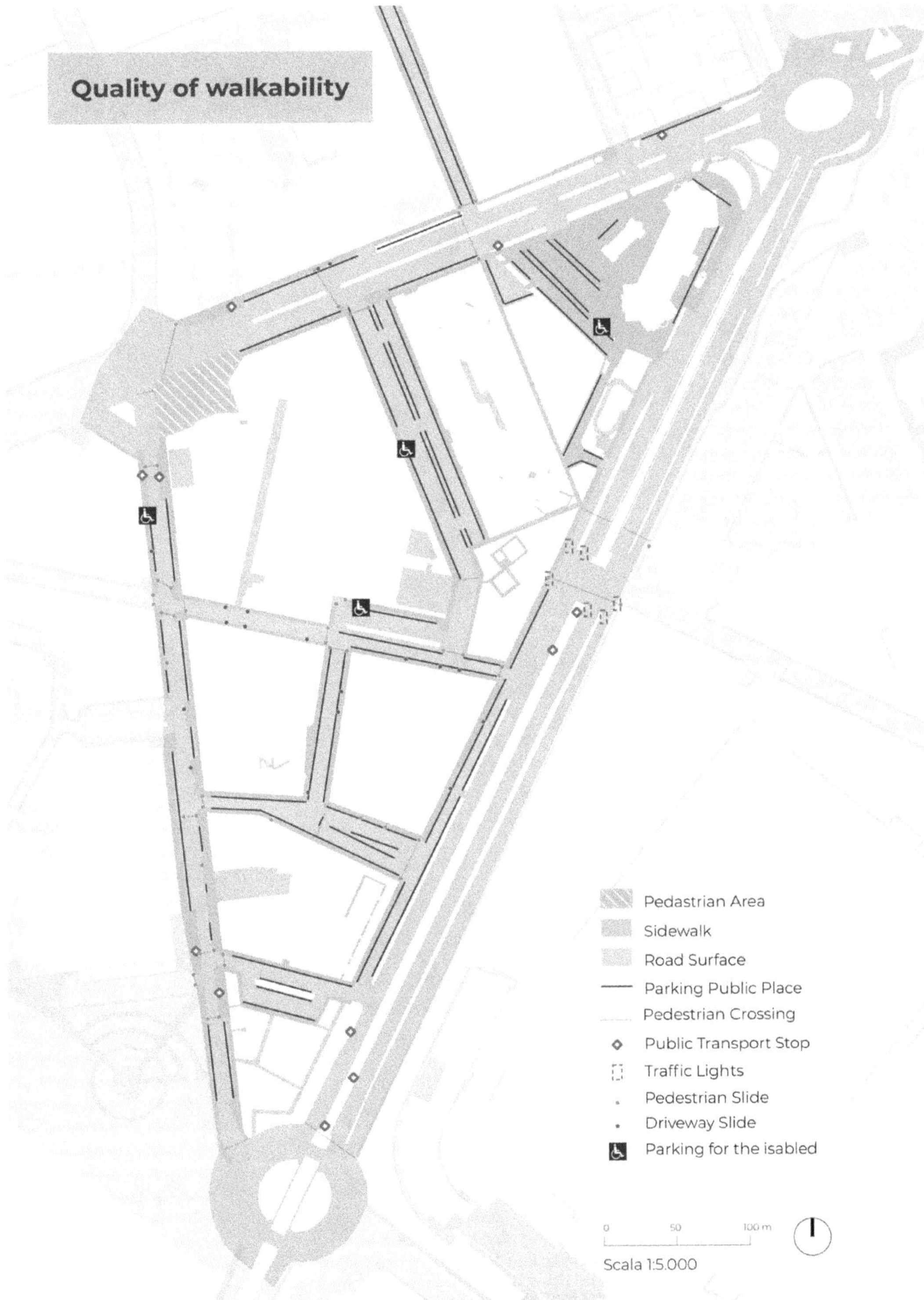

Figure 7. Waterproof Mante - Map of the Pietra Alta area with representation of the Waterproof Mantle

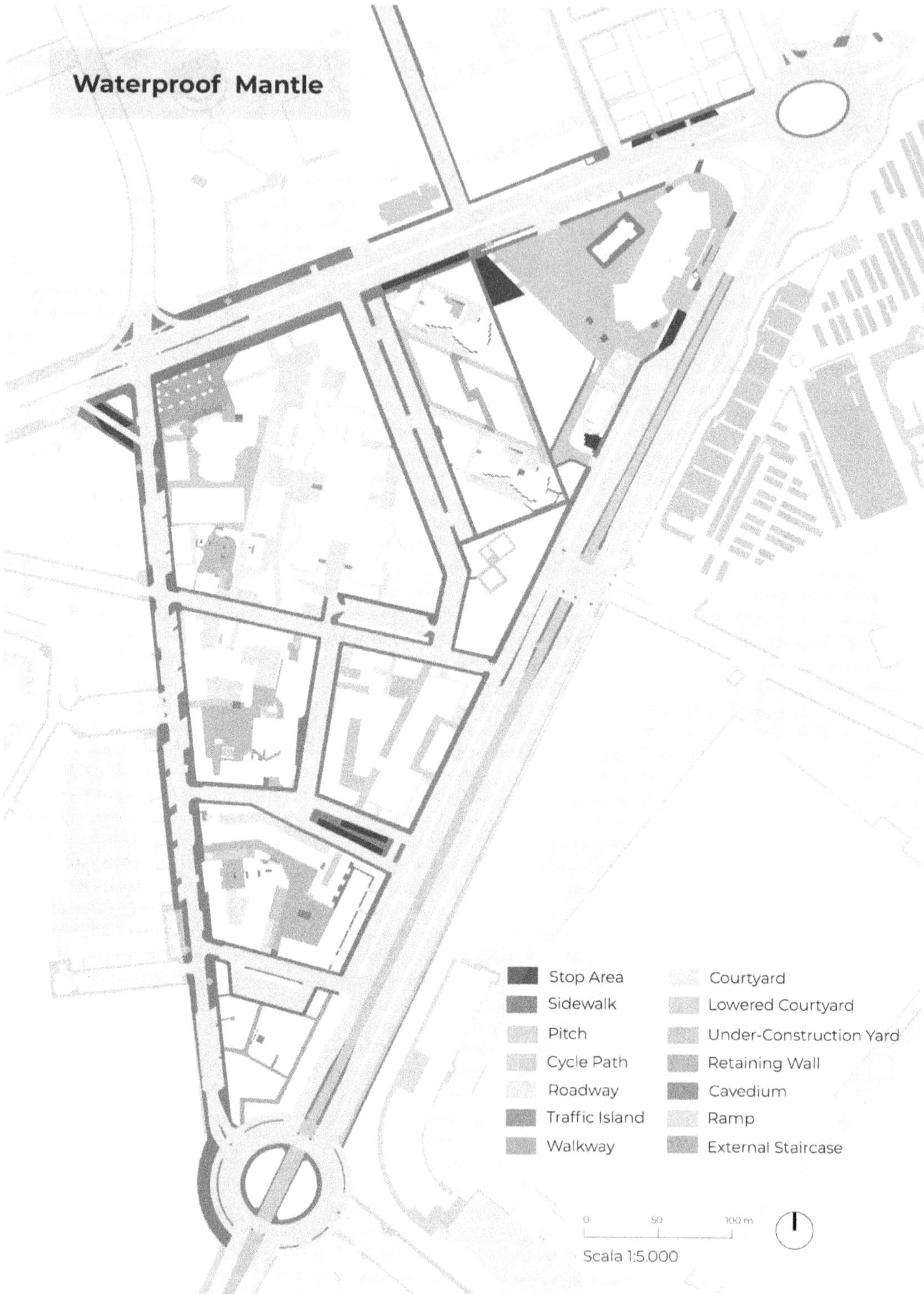

Figure 8. Green space - Map of the Pietra Alta area with representation of the Green space

Figure 9. Lighting - Map of the Pietra Alta area showing the distribution of street lighting

Figure 10. Distance. From the analysis of the uses of the buildings that make up the study area, the basins of interest for pedestrians were drawn up, using both the variable distance in metres and the variable time, i.e. distance covered in a given time interval

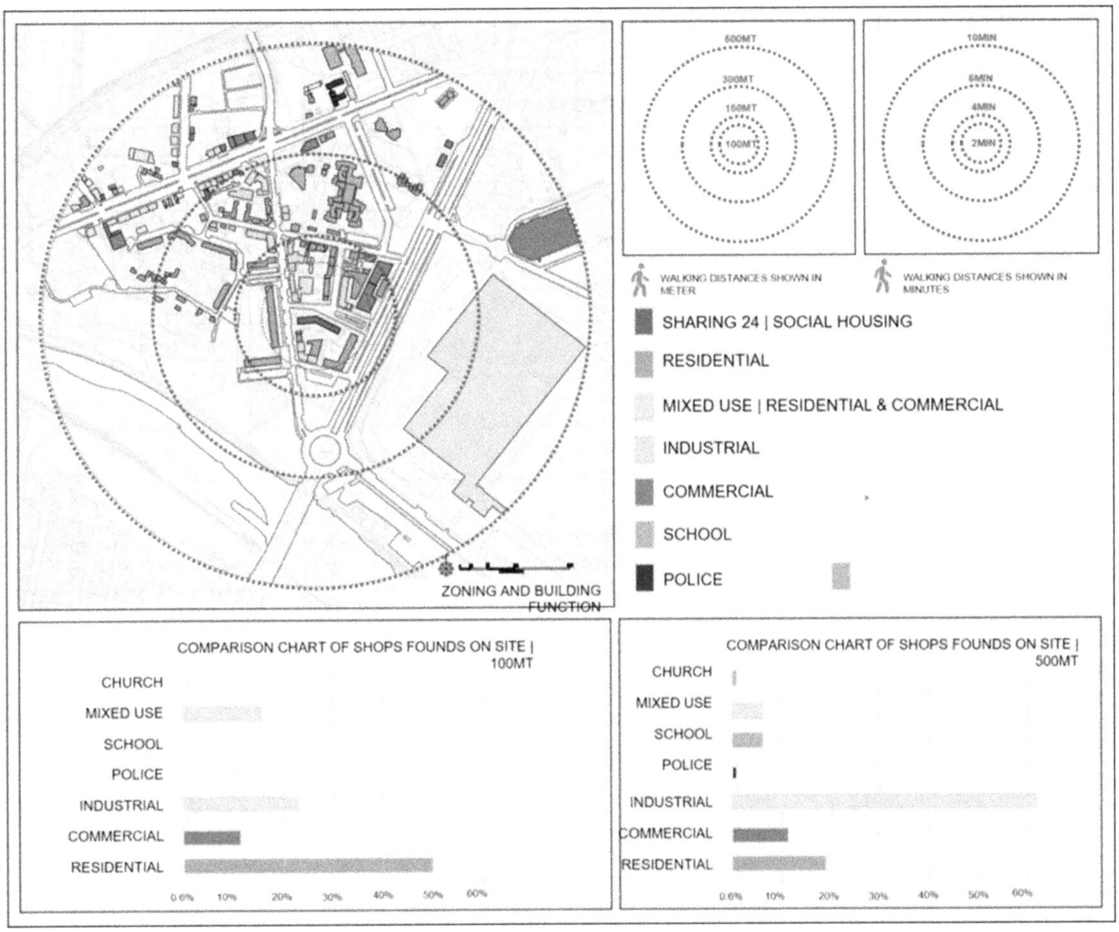

From a first analysis of the indices, it is immediately evident that there are different categories of indices, which can be classified according to the survey methods necessary for their determination:

- Direct assessment and survey, without subjective assessment of the surveyor: this category includes indices such as the presence of cycle paths, the number of lanes for vehicles, the speed - minimum and maximum, the presence of pedestrian crossings with traffic lights, the presence of chutes for overcoming height differences, the degree of maintenance of the road surface, the degree of maintenance of the pavement in pedestrian areas, the presence of public lighting on pedestrian and vehicle paths;
- Subjective evaluation of the detector: these are indices linked to the perception of the environment in which the pedestrian is inserted. These indices allow us to assess the comfort perceived during the walk: these indices include, for example, the assessment of the quality of buildings and their degradation and the perceived degree of safety - not only through the assessment of illuminance;

Figure 11. Percorsi. Within the map, the 4 different pedestrian routes were inserted, all originating from the Social Housing Sharing24 building

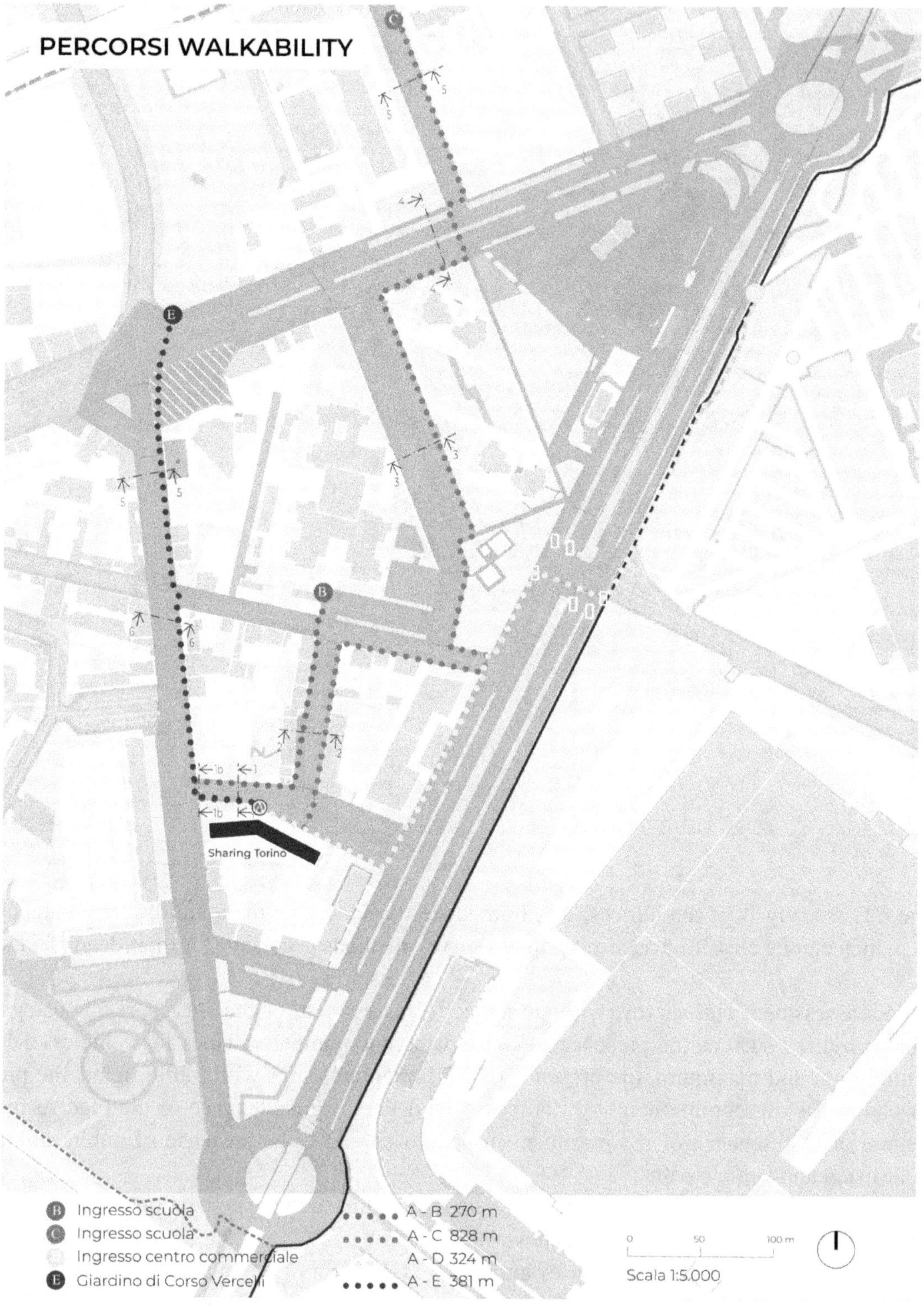

- Indirect evaluation, deduced from the survey of other parameters: for example, the survey and analysis of the variety of commercial activities allow us to indirectly evaluate the interaction between the surrounding environment and the pedestrian when choosing the route to take to reach a given point.

For the analysis and representation of the indices, the discipline of surveying plays a fundamental role as a necessary means both for the knowledge, management and representation of a large quantity of data and as a tool and key for simplifying the complexity of the relationships that exist between the various elements surveyed. All the data, collected and represented according to need using different methods and representation techniques considered most suitable depending on the expected output, were collected within a complex and articulated multi-scale and multi-content model, to make it easily interrogable during the phases of analysis and reading. The urban environment, and in particular the routes, are investigated not only as a representative place of geometric and aesthetic elements, but as a "container" place that contains elements that characterise and influence the behaviour of the user in the choice of a pedestrian route or in the choice of the use of a public or private means of transport. It was possible to analyse numerous data within a GIS model, from which numerous thematic tables were produced, to represent, in a first phase, the individual indices necessary for the assessment of the walkability of an area (Figure 06 Quality of Walkability - 07 Waterproof mante - 08 Green Space - 09 Lighting).

To obtain an exhaustive mapping of the area under study, all the individual indices and parameters identified as representative parameters and indices for determining the walkability assessment of the area were recorded and represented. Various representations will then be presented in order to define the individual parameters and to represent the real picture of the area. The CIM will have the possibility of being interrogated by single themes - or rather by single indices - returning single thematic tables or representing the overall assessment of the area in a single synthesis picture.

Case Study Area and Walkability Survey

At the micro-urban level, we have focused our attention on neighbourhood streets, and the direct relationship that is established between the individual who walks along with them and the context in which they are set. These relations are to be found within all those microelements that influence the sense of safety, perception, efficiency, and pleasantness of the route and, at the same time, the real possibility offered to the pedestrian to access and use the spaces. Within this context, the evolution of the use of IT tools, through advanced technologies, has allowed us to process large quantities of georeferenced data that can be shared on web platforms (Bultrini, 2017).

By carrying out a series of survey campaigns at different times of the day - morning, late afternoon and evening - the patterns that best represent the preferences and daily practices of residents were described, identifying the points of greatest potential and interest in the area. In order to determine the average walking catchment area, maps were made of the catchment areas that can be reached on foot with reference to both the minimum distance - from 100m to 500m - and the walking time - from 2 minutes to 10 minutes (Figure 10 Distance).

The results were summarised and represented within the route map. The four routes described have a single starting point (point of interest A) corresponding to the building in Via Ivrea n.24, which houses Sharing24, used for Social Housing (see map "Walkability Routes" - Figure 11). Along the four identified routes, a first mapping of the neighbourhood was carried out, identifying the landmarks on an urban scale that attract visitors from all over the city, such as the new shopping area (Yellow Route - point of interest D) or the points of greatest interest for the residents of the area, as areas dedicated to services, such as schools and public green spaces. The four routes have been represented by the use of different colours:

Red route I point of interest B – Entrance to Via Cavagnolo Secondary School
Green route I point of interest C – Entrance to Resi Marinotti Nursery School
Yellow route I point of interest D - shopping centre entrance
Blue route I point of interest E - Corso Vercelli Public Garden entrance

Through the mapping of the routes it was possible to identify other points of interest, such as the Parish of San Michele Arcangelo, with its pedestrian area in front of the entrance and the play area dedicated to the oratory, the green areas located near the schools, and the neighbourhood shops, located on Via Ivrea and Corso Vercelli, where the residents of the area can go using only pedestrian routes to find basic necessities. The results of the analysis of the main routes were synthesized within GIS and BIM models that allowed us to better represent the data collected and to read the urban fabric that characterizes this part of the city, which is very jagged and irregular.

In parallel to the direct surveys, the urban fabric of the first outskirts of Turin was analysed using the adoption and evaluation of indices, aimed at representing and mapping quantitatively and qualitatively the parameters characterising the urban scene studied. From the survey and the analysis of qualitative data related to the evaluation of walkability, we have identified, through a screening by levels, the characteristic attributes of the arcs of the graph of the road network. The values, whether qualitative and/or quantitative, of these attributes were collected through direct observation of the roads and urban scenes, in situ, between March and October 2020. As can be seen from the indices collected and represented in the various maps, the analysis was extended to many qualitative and detailed aspects of the urban environment, which are often not represented within studies focused on the assessment of walkability. Aspects that are determining factors for the evaluation of the relationship between the urban environment, characterised by road and pedestrian traffic, and the behaviour of pedestrians themselves, have been detected, focusing attention on urban elements and conditions that objectively and subjectively influence the choices and attitudes of people moving within the urban scene analysed. Some indices are easily and univocally detectable and assessable, such as the presence of a cycle path (assessable by detecting cycle lanes), or the number of lanes dedicated exclusively to road traffic, the maximum or minimum speed allowed, the presence of pedestrian crossings with the presence of timed traffic lights or characterised by the presence of chutes for overcoming height differences and the removal of architectural barriers (between pavement level and street level). Other parameters, however, are more related to the time of the survey and the subjectivity of the surveyor, such as the quality and degree of maintenance of the pedestrian pavement, the safety of the streets, given by the evaluation of adequate public lighting, the level of services available to pedestrians, such as the presence of benches or shaded areas for parking, etc. …

Data Processing, Analyses and Representation

For the representation of the selected routes, different types of analyses, evaluations and graphic representations will be reported below, each of which has the objective of representing and highlighting different indices and evaluations, through the use of suitable codes and representation techniques.

Figure 12. Route A was analysed by developing the fronts along the sections of road travelled by users to reach the entrance to the gardens on Corso Vercelli. This plan shows the road sections developed to analyse the different characteristics of the road sections travelled.

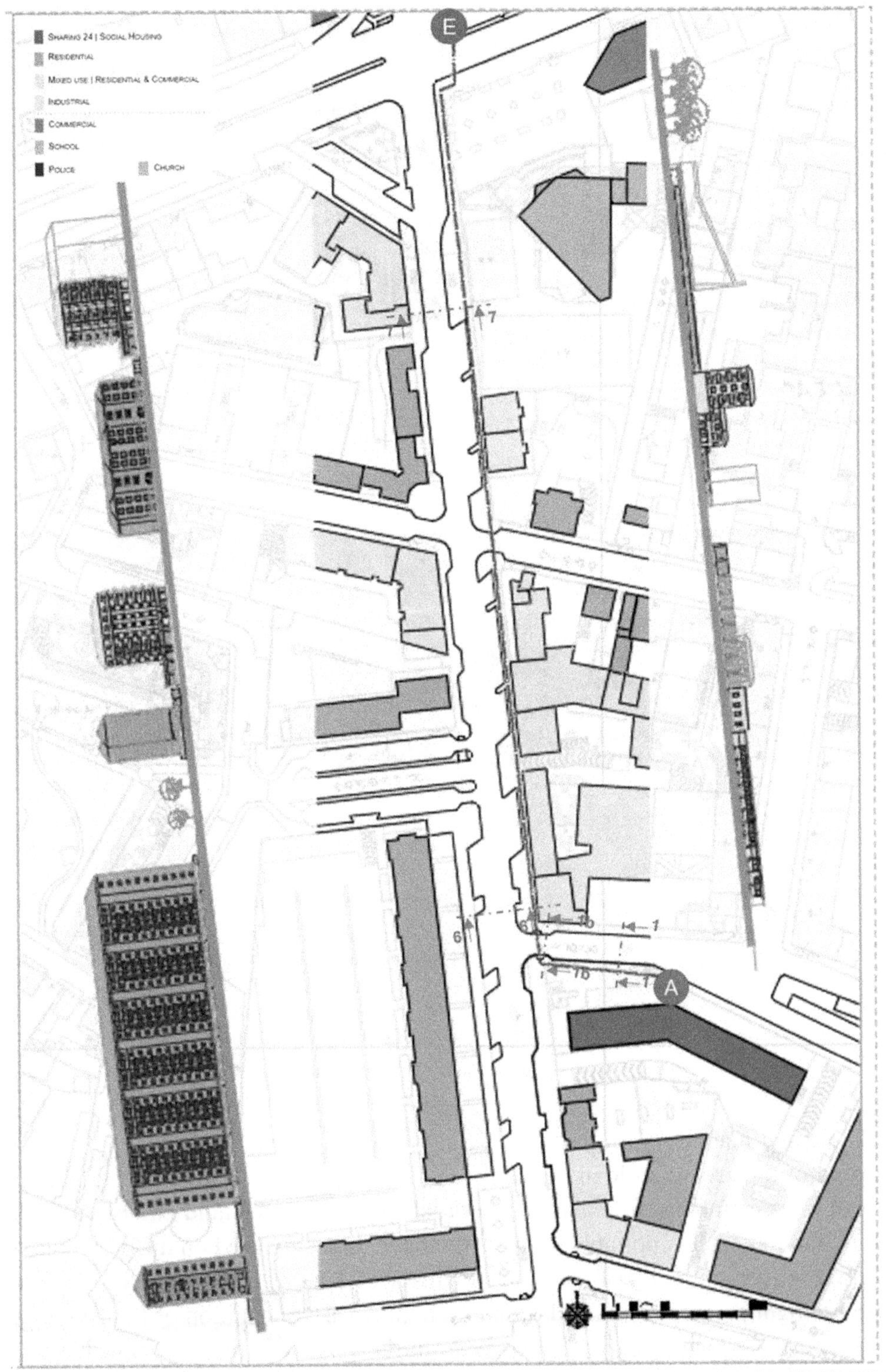

Figure 13. The four road sections characterise four nodal points of route A: the elaborated sections represent pedestrian crossings, road sections with parking spaces near the pavement, bus stops, street lighting, pedestrian lighting

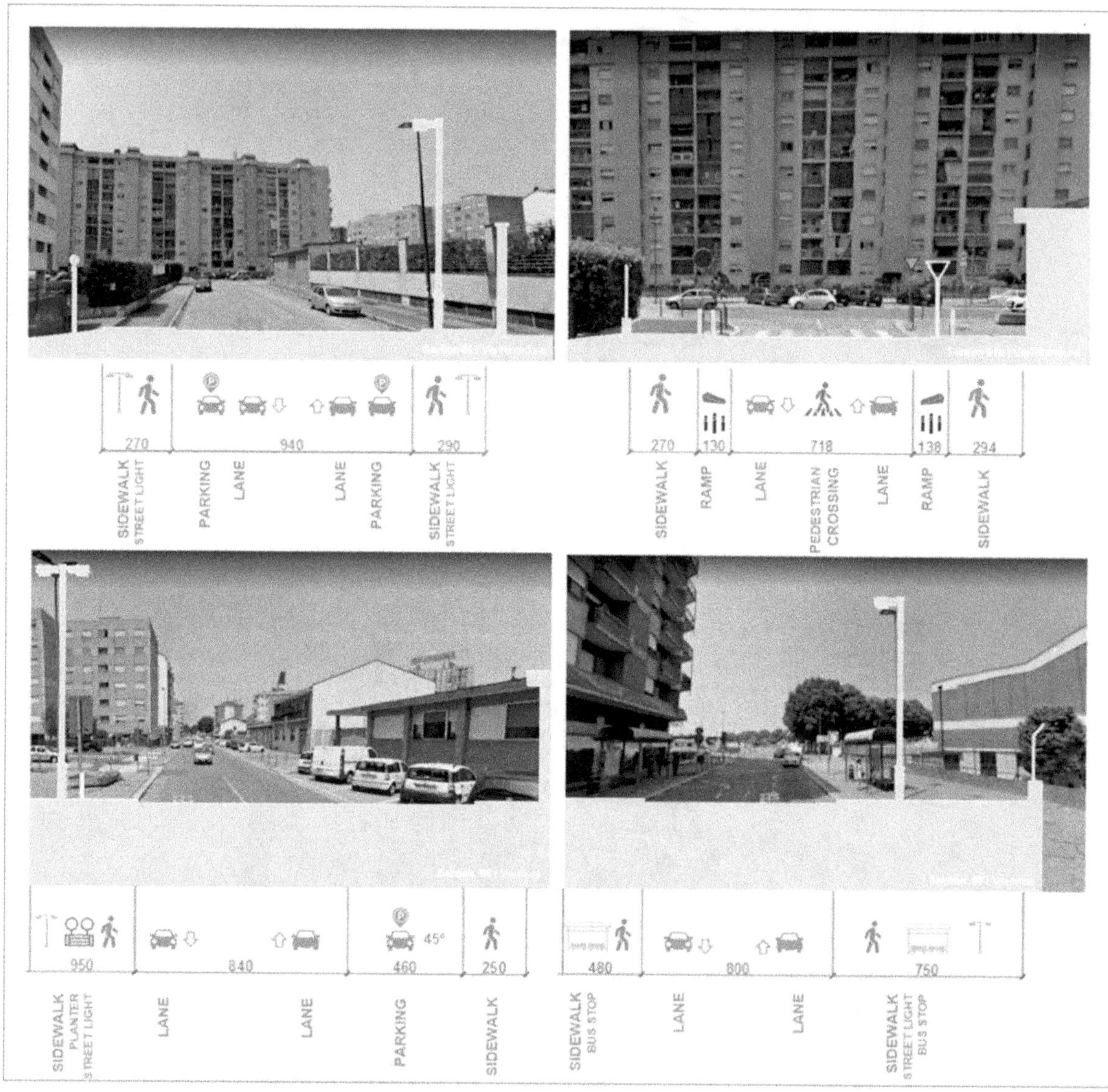

From the integrated reading of all the indices related to walkability it will be possible to have an integrated reading of the urban fabric in which the survey took place. Many indices will be described through different representations, for example, in order to better understand parameters related to the quality of pedestrian routes, it is possible to make thematic tables focused on different indices in order to highlight different characteristics of the routes themselves. The analysis of the typical road sections of a route will highlight parameters related to public lighting, fragmentation of the fronts, presence of car parks, etc… (*Figure 12* and *Figure 13*). By extracting the sections directly from the BIM model it was possible to highlight:

Figure 14. Planimetry with analysis of the uses of the buildings and indication of the Red and Green routes

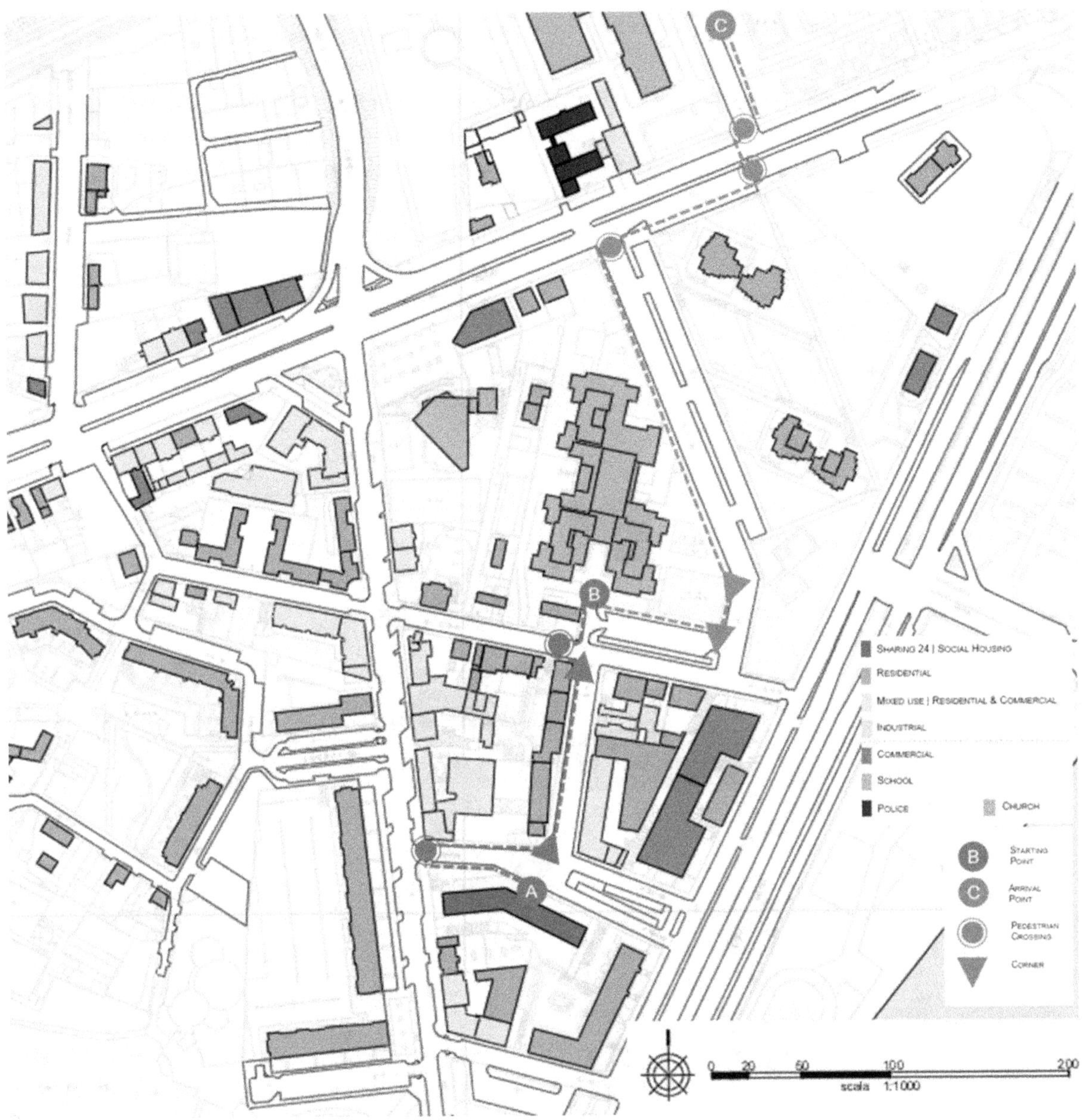

- the presence of different types of public lighting;
- the different elements that make up the pedestrian crossing: slide for overcoming height differences, bollards, traffic lights, pedestrian crossings;
- the presence of parking spaces along the pedestrian path: parking spaces parallel to the driveway/pedestrian path, parking spaces inclined at 45° concerning to the direction of travel and the pavement;
- the characteristics of the driveway: one-way, two-way, presence of road signs, possible speed limits (30km areas);

Figure 15. Development of the Red route - pedestrian route between point A, Social Housing Building and point B, Via Cavagnolo Secondary School

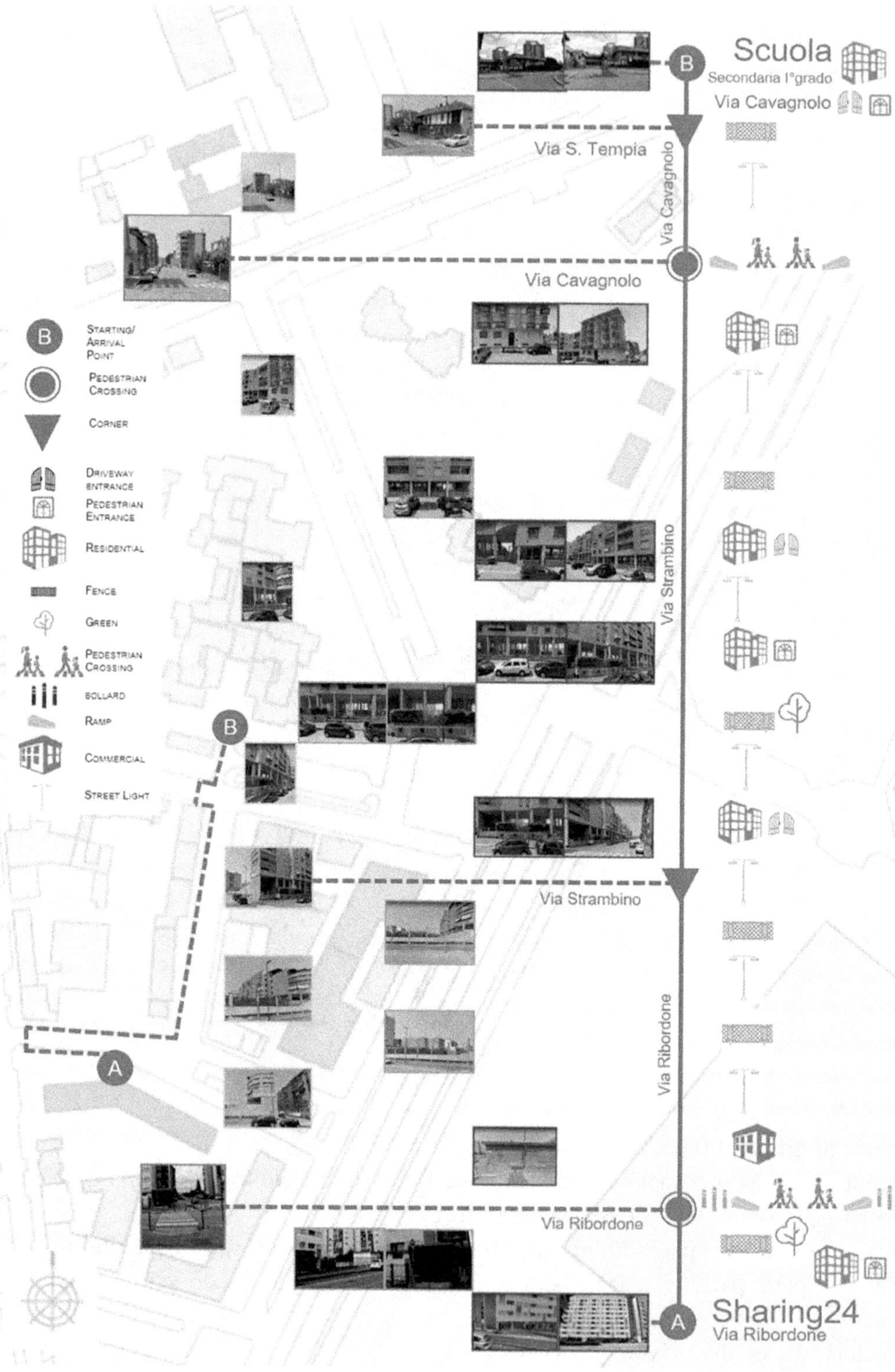

Figure 16. Development of the Green route - pedestrian route between point B, Secondary School of Via Cavagnolo and point C, Nursery School Resi Marinotti

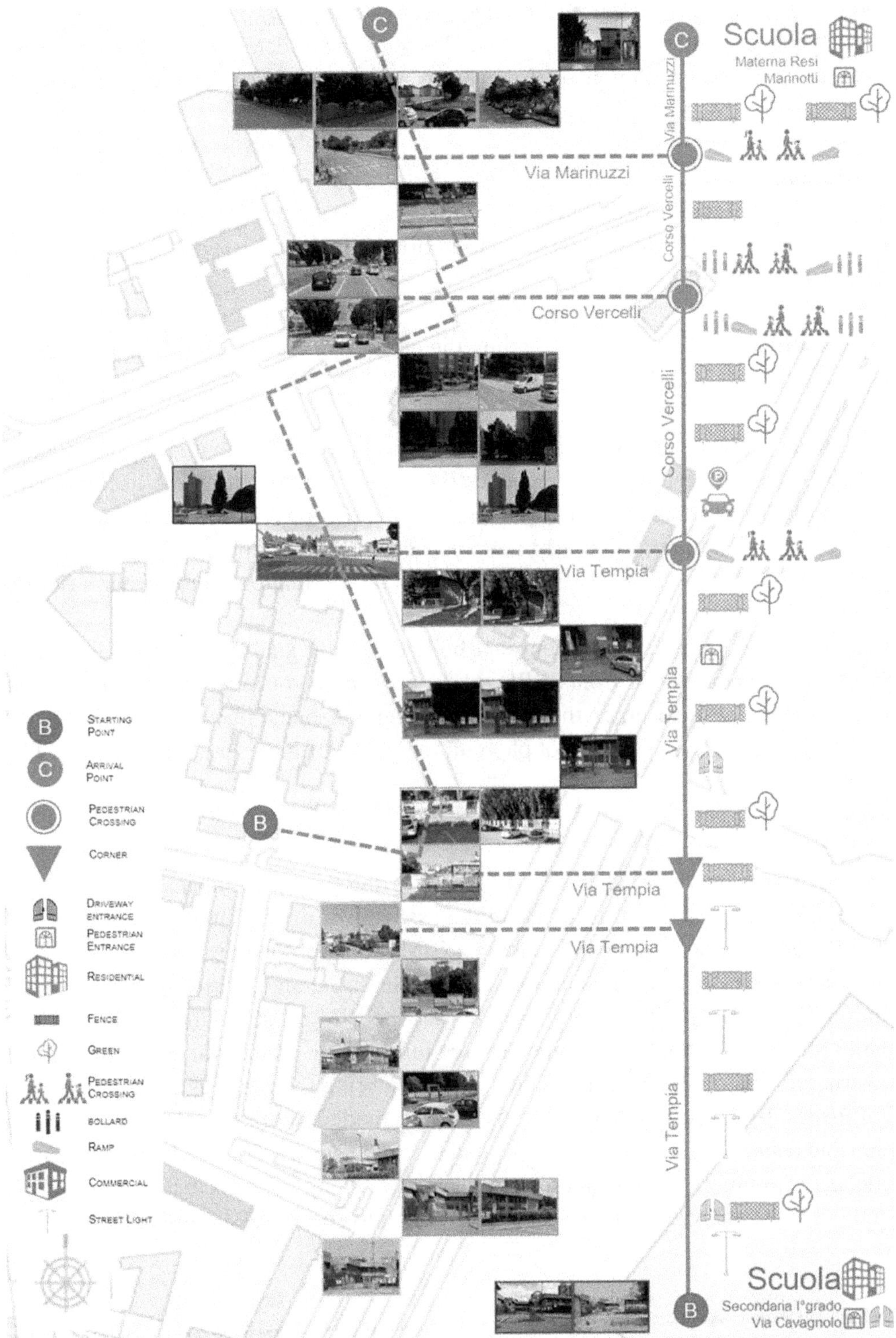

- the presence of public transport stops: type of stop, presence of bench for parking, presence of shelter;
- the type of fences and the presence of private green areas.
- similarly, to qualitatively read the urban fabric in which the selected pedestrian routes are inserted, and to understand the nodal points of the pedestrian routes (e.g. the Red route and the Green route), the routes were developed highlighting the presence of:
- residential buildings;
- commercial buildings;
- pedestrian entrances and driveways;
- fences with railings and/or low walls delimiting private and public space;
- private or public green areas;
- pedestrian crossings;
- bollards - both for crossings and for dividing the pedestrian area from the driveway area;
- ramps for overcoming height differences.

From the analysis of the development of the routes it will be possible to understand qualitatively the perception of the urban environment by the pedestrian walking along the route itself (Figure 14): the fragmentation of the fronts, due to the backwardness of the buildings, will most probably communicate to most users a sense of loneliness; similarly, the presence of a high number of public lighting elements and the presence of buildings characterised by a commercial destination on the ground floor will communicate a sense of security.

The same perception of the environment can be detected by analysing the images of the fronts: qualitatively a homogeneous set of fronts will make a pedestrian route more pleasant, since it will be more pleasing to the eye of the user, on the contrary, a fragmented and uneven set of fronts will be read by the pedestrian as a disorderly set and not pleasant to the eye of the user. (Figure 15 and Figure 16)

Using Maps / Using Models

In order to properly manage the amount of data that describe the urban fabric under analysis, and the plurality of users, technical and not, involved in the assessment operations, it is possible to make the BIM model of the area, to the GIS model, making a CIM model of the neighbourhood. Inside the CIM model will be represented the 3D model at the urban scale of the area, which inside collects: the BIM models of individual buildings - and in our specific case the BIM model of the building that houses the Social Housing Sharing 24, the data and open source information made available by the Municipality, such as infrastructure and roads, and information deduced during the campaigns of the direct survey, on the quality of places. By linking BIM to CIM, users are provided with an interactive environment, rich in information and data, which can be communicated through ad hoc elaborated representations, such as 3D city models, alphanumeric graphs, thematic tables, etc., where the information in addition to being accessible to all, can be analyzed, queried and shared in real-time (S. Hisham, 2018).

RESULTS AND DISCUSSION

An in itinere part of a larger work that will be completed in the coming months has been proposed here. The pretext is the update of the evaluation of the social return on investment index carried out in 2015 by the Fondazione Sviluppo e Crescita CRT of Turin. In 2021 it will be ten years of activity of Sharing and therefore we thought to give an updated reading of the social return by defining an evaluation model that on the one hand makes the process of data collection and analysis more expeditious, on the other hand, includes in the study variables of a wider context, at the scale of the neighbourhood, incorporating parameters and indicators of urban quality and resilience, always read through the filter of social impact.

The evaluation model we are thinking of wants to be light and easy to apply, maintaining transparency, representativeness and significance.

The objective is twofold: on the one hand to read the evolution of this node within the urban fabric in which it lives. On the other hand, to define an evaluation tool that can operate in the monitoring of different interventions (which obviously will also need variables sewn case by case), but that can also provide elements to identify in advance, ex ante, those social and territorial realities that to a greater extent could support the success of the investment in terms of social impact.

This is a field where social impact must be tied together with local actions and urban infrastructure triggers, involving public and private actors.

Walkability is one of the fundamental pillars of a sustainable city (Southworth 2005, Ariffin and Zahari 2013 and Rafiemanzelat et al. 2017) and the benefits of a walkable environment comes from an environmental, social as well as economic dimension of sustainable development. Currently, the most recent and significant investigations are directed towards further complex work aimed at defining processing equipment to manage information assets of various kinds, designed to make more effective and efficient multidisciplinary research, in order to exploit the information potential of the graphical representation for those environmental and territorial systems of which you want to know and improve the degree of resilience, or the measurement of the ability to absorb and mitigate the effects of critical anthropogenic and natural events related to different levels and scales of influence. The path has therefore been enriched by making new supports for the path of Drawing. This highly collaborative field of work has found institutional expression at the Politecnico di Torino through the recent establishment of an Interdepartmental Research Centre specifically dedicated to in-depth studies on the theme of resilience, responsibility for interventions and risk (R3C), to which many researchers from different scientific sectors, including our Drawing area, have adhered. Increasingly, the evolution of information systems is conforming tools and applications towards structures that are easily portable and interconnected at all times. The storage and management of data for their graphic re- presentation follow consolidated criteria and forms of knowledge, but in a more efficient way and towards a wealth of increasingly numerous and heterogeneous information. Technological evolution also offers innovative elements and confirmations of some methods and techniques or the abandonment of others, in favour of more efficient and secure processes. The path that the contribution has taken through a number of fundamental stages, critically analyzed precisely because it highlights elements that have remained compared to those that have been abandoned, has been enriched by making new areas of research and renewed ideas. Changes and permanences characterize the history of drawing; we wanted to give a personal testimony of the aspects of method contained in experiences that we can consider distant, in the face of technical progress that proceeds at high speed, and how these aspects remain in the current developments, thus being able to consider them, reasonably, immutable even with respect to the frontiers that are expected.

From the need to move around independently on foot, given a safe and comfortable environment, people seek utility, a sense of belonging, and pleasantness as additional and distinct needs to enhance their experience of living in city neighbourhoods. Using survey data and assuming relief given by geo-referenced interviews, the proposed work highlights the integration of residents' perceptions and objective measures to understand the impact of environmental features on the views and behaviour of neighbourhood residents.

We mentioned the sensitivity of knowledge to innovation in survey methods and tools. The naked eye observation tells of the elements, the use of microscopes and telescopes or X-rays reveal, make visible knowledge in ways never possible before. The active involvement of citizens is the development we want to give to the work, even with new tools for collecting and analyzing the feelings of those who live in the places. It remains open the knot of solutions, an abacus, a handbook that somehow provides us with a vocabulary to build the discourse of possible interventions to improve the quality of public space and urban well-being.

CREDIT AUTHORSHIP CONTRIBUTION STATEMENT

The contribution is part of wider research work on the study and representation of urban and environmental quality and related resilience aspects. The research group is coordinated by professors Giorgio Garzino and Maurizio Bocconcino and is composed by ing. Mariapaola Vozzola, all belonging to the Department of Structural, Building and Geotechnical Engineering of the Politecnico di Torino, and arch. Anna Rabbia of the Fondazione Sviluppo e Crescita CRT. In particular, they have edited the following parts: introduction and methodological aspects Maurizio Marco Bocconcino, state of the art Mariapaola Vozzola, Empirical analysis and Metrics Maurizio Marco Bocconcino, Mariapaola Vozzola and Anna Rabbia, Proposal of Indexes Mariapaola Vozzola e Anna Rabbia, Case study and analytical elaboration (digital informative model and system) Mariapaola Vozzola (bim and cim) and Anna Rabbia (gis), Results and discussion Maurizio Marco Bocconcino.

Acknowledgement

The Authors thanks Fondazione Sviluppo e Crescita CRT (president Cristina Giovando) Urban and Social Regeneration group that finances part of the research (project manager Bianca Viarizzo, project coordinator Anna Rabbia, scientific coordinators Franco Prizzon and Maurizio Marco Bocconcino of Politecnico di Torino, research collaborators Mariapaola Vozzola and Giada Mazzone).

The research is developed in collaboration with the R3C - Responsible Risk Resilience Centre of the Politecnico di Torino, an interdisciplinary research centre focused on Urban Resilience.

The Authors also thanks to the Reviewers for their suggestions and integrations.

REFERENCES

Ariffin, R. N. R., & Zahari, R. K. (2013). Perceptions of the urban walking environments. *Procedia: Social and Behavioral Sciences*, *105*, 589–597. doi:10.1016/j.sbspro.2013.11.062

Aultman-Hall, L., Roorda, M., & Baetz, B. W. (1997). Using GIS for Evaluation of Neighborhood Pedestrian Accessibility. *Journal of Urban Planning and Development*, *123*(1), 10–17. doi:10.1061/(ASCE)0733-9488(1997)123:1(10)

Batty, M. (2005). 2005, Agents, cells, and cities: New representational models for simulating multiscale urban dynamics. *Environment & Planning A*, *37*(8), 1373–1394. doi:10.1068/a3784

Batty, M. (2013). *The New Science of Cities*. The MIT Press. doi:10.7551/mitpress/9399.001.0001

Blečić, I., Cecchini, A., Fancello, G., Talu, V., & Trunfio, G. A. (2015). Camminabilità e capacità urbane: valutazione e supporto alla decisione e alla pianificazione urbanistica. Territorio Italia, 1, 49 - 65.

Börner, K., Sanyal, S., & Vespignani, A. (2007). Network Science. In B. Cronin (Ed.), *Annual Review of Information Science & Technology* (Vol. 41, pp. 537–607). Information Today, Inc./American Society for Information Science and Technology.

Bultrini, M. (2017). Il walkability index: un indice di camminabilità urbana. In *Qualità dell'ambiente urbano. XIII Rapporto. Edizione 2017. Focus su Mobilità pedonale in città. ISPRA, Stato dell'ambiente 75/2017*. Available on line: https://www.isprambiente.gov.it/files2017/pubblicazioni/stato-ambiente/Focus2017.pdf

Calvano, M., Cirelli, M., & Lo Turco, M. (2020). Display the Invisible. Automated Algorithms to Visualize Complex Phenomena. *Proceedings of the 2nd International and Interdisciplinary Conference on Image and Imagination*.

Caperna, A., Giangrande, A., Mirabelli, P., & Mortola, E. (2013). Partecipazione e ICT. Gangemi.

Castells, M. (1996). *The Rise of the Network Society*. Blackwell.

Cavallari Murat, A. (1968). Forma urbana e architettura nella Torino Barocca. Dalle premesse classiche alle conclusioni neoclassiche, Unione tipografico-editrice torinese, Torino.

Cervero, R., & Kockelman, K. (1997). Travel demand and the 3Ds: Density, diversity, and design. *Transportation Research Part D, Transport and Environment*, *2*(3), 199–219. doi:10.1016/S1361-9209(97)00009-6

Clifton, K., Ewing, R., Knaap, G. J., & Song, Y. (2008). Quantitative analysis of urban form: A multidisciplinary review. *Journal of Urbanism*, *1*(1), 17–45. doi:10.1080/17549170801903496

Cullen, G. (1971). *The Concise Townscape*. Architectural Press.

Cullen, G. (1976). *Il paesaggio urbano: morfologia e progettazione*. Calderini Editore.

Duncan, D. T., Aldstadt, J., Whalen, J., & Melly, S. J. (2012). Validation of Walk Scores and Transit Scores for estimating neighborhood walkability and transit availability: A small-area analysis. *GeoJournal*, *78*(2), 407–416. doi:10.100710708-011-9444-4

ESUM. (2017). *Human perception of urban environment.* https://www.esum.arch.ethz.ch

Falcinelli, R. (2011). *Guardare, pensare, progettare. Neuroscienze per il design.* Stampa Alternativa & Graffiti.

Frank, L. D., Saelens, B. E., Powell, K. E., & Chapman, J. E. (2007). Stepping towards causation: Do built environments or neighborhood and travel preferences explain physical activity, driving, and obesity? *Social Science & Medicine, 65*(9), 1898–1914. doi:10.1016/j.socscimed.2007.05.053 PMID:17644231

Frank, L. D., Sallis, J. F., Conway, T. L., Chapman, J. E., Saelens, B. E., & Bachman, W. (2006). Many pathways from land use to health: Associations between neighborhood walkability and active transportation, body mass index, and air quality. *Journal of the American Planning Association, 72*(1), 75–87. doi:10.1080/01944360608976725

Frank, L. D., Schmid, T. L., Sallis, J. F., Chapman, J., & Saelens, B. E. (2005). Linking objectively measured physical activity with objectively measured urban form. *American Journal of Preventive Medicine, 28*(2), 117–125. doi:10.1016/j.amepre.2004.11.001 PMID:15694519

Girardin, F., Calabrese, F., Dal Fiore, F., & Ratti, C. (2008, October-December). Digital Footprinting: Uncovering Tourists with User-Generated Content. *IEEE Pervasive Computing, 7*(4), 36–43. doi:10.1109/MPRV.2008.71

Girardin, F., Del Fiore, F., Ratti, C., & Blat, J. (2008). Leveraging Explicitly Disclosed Location Information to Understand Tourist Dynamics: A Case Study. Journal of Location-Based Service, 2(1). doi:10.1080/17489720802261138

Girardin, F., Vaccari, A., Gerber, A., Birderman, A., & Ratti, C. (2009). Quantifying Urban Attractiveness from the Distribution and Density of Digital Footprints. *International Journal of Spatial Data Infrastructures Research*, 4.

Handy & Clifton. (2000). *Evaluating Neighborhood Accessibility: Issues and Methods Using Geographic Information Systems.* The University of Texas - Austin. SWUTC/00/167202-1.

Heidi, S. (2018). *Non-Spatial Statistic for Analysing Human Perception of the Built Environment* (Master Thesis). ETH Zurich.

Hollenstein, D., & Bleisch, S. (2016). *Walkability for different urban granularities.* In The International Archives of the Photogrammetry, Remote Sensing and Spatial Information Sciences, Prague, Czech Republic.

Jacobs, J. (1961). *The Death and Life of Great American Cities.* Random House.

Kocher, J., & Lerner, M. (2007). *Walkscore.* Available: https//walkscore.com/

Kovács-Győri, A., Resch, B., & Zeile, P. (2019). Defining and assessing walkability: An concept for an integrated approach using surveys, biosensors and geospatial analysis. *Urban Development Issues, 62*(1), 5–15. doi:10.2478/udi-2019-0008

Kuznetsov, S., & Paulos, E. (2010). Rise of the Expert Amateur: DIY Projects, Communities, and Cultures. *Proceedings of the 6th Nordic Conference on HumanComputer Interaction: Extending Boundaries*, 295–304.

Lazer, D., Pentland, A., Adamic, L., Aral, S., Barabási, A., Brewer, D., Christakis, N., Contractor, N., Fowler, J., Gutmann, M., Jebara, T., King, G., Macy, M., Roy, D., & Van Alstyne, M. (2009). Computational Social Science. *Science, 323*(5915), 721-723. . doi:10.1126cience.1167742

Leslie, E., Coffee, N., Frank, L., Owen, N., Bauman, A., & Hugo, G. (2007). Walkability of local communities: using geographic information systems to objectively assess relevant environmental attributes. *Health & Place, 13*(1), 111-22. .healthplace.2005.11.001 doi:10.1016/j

Lewin, K. (1961). *1936, Principles of topological psychology. Trad. it.: Principi di psicologia topologica. Traduzione a cura di A.Ossicini*. Edizioni OS.

Li, X., Hijazi, I., Koning, R., Lv, Z., Zhong, C., & Schmitt, G. (2016). Assessing Essential Qualities of Urban Space with Emotional and Visual Data Based on Gis Technique. *ISPRS International Journal of Geo-Information*, *5*(11), 218. doi:10.3390/ijgi5110218

Lurija, A. (1973). *The working brain. Trad. it.: Come lavora il cervello. Introduzione alla neuropsicologia*. Edizione Il Mulino.

Lynch, K. (1960). L'immagine della città. Versione italiana a cura di Paolo Ceccarelli, Marsilio Editore, 2001-2006, Venezia.

Lynch, K., & Appleyard, D. (1964). The View from the Road. MIT Press.

Lynch, K., & Rivkin, M. (1990). A Walk around the Block. In T. Banerjee & M. Southworth (Eds.), *City Sense and City Design, Writings and Projects of Kevin Lynch* (pp. 185–205). MIT Press.

Marr, D. (1982). Vision. A Computational Investigation into the Human Representation and Processing of Visual Information. Freeman & Company.

Marr, D., & Nishihara, H. K. (1978, February 23). Representation and Recognition of the Spatial Organization of Three-Dimensional Shapes. *Proceedings of the Royal Society of London. Series B, Biological Sciences*, *200*(1140), 269–294. doi:10.1098/rspb.1978.0020 PMID:24223

McMillan, T. E. (2005). Urban Form and a Child's Trip to School: The Current Literature and a Framework for Future Research. *Journal of Planning Literature*, *19*(4), 440–456. doi:10.1177/0885412204274173

Mehta, V. (2008). Walkable Streets: Pedestrian Behavior, Perceptions and Attitudes. *Journal of Urbanism*, *1*(3), 217–245. doi:10.1080/17549170802529480

Mehta, V. (2013). Evaluating Public Space. *Journal of Urban Design*, *19*(1), 53–88. doi:10.1080/13574809.2013.854698

Moretti, M. (2016). *Senso e paesaggio. Analisi percettive e cartografie tematiche in ambiente GIS*. Franco Angeli.

Partnership for a Walkable America, Pedestrian and Bicycle Information Center. (2005). *Walkability Checklist: How walkable is your community?* https://cachampionsforchange.cdph.ca.gov/en/Documents/Let%27sWalk.pdf

Rafiemanzelat, R., Emadi, M. I., & Kamali, A. J. (2017). City sustainability: The influence of walkability on built environments. *Transportation Research Procedia*, •••, 24.

Randall, T. A., & Baetz, B. W. (2001). Evaluating Pedestrian Connectivity for Suburban Sustainability. *Journal of Urban Planning and Development*, *127*(1), 1–15. doi:10.1061/(ASCE)0733-9488(2001)127:1(1)

Rišová, K. (2020). Walkability research: Concept, methods and a critical review of post-socialist studies. *Geograficky Casopis*, *72*(3), 219–242.

Sacks, O. (1985). The Man Who Mistook His Wife For a Hat. Trad. it.: L'uomo che scambiò sua moglie per un cappello (trad. di Morena, C.). *Adelphi Edizioni, Milano*, *2001*, 318.

Saelens, B. E., Sallis, J. F., & Frank, L. D. (2003). Environmental correlates of walking and cycling: Findings from the transportation, urban design, and planning literatures. *Annals of Behavioral Medicine*, *25*, 80–91.

Schaeben, C. (2017). *People's Perception of Urban and Architectural Features*. Retrieved from: https://www.research-collection.ethz.ch/handle/20.500.11850/266634

Sherrington, C. S. (1906). The Integrative Action of the Nervous System. New Haven: Yale University Press.

Simpson, J., & Simpson, K. (2018). Visual engagement with urban street edges: Insights using mobile eye-tracking. *Journal of Urbanism*.

Sitte, C. (1953). L'arte di costruire le città, Officine Grafiche dell'Editore Antonio Vallardi, Milano.

Southworth, M. (1997). Walkable suburbs? An evaluation of neotraditional communities at the urban edge. *Journal of the American Planning Association*, *63*(1), 28–44. doi:10.1080/01944369708975722

Southworth, M. (2005). Designing the walkable city. *Journal of Urban Planning and Development*, *131*(4), 246–257. doi:10.1061/(ASCE)0733-9488(2005)131:4(246)

Southworth, M., & Ben-Joseph, E. (1995). Street standards and the shaping of suburbia. *Journal of the American Planning Association*, *61*(1), 65–81. doi:10.1080/01944369508975620

Southworth, M., & Owens, P. M. (1993). The evolving metropolis studies of community, street form at the urban edge. *Journal of the American Planning Association*, *59*(3), 271–287. doi:10.1080/01944369308975880

Talen, E. (2003). Neighborhoods as Service Providers: A Methodology for Evaluating Pedestrian Access. *Environment and Planning. B, Planning & Design*, *30*(2), 181–200. doi:10.1068/b12977

Talen, E., & Koschinsky, J. (2013). The Walkable Neighborhood: A Literature Review. *International Journal of Sustainable Land Use and Urban Planning*, (1), 42—63.

Taylor, M. C. (2002). *The Moment of Complexity: Emerging Network Culture*. University of Chicago Press.

The Prince's Foundation. (2020). *Walkability and Mixed-Use. Making Valuable and Healthy Communities.* https://content.knightfrank.com/research/2139/documents/en/walkability-and-mixed-use-making-valuable-and-healthy-communities-7667.pdf?fbclid=IwAR161EMGOJ4kszJPO-DgWJ2wLUWWKVw-G2ufSKaCs1BoL6H7jT73e5C77hEg

Tsiompras, A. B., & Photis, Y. N. (2017). What matters when it comes to "Walk and the city"? Defining a weighted GIS-based walkability index. *Transportation Research Procedia, 24*, 523–530. doi:10.1016/j.trpro.2017.06.001

KEY TERMS AND DEFINITIONS

Building Information Modeling or BIM: ISO 19650:2019 defines BIM as: Use of a shared digital representation of a built asset to facilitate design, construction, and operation processes to form a reliable basis for decisions.

City Information Modeling or CIM: City Information Modeling is to be understood as a digital representation of the city, focusing on city management, building design, planning, and monitoring. The City Model should be understood as and support for the management of a Smart City infrastructure.

Metrics: Measures of quantitative assessment commonly used for assessing, comparing, and tracking performance or production. Generally, a group of metrics will typically be used to build a dashboard that management or analysts review on a regular basis to maintain performance assessments, opinions, and design strategies.

Geodesign: The term geo can be simply defined as geographic space – space that is referenced to the surface of the earth (geo-referenced). In general, thinking of geographic space brings to mind a 2D geographic space (a flat map) or, for those who are a bit more advanced in their thinking, a 2.5D geographic space – that is, an undulating surface (a relief map). This thinking could also be extended to include 3D geographic space, providing the ability to geo-reference what lies below, on and above the surface of the earth, including what exists inside and outside buildings, as well as 4D geographic space, giving the added ability to geo-reference time-dependent information such as population growth or the migration of a toxic plume through a building. The word design, the second component of geodesign, can be defined as either a noun or a verb. As a noun, design generally refers to some object or other entity. As a verb, it usually refers to a process or series of activities (ESRI, 2018).

Geographic Information System or GIS: A geographic information system (GIS) is a framework for gathering, managing, and analyzing data. Rooted in the science of geography, GIS integrates many types of data. It analyzes spatial location and organizes layers of information into visualizations using maps and 3D scenes. With this unique capability, GIS reveals deeper insights into data, such as patterns, relationships, and situations—helping users make smarter decisions.

Walkability: Walkability is a measure of how friendly an area is to walking. Walkability is influenced by many factors, including the quality of paths, pavements or crossings, roads, lighting and accessibility of buildings, and perceptions of safety. Walkability is one of the basic concepts of sustainable urban design.

ENDNOTES

1 For more information on the development of the term walkability and its definition, see the research of Katarína Rišová, 2020.

2 Different sources were studied: from ICE - Indicators of the European Community (European Municipalities Indicators Towards a Local Sustainability Profile, 2003); Towards a Local Sustainability Profile - European Municipalities Indicators (ICE) Results from the Province of Turin (2003); Indices introduced by the Municipality of Cervia in its DPQU plan (2017).

Chapter 19
Combining BIM, GIS, and IoT to Foster Energy Management and Simulation in Smart Cities

Edoardo Patti
Politecnico di Torino, Italy

Francesco G. Brundu
https://orcid.org/0000-0003-2133-1426
Politecnico di Torino, Italy

Andrea Bellagarda
Politecnico di Torino, Italy

Lorenzo Bottaccioli
Politecnico di Torino, Italy

Niccolò Rapetti
Politecnico di Torino, Italy

Vittorio Verda
Politecnico di Torino, Italy

Elisa Guelpa
Politecnico di Torino, Italy

Laura Rietto
Politecnico di Torino, Italy

Enrico Macii
Politecnico di Torino, Italy

Andrea Acquaviva
Università di Bologna, Italy

Alexandr Krylovskiy
Fraunhofer Institute for Applied Information Technology, Germany

Marco Jahn
Fraunhofer Institute for Applied Information Technology, Germany

ABSTRACT

This chapter presents a novel distributed software infrastructure to enable energy management and simulation of novel control strategies in smart cities. In this context, the following heterogeneous information, describing the different entities in a city, needs to be taken into account to form a unified district information model: internet-of-things (IoT) devices, building information model, system information model, and georeferenced information system. IoT devices are crucial to monitor in (near-) real-time both building energy trends and environmental data. Thus, the proposed solution fulfills the integration and interoperability of such data sources providing also a correlation among them. Such correlation is

DOI: 10.4018/978-1-7998-7091-3.ch019

the key feature to unlock management and simulation of novel energy policies aimed at optimizing the energy usage accounting also for its impact on building comfort. The platform has been deployed in a real-world district and a novel control policy for the heating distribution network has been developed and tested. Finally, experimental results are presented and discussed.

INTRODUCTION

Urban population have reached more than 54% of the total global population (World Health Organization, n.d.) and buildings represent nearly the 40% of total energy consumption. New control policies to optimize energy consumption in cities that include buildings and grids (e.g. power and heating distribution networks) are needed in order to reduce the overall energy consumption. For example, smoothing thermal energy peaks includes information on building energy profiles as well as grid topology and features to determine their actual effect on network infrastructures. This knowledge can be made accessible by means of ubiquitous sensors installed in buildings as well as in energy distribution networks. In the case of district heating, it is important to identify building energy profiles as well as distribution networks topologies and characteristics to reshape thermal energy peaks. Such reshaping has an effect on buildings air temperature profiles which can be carefully assessed to maintain an optimal level of user comfort.

Exploiting the Internet-of-Things (IoT) (Ashton, n.d.) communication paradigms (Giusto, Iera, Morabito, & Atzori, 2010), monitored data can be made available at the utility control centers. With such a data it is possible to implement smart control policies on cloud or ICT cluster infrastructures. Furthermore, IoT eases the integration of heterogeneous data sources that can be exploited to develop smarter policies. In such a way, Building Information Models (BIMs) (Eastman, Eastman, Teicholz, & Sacks, 2011) and Geographical Information Systems (GISs) (Tomlinson, 1968) can be integrated with sensor data. BIMs are 3D parametric models that describe all the elements of a building reporting construction and physics characteristics of the element. BIM allow to perform energy simulations that can be used to evaluate the impact of thermal reshaping on indoor air temperature. Such a simulation can be compared with temperature profiles provided by indoor sensors to tune energy models that can be used to evaluate policy impact on buildings where environmental sensors are not deployed. Researchers have proposed many frameworks in the last decades to exploit IoT technologies at building and house level (Kamilaris, Pitsillides, & Trifa, 2011) (Patti E., et al., DOI: 10.1109/JSYST.2014.2302750) (Bonino, Castellina, & Corno, 2008), and software solutions have been proposed to enable interoperability among various data formats and protocols (Da Xu, He, & Li, 2014) (Athanasios, Gialelis, Alexakos, Georgoudakis, & Koubias, 2006) (He & Da Xu, 2014). However, the integration of district models with sensor data remains a challenge.

In this chapter, we present a distributed IoT platform able to collect, process and analyze energy consumption data and structural features of systems and buildings in a district. In particular, data from Building Information Models (Eastman, Eastman, Teicholz, & Sacks, 2011) (BIMs), System Information Models (Brundu, et al., 2015) (SIMs) and Geographical Information Systems (Tomlinson, 1968)

(GISs) are interconnected and enhanced by historical and (near) real-time data collection from heterogeneous IoT tools. These instruments are installed across the district to track and control the energy delivery systems (both heating and power). The proposed solution exploits data from sensors installed in heat exchangers (i.e. the connected "things") to determine buildings energy profiles, while network models are used to evaluate the peaks accounting for grid characteristics and topology. Moreover, user comfort inside the building is monitored with temperature sensors. Such a data are also exploited to tune building models. This enables to develop models that can be used to evaluate the impact on habitants' comfort on buildings that are not instrumented with environmental sensors. The proposed solution has been tested in a real-world case study with hundreds of devices connected to monitor and manage the heating distribution network. Moreover, a policy aimed smoothing peaks of district heating network by rescheduling buildings energy requests has been simulated and evaluated. In a nutshell, the IoT platform acts as a district information management system, in which different stakeholders playing in a smart city scenario can collaborate and provide novel energy management policies and simulations.

Background

Due to the need of optimizing energy consumption in a smart city domain, different ICT solutions have been proposed. In the following the most relevant in this field are presented.

In Kim et al. (Kim, Thottan, Kolesnikov, & Lee, 2010), a data-centric middleware, that exploits a publish/subscribe model (Eugster, Felber, Guerraoui, & Kermarrec, 2003), is presented. The solution allows decentralized monitoring and control in a Smart grid context, which is appropriate for delivering information. However, the presented solution is not yet sufficient to have data access that is independent of this model. Indeed, the request/response communication approach is also needed to provide novel services that can easily retrieve data without having to wait for new events.

Patti et al. (Patti, et al., DOI: 10.1109/TSG.2014.2375197), presented a distributed software infrastructure for general purpose services in power systems. The proposed solution enables the interoperability across heterogeneous devices by creating a secure peer-to-per network.

W.r.t these solutions, the IoT platform presented in this chapter aims at creating a virtual model of a city district for providing Smart City services. Therefore, it is not yet necessary to accept data coming from IoT devices deployed around energy distribution networks. Such data must in fact be integrated and correlated with information about buildings, often geo-referenced.

In the Smart City scenario, several middleware solutions have been proposed to integrate heterogeneous data sources. Presented in (Neves, Santos, & Pinto, 2015), the ReActOR system is characterized by a tiny footprint and can be used as a service. It consists of three layers: a facade layer (Web Service), a core layer and a layer of extensions (supporting various technologies). It enforces user authorization and maps each user to a set of devices that they can manage, control, and sense. Moreover, ReActOR provides identifiers only for the different agents in the system, hiding data about users and devices. This middleware is limited on integrating only hardware and it does not support the integration of different data sources, such as Database Management Systems. Cândido et al. (Cândido, Colombo, Barata, & Jammes, 2011) presented a solution focused on interoperability, modularity, uncomplicated management and adaptation, and composed by a set of different components. Each module provides its services to the network using open web standards. This enables the composition of interoperable modules. However, this approach is limited to industrial automation scenarios.

SemsorGrid4Env (SemsorGrid4Env, n.d.) is a service-oriented architecture that enables the design of open large-scale semantic-based sensor network applications for environmental management. Further, it eases a simple and fast development of thin applications (e.g. mashups) allowing the integration of real-time data with historical data from other heterogeneous data-sources. This solution is tailored to environmental management and cannot be applied seamlessly to city district setting.

Different research projects and initiatives contributed to the definition of models and guidelines for enabling interoperability across various application domains. For instance, in the IoT-A project (Nettstraeter, 2012) an IoT reference model is presented. Such a model allows the description of an IoT solution by using shared building blocks, a reference architecture and general advices to IoT architects. Furthermore, the OneM2M alliance (OneM2M, n.d.) aimed at developing detailed technical specifications, to address the need for a common M2M Service Layer, using existing IoT and Web standards. However, it lacks in covering and analyzing aspects of IoT platforms, such as scalability, availability and deployment. FI-WARE (FI-WARE, n.d.), another research project, is devoted to design a service infrastructure for the Future Internet vision. Such infrastructure would be composed by reusable components, which could be selected and complemented with additional specific components.

Barbierato et al. (Barbierato et al., 2020) presented a distributed multi-model co-simulation platform based on smart grid architecture model to foster general purpose services in smart grids. The platform is used to demonstrate a testbed environment, where they can simulate realistic scenarios to assess their algorithms and services. In (Chapman et al., 2014), a framework to couple a multi-agent stochastic simulation of occupants with a building performance simulation tool based upon Energy+ was presented. (Thomas et al., 2014) focused on a multi-scale coupling process to enable a co-simulation between i) an energy model of a building simulated with Energy+ and ii) an energy model of a city simulated with CitySim. Finally, Wang et al. (Wang et al., 2017) developed a proof-of-concept urban energy co-simulation framework, which couples two simulation tools, EnergyPlus and NO-MASS, showing that the results were as expected and accurate in comparison with the framework results obtained by (Chapman et al., 2014).

The following work contributes to bring novelty in the design and development of a distributed IoT platform, based on open standards of the Web. Such a platform creates a single virtual parametric district model that is in charge of providing a standard access to heterogeneous IoT devices and District Information Systems deployed across the city. The presented platform provides information that is correlated, enriched, as a shared open repository. This IoT platform is instrumental for energy simulations on alternative control policies for the energy distribution network. To highlight our contribution, *Table I* reports a comparison of our distributed software infrastructure with the reviewed literature solutions.

IoT SOFTWARE INFRASTRUCTURE

The development of a virtual parametric district model for energy simulation and management through the creation of a distributed IoT platform is a challenging activity. Hundreds of heterogeneous IoT devices and data sources, which can be leveraged for energy consumption monitoring and management, must be integrated transparently in such a platform. Moreover, the platform also must be reliable and scalable. Following these considerations, it is necessary to step back and consider a lower level of detail for the districts, or focus on specific entities (e.g. buildings, IoT devices), which required a suitable model. There are two communication paradigms implemented by the proposed infrastructure: i) REST-based

Table 1. Comparison among our distributed software infrastructure and literature solutions

	Publish/ Subscribe	Request/ Response	Decentralized Middleware	IoT	BIM	SIM	GIS	Simulation Engine
Kim et al.	ü	û	ü					
Patti et al.	ü	ü	ü	ü				
ReActOR	ü	ü	ü	ü				
Cândido et al.	ü	ü	ü	ü				
SemsorGrid4Env	ü		ü	ü				
Barbierato et al.	ü	ü	ü	ü		ü		ü
Chapman et al.					ü			ü
Thomas et al.					ü	ü		ü
Wang et al.					ü			ü
Proposed Solution	ü	ü	ü	ü	ü	ü	ü	ü

non-real-time request/response (Fielding, 2000) and ii) MQTT-based (near-) real-time publish/subscribe (MQTT, n.d.).

LinkSmart® OpenSource Middleware (Linksmart, n.d.) is the basis for the infrastructure of this IoT software, which has been extended addressing requirements for energy management in a Smart City context. The platform includes the following tasks: i) modeling of real-world IoT devices and ICT systems, ii) providing their search and discovery by applications, and iii) exposing their data via common application protocols. This section describes several abstraction models and APIs implemented by a number of integration components and services defined by the platform in order to implement these tasks. The proposed software infrastructure is made up of three layers: i) *IoT Devices and Technologies Integration layer*, ii) *Services layer* and iii) *Applications layer*. The architectural schematics are shown in Figure 1, and each layer is described in more detail in the rest of this section.

IoT Devices and Technologies Integration Layer

ICT infrastructure made of heterogeneous devices and technologies, exploiting different communication standards and protocols, are leveraged by the proposed IoT platform. More specifically, in order to allow the interoperability among heterogeneous devices, *Device Connectors* are exploited by the *IoT Devices and Integration Layer* (bottom layer in Figure 1).

Device Connectors are integration components that describe real-world IoT Devices and ICT systems and integrate them in the middleware by using a common abstraction model. The Device Connector is a middleware-based software component that acts as a bridge between the middleware network and the underlying technologies, devices or subsystems (Patti, Acquaviva, & Macii, 2013). Web Service APIs for discovering and querying ICT data sources for each integrated system are exposed by such Device Connector, allowing the applications and other services seamless data access. In order to access data from all integrated systems, the middleware clients use the same Web Service API, instead of implementing system-specific and proprietary APIs of the integrated systems. Further functions of the Devicc Connector include its ability to provide a (near-) real-time communication by acting as a publisher and/or subscriber for the integrated IoT devices and ICT systems (see next Section B).

Figure 1. Architectural schema for the proposed IoT software infrastructure for Energy Management and Simulation

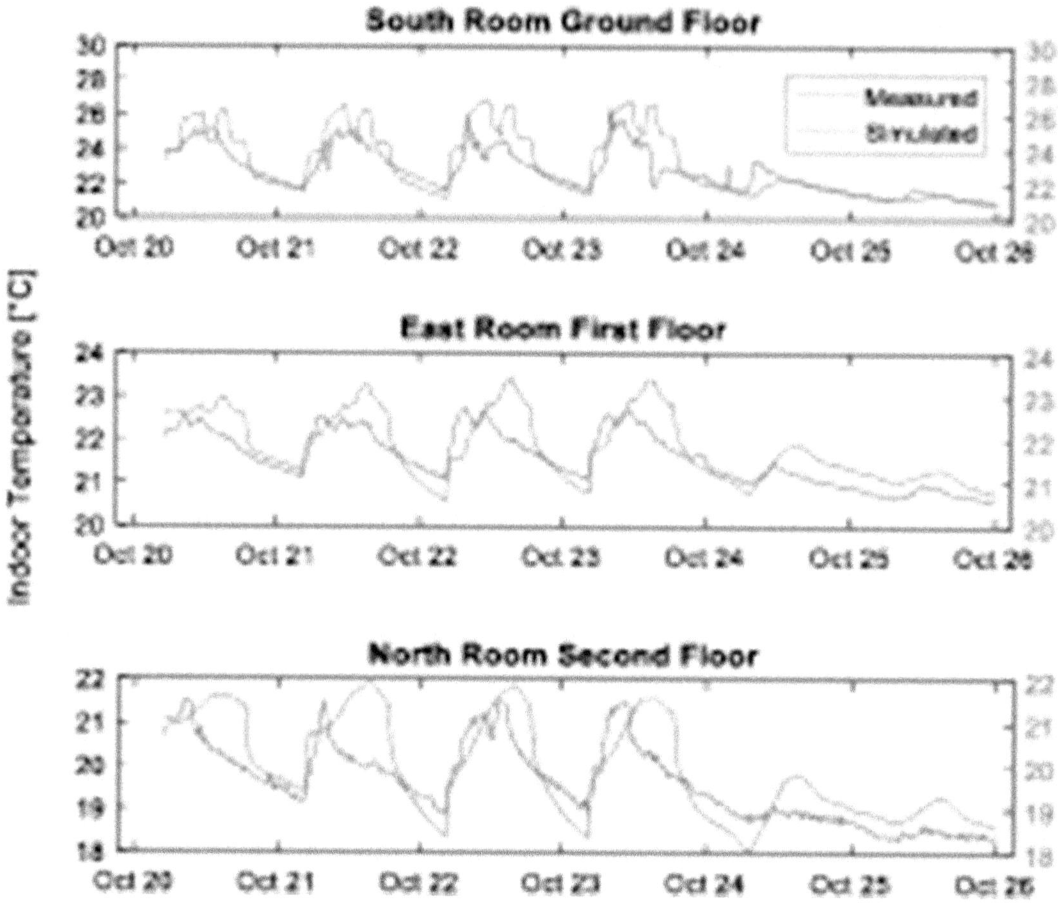

To transmit sensor measurements the Device Connector exploits the SenML (Sensor Markup Language) data format, defined as draft by *The Internet Engineering Task Force* (IETF) (SENML Draft, n.d.).

Services Layer

The founding pillar of the proposed IoT software infrastructure is the *Services Layer*. In order to access and manage information coming from heterogeneous IoT devices and technologies by leveraging a Web Service approach, this layer provides specifically designed components. The layer is also responsible for taking data from different data sources and correlating them by creating a virtual District Information Model. Furthermore, it provides features for control policies simulation for energy consumption optimization (e.g. electric, heating). The following section describes the main components of this layer.

1. Service Catalog: The network available services are described through the exposure of a JSON-based RESTful API. It is the entry point for applications, used to discover available services in the network. Entries of services such as the Resource Catalog and the MQTT Broker are contained here.
2. Resource Catalog: It constantly updates the endpoints registry for IoT devices and resources which are available in the infrastructure. Through the Resource Catalogue, a JSON-based RESTful API is exposed to: i) Device Connectors, which register the endpoints to access devices and resources, and ii) Applications, which discover such devices and their access endpoints to access information. The Resource Catalog allows new IoT devices and ICT systems can be transparently added, removed or replaced in the system, therefore providing the whole infrastructure with flexibility.
3. Message Broker: By applying the MQTT (Message Queue Telemetry Transport) (MQTT, n.d.) communication protocol, the different components of the infrastructure are provided with (near-) real-time asynchronous communication. This protocol allows loosely-coupled event-based systems to be developed, by following the publish/subscribe paradigm (Eugster, Felber, Guerraoui, & Kermarrec, 2003). Each device or component in the platform can i) publish data, and ii) subscribe to specific event notifications. Receiving data from publishers and forwarding data to subscribers is therefore responsibility of the Message Broker. Both internal components or client applications can be used to exploit this module. The dependencies between interacting entities is removed by this approach, thus increasing the infrastructure scalability.
4. Historical Datastore: IoT applications and services require, as a basic feature, the querying of historical data. However, in order to access historical sensor data a unified API is mandatory, especially considering a typical Smart City platform and the heterogeneity of its data-sources. A Historical Datastore API with standalone service implementation is defined by the platform, as well as for each integrated ICT system that already provides a database with historical data and an API to access it. The integration of new ICT or already existing data sources can thus be achieved by using the standalone service implementation. In addition, it allows high-frequency sensor measurements to be down-sampled and basic aggregates on them to be calculated by providing an optional aggregation functionality. It is possible to access historical sensor data across all integrated systems without data duplication by providing a unified API, which is allowed by the integrated system-specific implementations of the Historical Datastore API. Because of large volume and of storage and access policies, it can be problematic to manage the integration of historical data from a large ICT system. By having data requests forwarded to the integrated system and access policies between the middleware and the data owners being enforced, both these issues can be solved. This can be done through system-specific implementations of the Historical Datastore API. Huge amount of data can be stored and managed by the Historical Datastore. Historical data analysis can therefore support energy managers to develop control policies, through the availability of the above-mentioned data to Big Data Analytics tools thanks to its API.
5. Semantic Metastore: Another important feature of the platform is semantic interoperability, which allows the annotation and interlinking of integrated data sources and the query of the latter using the semantic attributes, through the use of Semantic Web technologies. Furthermore, higher-level services and applications can be described by using such annotations, and real-world entities they operate with that are not directly managed by the platform can also be modelled. The Semantic Metastore service implements this feature, to provide higher-level, developer-friendly APIs for populating and querying the semantic knowledge base through REST Web Services by building

on the off-the-shelf Semantic Web technologies such as Apache Jena (Jena, n.d.). Both internal components or client applications can query the Semantic Metastore.

6. DIM Services: There are different entities which compose a District Information Model (DIM). More specifically, there are three specific data sources that can be defined, which are represented using relational databases, and can be integrated to get a comprehensive view of a district:
    a. ***Building Information Models*** (BIMs (Eastman, Eastman, Teicholz, & Sacks, 2011)) are parametric 3-Dimensional models, where each model describes a building - both structurally and semantically (e.g. by defining materials and costs (Osello, 2012)). Decisions for the whole lifecycle of the building are supported by BIM;
    b. ***Geographical Information Systems*** (GISs (Tomlinson, 1968)) map the geographical location and topology of district entities, such as district buildings or energy distribution networks: data management and modelling of data for advanced cartography is provided by GIS;
    c. ***System Information Models*** (SIMs (Brundu, et al., 2015)) describe size and structure of energy distribution networks. The purpose of SIM is to exploit parametric and topological data to provide information for both visualization and simulation.

Currently, the integration and correlation of all these data-sources from the district point-of-view cannot be carried out by any commonly agreed method, even if they could all be represented using a relational database. All these heterogeneous data-sources cannot be collected under a single database due this fragmentation, because i) there is incompatibility between most Database Management System (DBMS) technologies, and ii) different databases could have conflicting semantics for the same value. On the other hand, RESTful Web Services and distributed deployment are the pillars of the following subsystem. Each component of the system is provided by a uniform interface through the use of RESTful Web Services. More specifically, no matter what data source is actually queried, the return is always

Figure 2. DIM Services schema

the same response format. Therefore, the result of a same query returns the same response to the client application even if there is a change in a component technology (e.g. a DBMS). Figure 2 shows a schematic representation of the DIM Services subsystem.

The subsystem is built in a micro-service fashion (Fowler, n.d.), i.e. it consists of a suite of small, autonomous, lightweight services. From a logical point-of-view, two independent sub-layers make up each component in DIM services: i) *Data layer* and ii) *Interface layer*. The following section presents the BIM Service Provider structure, which can be extended easily to GIS and SIM Service Providers. For the first sub-layer, each building in the district is represented by a database, and the set of several databases represents the Data layer. The internal representation of BIM models (which have been created with proprietary software Autodesk Revit (Autodesk Revit, n.d.)) can also be stored in a relational DBMS, which is used to manage the databases. For the second sub-layer, a RESTful Web Service interface is provided to the underlying databases by the Interface layer. This sub-layer uses SQL queries and prepares JSON results to interact with the Data layer, and also provides building resources, such as proprietary (Revit files) and public BIM standards (IFC and gbXML).

To complement the BIM information, other kinds of data sources are needed within the District Information Model, as depicted in Figure 1. In particular, GIS and SIM have been identified. A Web Service interface is required for both these data-sources, which can use DIM context layer to integrate into DIM Web Services. Through the REST API, JSON results are returned by both the GIS and the SIM Service provider. The main interface to district information, which is the DIM Context layer, is above each Service provider. The DIM Context layer creates a virtual parametric model of the district that can be queried as a whole by client applications, and also acts as a translator from the district context (district-based queries) to a specific data-source context (e.g. building-based queries). On the other hand, it also integrates back responses from the different Service Providers to a single response. The DIM Context layer also behaves as a proxy to the Service Provider Interface layer, and therefore works independently from each of the latter. It is also able to compose complex queries, which may span different information domains (e.g. building and geographical information), through the use of inter-provider queries. For example, it offers the opportunity to retrieve information about buildings composed by a specific number of stories, by querying both the BIM and the GIS Service Providers.

7) Simulation engine: It is a specific component of the proposed IoT software infrastructure, aiming to support the development of new control policies and the execution of energy simulations in a district environment. It provides a standard entry-point to the district simulator through a Web Service enabled interface.

Moreover, the Simulation engine exploits both publish/subscribe and request/response communication approaches (MQTT and REST, respectively) in order to retrieve information from all the other components of the platform, by providing features to develop control policies. District Information Models, and also historical or (near-) real-time information collected from the heterogeneous IoT devices can therefore be taken into account by such simulations aimed to optimize the energy consumption. For example, the environmental information from the Historical Datastore that collects sampled data sent by IoT devices can be gathered by this module. (Near-) Real-time data from IoT devices can also be received by the Simulation engine, through the Message Broker, since it can also work as a subscriber. Information about BIM, GIS and SIM are correlated by the DIM Services, which can also retrieve information about a district as a whole. Semantic information retrieved from the Semantic Metastore is also exploited by the DIM Services in order to perform such correlations.

It is important to highlight that simulations for the different energy flows in the district (e.g. electrical and heating) can be performed by the Simulation engine module. A more in-depth description of control policies for the district heating network are provided in next Section *ENERGY OPTIMIZATION RESULTS*.

A. Applications Layer

The data coming from the underlying layers of the proposed IoT infrastructure can be managed and post-processed by developing specific applications, with the support of a set of APIs and tools provided by the Applications Layer (the highest layer in Figure 1). Such applications may be web-based, mobile, desktop and semantic web-clients. The Web Services approach enables the interoperability among different IoT devices and technologies at this level. Example of applications are:

- the *Benchmarking tool* (see Figure 3) compares different buildings' energy behavior. End-user for this tool are building managers, households and energy distributors. By providing some input parameters analysis (e.g. building typology, building use, etc.), the application allows end-user to filter buildings that are compared and benchmarked together. The benchmarking tool retrieves data from DIM Services (i.e. BIMs, GISs and SIMs). Due to privacy concerns, each building is represented by a virtual building which summarize the features of all the similar buildings under review.
- the *WebGIS Dashboard* (see Figure 4) is a web application which integrates different information layers. This application exploits both 2D and 3D maps to show heating and electricity consump-

Figure 3. Benchmarking Tool

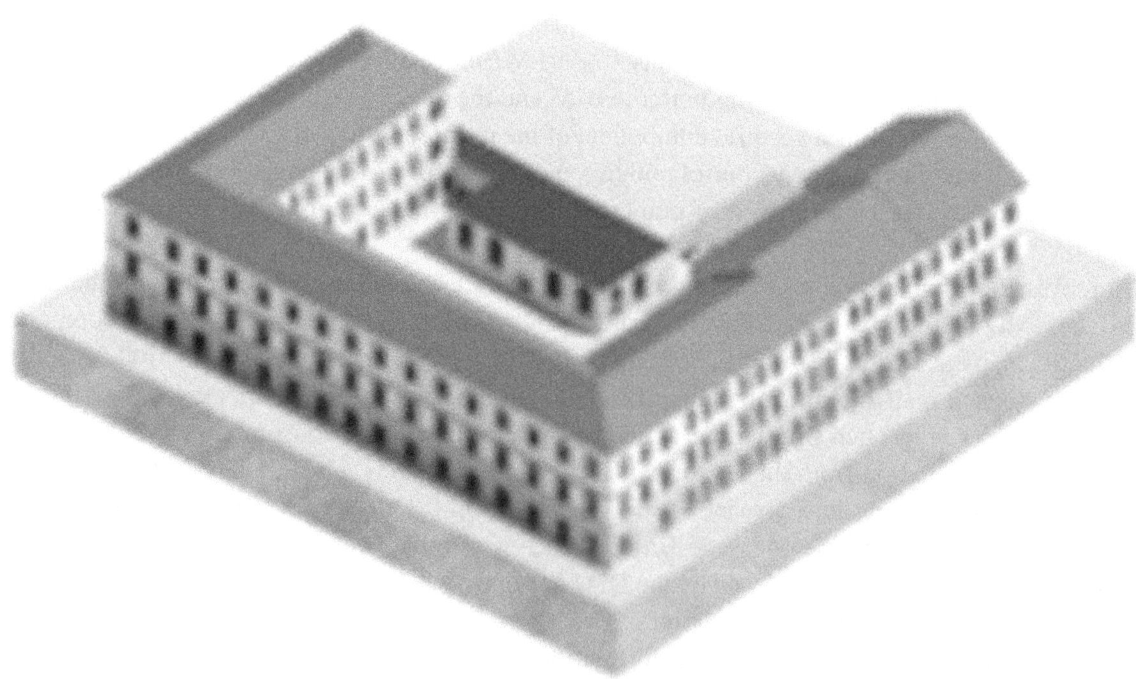

tion around the district, both to provide monitoring of sensor networks and advanced data analysis through the integration of heterogeneous data. It retrieves data from DIM Services (i.e. BIMs, GISs and SIMs). Proposed end-users for this dashboard are building managers, households, energy distributors, energy managers and public administrators.

Figure 4. WebGIS Dashboard

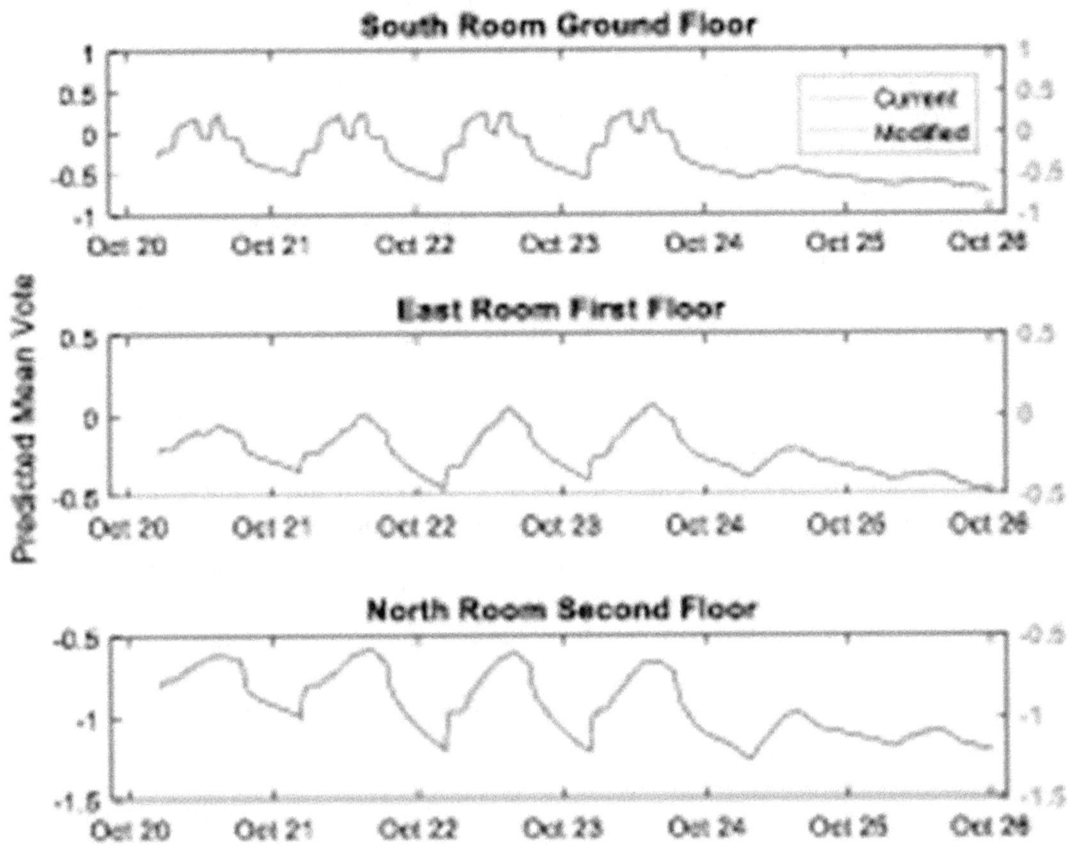

CASE STUDY

The proposed platform is implemented to provide services for energy management in a smart city context. To test the capabilities of the presented solution a case study in a real district in Turin, north-west of Italy, has been deployed. In the case study, the district heating (DH) provides thermal energy to nearly 50% of public and private buildings. Moreover, to collect indoor data about air temperature and relative humidity on seven representative buildings, IoT devices forming different Wireless Sensor Networks (WSN) have been deployed.

The DH in Turin is one of the largest in Europe, being composed by about 580 km of double pipeline, with more than 5700 buildings connected (over 55% of the town). The thermal request presents a high variability during the 24-h of a day and during the heating season. In particularly daily requests are composed of a morning peak of about 1300-1400 MW while the afternoon request ranges from about 850 MW in typical winter days to about 400 MW in typical middle season days. The thermal requests are satisfied with three efficient cogeneration systems, able to produce a total thermal power of about 740 MW. The exceeding request is covered using thermal storage units and auxiliary boilers.

In the selected district nearly 4000 IoT devices are deployed for monitoring and managing the DH. They collect data (e.g. instantaneous power, cumulative energy consumption, water flows and temperatures) every five minutes. With such a number of sensors, about 70% of the buildings connected with the DH network are monitored. The remaining buildings have been characterized assuming a behaviour equal to the average profile of the buildings in the network. This approach has been validated through comparison of the calculated and measured thermal request profiles at the plants. Collected data are sent to a remote *Historical Datastore* unit by means of embedded gateways. Each gateway is accompanied by a GPRS modem and executes the following tasks: i) scheduling of data collection, ii) management of IoT sensors, iii) communication via GPRS, and iv) data transfer to remote *Historical Datastore*. The *Historical Datastore* for DH consists of four dispatchers, to improve reliability and scalability. Each measurement is sent to a different dispatcher, which forwards it to a database.

Differently, a significant part of the infrastructure consists of heterogeneous data-sources, e.g. BIMs. Such models have been prepared for selected buildings and, the corresponding SIM, for the district heating network.

This chapter assesses how this infrastructure is beneficial: i) to optimize the energy demand, ii) to manage a large number of IoT devices, and iii) to exploit of parametric models for energy simulations. Thanks to the integration of BIMs the platform it is able to perform fine-grained energy simulations, down to the room level, and are useful to plan retrofitting of existing buildings.

The most popular construction types in the selected district are load-bearing masonry (mostly used until the 1950) and reinforced concrete. Many cases present a mixed solution between the two technologies. For this reason, seven representative buildings have been considered and IoT devices for indoor air temperature and relative humidity have been deployed on them. They exploit the *Device Connector* component to be integrated into the proposed software infrastructure. In particular, such sensors communicate using MQTT protocol (Eugster, Felber, Guerraoui, & Kermarrec, 2003) (MQTT, n.d.) and send collected data to a *Historical Datastore*.

For this case study, a Primary School in the selected district has been chosen (see the BIM in Figure 5). This primary school was originally built in 1902 and almost totally rebuilt at the end of the World War II. The building structure is a three-storey load-based masonry with a heavy massive envelope. The main heating supply is based on radiators. Such rooms have been selected considering the orientation. Three representative rooms have been selected to better describe the air temperatures trends of the case study and the building shape: the first room (S) is oriented to South, the second room (E) to East and the third room (N) to North-West. Further, the room N is also located in the highest floor of the North-West corner of the primary school.

Figure 5. BIM of the Primary school selected as case study building

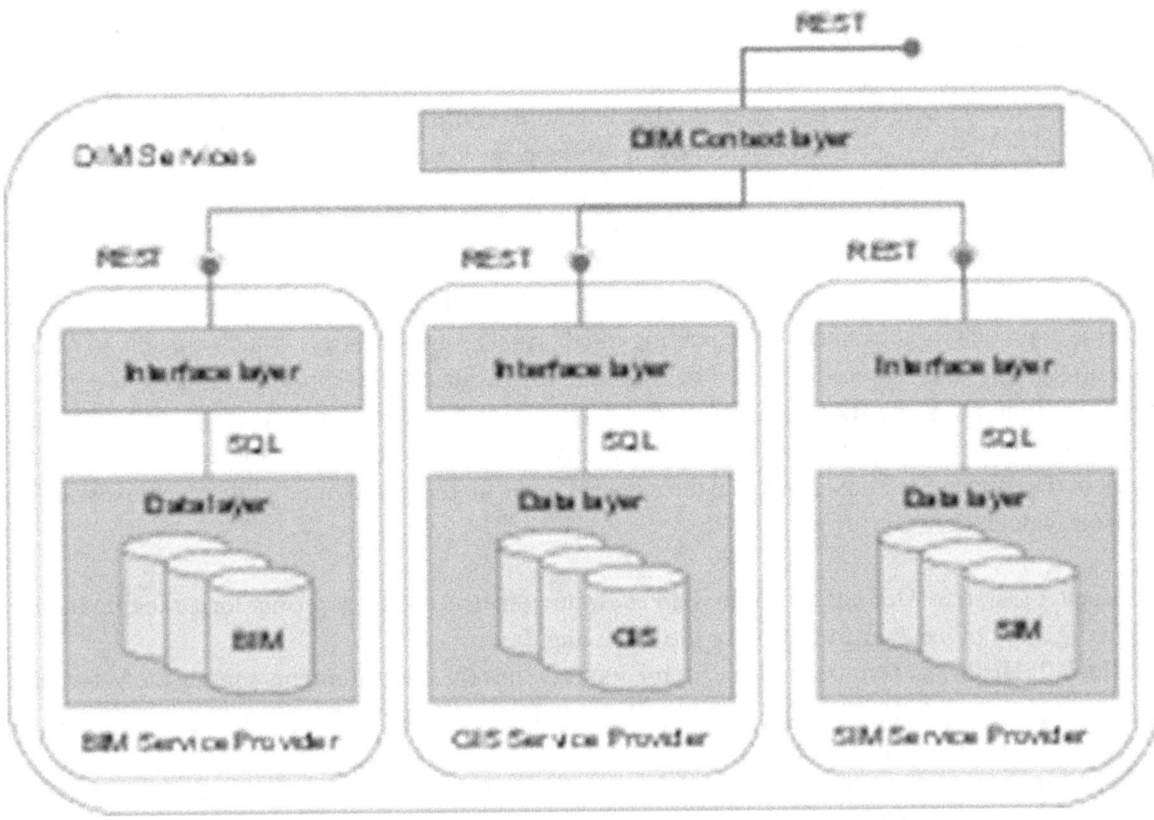

ENERGY OPTIMIZATION RESULTS

Figure 6. Procedure for the analysis of alternative thermal request profiles

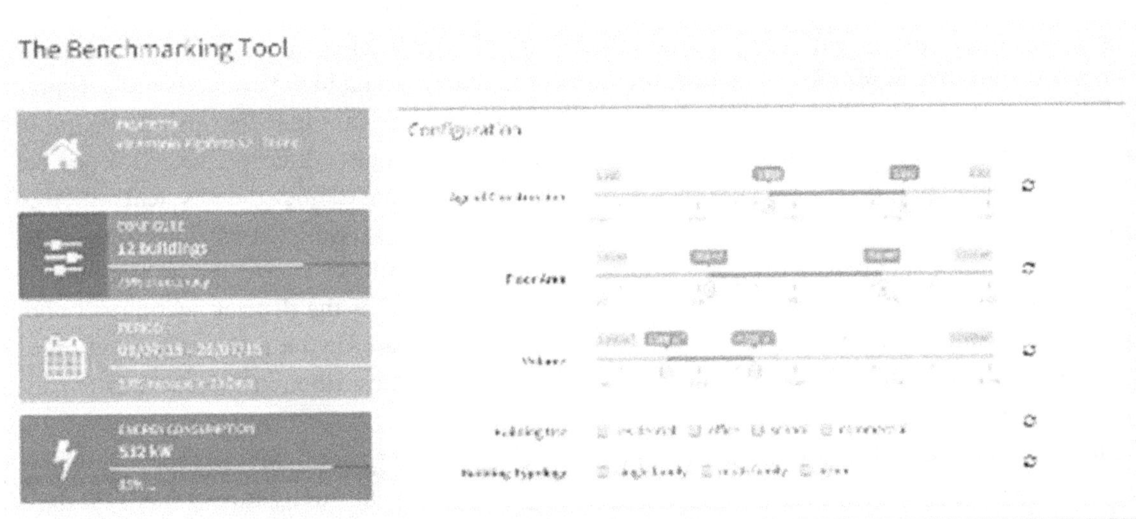

Methods and Policies

The use of cogeneration has to be maximized in DH to reduce the primary energy consumption in the supply of heating to the connected buildings. Such a goal can be reached through thermal storage (Verda & Colella, Primary energy savings through thermal storage in district heating networks, 2011). However, other improvements can be obtained by optimizing the thermal request profiles of the buildings acting at district level. This strategy plays a crucial role considering that the district heating network is in continuous evolution, with increasing number of buildings connected to the network. Meanwhile, the construction of new storage units has to face the issues related with land occupation and also the investment costs. Exploiting the IoT infrastructure previously described it is possible to implement an innovative approach for the analysis of changing the thermal request profiles, which are able to produce similar effects than thermal storage units. In particular, by means of the *Simulation Engine* presented in previous Section *IOT SOFTWARE INFRASTRUCTURE*, it is possible to run thermal request simulations following the approach described in Figure 6, that involves all the buildings connected to the heating network in the district. Red boxes indicate the external input for the procedure, while blue boxes perform the various calculations.

- **Block 1** (Strategy Definition), defines an alternative thermal request profile for a specific building on the basis of weather forecasts for the following day.
- **Block 2** (Historical Datastore - measurements) provides the measurements sent by the IoT-enabled heat exchangers in the buildings through *Message Broker* from the *Historical Datastore* of the infrastructure. After this, the algorithm computes the daily thermal consumption associated with the actual request profile, as well as it evaluates: i) the status of the heat exchanger, and ii) the values of parameters characterizing the heat losses and the dynamic behavior of the building (Verda & Baccino, Primary energy reductions in district heating networks through variation of the thermal load profile of the users, 2014). The other quantities are calculated every day by exploiting almost steady state behavior reached in the afternoon and transient evolution when the heating system is switched off or attenuated.
- **Block 3** (Building compact model) verifies the applicability of the calculated strategy for the building. In particular, it compares the expected average internal temperatures resulting from the proposed rescheduling with the original scheduling. The internal comfort of the building is imposed as a constraint for the new scheduling, which can be adjusted in the case it is not acceptable.
- **Block 4** (DIM (SIM + GIS) model of the network) is an additional input for this tool. Exploiting the DIM Services, this block retrieves the SIM model of the network correlated with geo-referenced information (GIS) of the various sub-networks. This model defines the physical structure of sub-networks, and it is updated every year to properly consider possible expansions of the district heating to new areas or possible connection of additional buildings.
- **Block 5** (Network model) calculates the thermo-fluid dynamic of the district heating network to estimate from the thermal request of the users the thermal load for the full district and then for the thermal plants. The physical model of the network is fully described in (Sciacovelli, Verda, & Borchiellini, 2013).
- **Block 6** (Plant model) evaluates the KPIs used to estimate energy and environmental impacts of the new strategy w.r.t the current request profiles. Main KPIs are the reductions in primary energy consumption and CO_2 emission. Set points and heat scheduling for each heat exchanger are

produced by the thermal request simulation, to allow the simulation at building level to proceed. The proposed platform provides fine-grained monitoring which, together with the IoT devices deployed in the buildings, allow the feasibility of such a control policy. Detailed information about energy consumption of buildings served by the heating network are available. Thanks to this capillary IoT platform, it is actually possible to quantify both i) the cumulative heating request profile at the energy power plant, and ii) the energy request profile.

Simulations at building level were also carried out in order to assess the impact of the new district heating control policy. The BIM for the building, developed with Autodesk Revit 2015, is used as an input for such simulations. Each building feature is used when compiling the BIM, and the building typology is also taken into account. In order to calculate the evolutions of the internal temperature, the BIM model is then imported into Energy+ environment by means of the DIM Services component of the infrastructure. The room occupancy is then set, and heating set points and schedules provided by the district thermal request simulation are used to run the Building Simulation Engine

Thermal parameters, such as the indoor temperature and thermal comfort for the building, are then estimated by the Building Simulation Engine. For the building selected as case study, simulated indoor air temperature trends are compared with real measurements operated in selected rooms, where IoT sensors have been deployed. Such real-world measurements are sent by IoT devices to the *Historical Datastore* through the *Message Broker* every 15 minutes. This study investigates the results of simulations performed for the selected building on working days, when the primary school is occupied by users.

A. Experimental Results

For the city under investigation two different scenarios have been considered, corresponding with a typical winter day, in order to evaluate how proper change of the thermal request profiles of the buildings can generate possible benefits. The current request profile of the buildings is considered in the reference scenario (solid line), with the thermal loads of the plants being reported by Figure 7. Then the time the heating system is switched on is anticipated and postponed in a second scenario (dashed line), therefore modifying the request profiles of 50% of the buildings. The request profile curve of each building is unmodified, even though the alteration occurs within less than a 30 min window.

For a large percentage of the buildings the start-up time is anticipated, which is the main content of the strategy optimization. From about 4:30 a.m. to 6:00 a.m., the total thermal load increases as a consequence. About 0.14 GWh of heat is moved between periods, more specifically from periods where the total capacity of the cogenerators is exceeded to periods characterized by a total load smaller than the capacity of cogenerators, while more than 120 MW are also reduced from peak request, as shown in Figure 7. As a consequence, auxiliary boilers can be substituted by the cogenerators to produce this amount of heat. In order to produce a unit of heat, 1.11 units of fuel are requested, therefore the heat produced through boilers has a primary energy factor of about 1.11. On the other hand, the combined cycles operating in cogeneration mode which produce heat have a primary energy factor of about 0.36. When these combined cycles operate in full electricity production, their electrical efficiency is about 0.58, which justifies the above small value. When they operate in cogeneration mode, for each 10 MW of heat produced, the electricity production decreases of about 2.1 MW (Verda, Chirio, & Ponta, 2006). The primary energy consumption achieved a corresponding reduction of 3%. This reduction depends in a non-linear way on the percentage of buildings which heating schedule can be modified: if a reduction

Figure 7. Thermal loads in a typical winter day obtained by using the current request profiles and the modified request profiles

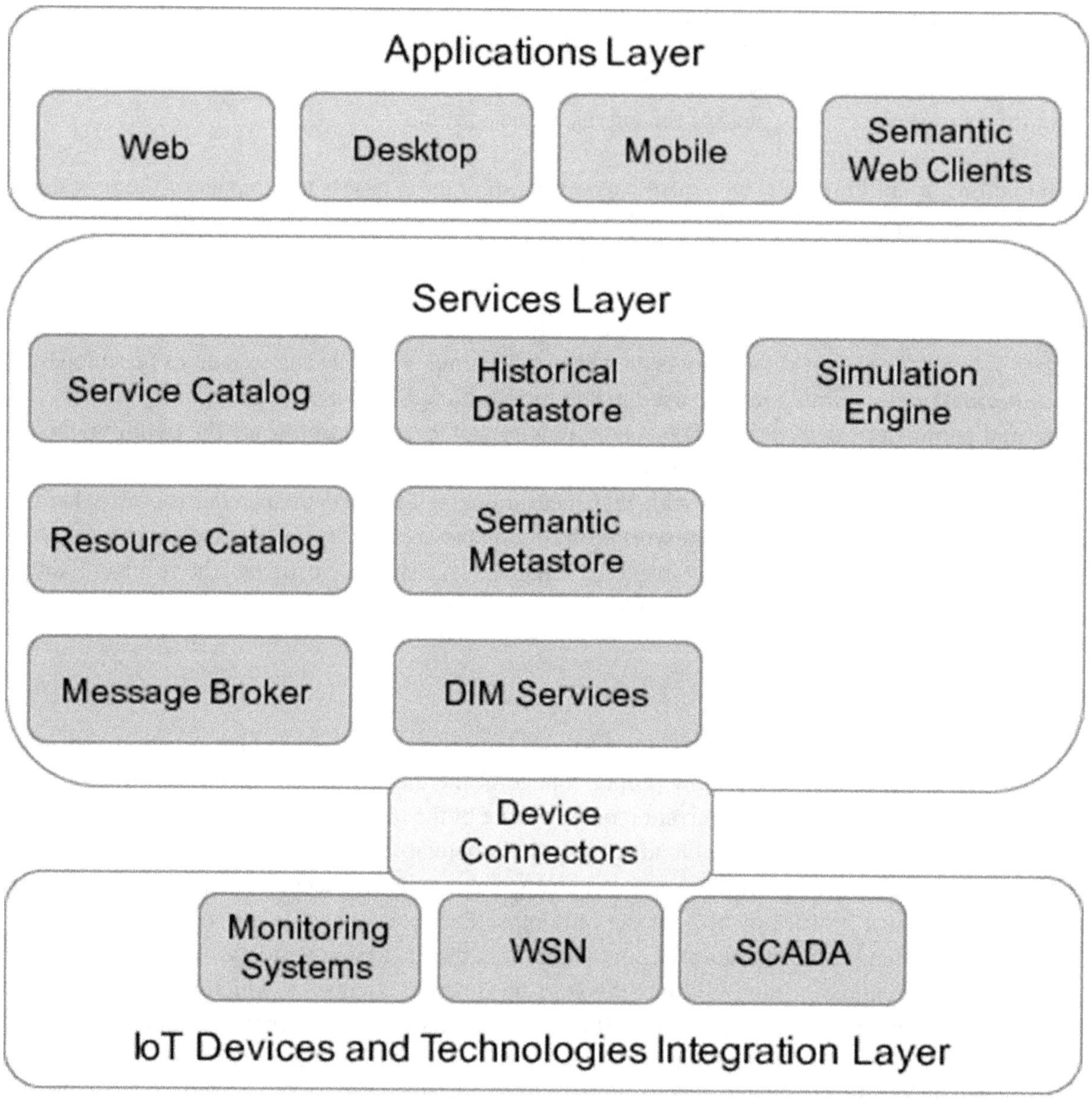

of 25% occurs to the percentage of buildings, a reduction to about 2% occurs to the primary energy saving, while if an increase of 75% occurs to the percentage of buildings, an increase to about 3.5% occurs to the primary energy savings.

It is important to highlight that an even larger reduction in the primary energy consumption would result from the application of a strategy aiming at optimizing the shape of the thermal request profiles of the buildings.

The acceptability of the changes in the thermal profiles in the various buildings is evaluated by applying the building model once the new schedule is defined. More specifically, acceptable new comfort

Table 2. Energy schedules for selected building

	Start	Pause	Restart	Stop
Current	5:30 am	---	---	10:00 pm
Proposed	5:00 am	3:00 pm	3:30 pm	10:00 pm

Table 3. Difference between simulated and measured temperature and humidity (current schedule)

ΔT sim. - meas.		**Room S**	**Room E**	**Room N**
Max	[°C]	0.44	0.71	0.38
Min	[°C]	-0.09	-0.03	-0.25
Average	[°C]	0.37	0.23	0.32
St. Dev.	[°C]	0.98	0.46	0.69
Δφ sim. - meas.		**Room S**	**Room E**	**Room N**
Max	[%]	6.18	5.32	5.28
Min	[%]	-6.02	-5.01	-4.89
Average	[%]	1.28	0.76	0.52
St. Dev.	[%]	4.76	4.02	3.56

conditions for the end-users are achieved by checking internal temperatures when the thermal request profile of a building is modified. *TABLE II* presents the current and the proposed (anticipated) schedules for the primary school. An estimation of the comfort has been carried out at building level by using the temperature trends and by exploiting the Predicted Mean Vote (PMV) methodology (Fanger, 1970), which is a measure of thermal comfort of the building. PMV has been estimated for i) the current energy distribution schedule and for ii) the schedule proposed as output from the district model.

In order to assess whether a correct simulation of a real-world building could be carried out by the model, a comparison between simulated and measured temperatures from IoT devices has been completed. Simulation models can actually be validated by using an unprecedented fine-grained measurement of indoor conditions, which is allowed through the use of IoT devices, which can be exploited through the infrastructure.

Table 4. PMV current schedule vs proposed schedule

		Room S	**Room E**	**Room N**
Current	**Min**	-0.06	-0.32	-0.80
	Max	-0.34	0.09	-0.30
	Avg	0.16	-0.09	-0.48
Proposed	**Min**	-0.09	-0.37	-0.82
	Max	0.31	0.04	-0.36
	Avg	0.13	-0.12	-0.52

The impact on comfort of the scheduling policy, shown in Figure 7, can be assessed by using the infrastructure, and most importantly the Simulation Engine, as explained by Figure 8 and Figure 9. By comparing the real data taken from installed sensors with simulated trends, we show how the latter is reliable. Also, we envision a potential usage of the infrastructure where sensor data can be used to tune the simulation model.

For the three selected rooms (S, E and N), the actual and simulated (with the current schedule) temperature trends are shown in Figure 8. The trends are estimated in four working days and during the weekend. *TABLE III* also provides a brief description of the evolution of temperature and relative humidity (φ). The maximum temperatures are tended to be overestimated and the minimum temperatures slightly underestimated by the simulation, in all cases. The maximum temperatures are experienced by the South room (S), which is constantly exposed to solar radiation during the day. Certain unforeseeable factors such as ventilation, which usually occurs by manually opening the windows, window shading and occupancy of rooms generate larger deviations in the central part of the day. Since the simulations should rely only on predictable data at a district level, they do not consider these effects, even if they have a significant impact on the temperatures measured by the IoT devices. In this case, the Mean Absolute Error (MAE) between simulated and measured temperature is 0.37 °C with a standard deviation of 0.98 °C, the Root Mean Square Error (RMSE) is 0.48 °C and the Mean Bias Error (MBE) is -0.18°C. On the other hand, other rooms do not fully benefit of solar exposure during the day, like the east room (E) and north room (N). The best model prediction is in Room E, where MAE is 0.23 °C, the standard deviation is 0.46 °C, RMSE 0.13 °C and MBE is -0.11 °C. Room N, which is located at the top floor of the primary school in the north-west corner, has a MAE equal to 0.32 °C with a standard deviation of 0.69 °C, RMSE equal to 0.27 °C and MBE equal to -0.16 °C. The evaluation of internal comfort conditions is not significantly affected by relative humidity, even if it is slightly overestimated in simulations.

It is important to underline how the dynamic behavior of the building must be captured, since meteorological forecasts for the following day are used by the model, with the objective to check the effects of variations in schedules. This can be analyzed when by comparing the average temperature degrease when the heating system is switched off (i.e. between 8:30 p.m. and 4:30 a.m.). The measured values for the three rooms are 0.05 °C/hour, 0.08 °C/hour and 0.10 °C/hour, respectively, while the simulated values are 0.06 °C/hour, 0.10 °C/hour and 0.10 °C/hour. The model is suitable for the analysis, since these differences are sufficiently small.

The current schedules for energy distribution and the schedules proposed by the output of the district simulation have been compared by using the PMV methodology, in order to better assess the degree of comfort. The simulation results have been used to perform this. As shown in Figure 9, the expected PMV for the proposed schedule is only slightly below that corresponding with the current schedule. As the last value representative of slightly cold comfort conditions, PMV varies from about 0.3 in the central hours of the day and -1.3 at night. Values below -1 are only experienced in the north room and outside the working hours. *TABLE IV* presents a comparison between the current and proposed schedules, where only the working hours (8 am - 5 pm) are considered. A 20 min pause is introduced for the heating system between 5:50 and 6:10, which during daytime causes a reduction of about 0.03, with slightly smaller PMV in the selected rooms (on average) caused by the propose schedule. The internal temperatures and the PMV corresponding with the current schedule are recovered in the early afternoon for South and East rooms and in the late afternoon for the North room through slightly longer operation of the heating system in the afternoon.

Figure 8. Simulated and actual temperature trends for selected rooms

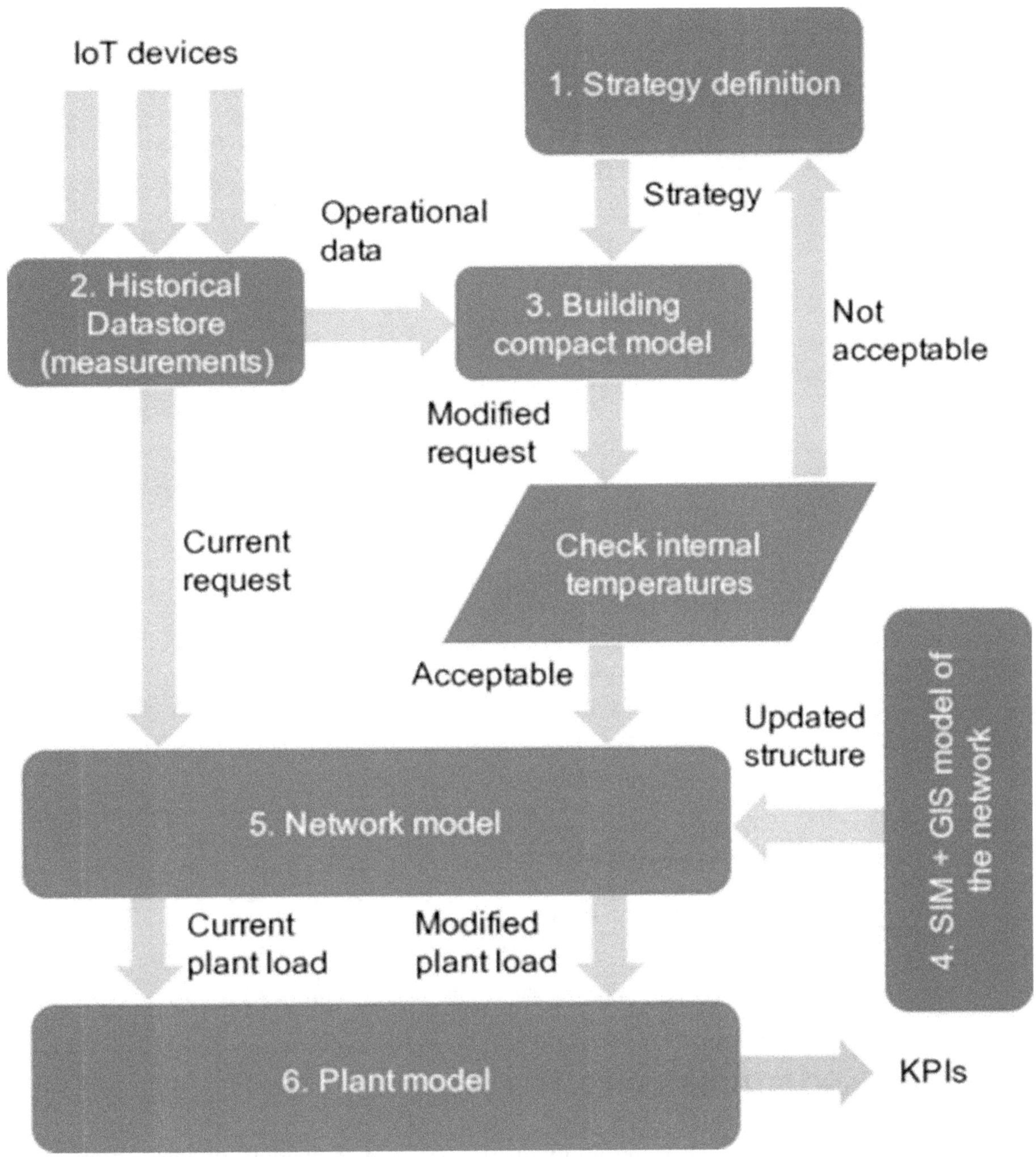

The adoption of a new energy distribution schedule by the energy provider is supported by the building simulation in the presented case, since it would promote energy saving (at the district level) without affecting the user comfort (which has been estimated using the building model).

Figure 9. PMV methodology for proposed schedule for selected rooms

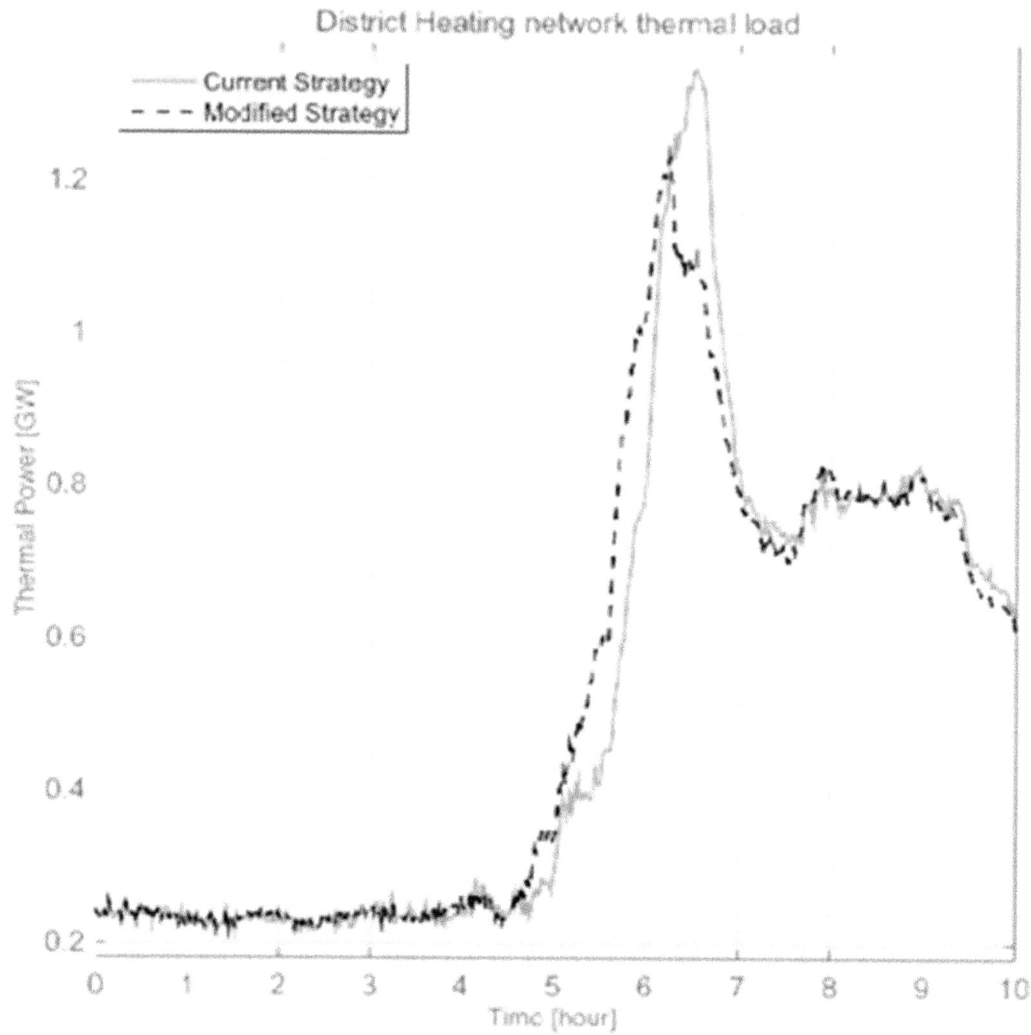

CONCLUSION AND FUTURE WORK

A novel IoT platform for the management of city district data and simulations of energy flow is presented in this work. More specifically, this platform is instrumental to i) monitor and manage a whole city district through the integration of heterogeneous IoT devices; ii) share building and energy network resources, both for visualization and simulation of energy policies, at building and district level; and iii) assess the quality of the energy model of buildings.

The micro-service paradigm has been followed in order to design and develop the present platform, which is also able to reliably be scaled up. Furthermore, the use makes it easy its integration into existing systems is facilitated by the use of shared open standards of the Web. The platform is ready to send actuation commands to the devices deployed in the district. For example, the DH scheduling control system is currently being tested with it. In this case, heat exchangers are controlled by sending scheduled on-off heating switching times to IoT devices.

Finally, a currently unmet need in state-of-art applications, the need for an energy management system for a city district, is satisfied by this platform. Simulations for energy distribution policies from district level down to building room level can be carried out by this platform, with a special focus on both savings of district energy and comfort level of the end-user, as the real-world case study showed. It is important to underline how the proposed IoT platform provided a fine-grained monitoring which made such policies feasible. As future works, the platform will be extended to integrate simulators like OPAL-RT and RTDS to unlock real-time simulation also with Hardware-in-the-Loop capabilities suitable to study in-depth electrical behaviors in smart grids.

ACKNOWLEDGMENT

This work was supported in part by EU, FP7 SMARTCITIES 2013 DIMMER project (District Information Modelling and Management for Energy Reduction).

REFERENCES

Ashton, K. (n.d.). *That 'Internet-of-Things' Thing*. Retrieved September 2015, from http://www.rfid-journal.com/articles/view?4986

Athanasios, K. P., Gialelis, J. V., Alexakos, C. E., Georgoudakis, M. J., & Koubias, S. (2006). Vertical integration of enterprise industrial systems utilizing web services. *IEEE Transactions on Industrial Informatics*, *2*(2), 120–128. doi:10.1109/TII.2006.875507

Autodesk Revit. (n.d.). Retrieved November 2015, from http://www.autodesk.com/products/revit-family/overview

Bonino, D., Castellina, E., & Corno, F. (2008). The DOG gateway: Enabling ontology-based intelligent domotic environments. *IEEE Transactions on Consumer Electronics*, *54*(4), 1656–1664. doi:10.1109/TCE.2008.4711217

Brundu, F. G., Patti, E., Del Giudice, M., Osello, A., Macii, E., & Acquaviva, A. (2015). DIMCloud: a distributed framework for district energy simulation and management. In Internet of Things. User-Centric IoT (pp. 331-338). Springer International Publishing. doi:10.1007/978-3-319-19656-5_45

Cândido, G., Colombo, A. W., Barata, J., & Jammes, F. (2011). Service-oriented infrastructure to support the deployment of evolvable production systems. *IEEE Transactions on Industrial Informatics*, *7*(4), 759–767. doi:10.1109/TII.2011.2166779

Da Xu, L., He, W., & Li, S. (2014). Internet of Things in industries: A survey. *IEEE Trans. Ind. Informat.*, *10*(4).

Eastman, C., Eastman, C. M., Teicholz, P., & Sacks, R. (2011). *BIM handbook: A guide to building information modeling for owners, managers, designers, engineers and contractors*. John Wiley & Sons.

Eugster, P. T., Felber, P. A., Guerraoui, R., & Kermarrec, A.-M. (2003). The many faces of publish/subscribe. *ACM Computing Surveys*, *35*(2), 114–131. doi:10.1145/857076.857078

Fanger, P. O. (1970). *Thermal comfort. Analysis and applications in environmental engineering*. Danish Technical Press.

FI-WARE. (n.d.). Retrieved September 2015, from https://www.fiware.org/

Fielding, R. T. (2000). *Architectural styles and the design of network-based software architectures*. University of California.

Fowler, M. (n.d.). *Microservices*. Retrieved September 2015, from https://martinfowler.com/articles/microservices.html

Giusto, D., Iera, A., Morabito, G., & Atzori, L. (2010). *The internet of things: 20th Tyrrhenian workshop on digital communications*. Springer Science & Business Media.

He, W., & Da Xu, L. (2014). Integration of distributed enterprise applications: A survey. *IEEE Transactions on Industrial Informatics*, *10*(1), 35–42. doi:10.1109/TII.2012.2189221

Jena, A. (n.d.). Retrieved January 2016, from https://jena.apache.org

Kamilaris, A., Pitsillides, A., & Trifa, V. (2011). The smart home meets the web of things. *International Journal of Ad Hoc and Ubiquitous Computing*, *7*(3), 145–154. doi:10.1504/IJAHUC.2011.040115

Kim, Y.-J., Thottan, M., Kolesnikov, V., & Lee, W. (2010). A secure decentralized data-centric information infrastructure for smart grid. *IEEE Communications Magazine*, *48*(11), 58–65. doi:10.1109/MCOM.2010.5621968

Linksmart. (n.d.). Retrieved December 2015, from http://linksmart.eu

MQTT. (n.d.). Retrieved November 2015, from http://mqtt.org

Nettstraeter, A. (2012). Architectural reference model for IoT. *EC FP7 IoT-A (257521) D, 1*, 2.

Neves, D. T., Santos, M., & Pinto, M. (2015). ReActOR: A middleware as a service to interact with objects remotely. In *Proc. ICIT 2015* (pp. 2433-2439). IEEE. 10.1109/ICIT.2015.7125456

OneM2M. (n.d.). Retrieved September 2015, from https://onem2m.org/

Osello, A. (2012). *The Future of Drawing with BIM for Engineers and Architects*. Dario Flaccovio Editore.

Patti, E., Acquaviva, A., Jahn, M., Pramudianto, F., Tomasi, R., Rabourdin, D., Virgone, J., & Macii, E. (2016, September). Event-driven user-centric middleware for energy-efficient buildings and public spaces. *IEEE Systems Journal*, *10*(3), 1137–1146. Advance online publication. doi:10.1109/JSYST.2014.2302750

Patti, E., Acquaviva, A., & Macii, E. (2013). Enable sensor networks interoperability in smart public spaces through a service oriented approach. *IEEE International Workshop on Advances in Sensors and Interfaces (IWASI)*, 2-7. 10.1109/IWASI.2013.6576081

Patti, E., Syrri, A. L., Jahn, M., Mancarella, P., Acquaviva, A., & Macii, E. (2016, March). Distributed Software Infrastructure for General Purpose Services in Smart Grid. *IEEE Transactions on Smart Grid*, *7*(2), 1156–1163. Advance online publication. doi:10.1109/TSG.2014.2375197

Sciacovelli, A., Verda, V., & Borchiellini, R. (2013). *Numerical design of thermal systems*. Clut.

SemsorGrid4Env. (n.d.). Retrieved September 2015, from https://www.semsorgrid4env.eu/

SENML Draft. (n.d.). Retrieved from https://tools.ietf.org/html/draft-jennings-senml-10

Tomlinson, R. F. (1968). A geographic information system for regional planning. In Symposium on Land Evaluation, Commonwealth Scientific and Industrial Research Organization. Melbourne: MacMillan of Australia.

Verda, V., & Baccino, G. (2014). Primary energy reductions in district heating networks through variation of the thermal load profile of the users. *Proceedings of ECOS*, 15-19.

Verda, V., Chirio, M., & Ponta, A. (2006). A Thermoeconomic Based Tool for the Real-time Management of a Cogenerative Combined Cycle. *ASME 2006 International Mechanical Engineering Congress and Exposition.*

Verda, V., & Colella, F. (2011). Primary energy savings through thermal storage in district heating networks. *Energy*, *36*(7), 4278–4286. doi:10.1016/j.energy.2011.04.015

World Health Organization. (n.d.). *Urban population growth*. Retrieved February 2016, from https://www.who.int/gho/urban_health/situation_trends/urban_population_growth_text/en/

Chapter 20
Customized Data Capture for BIM:
Using APIs and Visual Programming

Kjartan Gudmundsson
KTH Royal Institute of Technology, Sweden

Giuseppe Digregorio
KTH Royal Institute of Technology, Sweden

Jiayu Cui
KTH Royal Institute of Technology, Sweden

ABSTRACT

Building information models (BIMs) make it possible to share structured data in the planning and building process and during the live cycle of the building. The models can be enriched with information in all stages from planning to end-of-life. Predefined information requirements, data structures, and file formats allow the model to be used for design, cost planning, and time scheduling as well as analysis of building performance, and ideally also for facilities management. This chapter describes the possibilities of using application programming interfaces (APIs) to enhance the usefulness of BIM models. This provides the user with the possibility of creating own applications with real time bidirectional data exchange. Of particular interest is the exchange of data with web-based data sources. The applicability is illustrated with examples of environmental analysis based on web-based data and the use of web-based forms to enrichen BIM models with data input from building users.

INTRODUCTION

The usefulness of sharing structured building data in accessible format and the applicability of that data in a well-defined context is undeniable. Ideally, the different actors in the domains of architecture, engineering and construction (AEC) can collaborate and share data without loss of information and without

DOI: 10.4018/978-1-7998-7091-3.ch020

the need for the manual repetition of data input. Consequently, much effort has been put into defining effective means of structuring and sharing data in BIM models as well as platforms for sharing documentation in building projects. This chapter focuses on the need of practitioners in the AEC domain to be able to develop custom-made tools that can be used together with design authoring software for BIM models. It is explained how APIs can be used to develop custom-made tools with real time bidirectional data exchange between BIM models and existing applications. It will also be explained how this can make useful the vast amount of existing program code and resources. Furthermore, it is described how APIs can be used to access and navigate in the data structure of BIM models and the most common formats used for web based data exchange. It is also explained how automated data capture can be used to enrichen BIM models with information. It is shown by examples how this can be used to capture data for the calculation of environmental mid-point impact as well as well as by using web-based forms for the gathering of real-time data from building users. This is carried out with the Dynamo visual programming plug-in for Autodesk Revit design authoring software. In a broader context, the authors hope to contribute with processes that can be of use for the linking of a physical asset with its digital representation or digital twin.

Background

According to BIM Alliance Sweden (2020), BIM is about creating and using digital models of construction work in the building process. The digital models that are created are called Building information models and the working process is called Building information modelling. Building information management is the handling and administration of the models and the process of creating them. All three concepts are abbreviated BIM.

The core of BIM is the individual object. The object carries information about the geometry, attributes, classification and relations to other objects. Together the elements constitute a BIM model. The information associated with the object is what separates BIM from CAD drawings, CAD models and 3D models. Thus, a wall is not merely a geometrical volume but it can also contain information about properties such as materials, width, height and type as well as the relationships to other elements in the model.

One outspoken aim of digitalization in the building sector has been to create continuous flows of information between all actors involved in planning and building. Data deliveries can be defined at different points from planning and further through the whole life cycle of a building. The definitions can include a definition of the data set, the reliability of the data, the structure of the data and the file format that should be used.

Ongoing efforts include the establishing of classification systems that include infrastructure and the possibilities to share data between BIM and GIS (Smart Built Environment, 2020) as well as the implementation of CIM or City Information Modelling. The classification of objects in BIM models means that the objects are sorted according to a structure that makes it possible for different actors to find the information needed. The classification system commonly agreed upon by the building industry is IFC or Industry Foundation Classes that is a platform neutral open file format and an official ISO Standard that architects and engineers can use to share data and use data in their areas of expertise. A large number of commercial tools use IFC files for the viewing BIM models and the sorting of BIM data. In addition, there is a vast number of regional classification systems.

Current research efforts are also concerned with processes and platforms for sharing project information through single sources of information in a common data environment (CDE) that can store BIM data as well as data in a more conventional format. Previous research has reasoned that this can make it possible to leave the traditional procedural flows of the building industry and allow for collaborative working processes that are supported by a single centralized CDE in which all the actors involved (owner, designer and planner, manufacturer and builder, operation manager, user) can operate through all phases of the building process and throughout the live cycle of the building (Daniotti et al, 2020).

The paper of Turk (2017) points out the importance of standards for the definition of concepts (not definitions of building specific concepts) and for syntax as well as standards for software interactions and data sharing with so-called APIs. Furthermore, the same author emphasizes the use of generic standards used in service-oriented architectures across industries. Recent efforts have been directed towards the application of semantic web technologies for the AEC domains, in particular the collections of information, called ontologies. According to Berners-Lee et al (2001) the typical kind of web ontology has a taxonomy, that defines classes of objects and relations among them, as well as a set of inference rules. Pauwels et al (2017) did an extensive study of the development and application of, the notably increased, semantic web technologies in the AEC domains. Referring to previous research, the same authors argue that the use of semantic web technologies can improve interoperability in the AEC industries by providing a way to describe information in a computer-understandable manner. Furthermore, the authors point out that information from a variety of heterogeneous sources and diverse domains can be linked into one web of linked building data. Finally, the authors mention the possibility to use the logical basis of these technologies to deduce extra information from the originally represented information.

The aim of this chapter is to illustrate the vast possibilities to develop custom-made tools for the enrichment of BIM with web-based data on common formats. It is of particular interest to show how this can be done with well-known design authoring tools in combination with low barrier of entry visual programming tools. Furthermore, it is of interest to show how actors without any knowledge of BIM can contribute to the gathering of data as well as to illustrate some of the possibilities of providing accessible visual information about a building.

APIs and Visual Programming

API is the acronym for Application Programming Interface. APIs are central to automated data exchange and offer the possibility of real-time bidirectional data exchange with BIM models. APIs can therefore contribute to an increased interoperability and allow for the development of customized tools in the AEC domain. Furthermore, they allow for the automated use of web-based data. While some fundamental knowledge of coding is required, visual programming tools lower the threshold in terms of programming experience.

APIs and Automation for BIM

APIs make it possible for software to interact and share the data and functionalities of an existing software library (while not limited to that). An API is a specification of how to write a program code that can be used to call for the use of functions and the reading or modifying of data outside the application. One example is the Autodesk .NET API that gives the user the possibilities to create own tools instead of being limited by existing ones. The API gives the possibility to access the database that Autodesk Revit creates for each project as well as the Revit functions. In short, .NET is an open source developer platform for building applications developed by Microsoft (2020). The Windows version of .NET is the

.NET Framework that is a part of the Microsoft Windows operative system and handles the execution of programs while giving the possibility to access different libraries of previously coded solutions for frequently occurring programming tasks, including file reading and writing, XML document manipulation and network communication. The programs are executed in the Common Language Runtime software environment (CLR).

APIs have immediate practical purposes in BIM and the AEC sector. One example is the .NET API for the Autodesk Revit design authoring software that can be used to automate repetitive tasks as well as to extend the functionality of Revit by own applications for simulations or conceptual to give just a couple of examples. The Revit .NET API allows programming with any .NET compliant language such as VB.NET, C# or IronPython. This can for instance be done, from within the Revit, using the Macro Manager in Revit or using Visual Studio, an integrated development environment (IDE) from Microsoft. For those with lesser experience with programming, however, visual programming offers a more accessible opportunity to extend the functionality of Revit with the .NET API. Dynamo is tool for visual programming that uses the API for Revit. Dynamo has a visual user interface that makes the programming easier in comparison with textual programming. Dynamo can be used as a stand-alone application or a plug-in that starts from within Revit.

Particular design authoring tools such as Revit have an own hierarchical structure for organizing data. Everything in the model is an object, the properties and methods of which are defined by a template, i.e. the class of the object. On the top of the hierarchy, are the categories, the categories have families and the families have one or many types and finally we have instances of the types. The families share a common set of type properties while the values of these properties are not necessarily the same for each element and changing the value of an instance property affects only the elements selected. A knowledge of this hierarchy is useful when other applications are used to select and manipulate Revit elements such as when using Dynamo together with Revit. This also allows for effective export of data as well as the writing of code with the purpose of extracting data.

In the context of this chapter, it is of particular interest how to use APIs to populate element instances in a BIM model with values that are retrieved from web-based resources. The most common data formats found in modern APIs are JSON (JavaScript Object Notation) and XML (Extensible Markup Language) (Breje et al, 2018). Both are direct data formats in the sense that they are designed to handle data directly between machines. JSON is the most widely used data format for data interchange on the web (Infoworld, 2020). It is an open standard data exchange format that both humans and machines can read and write. JSON is a text format that is completely language independent but uses conventions that are familiar to programmers of several languages. JSON contains a collection of name-value pairs. The value can be a string, a number, a Boolean, null, an object or an array (Ecma International, 2020). XML defines a set of rules for encoding documents in a format that is both human-readable and machine-readable. XML creates a hierarchical tree-like data structure in which the elements are defined by tags and the top-level element is the root element. Elements can contain other elements. The elements have one or more attributes that define the properties of the elements by name-value pairs. XML supports not only text and number data types but also other data types such as images, charts, and graphs (W3C, 2020).

In this context, it must be mentioned that while an IFC data model is usually stored as a STEP file, IFC data can also be shared on the XML format as IFCXML files. Further worth mentioning is the Green Building XML schema, or gbXML, that is based on XML and can be used to create files in an open format that can be used for the transfer of building information to and from architectural and engineering models. GbXML is supported and adopted by many of the leading BIM vendors (Green Building XML, 2020).

The applicability of APIs for getting data from web services can be illustrated with the Python code below, in which a call is made to https://openweathermap.org/ with a request for the weather data on JSON format for the city of Torino, Italy on the 6th of July 2020.

The computer code in Python:

```
import requests

url = 'http://api.openweathermap.org/data/2.5/weather?id=3165524&units=metric&
appid=my API key'

json_data = requests.get(url).json()

formatted_data = json_data ['main']['temp']

print(formatted_data)
```

The API response when run:

31.54

This request uses the id for the city of Turin, that is equal to 3165524, as well as the appid of the the user that makes the call (in this example it is substituted with "my API key" – the real user key is not shown here). To get the value for the temperature we look for the key "temp" that is nested within the array of the values for the key "main".

Visual Programming With Dynamo

Dynamo is an open source visual programming environment that can be used as a standalone tool for the testing and the development of features or as a programming platform accessed from within Revit. In Dynamo, the relationships and instructions of a program are defined through a graphical user interface. The main components of Dynamo are Nodes and Wires. The Nodes perform an operation, such as the storing of a number or a string, querying from the Revit database, creating Revit elements, managing of data and computations. Dynamo has a library of built in Nodes but that library can be extended by importing Nodes and collections of Nodes or packages that have been previously created by other users. The Wires connect the Nodes and govern the flow of information into nodes and out of Nodes through Ports.

Figure 1 illustrates the anatomy of a simple set of operations for modelling a cuboid using built in Nodes from the Dynamo library. The Node "Cuboid.ByLenghts" is fed with the point of origin from the Node "Point.ByCoordinates" through a wire from the Port named "Point" to the Port named "origin". The "x", "y" and "z" ports all have a default value of zero that is used when the Port is not fed with another value.

A node can include a program code, written for instance in IronPython that may in some instances provide more efficient programming. Dynamo also makes it possible to import and export data on different formats and to and from different software platforms. The resources can be local as defined by a file path or a remote server as defined by a URL. In this manner, the user can for instance retrieve spreadsheet data from a file on a local computer or use an API to make a call for real real-time or historical weather data at a particular location as shown in the previous example.

Figure 1. An example of Dynamo Nodes and Wires

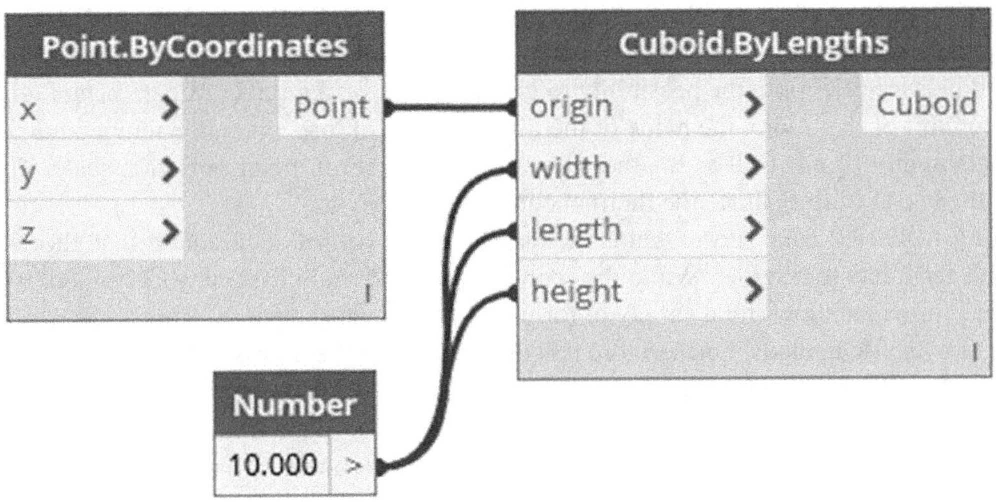

Examples of Implementation

By the means of APIs, the functionality of BIM models and BIM authoring tools can be extended. This includes the possibilities of developing custom-made applications for analytical work as well as data exchange with BIM models and web-based resources. The authors wish to illustrate those possibilities by showing some examples of possible workflows. This work includes the testing of every step of the workflows. The first example shows how user input, through web-based forms, can be used to collect information about building performance and how a BIM model together with other applications can be used to visualize the results of the following assessment in a virtual environment. The second example shows how web-based data, concerning the environmental properties of materials and components, can be used together with a BIM model to compare the environmental performance of different design alternatives.

Feeding BIM Model with Real Time User Data

The aim of this study is to configure a protocol for feeding a BIM model with real-time user data input. Real time data is here used to denote data that is not kept or stored but sent directly to the application in which it is used as soon as it comes in. Of particular interest is to demonstrate a protocol that allows the user to insert data using a QR-code that redirects the user to a web-based form and to find effective means of collecting and elaborating on that data. Furthermore, it should be possible to visualize the data as well as calculations based on that data in a manner that is understandable to experts and non-experts alike. A pilot study is used to illustrate the potential of the proposed protocol. The immediate practical purpose of such as protocol is for instance the possibility of providing the building services engineer feedback on the quality of the indoor environment.

The building used in this study is the "Undervisningshuset" (The Teaching House) situated at Brinellvägen 28a on the central campus of the KTH university in Stockholm, Sweden. The intention is to give a person situated in a specific room the possibility to give feedback on a number of parameters related to the thermal comfort of that room: the name of the user, the date of inspection, the temperature (°C) and the relative humidity (%) as well as the thermal comfort on a seven-point sensation scale. Those parameters can then be used to evaluate the thermal comfort of the room.

The user data is collected with a form that is the user interface part of web request that allows the user to insert and send data to a server. While the scope of the pilot study is somewhat limited, it was of interest to use a form that allows for a variety of forms for user input, such as text input, check boxes, multiple choice or even file uploads. Furthermore it is of interest that the user is able to access the web form by using a QR code that can be natively scanned without downloading external app on various mobile device operating systems such as iPhones with iOS11 or Android devices, for instance by using the camera app to scan the QR code. The work of Digregorio (2020) contains an extensive study of the various methods available to create web-based forms for this purpose. In this particular study, however, it was decided to work towards a protocol that relies on commonly available and well-known software and platforms from Google and Microsoft.

The QR-code in the room (shown in figure 2) was generated with a free online QR generator. Using the camera of a mobile phone the QR code is then used to redirect the user to a Google Forms form, asking the user to fill in the information requested. One advantage of Google Forms is that the URL of the Google form can be used directly and the form must not be embedded in a website.

The form is linked to a Google Sheets spreadsheet from which the online tool Zapier is used to transfer the data to an Excel spreadsheet. This is done through an automated workflow in which an action such as the updating of an Excel spreadsheet is triggered by a pre-set particular event such as is the addition of a new row in Google Forms.

Having a protocol for gathering input data, the next step is to store and manage the data in the BIM model of the building. By choice, this is done with the Dynamo visual programming software in combination with Revit using the native file format of the BIM model of the building. This is done by activating Dynamo from within an open Revit document. It is then possible to use Dynamo to read data from the Excel spreadsheets (see figure 3.) and to manipulate that data with built-in or imported Nodes or Python scripts as well as the possibility to send and get information from the Revit model.

With Dynamo, the user data can serve as an input for any kind of calculations or assessments, the results of which can be used as values of the instance properties of Revit elements. Those operations can be carried out with Dynamo, using the functionalities of built in Nodes or imported packages or by writing own programming code. The assessment can for instance be the comparison of data with a predefined set of criteria for validating the performance.

The next step is to calculate the predicted mean vote or PMV index that describes the indoor thermal environment in terms of the mean value of votes of a group of occupants on a seven-point thermal sensation scale. In this simplified study, the calculation is based on the data from the web form together with assumed values concerning, among other, the clothing and the level of physical activity. The PMV is then used to calculate the PPD (Predicted Percentage of Dissatisfied) or the ratio of people that are not satisfied with the thermal comfort of the room. The calculations are carried out with an IronPython script Node in Dynamo.

Figure 2. QR code linked with the form

When the user input data from the web-based form has been collected and assessed with Dynamo the next step is to populate the BIM model with data. This is done by creating a new Revit family, called "Totem" that consists of a cube shaped geometry and by defining a set of parameters for that family. This set of data must match the set of data that is to be collected and visualized in the BIM model. An element instance of the "Totem" is placed in the room in which the user makes the observations. The element instance is then populated with values from Dynamo as shown in figure 4. Those values include not only the user input but also the calculated values such as the PPD. Furthermore, Dynamo is used to set the color of the "Totem" in Revit in accordance with the resulting PPD value.

The next step of the protocol is, however, to allow the sharing of the results with other users. While this can obviously be done by sharing the Revit file or even by sharing an exported IFC version of the BIM model; it is of particular interest to find a process that allows for sharing with users with limited knowledge, experience and access to BIM software. In this study this is accomplished with the Unity Reflect virtual environment software. In principle, Unity Reflect allows the user to visualize real-time transfer of BIM data on the screen of a PC (or any devices supporting iOS and Android) or using Augmented Reality or Virtual Reality devices. The BIM data is stored on a computer that runs Unity Reflect or on a network server while a cloud-hosted project management server stores the project names and permissions in order to determine the access right of users. One immediate practical application is the real time visualization of changes made by project members in an AEC team; another is the possibility

Figure 3. Using Dynamo to retrieve data from Excel

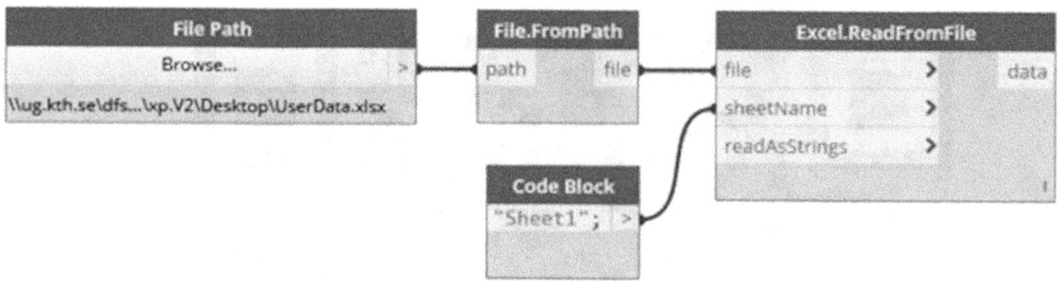

to convey information to non-experts. The strength of Unity Reflect, compared to other BIM model viewers is the possibility to observe changes in real time.

The game-like user interface of the software allows the user to travel in the model and allows for the simultaneous access of multiple stakeholders as well as the possibility to filter the visibility of the model content in the view. Figure 5 shows the “Project View” mode of “Undervisningshuset” in Unity Reflect.

While every change in the Revit model will also happen in real-time in the shared 3D model, there are still some challenges when the Revit model is connected with Dynamo. It was therefore necessary

Figure 4. Populating the Revit Elements with values from Dynamo

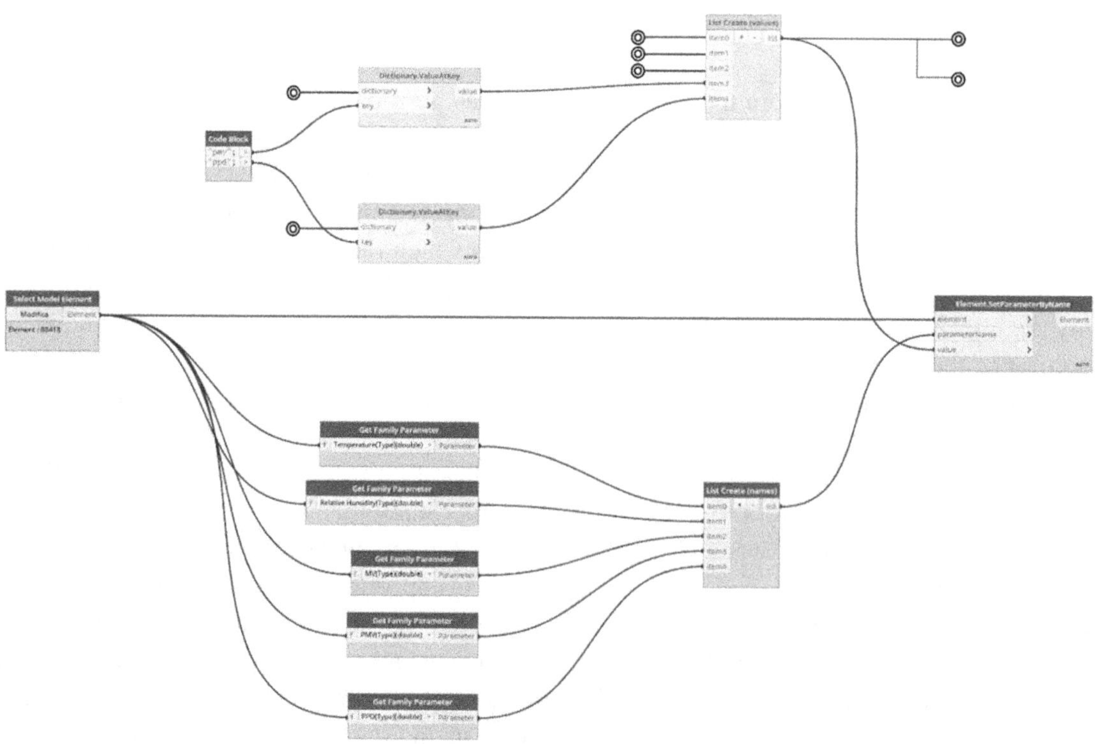

Figure 5. Unity Reflect 3D model of "Undervisningshuset"

to re -synchronize the model when the color of the Totem is changed changes but this process takes only few minutes and is not considered a serious issue.

Figure 6 shows a virtual view of the room from Unity Reflect. The colored box on the floor is the "Totem" and the color illustrates the value of the assessment, with a value of PPD equal to or above 10 the color is green whereas the color will become red when the PPD value is lower than 10.

Figure 6. Visualization of the results with Unity Reflect

Further account of this work can be found in the publication of Digregrorio (2020).

BIM Models Together With XML Data From Environmental Product Declarations

Earlier research has pointed out the applicability of LCA as a tool to improve the quality, the energy efficiency and the sustainability of new buildings and the usefulness on different system levels (Ingrao et al, 2018). Other studies have shown how LCA can be used for the environmental evaluation of buildings that have been designed as low energy or sustainable buildings, especially in urban areas and developed countries (Cabeza et al, 2014). Meex et al (2018) discuss the challenges concerned with applying LCA as a tool for supporting or optimizing design decisions during early design stages and argue that the due to its extensiveness and complexity, as well as lack of knowledge amongst architects, the LCA methodology is most often applied as a post-design evaluation tool. Furthermore, the authors present a framework with requirements for applying LCA-based tools by architects in early design and stress the importance of meaningful feedback and suggestions for design alternatives. Previous research has also highlighted the need for tools that can integrate environmental, social and economic criteria into the design framework (Antón & Díaz, 2014a), stating that the aim must be to integrate important tools like BIM and LCA at an early stage in order to make full use of their potential. The same authors, however, point out the lack of interoperability between BIM and LCA software. In another paper, Antón & Díaz (2014b) propose two approaches to perform to integrate BIM and LCA. The first approach is the export of BIM data to LCA software. The main drawbacks of this process is that any alterations in the BIM model have to be modified in BIM tools and then re-export to LCA software again and the interoperability issues between BIM models and LCA software. The second approach is the automatic linking of BIM models with environmental information in LCA databases through an unknown process that provides the BIM objects with environmental data properties. In more recent work, Tsikos et al (2017) have evaluated the potential applicability of using Dynamo for integrating LCA tools with BIM and compared the results from case studies with the results obtained from established LCA software; the results show that the integrated method can deliver results of equal quality and precision while demanding much less effort.

The aim of this study is to illustrate a possible workflow for using BIM models together with data from Environmental Product Declarations, EPDs that can be used to calculate the environmental midpoint impact of different design alternatives. An EPD is an independently verified and registered document that gives transparent and comparable information about the life-cycle environmental impact of products and can provide a valuable input to the LCA of buildings. Until recently EPDs have been published in a static pdf format but ongoing efforts include the transfer of some of the information of an EPD into a machine-readable XML format (EPD: The International EPD System, 2020). In previous research (Karami et al, 2015), quantity take-off from a BIM model together with EPDs were used to do a comparative study of the environmental impact of different design options for residential buildings while relying on EPDs on PDF format and a manual process for entering EPD data into the calculations.

This study gives an example of how the different impact assessment indicators can be retrieved from an EPD on the XML format. The purpose is to facilitate the effective evaluation of the environmental performance in the early stages of the design process. The indicators of the impact assessment are the following: the abiotic depletion potential for non-fossil resources (ADPE), the acidification potential (AP), the global warming potential (GWP), the eutrophication potential (EP), the photochemical ozone creation potential (POCP) and the ozone Depletion Potential (ODP). The data used in this study is from the publicly available EPD-Norway's Digital EPD database of the Norwegian EPD Foundation (2020) that contains LCA-based data from Environmental Product Declarations (EPDs) by manufacturers, digitalized in XML format. Another important aspect of the workflow is to present the results in a manner that provides the user the possibility to compare different design alternatives.

A simple model of a two storey single family house was created in Revit for the purpose of this study (Cui, 2020). The building is depicted in figure 7.

Figure 7. A view from a Revit model of the building used in the study

The first step is to find a method for extracting data from the XML documents containing the EPDs. In this study, the "Bakery" package for Dynamo is used to import the XML file and to navigate in the XML element tree structure of the EPD. The Bakery Node "Read File Path to Element Tree with UTF8 roundtrip" is used to extract the element tree from the XML file. The XML Path language, or XPath, is then used to extract the required information. XPath is a well-known query language for finding and extracting information from XML documents that can be used to navigate and find the target elements, attributes and text throughout the XML document. XPath uses path expressions to select the information wanted in an XML document. In short, the expressions resemble file paths in a computer system while the reader can find an accurate account of the syntax for location paths at the W3C website (2020b). The XML document is human-readable and in this study the XML file is first opened with a browser to identify the location of the required data. This reveals that the LCA results are located in a children element called "LCIAResults". Dynamo is then used to put together the XPath expressions that are used to locate the required values; figure 10 in the appendix gives an example of the path for locating the environmental impact values. With the element tree and the XPath expressions the Node "XPATH Expression Find for Element Tree" is used to find elements in an element tree and the Node "Get XML Children Texts and Tags from XML Elements" is used to get the impact values and the names of the categories from the XML elements as shown in figures 11 and 12 in the appendix. The impact values extracted from the XML file are written, however, in string format and not as numbers. While Dynamo has a built in Node for converting strings to numbers, this Node cannot recognize the scientific notation used in the XML file of the EPDs. For this reason, a python script was used to convert the scientific notation to number that can be recognized and used to calculate the environmental impacts.

The IronPython script used in Dynamo:

```
import clr
clr.AddReference('ProtoGeometry')
from Autodesk.DesignScript.Geometry import *
#The inputs to this node will be stored as a list in the IN variables.
dataEnteringNode = IN

#Assign the output to the OUT variable
OUT=[]
#Loop through the list of items
for items in IN[[REMOVED EQ FIELD]]:
      tmp=[]
      for item in items:
            item=eval(item)
            tmp.append('{:.8f}'.format(item))
      OUT.append(tmp)
```

The code imports the clr module, i.e. the environment in which the .NET applications run. The AddReference function in the clr module is then used to load an assembly called "ProtoGeometry" from the Dynamo Core library. Everything from the Autodesk.DesignScript.Geometry library is then imported. The remaining parts of the script collects input as strings from the XML file. The format function is then used to convert the number in decimal form with 8 places following the decimal point and the value is added to the end of the list using the append method and then returned to Dynamo by assigning them to the OUT variable.

The next step is to use Dynamo to do a quantity take-off from the BIM model. There are numerous ways to use Dynamo for selecting elements from the active Revit document, such as filtering by category, family or element type or by picking the element directly from the active Revit document with the "Select Model Element" Node. Subsequently the required element parameter can be obtained by feeding the built in Node "Element.GetParameterValueByName" with a string containing the name of the desired parameter as shown in figure 8.

An obvious but important aspect of the quantity take-off is that the parameter used in the quantity take-off must match that of the functional unit of the EPD that may be a mass, a volume, a number or an area.

Figure 8. An example of using Dynamo to retrieve elements and their properties

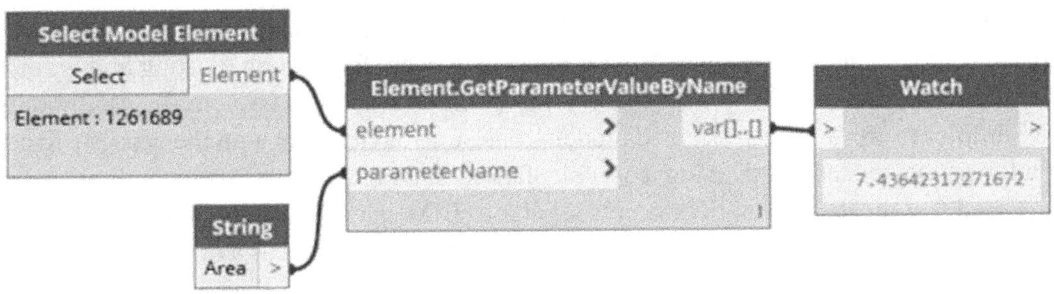

In the next step, the EPD data together with the quantity takeoff is used to calculate the environmental impact of the different materials and components. This give the resulting values for the different impact categories such as the Abiotic depletion potential and the Acidification potential.

The final step is to visualize the results with the primary aim of comparing different design alternatives. This can be done with the Bar Chart Node in Dynamo that is fed with the necessary values. The results shown in figure 9 are an example of such output. The input data of this example is from the XML files containing the EPDs of two different types of doors from the Norwegian EPD Foundation.

Figure 9. A diagram in Dynamo showing a comparison of the environmental impact of design alternatives

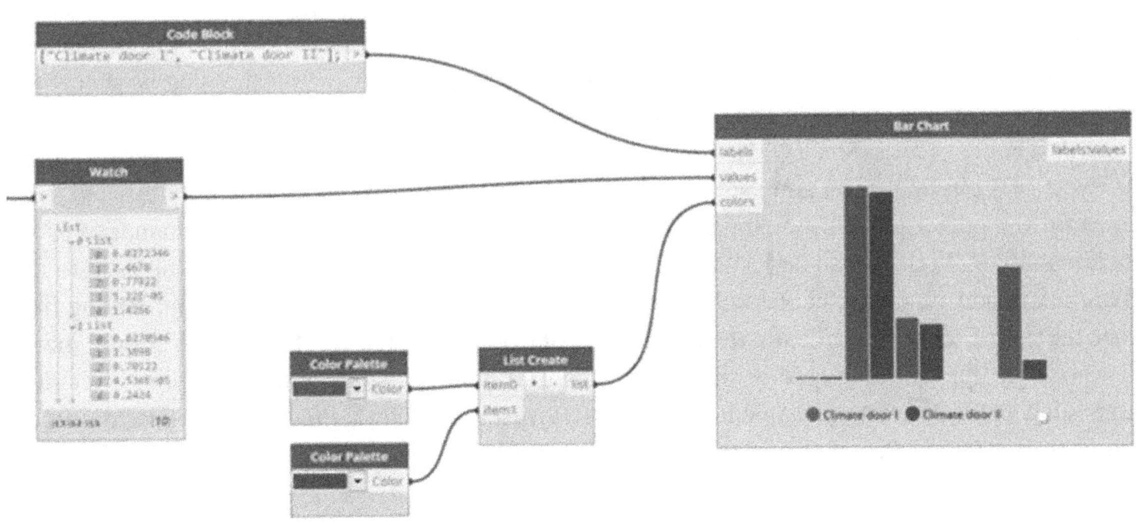

A more detailed account of the use of BIM models together with XML data from Environmental Product Declarations can be found in the work of Cui (2020).

FUTURE RESEARCH DIRECTIONS

Professional in the AEC sector have since decades had the opportunity to create digital models of buildings and building components and systems. Earlier CAD tools were focused on drafting and graphical representation while the arrival of BIM in the first years of the 21st century (Autodesk, 2002) brought about the possibilities to create object-oriented models consisting of elements with the capacity to carry a variety of information needed for planning, cost calculations and facilities management. It must, however, be mentioned that the discussion of concepts similar to BIM started much earlier.

A much more recent concept is that of a Digital Twin that may be defined as a digital representation of a physical asset. The additional value is that of linking the physical and digital twin for a regular exchange of data. In short, the digital twin can simulate and provide real-time feedback on performance of the physical object or serve as a prototype to the physical object (networkworld, 2020). A recent publication by Building Smart (2020a) points out that dynamic data gathering and exchange can be done by the means of AI, machine learning, sensors and Internet of Things while the key problem lies in the integration of data coming from different sources in different formats. The authors of this chapter argue that digitalization of the AEC domain must therefore stretch beyond the use of BIM and must be aligned with the possibilities offered by current data exchange technologies. Building Smart also emphasizes the connection to social and economic infrastructure as well as the involvement of users and citizens. The active participation of users and citizens is in line with the emerging trends of using new tools and user interfaces to make technology available to people as opposed to being confined to the specialist user (Gartner, 2020).

The purpose of this chapter is to explain and illustrate by examples how the applicability of BIM models can be extended to the purposes of real-time data exchange. The possibilities of engaging building users and other non-experts in the gathering of data is of central interest, as is to make the results understandable and useful to actors outside the AEC domain. Furthermore, the value of data generated by experts outside the AEC domain must be appreciated. With the intention of engaging the actors in the AEC domain, this work is done with common design authoring tools while showing how new tools for programming and visual programming allow for the manipulation of BIM models through bidirectional data exchange. This offers a wide variety of possibilities in the design phase as well as throughout the life cycle of a physical asset and its digital twins. This chapter gives examples of automated digital processes for the gathering of user data and web based data and how the data can be managed as well as how the analytical results can be visualized, but the same or similar workflows can be used to manage, manipulate and visualize the data of various phenomena. One possible application, among many, would be the use of sensor data, i.e. for the monitoring of bridges and the validation of structural models. Another possible application would be the monitoring of room occupancy and the indoor air quality of buildings such as for detecting possible hazards due to viruses or bacteria in which real-time data would be used to activate measures of sanitation.

CONCLUSION

It is clear that visual programming and APIs provide the actors in the AEC domain with the opportunities to enhance the applicability of BIM and BIM models. This work highlights the possibilities to develop custom-made applications with real time bidirectional data exchange with BIM models.

The possibility of data exchange between BIM models and web based data sources on common formats, such as JSON and XML is of particular interest. Likewise, it is compelling to see the possibilities to involve actors without BIM expertise in the gathering of data for BIM and the possibilities to present data and analytical finding to experts and non-experts alike.

The applicability is illustrated with examples of workflows for automated processes. The first example shows the use of web based form to gather data on the indoor climate in a building and how building performance can be visualized with a user interface based on a 3D model of the building. The second example illustrates the use web based environmental data and how that data together with information from a BIM model can be used to compare the environmental performance of different design alternatives. The scopes of the examples are somewhat narrow and simplified but the workflows can easily be adapted to a variety of different applications.

ACKNOWLEDGMENT

This research received no specific grant from any funding agency in the public, commercial, or not-for-profit sectors.

REFERENCES

W3C. (2020a). *XML Essentials*. https://www.w3.org/standards/xml/core

W3C. (2020b). *XML Path Language (XPath)*. https://www.w3.org/TR/1999/REC-xpath-19991116/#location-paths

Antón, L. A., & Díaz, J. (2014a). Integration of LCA and BIM for Sustainable Construction. *International Journal of Civil and Environmental Engineering*, *8*(5), 1378–1382.

Antón, L. A., & Díaz, J. (2014b). Integration of Life Cycle Assessment in a BIM Environment. *Procedia Engineering*, *85*, 26–32. doi:10.1016/j.proeng.2014.10.525

Autodesk. (2002). *Building Information Modeling in Practice*. https://images.autodesk.com/apac_grtrchina_main/files/aec_bim.pdf

Berners-Lee, T., Hendler, J., & Lassila, O. (2001). The Semantic Web: A New Form of Web Content That is Meaningful to Computers Will Unleash a Revolution of New Possibilities. *Scientific American*, *284*(5), 34–43. doi:10.1038cientificamerican0501-34 PMID:11681174

BIM Alliance Sweden. (2020). *Vad är BIM?* https://bimalliance.se/vad-aer-bim/

Breje, A.-R., Gyorödi, R., Gyorödi, C., Zmaranda, D., & Pecherle, G. (2018). Comparative Study of Data Sending Methods for XML and JSON Models. *International Journal of Advanced Computer Science and Applications*, *9*(12), 198–204. doi:10.14569/IJACSA.2018.091229

Building Smart International. (2020a). *Enabling an Ecosystem of Digital Twins*. https://www.buildingsmart.org/wp-content/uploads/2020/05/Enabling-Digital-Twins-Positioning-Paper-Final.pdf

Cabeza, L. F., Rincón, L., Vilariño, V., Pérez, G., & Castell, A. (2014). Life cycle assessment (LCA) and life cycle energy analysis (LCEA) of buildings and the building sector: A review. *Renewable & Sustainable Energy Reviews*, *29*, 394–416. doi:10.1016/j.rser.2013.08.037

Cui, J. (2020). *Integration of Life Cycle Assessment within Building Information Modeling Environment* [Unpublished master thesis]. KTH, Stockholm, Sweden.

Daniotti, B., Pavan, A., Lupica Spagnolo, S., Caffi, V., Pasini, D., & Mirarchi, C. (2020). Collaborative Working in a BIM Environment (BIM Platform). In BIM-Based Collaborative Building Process Management. Springer.

Digregrorio, G. (2020). *Real-time data and BIM: automated protocol for management and visualization of data in real time* [Unpublished master thesis]. KTH, Stockholm, Sweden.

Ecma International. (2020). *The JSON Data Interchange Format, Standard ECMA-404*. http://www.ecma-international.org/publications/standards/Ecma-404.htm

EPD. The International EPD System. (2020). *What is an EPD?* https://www.environdec.com/What-is-an-EPD/

Gartner. (2020). *Gartner Top 10 Strategic Technology Trends for 2020*. https://www.gartner.com/smarterwithgartner/gartner-top-10-strategic-technology-trends-for-2020/

Green Building, X. M. L. (gbXML) Schema. (2020). *gbXML*. https://www.gbxml.org/index.html

Infoworld. (2020). *What is JSON? A better format for data exchange*. https://www.infoworld.com/article/3222851/what-is-json-a-better-format-for-data-exchange.html

Ingrao, C., Messineo, A., Beltramo, R., Yigitcanlar, T., & Ioppolo, G. (2018). How can life cycle thinking support sustainability of buildings? Investigating life cycle assessment applications for energy efficiency and environmental performance. *Journal of Cleaner Production*, *201*, 556–569. doi:10.1016/j.jclepro.2018.08.080

Karami, P., Al-Ayish, N., & Gudmundsson, K. (2015). A comparative study of the environmental impact of Swedish residential buildings with vacuum insulation panels. *Energy and Building*, *109*, 183–194. doi:10.1016/j.enbuild.2015.10.031

Meex, E., Hollberg, A., Knapen, E., Hildebrand, L., & Verbeeck, G. (2018). Requirements for applying LCA-based environmental impact assessment tools in the early stages of building design. *Building and Environment*, *133*, 228–236. doi:10.1016/j.buildenv.2018.02.016

Microsoft. (2020). *What is. NET?* https://dotnet.microsoft.com/learn/dotnet/what-is-dotnet

Networkworkworld. (2020). *Internet of Things*. https://www.networkworld.com/article/3280225/what-is-digital-twin-technology-and-why-it-matters.html

Pauwels, P., Sijie Zhang, S., & Lee, Y.-C. (2017). Semantic web technologies in AEC industry: A literature overview. *Automation in Construction*, *73*, 145–165. doi:10.1016/j.autcon.2016.10.003

Smart Built Environment. (2020). *Projekt DigSam - Digital Samhällsbyggnadsprocess*. https://www.smartbuilt.se/digsam

The Norwegian EPD Foundation. (2020). *EPD-Norge Digi*. https://digi.epd-norge.no/

Tsikos, M., & Negendahl, K. (2017). *Sustainable design with respect to LCA using parametric design and BIM tools* [Paper presentation]. World Sustainable Built Environment Conference, Wan Chai, Hong Kong. https://backend.orbit.dtu.dk/ws/portalfiles/portal/133787517/Sustainable_Design_with_Respect_to_LCA_Using_Parametric_Design_and_BIM_Tools.pdf

Turk, Z. (2017). Ten questions concerning building information modelling. *Building and Environment*, *107*, 274–284. doi:10.1016/j.buildenv.2016.08.001

ADDITIONAL READING

Building Smart International. (2020b). *Industry Foundation Classes (IFC)*. https://www.buildingsmart.org/standards/bsi-standards/industry-foundation-classes/

Deutsch, R. (2015). *Data-Driven Design and Construction: 25 Strategies for Capturing, Analyzing, and Applying Building Data*. Wiley.

Dynamo. (2020). *The Dynamo primer*. https://primer.dynamobim.org/

Eastman, C. M., Eastman, C., Teicholz, P., Sacks, R., & Kathleen Liston, K. (2008). *BIM Handbook: A Guide to Building Information Modeling for Owners, Managers, Designers, Engineers and Contractors*. John Wiley & Sons. doi:10.1002/9780470261309

GeeksforGeeks. (2020). *JSON Data Types*. https://www.geeksforgeeks.org/json-data-types/

Hauschild, M. Z. (2018). Introduction to LCA Methodology. In *Life Cycle Assessment: Theory and practice* (pp. 59–66). Springer., doi:10.1007/978-3-319-56475-3_6

International Organization for Standardization. (2005). Ergonomics of the thermal environment - Analytical determination and interpretation of thermal comfort using calculation of the PMV and PPD indices and local thermal comfort criteria. *ISO Standard*, (7730), 2005.

Kilkelly, M. (2020, May 1). *What Is Dynamo and 5 Reasons You Should be Using It*. https://archsmarter.com/what-is-dynamo-revit/

Python. (2020). *Python*. https://www.python.org/

KEY TERMS AND DEFINITIONS

.NET: An open source developer platform that can be used to develop applications. .NET Framework is a part of the Microsoft Windows operative system and is a collection of components for the execution of programs. It also contains a class library with code that can be used for frequently occurring tasks.

API: Application Programming Interface, a predefined specification of how applications use and communicate with a library of information and functions.

BIM: Building Information Model/Modelling/Management.

EPD: Environmental Program Declaration. A third party verified document that describes some key environmental properties of goods and services.

XML: Extensive Markup Language, a set of rules for encoding documents in a format that cand be read by machines while readable for humans.

Web Form: Pages requested by users with browser.

APPENDIX

VISUAL PROGRAMMING FOR READING XML FILE

Figure 10. Dynamo Nodes for constructing the XPath expression used to locate the impact values

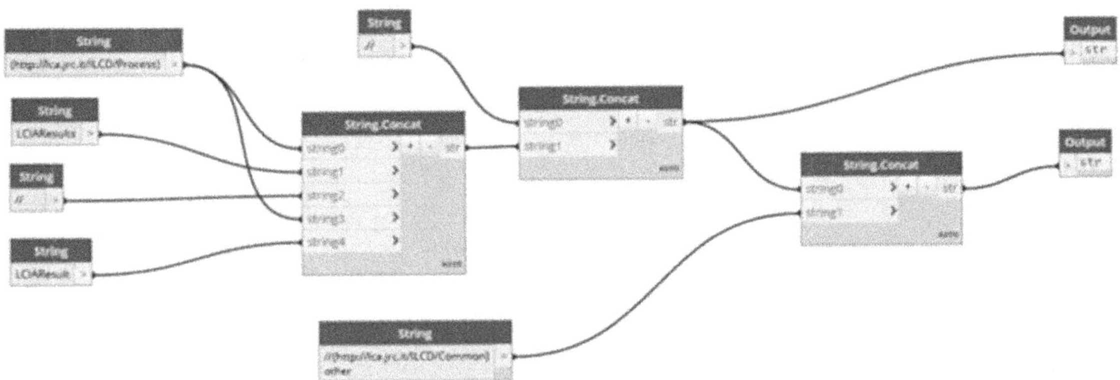

Figure 11. Dynamo Nodes for getting impact values

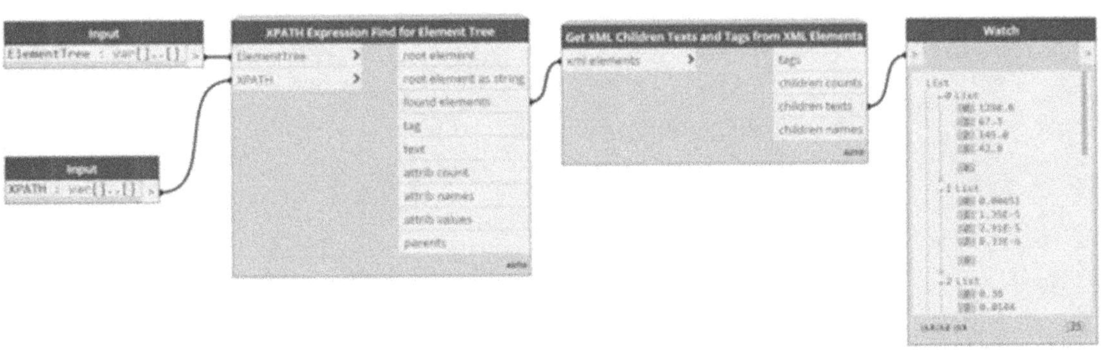

Figure 12. Dynamo Nodes for getting impact categories

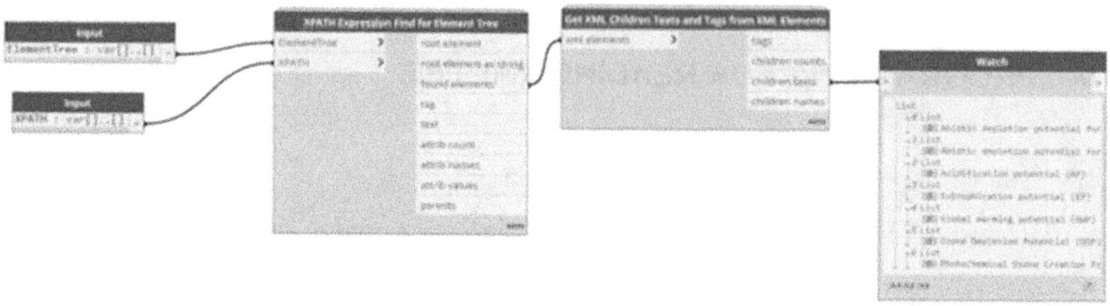

Chapter 21
Participatory Design of Use Cases for an IoT Open Platform to Support Smart Urban Development: Approach and Method

Alice Schweigkofler
Fraunhofer Italia Research, Italy

Katrien Romagnoli
Fraunhofer Italia Research, Italy

Gabriel Sanz Salas
Systems s.r.l., Italy

Dieter Steiner
Fraunhofer Italia Research, Italy

Michael Riedl
Fraunhofer Italia Research, Italy

Dominik Matt
University of Bolzano-Bozen, Italy

ABSTRACT

The chapter describes the approach for the South Tyrolean city of Meran in the creation of use cases and the implementation of an urban agenda (roadmap) for the development of the city from a smart city perspective, with the involvement of citizens, experts, and local administrators. A list of key services, based on a technical and economic pre-feasibility study and social impact assessment, has been developed and will be able to be provided through a digital platform. In particular, the example of the concrete

DOI: 10.4018/978-1-7998-7091-3.ch021

development of a use case about public lighting is presented in detail: starting from the identification of the use case to the execution of the installation of 100 intelligent streetlights and 5 test-sites for the monitoring of water consumption up to the visualization of the collected data.

INTRODUCTION

The topic of smart cities is achieving more and more relevance locally and internationally. Nowadays, hundreds of cities around the globe are giving rise to many initiatives and the concept of smart cities is rapidly gaining a worldwide attention (Bibri et al. 2017). The agenda for smart urban renovation is expected to solve a multiplicity of challenges (Haarstad 2019), with the key aim of significantly increase the urban quality of life. Many cities are developing smart city roadmaps in order to ensure a successful and effective outcome facilitated by a smooth transition. According to the study "The smart City Breakaway 2019", conducted by the international consultancy company Roland Berger, there are 153 cities in the world that have published an official smart city strategy. Just 15 of them have projects that demonstrate a comprehensive strategic approach and only 8 cities are at an advanced stage of implementation. Among the top three smart cities in the world there are two European capitals i.e. London and Vienna. Considering other sources, it emerges that one third of UK cities with a population over 100.000 people have smart city ambitions, while two-thirds of US cities are investing in some form of smart technology. The national governments of India, China and Singapore are promoting smart cities through competitions, funding programs, policy agendas and pilot projects with support of transnational organizations (e.g. Joint Programming Initiative Urban Europe, Bloomberg, Philanthropies) and technology providers (IBM, Cisco, Google). The European Union (EU) has been a particularly strong promoter of smart cities through the European Commission´s Horizon 2020 program on Smart Cities and Communities.

Smart services cover many different areas and should therefore be flexible and able to adapt to the market needs. Considering the different applications and services of a smart city, we can identify the following areas: mobility, school, tourism, government, health. However, areas of intervention can expand as cities are considered 360°. In this wide scenario, most of the methods for becoming "smart" for a city imply the use of a strategic approach (Angelidou 2015). Despite this, a standard approach still needs to be defined and therefore, the ability of smart city initiatives to provide an integrated and systematic answer to urban challenges is constantly questioned (Fernandez-Anez et al. 2018). It appears clear that a holistic approach is still missing if we consider that apparently, a city could be qualified now as "smart" only introducing single, extemporaneous and non-coordinated actions. Therefore, one of the clearest emerging research questions is the following: how should decision makers such as local authorities, small and medium enterprises as well as utility providers, approach the transition towards a smart city? This research work gives a contribution to this issue, providing a methodology for the drawing of an urban agenda in terms of smart city, which is based on a participatory approach.

Another significant aspect is the definition of "smart" itself, that has been in the center of an intense debate for several years. Despite the ongoing discussion, there is no agreed and univocal definition of smart city (Angelidou 2015). In fact, there has been a lot of talk about smart city in recent years, but the term itself is still often used vaguely or inappropriately, sometimes with an exclusively technological focus. From time to time Information and Communication Technologies (ICT) solutions appear as the primary facilitators for the integration of technology in every aspect of the urban environment (Bifulco et al. 2015). However, this identification of "smartness" with "technology" can be highly misleading. Indeed, even if technological solutions are already available, the biggest difficulty in their concrete adoption is about the translation of the available technological solutions in complex and sustainable urban management plans. For this reason, the authors rather assume as a good starting point the definition of smart city given by the "Observatory Smart City" (2018) of ANCI (National Association of Italian Municipalities). According to it, a smart city is defined as a city which uses ICT tools as an innovative support in the management areas and for the provision of public services, following a strategic vision in an organic way, thanks also to the help of public-private partnerships, to improve the livability for its citizens; a city that uses information from various fields in real time, and exploit both tangible resources (for example transport, energy infrastructures, natural resources) and non-tangible resources (human capital, education and knowledge and intellectual capital of companies); a smart city is able to adapt itself to the needs of users, promoting their own sustainable development.

Considering this complexity, research activities are mostly aiming at studying models for a smart city definition and their evaluation. Indeed, the impact assessment of various measures in the short, medium and long term on the urban environment will have to be carried out in a holistic way, given that the rapid evolution of technologies is increasing the risk of services to become rapidly obsolete. In particular, four different levels of implementation, structured in "layers" need to be addressed (Ernst & Young, 2016):

Layer 1 - Infrastructure: Enabling networks and technological equipment (TLC Networks, transports, energy, territory and environment).

Layer 2 - Sensors: Internet of Things (IoT) to collect big data from the connected objects from the city, which collect data about the infrastructure, the environment and on user behaviors.

Layer 3 - Service delivery platform: Platform capable of processing and enhancing the territory's big data acquired from the sensors-layer, in order to improve existing city services and to create new ones.

Layer 4 - Applications and services: The service provision, provided via mobile and web applications, which represents the contact point with end users.

With this concept in mind, the approach described in this chapter is based on the implementation of a layered structure. The major urgency is represented in this context by the development of standardized processes and decision-making tools, which should be able to help in the definition of priorities for urban design and planning in order to increase the quality of public decisions (Pereira et al. 2017). Starting from these observations, the assumption is taken that a holistic methodology to support the smart urban development is more than ever crucial and its adoption is necessary to establish new smart city services. The focus of the present study is about the development of a methodology and the adoption of a strategic decision-making tool to support cities in defining their transition towards a smart city, including the major steps in this transition and considering the complex technological interaction of infrastructure, sensors, data management and services.

Background

This work is part of the research project "OpenIoT4SmartCities", funded by the Operational Program for European Regional Development Fund ERDF 2014-2020 (CUP: B11B17000720008) in Südtirol/Alto Adige. The main aim of the project is to state a methodology for the design and development of a platform for the supply of new urban services, called OpenIoT platform. An IoT platform is a multi-layer technology that provides a set of ready-to-use features to speed up the development of IoT projects. It is an important piece in the IoT that enables the communication between objects (Uwiringiyimana et al., 2021). A high number of IoT platforms are available at the market today, trying to answer this issue. Research into platforms is at an early stage with many questions and issues (Ng et al., 2015). There are different IoT platforms, which work as mediators between the varying kind of hardware and the different connectivity options. Every platform has at least one better specific thing to offer, therefore, it is not easy to choose the right platform for a specific issue. Within the project, there is the interest of the company Systems to develop its own platform, which should represent an enabling tool for the smart urban development of a city i.e. will be able to support and encourage the growth of new services for - and with - citizens, companies, tourists, and public administrations. It should be open, flexible and scalable and will include the necessary features (a) to permit the integration of smart services for a digitalized city; (b) to be 360° "open" to allow a high degree of flexibility in terms of infrastructures and technologies, user rights and proposed functionalities; (c) to allow its adoption in different contexts and cities in a modular and scalable way. The purpose comes from the desire of combining the possibilities offered by the available ICT infrastructure and services, with the modern urgencies emerging from urban contexts, applying new technologies to new city's issues.

Once defined the strategic platform's features, it´s important to list specific requirements. Considering that today the IoT is involved in all aspects of our lives (Based, 2021), one of the main challenges that IoT companies face is to select the appropriate business model (Escribano et al., 2021) for platform development. In the end, each organization is different, therefore, the set of different architectural components is very extensive. This leads to a situation where companies have different IoT projects and replication of IoT solutions is hard or even impossible (Escribano et al., 2021). For this reason, the first aim of this study is to present a methodology to support and contextualize the development of an IoT platform, to undertake a realistic and coherent path towards smarter cities.

A solid methodological basis is essential for the implementation of smart city technologies in an urban context. As already mentioned, in fact the context complexity, together with the strong territorial specificities, make difficult to generate an overall and unitary vision. For this reason, many smart city projects have specific and less transversal implementations. As a support to the technical development, different smart city implementation methodologies and frameworks have been analyzed, among other Morgenstadt City Lab (Fraunhofer Gesellschaft, 2019), the Smart City Guidance Package of the Norwegian University of Science & Technology (Borsboom-van Beurden et al. 2019) and the Smart City Framework developed by the British Standard Institution (2018). The suggested steps, but also the attention to the most critical points, such as the interdisciplinary project team and the determination of relevant urban stakeholders were considered. Finally, a methodology has been proposed, based on three fundamental elements:

- The analysis of the digital needs of a city as a "reference".
- The setting of a strategic, shared "vision" of the future of the city.

- The definition of a "roadmap" of services development for a sustainable planning of activities.

These elements are settled as the methodological, theoretical support for the development of the platform and its adoption and utilization. The developed methodology meets the requirement as well to be scalable on several urban contexts and provides decision-making support for a participatory planning in the transition towards a smart city.

Since the definition of a methodology cannot disregard a testing phase, a real city has been considered for this purpose. Therefore, this work gives a look at Meran, a city located in the mountainous region of South Tyrol, in Italy, which presents unique challenges and opportunities. The ongoing experience of the city of Meran in its transition towards a smart city is in line with the city´s objective of becoming one of the best regional examples of smart city. Its long-term challenge is to significantly increase its overall quality of life, for the benefit of all the inhabitants and visitors. This paper will present the results of previous analysis, consultation meetings and workshops that have been conducted by the research team together with the public administration and local service providers of the city of Meran in a participatory way.

The Vision of Meran Smart City

A vision is a crucial step in the transition towards a smarter city for various reasons and for multiple stakeholders. A vision puts the transition of the city into a broader, strategic long-term plan. It motivates and helps focusing on a common objective. To achieve these benefits, a vision should be concise, future-oriented, stable, challenging, abstract, inspiring and needs a time-horizon. To create the smart city vision for Meran a structured workshop was organized to gather all stakeholders with a vision of the city and/or a significant impact to make it a reality: mayor, councilors, incubators, youth representors, key service providers etc. The workshop consisted of three parts. First, the project context and its objective, timeline and funding were shared together with the objective of the workshop itself. Second, the importance and the characteristics of a well-structured vision were explained and compared with those of cities around the world. Lastly, the participants followed a structed methodology to give input for the Vision of Meran smart city.

With the input from the workshop, Fraunhofer Italia developed the vision concept for Meran in 2030, which is shown below. This consists of three core pillars: "Innovation meets authenticity", "Meran cares" and "Environment awareness".

- The first pillar refers to the intention of the city to become a European inspirational example for an innovative city which highly respects its culture and at the same time promotes tolerance.
- The second pillar emphasizes the importance of digitalizing a city without losing sight of human connection and the citizens' needs. The city cares by remaining accessible to all, without excluding minorities.
- The third pillar envisions a city close to nature at human pace. The environmental awareness is reflected in a peaceful city which allows a Zen way of living.

The four fundaments of the city are presented in four key values for Meran as a smart city, which are inclusion, culture, sustainability and innovation. The vision itself is summarized in a slogan which is

"Walk through Meran where authenticity and innovation meet for better life" and in a picture, as shown in Figure 1, for a fast and immediate communication.

Figure 1. Meran 2030: where authenticity and innovation meet for better life; a city life close to nature in which all people are environmentally aware; Meran cares for all

USE CASES DESIGN WITH A PARTICIPATORY METHOD

The main focus of this section is to present the approach and methodology adopted for the processing of use cases for the development of the OpenIoT platform. In fact, despite technical challenges, companies need to face a number of barriers coming from stakeholders including end-users (Based, 2021). This challenges with respect to use case development include stakeholders perspective understanding (Escribano et al., 2021) and customer awareness and permissions achievement. A tight communication between companies as IoT system builders and consumers of these systems is crucial. The developed methodology includes therefore the involvement of citizens, technicians and local employees for the elaboration of the use cases, together with their placement into a "roadmap". In this way, it was possible to develop a service culture within the city actors and to create a business ecosystem thinking with multiple actors who interact with each other.

Here, "use case" is defined as a "smart" urban service (i.e. requires ICT infrastructure, IoT or data analysis), related to the information about its elementary utilization by users, which can be provided through a digital platform. Moreover, the authors define here "roadmap" as a strategic plan which contains the key steps and milestones needed to reach a defined goal or a desired outcome. The methodology adopted brought in a first step to the generation of a list of key urban service ideas, designed on the specific needs of Meran, and to a second step, where the service ideas have been evaluated in a pre-feasibility study for their real implementation as use cases in the city, with regard to technical and economic aspects, together with a social impact assessment.

The service ideas were generated as a result of a series of workshops held together with the city's key-stakeholders, during which brainstorming was done based on the most important issues linked to urban development. The following workshop themes were considered: (1) security and illumination, (2) mobility: flows, (3) mobility: means, (4) environment and waste, (5) accessibility and inclusion, (6) energy and (7) e-governance, as a transversal topic. The thematic areas covered during the various workshops were identified based on the following criteria:

- The theme is of strategic importance to realize the smart city vision.
- The theme has an important role in other projects the city is engaged in.

The objective of each workshop was to draft service ideas that respond to city needs and that can be implemented in the short, medium but also long term. These respond to very general needs and only part of them have been evaluated as "smart" i.e. have been selected for further implementation for the smart city roadmap of Meran. The next sections will give an idea of the workshop outputs, especially explaining the methodology adopted for their brainstorming though a participatory design process, together with the explanation of the output evaluation criteria for the smart city use cases development.

Involved Stakeholders

An important step in this process is about the definition and the active involvement of stakeholders, which are those people that have a strategic role in the management and planning of the city, and typically belong to the following entities:

- Public administration.
- Public services for citizens (tourism, merchants, health, museums, retirement homes, trade associations, etc.) and service companies (public transport, energy providers, etc.).
- Social cooperatives.
- Representatives of civil movements or social groups.
- Start-ups or technology providers.

The stakeholders identified are invited to participate in all the workshops that are related to their field of interest. During the pilot project in Meran 45 stakeholders in total were involved, some of which participated to more than one workshop. In the Table 1 below the workshop series and the relative participation are listed. In each workshop three moderators were present.

Table 1. Overview of the workshops organized for the development of Smart City Services in Meran

Nr.	Name	Objective	Participants
1	Security and lighting	**How to make Meran safer?** Development of smart services and applications that will make Meran safer or increase the perception of security. The topic of lighting can be linked to that of security for some solutions.	8
2	Mobility: fluxes	**How to manage flows and make traffic flow more fluid?** Road traffic; parking; signalling; hub freight delivery.	13
3	Environment and waste	**How to make Meran more connected to nature and cleaner?** Waste management; circular economy; environmental monitoring; water, soil and air quality; light and noise pollution; green management; smart irrigation.	7
4	Mobility: means of transport	**How do I move around Meran and the surrounding area?** Means of transport; (e)-bike/car sharing; public transport; intermodal transport; pedestrian zones.	9
5	Inclusion and accessibility	**How to make life and Meran's city-services more inclusive and accessible to all?** Peer-to-peer accommodation; Ambient Assisted Living (AAL); architectural barriers; facilitation of accessibility, buildings and points of interest; guidance and dedicated parking; smart watch.	9
6	Energy	**How to reduce emissions through better energy management?** Smart grid and smart meters; telemetering; Demand Side Management (DSM); electricity; district heating; hydrogen; building consumption; mobility and tertiary sector.	9
7	E-governance	**How to make services to citizens more accessible and digital? What new services can be offered in Meran?** Registry office; online registration; benefits; health services; online school services; digital information and help desks; online payments; information portals.	13

Workshop Methodology

The methodology adopted for the generation of the service ideas is based on the Design Sprint methodology and the Design Thinking, 6-3-5 Method. They have been adapted to the context of the workshop. The original Design Sprint is a unique five-day process for validating ideas and solving challenges through prototyping and ultimately testing ideas with customers. The Design Sprint methodology was developed by Google (Knapp et al. 2016). Multiple teams within Google experimented with different methods and applied them to support divergent and convergent thinking with teams. The resulting framework and set of methods are flexible, and teams are continuing to adapt it based on different goals and organizational cultures. The original process is sketched as follows:

Day 1: Understand. Map out the problem and pick an important area to focus on.
Day 2: Ideate. Sketch out competing solutions on paper.
Day 3: Decide. Make decisions and turn your ideas into a testable hypothesis.
Day 4: Prototype. Hack together a realistic prototype.
Day 5: Test. Get feedback from real live users.

On this basis, each workshop was structured in three steps. The first phase of the workshop methodology corresponds to the first day of the Design Sprint methodology. In this phase the conditions and context related to the topic are discussed. Firstly, moderators and stakeholders give an insight into the cities' existing initiatives and reflect on the ongoing relevant projects. In a second stage, participants compile a Strengths – Weaknesses – Opportunities - Threats analysis (SWOT analysis), by writing key-words on post-it in short time. The contributions are collected by the moderators and posted on a panel, which are then commented and together discussed.

The second phase of the workshop corresponds to the second and third day of the Design Sprint methodology. In this phase as many service ideas as possible are generated on the thematic areas of the workshop. Before the generation of the ideas starts, the moderators define and clarify the scope and boundaries of the topic and describe the cities' smart vision. The vision acts as the common ground from which the ideas should be generated. Smart city service ideas are generated based on the Design thinking methodology. The adopted approach allows participants to co-create and at the same time to work autonomously. The original methodology used is referred to the 6-3-5 Method, but it is adapted to fit the context. Each participant writes a proposal of solution to the issue and a correlated vision on a canvas. After every 3 minutes, the paper is passed on to the neighbor. Each participant follows the proposed idea already listed and adds new contents, in order to further iteratively develop the proposals. In other words, each participant writes a further solution to each basic idea, and this step is repeated two times. In the third round, the participant summarizes all contributions into one single idea and presents it to the group. This entire process is repeated twice during the workshop. At the end, the total number of ideas generated equals two times the number of participants. In addition, four people contributed to the development of each idea. After the presentations, all papers are physically gathered on a pin-wall where identic or very similar ideas are clustered.

The third phase of the workshop corresponds to the third and fourth day of the methodology. In this phase the best ideas are chosen and further developed. Participants are given stamps to vote, the number of votes equals the number of ideas each generated plus one. For example, in Meran two rounds of 6-3-5 Method were done, resulting in 2 ideas and 3 votes per participant. The most voted ideas are selected for further implementation and participants are free to choose which idea they prefer to work on and further develop. As a result, the working groups vary in size and each moderator supports one or two working groups, leading the participants through the implementation of the idea according to the following points: Story board; Service objectives; Advantages; Obstacles; Stakeholders to be involved during the development; Users and stakeholders to involve during the deployment and use phase; Urgency and need for the implementation of the service; Service requirements; Possible pilot projects.

Intermediate Output: The Stakeholder´S Service Ideas

In the pilot city of Meran, a total of 131 ideas for urban services were generated by the 6-3-5 brainstorm method, after which similar ideas were merged, resulting in 100 unique ideas. In Table 2 all use cases are listed at a glance:

This list serves as the basis for planning and structuring the smart city roadmap for the city of Meran. All input collected are analyzed and evaluated. This way each positively assessed input can be developed into a use case and included within the timeline of the roadmap, according to its consistency with the smart city concept and its technological feasibility. In the table above is indicated in the column "use case" if the service idea has been further developed in a use case, as a potential smart service for the city that can be provided totally or partially by the OpenIoT platform. The added number refers to the score assigned to them during the evaluation process shortly described in the next section.

Table 2. Workshops first output: list of urban service ideas

Name	Service ideas short description	Use case	Votes
	Workshop 1. Security and lighting		
Meran4Citizen	App that collects information from police and security forces about requested interventions and enhances communication with them to optimize their resources across the territory based on predictive police interventions.	UC 11	5/24
Intellipark	Advanced public parks with a focus on smart-lighting, crime-prevention and automated unsafe-situation detection.	UC 27	2/24
Bike Light	Improved quality of the public lighting of bike lanes and identification of un-even surfaces through a raster which is projected on bike lanes.	UC 30	0/24
Neighborhood watch digital	Safety communication app used to report potential threats and dangerous places or situations, which alerts the police and can launch a new intervention.	UC 31	8/24
Meran Safe Cycle	App for tracking and safeguarding parked bicycles and a protection system that alerts (the owner) in case someone is trying to steal a bike.	UC 38	1/24
Safer Territory	Centralized system analyses security alerts and serves as indicators for better decisions about appropriate intervention.	UC 65	1/24
Territory Control	Improved police coordination based on predictive analytics for better distribution of forces on high-risk locations of criminal acts.	No	7/24
Mobility optimization	Creation of an efficient city plan for safe cycling and pedestrian routes.	No	0/24
"Conditioned freedom"	Light signs on the roads before/after curve to inform about cars coming in the driver's direction.	No	0/24
"Divine Comedy"	Installation of high definition cameras for monitoring of all access points to the city.	No	0/24
Smart Watch	Cameras that identify suspicious movements with the help of a software that analyses the images in real-time and can control camera movements.	No	0/24
School information day	Teaching children how to communicate effectively in a multi-ethnic society with appropriate use of technology and attention towards conflict prevention, through group work.	No	0/24
	Workshop 2. Mobility: fluxes		
Priority Traffic	Giving priority to specific type of fluxes based on real-time traffic data.	UC 1	5/39
Intelligent Bike	Collection of bike traffic data to improve public bike lanes through a corporate biking incentive program for commuters.	UC 6	4/39
Auto-Stop	Uninstalling traffic lights by introducing smart cars which give priority to pedestrians and bikes.	UC 7	5/39
Flex-Bus	Optimization of bus capacity by monitoring the number of public transport users in real-time and favoring prior seat reservation.	UC 8	2/39
Park and Bike	Traffic is guided to a public parking outside the city center from which bicycles and public means of transport can be easily used to reach the city center.	UC 9	6/39
Traffic Alert	Real time traffic and road condition monitoring for individual route optimization including all transport means.	UC 10	3/39
Bike at lighting pole	Network of e-bike charging station at illumination poles, indicating availability through an app and offering individual digital locks.	UC 12	1/39
City Toll	Toll for people using cars in the city center, especially single-car users, and benefits (e.g. shopping vouchers) for those who do not.	UC 15	0/39
Green Coins	Incentive for users of sustainable transport means by using green coins (exchangeable shop vouchers etc.) and penalize those who do not by monitoring their use.	UC 17	3/39
Access Trader Meran	Cars receive an amount of tradeable credits to access the city, related to the type of car, and rewards are given for not using all tickets.	UC 40	4/39
Smart parking	Application which leads cars to closest available parking lot, including garages.	UC 41	2/39
Go&Go	Minimization of stopping at traffic lights by informing drivers about best speed and adapting traffic light duration to real-time traffic.	UC 43	0/39
Bikes first!	Use of Süd-Tirol pass (i.e. local public transport ticket system) for bike-rental services and priority to bike traffic.	UC 46	1/39
Shared on demand	Zero emission public transport which can be requested on demand.	UC 58	1/39
E-car Share	E-car sharing stations also installed in private garages.	UC 59	1/39
Seat Share	App for car-pool community allowing seat-reservation, for commuting purposes.	UC 60	0/39
Slow mobility days	Car-free city 2 days per month and compensated with free public transport and bike or e-car sharing.	No	1/39
"Don't enter by car" City planning	Discouragement of cars entering the city through smart traffic lights and street directions, guiding towards surrounding Park and Ride (P+R) parking.	No	0/39
Hop On	Ropeways bring people back and forth on a regular basis across Meran, better connecting the center with the higher located parts of the city	No	0/39
	Workshop 3. Environment and waste		
Intelli-Bin	Optimization of waste collection services through sensors on waste bins which help optimise (minimise) garbage truck-routes and avoiding waste bins that are not full yet.	UC 3	5/21
Ambient Aware	Sensibilization of citizens on energy consumptions, through the possibility to visualize real time data from sensors that monitor environmental impacts (waste, water, energy, and emissions).	UC 14	0/21
Waste? No, thanks	No spilling water while collecting rainwater for public use and irrigation. Collectors are equipped with sensors to optimise water consumption.	UC 24	5/21
Waste tracking	Individual fee on waste based on produced waste volume and type.	UC 34	1/21
Green Fields	Designation of a specific green area to a school on annual basis for which they need to take care and monitor its public use, water consumption, etc.	UC 49	3/21
Circular Economy	Facilitating circular economy (e.g. food waste, second hand) through an information sharing platform application.	UC 62	0/21
No traffic Meran	Reduction of motorized traffic and creation of a maximum level of allowed traffic, optimize traffic fluxes, and encourage citizens to use bikes.	No	3/21
Environment and waste	Enhancing cleaning of public spaces through involvement of citizens and sharing information about waste for citizens sensibilization.	No	2/21

continued on following page

Table 2. Continued

Name	Service ideas short description	Use case	Votes
Zero Waste	Activation of zero waste initiatives and awarding excellent users under a brand called Meran cares.	No	0/21
Green city	Introduction of more green areas in the city on roof, facades, and spots.	No	2/21
Workshop 4. Mobility: means of transport			
Info Route	Real time information on public transport means at bus stops.	UC 5	4/27
Share Meran	Forbidding private cars in the city center and replace them by autonomous shuttles, bike sharing and e-taxi.	UC 18	3/27
RAIL-net	Reserving lanes for electric (autonomous) vehicles using induction for charging while driving.	UC 21	1/27
Moving tourists	Mobility info-app for tourists which collects data on tourist fluxes and triggers flexible shuttles in case of crowded situations.	UC 35	1/27
Share e-bike	E-bike sharing service giving information on availability and location of bikes when parked both in- and outside their docking station.	UC 36	1/27
Cargo-bike	Central delivery deposits for cargos and parcels from merchants and online shoppers. Re-distribution (last-mile) happens through cargo bikes or by storing the parcel in lockers for self-pick-up.	UC 42	4/27
LTZ 4.0	Extend limited traffic area but assure access to merchants through effective public transport or parking corridors.	UC 44	0/27
Intelligent City	Smart navigation systems for automatically redistributing traffic, based on real-time data and avoiding e.g. school zones.	UC 45	0/27
Public transportation as data hub	Smart public transport vehicles collecting data on traffic.	UC 61	0/27
Free ride for busses	E-/H_2 busses are given a green wave at traffic lights.	No	3/27
Tram	Introduction of a tram connecting Meran and surrounding sub-urban areas.	No	0/27
Moving walkway	Moving walkway with hop on, hop off possibility- linking outskirt parking areas with the city center.	No	3/27
H2	Hydrogen transport systems.	No	2/27
RopeBusWay	Introduction of ropeways for people and goods.	No	1/27
Nodal Points	Establishing nodes where different transport systems cross.	No	4/27
Hover Boards	Introduction of E-skateboards which can automatically drive towards the destination.	No	0/27
Workshop 5. Inclusion and accessibility			
MeranUber	Private car, bike and garage sharing at a lower cost than car sharing service companies and with a dedicated insurance model that allows sharing a private property with third persons.	UC 19	4/27
Safe2Go	Smart crosswalks guiding those in need (through lights, sound etc.) safe to the other side of the street and communicating dangerous situations to surrounding cars.	UC 20	1/27
Look beyond	Application for sharing information about e.g. accessibility of bus stops and facilitating the use of public transportation by disabled people.	UC 25	3/27
Ultra-yellow parking	Sensors automatically detect if a parking lot for disabled people is fraudulently used and alarm the police.	UC 39	1/27
Meran inclusive	Hotspots across the city for sharing information that is accessible to all, including disabled people and elderly.	UC 47	0/27
Info desk	Public administration and info platform that is accessible throughout Hubs in the city but also via app to complete administrative tasks	UC 54	0/27
The Fun Space	One integrated platform for all events, fed by organizations and including taxi reservations. and event participants can offer a shared ride to and from the event location.	UC 63	7/27
Monitoring	Tracking platform for elderly, children and disabled people.	UC 64	1/27
Open City	Creation of more public spaces (e.g. parks with mini libraries) where all people can relax and meet other citizens.	No	0/27
House of colors	Common living space for intercultural and inter-generational exchange.	No	3/27
Smart4Kids	Program to stimulate the development of children's´ affinity with nature e.g. public greenhouse garden for children.	No	4/27
No- or low-cost	Using public facilities to promote sustainable choices e.g. free public transport, bonus for non-motorized vehicle usage, etc.	No	0/27
Barrier Free	Integration of information, relevant for wheelchairs users, in the public space in an application.	No	0/27
Meran Rocks	Organization of events dedicated to all citizens	No	2/27
Me-Bo (Meran-Bolzano) train	Magnetically induced high-speed train linking Meran with the capital of the province and facilitating use for tourists.	No	1/27
Workshop 6. Energy			
Meran eco-mobility plan	Electric vehicles as substitutes of motorized transport including commercial goods transportation.	UC 2	1/27
Public temperature monitoring	Monitoring buildings' energy consumption and generate in a semi-automatic way suggestion of improvements. Award for best practice.	UC 4	6/27
Kick your green!	Award for the greenest district after sharing the amount of green energy produced and the amount of energy consumed in every district in real-time on a public portal.	UC 13	2/27
District heating network	Predictive analysis of the expected energy demand for a District Heating Network (DHN) and simultaneous calculation of a decarbonization strategy of the DHN. Portal with information about the benefits of DHNs for new users in the form of a decision-making tool.	UC 16	3/27
Web App "Improve your building"	Web app giving advice, based on technical data, on how to prioritize retrofitting actions for energy efficiency purposes.	UC 50	3/27
Charging e-mobility	Balancing the grid based on V2G-technology (Vehicle to Grid) with public e-cars and bikes sharing batteries.	UC 53	2/27
Easy monitoring	Informing elderly people through their smart TV about their individual energy consumption.	UC 66	0/27
Sunshine: energy source, life source	Decision-making tool for citizens offered by the city to simulate technological/commercial benefits from solar panels.	No	1/27
Energy&CO2	Roadmap with actions to decarbonize Meran by 2040 with a group of organizations and testimonials from best practices.	No	2/27

continued on following page

Table 2. Continued

Name	Service ideas short description	Use case	Votes
Renewable Energy Meran	Install solar (thermal) panels on every roof in the city.	No	1/27
Fresh – how much does it cost me?	Identification of energy efficiency opportunities at level of cooling buildings e.g. sensors giving immediate feedback to tourists in hotel rooms.	No	4/27
Meran 100% renewable	Database of all renewable energy systems to use for forecasting, incentives and maintenance purposes.	No	2/27
Energy storage	Combining solar panels with battery systems to facilitate an optimized auto-consumption of produced energy.	No	0/27
Workshop 7. e-governance			
My smart platform	Integration of all existing public services in one unique portal and extend the offer of intelligent public digital services.	UC 22	7/42
Digital democracy	Platform for citizens, where topics of interest and ideas can be discussed and voted. The top 3 ideas are discussed during the city council together with experts and the idea-giver.	UC 23	7/42
Ladies 'taxi	Taxi dedicated to women which can be paid and accessed with the Süd Tirol pass.	UC 26	4/42
No waiting room	Online booking of time slot for an appointment at the municipality to avoid lining-up.	UC 28	3/42
Volunte@r	Platform for requesting and offering all possible volunteering activities (doing groceries, walking a dog, replacing a light bulb, etc.) and helping to settle the administrative parts of volunteering.	UC 29	3/42
Alexa from Meran	Robotic assistant which guides citizens through all online public services e.g. voice-controlled online new ID-card request.	UC 32	2/42
Love for my town	Forum where citizens can ask direct questions to the public administration and FAQ are available for everyone.	UC 33	3/42
Meran feedback system	Chat-bot which helps citizens in 95% of the cases in an automatic way.	UC 37	0/42
Transparent process	Citizens and associations can inform themselves online about their administrative requests and view the progress and its status in detail at any given moment.	UC 48	7/42
Digital Kiosk	Ticketing service for online portal in which information is shared.	UC 51	1/42
Digi-doc	Online system to make doctor's appointment and communicate medical certificates to employer and INPS (i.e. the national social welfare institution)	UC 52	1/42
"Merenergy" app	Managing energy, water, waste consumption in a single application, cooperation of the various utility providers.	UC 55	0/42
Inter-operability	Open source API's to favor inter-operability.	UC 56	2/42
Digitalize the elderly	Facilitating the use of digital services by elderly through vocal assistance and other smart systems.	UC 57	2/42
Digital Divide	Encouraging the use of digital media with the dissemination of information and courses. Giving all categories of citizens access to digital infrastructure.	No	0/42

From Service Ideas to use Case

In a first screening, all service ideas were evaluated according to (i) their importance for the cities' stakeholders, (ii) their contribution to the city vision, (iii) their estimated potential impact and (iv) the service's smartness degree. This last criterion evaluates the smartness of an idea in terms of IoT needs i.e. if the service needs a platform, sensors, digital infrastructure, only internet connection or even if it does not need IoT technology. After this first evaluation process, from a total of 100 ideas, only 5 did not fit in with the Meran city vision, whereas 40 fit in only partially and the majority totally fit in. With regard to the "smartness" criterion, 31 ideas were classified as "No IoT needed", which means that they are not considered as "smart". Ideas which did not fit in with the vision or did not need any IoT technology were excluded from the use case development process. Consequently, the list of service ideas generated during the workshop has been shortened from 100 to 66 items, consisting in the only ideas which contribute to the city vision and which are considered to have a positive impact on the city's inhabitants.

These selected ideas can be used for the generation of smart city use cases. For each use case a pre-feasibility study has been conducted, in which the technical, economical and legal feasibility were evaluated. The smart city service has been further implemented and synthesized in a storyboard, which gives an immediately explanation of the service contents. The pre-feasibility study gives an indication about the complexity of realizing a smart service while considering the context of Meran. This complexity is translated into a time estimation for the implementation of the service and into an estimation of the preparatory phase duration before the service launch, where the legal context, the founds, the basis for the social acceptance and communication are given. The time for the implementation mainly depends on the technical and infrastructural difficulties for the service development. To guide this process, an operating tool in Excel has been developed. This tool allows to make evaluations in a standardized and

Figure 2. The Meran Roadmap 2040: thematic visualization

Figure 3. The OpenIoT conceptual platform design

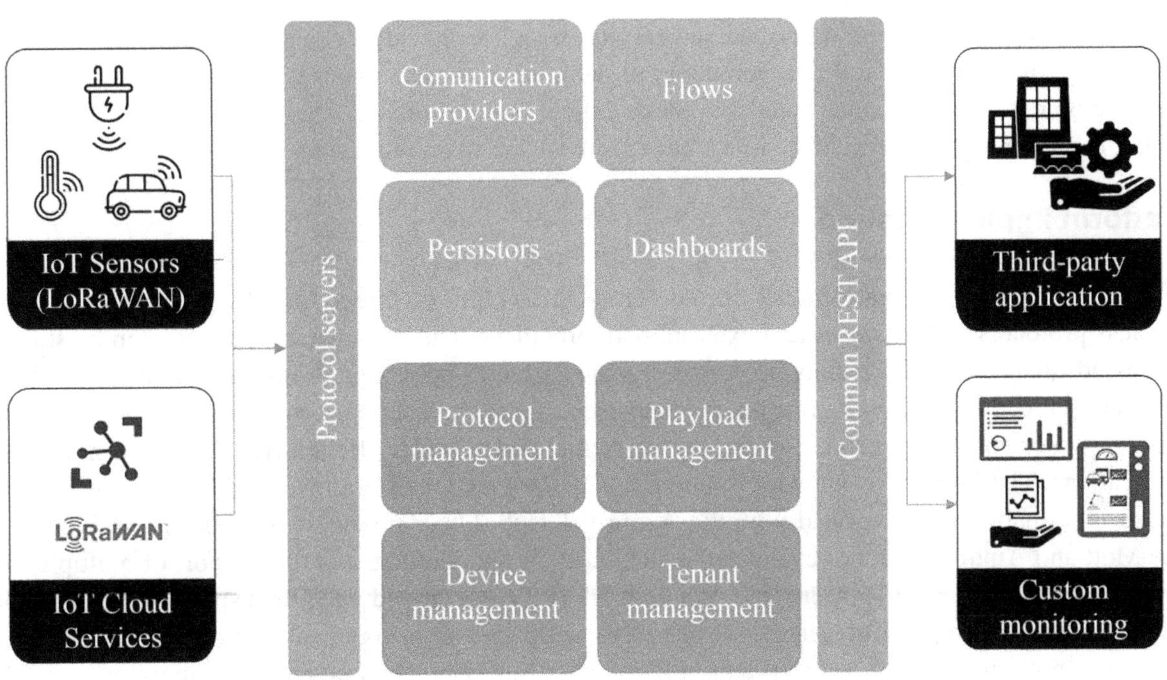

semi-automatic way. A detailed description of the use case evaluation methodology and the pre-feasibility analysis is described in Schweigkofler et al. (2020).

Results consists in an organized list of use cases, which represent the fourth level of smart city implementation i.e. "Layer 4 - Applications and services" (Ernst&Young, 2016). These are presented in the form of a classification of urban services in a smart city roadmap 2040 for the city of Meran. Through the operating tool, it is possible to visualize the roadmap in three different modes: (1) Chronological: use cases that can start immediately with the implementation appear first in the roadmap; (2) Thematic: use cases are grouped in thematic areas depending on the topics covered during the workshops (see Figure 2 where compare the first 46 use cases); (3) Ranking: use cases are visualized on the basis of the score assigned to them during the evaluation process.

PILOTS FOR THE OPENIOT PLATFORM DEVELOPMENT

After evaluating the outcome of the workshops, the following actions were implemented. The present section gives an example of the technical implementation of the first three smart city levels for a specific use case i.e. "Layer 3 - Service delivery platform", "Layer 2 – Sensors" and "Layer 1 – Infrastructure", as described in the introduction section of this chapter. The first steps in technical development of the OpenIoT platform were made within two pilot projects selected from the most suitable and elementary use case. This use case has been selected in cooperation with the municipality of Meran, based on specific problems, wishes and needs. The pilot actions were chosen in the field of monitoring public lighting as well as on reading water meters, according to the service idea "Ambient aware" proposed during the workshop "Environment and waste". Information about the implementation of this use case were given, i.e. the selection of sensor technology, the installation of devices, the first tests and the implementation process of a Long Range Wide Area Network (LoRaWAN) based communication system, planned for the total coverage over the city´s area. Furthermore, the creation of a dashboard for data visualization implementation process were described and results about conducted tests and measurements were explained together with a final discussion about possible next steps. In this section, only an overview of the development is given, since details on the technical implementation remain Intellectual Properties (IP) of the company Systems. More information is available upon request by contacting the authors.

Platform Implementation

An open data processing platform has been designed. This platform integrates different sensor systems, wireless protocols and networks and external data providers. The aggregated data can then be stored, processed, analyzed and the results are provided using an open Application Programming Interface (API) to other applications, including third-party ones (see Figure 3). The platform was implemented using C# and python for the backend, running on the Azure cloud, which offers a rich environment for IoT applications. The LoRaWAN network is integrated by the use of the Loriot Network Server, the leading enterprise capable tool integrated with the Azure IoT Hub. Other protocols have been also included, like Mqtt and Amqp in order to evaluate the flexibility of the platform and the support of multiple IoT devices. LoRaWAN is low-power wide area network (LPWAN) based on spread spectrum modulation techniques and it is implemented in the IoT networks worldwide for energy management, natural resource reduction, pollution control, infrastructure efficiency, disaster prevention, etc (Uwiringiyimana et al.,

Figure 4. The OpenIoT platform: overview network and device analysis

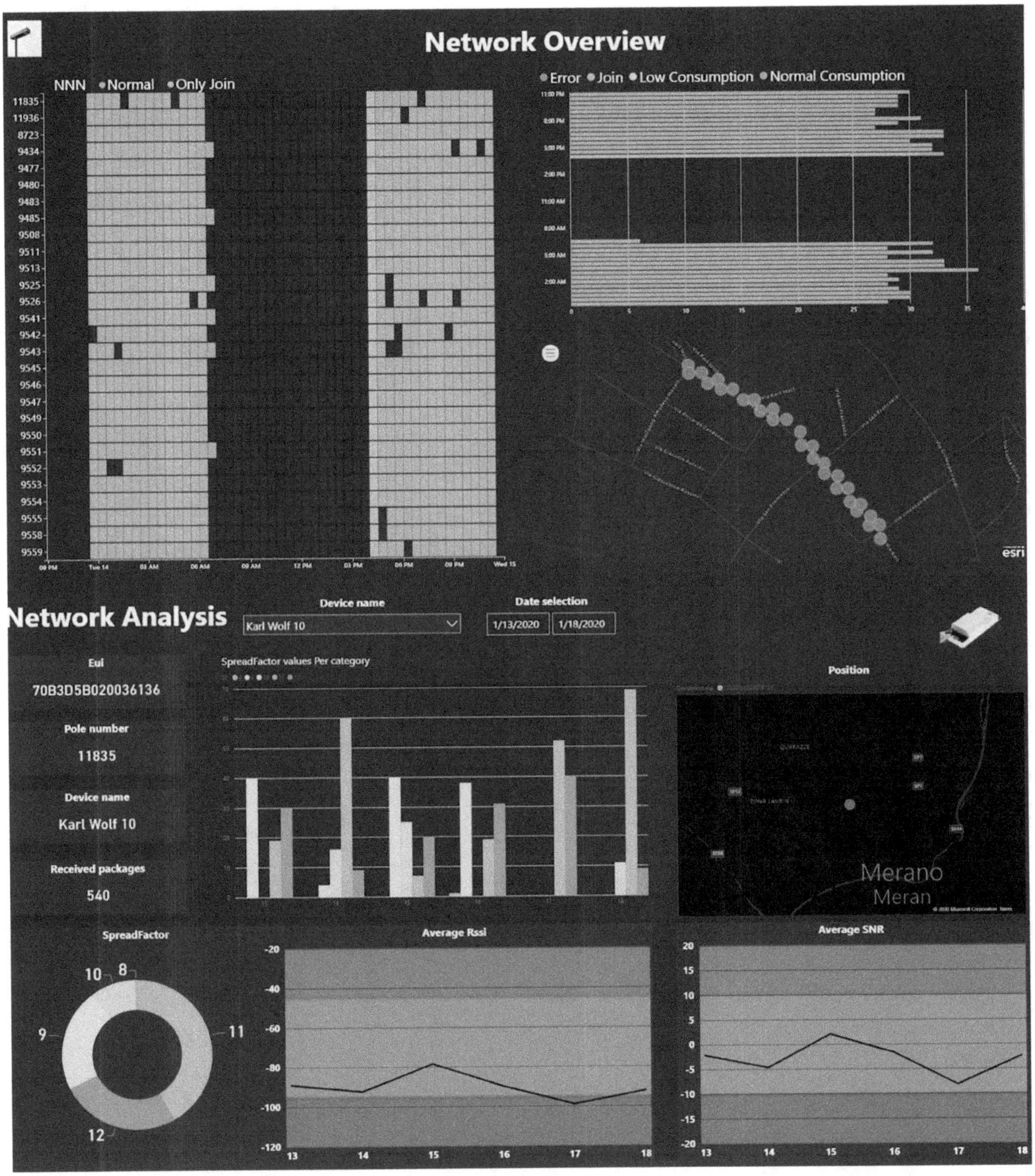

2021). Considering this and the actual infrastructure availability in the city of Merano, LoRaWAN has been chosen as the best radio technology to develop the two selected pilot actions in the field of monitoring public lighting and water meters.

Figure 5. The OpenIoT dashboard for public light (top) and water consumption monitoring (bottom)

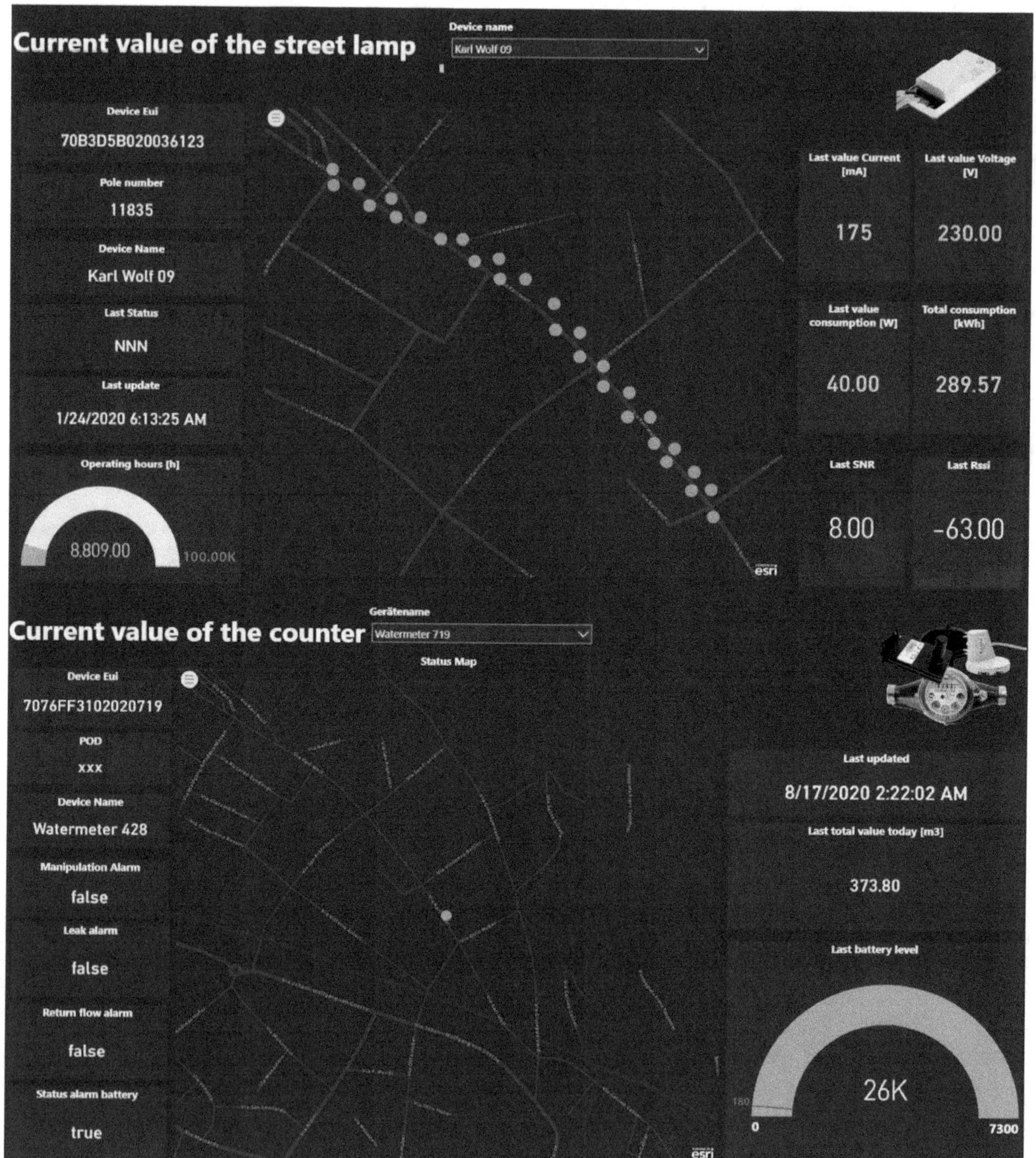

Based on the requirements elaborated in the workshops and the experience provided by the stakeholders, it was decided to implement a LoRaWAN-based communication system. This allows each sensor to be wireless and low-power, but to still communicate in a secure encrypted way. For testing purposes, three gateways connecting the LoRa-network to an IP based network were installed on premises belonging to the municipality of Meran. These gateways cover the current test sites. Nevertheless, due to the mountainous landscape in the city, they do not cover 100% of the city's territory. For a permanent installation, two of these temporary gateways will be replaced in the future by four pieces, which will be then permanently installed in- and outside of Meran's territory, providing complete coverage.

Public Street Lighting and Water Management

The public lighting provider in the city of Meran is currently replacing over 6000 light points with modern LED lighting systems. As a service provider, the purpose is to replace all the light points in the following years, and this work is expected to be completed in 2023. The public lighting provider decided to install sensor-ready light fixtures, which include Zhaga Book 18 Luminaire Extension connectors, allowing external sensor / controller modules to access a 24V power supply and the luminaires Digital Addressable Lighting Interface (DALI).

For the test site, 100 luminaires in the city have been featured with a controller interfacing the lamps driver with the LoRaWAN network, allowing monitoring (defect detection, power monitoring, etc.) and control (dimming) of the attached light. For the monitoring of water consumption, residential multi-jet water meters, featured with static pulsars and a LoRaWAN-based wireless communication module, have been installed at five selected test-sites. The installed communication module is external to the meter itself. This allows a) the installation of the communication module up to a 10m distance from the meter, thereby allowing flexible positioning of the antenna and b) the independence of the installed meter, as the installed meter can be replaced by a different make/model without replacing the communication module. The downsides of the chosen approach are both the increasing of complexity, due to one additional component, and the wiring effort together with the susceptibility of wires to be damaged by rodents.

Another relevant aspect is that the platform has been designed to provide a standard solution for device management. Therefore, each user without technical expertise could perform the customizations and control of each component. This requirement needs standardization, also included settings, measurement, and commands definition, which permit to adopt the same control logit for each component, no matter what specific characteristics each device manufacturer requires.

Dashboard

As shown in Figure 4, a dashboard application has been implemented in Microsoft Power BI on top of the OpenIoT platform stack. This dashboard application allows the real-time monitoring of all connected sensors during the testing phase. The graphical user interface runs in a state-of-the-art web browser and is thereby accessible independently from the used operating system or hardware platform, enabling user access from every kind of computer or smartphone. For a better understanding of the LoRaWAN network response in the context of Meran city, it was necessary to generate different dashboards that would have summarized the situation of the network on different periods of time.

Day by day the network can be monitored to assure the quality of the device's communication, considering time boxing intervals of 30 minutes as a successful operation time. To ensure that the quality of the communication response to the demands of a smart city, a dashboard that monitors the network quality has been designed. Considering Rssi (Strength signal) levels on all received payloads, and the SNR (noise) surrounded the communication.

The current version of the application allows the monitoring of operating hours, current power consumption, total energy use, voltage and dimming level of the installed lights (see Figure 5 top). In the case of water meters, the required dashboards focused on consumption as well as different alarms: manipulation, leak, return flow and battery (see Figure 5 bottom).

RECOMMENDATIONS AND FUTURE DEVELOPMENTS

Whit regard to the OpenIoT platform, the following remarks for further implementation are given:

- Currently, the implemented application is merely a showcase of the possibilities offered by the platform. As the implemented system is open by design, the collected data, analysis results and alerts are also available for future third party applications and services. In particular, the platform can be further developed by selecting and integrating more strategic use cases. Especially in the thematic area of mobility, the platform would benefit from further testing and validation.
- The application was implemented with a microservice oriented architecture, allowing deep customization for each device model, protocol, and smart city requirements. This feature has been implemented to allow the progressive size scaling of each dedicated microservice, so that the requirements of a growing smart city can be fulfilled (Uwiringiyimana et al., 2021).

Some recommendations regarding the future adoption of the methodology for the use case development are listed below:

- Both the methodology for generating use cases for a smart city roadmap and the operating tool itself have only been tested on Meran, a medium-sized, alpine city. Even though the methodology and tools are developed to be adaptable to any kind of city, they have not been tested on other cities yet. It is therefore recommended to test and validate the operating tool and its methodology in different contexts.
- The operating tool is currently an excel-based instrument, and it lacks a user-friendly interface. Especially, considering that the tool is designed to be used and updated by public administrators as a support in their transition towards a smarter city, a more intuitive interface would be needed. A web-application of platform is therefore advised.
- To make possible a participatory approach, it appears crucial to facilitate a convenient time, location and duration for the workshop and gives incentives in order to involve the highest number of stakeholders and especially to have them participate on a voluntary base. Since the methodology described in this chapter depends heavily on the input from the cities' stakeholder, the phase of their involvement is crucial for the quality of the outcomes.

CONCLUSION

In this chapter, a methodology for the participatory design of use cases for the smart city has been described, together with the implementation of a use case for the development of an open platform for data visualization. This final section will conclude the chapter with the discussion regarding the developed methodology and approach for the transition towards smart city adopted in Meran.

The goal of realizing a flexible and scalable model, designed to be replicated also in different contexts, has been achieved. This is guaranteed by the strict methodology developed for all phases, from the stakeholder involvement, to the vision sketch, to the urban service ideas generation and evaluation. In addition, the developed tool that supports the evaluation phase of the use cases is in facts a holistic and operative tool that can help public administrations to get a clear direction.

With regard to the OpenIoT platform, two pilot applications have been developed and the OpenIoT platform is actually in a prototype stage. Further development of the platform will be done, according to the use case list designed within the project. The platform prototype reflects the planned strategic features described in the introductive section of this chapter: (a) it can potentially provide a high number of smart services according to the developed use cases; (b) it permits the integration of different devices, communication protocols as well as users and third-party developers; (c) it permits to scale itself progressively for further application in other contexts.

The coordinated adoption of such a holistic methodology and technology should boost the development of new services for and with citizens, businesses, tourists and public administrations.

ACKNOWLEDGMENT

This paper has been supported by funding from the European Regional Development Fund ERDF 2014-2020 Südtirol/Alto Adige, in the project OpenIOT 1069 (CUP: B11B17000720008). The authors would like to thank the City of Meran, as well as the involved company Systems SRL for their contribution.

REFERENCES

Angelidou, M. (2015). Smart cities: A conjuncture of four forces. *Cities (London, England)*, *47*, 95–106. doi:10.1016/j.cities.2015.05.004

Based, W. H. E. (2021). *IoT Platforms, Use Cases, Privacy, and Business Models*. doi:10.1007/978-3-030-45316-9

Bibri, S. E., & Krogstie, J. (2017). Smart sustainable cities of the future: An extensive interdisciplinary literature review. *Sustainable Cities and Society*, *31*, 183–212. doi:10.1016/j.scs.2017.02.016

Bifulco, F., Tregua, M., Amitrano, C. C., & D'Auria, A. (2015). ICT and sustainability in smart cities management. *International Journal of Public Sector Management*, *29*(2), 132–147. doi:10.1108/IJPSM-07-2015-0132

Borsboom-van Beurden J., Kallaos J., Gindroz B., Costa S. and Riegler J. (2019). *Smart city guidance package*. Norwegian University of Science and Technology/ European Innovation Partnership on Smart Cities and Communities. Faculty of Architecture and Design.

Ernst & Young. (2016). *Rapporto Smart City Index 2016*. www.comune.bologna.it/iperbole/ piancont/archivionov/tabelle_grafici/EYsmartindex/2016-EY-smart-city-index.pdf

Escribano, C. P., Theologou, N., Likar, M., Tryferidis, A., & Tzovaras, D. (2021). Business Models and Use Cases for the IoT. *IoT Platforms, Use Cases, Privacy, and Business Models*, 51–80. . doi:10.1007/978-3-030-45316-9_3

Fernandez-Anez, V., Fernández-Güellb, M. J., & Giffingerc, R. (2018). Smart City implementation and discourses: An integrated conceptual model. The case of Vienna. *Cities (London, England)*, *78*, 4–16. doi:10.1016/j.cities.2017.12.004

Fraunhofer Gesellschaft. (2019). *Morgenstadt - City of the future Initiative*. https://www.morgenstadt.de/en.html.Accessed

Haarstad, H., & Marikken, W. W. (2019). *Are smart city projects catalyzing urban energy sustainability?* Elsevier. *Energy Policy*, *129*(June), 918–925. doi:10.1016/j.enpol.2019.03.001

Knapp, J., Zeratsky, J., & Kowitz, B. (2016). *Sprint: How to Solve Big Problems and Test New Ideas in Just Five Days*. Simon and Schuster.

Ng, I., Scharf, K., Pogrebna, G., & Maull, R. (2015). Contextual variety, Internet-of-Things and the choice of tailoring over platform: Mass customisation strategy in supply chain management. *International Journal of Production Economics*, *159*, 76–87. doi:10.1016/j.ijpe.2014.09.007

Osservatorio Nazionale Smart City. (2018). *Vademecum per la cittá intelligente*. www.osservatoriosmartcity.it

Pereira, G. V., Cunhab, M. A., Lampoltshammera, T. J., Paryceka, P., & Testac, M. G. (2017). Increasing collaboration and participation in smart city governance: A cross-case analysis of smart city initiatives. *Information Technology for Development*, *23*(3), 526–553. doi:10.1080/02681102.2017.1353946

Schweigkofler, A., Romagnoli, K., Steiner, D., Riedl, M., & Matt, T. D. (2020) *Methodology and operating tool for urban renovation: the case study of the Italian city of Meran*. SSPCR 3rd International Conference on Smart and Sustainable Planning for Cities and Regions.

The British Standards Institution. (2018). *Smart city framework – Guide to establishing strategies for smart cities and communities*. https://www.bsigroup.com/en-GB/smart-cities/Smart-Cities-Standards-and-Publication/PAS-181-smart-cities-framework/

Think: Act Magazine. (2019). *The Smart City Breakaway*. Roland Berger GMBH.

Uwiringiyimana, M. M., Nandagopal, G., Guan, Y., Vinkovič, S., Kölsch, J., & Heinz, C. (2021). IoT Platforms. *IoT Platforms, Use Cases, Privacy, and Business Models*, 21–49. . doi:10.1007/978-3-030-45316-9_2

KEY TERMS AND DEFINITIONS

Dashboard: An information management tool that visually tracks, analyzes, and displays key performance indicators (KPI), metrics and key data points to monitor the progress towards a desired outcome. The monitored data are customizable to meet specific needs. Behind the scenes, a dashboard connects to files, services, and APIs, but on the surface displays all data in the form of tables, line charts, bar charts and gauges.

Design Thinking: A non-linear, iterative process that teams use to understand users, challenge assumptions, redefine problems and create innovative solutions to prototype and test.

IoT (Internet of Things): Is a system of interrelated computing devices, mechanical and digital machines, objects, animals or people that are provided with unique identifiers (UIDs) and the ability to transfer data over a network without requiring human-to-human or human-to-computer interaction.

Roadmap: Any plan or guide to show how to progress towards a pre-defined objective.

Smart City: A well-defined geographical area, in which high technologies such as ICT, logistic, energy production, and so on, cooperate to create benefits for citizens in terms of wellbeing, inclusion and participation, environmental quality, intelligent development; it is governed by a well-defined pool of subjects, able to state the rules and policy for the city government and development.

Use Case: A specific situation in which a product or service could potentially be used.

Vision: The ability to imagine how something could develop in the future, or the ideas that come from imagining in this way.

Chapter 22
Building Stock Energy Models and ICT Solutions for Urban Energy Systems

Ilaria Ballarini
Politecnico di Torino, Italy

Vincenzo Corrado
https://orcid.org/0000-0001-9391-5602
Politecnico di Torino, Italy

Matteo Piro
Politecnico di Torino, Italy

ABSTRACT

The existing building stock presents a high potential of energy savings and CO^2 emissions reductions. To this purpose, literature provides novel city-scale building-oriented studies, aimed at developing suitable tools for stakeholders, city planners, and decision-makers. To achieve an effective urban energy planning, urban energy systems (UES) models are developed; they employ a multi-domain approach, embracing the complex interactions in urban areas, such as energy flows, environmental indicators, social and economic factors. To perform an advanced modelling and to simulate the complexity of the UES, ICT (information and communications technology) represents nowadays the right answer to the needs of integration of data, tools, and actors in different domains. The chapter investigates the current studies in the field of building stock energy modeling and the application of advanced technologies to develop UES models. As an exemplification, the technological approach followed in the SEMANCO project to support urban scale energy modelling is presented.

DOI: 10.4018/978-1-7998-7091-3.ch022

INTRODUCTION

The world population is currently about 7.8 billion and it is expected to increase to 10 billion in 2050. The current share of urban and rural population at world level is 56% and 44%, respectively, but it is foreseen a growing of the population share living in urban areas to 68% in 30 years. In Europe, despite 3.7% decrease of total population is expected in 2050 compared to 2020, the urban population share will move from 75% in 2020 to 84% in 2050 (Ritchie & Roser, 2018). The increase of urbanization determines a corresponding intensification of human activity and the related energy consumption; thus, it is necessary to carry out effective urban energy planning that takes into account sustainable development plans and quantifies the effect of related policies.

The building stock represents a significant share in energy use and greenhouse gas emissions; in 2017, it accounted for 42% of final energy consumption and 17% of CO_2 emissions from fuel combustion in Europe (European Commission [EC], 2019). Therefore, it offers a great potential for energy efficiency and integrated sustainable energy solutions. The key-role of buildings in the energy balance of cities has boosted novel city-scale building-oriented studies, aimed at developing suitable tools for stakeholders, city planners and decision-makers. These tools will enable to understand urban energy systems and to formulate energy plans, suggest sustainable initiatives and decide on constructive policies (Torabi Moghadam et al., 2017).

To develop urban energy planning tools, models of Urban Energy Systems (UES) should be conceived through an integrated multi-domain approach, which embraces complex interactions in urban areas, such as energy flows, environmental indicators, social and economic factors. The increase of model data and complexity requires more advanced technologies. In this field, Information and Communications Technology (ICT) represents the right answer to the need of integration of data, tools and actors, spanning across different domains and scales of analysis (i.e. building, neighborhood, city, and region). ICT is a very wide term; sometimes "digitization" and "datafication" are used to refer to the issues derived from data retrieval, modelling and analysis in urban contexts. This technology grants automatic access to data dispersed in different sources, interoperability of data and tools for energy simulation, optimization and decision support through multi-criteria analysis, and visualization in a 3D environment. To these objectives, ontology and semantics form the basis to develop UES models using ICT (Pauwels et al., 2017). Future UES models will take advantage from the nascent era of digitalization, through the inclusion of the Internet of Things (IoT) and Artificial Intelligence (AI) agents (e.g. data analytics, machine learning) (Boje et al., 2020), as to boost the development of Smart Cities.

In this context, the present chapter aims to investigate recent studies on integrated urban energy models, starting from building stock models and related tools, highlighting their strengths and limits, up to multi-domain energy models. In the field of UES, the chapter specifically focuses on projects and initiatives addressed to the creation of effective urban energy planning tools. As an exemplary case, the experience of SEMANCO (*Semantic Tools for Carbon Reduction in Urban Planning*), a European project part of the EC 7th Framework Programme, as well as the peculiarities of the UES tool developed in the project (Corrado et al., 2015) are presented. SEMANCO aimed to create an effective decision support system to reduce carbon emissions. Its approach was based on the interrelation of different actors – policy makers, planners, engineers, consultants, and inhabitants – to correlate a diversity of problems, spanning across distinct domains and geographic scales.

The chapter has been structured as follows:

- Background analysis on the approaches for energy and environmental modelling of building stocks.
- Presentation of urban building energy models (UBEMs), as recognized in literature as the most used techniques for building stock energy analysis, and urban scale energy models (USEMs), that integrate multi-domain energy assessment at city scale.
- Specification of Urban Energy Systems (UES) for the development of urban energy planning tools, and characterization and application of ICT systems for the integration and interoperability of data and tools.
- Description of the technological approach followed to create an effective tool for carbon reduction in urban planning, through the experience and the outcomes of the SEMANCO project. Presentation of the SEMANCO platform and of EECITIES (Energy Efficient Cities), an energy analysis service provider developed in the project.
- Conclusions and suggestions for future research in the context of Smart Cities development.

Background

Integrated urban energy modelling is a complex approach that concerns the energy flows coming into, passing throughout, and going out of cities. To understand the drivers and patterns of energy flows, the element of the urban metabolism should be analyzed. In literature, urban metabolism has been discussed in social, natural, historical and political sciences. Urban metabolism is either recognized as an urban metaphor to describe the transformation of nature by society, or discussed through an ecological approach where the city is considered as an ecosystem. Urban metabolism also concerns the transformation that energy and materials undertake, and the effects of this transformations on the city. This last aspect matches with the engineering approach of urban metabolism defined by Carreón & Worrell (2018) as "a process that involves the exchange of energy and materials within a specific environment". According to Carreón and Worrell (2018), the drivers of energy flows are recognized as factors that directly or indirectly cause a change in the services provided by the energy systems, and can be related to environment (e.g. climate, energy sources), technology, economics, society, demography and politics. Devising an integrated urban energy modelling means developing an Urban Energy System (UES), i.e. "a formal system that represents the combined processes of acquiring and using energy to satisfy the energy service demands of a given urban area" (Keirstead et al., 2012). In this context, urban metabolism is a paradigm that overlaps with urban energy system.

Among the elements of urban metabolism, the influential role of buildings is well recognized. In literature, several studies concern building stock models, including both buildings, or demand-side, and districts, or supply-side, energy models, as well as their interactions. Within the demand-side energy models, the urban building energy models (UBEMs) are the most widely applied and well recognized by the research community (Johari et al., 2020). Despite UBEMs are largely included in the UES models, the latter are still scarcely investigated and not widespread.

Building Stock Energy Models

According to the scientific literature, the building stock energy models are usually classified in function of two different methodological approaches: the top-down and the bottom-up models (Johari et al., 2020; Abbasabadi & Ashayeri, 2019; Kavgic et al., 2010; Swan & Ugursal, 2009).

The applicability of one technique compared to the other depends on the final aim to be pursued, on the correlation with different variables and on the diverse levels of detail of the input parameters. Specifically, the important differences between these two approaches concern: the different degrees of aggregation of data, the different computational capacity required by the computer, and the different purposes for which they are conceived and used.

The top-down model is mainly used to predict a future scenario, based on situations from the past time series. The bottom-up, instead, is used to define the current circumstances and ultimately to investigate future possibilities, such as improving energy performance of certain buildings. In general, the top-down model determines a broad-spectrum view of the system without venturing too far into detail. On the contrary, the bottom-up model specifies each component, and as a result provides a deeper knowledge of the phenomenon to be studied. In the following subsections, top-down and bottom-up building stock energy models are presented in terms of main features and applications, strengths and limits, as pointed out in literature.

Top-down Models

The top-down model in general concerns the resolution of complex systems, in which the final objective is to divide the output into a finite number of parts, in order to more easily identify the existing interconnections between them. It consists in associating a predefined number of black boxes to the problem, in order to mathematically demonstrate the possibility of obtaining that result without necessarily having to depend on the knowledge of the physical laws that govern the phenomenon itself. Through the analysis of the final solution, it is possible to identify the missing resources so as to complete the codification of the problem.

The top-down models, typically based on historical data series, are developed to investigate the macro-relationships between the energy, technological, and economic sectors. Usually, the initial parameters from which building consumption is calculated are a function of economic variables (gross domestic product, unemployment, inflation, per capita income), energy price, and/or climate data. Since it must work with aggregated data, the top-down model does not contain a direct and explicit representation of the technologies adopted.

The econometric model is an example of a top-down approach, which is developed to determine the influence of economic variables on the effects of energy policies adopted by following a statistical calculation. The difficulties in applying this methodology depend on the energy-economic interrelation of the parameters considered. The main purpose in the adoption of this model, implicitly based on the microeconomic theory, is the prediction of the energy demand of a building stock in a statistical way, generally adopting linear regression equations according to per capita income.

In literature, it is often possible to come across the distinction between econometric and macro-econometric models, whose diversity is mainly based on different applied economic theories: microeconomics in the first case, and macroeconomics in the second. In the macro-econometric model, the aggregated data are based on national income and not on per capita income. Many examples of econometric top-down models can be found in literature, like the work of Bianco et al. (2009), in which an energy model able to predict the urban energy use has been developed through regression analyses that correlate gross domestic product and population to yearly electricity use in Italy. Another example of a top-down approach is the one proposed by Haas & Schipper (1998), who developed several econometric models, thus correlating energy variables to economic ones, for countries belonging to the OECD (Organisation for

Economic Co-operation and Development). Specifically, the authors performed an analysis to estimate the energy demand of a residential building stock by correlating it to the price of energy, to the related consumption expenditure, and to the value of degree days in the reference country.

Having to do with multi-collinearity, one disadvantage of the top-down building stock energy model lies in the impossibility of evaluating the influence of a precise and defined technological measure, but the overall trend would be established in any case.

Strengths and limits of the top-down energy models are summarized in Table 1.

Table 1. Strengths and limits of top-down building stock energy models

Strengths	Limits
• Simplicity linked to reliance on aggregated historical data • Capability to extract relationships between factors that indirectly determine urban energy use • Ability to predict long-term energy consumption • Independence from detailed technological description	• Dependency on historical trends data • Generalization of the existing conditions • Absence of determination of individual end-uses • Unsuitability to examine changes in technology

Bottom-up Models

A bottom-up model, unlike the top-down one, builds an ordered and sequential action strategy to achieve the final result, providing a detailed explanation of each subsystem and of how it is connected to the others. The bottom-up building stock energy model is based on disaggregated and extensive data. These are acquired by the developer beforehand through analytical, statistical or empirical means, in order to be able to proceed with the subsequent calculation phases. Disaggregated data are applied to estimate the end-use energy of individual buildings or group of buildings, and then the aggregated energy demand is extrapolated at a wider territorial scale. This way, it is therefore possible to individually evaluate the behavior of potential energy refurbishment interventions aimed at improving the energy performance of predefined building stock in detail. Consequently, a major skill required to the developer is to know how to manage the complexity and uncertainty of the input parameters needed to complete the model.

In literature, bottom-up building stock energy models are commonly divided into three categories, namely *engineering* (or *physical*) *models*, *data-driven models*, and *hybrid models*.

The *engineering* (or *physical*) *models* are based on an analytical calculation; the methodological approach entails the solving of the energy balance equation of the system. To pursue this objective, a complete knowledge of the input parameters is necessary. It can be obtained from in-field measurements or from empirical references, i.e. based on the observer's experience. To this regard, it is necessary to know the climatic data of the site of interest, the geometrical properties of the buildings and of the urban environment (surfaces, volumes, heights, window-to-wall ratio), the thermo-physical characteristics of opaque and transparent envelope components, information on the user behavior and technical building systems data. The engineering model is the most used approach in literature to assess the energy use of a building stock; it is usually recognized as *urban building energy model* (UBEM). The following section of the chapter is specifically dedicated to present UBEMs and related lines of research.

Starting from a large sample of data, the *data-driven models* allow associating the building characteristics and other parameters, such as user features, to the energy use by means of statistical analysis or models based on artificial intelligence. The mathematical relationship identified is then used to estimate the energy use related to specific end-uses (Abbasabadi & Ashayeri, 2019). The data-driven statistical models involve regression techniques, such as linear or polynomial regression. These methods are widely applied in estimating the energy use due to very simple structure and fast processing time. Anyway, the regression model is not able to take into account complex patterns related to the real world. This limitation can be overcome by the data-driven models based on artificial intelligence, which commonly apply Machine Learning (ML) techniques. These allow identifying statistical patterns in data, fitting the model, and then finding the mathematical law to correlate the energy use to building parameters. The ML techniques include both un-supervised and supervised methods. The former relies on clustering approach, while the latter comprises neural network techniques. An example of a data-driven approach is the one proposed by Pasichnyi et al. (2019), which is based on the creation of a building stock model, referring to the urban reality of Stockholm, to investigate what interventions can be made to the existing building stock. Three improvements are proposed, and the best combination ensures a decrease of 18% in annual demand for the heating of the building stock involved.

A common issue of the building stock energy models is the lack of a complete picture of input data that can be discussed in terms of "data availability", "data accessibility", and "data accuracy" (Goy et al., 2020). The advantage of using a data-driven model compared to an engineering model consists in getting results that are more accurate. However, the accuracy of data-driven models can be easily compromised whereas the available dataset is not sufficiently representative or the data are aggregated. On the other hand, the engineering models do not suffer from the lack of historical data (e.g. available energy use data), differently from the data-driven ones, but are more sensitive to the estimation uncertainties of energy use due to the simplification of building characteristics and user behavior data.

In agreement with the above approaches, an example is the research carried out by Nageler et al. (2018). The authors performed an energy simulation through the construction of two models. The first one involves the creation of a dynamic physical-based model. The second one, instead, defines a mathematical algorithm of the data-driven type. These digital objects have been created to estimate the thermal behavior of a residential area composed of 34 buildings, and show good agreement against measured data. In addition, the application fields of these approaches have been discussed.

Table 2. Strengths and limits of the engineering and of the data-driven bottom-up energy models

Engineering models		Data-driven models	
Strengths	**Limits**	**Strengths**	**Limits**
• Independence from historical data • Easy assessment of efficiency measures • High adaptability with BIM tools	• Computationally time consuming • Loss of accuracy due to simplifications • Need of detailed input data	• Efficiency of processing time • Capability to include occupancy factors • Accurate prediction of energy use	• Relying on historical data • Limited adaptability in case of design variations • Loss of accuracy in case of data unavailability

The *hybrid models* are designed to overcome the limits of both engineering and the data-driven models, and to meet the need to complete the initial input data picture. Indeed, the stochastic distribution of certain parameters (e.g. those concerning occupant's use) produces a certain degree of uncertainty in

Figure 1. Classification of building stock energy models

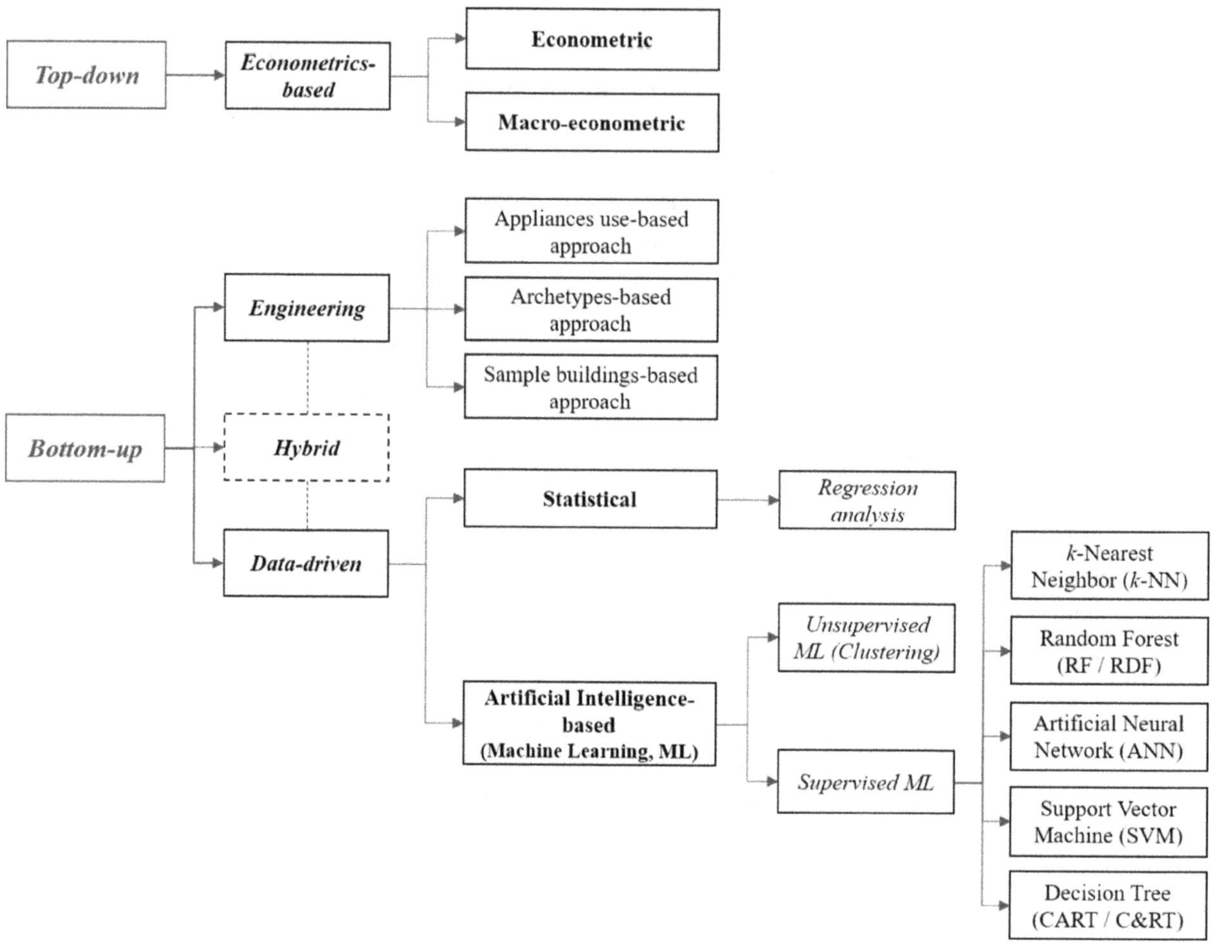

the creation of the engineering model. This issue can be overcome by deriving the necessary input data through statistical (data-driven) analysis of energy use historical data. This will then lead to the definition of a sophisticated bottom-up approach, whose aim is to determine the energy needs of a building stock, by extracting data from different information databases.

Strengths and limits of the engineering and of the data-driven bottom-up energy models are summarized in Table 2.

Figure 1 provides the classification schema of the building stock energy models.

URBAN BUILDING ENERGY MODELS

Classification and Features

Urban building energy models (UBEMs) are usually classified as bottom-up engineering models, despite different keys of interpretation attributable to the definition of UBEMs are provided in literature. For Sola et al. (2020) and Hong et al. (2020), the concept of UBEM is independent from the bottom-up approach adopted, engineering or data-driven. Reinhart & Cerezo Davila (2016) and Johari et al. (2020), on the other hand, refer the concept of UBEM to the engineering bottom-up model implemented to constitute an energy model, suitable for estimating the operational energy demand of a certain building stock. The energy demand, based on the energy services considered (space heating, space cooling, domestic hot water production, ventilation, artificial lighting, etc.), depends on the resolution of the mass and energy conservation equations.

In agreement with Swan & Ugursal (2009), which carried out a classification of the building stock energy models for the residential sector, the engineering models – and hence the UBEMs – can be further classified into three different modeling approaches:

- the *appliances use-based approach*, that involves the estimation of energy demand through the study of the appliances used at regional or national level within housing units according to parameters, such as type, use, efficiency and consumption,
- the *archetype-based approach*, based on the definition of archetypes, i.e. buildings considered representative of subsets of the building stock in function of energy-related parameters (construction period, building size, use category, etc.), the subsequent assessment of the energy behavior of each archetype, and then the upscaling of the results to a wider territorial scale, and
- the *sample buildings-based approach*, based on the assessment of a wide variety of buildings with actual available data. If the buildings constitute a representative sample of the national or regional building stock, thanks to a large database of information, the energy demand of the stock can be derived through weighted factors.

A schematic representation of the three UBEM approaches is shown in Figure 2.

The first approach, which is less used, can be applied only for specific cases, for instance, if the electrical needs of a building should be estimated. In this case, it is necessary to know the type of appliances, the power absorbed and their use profile. This methodological approach falls within the bottom-up methodology, as it starts from a series of disaggregated data. Anyway, the most applied approaches are the last two, whose difference is substantially based on the different number of buildings concerned in the model. Whereas the number of archetypes is limited, despite they can be constructed using a large dataset of real buildings, in the third case, real data of hundreds or thousands of buildings have to be collected. In agreement with the methodologies presented, an example is provided by Cerezo et al. (2017) that proposed four archetype-base modeling methods to compare the simulated Energy Use Intensity (EUI) applied on 336 residential buildings located in a district of Kuwait City. Two of these methods are based on the deterministic approach, one on the probabilistic approach and the fourth introduces a new Bayesian calibration, which is significantly closer to the measured data.

The UBEM is a useful tool for urban planners, designers, or public administrations who are willing to analyze the energy performance of a city's environment, and to support and encourage programs aimed at improving the energy performance of the reference building stock. The main difficulty that the developer of an UBEM must be able to govern concerns the high degree of uncertainty in determining some of the energy data, especially when they refer to an existing building stock. Some information, available from statistical or empirical documents, can be obtained from municipal, regional and national databases, standard and legislative references, and scientifically recognized publications (Piro et al., 2020). However, the degree of detail of the input parameters depends on the purpose and on the calculation engine

Figure 2. Types of urban building energy models (UBEMs): a) appliances-use based approach, b) archetypes-based approach, c) sample buildings-based approach

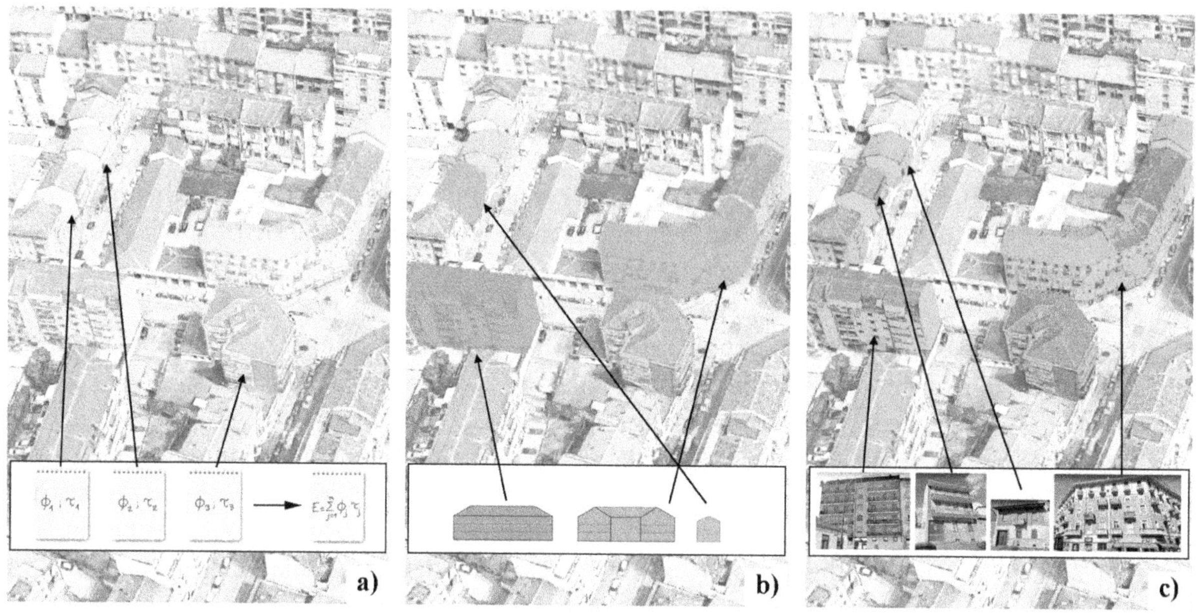

used. The energy performance assessment in an UBEM can be based either on steady-state monthly calculation methods, or on dynamic models with hourly or sub-hourly time steps. Furthermore, once the model has been generated, it is necessary to proceed with the calibration steps. Therefore, in addition to the input data sets, a considerable number of terms of comparison (e.g. real energy consumption data) are also required to establish a good quality.

According to Reinhart & Cerezo Davila (2016), the accuracy of a building stock energy model increases from individual building to urban level to regional level. On average, the deviation between estimated annual energy use and measured annual energy consumption amounts to 7-21% for heating loads in case of UBEMs involving multiple buildings. An example of an accurate UBEM is the one developed by Hedegaard et al. (2019), defined to estimate the energy consumption of a Danish neighborhood, consisting of 159 single-family houses, served by a district heating system. The described model aims to decrease the daily overloads required from the network by predicting a consumption pattern comparable to the real situation.

In the scientific context, a significant amount of tools for UBEMs development has been created, the majority of them within academic contexts. They usually employ BIM (Building Information Model) to store, process and visualize data. In such tools, the UBEM is usually composed of a three-dimensional geometric model, climatic and topographic information from the site of interest, opaque and transparent thermo-physical characteristics of the envelope components, data on the building use, and data about the technical building systems.

To represent the geometric reality of the city context, the Open Geospatial Consortium (OGC) has conceived a standard model, called CityGML (Gröger et al., 2012), with different Level of Details (LoD). The geometric and informative precision of an UBEM is a function of the purpose of the analysis to be pursued. For example, a three-dimensional representation of masses with planar surfaces, represented by a LoD1 (according to the standard provided by the OGC), may be sufficient to calculate the requirements of a building stock. LoD2, on the other hand, also allows the representation of oblique surfaces, so that sunlight studies can be carried out with a better degree of precision. Finally, LoD3 and LoD4 complete and improve the quality of the previous geometrical model, including external window frames, and potentially internal building structures. Therefore, it is established that the level of detail of a simulation at a building and at a district scales would be different from one carried out on an urban or regional context.

From UBEM to USEM - Urban Scale Energy Model

The urban building energy models, which only deal with building energy demand during operational phase, are often accompanied by other models, to take into account also supply and distribution energy in cities (Allegrini et al., 2015). The urban environment can be considered like a box, in which a series of energy conversion processes take place, which includes several areas, such as buildings, industry, transport, and city infrastructure in general. The approach to understand this urban thermodynamic system must be holistic and integrated, by including the exchange of different energy flows from different perimeters. In agreement with studies found in literature, this concept can be included in that of the *urban scale energy model* (USEM), as specified by Sola et al. (2020).

The USEM is a virtual model that involves a spatio-temporal resolution of energy flows in cities; it describes a certain energy system and proposes possible strategies to improve its urban energy performance. The virtual modeling of the USEM allows the evaluation of system losses, load peaks, unconsumed energy shares, and the relationship between consumer and supplier. In agreement with Basu et al. (2019), the USEM is not only distinguished by its physical elements, but also by the people who are part of the distribution process (agents, suppliers, consumers), as well as non-material entities (consumer feedback, incentives, energy policies, etc.).

About the incorporation of different urban energy components, models for building operational energy, models for transportation energy, and models for building embodied energy commonly coexist and usually operate independently from one to another. Instead, integrated energy models are scarcely developed, as also stated by Abbasabadi & Ashayeri (2019).

In the domain of USEMs, taking into account the different assessment models, both of the energy flows in cities (i.e. generation, distribution, and final energy demand) and of the energy consuming domains in cities (i.e. buildings, transport, industry), a detailed classification of the main existing tools can be drawn, as provided in Table 3.

URBAN ENERGY SYSTEMS

Overview

An energy efficient urban planning should be conceived by taking into account that urban elements, as energy consumers (e.g. buildings, transport) and energy suppliers, operate simultaneously at different levels and dimensions. This implies that models and related tools for the energy assessment of specific elements of cities should be integrated each other, as to mirror the complexity of the urban system with a hierarchical approach. To this purpose, the concept of *Urban Energy System* (UES) has been introduced to define "the combined process of acquiring and using energy to satisfy the demands of a given

Table 3. Classification of the main existing tools for urban energy use assessment

Tool name	Reference	Model category [(a)]	Modelling approach [(b)]	Calculation time step[(c)]	Modelling object		
					Energy demand		Energy supply and networks
					Buildings	Transport	
UMI	Reinhart et al. (2013)	UBEM	EN	H	•		
CityBES	Chen et al. (2017)	UBEM	EN	SH	•		
SimStadt	Nouvel et al. (2015)	UBEM	EN	M	•		
TEASER	Remmen et al. (2018)	UBEM	EN	H	•		
CHREM	Swan et al. (2009)	UBEM*	EN, DD	SH	•		
BEM-TEB	Bueno et al. (2012)	UBEM	EN	H	•		
CitySim	Robinson et al. (2009)	USEM	EN	H	•	•	•
UrbanOpt	Polly et al. (2016)	USEM	EN	SH	•		•
MESCOS	Molitor et al. (2014)	USEM	EN	H	•		•
HUES	Bollinger & Evins (2015)	USEM	EN	SH	•		•
CEA	Fonseca et al. (2016)	USEM	EN, DD	H	•	•	•
SynCity	Keirstead et al. (2010)	USEM	EN	D	•	•	•

[(a)] UBEM = urban building energy model, UBEM* = engineering model with additional statistical analysis, USEM = urban scale energy model
[(b)] EN = engineering, DD = data-driven
[(c)] M = monthly, D = daily, H = hourly, SH = sub-hourly

urban area" (Keirstead & Shah, 2013). In addition, since the urban energy system is a dynamic model, it should be constructed as to be able to both understand the current state of the system and simulate its evolution (Shah, 2013). The evolution of the UES will help stakeholders to get answers on the future urban energy efficiency and, subsequently, to take decisions.

Compared to an UBEM or to an USEM, an UES is not limited to the urban energy model itself, which is still the object of analysis, but the UES rather concerns the process to conceive and develop the model in a dynamic and multidisciplinary view. Although the multidisciplinary approach between UES and USEM is shared, what differentiates them is the simultaneous consideration of energy issues characterizing the urban environment. The description of Urban Energy Systems can be assimilated to that of a metabolic system (Keirstead & Shah, 2013), developed to assess the social, economic, and environmental impacts. This concept consists of the analysis of all the transformation processes of material and energy, dependent on external inputs, which occur within a given urban context. Therefore, although in USEM the simultaneous consideration of different energy models is attempted as much as possible, in UES the holistic assessment becomes a necessary condition. From this point of view, both UBEM and USEM can be defined as subsystems of UES.

In order to assess the temporal and spatial dynamism of the urban energy flows, the models of urban energy systems should take into account that, in the development of the UES, multiple experts and knowledge domains are involved. It means that the development of an urban energy system has an impact not only on the improvement of the energy efficiency of a territory, but it also has economic, political, social, demographic and cultural implications. These implications, in turn, have effect of the urban energy flows, as in a circular process.

ICT Solutions for Developing Urban Energy Systems

To develop models of urban energy systems, a very high amount of input data coming from different disciplines is required. As the urban energy systems are complex and dynamic entities, the related information is usually heterogeneous and dispersed in numerous databases. In addition, UES models should rely on integrated energy assessment tools interoperable with reliable input data.

Nowadays, large amount of data are being generated by digitalization, thanks to the wide spreading of Information and Communication Technology (ICT). The introduction of ICT solutions into the market has ensured higher energy efficiency in the use of a building, through an effective control of the management and measurement processes, and comfort adapted to the needs of the occupants. Taking advantage from new technologies (e.g. smart meters), the collection and processing of consumption data is considered an effective driver for the increase of the energy efficiency and the improvement of the environmental footprint of the existing building stock. It has been quantified that, if properly used, the adoption of ICT within an inhabited space ensures energy savings of near 20% (Hannus et al., 2010).

In agreement with Jáñez Morán et al. (2016), in the context of energy management of buildings and energy communities, ICT consists of three fundamental pillars: the measurement systems, the sharing structure through which the exchange of information takes place, and the Building Energy Management Systems (BEMS), which ensure the operator's control over his own thermal, electrical or lighting system. ICT revolves around a huge volume of information that has to be collected, processed, and visualized by a precise computer structure in real-time.

In this regard, the importance of Information and Communication Technology should be discussed also with regard to the development of urban energy systems. In the work of Marinakis et al. (2020), a big data platform for smart energy services that integrates types of data from different sources and domains has been created. One of the main structural components of this platform is a "data interoperability and semantification layer".

The application of semantic technologies is fundamental to access data and integrate them from different domains, besides assuring the interoperability between data and tools in the context of urban energy systems. Ontologies are used to create vocabularies, shared between experts from different knowledge domains, and establishing relationships between the different objects within urban energy systems. According to Gruber (1993), an ontology is "a description (like a formal specification of a program) of the concepts and relationships that can exist for an agent or a community of agents". Therefore, an UES supported by ontologies built by domain experts allows for an effective representation of a complex reality.

Some international initiatives and projects have developed methods and instruments to carry out sustainable urban energy planning through an effective energy assessment of cities and the optimization of energy use. Most of them take advantage of ICT systems and Decision Support Systems (DSS). Some examples of platforms and tools developed in this context are provided in the following list. The models proposed fall into the category of UES, since their objective is to provide an assessment of a predefined urban context according to social, environmental, and economic indicators. As a result of this analysis, it will be possible to identify inefficient urban systems in order to develop regeneration plans.

- TRACE (*Tool for Rapid Assessment of City Energy*) is a DDS tool that provides authorities with an assessment of the energy use in public sector, the identification of cost-effective and feasible energy efficiency measures, covering many knowledge domains, such as buildings, transport, water use, solid waste, public lighting and energy supply (World Bank, 2018).
- i-SCOPE (*Interoperable Smart City Services through an Open Platform for Urban Ecosystems*) is an open platform focusing on different domains and providing three main services for *Smart Cities*: improved personal mobility of aging and diversely able inhabitants, optimization of building energy consumption, and real-time environmental monitoring of urban noise (de Amicis et al., 2012).
- e-SCEAF (*Smart City Energy Assessment Framework*) is a web based DSS tool that has been developed to evaluate the energy performance of a city or an individual sector with the aim to optimize energy use, to reduce CO_2 emissions and to minimize energy costs. Through appropriate indicators, the tool is able to assess the performance over three main axes, that are: the "political field of action" (i.e. achievement of energy savings and environmental targets), the "Energy and Environmental Profile" (i.e. calculation of CO_2 emissions, energy consumption and renewable energy production), and the "Related Infrastructures and ICT" (i.e. analysis of ICT, automations, monitoring and forecasting systems in the buildings of the city) (Papastamatiou et al., 2017).
- DPL (*Sustainability Profile for Districts*, in German) is a tool aimed at assessing environmental, social and economic profiles of districts. It uses a limited number of indicators based on collected data from municipal registers. If data are not available, the model allows alternative methods to estimate the indicators. From these data, environmental, social and economic profiles for the district are calculated. In addition, if required by the local authorities, the tool allows new indicators to be included (Kortman et al., 2001; Jensen, 2009).

DEVISING AN EFFECTIVE URBAN ENERGY PLANNING

The Experience of the SEMANCO Project

In the context of the urban energy systems and advanced technologies for their modelling, the following section is aimed at presenting the experience carried out in SEMANCO (*Semantic Tools for Carbon Reduction in Urban Planning*), a project co-founded by the European Commission within the Seventh Framework Programme, under the theme "ICT systems for energy efficiency" (SEMANCO, 2015).

An integrated platform to model multiple energy systems using a variety of tools was developed in SEMANCO. The tools operate with semantically modelled data located in different data sources, and include instruments for visualizing, simulating and analyzing the multiple aspects that determine the energy and environmental performance of cities. The framework of data and tools allows different users (e.g. planners, policy makers, consultants) to take decisions about the improvement of the energy efficiency in cities, or part of them (e.g. even single buildings).

The main innovative aspect of SEMANCO consists in the close collaboration between experts in different knowledge domains aimed at creating the ontology and the platform. In such a way, the knowledge of each expert about a specific part of the overall system has been usefully taken into account; his/her knowledge is determined by the tools and method used in the discipline, by experience and by the information that he/she has at any given moment (Madrazo et al., 2014a).

To develop the models of the urban energy systems integrated in the platform, a *use-case* approach was assumed in SEMANCO. The use-case was defined as "a pre-conceptualization of a model which represents an urban energy system, as thought by experts within a particular context" (Madrazo et al., 2014a). To solve the problem described in the use-case, a series of actions, named *activities* in the project, were identified. An example of use-case, related activities and data needed to solve the problem, are shown in Figure 3. The use-case approach identifies a complex network in which the same activities can be shared by different use-cases. Each use-case and/or activity can be performed at different scales of analysis, such as micro-scale (i.e. building), meso-scale (i.e. neighborhood), macro scale (i.e. district/city).

The terminology used in the use-cases, in the activities and in the related data was set in accordance with standards and official references as to create the universally shared vocabulary (semantics) for the development of the SEMANCO ontology. The ontology created in SEMANCO constitutes the SEIF (Semantic Energy Information Framework) that forms the connection between the distributed data sources and the tools that use the data modelled semantically.

The development of the ontology in SEMANCO concerned the following six steps (Nemirovski et al., 2013):

Figure 3. The use-case approach in SEMANCO to develop UES models: an example of use-case, related activities and data

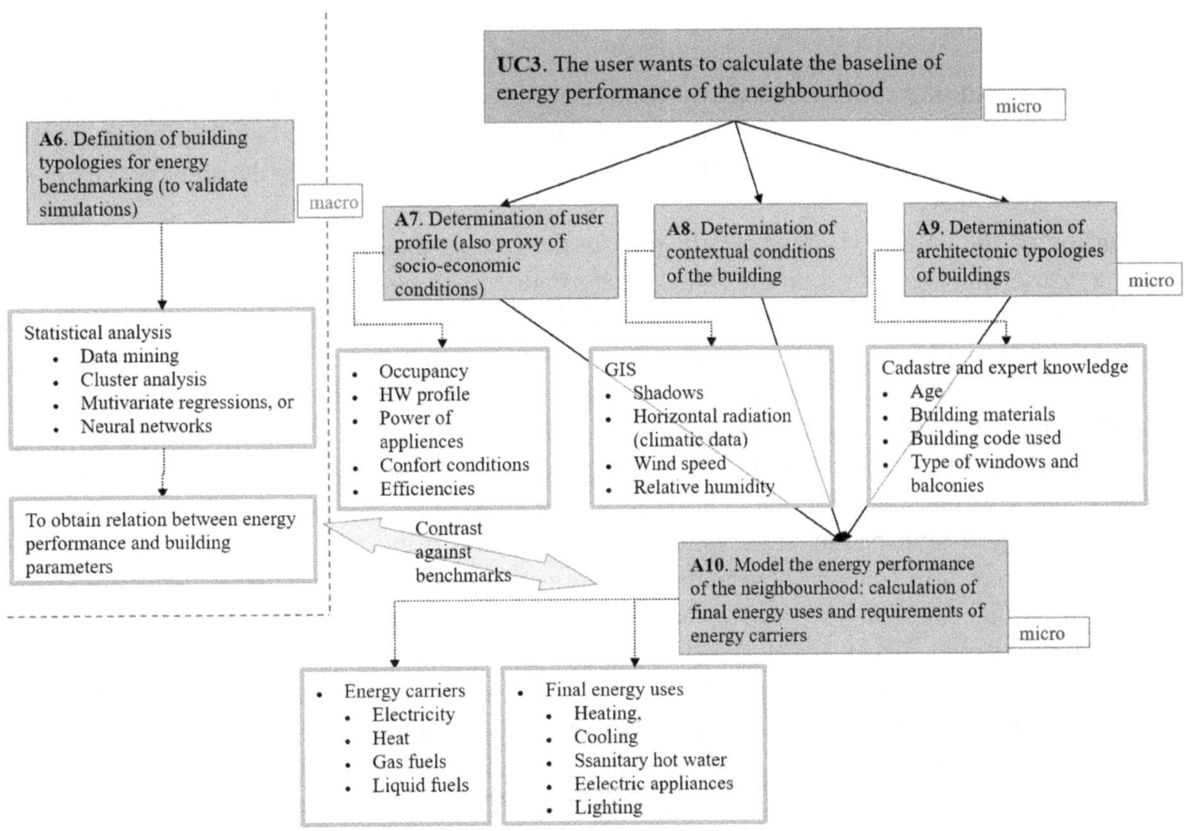

1. **Vocabulary capture:** the knowledge that domain experts have about the problems addressed in the use-cases is made explicit. The experts describe how actors, tools and data can solve a specific objective under a particular policy framework. In this step, the activities expressed in the different use-cases of SEMANCO were described by means of competency questions (Suárez-Figueroa et al., 2012), and then the data sources that contain the information required were identified (Madrazo et al., 2014a).
2. **Construction of the vocabulary:** the terminology provided through the use-cases was set up in form of *Energy Standard Tables*. According to Corrado et al. (2015), these tables constitute an informal vocabulary that precedes the construction of the formal vocabulary (i.e. the ontology), and represent an effective instrument to support collaboration between ontology engineers and experts from different domains. The *Energy Standard Tables* were implemented as a set of spread-sheets, where terms, descriptions, units of measure, and relationships between concepts are provided. These tables stand for a conceptualization of an UES, which is an inter-disciplinary domain, by providing the definition of the terms that experts consider relevant because related to the issue expressed in the use-case. The *Energy Standard Tables* encompass categories of energy data and energy-related data, such as energy cost data, climatic data, environmental data, building technical data, legislative constraints, geographical data, land and buildings registry data, urban planning data, socio-economic data, and demographic data. A total number of 25 *Energy Standard Tables* were created in the SEMANCO project, encompassing about 1000 concepts (Corrado & Ballarini, 2013). An example of *Energy Standard Table* for the *building* concept, included in the building technical data category, is shown in Figure 4.

3. **Mapping data sources to vocabulary:** the names of the data items, included in the data sources, which have been identified in the activities of the use-cases, are mapped on the vocabulary (i.e. concepts of the *Energy Standard Tables*). The mapping was carried out by the domain experts and by the data owners.
4. **Coding of the ontology:** the vocabulary set in the *Energy Standard Tables* was then translated in a formal ontology based on the DL-LiteA formalism. An ontology editor, in accordance with Wolters et al. (2013), performed the coding of the semantic energy model in SEMANCO. The SEMANCO ontology editor presents two possible simulation views of the developed ontology: (i) one view for the editing of the concepts taxonomy, (ii) another view for the editing of the graph of non-subsumption relations (Madrazo et al., 2014a).
5. **Integration of the data sources:** the outputs generated in steps 3 and 4 were used to transform the contents of the data sources into RDF (Resource Description Framework) resources. To do so, the mappings established in step 3 were coded as relations between a relational database and the ontology created in step 4. These mappings are usually implemented with declarative mapping languages that offer rich expressive features to bring the rigid relational schemas to real cases (Nemirovski et al., 2013).
6. **Evaluation:** the last step concerns the evaluation of the ontology created in the previous steps. Three elements were evaluated: (i) the ability of users to understand the ontology structure, (ii) the mapping compliance, and (iii) the computational efficiency, consisting in the capability of the ontology to support query with a short response time.

Figure 4. Example of Energy Standard Table for structuring the terminology related to the building concept. Extract from the SEMANCO project (Corrado & Ballarini, 2013)

Name/Acronym	Corresponding Name in D3.1	Description	Reference	Type of data	Unit	Reference to other sheets
Building	-	construction as a whole, including its envelope and all technical building systems, for which energy is used to condition the indoor climate, to provide domestic hot water and illumination and other services related to the use of the building	EN 15603	-	-	-
has **Building_Name**	-	name (ID) of the building	-	string	-	-
has **Age**	*building age*	construction period of the building	-	string	-	-
is **Year_Of_Construction**	-	year of construction of the building	-	string	-	-
is **Age_Class**	*building age class* [new]	period of years to be defined according to typical construction or building properties (materials, construction principles, building shape, ...)	TABULA	string	-	-
has **From_Year**	-	first year of the age class	TABULA	string	-	-
has **To_Year**	-	last year of the age class	TABULA	string	-	-
has **Allocation**	-	specification of the region the age class is defined for	TABULA	string	-	-
has **Identifier**	-	-	SUMO	A,B,C,D	-	-
has **Building_Typology**	*building typology*	building typology	-	string	-	-
is **Flat**	-	apartment in a building	-	string	-	-
is **Detached_Building**	-	small building, without attached buildings	TABULA	string	-	-
is **Semi-Detached_Building**	-	small building, with an attached building	TABULA	string	-	-
is **Terraced_Building**	-	small building, with two attached buildings	TABULA	string	-	-
is **Row_Building**	-	big building, with prevalent horizontal extension	TABULA	string	-	-
is **Tower_Building**	-	big building, with prevalent vertical extension	TABULA	string	-	-
is **Courtyard_Building**	-	big building having "L" or "U" shape	TABULA	string	-	-
has **Internal_Courtyard_Orientation**	-					
has **Conservation_State**	*conservation state*	conservation state of the building	-	string	-	-
is **New_Building**	-	building to be designed	-	string	-	-
is **Existing_Building**	-	existing building	-	string	-	-
is **Refurbished_Building**	-	building to be refurbished	-	string	-	-
has **Building_Use**	*building use*	use of the building	-	string	-	"b_use"
has **Building_Geometry**	-	geometry of the building	-	-	-	-
has **Building_Floor_Area**	*building floor area* [new]	sum of the areas of the building storeys	-	real	m^2	-

The methodology and the technological structure of SEMANCO were developed and tested in the project by means of three case studies, Manresa (Spain), Newcastle (UK) and North Harbour (Denmark).

The SEMANCO Integrated Platform and EECITIES

The SEMANCO integrated platform is the front-end for users; it grants the access to tools and data, and provides services to model urban energy systems. The combination of experts' knowledge, captured through the use-cases, and the data linked to external data sources, accessible through the SEIF, constitute the basis to create models of UES. For a particular urban area – a city, a district, or a neighborhood – multiple urban energy models can be created combining data and tools in various ways. Within a given urban model, users can use the tools to assess the energy performance of the urban area, and to carry out plans to improve it. The open structure of the platform enables an urban energy model to be enhanced when new tools and data – either from existing data sources or from the data generated by the different applications – become available.

Different layers are provided for in the SEMANCO platform, such as:

- *urban energy system* that refers to a particular urban area delimited by the available data and by objectives to be achieved applying specific tools,

- *urban energy model* that is the conceptualization of the UES,
- *plans* that are defined to improve the energy performance of the urban area (see, for instance, Figure 5). The baseline, which represent the current condition of the analyzed area in terms of energy performance, can be estimated by using the energy assessment tools and the available data,
- *projects* that can be proposed as alternatives to improve the energy efficiency considering different measures. The improvements obtained by applying the proposed measures are then compared with the baseline in order to identify the optimal solution.

For example, a *plan* can be proposed to refurbish the old town of Manresa (one of the case studies implemented in the SEMANCO platform). This can be achieved in different ways, each one represented by a different *project*: for instance, the energy refurbishment of the external vertical walls, the thermal insulation of roof, or the replacement of technical building systems (e.g. heat generator). The results of implementing these measures are determined using the same tools already applied to calculate the baseline. Then, a multi-criteria decision-making tool is used to rank the performance of the different *projects* (Madrazo et al., 2014b).

The platform provides different types of tools, embedded (i.e. part of the platform), interfaced (i.e. existing tools interacting with other tools and services in the platform), and external (i.e. accessed through the platform, operating separately and exporting output to be imported in the platform). The tools developed in SEMANCO for the energy analysis of the three case studies included SAP (*Standard Assessment Procedure,* used in the UK), UEP/UEO tool (*Urban Energy Planning/Optimization*, used in Denmark) and UsiT-*Improvements tool* (used in Spain). In addition, the platform allows performing multi-criteria decision analysis (MCDA), cluster analysis and data-mining analysis.

One of the main outcomes of SEMANCO is EECITIES (*Energy Efficient Cities*, http://www.eecities.com/), a provider of energy analysis services to support the planning of energy efficient cities (Madrazo et al., 2015). The services represent the exploitation of the technologies created in the project, while the service providers are the domain experts involved in SEMANCO. A specific web portal is dedicated to EECITIES (EECITIES, 2015). It has been structured into: technologies (i.e. a summary of the tools provided), services that experts can facilitate, applications (i.e. objectives to be achieved by using the platform) and learning, through access to tutorials and other information. The EECITIES members are experts involved in the development of the SEMANCO platform, as follows:

- ARC Engineering and Architecture La Salle, Barcelona, Spain (developer of the SEMANCO integrated platform, ontology design and data integration)
- University of Teesside, Middlesborough, United Kingdom (developer of the energy assessment and improvement tools SAPMAP)
- CIMNE BeeGroup, Barcelona, Spain (developer of the USiT tool, multicriteria evaluation)
- Politecnico di Torino, Department of Energy, Torino, Italy (energy data structuring)
- Hochschule Albstadt Sigmaringen, Germany (ontology design, energy data analytics)
- Agency 9, AB Lulea, Sweden (3d visualization, developer of the software 3dMaps)
- Ramboll, Copenhagen, Denmark (developer of the UEP evaluation tool)
- National Energy Action, Newcastle, UK (energy experts)
- Fòrum, Manresa, Spain (building energy experts)

Figure 5. Example of a plan in the SEMANCO platform
(Source: EECITIES, 2015)

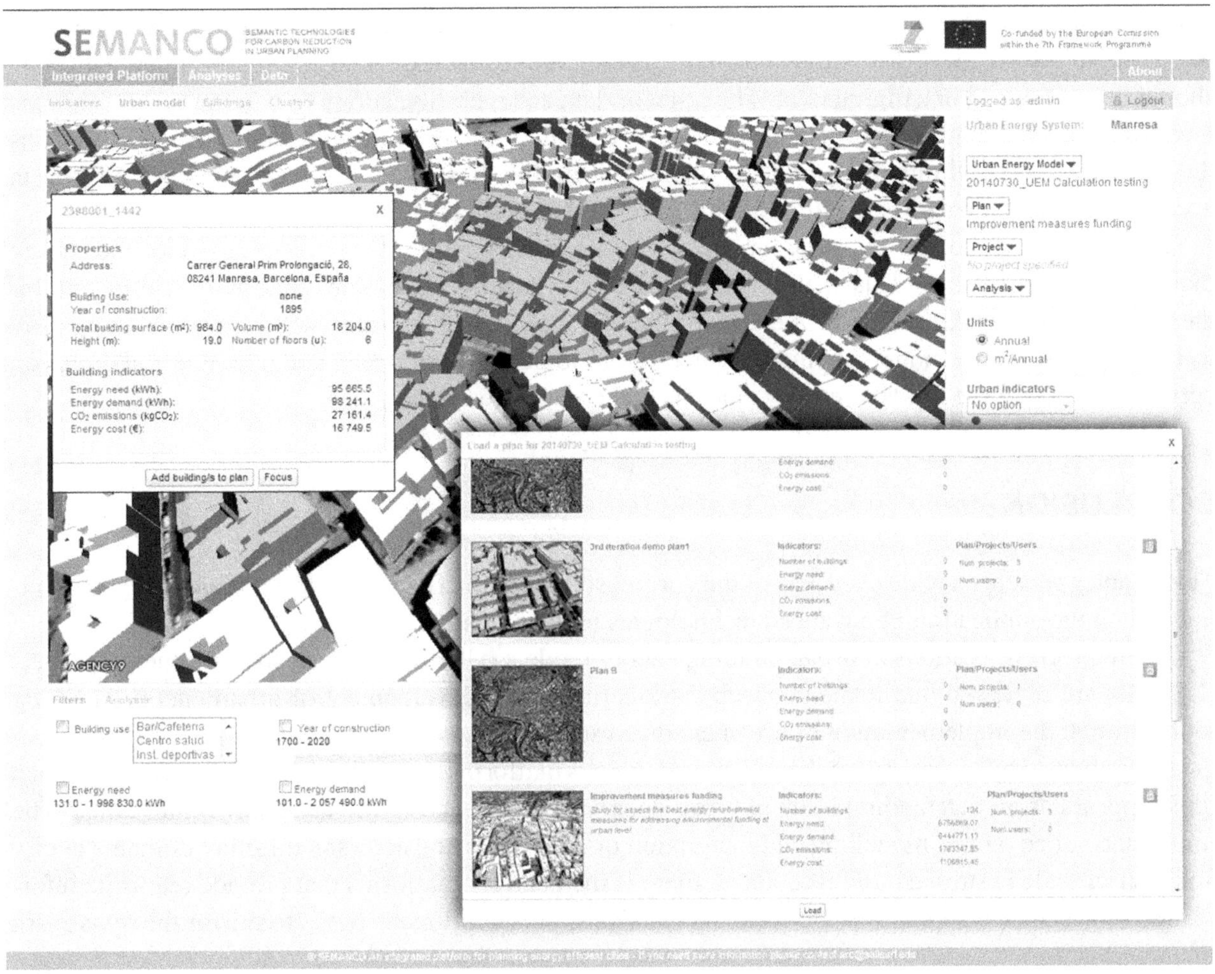

FUTURE RESEARCH DIRECTIONS

Future research in the field of efficient urban energy planning should be based on the lines of research provided in literature and the methodologies highlighted in the present chapter, and finalized at solving some still open issues. One of these is related to the presence of inaccuracies in the building stock energy models. Accurate urban energy performance modelling is still a challenge, due to the uncertainties of input data and the use of assumptions in the engineering models, on the one hand, and the use of aggregated data in the data-driven models, on the other hand. An effective mean to give robustness and reliability to building stock energy models would be the advancement in the use of hybrid models, got by coupling the strengths of engineering methods (UBEMs) with those of the data-drivel models. For instance, the widespread archetype approach can be enriched with probabilistic input data derived from machine learning techniques.

Another important aspect to be investigated is the dynamic integration of the different parts of an urban scale energy model (USEM), as to be conceived more and more as an urban energy system.

In the field of urban energy systems, the experience of the SEMANCO project allowed the creation of an effective tool to develop UES. Anyway, after the completion of SEMANCO, the EECITIES integrated platform did not work as expected. This was mainly due to the need to manually upgrade the input data, with the consequent loss of dynamism that is required from an UES. A continuous input data flow should be guaranteed in a digital twin. The issue of data accessibility, curation, continuity over time, is a cornerstone of integrated data platforms. These aspects originate from inherent difficulties, such as the data formats (i.e. limits related to data interchanges), the data ownership (i.e. limits related to data accessibility), and the data reliability (i.e. limits related to accuracy).

The enhancement of the integrated models of urban energy systems is a precondition for the energy planning of Smart Cities. This will be supported by advanced technologies that make use of ontologies, and encouraged by the nascent era of digitalization, in which big data will be available from IoT technologies. The new models should be conceived as to guarantee a strong and effective collaboration between experts in the fields of energy analysis and computer science.

CONCLUSION

This chapter presents a wide overview of the current studies in the field of building stock energy modeling and on the application of advanced technologies to develop urban energy systems.

The main classifications of urban building energy models (UBEMs) and urban scale energy models (USEMs) are reviewed and related concepts are defined. The development of urban energy planning tools through the implementation of urban energy systems is also discussed.

As confirmed by several research projects and various platforms and tools implemented, the addressed topic appears as an outstanding issue in urban energy planning and in the design of future cities. The importance of advanced modelling and integration of data, tools and actors in different domains at city and district scale is stressed. In this context, there is the need for platforms that provide real-time information (e.g. sensors, appliances, etc.) and data analysis, on a continuous basis, to mirror the dynamism of an UES.

In addition, difficulties related to access and management of data should be overcome by means of standardized procedures for data collection and integration. Despite significant progress has been achieved in the field of computer science in terms of data exchange protocols, further research is needed to increase energy data availability and quality. A universally recognized ontology of energy data and related-energy data (e.g. building-physics properties, user behavior, climatic data, etc.) is still lacking; a step toward this direction has been done by the SEMANCO project. International technical standards represent an effective mean to provide a common structure for information exchange between urban energy models in the next future. In this way, a development of digital twins more oriented to urban energy systems can be attained.

REFERENCES

Abbasabadi, N., & Ashayeri, M. (2019). Urban energy use modeling methods and tools: A review and an outlook. *Building and Environment*, *161*, 106270. doi:10.1016/j.buildenv.2019.106270

Allegrini, J., Orehouning, K., Mavromatidis, G., Ruesch, F., Dorer, V., & Evins, R. (2015). A review of modelling approaches and tools for the simulation of district-scale energy systems. *Renewable & Sustainable Energy Reviews*, *52*, 1391–1404. doi:10.1016/j.rser.2015.07.123

Basu, S., Bale, C. S. E., Wehnert, T., & Topp, K. (2019). A complexity approach to defining urban energy systems. *Cities (London, England)*, *95*, 102358. doi:10.1016/j.cities.2019.05.027

Bianco, V., Manca, O., & Nardini, S. (2009). Electricity consumption forecasting in Italy using linear regression models. *Energy*, *34*(9), 1413–1421. doi:10.1016/j.energy.2009.06.034

Boje, C., Guerriero, A., Kubicki, S., & Rezgui, Y. (2020). Towards a semantic Construction Digital Twin: Directions for future research. *Automation in Construction*, *114*, 103179. doi:10.1016/j.autcon.2020.103179

Bollinger, L. A., & Evins, R. (2015). HUES: A holistic urban energy simulation platform for effective model integration. In *Proceedings of International Conference CISBAT 2015* (pp. 841-846). Academic Press.

Bueno, B., Pigeon, G., Norford, L. K., Zibouche, K., & Marchadier, C. (2012). Development and evaluation of a building energy model integrated in the TEB scheme. *Geoscientific Model Development*, *5*(2), 433–448. doi:10.5194/gmd-5-433-2012

Carreón, J. R., & Worrell, E. (2018). Urban energy systems within the transition to sustainable development. A research agenda for urban metabolism. *Resources, Conservation and Recycling*, *132*, 258–266. doi:10.1016/j.resconrec.2017.08.004

Cerezo, C., Sokol, J., AlKhaled, S., Reinhart, C., Al-Mumin, A., & Hajiah, A. (2017). Comparison of four building archetype characterization methods in urban building energy modeling (UBEM): A residential case study in Kuwait City. *Energy and Building*, *154*, 321–334. doi:10.1016/j.enbuild.2017.08.029

Chen, Y., Hong, T., & Piette, M. A. (2017). City-Scale Building Retrofit Analysis: A Case Study using CityBES. In *Proceedings of BS 2017: 15th Conference of IBPSA* (pp. 259-266). International Building Performance Simulation Association.

Corrado, V., & Ballarini, I. (2013). *Guidelines for Structuring Contextual Data.* SEMANCO Deliverable 3.3. http://www.semanco-project.eu/deliverables.htm

Corrado, V., Ballarini, I., Madrazo, L., & Nemirovskij, G. (2015). Data structuring for the ontological modelling of urban energy systems: The experience of the *SEMANCO* project. *Sustainable Cities and Society*, *14*, 223–235. doi:10.1016/j.scs.2014.09.006

De Amicis, R., Conti, G., Patti, D., Ford, M., & Elisei, P. (2012). I-Scope–Interoperable Smart City Services through an Open Platform for Urban Ecosystems. In M. Schrenk et al. (Ed.), *Proceedings of REAL CORP 2012* (pp. 1219-1226). Academic Press.

EECITIES. (2015). *EECITIES - Energy Efficient Cities. Service platform to support planning of energy efficient cities.* http://www.eecities.com/

European Commission (EC). (2019). *EU Energy in figures, Statistical Pocketbook 2019.* Publications Office of the European Union.

Fonseca, J. A., Nguyen, T., Schlueter, A., & Marechal, F. (2016). City Energy Analyst (CEA): Integrated framework for analysis and optimization of building energy systems in neighborhoods and city districts. *Energy and Building*, *113*, 202–226. doi:10.1016/j.enbuild.2015.11.055

Goy, S., Maréchal, F., & Finn, D. (2020). Data for Urban Scale Building Energy Modelling: Assessing Impacts and Overcoming Availability Challenges. *Energies*, *13*(16), 4244. doi:10.3390/en13164244

Gröger, G., Kolbe, T. H., Nagel, C., & Häfele, K. H. (2012). *OGC City Geography Markup Language (CityGML) Encoding Standard*. Open Geospatial Consortium.

Gruber, T. R. (1993). A translation approach to portable ontology specifications. *Knowledge Acquisition*, *5*(2), 199–220. doi:10.1006/knac.1993.1008

Haas, R., & Schipper, L. (1998). Residential energy demand in OECD-countries and the role of irreversible efficiency improvements. *Energy Economics*, *20*(4), 421–442. doi:10.1016/S0140-9883(98)00003-6

Hannus, M., Samad, A., & Zarli, A. (Eds.). (2010). *ICT Supported Energy Efficiency in Construction. Strategic Research Roadmap and Implementation Recommendations*. REEB Project Consortium.

Hedegaard, R. E., Kristensen, M. H., Pedersen, T. H., Brun, A., & Petersen, S. (2019). Bottom-up modelling methodology for urban-scale analysis of residential space heating demand response. *Applied Energy*, *242*, 181–204. doi:10.1016/j.apenergy.2019.03.063

Hong, T., Chen, Y., Luo, X., Luo, N., & Lee, S. H. (2020). Ten questions on urban building energy modeling. *Building and Environment*, *168*, 106508. doi:10.1016/j.buildenv.2019.106508

Jáñez Morán, A., Profaizer, P., Herrando Zapater, M., Andérez Valdavida, M., & Zabalza Bribián, I. (2016). Information and Communication Technologies (ICTs) for energy efficiency in buildings: Review and analysis of results from EU pilot projects. *Energy and Building*, *127*, 128–137. doi:10.1016/j.enbuild.2016.05.064

Jensen, J. O. (2009). Sustainability Profile for Urban Districts in Copenhagen. In *Proceedings of Sustainable Cities and Regions: Enabling Vision or Empty Talk?* Aalborg University. doi:10.4337/9780857937162

Johari, F., Peronato, G., Sadeghian, P., Zhao, X., & Widén, J. (2020). Urban building energy modeling: State of the art and future prospects. *Renewable & Sustainable Energy Reviews*, *128*, 109902. doi:10.1016/j.rser.2020.109902

Kavgic, M., Mavrogianni, A., Mumovic, D., Summerfield, A., Stevanovic, Z., & Djurovic-Petrovic, M. (2010). A review of bottom-up building stock models for energy consumption in the residential sector. *Building and Environment*, *45*(7), 1683–1697. doi:10.1016/j.buildenv.2010.01.021

Keirstead, J., Jennings, M., & Sivakumar, A. (2012). A review of urban energy system models: Approaches, challenges and opportunities. *Renewable & Sustainable Energy Reviews*, *16*(6), 3847–3866. doi:10.1016/j.rser.2012.02.047

Keirstead, J., Samsatli, N., & Shah, N. (2010). SynCity: An integrated tool kit for urban energy systems modelling. *Energy efficient cities: Assessment tools and benchmarking practices*, 21-42.

Keirstead, J., & Shah, N. (Eds.). (2013). *Urban Energy Systems: An Integrated Approach*. Routledge. doi:10.4324/9780203066782

Kortman, J., van Ewijk, H., van Konijnenburg, P., Lanting, R., de Groot-van Dam, A., Kleefman, R., & Timmermans, F. (2001). *Duurzaamheidsprofiel van een locatie, Ontwikkeling en est van het DPL instrument versie 1.0*. IVAM Environmental Research, TNO - Building and Construction Research and TNO - Environment, Energy and Process Innovation.

Madrazo, L., Nemirovski, G., & Sicilia, Á. (2014a). Shared Vocabularies to Support the Creation of Energy Urban Systems Models. In European Commission (Ed.), ICT for a Low Carbon Economy - EEBuilding Data Models - Energy Efficiency Vocabularies & Ontologies (pp. 130-150). Publications Office of the European Union.

Madrazo, L., Sicilia, Á., Ortiz, D., Oliveras, J., Ronn, T., Lynch, D., Crosbie, T., & Carpenter, M. (2015). *Energy service platform web portal / Energy service platform web portal contents*. SEMANCO Deliverables 5.7 / 7.6. http://www.semanco-project.eu/deliverables.htm

Madrazo, L., Sicilia, Á., Ortiz, D., & Pleguezuelos, J. (2014b). *Operative integrated platform*. SEMANCO Deliverable 5.6. http://www.semanco-project.eu/deliverables.htm

Marinakis, V., Doukas, H., Tsapelas, J., Mouzakitis, S., Sicilia, Á., Madrazo, L., & Sgouridis, S. (2020). From big data to smart energy services: An application for intelligent energy management. *Future Generation Computer Systems*, *110*, 572–586. doi:10.1016/j.future.2018.04.062

Molitor, C., Groß, S., Zeitz, J., & Monti, A. (2014). MESCOS - A multienergy system cosimulator for city district energy systems. *IEEE Transactions on Industrial Informatics*, *10*(4), 2247–2256. doi:10.1109/TII.2014.2334058

Nageler, P., Koch, A., Mauthner, F., Leusbrock, I., Mach, T., Hochenauer, C., & Heimrath, R. (2018). Comparison of dynamic urban building energy models (UBEM): Sigmoid energy signature and physical modelling approach. *Energy and Building*, *179*, 333–343. doi:10.1016/j.enbuild.2018.09.034

Nemirovski, G., Nolle, A., Sicilia, Á., Ballarini, I., & Corrado, V. (2013). Data Integration Driven Ontology Design, Case Study Smart City. In *Proceedings of 3rd International Conference on Web Intelligence, Mining and Semantics, WIMS'13*. Association for Computing Machinery (ACM). 10.1145/2479787.2479830

Nouvel, R., Brassel, K. H., Bruse, M., Duminil, E., Coors, V., Eicker, U., & Robinson, D. (2015). SimStadt, a new workflow-driven urban energy simulation platform for CityGML city models. In *Proceedings of International Conference CISBAT 2015* (pp. 889-894). Academic Press.

Papastamatiou, I., Marinakis, V., Doukas, H., & Psarras, J. (2017). A decision support framework for smart cities energy assessment and optimization. *Energy Procedia*, *111*, 800–809. doi:10.1016/j.egypro.2017.03.242

Pasichnyi, O., Levihn, F., Shahrokni, H., Wallin, J., & Kordas, O. (2019). Data-driven strategic planning of building energy retrofitting: The case of Stockholm. *Journal of Cleaner Production*, *233*, 546–560. doi:10.1016/j.jclepro.2019.05.373

Pauwels, P., Zhang, S., & Lee, Y. C. (2017). Semantic web technologies in AEC industry: A literature overview. *Automation in Construction*, *73*, 145–165. doi:10.1016/j.autcon.2016.10.003

Piro, M., Ballarini, I., & Corrado, V. (2020). Use of Energy Performance Certificates data repositories in Urban Building Energy Models. *AiCARR Journal*, *64*(5), 44–48.

Polly, B., Kutscher, C., Macumber, D., Schott, M., Pless, S., Livingood, B., & Van Geet, O. (2016). From zero energy buildings to zero energy districts. ACEEE Summer Study on Energy Efficiency in Buildings.

Reinhart, C. F., & Cerezo Davila, C. (2016). Urban building energy modeling – A review of a nascent field. *Building and Environment*, *97*, 196–202. doi:10.1016/j.buildenv.2015.12.001

Reinhart, C. F., Dogan, T., Jakubiec, J. A., Rakha, T., & Sang, A. (2013). UMI - An urban simulation environment for building energy use, daylighting and walkability. In *Proceedings of BS 2013: 13th Conference of IBPSA* (pp. 476-483). International Building Performance Simulation Association.

Remmen, P., Lauster, M., Mans, M., Fuchs, M., Osterhage, T., & Müller, D. (2018). TEASER: An open tool for urban energy modelling of building stocks. *Journal of Building Performance Simulation*, *11*(1), 84–98. doi:10.1080/19401493.2017.1283539

RitchieH.RoserM. (2018). *Urbanization*. https://ourworldindata.org/urbanization

Robinson, D., Haldi, F., Kämpf, J., Leroux, P., Perez, D., Rasheed, A., & Wilke, U. (2009). CitySim: Comprehensive micro-simulation of resource flows for sustainable urban planning. In *Proceedings of BS 2009: 11th Conference of IBPSA* (pp. 1083-1090). International Building Performance Simulation Association.

SEMANCO. (2015). *Semantic Tools for Carbon Reduction in Urban Planning*. http://www.semanco-project.eu/

Shah, N. (2013). Modelling urban energy systems. In J. Keirstead & N. Shah (Eds.), *Urban Energy Systems. An Integrated Approach*. Routledge.

Sola, A., Corchero, C., Salom, J., & Sanmarti, M. (2020). Multi-domain urban-scale energy modelling tools: A review. *Sustainable Cities and Society*, *54*, 101872. doi:10.1016/j.scs.2019.101872

Suárez-Figueroa, M. C., Gómez-Pérez, A., Motta, E., & Gangemi, A. (2012). *Ontology Engineering in a Networked World*. Springer. doi:10.1007/978-3-642-24794-1

Swan, L. G., & Ugursal, V. I. (2009). Modeling of end-use energy consumption in the residential sector: A review of modeling techniques. *Renewable & Sustainable Energy Reviews*, *13*(8), 1819–1835. doi:10.1016/j.rser.2008.09.033

Swan, L. G., Ugursal, V. I., & Beausoleil-Morrison, I. (2009). Implementation of a Canadian residential energy end-use model for assessing new technology impacts. In *Proceedings of BS 2009: 11th Conference of IBPSA* (pp. 1429-1436). International Building Performance Simulation Association.

Torabi Moghadam, S., Delmastro, C., Corgnati, S. P., & Lombardi, P. (2017). Urban energy planning procedure for sustainable development in the built environment: A review of available spatial approaches. *Journal of Cleaner Production*, *165*, 811–827. doi:10.1016/j.jclepro.2017.07.142

Wolters, M., Nemirovski, G., & Nolle, A. (2013). ClickOnA: An Editor for DL-LiteA based Ontology Design. *Proceedings of the 26th International Workshop on Description Logics.*

World Bank. (2018). Tool for Rapid Assessment of City Energy (TRACE 2.0): A Manual for Experts and City Officials. Energy Sector Management Assistance Program. World Bank.

ADDITIONAL READING

Cajot, S., Peter, M., Bahu, J.-M., Koch, A., & Maréchal, F. (2015). Energy planning in the urban context: Challenges and perspectives. *Energy Procedia*, *78*, 3366–3371. doi:10.1016/j.egypro.2015.11.752

Cuenca, J., Larrinaga, F., & Curry, E. (2020). DABGEO: A reusable and usable global energy ontology for the energy domain. *Journal of Web Semantics*, *61-62*, 100550. doi:10.1016/j.websem.2020.100550

Ellis, K. A., Pesch, D., Klepal, M., Look, M., Greifenberg, T., Boudon, M., Kelly, D., & Upton, C. (2017). Information and Communication Technology for EPN. In A. Monti & ... (Eds.), *Energy Positive Neighborhoods and Smart Energy Districts* (pp. 101–157). Elsevier. doi:10.1016/B978-0-12-809951-3.00005-3

Naber, E., Volk, R., & Schultmann, F. (2017). From the building level energy performance assessment to the national level: How are uncertainties handled in building stock models. *Procedia Engineering*, *180*, 1443–1452. doi:10.1016/j.proeng.2017.04.307

Oregi, X., Roth, E., Alsema, E., van Ginkel, M., & Struik, D. (2015). Use of ICT tools for integration of energy in urban planning projects. *Energy Procedia*, *83*, 157–166. doi:10.1016/j.egypro.2015.12.206

Shabani, A., & Zavalani, O. (2017). Predicting Building Energy Consumption using Engineering and Data Driven Approaches: A Review. *European Journal of Engineering Research and Science*, *2*(5), 44–49. doi:10.24018/ejers.2017.2.5.352

Sola, A., Corchero, C., Salom, J., & Sanmarti, M. (2018). Simulation Tools to Build Urban-Scale Energy Models: A Review. *Energies*, *11*(12), 3269. doi:10.3390/en11123269

Sousa Monteiro, C., Costa, C., Pina, A., Santos, M. Y., & Ferrão, P. (2018). An urban building database (UBD) supporting a smart city information system. *Energy and Building*, *158*, 244–260. doi:10.1016/j.enbuild.2017.10.009

KEY TERMS AND DEFINITIONS

Bottom-Up Building Stock Energy Model: Energy model of the building stock, in which disaggregated data are used to estimate the end-use energy of individual buildings or group of buildings, and then the aggregated energy demand is extrapolated at a wider territorial scale.

Ontology: An explicit specification of a conceptualization, in which a set of objects is described through relationships among them to form the universe of discourse.

Top-Down Building Stock Energy Model: Energy model of the building stock, in which aggregated data of energy consumption are correlated to economic or other parameters to derive relationships for the prediction of the energy demand at a smaller territorial scale.

Urban Building Energy Model: Type of bottom-up building stock energy model that derives the energy demand through an engineering or physical-based approach, i.e. applying building heat balance equations.

Urban Energy System: The combined process of acquiring and using energy to satisfy the demands of a given urban area that is modelled using integrated tools, as to mirror the dynamism and the complexity of cities with a hierarchical approach.

Urban Metabolism: Process that involves the exchange of energy flows and materials (e.g. water, food, waste) within an environment.

Urban Scale Energy Model: Energy model that involves a spatio-temporal resolution of energy flows in cities, allowing the evaluation of system losses, load peaks, unconsumed energy shares, and the relationship between consumer and supplier.

Use-Case: The pre-conceptualization of a model that represents an urban energy system, as thought by experts within a particular context.

Chapter 23
Semantic Data-Driven Models to Improve Energy Efficiency in Buildings and Cities

Álvaro Sicilia
ARC, Engineering and Architecture La Salle, Ramon Llull University, Spain

Gonçal Costa
ARC, Engineering and Architecture La Salle, Ramon Llull University, Spain

Leandro Madrazo
ARC, Engineering and Architecture La Salle, Ramon Llull University, Spain

ABSTRACT

The assessment of building energy performance requires data from multiple domains (energy, architecture, planning, economy) and scales (building, district, city) to be processed with a diversity of applications used by experts from various fields. In order to properly assess the performance of the building stock, and to develop and apply the most effective energy efficiency measures, it is necessary to adopt a comprehensive, holistic approach. In this chapter, three research projects are presented which apply Semantic Web technologies to create energy data models from multiple data sources and domains in order to support decision making in energy efficient building renovation projects: SEMANCO, OptEEmAL, and OPTIMUS. A final reflection on the results achieved in these projects and their links to ongoing research on digital twins is presented.

INTRODUCTION

In order to improve the energy efficiency of the building stock, data from multiple domains (e.g. energy production, user behaviour) and scales (e.g. building, district, city) must be accessible to all the stakeholders involved in the process. The measures to improve building energy performance, both in building retrofitting and new designs, rely on the decisions adopted by experts from various realms (e.g. policy

DOI: 10.4018/978-1-7998-7091-3.ch023

makers, planners, architects, developers, managers, owners, residents). In order to take well-informed decisions, these experts need accurate and reliable data about buildings and their performance to make well-informed recommendations.

Addressing the problem of building energy efficiency in a comprehensive manner, encompassing all of its dimensions and scales, requires a systems approach. Accordingly, buildings need to be seen as organisms that produce and consume energy and whose performance is driven by physical and non-physical factors; systems which are part of larger systems (e.g., a district, a city). In order to understand the functioning of these systems and sub-systems, models can be created with the information available (e.g. energy consumption, energy production, building envelope, etc.). These models need to identify the elements of the systems and their relationships. Then, based on these models, decision support systems can be created to help the stakeholders involved to better understand the energy performance of buildings and to adopt the most appropriate measure to optimize energy consumption and costs in their respective decision-making realms.

BACKGROUND

Nowadays, energy related data continue to be spread in numerous proprietary and open data sources whose quality levels and reliability are often questionable. Data are continuously changing, since cities and buildings are dynamic entities in continuous transformation. Moreover, energy related data are heterogeneous since they are generated by different applications used by experts in various fields. However, having continuous access to reliable data sources from multiple domains and applications is essential for the creation of decision support systems based on the integration of such data.

In fact, the lack of reliable information on building energy performance is hindering the widespread application of renovation programmes in cities. To meet the goals set by the Energy Performance of Buildings Directives (EPBD, 2002; EPBD, 2010 & EPBD, 2018) concerning nearly-zero energy buildings (nZEB), not only do all new buildings have to be energy efficient but existing ones must be renovated in order to reach similar standards. However, it is estimated that only 1% of the building stock is renovated each year in Europe (EC, 2020).

Identifying the most appropriate combination of measures to assure a balance between energy performance and investment costs is a key factor for the success of any building renovation project. The renovation process of building stock involves multiple stakeholders – city planners, architectural offices, specialists in various fields, product manufacturers, ESCOs, owners and residents – who operate at different decision realms and interact with each other in various ways during the different stages of the lifecycle of the building (e.g., design, construction, operation). Currently, the lack of decision support systems makes it difficult for stakeholders to make well-informed decisions based on an understanding of the whole complexity involved in the different stages of the building renovation process and follow-up on the consequences of their adoption. In particular, the absence of data on the costs and benefits has been recognized as a major obstacle to overcome in large scale building renovation programmes (Staniaszek et al., 2013). Moreover, the lack of information about the current performance of renovated buildings makes financial institutions, investors and owners reluctant to engage in retrofitting projects (Menassa & Ortiz-Vega, 2013) while private owners do not see the benefits of investing in the renovation of their homes due to the lack of information on the potential benefits in terms of energy costs.

The development of decision support systems which provide multiple stakeholders with the information required to assess and improve building energy performance – for existing and new designs – in an urban context involves:

- the integration of data heterogeneous data from multiple domains, scales and applications by means of formal vocabularies
- a representation of the knowledge provided by experts from multiple fields (architects, planners, energy consultants, energy providers) about a complex multidisciplinary problem such as energy efficiency
- the creation of models that help to understand building performance in an urban context considering multiple scales.
- interoperability between data generated by multiple applications, and between applications used in the evaluation of energy systems performance (e.g. energy simulation tools, computer-aided design tools).
- the automatic integration of knowledge from existing data sources and formal vocabularies.

These issues can be separately addressed with ad hoc programming modules for each case. However, if we want to share and reuse the data models (to create new ones) and the data integration processes (to include new data sources), then the use of standard languages and methods is compulsory. The application of Semantic Web technologies can help overcome some of the difficulties which are intrinsic to the development of data-driven decision support systems. In particular, ontologies can play a key role in achieving semantic-based interoperability by providing an unambiguous interpretation of the data to be exchanged between computers and between computers and humans.

In this chapter, we provide examples of the application of Semantic Web technologies to develop data-driven decision support systems to improve the energy efficiency of buildings. The development of these systems has been carried out in three research projects in which Semantic Web technologies have been used to integrate diverse data sources from different domains and to formalise the knowledge from domain experts. In the SEMANCO project, a semantic data integration process was devised to build an urban energy system to support decision making for the renovation of buildings at a city scale using simplified data models. In the OptEEmAL project, a semantic data transformation and integration process was developed to build a set of semantic data models to carry out detailed energy simulations at district scale. In the OPTIMUS project, a semantic framework was developed to capture data from distributed sources in real-time to propose short-term action plans for building energy management. Finally, the conclusion includes a critical reflection on the achievements made with the development of the semantic-based energy data models in these research projects, and on their further development within the current digital twins paradigm.

Urban Energy Systems and Models for Smart Energy Cities

Energy efficiency in cities is a systemic problem which involves multiple elements interacting over time. Accordingly, cities can be thought of as an urban energy system, that is, as "the combined process of acquiring and using energy to satisfy the demands of a given urban area" (Keirstead & Shah, 2013). Models of urban energy systems can be created with two purposes: to understand the current state of the system and to help take decisions to influence its future evolution (Shah, 2013). Urban energy systems and their models rely on data: the data which are necessary to reproduce the city's physical structure (e.g., GIS data) and the data generated by the activity of people, goods, and services.

With the participation of experts in multiple domains, models can be created to assess the performance of an urban energy system in a particular field (building, transport, energy, etc.), or in a combination of them. Such models include simplified representations of the city elements (e.g., buildings, energy grids) and of their interactions. Using the available data, models can be created to understand the interaction between the different elements and the general functioning of the system in order to support decision-making in planning (Shah, 2013). Thus, in the case of building energy planning, an urban energy model could show how much energy is consumed in an urban area, the type of energy that is, being consumed and how consumption can be reduced, as well as mapping the relationships between urban density and energy demand.

Applying Semantics to Urban Data Models

Semantic Web technologies can provide a plausible solution to address some of the issues involved in the development of urban energy systems and models, namely the integration of data from multiple domains and the formalization of the knowledge of domain experts. Ontologies, that is, formal descriptions of a term in a specific domain, are a fundamental component of the Semantic Web. As formulated by Gruber (1993), an ontology stands for "a description (like a formal specification of a program) of the concepts and relationships that can exist for an agent or a community of agents". Ontologies define objects and concepts that exist in a specific domain and the relationships between them. For example, an ontology for Smart Cities domain can have concepts such as Building, Citizen, and Car among others and the relations between them, a Citizen drives a Car, a Citizen lives in a Building, a car is parked in a Building. Ontologies can be used to help experts in different fields to establish relationships between data sources according to their knowledge and experience. In fact, the construction of a common vocabulary in a multidisciplinary field such building energy efficiency requires that the knowledge of the different experts be made explicit and formal. Therefore, an ontology can be thought of as a formalized knowledge about a particular domain carried out jointly by the various experts involved.

Ontologies can serve to foster links between the semantically modelled data and the various software applications used by experts to assess energy performance in cities. The connections between the tools they use and the data they handle can also be captured by the ontologies. This way, when a tool is used in a particular urban energy model, the data which the tool needs as input can be retrieved via ontologies. This makes it possible to create multiple urban energy models of an urban energy system, each one with its own set of tools and associated data. This way, Semantic Web technologies can facilitate the interoperability between the semantically modelled data and the variety of tools with which an urban energy model operates. Ontologies are used to share a view of a domain among the multiple stakeholders involved (typically, experts from multiple fields) and to represent the domain knowledge in a machine-readable format thus enabling automated processes to find inconsistencies in the data sources and to infer knowledge from the given data.

Some research work has already been done to address energy related data integration using Semantic Web technologies. The REEGLE project collected this kind of data from different sources such as the World Bank, International Energy Agency, and World Energy Agency among others (Bauer et al., 2011). The data integrated by REEGLE are basically statistics of building properties and building performance at the national level. These data are shared as Linked Data using light-weight ontologies.

Similarly, ontologies have been created for the domain of energy efficiency. One example is EEOnt (Vinagre et al., 2013), which provides a semantic representation of energy data for terms of the EPBD. It also provides tools for building energy assessment inventories, enabling the creation of formal, machine understandable and easily accessible certification schemes. Another example is DogOnt, firstly developed in 2008 and later reedited. DogOnt provides a bridge between devices, buildings and districts to develop smart city applications (Bonino et al., 2018).

In the Km4City project, an ontology to represent smart city data from multiple domains and a corresponding data ingestion system was developed (Bellini et al., 2014). The system can manage large volumes of data – static and dynamic – obtained from a variety of sources. This data is mapped to the km4City Ontology and stored into a Resource Description Framework (RDF) store making it possible for applications via SPARQL queries to provide new services to citizens and third-parties.

Within the RÉPENER project, a Semantic Energy Information System (SEIS) was developed to integrate energy-related data of buildings to provide quality, accurate information for building designers, facility managers, owners and energy experts to take informed decisions in building renovation projects (Madrazo et al., 2015). The system is based on an energy model formalised as an ontology which integrates multiple data sources (e.g., building energy certificates, building descriptions, simulation results, energy monitoring, and climate data).

A follow-up of RÉPENER is the ENERSI project, for which two applications – ENERHAT and ENERPAT – were created using Semantic Web technologies. These applications integrate building data, on-line building refurbishment assessment tools, and refurbishment policies to make it easier for building owners and city planners to take informed decisions to improve buildings' energy performance (Madrazo et al., 2018). The interconnection of the different components included in these applications was the collaborative work of ontology engineers, data providers and energy experts.

SEMANCO: An Integrated Platform to Support Building Energy Efficiency in Urban Planning

The SEMANCO project – developed during from 2011 to 2014 with the support of the 7th Framework Programme – is a prime example of the application of Semantic Web technologies to support the creation of urban energy systems and models. A semantic-based platform was developed in this project that integrates data from multiple sources and scales (e.g., micro, meso, macro) and tools which interoperate with the data. With this platform, users can create urban energy models which represent the combined knowledge of the various experts involved in the evaluation and planning of energy efficient cities.

A methodology was developed to design, implement, and apply urban energy models based on energy standards. The combined outputs of the project – methods and tools – gave rise to EECITIES[1], a provider of energy analysis services to support the planning of energy efficient cities. These services, based on the technologies developed in the project, provide quality information to enable key stakeholders (e.g., architects, engineers, building managers, local administrators, policy makers, and citizens) of a sustainable urban planning project to improve the decision-making process in their respective realms in order to reduce carbon emissions in cities.

In the SEMANCO project, Semantic Web technologies were applied:

- To formalise domain expert knowledge required for the creation of an urban energy model.

- To integrate data from multiple domains, sources and applications (e.g., census, climate, cadastre, building typologies, and socio-economic data).
- To make integrated data interoperable with energy simulation tools.

Semantic Energy Information Framework

The Semantic Energy Information Framework (SEIF) is one of the key technological components of the SEMANCO platform. The goal of the SEIF is to facilitate the retrieval of heterogeneous and distributed energy related data to be analysed by energy simulation tools and domain experts. Its main goal is to ensure data access interoperability so that tools and users can formulate data queries without knowing the underlying technical details concerning the access methods, the content of the sources, the data schemas or the specific semantics employed by each source.

The technological components of the SEIF are the SEMANCO energy model – formalized as an OWL DL-Lite$_A$ ontology – to represent energy efficiency in buildings (Nemirovski et al., 2013) and ELITE, a federated query engine to retrieve distributed data (Nolle & Nemirovski, 2013).

The ontology was created following a design process developed within the SEMANCO project (Nemirovski et al., 2013), whose purpose was to model a specific domain based on the integration of data sources in ways that support federated querying processes. A set of design patterns, document templates, tools and editors were developed to support these processes (Madrazo et al., 2013).

The Urban Energy Model

The Urban Energy Model (UEM) of the SEMANCO platform is a formal ontology created with Web Ontology Language 2 (OWL 2). This ontology includes concepts from diverse sources such as energy standards, use cases descriptions, and data sources related to the domains of urban planning and energy management. In particular, it contains the terms and attributes to describe regions, cities, neighbourhoods and buildings; energy consumption and CO_2 emission indicators, as well as climate and socio-economic factors related to energy consumption.

The ontology adheres to some European and international standards, among them:

- Energy efficiency and renewable energy sources (ISO/IEC CD 13273).
- Common terms, definitions and symbols for the overall energy performance rating and certification of buildings (ISO/DTR 16344).
- Assessment of overall energy performance of buildings (ISO/CD 16346).
- Presentation of real energy use of buildings (ISO/DIS 12655).
- Methods for expressing energy performance and for energy certification of buildings (ISO/CD 16343).
- Energy management systems – requirements with guidance for use (ISO 50001:2011).

The UEM is built on top of the Suggested Upper Merged Ontology (SUMO) which is a foundational ontology which ensures generality and expressiveness for a wide range of domains (Niles & Pease, 2001). SUMO includes several domain ontologies and contains over 20,000 terms and 80,000 logical statements. SUMO ontology was selected over other foundational ontologies (such as DOLCE, PROTON, General Formal Ontology, and Basic Formal Ontology) because is simple to understand and it supports reasoning. The UEM ontology concepts are subsumed in SUMO. For example, the concept *Building_Geometry* from UEM is subsumed by UEM's *Geometry* which in turn is subsumed by the SUMO's concept *Attribute*. This way, the UEM taxonomy can have UEM's concepts subsumed by UEM's concepts or by SUMO's concepts.

The UEM ontology was coded in DL-Lite$_A$ formalism, which was selected because it outperforms most other description logic formalisms when it comes to managing data distributed in heterogeneously structured sources (Poggi et al., 2008). The UEM contains numerous of values, most of which are measures to specify physical quantities such as area, volume, and energy consumption among others. UEM ontology adopts the measurement units of the SUMO ontology.

The UEM ontology is annotated to enable users to understand its structure and key concepts. Four annotation properties are used for each concept:

- **Label**: It is the formal name of a concept, one that humans can understand. For example, the concept of *Electrical_Appliances_Power_Installed* has the label of *Power installed on the electrical appliances*.
- **Comment**: It is a textual description of the concept provided by domain experts. The basic description stems from the existing standards and related domain publications. However, it can be written ad-hoc if it is not covered by the standards.
- **Reference**: It is the standard used to name and describe the concept.
- **Author**: It is the name of the person who created the concept. It helps to track changes since the ontology is created by different people.

Urban energy model construction methodology

The adoption of existing ontologies to create urban energy models avoids unnecessary work, reduces design errors and facilitates replicability in other contexts. Design ontology methodologies have been facilitated by Gruber (1995) and Uschold and King (1995). More recent approaches have been provided by Fernandes el al. (2011) and Suárez-Figueroa et al. (2012).

These methodologies are mainly concerned with the creation of the conceptual models of a specific domain, rather than the integration of the data sources in ways that facilitate querying through federated access. Therefore, it was necessary to develop an ontology design process specific for the SEMANCO project, which comprises three main phases: requirements capture, informal knowledge building, and knowledge formalization (Nemirovski et al., 2013).

Firstly, the **requirements capture** is carried out through a methodology based on case studies and use cases. A case study refers to the delimitation of the research scope to a geographic location, to the factors that influence the problem of carbon reduction in a particular urban area, and to the stakeholders involved in the planning issues at stake and the energy policy agenda (Madrazo, 2012). A use case connects data, tools and users and the interactions between them in order to fulfil a specific goal within an urban energy system; for instance, reducing carbon emissions at the district level. Use cases are composed of activities which describe the steps to be carried out in order to achieve specific goals. Activities encompassed by a use case are described as requirements and competency questions following the NeOn methodology (Suárez-Figueroa et al., 2012). Following this methodology, the data sources required to carry out the activities are identified and briefly described, and a set of templates are used to define case studies, use cases and activities. In the SEMANCO project, 3 case studies, 15 use cases and 44 activities were defined using such templates. Figure 1 is an example of a use case and an activity implemented in SEMANCO project. In this first stage, users (building owners, energy consultants, city planners), data (cadastre, energy consumption, building typologies), related standards (ISO/CEN), and tools (energy simulation tools) were identified. The activities described deal with a wide range of issues including the identification of areas with high instances of fuel poverty, the calculation of the potential of local solar gains, and the calculation of the CO_2 emissions of buildings and urban areas.

Figure 1. A use case to calculate indicators for new or existing urban development. An activity to define different alternatives for urban planning.

Acronym	UC10
Goal	To calculate the energy consumption, CO2 emissions, costs and /or socio-economic benefits of an urban plan for a new or existing development.
Super-use case	None
Sub-use caseUC9	
Work process	Planning
Users	1. Municipal technical planners 2. Public companies providing social housing providers 3. Policy Makers
Actors	4. Neighbour's association or individual neighbours: this goal is important for them to know the environmental and socio-economic implications of the different possibilities in the district or environment, mainly in refurbishment projects. They use to ask these questions to the municipality 5. Mayor and municipal councillors: In order to evaluate CO2 emissions impact of different local regulations or taxes
Related national/local policy framework	6. Sustainable energy action plan (Covenant of Mayors) 7. Local urban regulations (PGOUM, PERI, PE in Spain) 8. Technical code of edification and national energy code (CTE, Calener in Spain)
Activities	9. A1.- Define different alternatives for urban planning and local regulations 10. A2.- Define systems and occupation (socio-economic) parameters for each alternative 11. A3. Determine the characteristics of the urban environment 12. A4. Determine the architectural characteristics of the buildings in the urban plans 13. A5. Model or measure the energy performance of the neighbourhood 14. A6. Calculate CO2 emissions and energy savings for each proposed intervention 15. A7. Calculate investment and maintenance costs for each proposed intervention

Acronym	A1
Super-activity/use case	UC10
Sub-activities	A2, A3, A4
Goal	Define different alternatives for urban planning and local regulations
Urban Scale	Micro-Meso
Users	1. The municipality (councilors of urban planning, housing, environment and countryside, …) (stakeholder) 2. Urban planners 3. Public company of social housing 4. Owner/promoter of the building (stakeholder) 5. Neighbor's association (stakeholder) 6. Consultants and technicians from Engineering and consultancy companies 7. Supply companies (i.e. supply company of district heating)
Related national/local policy framework	• Sustainable energy action plan (SEP from Covenant of Mayors) • Local regulations 8. National energy codes (Código Técnico and certificación energética in Spain, DECC 2012 and HECA in UK, and Heat Planning Act, and danish Planning regulation in Denmark)
Issues to be addressed	1. To define the comparison of different CO2 emissions scenarios of urban planning, according to local energy requirements acts and/or Plans, in order to select the most efficient urban planning alternative in next steps. 2. To select a set of technologies, and local regulation in order to evaluate their CO2 impact • To select different scenarios to evaluate the socio-economic impact of different measures • To define alternative building performance levels in order to calculate scenarios of improvement of energy efficiency

Input Data

Name	Description	Domain	Format
Local regulations and requirements	Local regulations related to Energy Efficiency, RES, and CO2 emissions, as well as Local Urban regulation that can affect to de different proposals to implement	Energy efficiency Urban planning	Maps, and technical requirements
List of objectives from different users	List of scenarios of energy performance, energy supply, and/or urban planning	Energy efficiency Urban planning	Documents

Secondly, the **informal knowledge building** stage uses input the descriptions of the use cases and activities from the previous stage as input in order to create an initial vocabulary composed of a set of categorized terms connected by simple relations such as subsumption (is) and aggregation (has). Subsumption relations are used to define subtypes of classes creating a taxonomy. For example, *OfficeBuilding* is a *Building*, while *Building* is a *Construction*. Aggregation is a relation between two classes which is established between individuals of both classes. For example, the aggregation between the class *Building* and *City* might be established with the relation *location*. That is, the building *Casa Batlló* is located in *Barcelona*. The goal of this initial vocabulary is to identify the terms based on existing international standards for energy modelling and to provide a shared terminology of the urban energy model. The initial vocabulary is specified in the form of an energy standard table. Each category in this table contains the terms identified by the various activities (for example, *Energy_Need*, *Building_Volume*, and *Space_Heating_System*). The initial vocabulary contains the description of the terms and the relations between them and, in this regard, it can be equated with a formal ontology specification. Building an initial vocabulary is an important intermediate step towards the design of a semantic energy model. It significantly simplifies formal ontology coding by using a formal language, such as OWL. This task was carried out following the methodology for structuring and semantically modelling energy and contextual data developed in the SEMANCO project (Corrado et al., 2015). The initial vocabulary is composed of 24 categories including building use, climate and building geometry. Around 1,000 terms were collected including descriptions, references, units, and type of data. 18 standards (e.g., ISO/IEC CD 13273-1, ISO/IEC CD 13273-2, EN 15603 and the EN ISO 15927-1) and 16 references (e.g., research project, public recommendations, European directives) were used to create the energy standard tables.

Finally, the **knowledge formalisation** stage focuses on the codification of the urban energy model as an ontology based on the DL-Lite$_A$ formalism which is a dialect of OWL which is an ontology language for the Semantic Web. DL-Lite$_A$ was selected because outperforms most of OWL formalisms in the case of managing data distributed in heterogeneously structured sources which is one of the challenges that has to address urban energy models. The coding of the ontology was carried out using Click-On, an ontology editor developed within the project (Wolters et al., 2013). This editor provides a user-friendly interface which facilitates the participation of domain experts in the ontology building process. Besides, the editor supports the coding of DL-Lite$_A$ axioms to represent domains and ranges of object properties which require the processing of reasoning. The need to have these two features was the reason for the development of a custom editor instead of using an existing one like Protégé or TopBraid Composer. The formalisation of the ontology is carried out by energy experts and ontology engineers in a collaborative way thanks to the annotation properties of the ontology.

Urban Energy Model population

Populating an Urban Energy Model is part of the ontology design process devised within the SEMANCO project. The data sources – identified in the use case descriptions – are integrated in the model. To do so, a set of mappings between the data sources and the UEM ontology are needed. Therefore, it is necessary to identify the mappings by using an informal method that enables non-ontology engineers to participate in the knowledge construction process. A table is collaboratively filled by data owners, domain experts, and ontology engineers with the details of each data source. This way, each field in comma-separated values files or tables of a database is mapped to a concept of the ontology. For those fields in which the data source cannot be univocally mapped to the ontology, ontology experts need to deal with them by modifying the ontology or by implementing complex mappings.

Once the mappings between the data sources and the ontology are informally described, they can be coded in order to populate the UEM. This implies to transform the data sources into RDF following the structure of the ontology. The codification of the mappings is carried out using R2RML, a standard language to define mappings. The creation of such mappings is a complex process which involves experts from different domains with different skills. The process requires them to understand both the structure of the ontology and the data sources. To support their work, a mapping editor was created within the project, Map-On, which includes a graphical user interface to help non-technicians in the task of creating mapping using R2RML language (Sicilia et al., 2017). The editor hides the complexity of query generation from the user. 90% of the mappings could be coded thanks to Map-On, while the remaining 10% were manually coded since they required complex queries.

Exploitation of the Urban Energy Models

End users of the SEMANCO platform interact with the graphical user interface using a 3D map, tables, and charts which enable them to create plans and projects, and to assess their energy efficiency (Figure 2). An end user can create and operate an urban energy model without being aware of its underlying complexity. When the user zooms in or out, selects a building to visualize its data, or runs energy simulations, SPARQL queries are sent to the SEIF using the terms of the ontology. The SEIF uses the federation engine to retrieve data from the sources and translates the data into RDF following R2RML mappings. Therefore, the SEIF assures the interoperability between the data and the tools that use the data. The relation between data and tools is also handled by the SEIF (Sicilia et al., 2014).

The end user of the platform can select buildings from an UEM to create a renovation plan for a set of buildings selected in the city map. A renovation plan can include several projects, each one contemplating a series of retrofitting measures such as renovating existing windows, adding insulation to external walls, and replacing the boiler, among others. When the user calculates the energy performance for each plan and project, the SEIF retrieves the data needed by the energy simulation tools thanks to the semantic integration of different data sources. For each building, the energy simulation tools calculate a set of indicators (e.g., energy demand, energy need, costs). Those indicators can be aggregated at different scales: district, neighbourhood and city.

The projects created by the user can be evaluated using a multi-criteria decision tool included in the platform (Figure 3). Users can select the weight (importance) of the indicators calculated by the simulation tools and then choose other indicators, such as foreseen funding, to be included in the analysis funding. This way, the user can select the project that better suits the selected requirements.

The SEMANCO platform has been applied in demonstration scenarios in four cities:

Manresa, Spain. The platform was used to assess the effectiveness of the measures to refurbish buildings in two neighbourhoods. The users were architects, industrial engineers, engineers, and urban planners. The data sources integrated by SEIF were cadastre, census, socio-economic data, and building typologies (e.g., u-values, windows properties, systems). The energy simulation tool was URSOS simulation engine. The users created three projects: Upgrading the building envelope, improving the heating system, and installing energy generation systems fed with renewable sources.

Newcastle, UK. The goal was to install energy generation systems fed with renewable sources. The user was an energy consultant contracted by Newcastle City Council. The data sources accessed by SEIF were income, fuel poverty, and index of multiple deprivation. The method used to calculate the indicators was Simplified Assessment Procedure (SAP). Three projects were created by the users: Insulation based refit, renewables refit, and targeted fabric refit.

Copenhagen, Denmark. The purpose was to assess different strategies regarding energy supply– based on central and distributed solutions – in greenfield planning. The users were urban planners from the Environmental Department of the municipality. The data sources integrated by SEIF were building typologies (supply technologies, energy demand) and carbon emission coefficients. In this scenario, a custom urban energy planning tool provided by the energy consultant was used for the calculation. Us-

Figure 2. Visualization of energy need indicators of an urban energy model

ers created three projects for a new area: district heating projection, individual fossil fuel solutions, and ground source heat pump.

Torino, Italy. The goal was to assess the effectiveness of the measures to refurbish buildings in a city neighbourhood. The users of the platform were urban planners from the Environmental Department of the municipality. The data sources handled by SEIF were building typologies (supply technologies, energy demand) and carbon emission coefficients. Three projects were created by users: adding low emission windows, installing extra wall insulation, and installing photovoltaic panels.

BUILDING AND DISTRICT SCALE ENERGY DATA MODELS

In district level building renovation projects, it is important to take into account the interactions between the buildings; for example, the effect of shading of the surrounding buildings and the production and distribution of energy between the buildings. Taking into consideration data at the building and district levels, it is possible to create decision support systems that enable the analysis of the impact building retrofitting in a holistic manner (Romero et al., 2014).

Figure 3. The multi-criteria decision analysis (MCDA) tool enables users to select the best solutions by providing comparisons of multiple plans and projects for a given urban area

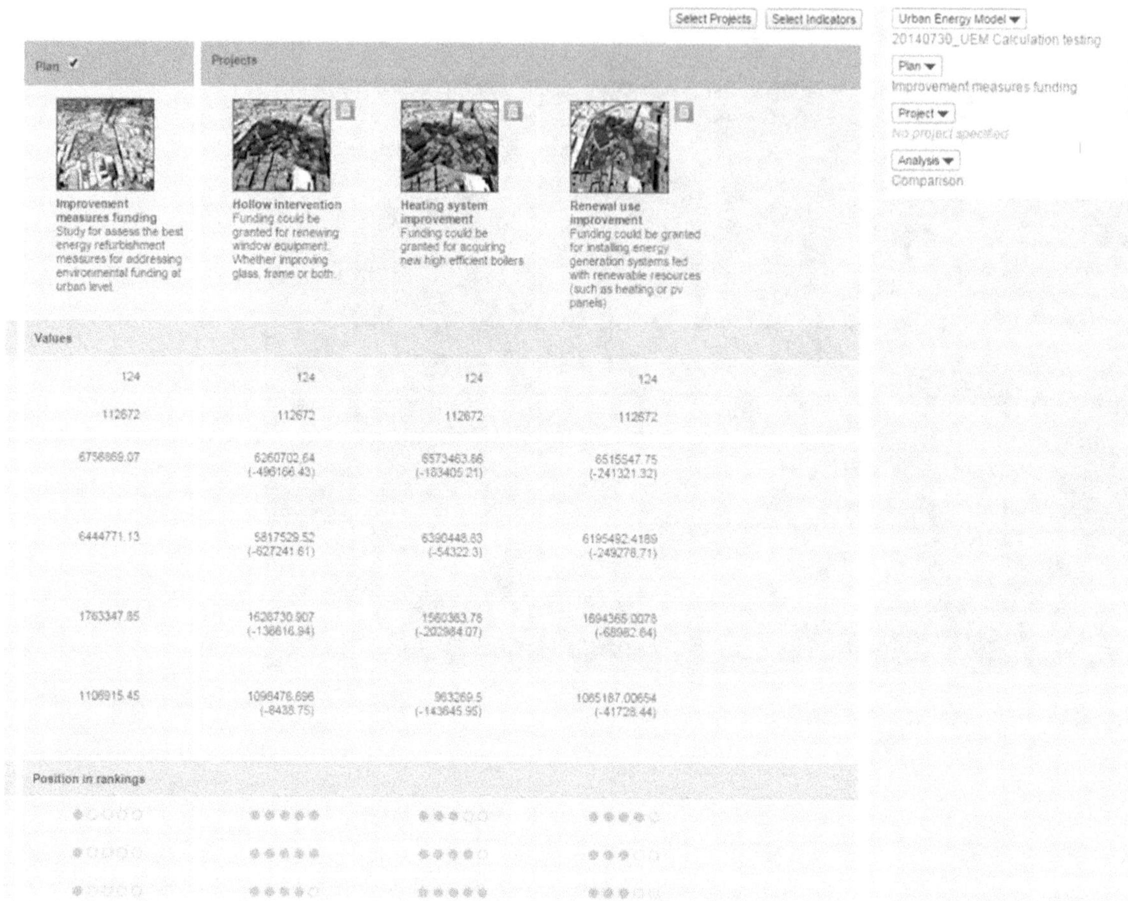

Typically, a building renovation project at the district level involves the evaluation of multiple design options for various buildings. This evaluation is carried out with energy simulation tools, and their ranking is based on constraints which affect single buildings (e.g., the type of building and its accessibility) and the whole district (e.g., lack of heating energy networks). Typically, the building energy models of the different designs are created from scratch using specialized tools (e.g., EnergyPlus, EcoDesigner). To perform simulations with these tools it is necessary to have a comprehensive energy data model which includes all the features of the buildings and their context (e.g., relative location to each other, shared energy sources) as well as the constraints specific of the renovation project (e.g. budget, used of materials and equipment, etc.). However, generating these models is a tedious task since it involves collecting data from multiple sources (e.g., BIM, GIS, weather data, socio-economic data) and in various formats (e.g., IFC and CityGML) and then processing these data so they can become inputs for the simulation tools. These difficulties can be overcome using Building Information Model / Modelling (BIM) in the development of the projects. With BIM, accurate virtual models of a building can be digitally constructed and their data analysed in multiple ways (Eastman et al., 2011). By combining a BIM model with other data sources, it is possible to generate energy data models which suit the requirements of the various simulation tools which might be used in a building retrofitting project.

The integration of BIM data with other data sources – among them the Web– using open standards and Semantic Web technologies can facilitate the automatic generation of comprehensive energy data models which can serve as input for subsequent analysis. In this line, Pont et al. (2015), explored the potential of Semantic Web technologies in order to acquire and reuse of AEC-FM domain data available on the Web, even if they are poorly structured. Their flexible, multi-objective design optimization system – developed as part of the SEMERGY project – was able to reuse these data to perform different kinds of building performance assessment and optimization.

Other alternatives to simplify the exchange, interlink and combination of data between different AEC-FM domains focuses on the use of ontologies. For example, Niknam and Karshenas (2017), proposed the use of a BIM shared ontology (BIMSO), a semantic bridge to exchange and integrate data from different AEC-FM domains. Its effectiveness was demonstrated through the implementation of a single search interface across these domains. Other examples of the use of ontologies to facilitate interoperability of building data in the AEC industry can be found in other research projects (Costa & Madrazo, 2015; Belussi et al., 2019; Bilal et al., 2019).

OptEEmAL: A Semantic District Data Model

The goal of OptEEmAL – a research project developed from 2015 to 2019 under the auspices of Horizon 2020 programme – was to develop an energy efficient design platform to improve the energy behaviour of buildings. The purpose of the platform was to optimize the process of generating and evaluating multiple design proposals for the building renovation projects included in a district. With this platform, project stakeholders would significantly reduce the time required to evaluate several thousands of design alternatives by means of evolutionary algorithms and parallel processing (Costa et al., 2016; Martin et al., 2019). The design alternatives were generated by applying different energy conservation measures (ECM) at the district level (Costa et al., 2020). An energy conservation measure is any kind of technological resource to improve the performance of buildings; for example, adding a layer of insulation to the façade of a building or add photovoltaic panels in all the buildings. Different types of input data (building data, urban data, energy conservation measures, etc.) were managed by the platform and further processed with external simulation tools to create design alternatives. For each alternative, their environmental impact was evaluated through a set of sustainability performance indicators (e.g., energy demand, global warning potential, energy use from district heating, and local thermal comfort).

District Data Model

The core of the platform is the District Data Model (DDM), a semantic-based framework to create a multidimensional, multiscale representation of the energy performance of buildings within a district. This framework facilitates the integration of the data from multiples sources (e.g., GIS, BIM, urban data, project briefs, etc.), in order to create energy data models from which a series simulation tools could retrieve the input data they need to perform the calculations. The semantically integrated data model includes detailed information about the geometry, materials, equipment, and indicators, at the building and urban scales. This way, the DDM facilitates the data required by the simulation tools in each of the domains (energy simulation, economic assessment, etc.) to obtain optimised energy efficient designs.

The input data models provided by users (IFC, CityGML, and contextual data) are transformed into semantic data models (an RDF representation of source data models using a standard ontology such as ifcOWL). In the DDM, these data models are represented in RDF according to various domain ontologies. Some of the ontologies have been reused (e.g., SimModelOWL) while others have been created from scratch. In turn, these models are transformed into simulation data models for specific domains (e.g., energy, urban, economic), while others – such as SimModel ontology (Pauwels et al., 2014) – have been extended to meet the requirements of the energy simulations tools.

A data integration and transformation process was devised to feed the DDM, consisting of three steps (Figure 4).

Figure 4. Data integration and transformation process

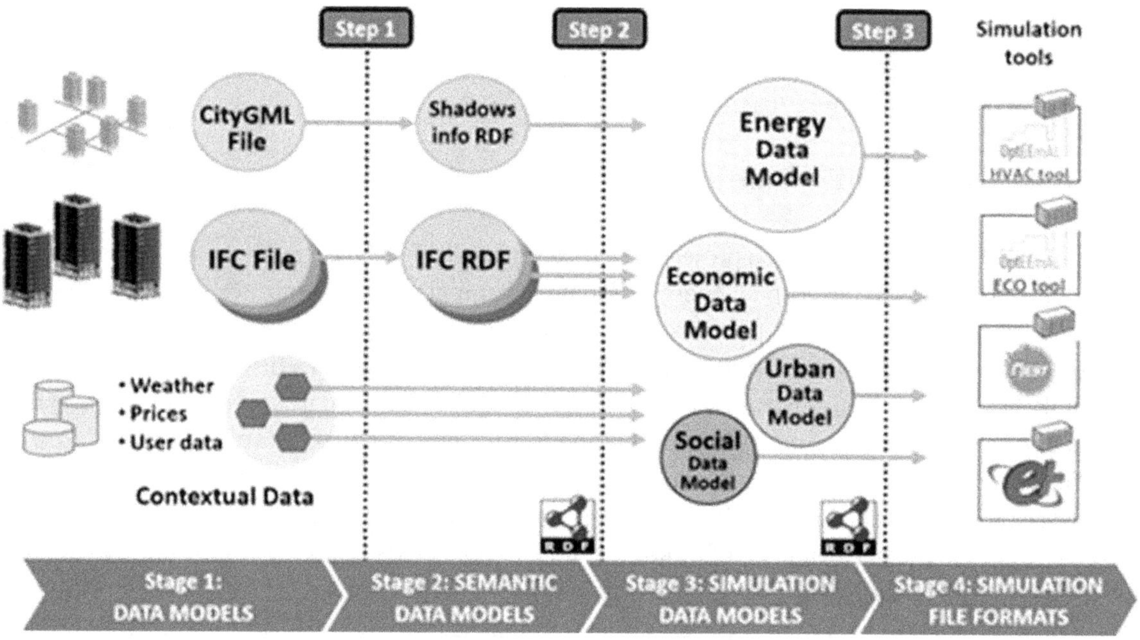

Firstly, the **input data models are transformed into semantic data models in RDF**. For example, BIM models represented according to the IFC standard and provided in the STEP format are transformed into RDF according to the ifcOWL ontology through a conversion tool (Pauwels, 2015). Similarly, GIS models represented in the CityGML standard are transformed into RDF using the District Neighbouring Surface tool (Lilis et al., 2018). This tool identifies all relevant surfaces of buildings and generates a list of shading surfaces necessary to perform energy simulations at district level.

Secondly, the **semantic data models are transformed into simulation data models.** The simulation data models are generic and representative enough to fulfil the data requirements to carry out different kinds of simulations. An example is SimModel (O'Donnell et al., 2011), which has been designed to represent energy simulation data. The simulation data models can get data from different semantic data models. For example, they can include geometry from IFC models and shadows from CityGML models. The transformation of the semantic data models into simulation data models is carried out using a set of RDF -to-RDF mappings. This way, the entities of the semantic data models (e.g., ifcBuilding) are mapped to the entities of the simulations data models (e.g., simBuilding). The mappings have been implemented by means of SPARQL construct queries in which the semantic and structure dissimilarities of the semantic and simulation data models are resolved. A total of 14 basic data mappings were used to implement a total of 35 SPARQL queries to transform all the data needed to generate the energy simulation data models.

Finally, the **simulation files are generated from simulation data models**. A simulation file format is a set of data required as input to a specific simulation tool. For example, in the case of EnergyPlus, IDF is the format for the input data. These simulation file formats are created by querying the simulations data models by means of SPARQL queries.

A more comprehensive description of this transformation and integration process can be found in Costa and Sicilia (2017) and Sicilia and Costa (2017).

The potential use of the OptEEmAL platform was demonstrated in three case studies of renovation projects:

Cuatro de Marzo district in Valladolid, Spain (Mediterranean climate zone). A building renovation project with a net building area of 210,000 m^2 encompassing five buildings (two towers and three blocks) located in a residential district. The target of the renovation was to reduce energy consumption and maximize investment.

Mogel district in Eibar, Spain (Atlantic climate zone). This is a residential neighbourhood with a net building area of 9,450 m^2 including fifteen buildings. The goal of the renovation was to improve the energy efficiency of the buildings. This was achieved primarily by improving the building envelope as well as with the installation of a solar panels for sanitary hot water production.

Polhem district in Lund, Sweden (Nordic climate zone). This is composed of five high school buildings and a library, all of them public buildings managed by the municipality of Lund. The total net area of the buildings is 26,987 m^2. The purpose of the renovation was to have a 10% decrease in energy consumption and using non-fossil energy sources.

ENERGY DATA MODELS FOR BUILDING ENERGY MANAGEMENT

In previous two projects, energy data models have been used for renovating existing buildings. However, energy savings can be achieved also at the operation stage. Typically, an energy manager controls the building energy systems to improve energy efficiency while occupant comfort is not compromised. Decision support systems can help energy managers to implement and monitor sustainable energy action plans to optimize the energy use in their premises and reduce CO_2 emissions (Papastamatiou et al., 2016).

Current decision support systems for energy efficiency use limited data sources, mainly data provided by a building management system (e.g., energy consumption, temperature, humidity and CO_2), while other data such as weather forecasting, and occupancy in most cases are not considered (Corry et al., 2013). Different research projects aimed at integrating multiple data sources for providing new business models and also services have been developed in recent years. With a more comprehensive vision of the building information, it is more feasible to address the management and maintenance of a building when multiple domains are involved. Under this vision, SEMERGY and KnoholEM have been focused on the use of Semantic Web technologies to create energy data models. In SEMERGY, these technologies were used to retrieval, restructuring, and enrichment of building product data aiming at semi-automated building performance assessment and optimization processes (Pont et al., 2015). A web-based decision support and an optimization environment was developed within the project including simulation methods for the building evaluation in different domains (e.g., energy, environment, finance). The interface of the decision-support system enables access to semantic data ontologies. In the KnoholEM project, the goal was to create a knowledge-based energy management for public buildings by means of holistic information modelling and 3D visualization (Anzaldi et al., 2014). Ontologies were created in accordance with current standards used in the construction (e.g., Industry Foundation Classes) and electrical field (e.g., OntoSensor). Other research works enhance building energy management systems through an analytics layer based on a reasoning module whose rules are extracted from historical monitored data by means of data mining methods. The building energy management system proposed by Howell et al. (2019), includes an ontology to make building data in IFC interoperable with monitored data from a sensor network.

OPTIMUS: A Data-Driven Decision Support System to Optimize Energy Building Management

OPTIMUS is a research project developed from 2013 to 2016 with the support of the 7th Framework Programme. The purpose of the project was to design, develop and deliver a Decision Support System (DSS) as an integrated web platform to help local authorities reduce energy consumption and CO_2 emissions in buildings under their premises (Marinakis et al., 2020). A data-driven DSS was developed to support building owners and facilities managers to improve energy efficiency in the daily use of their buildings. At the core of the DSS is a semantic framework which integrates distributed data sources in real time (e.g., sensors for monitoring temperature, light, humidity), prediction models (based on inverse models) to forecast building behaviour, short-term plans to support energy managers in building operation and a user interface which includes dashboards to display data, forecasts and short-term plans.

The prediction models are created with data obtained from five data sources: weather conditions, social, building energy management systems, energy prices, and renewable energy production. These data were obtained from multiple sources such as physical sensors installed in buildings (e.g. energy consumption), national agencies (e.g., weather forecast, energy prices) and web services (e.g., social media networks) (Sicilia et al., 2015). The data sources are interconnected in the DSS using Semantic Web technologies in order to propose building managers short-term actions plans to reduce building energy consumption while keeping comfort level. Then, managers could select the most suitable plans and applied to their management systems.

The semantic framework is composed of: (1) the OPTIMUS ontology, a conceptualization of the urban and building domain including monitoring devices, (2) a semantic data integration process to capture and integrate the data from the different sources and create a real time energy data model, (3) a set of RDF templates for modelling real-time information items according to the OPTIMUS ontology, and (4) a publish-and-subscribe system as a communication infrastructure based on Ztreamy system (Fisteus et al., 2014).

The data integration process is achieved by means of: (1) data capturing modules, to retrieve the raw data from each corresponding source and transform them into RDF, (2) a published-and-subscribe system, to ensure the connectivity between the modules and the DSS, (3) a semantic service, to contextualize the data sent by the modules, and (4) a triple store (OpenLink Virtuoso), to store the integrated data. Using these components, a data integration process is carried out in four steps (Figure 5).

First, a **data translation** is carried out to transform the data sources into RDF according to the OPTIMUS ontology. The translation is made by a set of RDF templates which includes the name of the sensor and the building in which the sensor is located. In this way, it is easy to reuse and deploy the RDF template in any kind of sensors.

Secondly, a **data communication** by which RDF data is sent to the DSS using Ztreamy, a publish-and-subscribe system whose main purpose is to publish streams of semantically-annotated data with special focus on scalability. The data between the sensor and the DSS is sent by means of channels provided by Ztreamy.

Thirdly, a **data contextualization** which adds context to the received data. That is, static information about sensor characteristics such as type of property observed (e.g., Solar_Radiation) and feature of interest (e.g., solar irradiation).

Finally, **data storage** by which the data modelled in RDF and contextualized in previous steps is uploaded to a triple store (i.e., OpenLink Virtuoso server). Once the data is in the triple stores, the other components of the DSS such as the prediction models, action plans, and management interface can access the data by means of SPARQL queries.

Figure 5. Semantic data integration process

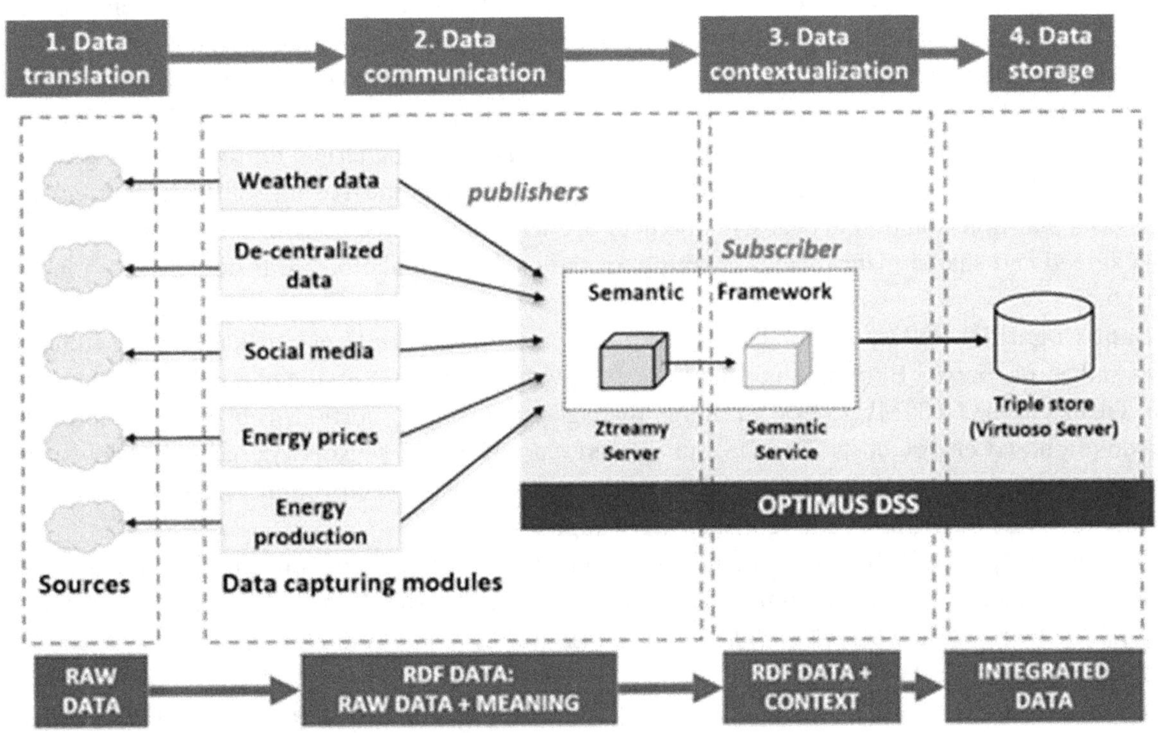

Finally, from the data models generated in real time using the OPTIMUS ontology, the DSS predicts the future behaviour of buildings, suggesting short-term actions to energy managers. The prediction models were developed based on data-driven methods (grey and black boxes). For example, a multiple linear regression (MLR) model was implemented using 13 regressors to predict the energy production of photovoltaic panels considering year, month, hour, solar radiation, humidity, dew point, wind speed, and their combinations. The description of the short-term plans proposed by the OPTIMUS DSS includes the following information:

Scheduling and management of the occupancy. Reduction of the building energy consumption by changing the location of building occupants. The schedule tells when the heating/cooling system can be turned off for some thermal zones.

Scheduling the set-point temperature. Adjustment of the indoor temperature set-point to optimise energy use, by taking into consideration the thermal sensation and comfort communicated by the building users through a web application.

Scheduling the on/off of the heating system. Optimization of the boost time of the heating system considering the forecasting of the indoor air temperature and the building occupancy levels.

Management of the air-side economizer. Schedule of the amount of outdoor air to be used for cooling the indoor environment, in order to reduce or eliminate the need for mechanical cooling when favourable conditions occur, using air-side economizer technology.

Scheduling the photovoltaic maintenance. Detection of the need for maintenance of the solar panels, alerting the user to check if corrective actions are necessary. This facilitates the identification of the panels malfunctioning.

Scheduling the sale/consumption of the electricity produced through the photovoltaic system. Optimization of the selling/self-consumption of electricity produced by a photovoltaic system considering different energy market scenarios (green strategy, finance strategy and peak strategy).

Scheduling the operation of heating and electricity systems towards energy cost optimization. Minimization of the energy cost by optimizing simultaneously the operating schedule of the heating (CHP and boilers) and electricity systems (grids, photovoltaic plants and batteries) for the upcoming week.

In OPTIMUS, modelling the energy data of buildings and their context using Semantic Web technologies was a plausible solution to deal with the heterogeneity of systems, domains, data and applications in an agreed and shared manner. The feasibility of the OPTIMUS platform was demonstrated in three case studies:

Sant Cugat del Vallès, Spain. The platform was used in two buildings managed by the city: the town hall and a theatre. Both buildings have an energy manager to set up the energy management system. Thanks to the OPTIMUS DSS, an energy manager could apply short-terms plans to reduce energy consumption and energy costs. The DSS integrated data from 204 sensors and actions plans such as *Scheduling sale/consumption of the electricity produced through the photovoltaic system* and *Scheduling the on/off of the heating and cooling system* were applied.

Savona, Italy. The DSS was used in this city to assess the effectiveness of the action plans in two areas: an elementary school and a university campus. The short-term plans applied were *Scheduling the set-point temperature* and *Scheduling the photovoltaic maintenance*. A total of 64 sensors were used to feed the prediction models of the DSS.

Zaanstad, Netherlands. The DSS was used to manage the town hall of Zaanstad. The DSS integrated data from 283 sensors and short-term plans such as *Scheduling and management of the occupancy* and *Scheduling the ON/OFF of the heating system* were applied.

CONCLUSION

In this chapter, three research projects have been presented in which energy related data have been modelled using Semantic Web technologies to support decision-making for improving energy efficiency in buildings, in different contexts: design for new and retrofitting projects, as single buildings and as part of districts and cities (SEMANCO, OptEEmAL), and building energy management (OPTIMUS). The adopted systems approach, common to the three projects, has facilitated the modelling of some of the complex issues involved in building energy efficiency – integrating data from multiple domains, bringing together knowledge from various experts, facilitating interoperability of data with a variety of tools –, although within the limitations of the data available in each project. Some of these limitations refer to the continuous access to real time data at the urban level and the creation of a virtual model of the building to monitor its energy performance over time, in line with the digital twins paradigm.

In the three projects, ontologies were created using methods based on well-stablished methodologies. One of the leitmotivs of using ontologies in a data integration process is to facilitate sharing and reusing of the outcomes (e.g., data models) and tools used in the process with the community (e.g., end users, third-parties, researchers). The ultimate goal is to extend the life of the data models and, more important, to not reinvent the wheel in each project. The creation of data models is not a unique process, but requires maintenance over time to get consolidated with the community. This maintenance work can increase the quality of a data model, by fixing errors and including new data sources, along with additional knowledge. To facilitate its utilisation by the community, it is necessary an exploitation plan of the data model either for public or commercial use.

The data models created within the three projects rely on external data sources. As the data sources are dynamic and can evolve over the time, the decision support systems should ensure the continuous access to them. This is a key issue on data-driven decision support systems because components such as prediction models, inference engine, energy simulation need to have reliable and updated data. Data access is not always guaranteed because, among other reasons:

- Data owners are reluctant to give access to data.
- Technology components such as sensors and networks can fail making gaps in the data streams.
- Formats in which data is shared becomes obsolete or are in a proprietary format disabling its use.
- Methods used for generate the data in the source change over the time, this makes difficult to compare data over the time.

In today's cities, massive streams of data are being produced every minute, hour, and day in different areas (transport, energy) captured by sensors, equipment and mobile devices. The incorporation of these data to the semantic models would enhance their complexity and open up new realms for analysis and decision-making for building energy systems. With the incorporation of larger, dynamic data streams would be possible richer energy data models that will enable reliable energy simulations of buildings, districts and cities. Indeed, those models could consider at the same time:

- Relationships between scales, as in the SEMANCO semantic data model.
- Detailed descriptions of the buildings to be renovated or energy efficient managed using standards, as proposed in the OptEEmAL project.
- Prediction models based on historical monitored data, as developed in the OPTIMUS project.

The three projects address different aspects of energy efficiency in buildings sharing the vision of solving data interoperability by means of Semantic Web technologies. Thus, the use of well-known and stablished ontologies such as ifcOWL (Pauwels & Terkaj, 2016) and Semantic Sensor Network (Compton, 2012) can help to address the interoperability between static data (virtual models of buildings, districts and cities) and the dynamic data (monitoring of the physical models).

Semantically enhanced data models can be created following the Digital Twins paradigm, in which data models are updated in real time enabling the access to integrated data (static and dynamic) and generating new knowledge by means of inferring processes which are native in the Semantic Web. Continuous access to data streams would help to increase the quality of the energy simulation outputs in the SEMANCO platform. The semantically enhanced data models developed in the OptEEmAL and OPTIMUS projects could be further developed to create a virtual model of the building, which would reflect its status and behaviour over time following the paradigm of the digital twin, that is, a virtual object that represents a physical object and the connection between the physical and virtual objects (Enders & Hoßbach, 2019). In this context, some of the future challenges is to continue with the development of semantic-based energy data models which support the objectives of the digital twin paradigm (Boje et al., 2020; Kupriyanovsky et al., 2020; Chevallier et al., 2020, Tagliabue et al., 2021).

ACKNOWLEDGMENT

The work described in this chapter was carried out with the support of the European Union Seventh Framework Programme 2007-2013 (SEMANCO, grant number 287534); European Union Horizon 2020 Programme 2014-2020 (OptEEmAL, grant number 680676); and European Union Seventh Framework Programme 2007-2013 (OPTIMUS, grant number 608703).

REFERENCES

Anzaldi, G., Corchero, A., Wicaksono, H., McGlinn, K., Gerdelan, A., & Dibley, M. J. (2014). Knoholem: Knowledge-based energy management for public buildings through holistic information modeling and 3d visualization. In I. González Alonso (Ed.), *International Technology Robotics Applications* (pp. 47–56). Springer International Publishing., doi:10.1007/978-3-319-02332-8_5

Bauer, F., Recheis, D., & Kaltenböck, M. (2011). Data. Reegle. Info – a new key portal for open energy data. In J. Hřebíček, G. Schimak, & R. Denzer (Eds.), *Environmental Software Systems. Frameworks of eEnvironment* (Vol. 359, pp. 189–194). Springer Berlin Heidelberg. doi:10.1007/978-3-642-22285-6_21

Bellini, P., Benigni, M., Billero, R., Nesi, P., & Rauch, N. (2014). Km4City ontology building vs data harvesting and cleaning for smart-city services. *Journal of Visual Languages and Computing*, *25*(6), 827–839. doi:10.1016/j.jvlc.2014.10.023

Belussi, L., Barozzi, B., Bellazzi, A., Danza, L., Devitofrancesco, A., Fanciulli, C., & Scamoni, F. (2019). A review of performance of zero energy buildings and energy efficiency solutions. *Journal of Building Engineering*, *100772*. Advance online publication. doi:10.1016/j.jobe.2019.100772

Bilal, M., Oyedele, L. O., Kusimo, H. O., Owolabi, H. A., Akanbi, L. A., Ajayi, A. O., Akinade, O. O., & Delgado, J. M. D. (2019). Investigating profitability performance of construction projects using big data: A project analytics approach. *Journal of Building Engineering*, *26*, 100850. doi:10.1016/j.jobe.2019.100850

Boje, C., Guerriero, A., Kubicki, S., & Rezgui, Y. (2020). Towards a semantic Construction Digital Twin: Directions for future research. *Automation in Construction*, *114*, 103179. doi:10.1016/j.autcon.2020.103179

Bonino, D., & De Russis, L. (2018). DogOnt as a viable seed for semantic modeling of AEC/FM. *Semantic Web*, *9*(6), 763–780. doi:10.3233/SW-180295

Chevallier, Z., Finance, B., & Cohen Boulakia, B. (2020). A Reference Architecture for Smart Building Digital Twin. *Proceedings of the International Workshop on Semantic Digital Twins co-located with the 17th Extended Semantic Web Conference (ESWC 2020).*

Compton, M., Barnaghi, P., Bermúdez, L., García-Castro, R., Corcho, O., Cox, S., & Huang, V. (2012). The SSN ontology of the W3C semantic sensor network incubator group. *Journal of Web Semantics*, *17*, 25–32. doi:10.1016/j.websem.2012.05.003

Corrado, V., Ballarini, I., Madrazo, L., & Nemirovskij, G. (2015). Data structuring for the ontological modelling of urban energy systems: The experience of the SEMANCO project. *Sustainable Cities and Society*, *14*, 223–235. doi:10.1016/j.scs.2014.09.006

Corry, E., Coakley, D., O'Donnell, J., Pauwels, P., & Keane, M. (2013). The role of linked data and the semantic web in building operation. *Proceedings of the 13th annual International Conference for Enhanced Building Operations. In 13th annual International Conference for Enhanced Building Operations.*

Costa, G., & Madrazo, L. (2015). Connecting building component catalogues with BIM models using semantic technologies: An application for precast concrete components. *Automation in Construction*, *57*, 239–248. doi:10.1016/j.autcon.2015.05.007

Costa, G., & Sicilia, Á. (2017). Methodology for data integration using SPARQL Constructs in the AEC industry. In *Proceedings of the 5th Linked Data in Architecture and Construction Workshop* (pp. 13-15). Retrieved from https://linkedbuildingdata.net/ldac2017/files/Presentations/LDAC2017_ARC.pdf

Costa, G., Sicilia, Á., Lilis, G. N., Rovas, D. V., & Izkara, J. (2016). A comprehensive ontologies-based framework to support the retrofitting design of energy-efficient districts. In *eWork and eBusiness in Architecture, Engineering and Construction-Proceedings of the 11th European Conference on Product and Process Modelling* (pp. 673-681). CRC Press, Taylor & Francis Group. 10.1201/9781315386904

Costa, G., Sicilia, Á., Oregi, X., Pedrero, J., & Mabe, L. (2020). A catalogue of energy conservation measures (ECM) and a tool for their application in energy simulation models. *Journal of Building Engineering*, *29*, 101102. doi:10.1016/j.jobe.2019.101102

Eastman, C. M., Eastman, C., Teicholz, P., Sacks, R., & Liston, K. (2011). *BIM handbook: A guide to building information modeling for owners, managers, designers, engineers and contractors*. John Wiley & Sons. doi:10.1002/9781119287568

Enders, M. R., & Hoßbach, N. (2019). Dimensions of Digital Twin Applications-A Literature Review. *Americas Conference on Information Systems*. Retrieved from https://aisel.aisnet.org/amcis2019/org_transformation_is/org_transformation_is/20/

Fisteus, J. A., García, N. F., Fernández, L. S., & Fuentes-Lorenzo, D. (2014). Ztreamy: A middleware for publishing semantic streams on the web. *Journal of Web Semantics*, *25*, 16–23. doi:10.1016/j.websem.2013.11.002

Gruber, T. R. (1993). A translation approach to portable ontology specifications. *Knowledge Acquisition*, *5*(2), 199–220. doi:10.1006/knac.1993.1008

Gruber, T. R. (1995). Toward principles for the design of ontologies used for knowledge sharing? *International Journal of Human-Computer Studies*, *43*(5-6), 907–928. doi:10.1006/ijhc.1995.1081

Howell, S. K., Wicaksono, H., Yuce, B., McGlinn, K., & Rezgui, Y. (2019). User centered neuro-fuzzy energy management through semantic-based optimization. *IEEE Transactions on Cybernetics*, *49*(9), 3278–3292. doi:10.1109/TCYB.2018.2839700 PMID:30028719

Keirstead, J., & Shah, N. (Eds.). (2013). *Urban energy systems: an integrated approach*. Routledge. doi:10.4324/9780203066782

Kupriyanovsky, V., Klimov, A., Voropaev, Y., Pokusaev, O., Dobrynin, A., Ponkin, I., & Lysogorsky, A. (2020). Digital twins based on the development of BIM technologies, related ontologies, 5G, IoT, and mixed reality for use in infrastructure projects and IFRABIM. *International Journal of Open Information Technologies*, *8*(3), 55–74.

Lilis, G. N., Giannakis, G., Katsigarakis, K., & Rovas, D. (2018). District-aware Building Energy Performance simulation model generation from GIS and BIM data. In *Proceedings of BSO 2018: 4th building simulation and optimization conference*. BIPSA. Retrieved from https://discovery.ucl.ac.uk/id/eprint/10056593/

Madrazo, L., Massetti, M., Sicilia, Á., Wadel, G., & Ianni, M. (2015). SEíS: A semantic-based system for integrating building energy data. *Informes de la Construcción*, *67*(537), e060. Advance online publication. doi:10.3989/ic.13.048

Madrazo, L., Nemirovski, G., & Sicilia, Á. (2013, September). Shared vocabularies to support the creation of energy urban systems models. *Proceedings of 4th ICT for Sustainable Places Conference, EEBuilding Data Models, Energy Efficiency Vocabularies and Ontologies. Nice, France.*

Madrazo, L., Sicilia, Á., & Gamboa, G. (2012). SEMANCO: Semantic tools for carbon reduction in urban planning. In *Proceedings of the 9th European Conference on Product and Process Modelling*. CRC Press. 10.1201/b12516-143

Madrazo, L., Sicilia, Á., Massetti, M., López Plazas, F., & Ortet, E. (2018). Enhancing energy performance certificates with energy related data to support decision making for building retrofitting. *Thermal Science*, *22*(3, Suppl. 3), 957–969. doi:10.2298/TSCI171005028M

Marinakis, V., Doukas, H., Tsapelas, J., Mouzakitis, S., Sicilia, Á., Madrazo, L., & Sgouridis, S. (2020). From big data to smart energy services: An application for intelligent energy management. *Future Generation Computer Systems*, *110*, 572–586. doi:10.1016/j.future.2018.04.062

Martín, S., Serna, V., Álvarez, S., García, M. A., Hernández, G., & Sicilia, Á. (2019). OptEEmAL: it-supported design tool for the generation of optimised energy retrofitting scenarios at district level. *European Conference on Computing in Construction*. 10.35490/EC3.2019.169

Menassa, C. C., & Ortiz-Vega, W. (2013). Uncertainty in Refurbishment Investment. In F. Pacheco Torgal, M. Mistretta, A. Kaklauskas, C. Granqvist, & L. Cabeza (Eds.), *Nearly Zero Energy Building Refurbishment* (pp. 143–175). Springer. doi:10.1007/978-1-4471-5523-2_6

Nemirovski, G., Nolle, A., Sicilia, Á., Ballarini, I., & Corrado, V. (2013). Data integration driven ontology design, case study smart city. In *Proceedings of the 3rd International Conference on Web Intelligence, Mining and Semantics* (pp. 1-10). 10.1145/2479787.2479830

Niknam, M., & Karshenas, S. (2017). A shared ontology approach to semantic representation of BIM data. *Automation in Construction*, *80*, 22–36. doi:10.1016/j.autcon.2017.03.013

Niles, I., & Pease, A. (2001). Towards a standard upper ontology. In *Proceedings of the international conference on Formal Ontology in Information Systems* - Volume 2001, (pp. 2-9). 10.1145/505168.505170

Nolle, A., & Nemirovski, G. (2013). ELITE: An Entailment-Based Federated Query Engine for Complete and Transparent Semantic Data Integration. In *Description Logics* (pp. 854-867). Retrieved from http://ceur-ws.org/Vol-1014/paper_50.pdf

O'Donnell, J., See, R., Rose, C., Maile, T., Bazjanac, V., & Haves, P. (2011). SimModel: A domain data model for whole building energy simulation. In *Proceedings of Building Simulation 2011: 12th Conference of International Building Performance Simulation Association*, (Vol. 12). Retrieved from http://www.ibpsa.org/proceedings/BS2011/P_1223.pdf

Papastamatiou, I., Doukas, H., Spiliotis, E., & Psarras, J. (2016). How "OPTIMUS" is a city in terms of energy optimization? E-SCEAF: A web based decision support tool for local authorities. *Information Fusion*, *29*, 149–161. doi:10.1016/j.inffus.2015.10.002

Pauwels, P. (2015). *IFCtoRDF converter*. Available online: https://github.com/pipauwel/IFCtoRDF

Pauwels, P., Corry, E., & O'Donnell, J. (2014). Representing SimModel in the web ontology language. In Computing in Civil and Building Engineering (pp. 2271-2278). doi:10.1061/9780784413616.282

Pauwels, P., & Terkaj, W. (2016). EXPRESS to OWL for construction industry: Towards a recommendable and usable ifcOWL ontology. *Automation in Construction*, *63*, 100–133. doi:10.1016/j.autcon.2015.12.003

Poggi, A., Lembo, D., Calvanese, D., De Giacomo, G., Lenzerini, M., & Rosati, R. (2008). Linking data to ontologies. In *Journal on data semantics X* (pp. 133–173). Springer., doi:10.1007/978-3-540-77688-8_5

Pont, U. J., Ghiassi, N., Fenz, S., Heurix, J., & Mahdavi, A. (2015). SEMERGY: Application of semantic web technologies in performance-guided building design optimization. *Journal of Information Technology in Construction*, *20*(8), 107–120. Retrieved August 21, 2020, from https://itcon.org/papers/2015_8.content.00001.pdf

Romero, A., Egusquiza, A., & Izkara, J. L. (2014). Integrated decision support tool in energy retrofitting projects for sustainable urban districts. In *Proceedings of the SB14 World Sustainable Buildings*, Vol. 3, (pp. 268-274). Retrieved from https://gbce.es/archivos/ckfinderfiles/WSB14/CreatingNewResources_volume3.pdf

Shah, N. (2013). Modelling urban energy systems. In J. Keirstead & N. Shah (Eds.), *Urban Energy Systems. An Integrated Approach* (pp. 151–175). Routledge.

Sicilia, Á., & Costa, G. (2017). Energy-Related Data Integration Using Semantic Data Models for Energy Efficient Retrofitting Projects. In Multidisciplinary Digital Publishing Institute Proceedings, (Vol. 1, No. 7, 1099, pp. 1-5). doi:10.3390/proceedings1071099

Sicilia, Á., Costa, G., Corrado, V., Gorrino, A., & Corno, F. (2015). A Semantic Decision Support System to optimize the energy use of public buildings. In *CIB W78 Conference* (pp. 676-685). Retrieved from https://itc.scix.net/pdfs/w78-2015-paper-070.pdf

Sicilia, Á., Madrazo, L., & Pleguezuelos, J. (2014). Integrating multiple data sources, domains and tools in urban energy models using semantic technologies. In *Proceedings of the 10th European Conference on Product and Process Modelling* (pp. 837-844). 10.1201/b17396-134

Sicilia, Á., Nemirovski, G., & Nolle, A. (2017). Map-On: A web-based editor for visual ontology mapping. *Semantic Web*, *8*(6), 969–980. doi:10.3233/SW-160246

Staniaszek, D., Rapf, O., Faber, M., Nolte, I., Bruel, R., Fong, P., Lees, E., & Verheyen, L. (2013). A Guide to Developing Strategies for Building Energy Renovation–Delivering Article 4 of the Energy Efficiency Directive. *The Buildings Performance Institute Europe (BPIE)*. Retrieved from http://www.bpie.eu/documents/BPIE/Developing_Building_Renovation_Strategies.pdf

Suárez-Figueroa, M. C., Gómez-Pérez, A., & Fernández-López, M. (2012). The NeOn methodology for ontology engineering. In *Ontology engineering in a networked world* (pp. 9–34). Springer. doi:10.1007/978-3-642-24794-1_2

Tagliabue, L. C., Cecconi, F. R., Maltese, S., Rinaldi, S., Ciribini, A. L. C., & Flammini, A. (2021). Leveraging Digital Twin for Sustainability Assessment of an Educational Building. *Sustainability*, *13*(2), 480. doi:10.3390u13020480

Uschold, M., & King, M. (1995). Towards a methodology for building ontologies. *Workshop on Basic Ontological Issues in Knowledge Sharing, held in conjunction with IJCAI-95.*

Vinagre Díaz, J. J., Wilby, M. R., Rodríguez González, A. B., & García Muñoz, J. (2013). EEOnt: An ontological model for a unified representation of energy efficiency in buildings. *Energy and Building*, *60*, 20–27. doi:10.1016/j.enbuild.2013.01.012

Wolters, M., Nemirovski, G., & Nolle, A. (2013). ClickOn_A: An Editor for DL-Lite_A Based Ontology Design. In *Description Logics* (pp. 1000-1010). Retrieved from https://www.semanticscholar.org/paper/ClickOn_A%3A-An-Editor-for-DL-Lite_A-Based-Ontology-Wolters-Nemirovski/b449eea3c1d57ecdf55f1d71d9dc13b844e553c2

ENDNOTE

1 www.eecities.com

Chapter 24
BIM Tools for the Energy Analysis of Urban Transformation Projects and the Application to the Development of Healthcare Infrastructures

Ezio Nicolas Bruno Urbina
Istituto Giannina Gaslini, Italy

Elisa Spallarossa
Archimede s.r.l., Italy

ABSTRACT

The aim of INDICATE was the creation of an innovative interactive software capable of providing designers, urban planners, and companies a "decision support system" in all urban development phases of a city: from the construction of a single building to the design of a master plan. Galliera test site experience was developed following the idea of using the INDICATE tool on Genoa site to understand how the tool can work in the planning stages, as if the preliminary design of the new hospital were not already defined, to understand the optimal solution to select from the energy point of view. The chapter shows how a tool such as INDICATE would have proven absolutely useful to comprehend the different energetic and economic impacts of different options and of different new building shapes. The experience gained with the INDICATE project and BIM implementation within other projects and realities of the authors could also be adopted to develop the implementation of the BIM in another important hospital in Genoa, the Giannina Gaslini Pediatric Hospital.

DOI: 10.4018/978-1-7998-7091-3.ch024

INTRODUCTION

The construction industry is facing a period connected to the digital transition. In the era of climate change, humanity is observing phenomena such as global warming and the melting of glaciers. In this context, the world of construction is becoming aware of the concepts of smart cities and digital twin to develop smart cities and buildings aimed at energy saving, reduction of CO_2 emissions.

Within this framework, The European Project Indicator-based Interactive Decision Support and Information Exchange platform for Smart Cities (INDICATE) proposes a novel city-wide decision support system which accounts for all major systems and activities relevant to developing energy-efficient cities. To create a 'Smart Economy', integrated smart urban planning tools including architectural masterplanning and detailed energy optimization and environmental analysis will be required. If this is not addressed, we will be unable to deliver the Millennium Development Goals as agreed by each of the UN member states. (INDICATE, 2016).

The aim of INDICATE was the creation of an innovative interactive software capable of providing designers, urban planners and companies a "decision support system" in all urban development phases of a city: from the construction of a single building to the design of a master plan. Galliera test site experience was developed following the idea of using the INDICATE tool on Genoa site, to understand how the tool can work in the planning stages, as if the preliminary design of the new hospital were not already defined, to understand the optimal solution to select from the energy point of view. The experience gained with the INDICATE project and BIM implementation within other Projects and realities of the authors could also be adopted to develop the implementation of the BIM in another important hospital in Genoa, the Giannina Gaslini pediatric hospital.

A considerable number of projects developed at urban scale aims at creating platforms/tools, useful to visualize energy-related information and perform energy simulations. The international literature, offers numerous examples of research related to the management of data at the urban scale aimed at saving energy. Some of these are reported below:

- The Optimising Energy Use in Cities through Smart Decision Support Systems (OPTIMUS) project, funded under the EU FP7 between 2013-2016, that aims to support the decision making process, improving the energy efficiency of buildings, by optimizing the energy use in their premises, and reducing CO_2 emissions. (Sicilia, 2015);
- The Optimised Energy Efficient Design Platform for Refurbishment at District Level (OptEEmAL) project is funded under the European Union's Horizon 2020 research and innovation programme. The platform delivers an optimized, integrated and systematic design based on an Integrated Project Delivery (IPD) approach and supported by the utilization of BIM models for an integrated, optimized and systemic energy oriented refurbishment at district level. (Romero, 2016);
- The District Information Modelling and management for Energy Reduction (DIMMER) European 7th Framework Programme (EU FP7) project, focuses on extending BIM method at district scale, managing all district-related data through the DIM platform, available to all the stakeholders involved at the district level. (Del Giudice, 2017).

Comparing to the above projects presented, INDICATE is focused on the development of a tool capable of optimizing interoperability between BIM GIS, thanks also to the stakeholders that are part of the consortium including world leaders in Dynamic Simulation Modelling & GIS, experts in urban planning, sustainable indicators, algorithm development and demand side management and Public Authority end users.

THE INDICATE PROJECT

The INDICATE project was funded with approximately four million euros by the European Commission as part of the 7th framework program for three years during 2013-2016. The test software provided suggestions regarding the best technologies on the market, infrastructure that needs improvement and load balancing of users, offering the most suitable financial plan. It contained a set of urban sustainability indicators, with a chain effect on the changes made to buildings and infrastructure and in the introduction of new technologies in the urban context. The INDICATE solution integrated several Information Technology (IT) tools, such as dynamic modeling systems, GIS (Geographic Information System) systems that exploit geolocated data, Computer Added Design (CAD) systems for 3D modeling and visualization, urban sustainability indicators, algorithms for managing energy demand and the corresponding automatic balancing of demand with energy from the grid (Grey, 2016).

The project involved two other pilot sites in addition to the Galliera hospital in Genoa (IT): the Irish city of Dundalk, as an example of urban planning plan, and a University campus in Dundalk The INDICATE consortium was coordinated by the IES (Integrated Environmental Solution) company of Glasgow (UK) (Melia, 2015).

Starting from the study of a typical modern hospital, which minimal specifications are presented in Italian law and regulations, the study of New Galliera Hospital aimed to evaluate the introduction of efficient and modern technologies and their impact on operative costs and energy efficiency of the Hospital. The success of the implementation of digitization in infrastructure complexes such as hospitals is based on knowledge in the various sectors, on sharing of skills and on identification of the great advantages of digitization but also any implementation difficulties shall be identified and addressed.

The new international (UNI EN ISO 19650 -1, 2019) and national regulations (DM 560/2017) also underline the key importance of developing digital methods and tools for managing existing assets.

Starting from the principles of these standards, from the experiences gained in the hospital and infrastructures BIM implementation, the authors of this chapter will describe the how was the development of the BIM project of the new hospital (lot 1) managed from an asset management point of view, the perspectives of the implementation of BIM methods and tools at the pediatric hospital Giannina Gaslini and how the dissemination of BIM methods can be carried out in various public organizations beyond the mandatory regulatory obligations.

INDICATE project proposed a city-wide decision support system which accounts for all major systems and activities relevant to developing energy-efficient cities (Musetti, 2016). To create a 'Smart Economy', integrated smart urban planning tools including architectural master planning and detailed energy optimization and environmental analysis were required. The project addressed these issues through the development of a decision support tool that can be used in all stages of urban development of a city. The tool main topics are:

(i) inform masterplanning at early stages of urban development;
(ii) (ii) help make decisions with respect to best technologies to integrate and their economic and environmental impact to the urban environment; and

(iii) (iii) optimize existing Smart Technologies to further reduce energy consumption and carbon emissions. This is achieved through the integration of Dynamic Simulation Modelling, Geographic Information Systems (GIS), 3D Urban CAD Modelling Tools, Sustainable Urban Indicators and algorithms for Demand Side Management and local balancing of energy use into a single software package.

At the date of the project existing tools tend to address single sub-systems and don't examine the city in a holistic manner with respect to all city sub-systems and their interactions. The few tools that address this do not have the Dynamic Simulation Modelling capability to simulate the environment and show the 'ripple effect' that a decision will have in the overall urban context.

CONCEPT AND OBJECTIVES

The INDICATE project wants to support decision makers and other stakeholders towards transforming their cities to a 'Smart' City. This has been achieved through the development of an interactive cloud-based tool, providing dynamic assessment of the interactions between buildings, the electricity grid, the EV grid and Renewable and ICT technologies. Recommendations were provided with respect to the best technologies to install, the infrastructure that requires improvement to enable local balancing and the utility services that offer the best financial plan. To understand how the decisions made might affect the overall urban environment, a set of Sustainable Urban Indicators has been developed, taking account for the knock on effects from changes made to the buildings, infrastructure changes or introduction of new technologies, within the urban context. The decisions taken for the buildings are therefore influenced by the potential impact on the city. The interactive cloud based tool developed has the following goals:

- PLAN development through a dynamic simulation, energy-based decision support tool, which takes into account the buildings and their interaction with the urban environment
- REDUCE energy consumption and carbon emissions through an indication of the impact of best practice Energy Conservation Measures via Dynamic Simulation Modelling
- INTEGRATE new technologies and services in the city to better manage supply and demand, via Dynamic Simulation Modelling, GIS and 3D urban modelling which will reliably inform the impact of the integrated technologies
- OPTIMISE existing installed systems, to enable local balancing through demand response analysis and tariff analysis via Dynamic Simulation Modelling, which will model the interactions between the buildings the installed systems and the electricity grid, across multiple buildings in the urban environment

As a result, the INDICATE cloud-based interactive decision support tool, has been developed to be used at any stage of development; it can be used in a city where the buildings and their urban environment are at an early stage of the sustainable development or in contrast where the city has already installed smart grid infrastructure and energy efficient technologies are present in many buildings. The project outcome is the delivery of a cloud-based interactive decision support tool that can be used by City Architects/Planners, Urban Planning Consultancy Firms, the Public and Business Community, ICT system suppliers and potential investors and developers in a city.

INDICATE Demonstration Sites

The INDICATE project has created a decision support tool that helps with the development of urban master plans; supports the decision maker with respect to the integration of technologies and renewables in their neighbourhood which will help reduce consumption and demand on fossil fuels; and supports existing systems optimization in cities that are already reaching 'Smart' City status. As such the tool can be used to support the decision maker regardless of the stage of development of the city or its technological advancement. To demonstrate each of these aspects of the tool, 2 cities demonstrators with detailed information about the site and the ability to make improvements and demonstrate the functionalities of the tool throughout the course of the project were chosen; these were Dundalk in Ireland and Genoa in Italy. There were 2 test demonstrations undertaken in Dundalk and 1 in Genoa. Both demonstrators included direct end user involvement through project partners Ente Ospedale Galliera (EOG - Galliera Hospital), Louth County Council (LCC) and Dundalk Institute of Technology (DKIT).

Integration-Optimisation of NEW ICT and Renewable Technologies, the Genoa Test Site

Genoa is the capital of Liguria and the sixth largest city in Italy, with a population of 600,000 within its administrative limits on a land area of 240 km^2. Genoa, which forms the southern corner of the Milan-Turin-Genoa industrial triangle of north-west Italy, is one of the country's major economic centers. Genoa is currently part of the Genoa Smart City initiative, which is aimed at improving the city quality of life through the sustainable economic development based on research and technology guided by the Municipality in a project of integrated planning; it also signed the Green Digital Charter at the end of 2009 to encourage cities to reduce the carbon footprint of their ICT and roll-out ICT solutions which lead to more energy efficiency in areas such as buildings, transport and energy; and has an SEAP (Action plan for sustainable energy), which will reduce CO_2 emissions by 23% in 10 years by increasing the use of public transportation, the use of renewable energies and by reducing energy consumption. The project will examine a test site within Genoa city that covers an area of 60 km^2. The area is located in downtown Genoa, in a district with dense buildings including two high schools, a sports centre with a pool, the police headquarters, a number of private residential dwellings and the Galliera hospital. The Galliera hospital is formed by a number of buildings across the site, totalling 110,000 m^2 of Gross Floor Area. Also included with the Hospital buildings is a Historical Building built at the end of 19thcentury. Currently it is not compliant to a modern hospital concept and as such, the municipality has started an ambitious renovating project of the site with the aim of minimizing energy consumption through designing a Green Hospital in line with the Green Guide for Health Care. Data available from the site includes metered information from the buildings belonging to the Galliera Hospital, ordinance survey information from the surrounding neighborhood and information with respect to the building typologies. Further to this, the existing hospital buildings were mapped by BIM tools by using the Industry Foundation Classes (IFC) format (BuildingSMART, 2021), which were imported directly to the Virtual City Model. The project in addition collected the necessary information to create a GIS of the area to be used with ESRI's ArcGIS platform and LIDAR information to create the CityEngine model to be used with the Virtual City Model (VCM). Existing Renewables available on the site include photovoltaic panels for electricity, water solar panels to produce warm water and energy riser systems to re-use the heating generated by technological and diagnostic medical machines (e.g. TAC, RM, PET-TC, LINAC);

the relevant information from these systems is available for integration with the VCM. Furthermore, the existing buildings include an installed a CHP station which will be updated to include the latest of the state of the art CHP technologies for neighborhoods energy systems.

The VCM will be used to provide the decision support with respect to what technologies should be integrated to help reduce energy consumption and what renewable technologies would be most appropriate for the region and the building typologies. Modelling of the electricity grid will be performed in order to provide decision support with respect to the infrastructure improvements that are required so that the new Galliera buildings can optimise the heating distribution to the hospital facilities and corresponding neighborhoods facilities, such as two high schools, a sport centre with a pool, offices and private houses; this is one of the main aims of the current development plans for the area. The algorithms will be applied here to ensure local balancing and optimization of energy and cost across these buildings. Recommendations for the CHP upgrade and its optimisation will also be specifically provided.

EXECUTIVE SUMMARY

In the following the authors provide an overview of the INDICATE first prototype application to Genoa test site, testing the system from the very beginning phase, creating the virtual city model of Galliera site and performing simulations with the first prototype of the INDICATE tool.

Before the actual preliminary design was developed, many other studies were developed, comprising both different new buildings concept design, both different function allocation for the existing buildings.

Currently the project of the whole New Hospital has been divided into different lots:

- Lot 1: consists in the construction of a new structure that will rise on the same site of the existing hospital;
- Lot 2 and Lot 3: consists in restoration and repair of the historical part of hospital halls.
- Sector 3: existing buildings to be deconstructed and new residential buildings to be built.

Galliera Hospital is going therefore to manage a very complex operation, both from the engineering and design point of view, both from the managing point of view.

Lot 1 (NEW Hospital) amount of works value is about 145 million euro and consists of a new building (three basement floors and three floors above ground) construction, in a highly urbanized area, to be realized after four existing buildings deconstruction. The planned restructuring of Lot 2 involves monumental buildings, currently used for hospital activities that will be converted to host logistical and organizational activities.

For what concerns the existing historical hospital buildings, BIM models were used to provide INDICATE's partners with energy simulation data needed, such as buildings location and MEP zones volumes calculations, and provide with the 3D models with the latest technology widely used in Europe. This way the historical buildings were mapped by BIM.

In the past years (since 2000) many options have been studied to identify the "optimal" options for either Galliera historical hospital refurbishment, or a new building hospital construction. The new building hospital has been designed with different shapes.

Galliera existing Hospital complex is composed by eighteen pavilions. Each pavilion is characterized by different use, structure, period of construction and original use. Main uses of buildings may directly concern healthcare (in case of hospital wards, operating rooms, surgeries, laboratories), or be characterized by traditional tertiary activities, such as administration offices. The construction period of buildings varies from mid-nineteen century to last decades of twentieth century. Moreover, some buildings that were initially built for purposes different from healthcare, where in a second time converted. The complexity and diversity of the installation drives to consider Galliera Hospital as a sub-district inside Genoa macro-district. Galliera sub-district is in turn composed by pavilions and aggregations of pavilions.

Together with Lot 2 and Lot 3 traditional bidimensional design development, a digitization campaign of historical buildings has begun. BIM models of the monumental complex have been developed. Creating Mep zones within BIM models and using Solibri Model Checker® Quantity Takeoff functionality, it was possible to collaborate with other partners providing different zones mechanical flow information details.

From public employers or real estate managers point of view, the usage of BIM tools like INDICATE makes it easier and faster to analyze and compare between different options in district development projects, in terms of energy consumption and environmental impact.

The overall INDICATE software was developed in order to have the following functionality four levels:

- Low level city-scale masterplanning for decision making regarding the supply and use of energy in buildings.
- More detailed district level simulations with higher level of data granularity.
- In-depth simulations for blocks of buildings, to enable local balancing across a number of buildings.
- Detailed building level simulation, for individual buildings.

INTRODUCTION

Galliera hospital technicians did not not have particular knowledge of either Esri City Engine, nor Virtual Environment tools; this way the creation of Galliera site and subsequent real installations simulation based on the decision support tool (in its first prototype edition), their integration with the buildings in the area, and the results of the improvement and return of investment, have been used to test the complete procedure a new user should perform to use the tool, from the very initial stage.

GALLIERA TEST SITE POINT OF VIEW AS INDICATE USER

Galliera test site experience was developed following the idea of using the INDICATE tool on Galliera site, to understand how the tool could work in the planning stages, as if the preliminary design of the new hospital were not already defined, to understand the optimal solution to select form the energy point of view.

Starting from the study of a typical modern hospital, which minimal specifications are presented in Italian law and regulations, the study of New Galliera Hospital aimed to evaluate the introduction of efficient and modern technologies and their impact on operative costs and energy efficiency of the Hospital.

Before the actual preliminary design was developed, many other studies were developed, comprising both different new buildings concept design, both different function allocation for the existing buildings. In the following the different ideas studied from year 2000 to our days are briefly described.

Sub Questions Detailing What Galliera Technicians Would Like to be Able to do With Regard to the Hospital

As stated above, in the following the questions that Galliera technicians would have liked to answer using INDICATE tool are provided, and in the next chapters, the results obtained by INDICATE tool application to Galliera site are described.

- What is the best scenario, from an energy point of view. Beginning from the actual hospital layout, we would like to compare different new building shapes and functional layouts (envisaged in different design proposals) to select the scenario that grants the highest energy saving. This can be done running benchmark simulations, changing building usage type and creating different shapes for the new hospital building. This simulation was performed with the first INDICATE prototype.
- In second way compare the result obtained using Italian benchmark data in INDICATE to values resulting from hospital energy bills to evaluate the tool's results.
- Once the best scenario has been selected, get in depth simulations, selecting different future interventions and looking at the impact of renewables on energy savings.

For Galliera hospital, different scenarios were developed, and are described in the following paragraphs. One of these scenarios (Scenario 4) contains the New Hospital building, with all new technologies implemented in the concept design, defined as future interventions.

Galliera new Hospital Design State of the art

Since May 2015, Galliera Hospital restarted the procedure for the design of the New Hospital. After a two years stop, the project of the whole New Hospital has been divided into different lots:

- Lot 1: consists in the construction of a new structure that will rise on the same site of the existing hospital;
- Lot 2 and B1, B2, B3 refurbishment: consists in restoration and repair of the historical part of hospital halls.
- Sector 3: existing buildings to be deconstructed and new residential buildings to be built.

Therefore the New Galliera Hospital realization does not consist of just a "simple" new hospital building construction, but of a much more complex operation, that will be developed in phases and which requires that the existing Hospital in all its functions is kept working, before, during and after the new building construction phase.

This aspect involves the complex organization of the so called "transition phase", including logistics and management issues related to:

- maintaining the hospital functioning during the preparation of the area before the new building construction. This phase requires that the buildings to be demolished are emptied and that all the sanitary and hospital functions are transferred in other actually existing buildings (where some unallocated areas have been identified) or in 'ad hoc' temporary buildings to be provided just for the construction period.

- maintaining the hospital functioning during the construction of the new building.

The New Galliera design, born as a response to structural inadequacies of the current buildings, provides also an opportunity for radical cultural change for all operators involved. It connotes as the physical realization of various initiatives such as the intensive care hospital, the lean hospital, the digital hospital, the green hospital, all initiatives that demonstrate Galliera innovation commitment.

This commitment, along with the complexity of the work, led the Hospital Management and Technical office to identify Building Information Modelling (BIM) (Osello, 2012) as the fundamental tool in the design process of the third millennium hospital, essential choice for managing the complexity of the operation, further confirmed at international level too.

For what concerns the existing historical hospital buildings, BIM models were used to provide IES with energy simulation data needed, such as buildings location and MEP zones volumes calculations, and provide with the 3D models with the latest technology widely used in Europe. This way the historical buildings were mapped by BIM.

In the past years (since 2000) many options have been studied to identify the "optimal" options for either Galliera historical hospital refurbishment, or a new building hospital construction. The new building hospital was designed with different shapes. Having a tool such INDICATE available, would have proven useful to comprehend the different energetic and economic impacts of different options and of different new building shapes.

In the following paragraphs, a description of existing buildings and new hospital building is provided and the procedures followed to create the Galliera Hospital virtual city model are provided, followed by the different simulations performed with the first INDICATE prototype.

EXISTING HOSPITAL LAYOUT AND BIM MODEL

Galliera Hospital complex is made of eighteen pavilions. Each pavilion is characterized by different uses, structure, period of construction and original use.

Main uses of buildings may directly concern healthcare (in case of hospital wards, operating rooms, surgeries, laboratories), or be characterised by traditional tertiary activities, such as administration offices.

The construction period of buildings varies from mid-nineteen century to last decades of twentieth century. Moreover, some buildings that were initially built for purposes different from healthcare, where in a second time converted.

The complexity and diversity of the installation drives to consider Galliera Hospital as a sub-district inside Genoa macro-district. Galliera sub-district is in turn composed by pavilions and aggregations of pavilions.

Together with Lot 2 and Lot 3 traditional bi-dimensional design development, a digitization campaign of historical buildings was begun. BIM models of the monumental complex have been developed at a comparable to L0 level of detail: spaces and main structural elements and architectural features such as main walls, interior partitions, ceilings, windows, doors, stairs and roofs have been modeled.

Creating Mechanical, Electrical, Plumbing (MEP) zones within BIM models and using Solibri Model Checker® Quantity Takeoff functionality, it was possible to collaborate with IES providing different zones Mechanical flow information details such as: type, Volume, specified supply and return airflow.

GALLIERA HOSPITAL created MEP zones within the BIM models with the Autodesk Revit® authoring tool. MEP zones were populated with data needed for Virtual Environment simulations. Once GALLIERA hospital created MEP information takeoffs with Solibri Model checker, these models were viewed with the free tool Solibri model viewer and used for simulation purposes.

This way many spreadsheets were substituted by quantity takeoff organized data with BIM models.

Figure 1. Galliera Historical buildings BIM model in Solibri Model Checker® with Mechanical flow quantity takeoff

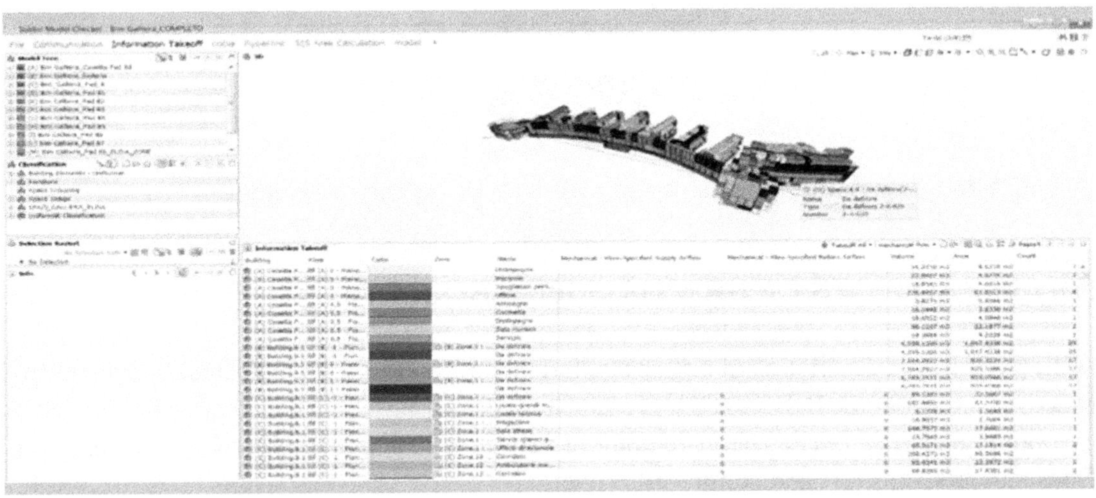

BIM models developed for both historical and new hospital buildings allow Galliera technicians to use Solibri Model Checker®, Quantity Takeoff functionality, for evaluating gross quantities and for immediately and very clearly visualize spaces and MEP zones.

Spaces, modeled with their hospital functions and placed in Solibri Model Checker®, have been mapped and grouped according to specific health and sanitary classifications. These classification groupings are strategical and functional for checking the correspondence of the hospital building to the Liguria Region health requirements.

MEP zones have been created to collect data required by IES for Virtual Environment energetic simulations. These data were initially recorded on excel spreadsheets, but with the creation of BIM models the same data were used to populate BIM database with different Air treatment zones, with all the correspondent data. The big help given by BIM models is that it is easier to locate zones, this aspect is very useful for complicated structures as Gallliera hospital. Having tridimensional models with HVAC data mapped helps information sharing and correspondent volumes and areas calculation is easier. Moreover this procedure is quicker and grants reduction of errors risk. Very complicated buildings such as building "A" have very complicated HVAC systems.

With BIM models Mep zones, every room was associated to its MEP system, for example, all air-handling units were associated to the group of spaces they serve, the same for zones served by fan coil systems. In the figure below the 3rd floor of building A is showed, with the highlighted MEP Zone called "FAN _COIL degenze piano 3 A", that is the zone made of hospital ward rooms, served by fan-coil system.

Figure 2. Galliera building A BIM model in Revit

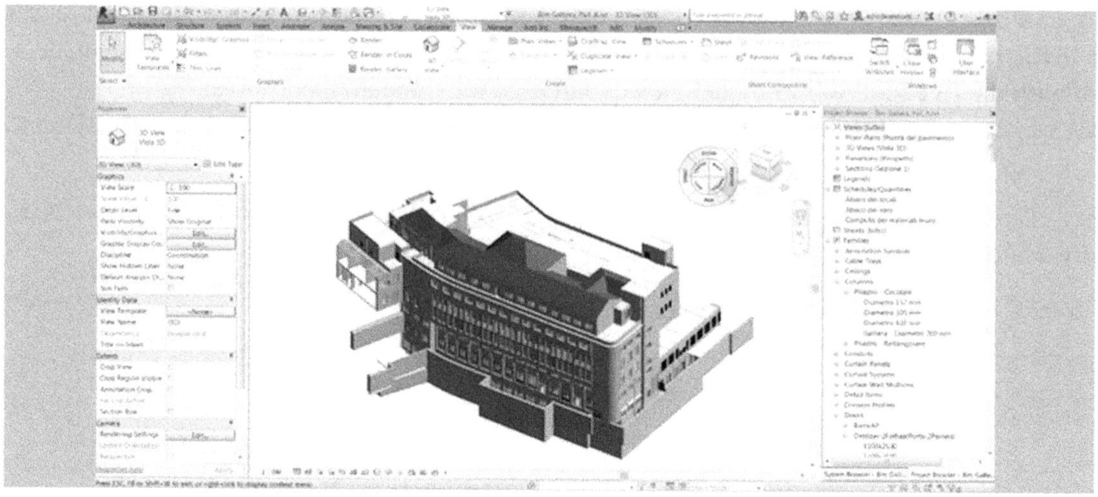

PREVIOUS PROJECTS OF NEW GALLIERA HOSPITAL

In the following, a brief description of different solutions studied for Galliera hospital from 2000 to 2008 is provided, to show how the INDICATE tool could have been used by Galliera technicians to understand the best option to select from an energetic point of view, and to show different solutions results to different stakeholders, directors and public authorities.

User testing of the first INDICATE prototype tool by Galliera has been made simulating different shapes of the new hospital building designed within different projects options.

In the following the different Galliera projects are described, and a note is added to identify the ones modeled in the INDICATE prototype.

Figure 3. Galliera building A BIM model in Revit – Mep zone definition to provide data for Virtual Envoronment energy models

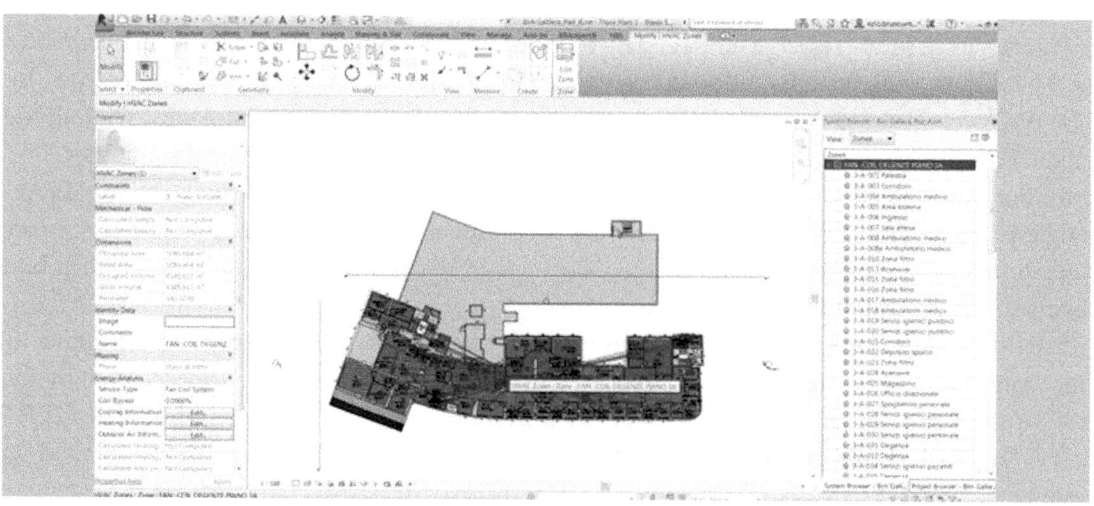

Galliera 2000

In 2000 a new project was developed, comprehending:

- Refurbishment of the existing monumental hospital. The historical buildings should have been renewed to continue maintaining hospital functions;
- New building creation. The new building, defined as "Nuovo Edificio Polifunzionale" (new polifunctional building) was designed as a 13 floors new building, with the shape showed in the following picture.

This project was not modeled in the INDICATE Prototype as its idea was too old.

Hypothesis Developed From 2000 to 2007

Three projects were developed in those years, two regarding the new hospital Galliera and one that envisaged also the creation of two new buildings for the police headquarters.

The three projects are in detail:

- "progetto Mauri" and "progetto Rocchi-Mor": two different solutions for a new hospital construction, see following paragraphs for description, only progetto Mauri has been developed as an INDICATE scenario (scenario 2);
- "progetto Bona e Galasso": New police headquarters construction within area actually occupied by buildings A, A1 and As, these areas should have been changed from actual hospital usage to police headquarters usage; this scenario has not been developed.

Figure 4. Galliera Mauri Project general framework

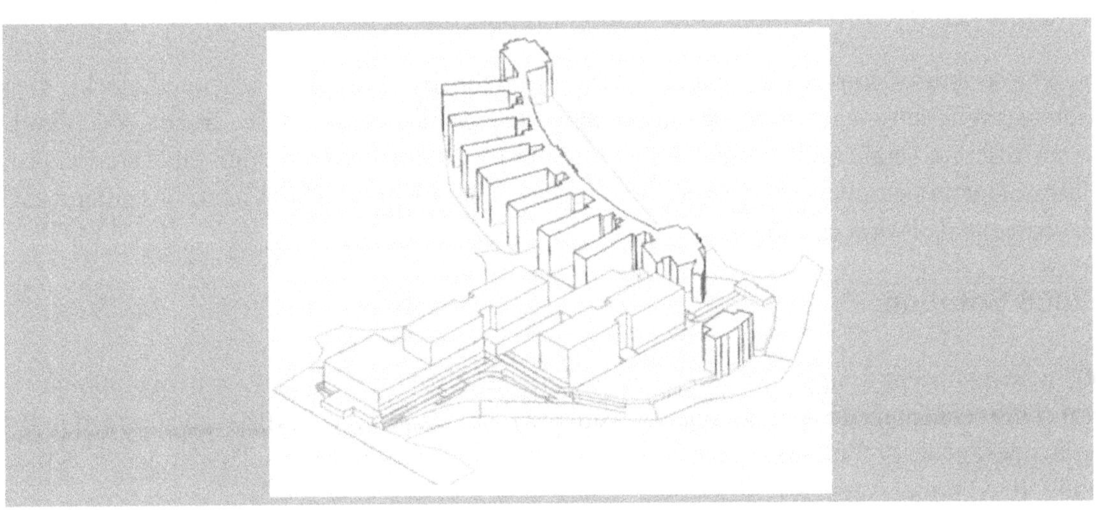

Mauri Project

This solution envisaged the creation of a very big new hospital building, as showed in the following picture.

Progetto Mauri Concept Design

This project was defined by its designers as a preliminary study of a base project. It was developed with the aim of giving the hospital administrators the main ideas of elements, indications and pathways to base the strategic evaluation and finally define the feasibility study. The project envisaged that hospital functions in the historical buildings should be replaced by public functions (libraries, museums, high schools, universities, congress halls, auditoriums, etc.), only buildings B6 and B7 would maintain the sanitary function, placing there laboratories and medical studies.

The new structure proposed by Mauri project had a main functional plate that supports the entire structure working and 4 hospital ward blocks. The total number of beds is 620, to be compared with the 417 ones existing at the project age. The hospital structural model envisaged is of blocks and plate structure, kind of structure that is not applicable nowadays. Moreover, the beds number is too big compared to the existing structure, consequently the structure could be designed with smaller dimensions.

Rocchi-Mor Project

This project reviews the Mauri one, envisaging a smaller structure, with new buildings for the police headquarters, designed by Galasso. The historical buildings maintain public functions. This project was not modeled in INDICATE tool.

The three main elements that constitute the Rocchi-Mor project are the following:

- New building project and construction, called "CORPO 1" for the police headquarters, located in place of the existing emergency and surgical units (buildings A/As).
- New building project and construction, called "CORPO 2" for the police headquarters, located nearby building "D", the direction offices buildings
- Refurbishment of the concentrical "comb shaped" historical buildings and construction of new buildings behind the hospital, in the zone between Via Vannucci and Corso Mentana.

The new structures proposed by Rocchi-Mor project consist of different separated blocks, disposed on different levels, and not connected to one another. This solution envisages the creation of 7 new buildings, with different height and, consequently, different roof levels. Even though this solution provides a less invasive volumetric layout, it causes both spaces and time growth both internal sanitary staff and hospital patients pathways lengthening.

The 2008 Solution

In 2008 a new concept design was developed to define the volumes needed for the creation of a new hospital. This conceptual design envisages the creation of a building with the shape defined in the following images, made of 3 underground floors and of 5 above ground floors. This project was modeled in the INDICATE tool (scenario 3).

Figure 5. A section of the 2008 design shape of New Galliera Hospital

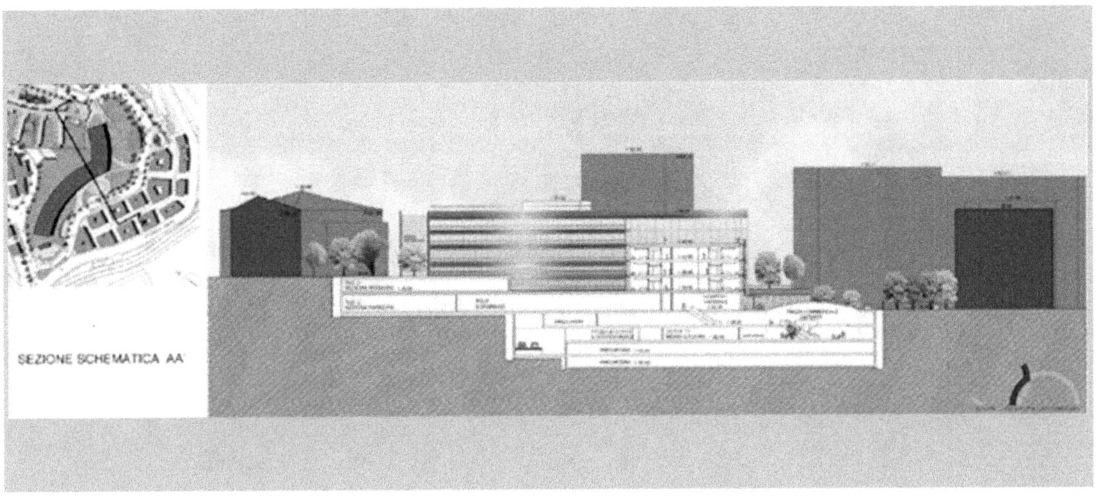

Besides defining the needed volumes for the new structure, based on more recent sanitary requirements respect to the ones of the previous projects, the 2008 solution provides the opportunity to reallocate actual sanitary functions of the existing historical hospital with different functions. The different functions envisaged by this project have been simulated within the INDICATE tool building.

The following figures show different functional layout for the existing historical buildings reorganization; the second hypothesis has been modelled in INDICATE to see different energy impacts of the different function distribution.

NEW GALLIERA HOSPITAL PRELIMINARY DESIGN CONCEPT AND LAYOUT

The preliminary design of the New Galliera hospital incorporates, through a holistic approach, the different aspects comprehending medical, organizational, functional aspects together with architectural, landscaping and urban ones.

The building is made of two main parts:

- A roof garden that connects the different lot elevations characterizes the partially underground part that organically follows the irregular lot size, made of three levels (emergency, surgery, medical surgery).
- Over the underground part the regular volumes of hospital wards develop, designed to guarantee both the maximum sanitary efficiency both the maximum urban integration

The new building has 7 floors: one basement parking, 3 partially underground, and 3 in elevation. The total intervention area is 26.000 m^2. The new building 50 m above sea level. Total hospital number of beds is 404.

Figure 6. Different functions distribution and corresponding legend

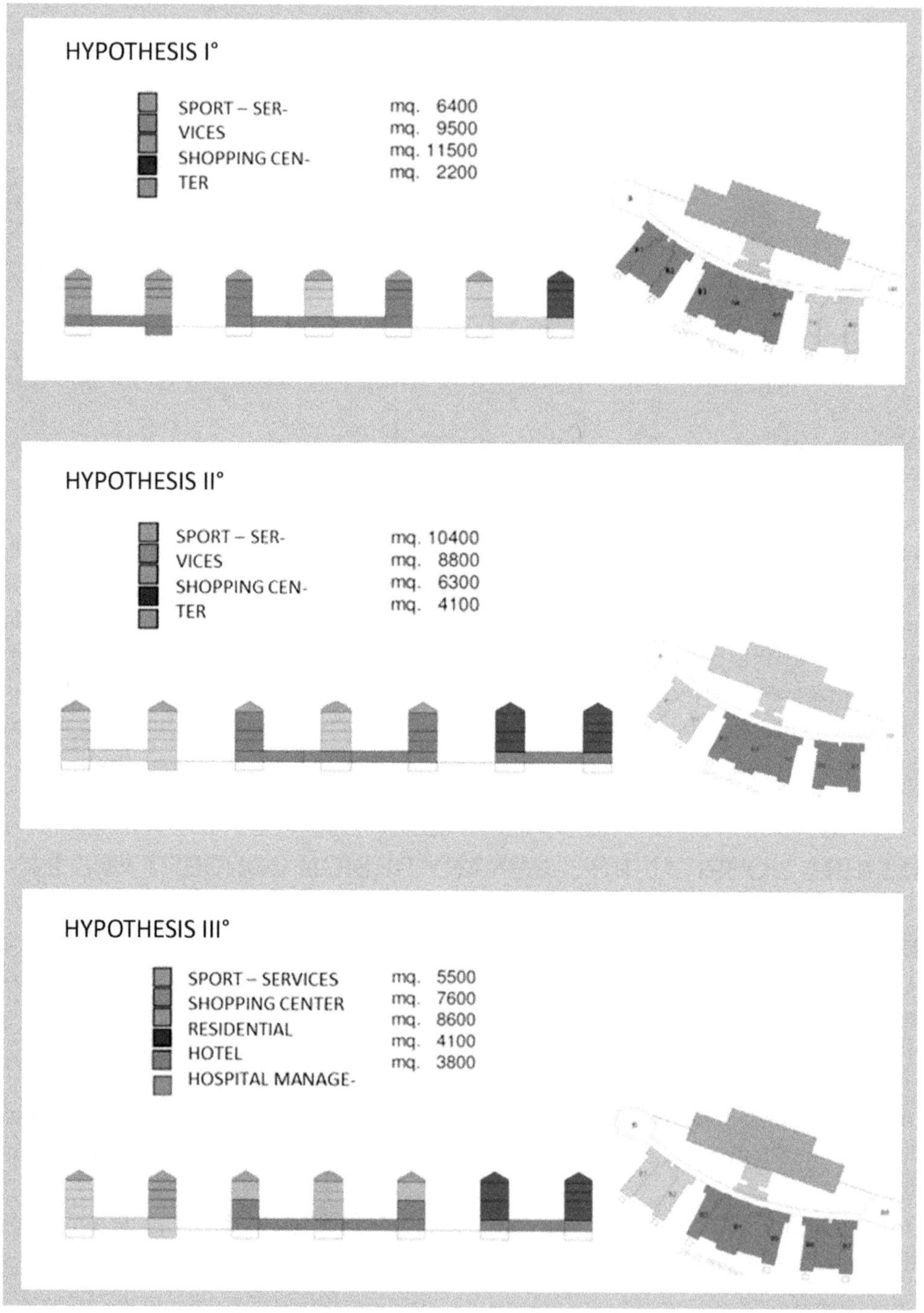

The particular site shape and the objectives of rationalization and multidisciplinary sanitary activities have conducted to the design of a unique shape, made of the conjunction of buildings network organized in a quintuple model.

Figure 7. New hospital layout

Distribution Layouts

In the following, the different hospital uses defined in each floor are described, comprehending both the new building (lot 1) and the refurbished historical buildings (lot 2). The design is thought divided into horizontal levels, that is one of the reasons the first approach for Galliera Building Modeling into INDICATE tool was developed with the same concept, dividing buildings into horizontal plans:

- Level -1: parking
- Level 0: emergency; surgery; technical installations
- Level +1: surgical unit; sterilization; intensive care unit; interventional radiology; ICU; outpatient surgery; pathological anatomy; technological center; installations; medical gas central
- Level +2: psychiatric services; free clinic; blood sampling center; dentists center; radiotherapy; technological center
- Level +3: day hospital; restaurant/coffee; shops; chapel; medical practice
- Level +4: hospital wards
- Level +5: hospital wards; childbirth; day surgery
- Level +6; mechanical and cooling installations

Historical Buildings Refurbishment (Lot 2 and 3)

Once the new hospital will be ready, all the historical buildings will be refurbished to allocate functions different from sanitary ones such as, for instance: offices, educational, museums, storage, changing rooms, elderly studies.

Urban Planning Variations

In May 28th, 2009 the Genoa city Council approved the urban planning zone variance for the New Galliera Hospital construction. This variance envisages the conversion of some existing hospital buildings, resulting unsuitable for new hospital functions, to residential usage, maintaining however buildings and gardens of the historical complex.

In this variation, buildings A, As and A1 will be refurbished or deconstructed and rebuilt, maintaining the same Net Internal Area (NIA), or Usable floor area (UFA), provided by appropriate parking and green areas around the residential buildings.

Building D will be refurbished or rebuilt and transformed into residential use, maintaining the same Net Internal Area (NIA), or Usable floor area (UFA), provided by appropriate parking and green areas around the residential building.

Building E will be refurbished and provided by appropriate parking.

Urban planning variations have been modeled in INDICATE, the corresponding scenario is called "Third Sector_NEW", that hypnotizes:

- For buildings A/As and A1: deconstruction and a reconstruction with different shapes, to realize residential buildings.
- For building E And D: refurbishing with usage change from hospital to dwellings.

GALLIERA HOSPITAL SCENE CREATION WITHIN INDICATE

First Approach for Galliera Building Modeling

As previously described, Galliera new hospital building concept design was developed working on different levels layouts and areas. The first approach in creating the model of Galliera site within INDICATE tool followed the same criteria, buildings were imported into INDICATE creating footprints obtained "slicing" into horizontal plans the different buildings. This way it was possible to create very well defined Galliera buildings models, in order to reproduce Galliera site for more detailed district level simulations with higher level of data granularity and for in-depth simulations for blocks of buildings, to enable local balancing across a number of buildings. Anyhow, from a check on behalf ESRI of the model produced, the procedure had to be modified and the scene to be built again. ESRI technicians explained Galliera Hospital technicians that modeling the buildings with horizontal layers does not work for ESRI CITY ENGINE because the latter doesn't know anything about how the shapes are logically connected. Therefore it is impossible from within City Engine to figure out that different shapes are actually the same building. To overcome this issue, it was suggested to have the shape of each "finger" of the hospital and then add the connections as separate shapes: doing so only small surfaces separate simulated buildings and the simulation is not that far off. On the other hand, with this modeling the analysis grain on each single building is highly reduced, for example in the NEW Hospital, where there are different usages in one same floor, it is needed to define just one usage per level. Galliera Hospital test was very focused on each building precise definition, while INDICATE is mostly about city scale analysis, so the user approach should be at city level, as many indicators are, too.

The Galliera INDICATE scene was developed with another approach, modeling buildings as unique shapes and assigning a main function to each building, losing the floor by floor detail. A great effort was made on behalf Galliera technicians to select the correct coordinate system (WGS84 UTM 32N) to join the different files necessary to build Genoa scene.

INDICATE Galliera Model

INDICATE Galliera Model First Approach

Figure 8. Galliera scene created with horizontal shape layers for each floor in each building

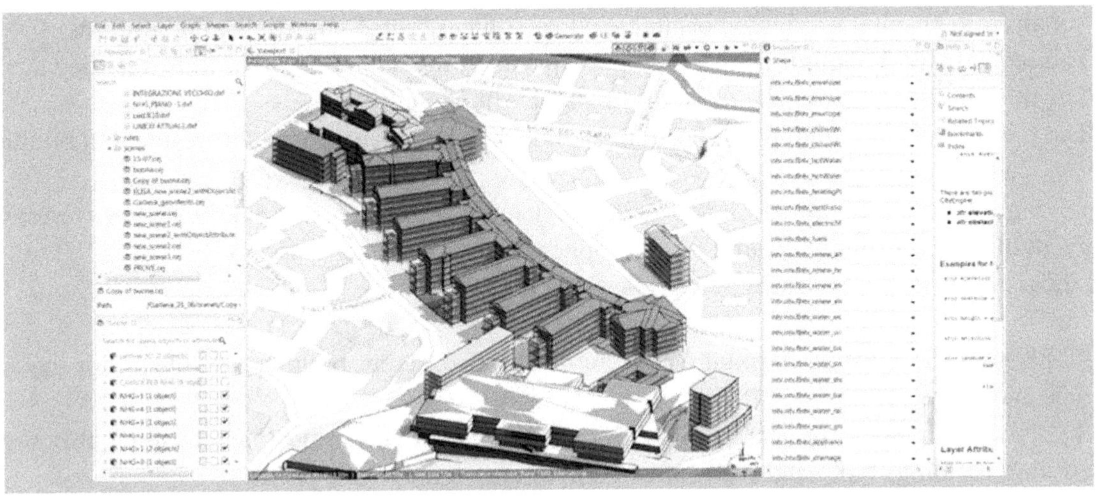

Correct Approach for Galliera Building Modeling

To create the Galliera site scene, the following procedure was followed.

- Create a new scene;
- Create the footprints (one for each finger and a connection shape);
- Create three different layers for: existing buildings, to be demolished buildings, new building;
- With the city engine trial version create the object attributes on all footprints;
- Assign values to the object attributes (INDICATE tool);
- Apply rule file;
- link rule attributes to object attributes with same name (click on drop down box arrow for a rule attribute -> connect attribute -> object attribute). Do this for all the rule attributes in the section "object attributes" with a non-bold heading;
- generate buildings;
- run simulations (INDICATE tool).

The attribute called OBJECTID must be a unique number for each shape.

First of all Galliera technicians downloaded the Open Street Map maps both with and without aerial photos.

Then the maps were uploaded to the City Engine tool and the coordinates were aligned. The maps were compared to the Auto Cad 3D (wireframe model) of that part of the city provided by Genoa's municipality and the terrain was shaped to represent the ground shape of this district with high detail level.

As previously stated, at the actual stage, a new building preliminary design has been performed; this new building will be built after some existing buildings deconstruction.

Simulations within INDICATE tool therefore were performed at the beginning, with the idea of creating the following layers within INDICATE tool:

- The first layer containing all the building footprints that will not change: historical buildings are heritage status, therefore they cannot be deconstructed, the only change within them will be functional: from hospital ward into hospital services (offices, nurse school, changing rooms)
- The second layer containing the building footprints of the buildings that will be tear down.
- The third layer containing the new building that will be built;
- The fourth layer for buildings that will be demolished and reconstructed as residential buildings.

Applying the tool to Galliera site, it was realized that more layers, or better, SCENARIOS, were needed to study all the possible options useful for the application of a Decision Support System as INDICATE, this is the reason why the following scenarios were created:

- SCENARIO 1: Existing hospital configuration, with different buildings defined with the actual type and usage
- SCENARIO 2: Mauri project configuration: creation of Mauri project new hospital building and existing buildings types and usage attribution as envisaged by the project
- SCENARIO 3: 2008 Project solution: creation of 2008 project new hospital building and existing buildings types and usage attribution as envisaged by the project
- Scenario 3 - RESIDENTIAL: 2008 Project solution of a new residential building close to pavilion D of the hospital
- SCENARIO 4: New Galliera hospital last design developed: existing buildings types and usage attribution as envisaged by the project
- SCENARIO 4 – NHG (only): New Galliera hospital last design developed: project new hospital building
- THIRD SECTOR_NEW: Residential buildings: urban transformation of actual hospital buildings into residential buildings after the new hospital be ready
- OTHERS BUILDING: other existing buildings in the area, mainly with residential usage
- Public building: other existing public buildings in the district besides the hospital

Required Object Attributes

ESRi provided the following list of required object attributes that must be first created for each footprint. This phase was very time consuming because the list had to be detailed for each building.

The objects attributes are recorded and modifiable within the "Inspector" tab of the INDICATE tool.

Figure 9. Scenario 1 and Scenario 4 in INDICATE tool – existing and new Hospital - building usage different colors visualization

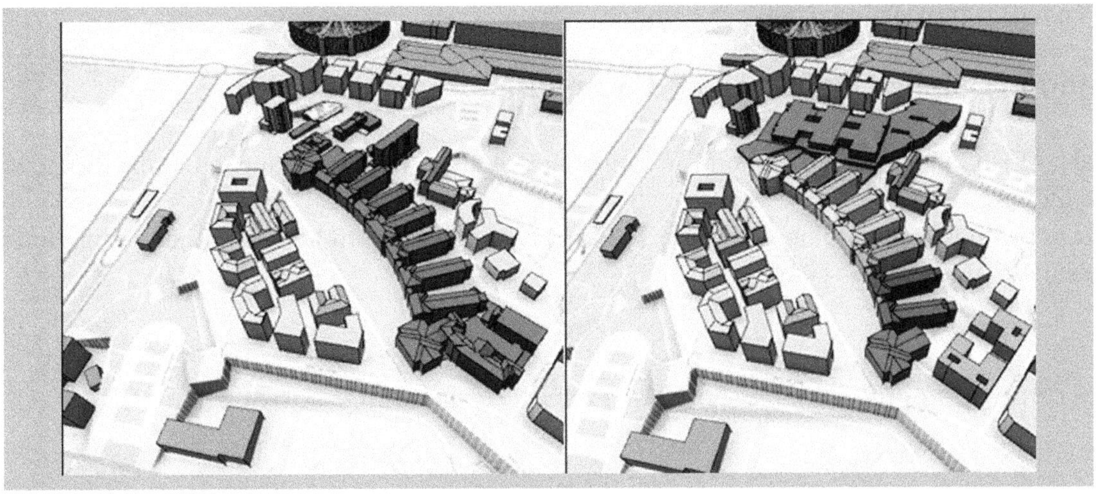

Observations for Modeling Approach

BIM models could be very useful to speed up and get easier the procedure of the scene creation. Moreover, it is widely recognized among engineering software, that interoperability should be granted by software houses. Interoperability is a characteristic of a product or system, whose interfaces are completely understood, to work with other products or systems, present or future, in either implementation or access, without any restrictions.

IFC Definition

The BIM format that guarantees interoperability is defined and IFC (industry foundation classes). At the time of the project, Esri City Engine does not support either IFC importing, or three dimensional file formats (.dxf 3D).

The implementation of these functions is highly desirable to be developed with INDICATE to avoid the long procedure of the tridimensional scene creations, that is error prone and time consuming.

Graphs Selection

To compare scenarios results the following graphs have been selected in the user interface:

- Energy by type
- Energy
- Energy after

Once the scenario has been selected, different simulations have been developed making different hypothesis on renewables applied to the buildings, and looking at their impact on energy. Interventions (existing and future) can be selected from the Inspector tab. As different complex hypothesis on interventions have been developed to compare Galliera scenarios, in the following the list of different interventions is provided.

Scenarios Detailed Definition and Comparison

With the INDICATE inspector, several "existing intervention" have been applied to the different Galliera buildings.

The actual hospital buildings (scenario 1) were characterized with high detail in existing interventions as in the spreadsheets below.

Figure 10. Scenario 1 interventions definition from Inspector tab

In the dashboard Energy as key number and Energy By Type as pie chart were examined.

Scenarios (from 2 to 6) have been used to compare as a first draft energy consumption on a district scale, using solely benchmark data, without defining existing interventions in the inspector

The Scenario 4 – NHG (only) in addiction with Scenario 4 (existing historical buildings with heritage status with future usage after refurbishment) gives the best solution from the energy saving point of view.

Galliera technicians also investigated how deep can the existing buildings be retrofitted keeping the hospital in activity: scenario 1 with future interventions applied.

With the INDICATE inspector, several "future intervention" have been applied to some buildings, characterizing the possible retrofitting (scenario 1) with high detail as in spreadsheets below.

The results in terms of energy of the best scenario (4), compared with scenario 1 with future interventions seemed not to be satisfactory, as scenario 4 needed to be better defined in terms of future interventions, as a green hospital. So several "future intervention" of a modern "green hospital" were applied to the new building as in spreadsheets below.

In the dashboard Energy and the EnergyAfter as key number were visualized.

Table 1. Scenario 1 existing interventions definition

SCENARIO	BUILDING SELECTIONS	EX. INTERVENTION	
Scenario 1	all	boiler	old Cop 0,75
		chiller	cop 3,8
		hvac system	central heating radiators
		lightining source	fluorescent T8
		lightining occ	300 lux
		appliance	30 W/m2
	padd. A-As, A1 (bunkers radiot.)	hvac system	CAV
		hvac fan	Average 7.0 ACH50
		hvac vent	mech. vent central extract
		hvac fresh air	high 15 l/s p.
	padd. C, D	hvac system	air cond. fan coil
		hvac fan	Average 7.0 ACH50
		hvac vent	windows
	padd. A1, B1, B2	hvac system	room split heat pump
		hvac fan	tight 6,0 ACH50
		hvac vent	windows
	padd. B9, D1	hvac system	air central term. Reheat
		hvac fan	very leaky 15 ACH50
		hvac vent	mech. vent central extract
		hvac fresh air	high 15 l/s p.

With this kind of data availability, decision makers have the opportunity to choose the strategy to apply: whether to retrofit of the old hospital or to construct a new building, on the basis of scenarios comparison and of objective and reliable data.

Table 2. Scenario 1 future interventions definition

SCENARIO	BUILDING SELECTIONS	FU. INTERVENTIONS	
Scenario 1	all	boiler	CHP 70 kWe
		chiller	excellent COP ..
		sys terminal	zone e thermostatic
		sys plant	BEMS
		hvac fan	Tight 6.0 ACH50
		hvac vent	mech. vent central extract
		hvac fresh air	moderate 8 l/s p.
		lightining source	LED 60 lm/w
		lightining control	networked swiching
		chilled waters pipes	insulated
		chilled waters tanks	insulated
		hot waters pipes	insulated
		hot waters tanks	insulated
		heating waters pipes	insulated
		heating waters tanks	insulated
		vet ducts	insulated
		electric motors	high eff.

Table 3. Scenario 4 future interventions definition – green hospital

SCENARIO	BUILDING SELECTIONS	FU. INTERVENTIONS	
Scenario 4 NHG (only) + scenario 4	NHG green	boiler	CHP 199 kWe
		chiller	CHP 199 kWe + adsober COP 1
	All	sys terminal	zone e thermostatic
		sys plant	BEMS
		hvac fan	passive house ACH50
		hvac vent	mech. vent central extract
		lightining source	BEST OLED 120 lm/w
		lightining control	networked swiching arvest
		chilled waters pipes	insulated
		chilled waters tanks	insulated
		hot waters pipes	insulated
		hot waters tanks	insulated
		heating waters pipes	insulated
		heating waters tanks	insulated
		vet ducts	insulated
		electric motors	high eff.
	NHG green	hvac system	VAV
		hvac fresh air	moderate 8 l/s p.
		lightining occ	300 lux
		envelope out	100 mm
		envelope floor	50 mm
		envelope doors	insulated
		envelope windows	1,4 W/m2K
		envelope glazing	Very reflective solar glazing (G=0.10)
		envelope finss	500 mm
		envelope roof	low albedo
		appliance	30 W/m2
	NHG slabs	hvac fresh air	high 15 l/s p.
	B-B8	boiler	CHP mini 20 kWe
		chiller	CHP mini 20 kWe + Ads.
		hvac system	air cond. fan coil
		hvac fresh air	moderate 8 l/s p.

At this stage, INDICATE users should keep in mind that data are collected at district or macro zone, in coherence with the tool usage, and cannot be detailed further considering, for example, the different systems provided in each different building floor.

However, it is possible to compare the impact of the transformation of some hospital buildings into modern and green dwellings. Applying several "future intervention" to the building of the third sector scenario, as in spreadsheets below, it is possible to compare energy consumption.

Visualizing in the dashboard Energy and EnergyAfter as key number, the different solution impact can be compared and applied to the whole district modelled, comprehending both public and other buildings in the area.

For the swimming pool ad sports center near the new hospital building a future intervention was also selected (change the boiler with a CHP 199 kWe), given the fact that it might be possible to link those buildings to the CHP power unit of the new hospital.

Figure 11. Scenario 4 (New Hospital) interventions definition from Inspector tab

INDICATE Prototype 1 Scenarios Energy Values and Comparison

The scenario 1 uses benchmark data to simulate the existing hospital. The value of energy consumption obtained with Italian benchmark data is similar to the one of the hospital energy bills that amounts in about 44 millions of kWh/year.

The scenario 2 uses benchmark data to simulate preliminary concept design 2000-2004 hospital. The value of energy consumption obtained with Italian benchmark data for the Mauri project configuration is higher than the actual hospital energy consumption.

The scenario 3 uses benchmark data to simulate preliminary concept design 2008 hospital. The value of energy consumption obtained with Italian benchmark data for the 2008 project configuration (higher than the actual hospital energy consumption). As previously stated, the 2nd hypothesis of the usage distribution was applied.

Scenario 4 analyzes benchmark data of a new hospital, therefore no interventions were selected. Energy consumption of the new hospital is lower than the existing one, even in case no future interventions are selected.

Table 4. Scenario "Third sector" future interventions definition

SCENARIO	BUILDING SELECTIONS	FU. INTERVENTIONS	
Third sector_new	Residential	hvac system	floor heating
		boiler	CHP mini 20 kWe
		chiller	CHP mini 20 kWe + Ads.
		envelope out	100 mm
		envelope floor	50 mm
		envelope doors	insulated
		envelope windows	1,4 W/m2K
		envelope glazing	Very reflective solar glazing (G=0.10)

Figure 12. Galliera scenario publication with City Engine web

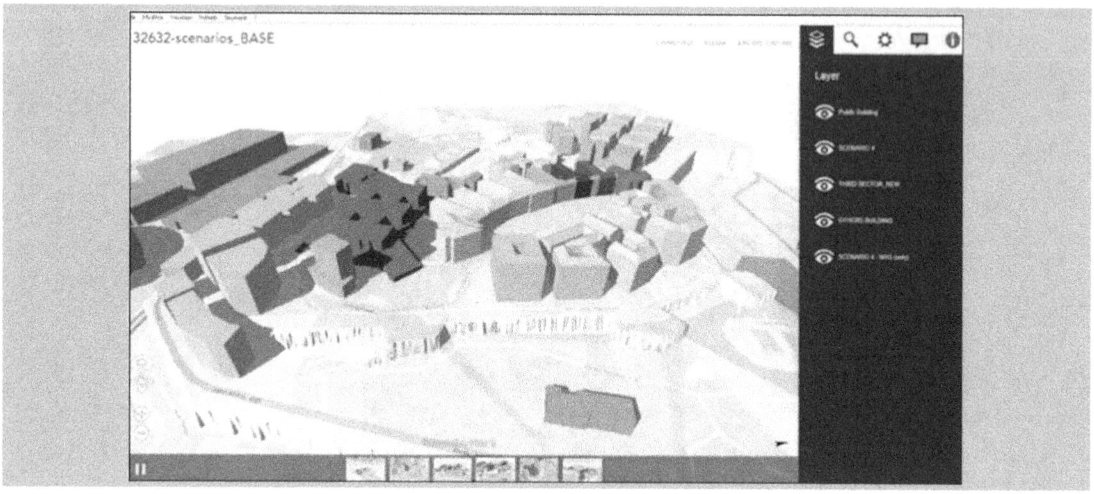

Therefore was analized the scenario 1, existing hospital with possibble optimal interventions. This scenario analyzes energy consumption of the existing refurbished hospital: future interventions are selected to see the energy savings possible in case existing buildings could be refurbished instead of the New Hospital construction. The historical hospital is heritage status, so interventions are focused on systems rather than envelope and glazing. The energy consumption of the refurbished hospital is lower that the new hospital analyzed with benchmark data and no future interventions.

Then was analized the scenario 4, New Galliera a s a green hospital. The new hospital is designed as a green hospital, and the corresponding future interventions are defined within this scenario. The energy consumption of the new hospital with "green" interventions is lower than the previous hypothesis.

Than was analyzed the energy consumption of the new hospital with "green" interventions, together with the existing refurbished hospital (lot2) and the "Sector 3" buildings.

Figure 13. The actual hospital scene in the first prototype and existing interventions definition

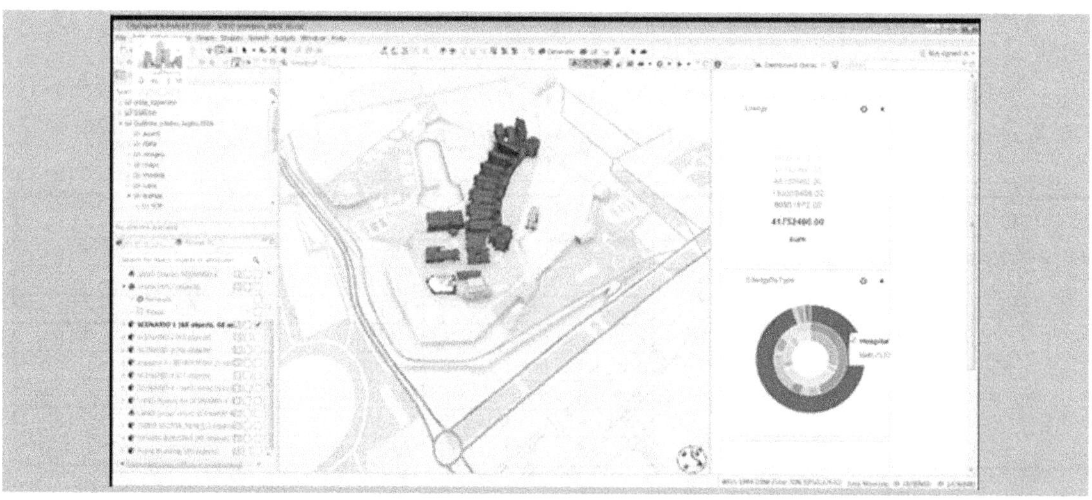

Figure 14. The Mauri project configuration

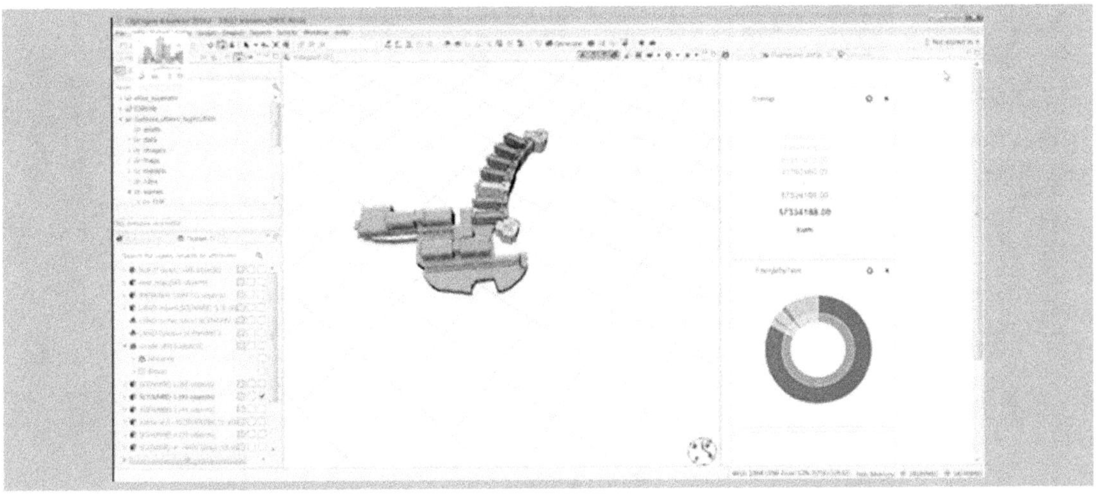

In conclusion was easy in INDICATE compare for any scenario the simulation with (Energy after) end without designed intervention (Energy):

- The energy consumption between the existing hospital (Energy) and the refurbished hospital (Energy after);
- The energy consumption evaluated for New Hospital design, with no interventions (Energy), and with actual designed interventions as Green building and corresponding future interventions selected (Energy after);
- The energy consumption evaluated for New Hospital design, with no interventions (Energy), and with actual designed interventions as Green building (Energy after), together with existing building refurbished (lot 2).

Figure 15. The 2008 project configuration

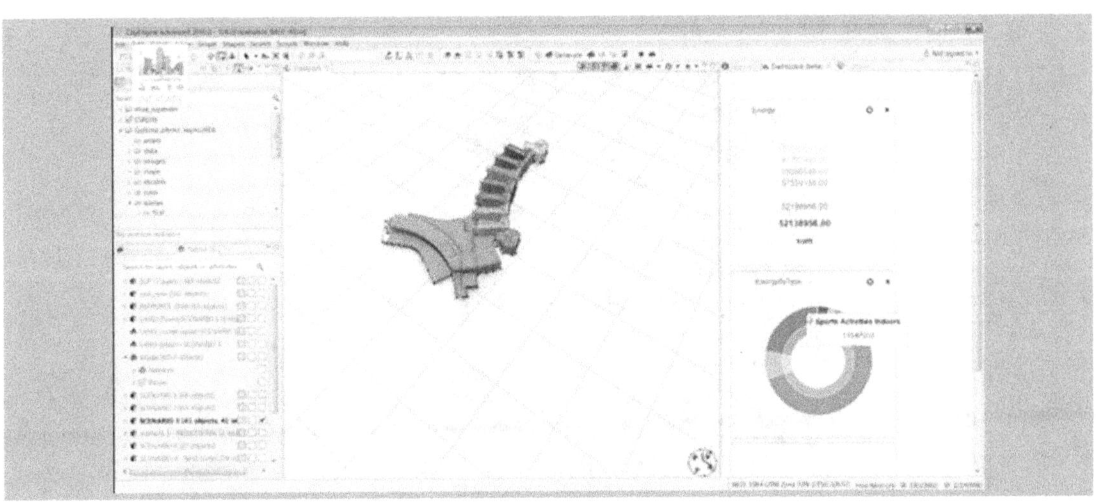

Figure 16. Scenario 4 analyzes benchmark data of a new hospital

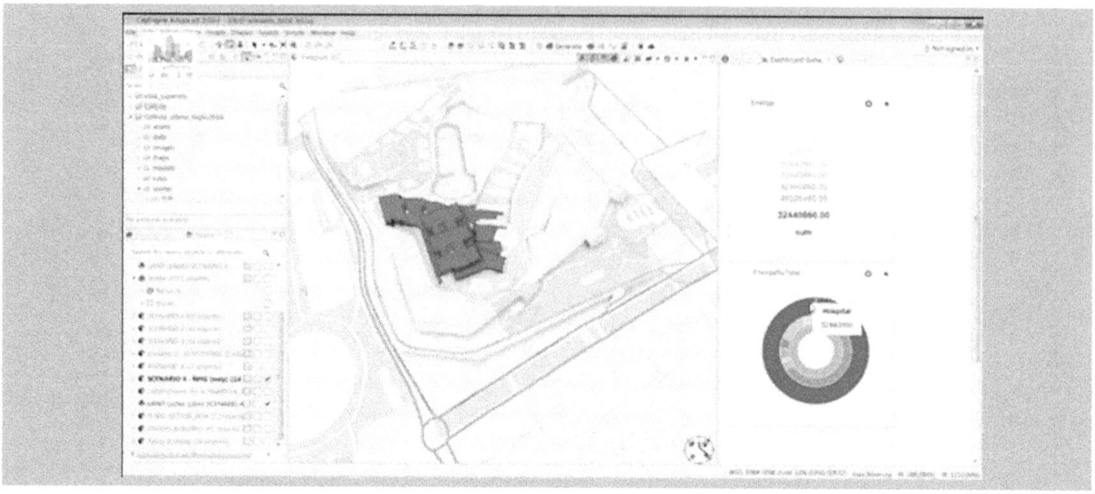

- The energy consumption evaluated for New Hospital design, with no interventions, (Energy) and with actual designed interventions as Green building, together with existing building refurbished (lot 2) and with SECTOR 3 (Energy after)

Figure 17. Scenario 1 with future interventions

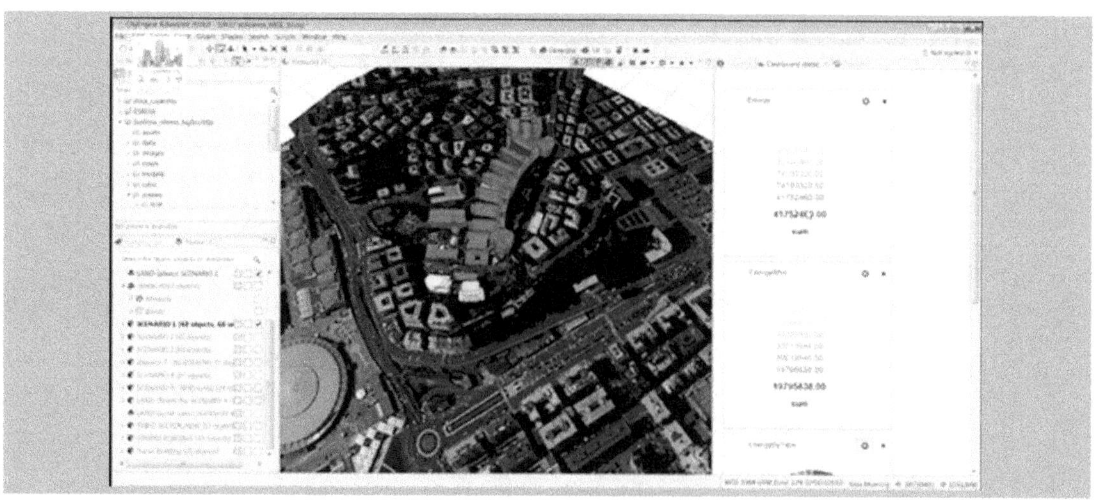

Figure 18. The new hospital is designed with "green" interventions

CONCLUSION

INDICATE

From Galliera point of view, the usage of the INDICATE tool makes easier and faster the analysis comparison between different options in district development projects, in terms of energy consumption and environment impact.

The outcomes of the simulations are clear and easy to explain to stakeholders and decision makers, specially using the second prototype of the INDICATE too that is clearer in terms of dashboard features, selections and options.

Figure 19. The energy consumption of the new hospital with "green" interventions, together with the existing refurbished hospital (lot2) and the "Sector 3" buildings

Figure 20. The energy comparation with no interventions, (Energy) and with interventions (Energy after), scenario 4 together with existing building refurbished (lot 2) and with SECTOR 3

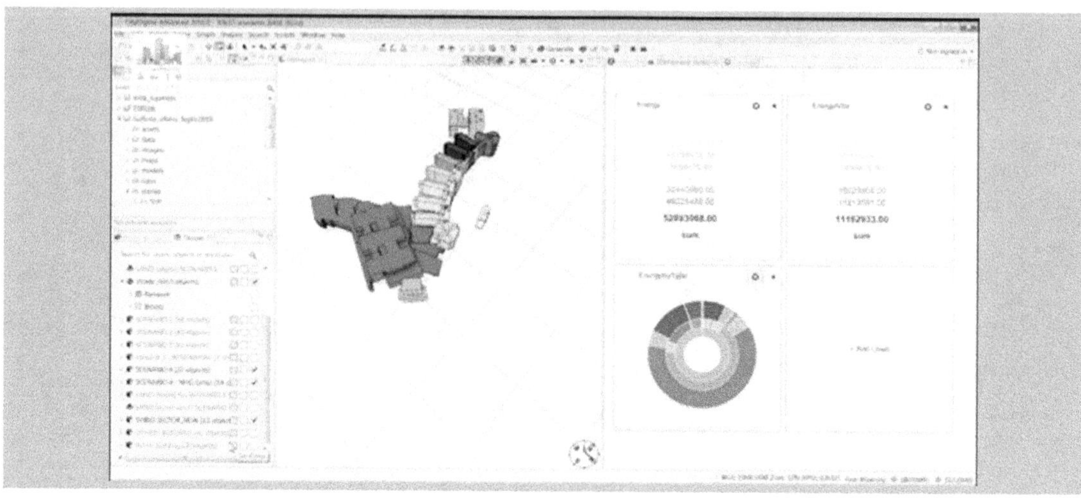

The first stage of the scenario creation requires more time than the intervention selection and results comparison. This first task requires an experienced modeler with very good technical skills and competences.

In the case of Galliera hospital this task was accomplished by an expert CAD 3D and BIM modeler, after a first trial time with the CityEngine tool of about 1 week, with some on line helps from ESRI. The objects modeled are less than three hundred, some of them created using copy and paste functions in different layers.

On behalf Galliera, we think that additional work can be done in terms of interoperability for the future. Both for BIM interoperability (IFC files) and for free on line 3D city models as Google Maps or similar. With those integrations probably the implementation of the basic scenario that a user needs to start using INDICATE tool will be easier and faster, becoming user friendly as the successive tasks.

Beyond INDICATE, the New Galliera BIM Tender

Thanks to the experience gained with the interoperable management of BIM models during the INDICATE project, the Galliera technicians have decided to found the subsequent phases of the New Galliera project on the use of the BIM methodology for the design, construction and maintenance of the new building.

An Employer Information Requirements (EIR), a BIM PLAN and an Asset Information Requirements were therefore prepared to form the basis of the tender for the subsequent phases of final design, executive, construction and management.

The tender was published in August 2017, and was the highest-amount BIM call for E&A activities that appeared that year.

During the preparation of tender documents in Italy, the first regulatory indications for the mandatory and progressive introduction of BIM in public procurement were being outlined, with the issue of Ministerial Decree 560/2017. At the same time, also in 2017, the guidelines for the BIM management of building and infrastructure projects were being defined in Italy, with the issue of the first UNI 11337 standards.

Due to the fact that the Italian regulatory framework was not yet complete at the date of publication, the Galliera technicians preferred to refer to the UK regulations, in particular the BS and PAS 1192 framework (PAS 1192-2:2013, 2013). This choice proved useful when the new ISO 19650 was published, as Galliera BIM tender documents were already improved with Asset Information Model (AIM), Asset information Requirement (AIR) and Project Information Model (PIM) implementation.

The documents prepared by the Galliera technicians thus flank the model of the new building prepared by the designers of the previous phase, Pinearq and associates, who developed a model that was also useful for the simulations performed with INDICATE during the project. The model was made available to competitors in IFC format and previously subjected to validation procedures with the use of the Solibri Model Checker®.

The call drawn up by Galliera was highly innovative for the Italian regulatory framework as the use of BIM was not left as optional but imposed as mandatory. Participants were asked to respond to the call with an information management offer, Building Execution Plan (BEP) to be developed on the basis of a guide model provided in the tender. The BEPs provided by the competitors were thus judged in the selection phase of the successful bidder on the basis of a substantial series of thirty-two criteria that can be grouped into eight categories:

- BIM contract / Contracting authority involvement
- BIM deliveries
- Data, verifications and validations
- Collaboration processes
- Viewing / stakeholder engagement
- Authoring (creation) of models according to different disciplines
- Construction
- Model-based evaluations / estimates and management of variants

The declared objective was to select not the best competitor in managing a BIM project but who was able to design complex hospitals such as the Nuovo Galliera, within an area full of buildings and constraints, including heritage, which knew how to use BIM processes to improve design processes and projected these processes already at the construction and maintenance management phase of the building.

The contract was awarded to the Politecnica association with other companies.

The project and its models delivered in September 2019 were subjected to a verification and validation process and then approved by the control bodies of the municipal administration of Genoa, the Fire Brigade and the Ministry of Cultural Heritage with the release of the building permit.

For the subsequent construction and maintenance phase, a similar body of documents is now being prepared consisting of AIR, EIR, BEP for the builder and maintenance engineer of the building based on PIMs models in IFC format created by Politecnica.

As a legal basis for the preparation of this documentation, we can now refer to the Italian standards (UNI 11337, 2017) and above all to the recently published new international ISO 19650 standard, also translated into Italian (parts 1 and 2).

Applications to Gaslini Children Hospital

What has been learned so far can also be applied to another hospital, even in the absence of a large new construction project such as that of the New Galliera, the authors aim to apply the BIM methodology to the Giannina Gaslini Children Research Institute in Genoa.

First of all, it can be highlighted how both the BIM approach and the parameterization of the models at the masterplan level typical of INDICATE project can be useful if they proceed in parallel.

In fact, the idea is to start facing the digitalization of Gaslini from two sides.

On the one hand, to create a masterplan with parametric models, on the other hand, for individual specific design and construction interventions, to require the designers in charge of using the BIM methodology on the basis of ISO 19650 requirements (consisting in preparation of AIR, EIR and BEP, for example). At present, the first steps moved towards digitalization within Gaslini hospital comprehend the two following renovation projects: the new gastro-surgery department and the expansion of the emergency room, that will be developed applying BIM methodology.

In addition to the already typical elements of INDICATE (type of walls, curtain walls and windows, Heating Ventilation and Air Conditioning (HVAC) systems, electrical systems, data transmission systems), other useful data for hospital management must be included in the masterplan, such as the intended use of healthcare in detail (departments, areas of high intensity of care and areas of medium and low intensity, outpatient and rehabilitation, laboratory services and clinical research, fire prevention, seismic adaptation). The aim is to model different simulation scenarios applied to entire buildings or to departments. These scenarios will be implemented with necessary information parameters to develop time schedule planning, in order to eventually highlight interferences and opportunities in the management of multiple construction sites within the hospital complex.

As required in ISO 19650, Gaslini Institute is equipping itself with BIM tools and training for its technicians to proceed with the creation of a parametric masterplan to evaluate future developments.

The BIM key Importance for Managing Existing Assets: the ISO 19650 Approach and Possible Implementation Towards the Actual Gaslini Hospital

The new international (ISO 19650 part 2 and part 5) and national (DM 560/2017) regulations underline the key importance of developing digital methods and tools for managing existing assets.

ISO 19650 is a high-level framework that can be applied to all types of assets and by all types and sizes of organizations, regardless of the procurement strategy used. Its flexibility allows it to be adapted to projects of a wide range of scale and complexity. A key principle of the standard is clearly defining information requirements at project inception. This provides the clarity and certainty all parties need to direct their efforts effectively with the minimum of confusion, cost and time overruns, reworks, waste and conflict. It also allows clients to verify whether their requirements have been satisfactorily met.

Iso 19650-5 sets out requirements for the security-minded management of sensitive information within building information modelling (BIM).

A BIM oriented appointing party, as Galliera hospital was, and as Gaslini hospital is going to be, should clearly identify requirements for the security-minded management of sensitive information within building information modelling (BIM). From this point of view, one of the most important requirement for existing assets BIM managements is the "Level Of Information Need" definition, that is directly connected with the appointing party objectives. In the past, with drawings, the 'granularity' of information was determined by scale or by a person's capacity to draw and to read a line on a drawing. Standards such as BS (British Standard) then specified the information to be attached. With digital information, however, there is no physical limitation. The purpose drives the information to be embedded in the model. The level of information need is therefore developed to assure that the information is implemented in the model at the time it is needed to fulfil its purpose.

For example, in the creation of the BIM model for space management of an existing hospital, such as Gaslini hospital, the Level of Information Need should be defined according to the following goals:

- Creation of models as space usage repositories, update and management
- Setting up a database as a single source of truth regarding geometrical dimensions
- Setting up a repository of point cloud surveys (when available)
- design for and validate spatial program requirements for Gaslini properties
- leverage the use of building information modeling (BIM) and the interoperability among BIM software to optimize and automate the checking of model integrity and design performance relative to the applicable standards
- assist design and construction teams in producing and updating BIMs as part of their submissions in case of new design and construction contracts for new construction and major modernization projects.

The purpose of having information requirements is to specify the information that the asset or project supply chain members need to deliver as part of their work. They will be stored in the asset information model or project information model respectively.

Information Requirements management represents one of the main concepts of ISO 19650: BIM is a process of information delivery in accordance with requirements management coming from the systems engineering approach, developed for other industries. It includes:

- Organisational Information Requirements (OIR) – those pieces of information needed to answer or inform high-level strategic objectives within the asset owner/operator in relation to the built assets owned, operated, used or managed by them.
- Asset information requirements (AIR) – those detailed pieces of information needed to answer the organisational information requirements.
- Project information requirements (PIR) – those pieces of information needed to answer or inform high level strategic objectives within the asset owner/operator or project client organisation, in relation to a particular built asset project.

A BIM oriented organization can make the most of BIM implementation developing its own OIR and AIR that can give birth to AIM, the asset information model that can be used as the single source of truth for asset management, only if specific requirements have been defined.

The implementation of BIM as a methodology gives the best result by far when the quality of the information deliverables is improved and when the collaboration is enhanced in a structured way and integrated into the project management process. It is essential that the project management process remains the same as always, with all necessary underlying knowledge and experience. BIM brings additional tools, metadata and concepts to the same fundamental traditional processes of design, planning, tendering, execution, communication and coordination. This naturally leads to improvements in reducing financial losses, saving time, improving quality and mitigating risks. The BIM approach supports the agile and lean principles for better planning of both inputs and deliverables. It improves communication and interoperability between different stakeholders and different project phases and leads to an integrated approach for changes and optimizations.

ISO 19650 explains processes in general terms. However, there are some key issues for the engineering and construction phases of BIM-enabled projects of which BIM appointing parties project managers should be aware that the required process must allow for developments in the project and any changes this creates to the deliverables, and must allow interoperable information flows.

Gaslini technical chief has also the role of the BIM manager that, associated to the appointed party in the contract, includes the processes of BIM management with the responsibility of the BEP, the information management, the information requirements management, the information workflow management and the information interfaces management. Considering this, even if it is not a full-time activity, a member of the appointing party team will be appointed for these responsibilities, especially concerning the importance of defining the process for exchanging information between all stakeholders

The frequency and quality of the information exchange and sharing is crucial to the success of a BIM implementation and should be brought in line with the expected deliverables. In order to ensure fast convergence of the BIM model towards a final design that complies with requirements, and ultimately with the reality of the asset, it is advisable to only share the information that is necessary to be shared at that particular moment, with other stakeholders. The BIM management is to organize the following tasks, and their implementation under the BEP (BIM Execution Plan), that is a crucial document for BIM implementation which focuses are:

- to merge all discipline models
- to check the compliance of the models according to the BEP or other agreement
- to check the evolution of the models and their compliance with the scope of work
- to check for clashes between the models
- to check the correct Level of information needs adoption
- to check the correct hardware and software required adoption

As it is well known, the BIM process creates greater communication and visibility around project progress for all stakeholders and so supports project management in a positive way. However, it requires a mind-set of sharing and transparency that is not always common in traditional projects, nor amongst many stakeholders across the whole construction sector, and especially within appointing parties. This is one of the reasons BIM implementation is hard to be developed within public administrations.

Open minded realities (like Gaslini one) recognizing BIM advantages can really take a great advantage in managing data and in growing professional experiences.

During the operational or maintenance phase, the appointing party need is getting more and more about enriching the BIM model with additional information and using this information for innovative approaches (Internet of Things, Smart Buildings, Smart Cities, etc.) and maintenance or for operational purposes.

In summary, effective BIM implementation requires a lean, start-with-the-end-in-mind approach, which involves project requirements, delivery and contract type, the number of stakeholders and existing infrastructure and buildings. Information which is entered and not used is considered waste, but too little information can inhibit some later BIM uses and analyses

Every participant in the supply chain should adhere to the information handover principles for the BIM model and related deliverables, to enable reuse of this data for participants further along the project pipeline and especially for Asset information management.

It is worth mentioning that this process also requires a much closer involvement of the appointing party during all project phases, and specifically regarding approval after each phase which is necessary for being able to converge towards a final design. This is why experienced BIM managers, BIM coordinators and specialist should be within the appointing party technical staff. With an agreed source of knowledge in the form of a BIM model, information is better communicated and remains within the knowledge base even after stakeholders who have implemented this knowledge have finished with their involvement in the project. Appointing party technical BIM staff should for this reason develop templates and BIM guides to uniform models and create repeatable processes.

This is also crucial for interoperability, that should be available for both the baseline data (BIM model, through the ISO 16739:2013 Industry Foundation Classes standard), analyses results (clash detection, cost simulations BIM5D, construction simulations BIM4D) and information management.

Interoperability can be used to share information among different hospital departments and different professionals. With well organized shared information, nurses, doctors and any other professional involved on hospital space management, for example, can use BIM tool for maintenance, asset, space management and to develop LEAN processes.

REFERENCES

BuildingSMART Website. (n.d.). https://technical.buildingsmart.org/standards/ifc/

Decreto Ministeriale n. 560, 01/12/2017

Del Giudice, M., Osello, A., Fonsati, A., De Luca, D., Musetti, A., & Marchi, F. (2017). District data management, modelling and visualization via interoperability. Proceeding of Building Simulation 2017.

Grey, T. (2016). INDICATE-Indicator-based Interactive Decision Support and Information Exchange Platform for Smart Cities- Project Outline. In Planning, Climate Change and Energy in Ireland - Irish Planning Institute Guide. Academic Press.

INDICATE. (2016). *Project website*. https://cordis.europa.eu/project/id/608775 retrieved at 8/02/2021

Melia, A., Nolan, E., & Kerrigan, R. (2015). INDICATE: towards the development of a virtual city model, using a 3d model of Dundalk city. *Proceedings of International Conference CISBAT 2015 "Future Buildings and Districts – Sustainability from Nano to Urban Scale.*

Musetti, A. (2016). *Deliverable 7.8. Third Dissemination Event*. INDICATE Project.

Osello, A. (2012). *The Future of Drawing with BIM for Engineers and Architects*. Dario Flaccovio Editore.

PAS 1192-2:2013 (2013). Specification for information management for the capital/delivery phase of construction projects using building information modelling

Romero, A., Izkara, J. L., Mediavilla, A., Prieto, I., & Pérez, J. (2016). Multiscale building modelling and energy simulation support tools. *Proceedings of the 11th ECPPM*, 315-322.

Sicilia, A., Madrazo, L., & Costa, G. (2015). *Building a semantic-based decision support system to optimize the energy use in public buildings: the OPTIMUS project. Sustainable Places 2015 event*. Savona.

UNI 11337 (2017). Edilizia e opere di ingegneria civile - Gestione digitale dei processi informativi delle costruzioni.

UNI EN ISO 19650 -1(2019). Organization and digitization of information about buildings and civil engineering works, including building information modelling (BIM) — Information management using building information modelling. Parte 1: Concetti e principi.

UNI EN ISO 19650 -2 (2019). Organization and digitization of information about buildings and civil engineering works, including building information modelling (BIM) — Information management using building information modelling. Parte 2: Fase di consegna dei cespiti immobili.

Chapter 25
Digital Twin for Maintenance Information Management:
Scenarios and Perspectives for Sustainable Smart Cities

Umberto Mecca
Politecnico di Torino, Italy

Giuseppe Moglia
https://orcid.org/0000-0002-5355-2154
Politecnico di Torino, Italy

Francesco Prizzon
Politecnico di Torino, Italy

Manuela Rebaudengo
Politecnico di Torino, Italy

ABSTRACT

BIM and the construction sector have long been an inseparable pair: in many European Countries it is a consolidated practice while in others there is a big debate about it but. Although there are important signs of a turning point, it still does not seem to be really feasible. In Italy, the leverage for transformation is certainly the public works sector, but this is not enough: to really talk about a whole digitization of the AEC sector, it is necessary to "attract" the private sector, where more than 50% of investments are invested. The chapter will attempt, starting from an Italian framework overview, to assess the main perceived obstacles of the applicability of a BIM model for facility management. The tool chosen by the authors as a preliminary approach to the problem, the SWOT analysis, allows an effective synthesis of the strengths and weaknesses resulting from such implementation.

DOI: 10.4018/978-1-7998-7091-3.ch025

INTRODUCTION

BIM and the construction sector have long been an inseparable pair: in many European Countries it is now a consolidated practice while in others (i.e. in Italy) there is a big debate about it but, although there are important signs of a turning point (Frontera, 2019), it still does not seem to be really feasible. As it is well known, digitalization in the construction sector - which now mainly impacts on public works - has to do with a shared approach of information management in the entire process of design and construction of works. Integration and collaboration are thus in opposition with the old (sectorial) model of production of goods and services: the contract was divided into competence areas and the specialists worked in their respective one, carrying out their task in substantial autonomy, to the detriment of the internal coherence of the project. BIM methodology represents an important paradigm shift in the construction sector and its benefits in terms of process optimisation have also been reflected in the legislation in many countries: in Italy from 1 January 2019 it is mandatory to use electronic modelling methods and tools for the design and construction of public works of EUR 100 million or more.

At the moment the regulation provides a voluntary introduction without any limits of amount, in accordance with the decision of the contracting authorities, but at the same time it establishes the obligation to introduce BIM for subsequent steps: from 1/1/2019, if the amount exceeds EUR 100 million; from 1/1/2020, if the amount exceeds EUR 50 million; from 1/1/2021, if the amount exceeds EUR 15 million; from 1/1/2022, if the amount exceeds EUR 5.2 million; from 1/1/2025 also for public works of less than EUR 1 million.

As the construction sector is very slow to take up innovations (for example, the prefabrication or the introduction of new non-traditional building systems), the perception of the market and the structure of the operators is partially inadequate with regard to the standards required by law. The costs of training personnel/professionals and of implementing (performing) BIM services are still too high compared to the actual current benefits, both for production and management: not all professionals are aligned to the same production standard, with a clear reduction in the competitiveness of the sector; not all contracting authorities are able to manage and implement BIM systems; few companies are equipped with internal structures and trained in order to interact with the other actors in the process. [...] Compared to a traditional design, BIM systems are not yet competitive if not in a long-term vision and particularly linked to the impact of maintenance, both from an economic point of view and in terms of prevention of negative events for the whole system (the work being planned). (Moglia et al., 2019).

The driver of transformation is - through a regulatory imposition - the public works, but this is not enough: to really talk about a whole digitalization of the AEC sector, it is also necessary to attract the private sector, where more than 50% of investments are directed (ANCE, 2019).

How can we spread the BIM culture and not only from a theoretical point of view? This chapter will attempt to take stock of the main perceived obstacles, such as the static nature of operators, start-up costs and the need for training towards a new technical level dedicated in particular to management.

Specifically, in paragraphs 2 and 2.1 it has been reported the analysis carried out on the construction sector in Europe to identify the average structure of production factors and those on the facility management sector to highlight how this represents a real opportunity to relaunch the AEC market. Paragraph 3 contains an analysis of the use of BIM in the construction sector to provide a framework for its potential development. Paragraphs 4 and 5 show an analysis of the literature on Building Information Modeling (BIM) and Digital Twin (DT). Section 6 discusses the strengths and weaknesses of using BIM for four key types of individuals: Professional, Building Manager, Owner or Tenant, Operator. Lastly, some concluding remarks as well as possible perspectives for the development of thc topic are given in section 7.

THE CONSTRUCTION SECTOR IN ITALY

Although the future effects of the pandemic COVID19 and the United Kingdom's exit from the EU (Brexit) are not yet known, one of the main components for the development of the European economy is certainly the Construction Sector (CS). Indeed, while it directly contributes to the creation of value (about 9% of GDP in 2016; European Construction Sector Observatory, 2019a), it also promotes the growth of other sectors, such as industry and the tertiary sector, and involves the consumption of raw materials, products, equipment and services to meet its own productive needs (European Construction Sector Observatory, 2019a). Besides being a driving force for the economy, it is also one of the main sources of employment, indeed, in 2015 it employed more than 12 million people, about 8.8% of total non-financial employment in the EU, most of which, about 63%, came from specialized construction activities. Analysing the average composition of companies operating within the sector, its fragmentary nature emerges: about 94% of them employ less than 9 employees, while large companies representing less than 0.5% of the total (European Commission, 2019).

It is worth highlighting that in 2017 61.5% of the total workforce arising from the CS in the 28 EU countries (before Brexit) came from the summing of people employed in Germany, UK, France, Spain and Italy (European Construction Sector Observatory, 2020d). However, examining in detail the evolution of employment in these countries, we can observe totally different behaviours and in particular two groups can be outlined: Germany and the United Kingdom on the one hand; France, Spain and Italy on the other. Generally speaking, it can be stated that in the first group the employment in the CS increased between 2010 and 2018. Indeed, an increase of +35.2% (from 2,938.001 to 3,973.029 employees) in Germany and of +12.3% (from 2,753.944 to 2,452.168 employees) in the UK has been assessed (European Construction Sector Observatory, 2019c, 2020b). In contrast, employment in the second group contracted, in some cases very sharply. France in 2018 suffered a decrease of 2.2% compared to 2010 levels, due to a decrease in the number of employees in the construction and manufacturing subsectors (-4.3% and -7% respectively). However, this situation was partially balanced by increases in employment in the sub-sectors of real estate (+4.7%), architecture and engineering (+6.5) (European Construction Sector Observatory, 2020a). During the same reference period (2010-2018), the employment in the CS literally collapsed both in Spain and in Italy. As for the former, the drop recorded was about 15.7% and the most affected subsectors were the manufacturing and construction sectors and the architectural and engineering activities, which respectively decreased of 23.7%, 22.9% and 4.5% in the same period of time. The real estate sub-sector is the only exception: an increase of 39.7% was recorded (European Construction Sector Observatory, 2020c). In Italy, there was a 7.2% loss in the number of construction companies between 2010 and 2018 and concerning the production, the number of employees decreased by 21.1% (European Construction Sector Observatory, 2019b). In particular, in 2019 the National Association of Building Constructors (in Italy: "ANCE, Associazione Nazionale Costruttori Edili") reported in its annual report that workers in the Italian CS fell by 642,000 compared to pre-crisis levels. The same document also reports a National Institute of Statistics' survey (in Italy: "ISTAT, Istituto Nazionale di Statistica") according to which there were about 500,000 companies in the sector in 2017, i. e. about 11% of the Italian industrial and service system, with an average size of 2.6 employees for each. The fragmentation of the Italian CS is also confirmed by the average composition of the enterprises, indeed 62% of the total is made up of only one person while 34.2% includes from 2 to 9 employees in total. The remaining percentage, i. e. 3.8%, is composed of more structured companies employing more than 10 people. It is worth mentioning that only 0.2% of the total employs more than 50 people (ANCE, 2020).

As already stated in the paper "BIM and maintenance: real or virtual scenarios?" we can see that "the world of professionals also reflects a similar, fragmented reality: many small firms (on average 1.5 employees) and a few large companies which, however, attract about 70% of the resources at stake (CNI 2019) for technical design and project management services" (Mecca et al., 2019).

The same fate awaits the contracting authority as well, indeed "on the entire national territory there are about 36,000 employees (Oreto, 2018): if we exclude the Central and State Administrations, a recent survey (Giachi, 2016) defines an average number of employees equal to 8.3 [...], with peaks of 20 employees in the National Health System Companies and minimum values of less than 4 employees, in small municipalities associated with unions but not common leaders." (Mecca et al., 2019).

The direct consequences of this extreme fragmentation of the EU's CS is that micro, small and medium-sized enterprises quite often lack the human and economic resources necessary for the training of their employees (European Construction Sector Observatory, 2020d). However, this is not the only factor that undermines the development of this sector; indeed, its instability and unattractive working conditions discourage young people from pursuing a career. Moreover, precisely because of the lack of turnover, there is still a worrying increase in the average age of the workforce employed. On the other hand, it is estimated that in the coming years there will be a high demand for qualified staff, around 1 000 000 by 2025, and the most requested professionals will be those qualified in energy efficiency and digital technology (European Construction Sector Observatory, 2020d).

Today, the digitalization of the CS is a challenging issue, but it is considered as being able to strongly contribute to the development of CS itself, with regard both to the creation of new jobs and to the savings (in terms of costs to be incurred at the time of the interventions) that would result. For example, it is estimated that digitalization of the non-residential construction subsector could lead to overall cost savings of between 13% and 21% in the engineering and construction phases and between 10% and 17% in the operational phase (Gerbert et al., 2016).

Therefore, since automation and computerization of processes through an increasing use of Building Information Modelling (BIM) in CS could lead to greater efficiency and productivity, the European Commission has supported and developed several policies to promote it. However, the implementation of such policies in the different EU countries has been and still is very different: in general, we can affirm that Northern European countries have been the first to implement BIM policies, while Western and Eastern European countries are slowly equipping themselves (European Commission 2019).

In Italy, the digitalization of the CS began with the adoption of the Ministerial Decree nr. 560 of 2017. With this document, a plan has been issued that defines, by increasingly broad ranges of amounts, the progressive mandatory adoption of electronic modelling methods and tools for the design and construction of public works. According to this plan, by 2025 the prescriptions indicated in the decree will also have to be extended to public works with tender-based amounts of less than €1,000,000, in other words effectively to all works (Ministero delle Infrastrutture e dei Trasporti, 2017).

THE FACILITY MANAGEMENT, AS KEY PART OF THE CONSTRUCTION SECTOR

To foster the sustainable development of our cities and communities (SDG 11) and to mitigate the impacts of climate change (SDG 13), we need to modify the way we build and manage our urban spaces (United Nations, 2015, 2020). As shown by the International Energy Agency (IEA) in its "Global Status Report for buildings and Construction 2019", the CS is the main culprit for greenhouse gas emissions, in addition to being the most energy consuming sector (Figure 1) (Global Alliance for Buildings and Construction, International Energy Agency, 2019)

Figure 1. Global share of buildings and construction final energy and emissions, 2018 (Global Alliance for Buildings and Construction, International Energy Agency, 2019)

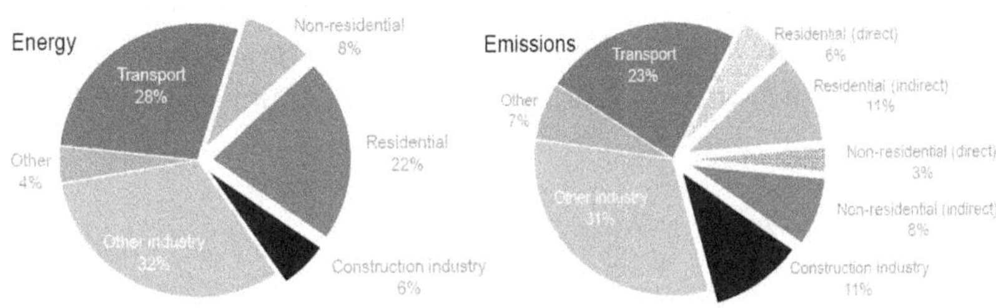

Therefore, in order to reduce emissions as much as possible and at the same time limit energy consumption, it is first necessary to act on the CS and in particular on its components.

A key role for this challenge is played by Facility Management (FM) services through the maintenance of buildings, neighbourhoods and cities. FM is defined by the International Facility Management Association (IFMA) as "a profession that encompasses multiple disciplines to ensure functionality of the built environment by integrating people, place, process and technology" (International Facility Management Association, 2020). Therefore, the FM includes all ordinary and extraordinary maintenance

Figure 2. Definition of FM services (GlobalFM, 2018)

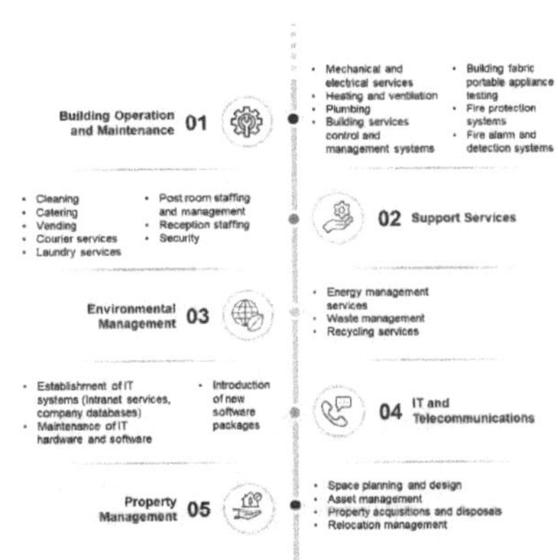

operations necessary to maintain, restore and/or improve the systems which buildings and infrastructures are composed of.

Forster & Sullivan, as shown in Figure 2, grouped FM services into five main spheres: Building Operation and Maintenance (BOM), Support Services (SS), Environmental Management (EM), IT and Telecommunications (ITT) and finally Property Management (PM) (GlobalFM, 2018).

Over the last few years, this sector increased significantly at a global level, as a result of the design of the different national strategies. They include, in particular, those that stimulate the creation of new green spaces within cities, those for the production of energy from renewable sources, those for the maintenance of buildings and those for the creation of routes and infrastructures for the development of alternative modes of transport such as public transport, electric cars and the use of "slow" vehicles. According to the "Global Facilities Management Market Report", globally, in 2017, the FM market was valued at $1.152 trillion and had recorded a growth rate of 7.4% between 2016 and 2017 (GlobalFM, 2018). This trend, over the same period of time, has also been confirmed at European level, for example the growth rates of the UK, Germany, France and Italy were respectively +0.8%, +1.5%, +1.2% and +0.7% [28]. These positive trends, as far as the housing sector is concerned, can be explained by observing the average age of European properties (Figure 3) (European Commission, 2020).

Figure 3. Breakdown of residential buildings by construction year (2014) (European Commission, 2020)

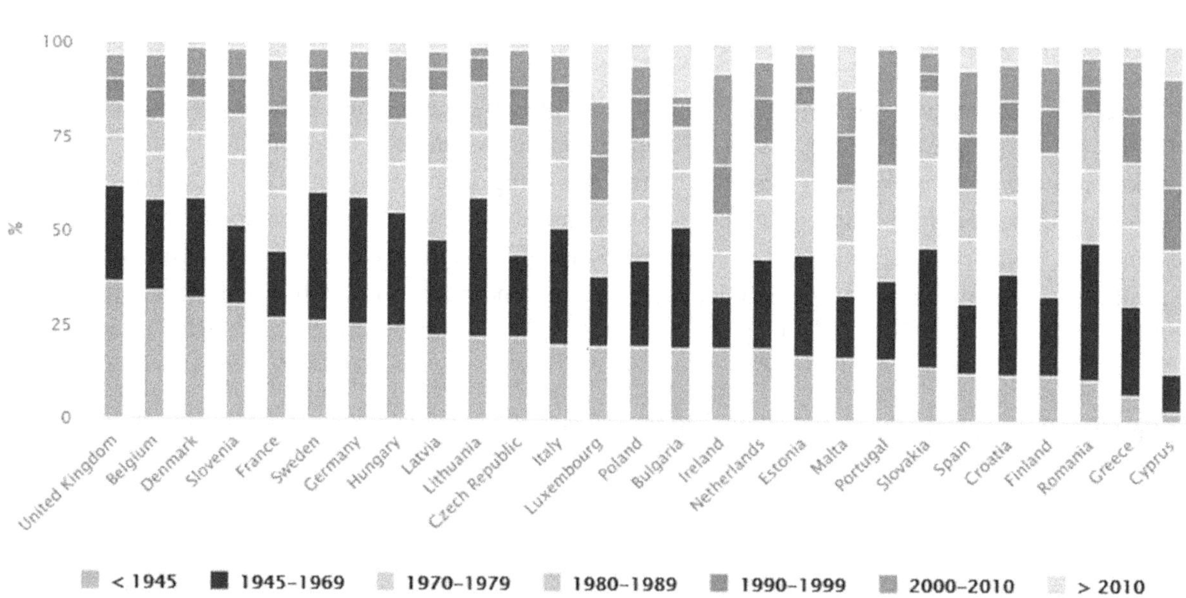

Indeed, more than half of these, in most European states, were built before 1970 and therefore potentially require extensive maintenance. In Italy there are about 60 million real estate units and 58% of them are for residential use (Mecca et al., 2020). Among the latter, about 55% are over 55 years old and, according to ISTAT data, 19.9% are estimated to be in poor state of conservation and 2.2% in very poor state (Mecca et al., 2019). For this reason, in the last few years the Italian government has been developing some incentive policies, establishing tax bonuses according to the type of interventions implemented, to stimulate structural and energy improvements in buildings. Before the pandemic due to the COVID-19 (years 2019-2020) in Italy, several incentive measures were active, including: the "renovation bonus" for restoration, rehabilitation and renovation works; the "ecobonus" for energy saving interventions; the "sismabonus" to ensure the safety of existing buildings in case of earthquake; the "green bonus" for the green upgrade of uncovered areas of existing buildings; the "façade bonus" for the restoration of the external facades of existing buildings (Agenzia delle Entrate, 2020a, 2020b, 2019b, 2019c).

These measures, have actually had a real impact on investment in housing upgrading, indeed, the latter accounting for about 37% of total investment in construction (ANCE, 2020).

It is worth mentioning that most of the interventions for the requalification of the operating heritage, in Italy, are classified as ordinary maintenance interventions (e.g. replacement of windows and doors, renovation of floors) or extraordinary (e.g. renovation of heating and cooling systems, structural consolidation, etc.) and therefore are included in the FM area. All maintenance interventions and FM in general could be considerably simplified if digital models of the buildings and infrastructures on which it is intended to intervene were available. Indeed, inside such models, all the information necessary for a detailed and real knowledge of the buildings would be available. Moreover, thanks to these models it would be possible to estimate a priori the state of a given asset and effectively plan its interventions. Since in Italy the sector of the requalification of the housing stock currently represents more than a third of the total investments in construction, it would be useful to start stimulating the use of BIM also in the private sector, starting with the drafting of guidelines (optional) to ensure a certain homogeneity and coherence of the data collected.

THE BIM MARKET FOR CONSTRUCTIONS

The birth and spread of BIM in the CS are relatively recent and are growing steadily, even though in different ways within individual national territories. A first global overview on the use of BIM was published in 2014, based on a survey carried out by McGrow Hill Construction in 2013 on a sample of 727 companies active in ten of the major construction markets including France, Germany, UK, USA and Canada, Brazil, South Korea and Japan, as well as Australia and New Zealand (European Commission, 2019; McGraw Hill Construction, 2014). The survey defined the degree of use of BIM methodologies according to the percentage of projects for which the interviewed companies had implemented them. In particular, four different levels of BIM implementation had been identified for that research. If the company had used BIM for less than 15% of the projects then it had implemented it lightly, if it had used it between 15% and 30% then it had implemented it on average, between 30% and 59% there was a heavy implementation and over 60% a very heavy implementation. The degree of use of BIM within the various countries investigated was determined according to this classification. In particular, it was determined that BIM had been used heavily or very heavily in 52% of the projects in North America, for 33% in European countries (France, Germany and the United Kingdom), for 30% in Australia and New Zealand, for 274% in Japan and finally between 25% and 30% in Brazil and South Korea (Figure 4) (European Commission, 2019).

According also to the 2013 survey, in almost all countries, more than 50% of companies said they used BIM for most of the commercial buildings, while the largest share of companies using BIM for the construction of institutional buildings was in the United States and the United Kingdom (Table 1).

Figure 4. Level of BIM implementation in the ten leading construction markets (European Commission, 2019)

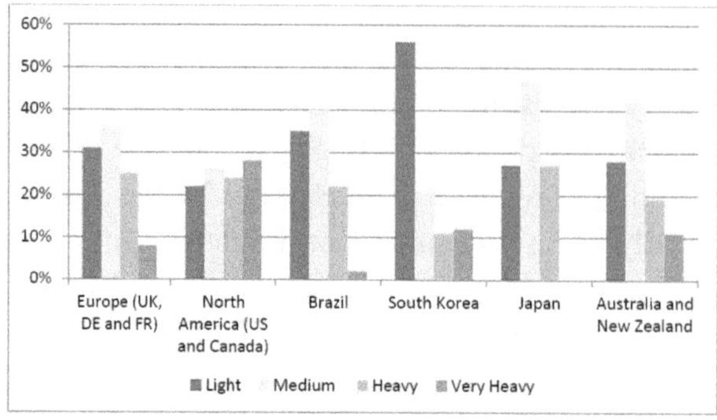

In 2016, the total value of the European BIM market was estimated at 1,8 billion euros and is expected to grow by a further 13% by 2023 (European Construction Sector Observatory, 2019a). Certainly, among the driving forces of BIM development we can find: the increasing complexity of projects and the increasing number of stakeholders involved in the processes; the need to effectively use the material and financial resources available, avoiding unnecessary waste; the strategies implemented by individual countries for the development of the digitalization of the construction sector. On the other hand, what limits the development of BIM is the structure of the construction market and its extreme fragmentation. The proof is given by the fact that the greater use of BIM is found among large companies, be they professional firms or contractors, rather than among small ones. There are many reasons for this, but the availability of human and economic resources of large companies, as well as the fact that they work more frequently on large complex works, are certainly a major factor (European Construction Sector Observatory, 2019a).

With regard to the Italian case, as mentioned above, the use of electronic modelling methods and tools for design and construction became mandatory only in 2019 and only for public works whose amount

Table 1. Share of companies using BIM per construction project type (European Commission, 2019)

Construction projects type	France	Germany	UK	US	Canada	Brazil	Japan	South Korea	Australia New Zealand
Commercial buildings (offices, retail, hotels)	68%	59%	69%	66%	54%	53%	63%	48%	70%
Institutional buildings (Education, healthcare, religious)	32%	31%	61%	77%	41%	31%	23%	35%	39%
Government / Publicly owned buildings (Courthouses, embassies, civic/sports and convention)	10%	22%	54%	68%	44%	12%	0%	51%	37%
Multifamily residential buildings	35%	44%	33%	18%	26%	19%	23%	20%	26%
Single-family residential buildings	19%	22%	17%	1%	10%	16%	0%	1%	4%
Infrastructure (roads, bridges, tunnels, dam, water/waste management)	19%	16%	33%	14%	31%	28%	13%	24%	25%

was over 100 million euros. Starting from 2019 the amounts of works for which the use of these methodologies is mandatory are expected to gradually decrease until 2025 when they will be compulsory for the whole public works sector (Ministero delle Infrastrutture e dei Trasporti, 2017). According to the association of architectural engineering and technical-economic consulting organisations (in Italy: “OICE, associazione delle Organizzazioni di Ingegneria di architettura e di Consulenza tecnico-Economica”), although 2018 was characterised by a decrease in calls for Architecture and Engineering Services (AES) and calls for design-only services, the AES calls for BIM have nevertheless registered a strong increase, with a peak towards the end of the year, when the State Agency (in Italy “Agenzia del Demanio”) published 90 calls. Despite a sharply upward trend in calls for tenders for BIM (+222.9% from 2017 to 2018), the ratio between these and the engineering and architecture services market recorded in 2018 sees BIM calls for tenders only reach 4.4% of the total, demonstrating that in Italy the development of BIM is only at its beginnings (OICE, 2019).

The activities entrusted with the 268 BIM calls for proposals in 2018 involved 46.3% design activities, 45.1% earthquake safety assessments and structural checks, 0.7% design verification services, 1.5% works supervision and 6.3% other technical services. In addition, analysing them by type of intervention, calls for tenders for renovation were 74.3% of the total (199 calls), while those for new projects were only 69, i.e. about 25.7% of the total (OICE, 2019). This trend is exactly similar to that of the housing sector and gives priority to maintenance work, renovation and redevelopment interventions over those of new construction. Finally, as far as the origin of the 2018 BIM calls for tenders is concerned, more than three-quarters of them come from State Administrations, Municipalities and Universities with different percentages, respectively 54.1%, 17.1% and 7.8% (OICE, 2019).

Figure 5. Number of indexed publications (source: Scopus Database; key: BIM, 2000-2020)

Year	Documents
2020	797
2019	1513
2018	1341
2017	1199
2016	992
2015	878
2014	800
2013	726
2012	570
2011	443

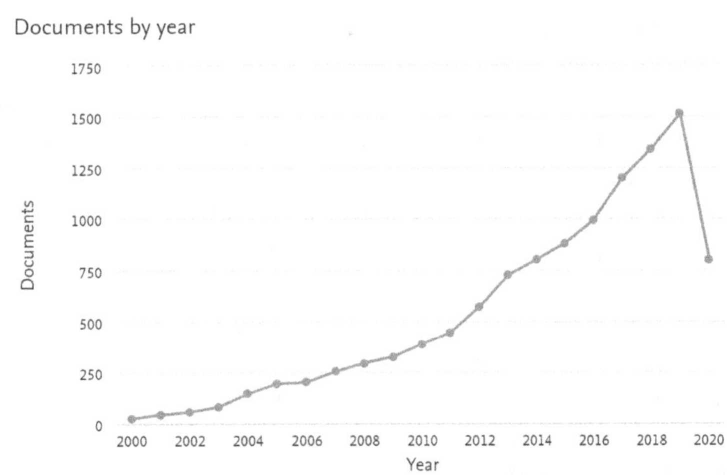

Figure 6. Countries where the authors of indexed publications are from (source: Scopus Database; key: BIM, 2000-2020)

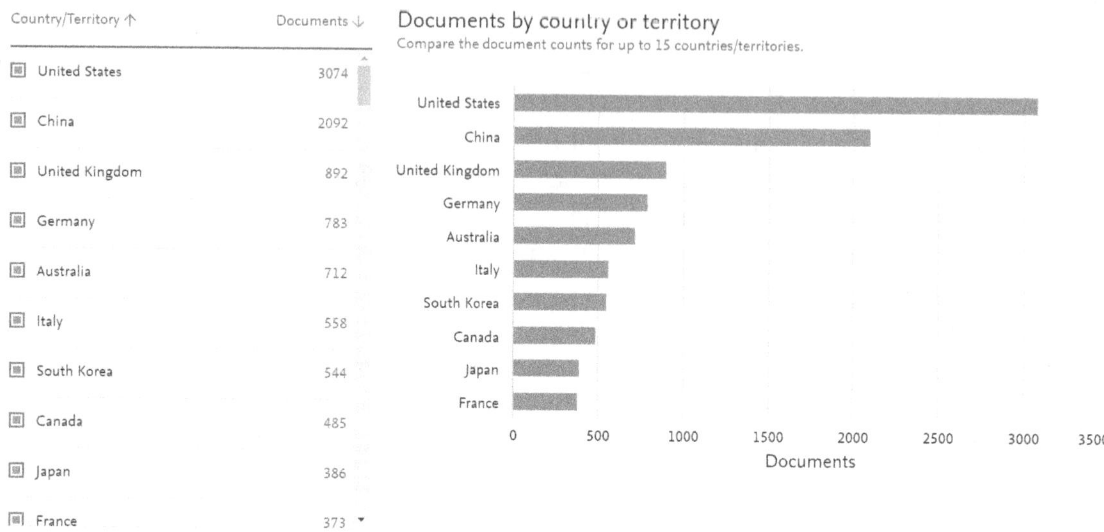

BIM TOPIC IN THE SCIENTIFIC RESEARCH

What dimensions does the scientific production on the topic of Building Information Modeling have? Usually the research anticipates by a few years the applications and operational practices, which is why it was decided to complete the framework characterization by introducing also this dimension. By simply searching for the acronym "BIM" among the keywords of all the indexed articles on Scopus database, 11,262 results are obtained (proceedings, articles in journals or book chapters), with a rapid evolution and diffusion of research from the early 2000s to today (Figure 5) and a clear prevalence of American researchers (Figure 6).

If we exclude the pharmacological/biochemical and related fields (in which we find the acronym BIM among the keywords, but to indicate a protein), the sample is little more than halved and still significant (over 6,200 articles), distributed in thematic areas with a prevalence in the engineering sector (almost 3/4 of the whole), followed by information technology (over 20%; Figure 7). Applications in less expected areas (environmental sciences, agriculture, earth sciences, art and humanities) cover less than 10% of cases and have focused on infrastructure (agriculture and earth sciences), historical-architectural-monumental heritage (art and humanities) and integration with the life-cycle approach (environmental sciences).

If we intend, instead, to cross-reference the theme with the maintenance of existing assets, we can repeat the same analysis by searching among the keywords both the acronyms "BIM" and "FM" or "Building Information Modeling" and "Facility Management"; in this way we obtain (only) 244 results distributed as follows (Figures 8 and 9):

The time frame is decreasing (since 2006/2008 and in a "consistent" way in the last 2-3 years) and Italy is the first European country with the most active authors: this appears to be in line with national policies to encourage the reduction of land consumption and the reuse of existing heritage, although another numerical impact would have been expected (13 cases out of the total, i.e. just over 5%). Why,

Figure 7. Indexed publications by subject areas (source: Scopus Database; key: BIM, 2000-2020, no pharma&chemistry)

Subject area ↓	Documents ↓
Agricultural and Biological Sciences	48
Arts and Humanities	190
Business, Management and Accounting	633
Computer Science	2187
Decision Sciences	237
Earth and Planetary Sciences	409
Economics, Econometrics and Finance	19
Energy	314
Engineering	4545

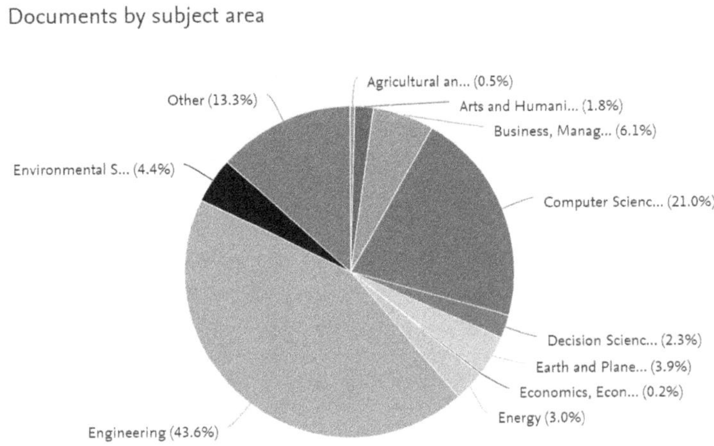

then, given the large number of existing buildings in our country, is this not yet the focus of the debate of the scientific communities?

Turning the field of observation to the thematic areas which the 242 publications refer, we note that - as expected - more than 2/3 of the publications address the Engineering, Computer science and Business and management sectors (Figure 10).

Figure 8. Number of indexed publications (source: Scopus Database; key: BIM&FM)

Year ↓	Documents ↑
2021	1
2020	22
2019	47
2018	38
2017	20
2016	25
2015	19
2014	18
2013	17
2012	9

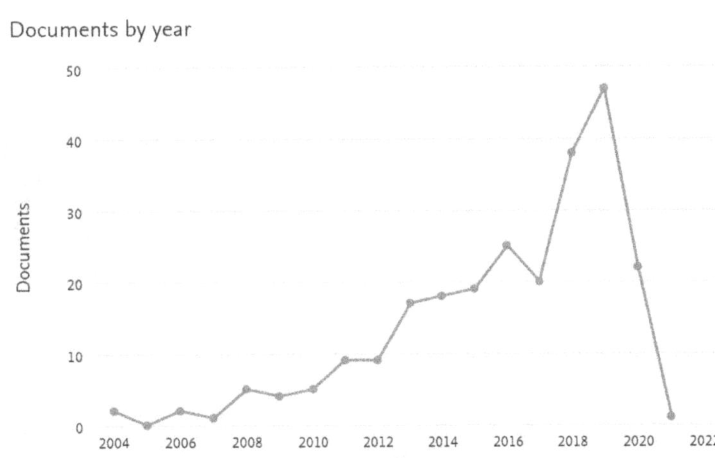

Figure 9. Country from which the authors of indexed publications are from (source: Scopus Database; key: BIM and FM)

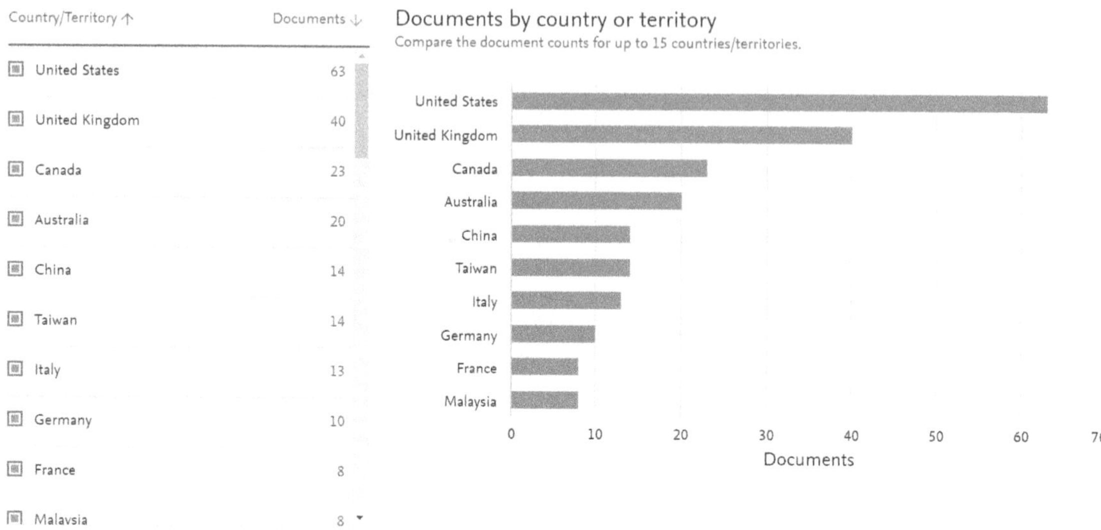

Another investigation with an interesting outcome is the one that associates the Digital Twin topic to BIM: repeating again the research, we obtain 47 results: a much smaller but rapidly growing number and in a short reference period (last 4 years) (Figure 11); this clearly indicates that the Digital Twin issue has only recently been added to the construction sector, even though it had been already widely present in other fields (aeronautics, industrial production, ...) since 2002 (Moglia et al., 2019).

Regarding the countries where the authors of indexed publications are from, Italy is second - in the EU - only to Germany (Figure 12) and once again the main field is Engineering, followed always by

Figure 10. Indexed publications by subject areas (source: Scopus Database; key: BIM and FM)

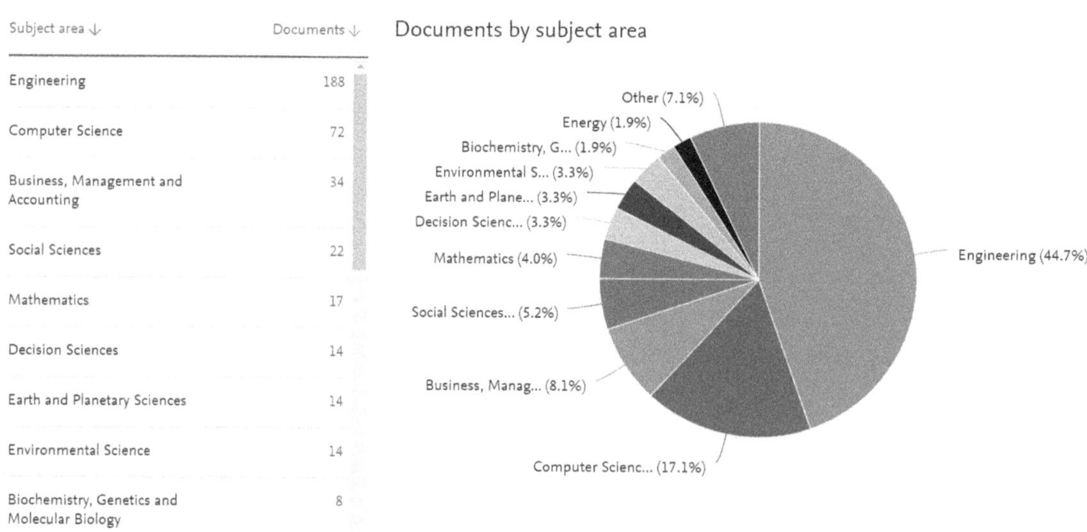

Figure 11. Number of indexed publications (source: Scopus Database; key: BIM&DT)

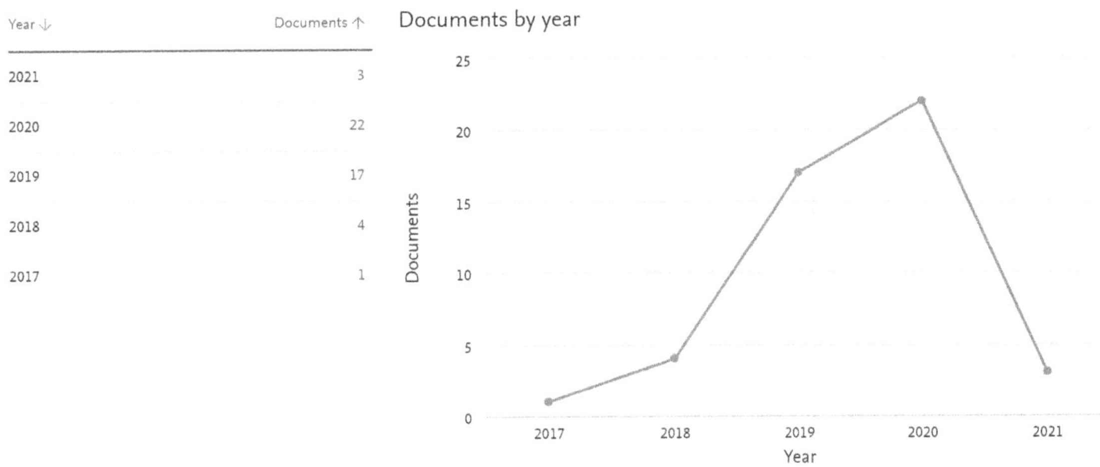

the IT area and then Social and Environmental Sciences (which strangely outnumber the economic/ management sector; Figure 13).

From this brief literature review, it is assumed that the use of BIM models as Digital Twin will increasingly animate the scientific communities, extending to other scientific fields, as it has already happened in the past for the topic of the Facility Management (FM). For now, the reduced reference sample (the 47 cases mentioned above) aims to enhance the positive aspects - as in all expansive phases - and not yet the critical aspects of application, as it perhaps requires a longer period to test data, reliable procedures of simulations. Moreover, although BIM is born and develops with different premises

Figure 12. Countries where the authors of indexed publications are from (source: Scopus Database; key: BIM and DT)

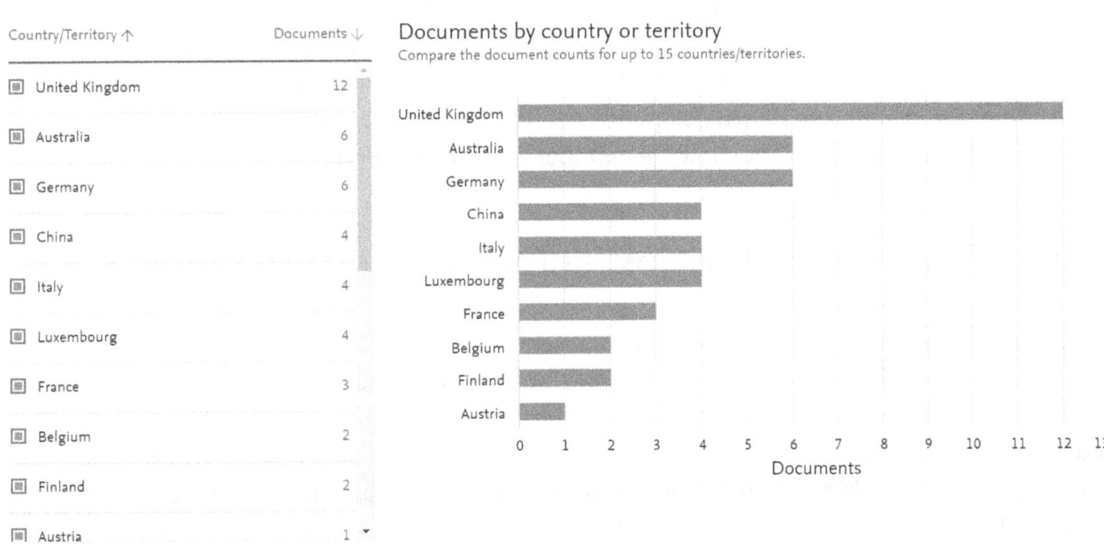

Figure 13. Indexed publications by subject areas (source: Scopus Database; key: BIM and DT)

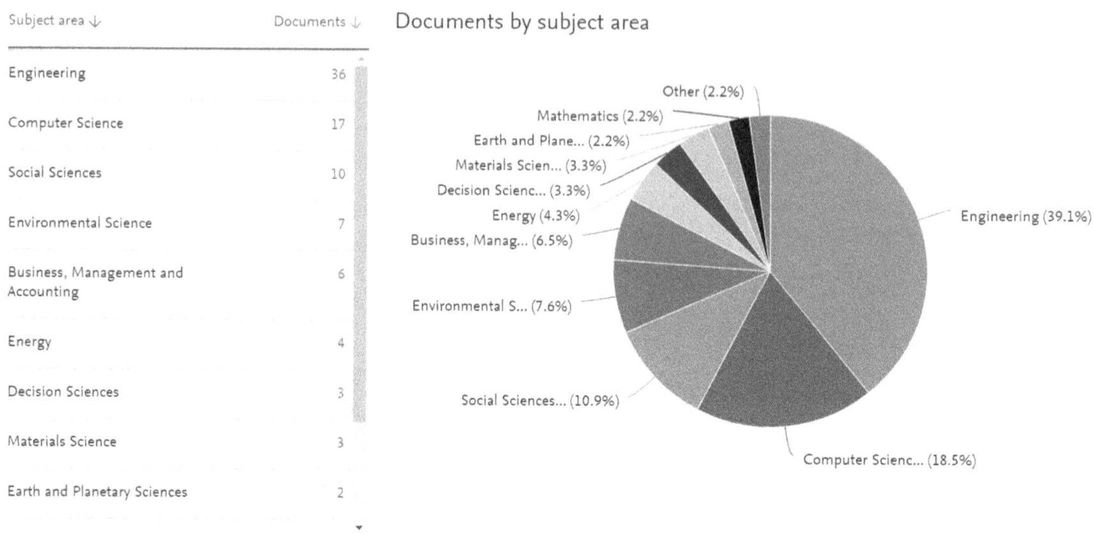

and objectives compared to DT, depending on the macro-phase of the building process, some points of convergence can be observed. During the design phase of a product, the DT constitutes the storage of all the information useful for its realisation and allows to carry out the necessary simulations to verify its correct functioning during its useful life: in the building sector these characteristics can be pursued by BIM tools and methodologies. The DT of an existing physical product, on the other hand, describes a specific correspondence between what has been achieved and its virtual version, moving towards the prevailing time horizon (that of management) (Moglia et al., 2019). Therefore, while waiting for a more numerous scientific debates on the DT, the following in-depth studies will focus on the general theme related to maintenance, i.e. BIM&FM. To understand how much the topic is still "operatively" to be explored, we have chosen to deepen the bibliographic research by looking in particular at all those open access publications that were published in the last three years (2020, 2019, 2018). We analyzed all these articles to understand if their main topic was new construction, facilities management, or was general in nature. Moreover, to understand the applicability level of the research, we classified the findings depending on the presence of a case study or not (Table 2). The Figure 14 shows four different graphs: the one on the left shows the percentage of articles published by topic (new construction, facility management, general), while the ones on the right show the number of articles by type.

DIGITAL TWIN & DATA COLLECTION: SCENARIOS AND KEY FIGURES

The life of an AEC product can be measured over centuries: how has it changed, how does it change and how will the modeling process, either analogic or digital? The search for the solution must be experimented starting from simple systems, for which the ideation, design, implementation, operation and end-of-life process can be modelled, with the aim of defining behavioural rules for each actor in the process in its relationship with the model that is formed and with the information that this model produces. In the AEC there is already a twin that is not digital but on paper and for this reason is not

Table 2. Review of articles published in the last three years (2020, 2019, 2018) and that are open access classified by topic and type (source: authors elaboration)

No.	Authors and Year	Topic	Type
1	Moretti, N. et al (2020)	Facility Management	Application
2	Lennartsson, M. et al (2020)	New Construction	Theoretical
3	Liu, Z. et al (2020)	Facility Management	Application
4	Kaewunruen, S. et al (2020)	General	Application
5	Wahbeh, W. et al (2020)	General	Application
6	Xie, X. et al (2020)	Facility Management	Application
7	Boje, C. et al (2020)	General	Application
8	Akbarieh, A. et al (2020)	General	Theoretical
9	Al-Saeed, Y. et al (2020)	New Construction	Application
10	Dou, S.Q. et al (2020)	General	Application
11	Lu, Q. et al (2020)	Facility Management	Application
12	Huynh, D. et al (2020)	Facility Management	Application
13	Poli, T. et al (2020)	Facility Management	Application
14	Khalil, A. et al (2020)	Facility Management	Theoretical
15	Kaewunruen, S. et al (2019)	Facility Management	Application
16	Roxin, A. et al (2019)	General	Theoretical
17	Schimanski, C.P. et al (2019)	New Construction	Application
18	Lu, R. et al (2019)	General	Application
19	Jouan, P. et al (2019)	Facility Management	Theoretical
20	Kaewunruen, S. et al (2019)	Facility Management	Application
21	Khajavi, S.H. et al (2019)	Facility Management	Application
22	Tchana, Y. et al (2019)	Facility Management	Theoretical
23	Antonino, M. et al (2019)	Facility Management	Application
24	Vivi, Q.L. et al (2019)	General	Application
25	Stojanovic, V. et al (2019)	General	Application
26	Lu, R.D. et al (2019)	General	Application
27	Tahmasebinia, F. et al (2019)	Facility Management	Application
28	Kaewunruen, S. et al (2018)	General	Application
29	Papadonikolaki, E. et al (2018)	General	Theoretical
30	Mayer, H. (2018)	General	Theoretical

less functional. For the existing heritage, the non-digital twin is also extremely important, because - as said before - construction often has a centuries-long life, and not a short useful life as it happens for technologies or production systems. Especially with a view to digitizing information on the existing heritage, it would be utopian to think about making all BIM models from scratch: it can be modelled by parts, as and when maintenance purposes or a "building booklet" is required.

Figure 14. Results of paper in-depth analysis (source: authors elaboration)

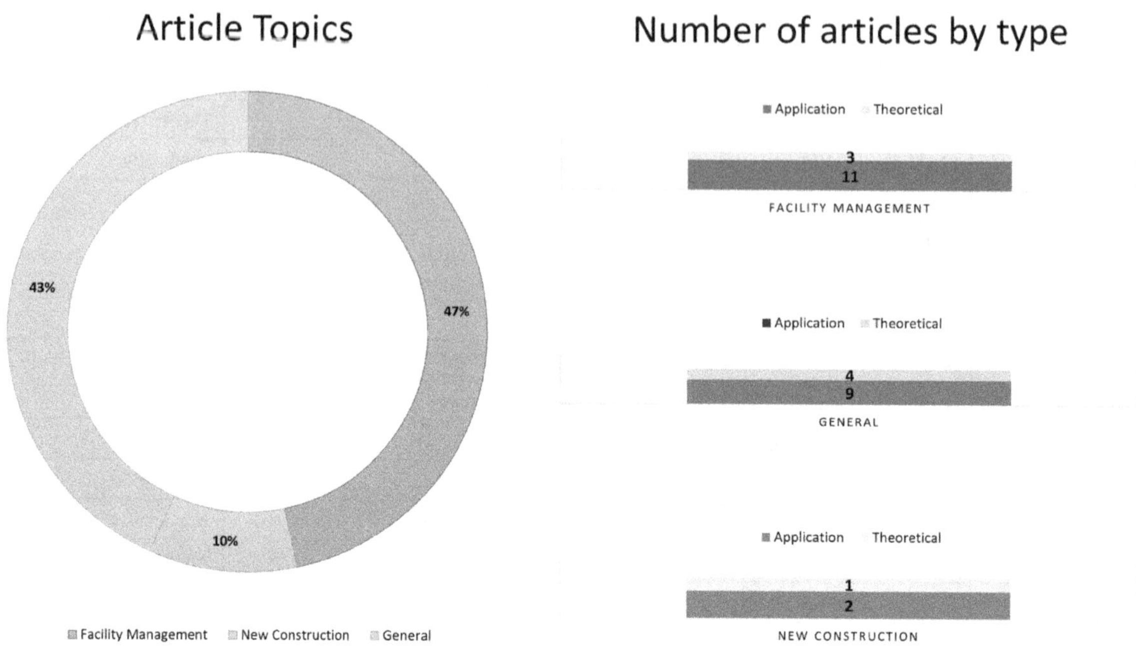

The model of the existing or the new one represents an artefact made up of components, which will be a future design and construction objective. In order to ensure that the BIM model functions as DT, i.e. it will be also predictive, it is necessary to define the priority components (safety, economic entity/damage ratio, ...), which we have to act on in terms of scheduled maintenance before "failure". Modeling by parts means considering each part, its replicability, its boundaries with the contact points/interface with the other parts: subsystems within systems. Both complex, consisting of various subsystems within the limit imposed by discretization. What should be asked from the BIM model to be able to say that it works, in the AEC industry, as a DT? In terms of maintenance, to prevent the occurrence of failures, especially for those priority components; in terms of risk, to identify which components to intervene on as a priority; in economic terms, to communicate in advance the amount of expenditure and/or the extent of damage related to non-intervention.

In order to become the DT of the existing building, the model should allow the drafting of a risk matrix based on the four categories of system behaviour (Grieves, 2017): desirable/ undesirable and expected/unexpected. In other terms, the technical information contained in the model, possibly derived from data sheets or list cost items, could produce a series of alerts for the most critical components, i.e. those whose failure has been classified as undesirable expected (priority) or undesirable not expected (secondary) behaviour. How to classify behaviours? Based on historical data, with fault occurrence rates, hazard levels (effects inside or outside the building; effects affecting a single or several people, potentially reaching the community) and economic impact of the damage (Moglia et al., 2019).

Collection of this information could be punctual, for individual interventions (e.g. building units up to the implementation of overall data per building), or on a larger scale (urban and/or territorial, when it comes to addressing the issue of security or resilience of territories to specific catastrophic events).

In this case, to represent maintenance at urban scale, moving from a City Model (CM) that contains dimensional information (shape, height, volume, impacts, ...) to one that shows the Buildings Conservation (BC), it is necessary to follow a simple logical path: from the city to the buildings, going back again to the city, representing maintenance information and, at the same time, implementing existing city model (Del Giudice et al., 2020). In the last years, several researches have been developed towards the smart city concepts, highlighting the value of using digital and telecommunication technologies for the benefit of its inhabitants and business. Moreover, the idea of smart city goes beyond the use of ICTs highlighting the meaning of a more interactive and responsive city administration, safer public spaces and meeting the needs of an ageing population. ICTs are becoming a key factor to connect the citizen with public administration, developing digital graphical interfaces able to represent data useful to arise smart city topics (United Nations, 2015, 2020). In this context, the creation of graphical databases such as at urban (GIS) and building (BIM) level allows to fill in and extract data to achieve the expected aims. Although BIM contains any/all information about the construction, the development of a detailed 3D parametric model requires a considerable effort depending on the established Level of Detail (LOD). On the other hand, a GIS dataset collects information on the urban and district scale useful for the development of analyses of macro systems for the evaluation of several policies such as energy saving, state of maintenance, through a simplified 2D/3D model. Obviously, parametric models created in different data domains have different characteristics because they refer to different kind of information. For example, GIS is adopted at urban scale while BIM is adopted at building scale and for both of them the ability to exploit geometric and alphanumeric information in a different way is requested.

If we consider the collection and use of systematic data on maintenance operations in buildings, "[…] a theoretical vacuum is created in the building sector [...]: just as, more generally, quality theories, so maintenance and especially scheduled maintenance, are now consolidated for household appliances rather than for buildings" (Agenzia delle Entrate, 2020a): there is no national database or extensive research on the subject; some recent data concern in particular the maintenance of infrastructures, which are the ones with the greatest negative impact in case of breakdown/incidental negative event […]. If there were an available database of the interventions carried out, it would certainly be useful for preliminary studies on scheduled maintenance. How can information be collected without imposing new regulatory requirements on the asset managers/owners? How can we make the built environment information available and "attractive" also from an economic point of view? It is therefore necessary to change the logic of the maintenance intervention, exploiting low costs and promoting data collection., also using –for example- UAV (Grosso et al., 2020).

KEY FIGURES CHARACTERIZATION

Given these promising prospects, what scenarios (real or utopian) do we foresee? As last in-depth examination, the chapter tries to analyse, for a standardizable case study (an existing residential building, built with traditional construction technologies), the strengths and weaknesses of the applicability of a BIM model for FM. The tool chosen by the authors as a preliminary approach to the problem, the SWOT analysis (Strengths-Strengths, Weaknesses -Weaknesses, Opportunities and Threats -Threats), allows an effective synthesis of the advantages and disadvantages resulting from such implementation (Mecca et al., 2019). To develop the construction of the SWOT matrices, a panel of experts (representative of the key figures potentially involved in the AEC sector digitalization process) were selected: user#1, a

Table 3. SWOT Analysis (#user1): the professional point of view

STRENGTHS	WEAKNESSES
1. reduction of survey time 2. complete documentation 3.increase project performance	1. need for professional updating 2.cost of licences
1. less responsibility due to increased asset knowledge	1. few contractors 2. low service utility perception by customers
OPPORTUNITIES	**THREATS**

member of a professional association (graduated, with more than 10-year experience in design on the existing building); user#2, a building manager (high school graduate, with 20-year experience in building asset management); user#3, an operator (secondary school graduate, with over 30-year experience in public and private contracts); user#4, an owner/tenant (a non-expert, living in a city apartment for more than 5 years). Experts were interviewed individually, through a remote communication platform. In the first part of the survey, after a brief description of the research, they were asked to provide a short overview of their skills in the AEC sector, in particular on BIM and DT topics. The second part of the interview, instead, took place in a dynamic mode: the interviewer showed, sharing his screen, a blank SWOT matrix and described its structure. Only one interviewer asked for clarification. Compilation of the matrices was done repetitive by the interviewer, at the respondent's direction. The following tables (3-4-5-6) show the results of the 4 interviews.

Here are brief summaries of the national framework, to highlight the significance of the 4 chosen figures/groups.

In terms of professionals, there are over 250,000 architects and engineers and 100,000 surveyors, 350,000 professionals in Italy who deal with the transformation of the territory: more than one for every 170 inhabitants. If we observe that in France and China the ratios are respectively 1/3,500 and 1/40 thousand inhabitants, in no country in the world is there such a density of professionals. (Gabbiani, 2018). This is a class of subjects fairly open to change -just over 40% are no more than 45 years old (InarCassa, 2020)- who work in a very fragmented reality: many small companies and few large ones, with an average gross income (freelance engineer or architect, only in 30% of cases) of about 27,500 euros/year. This widespread structure of small professional firms brings to light the first critical point in the approach to BIM: the cost of licenses and the need for professional updates (in the less young - about 60%-, who are presumed not yet trained on these issues) (Table 3).

Table 4. SWOT Analysis (#user2): the building manager point of view

STRENGTHS	WEAKNESSES
1. improvement of scheduled maintenance management 2. complete documentation 3.reduction of intervention time	1. need for professional updating 2. need for professional training 3. cost of licences
1. reduction of failure interventions 2. "aware" asset management	1 lack of economic resources 2. low perception service utility 3.need to update model
OPPORTUNITIES	**THREATS**

Table 5. SWOT Analysis (#user3): the owner/tenant point of view

STRENGTHS	WEAKNESSES
1. direct" control options, without giving up 2. complete documentation and testing opportunities	1. cost of service 2. cost of model upgrade
1. scheduled maintenance 2.reduction of intervention timing 3.reduction of overhead maintenance costs	1. low service utility perception 2. low willingness to spend
OPPORTUNITIES	**THREATS**

On the other hand, as far as building managers are concerned, the most recent data available online is not so up to date (2008) and refers to a research by the European Council of Real Estate Professions (CEPI), according to which they would be about 300 thousand in Italy, three times more than all those of the whole EU countries (CEPI, 2019). In this case, more than in the previous Table, there are important elements of strength and opportunity, which would favour a more effective and conscious maintenance of the built heritage. These elements, however, are opposed not only to a need for professional updating, but rather to a specific training on BIM, which is necessary because of the level of training linked to secondary school (Table 4).

The seventh edition of the "Real Estate in Italy", carried out by the Revenue Agency and the Department of Finance of the Ministry of Economy (Agenzia delle Entrate, 2019a) confirms that 75.2% of Italian families own the house they live in and are generally not young: 1.5 million owners under 35 years, compared to almost 14 million between 35 and 65 and 9.5 million over 65. The age factor weighs heavily on perceived weaknesses and criticalities, as is evident in Table 5.

Finally, the workers who carry out the maintenance. According to ISTAT, in 2018, 64% of construction companies in Italy have less than 10 employees, with a growing distribution in the 55-64 years-old bracket (a strong growth in the last 10 years; Vianello, 2019). For this key figure, the key balance weighs significantly more on the "positive" side (Table 6), directly perceiving all the benefits deriving from the availability of more information aimed at carrying out the intervention.

Table 6. SWOT Analysis (#user4): the operators point of view

STRENGTHS	WEAKNESSES
1. reduction of intervention time 2.reduction of risk of error 3. unexpected works reductions	1. need for professional updating
1. reliable intervention, based on reliable data	1. few contractors 2. low skill level category
OPPORTUNITIES	**THREATS**

The advantage of the operators (involved company and professionals), who are finally in the best working conditions is clear: having the avilability- in digital format – of the building materials. Less positive effects are perceived by the building manager (a more complete operator who, with the systematic data archive, favours more rapid alternations without loss of information) and even more by the owner, the only one involved in the new expenditure. The general impression is that the tool will really spread - and so the real breakthrough will be reached - when start-up costs will be reduced (software licenses and training courses for "BIM user", not for "BIM expert") and when users' perception will be linked to direct benefits in terms of reducing future maintenance costs. At the moment it is not yet possible to talk about coherence between digital model and real building condition: if you want to obtain an ex post modeling in a short time (also to reduce costs), it is necessary to proceed with a "simplified" modeling, to be implemented over time, distributing the costs to better align them with the benefits. The effective implementation of the final maintenance data could thus become an issue for building managers, who should be given "new technical training" as managers of the information model and not as service providers... In this way the model, produced by BIM experts, could work –for BIM users- as a "digital calendar" for inspections and maintenance operations (Mecca et al., 2019).

CONCLUSION

In brief, from the applications that have emerged from the state of the art (scientific publications), we can say that the experiments and studies in progress still concern realities that cannot be generalized both by type of intervention (hospitals, large infrastructures, university campuses, ...) and by user (at the moment large contracting stations or asset managers - such as insurance companies, real estate funds, ...- with spending capacity). As reported by Lennartsson et al. (2020), the "*diffusion of BIM use proceeds at a slower rate than the optimistic predictions*" and that is related to many factors. It is probable that BIM will really spread - and so it will reach the real turning point - when it is possible to reduce start-up costs and when users' perception is linked to direct benefits in terms of reducing future maintenance costs.

The attributes of the four actors (the professional, the manager, the owner, the operator) that characterize the maintenance process are therefore one of the focal points that condition and will condition the transformation of the maintenance of the objects (building, infrastructure) into one of the pillars of the smart city. These attributes are to be considered in a dynamic context, evolving both with regard to the skills of each actor and the actions that they are responsible for.

These ones should have cultural and language knowledge in the field of Digital Twin, although with limits suitable for the actions that he is responsible for pro tempore (which is the essence of Digital Twin for the specific field of maintenance of residential, non-residential buildings and infrastructure, of course for existing ones, but then also for those, in parallel, that will have to be realized with a design detail dedicated to scheduled maintenance, according to new rules, as already provided in Italy by the regulation on public works). Indeed, although the concept of Digital Twin is not clearly defined in literature (Wahbeh, 2020), as reported by Lu et al. (2020) this tool is able to integrate "*artificial intelligence, machine learning and data analytics to create dynamic digital models that are able to learn and update the status of the physical counterpart from multiple sources*". In any case, since the Digital Twin can give "*a more holistic and processual socio-technical characterization of the complex artifacts involved*" (Boje et al., 2020), and since it provides an "*accurate digital model of the equipment can be used for predictive maintenance*" (Khajavi et al., 2020). .

Each actor should, with individual responsibility, carry out what is foreseen, and foreseeable, through the role that has traditionally been built, or through the one that it has come to assume for himself; with collaborative initiative, he should develop and maintain contacts, exchange of information, operational synergies both within the four actors and globally between the four actors and the broader construction and urban planning sectors, also in order to foster the definition and training of a new actor that concentrates quality and skills of some of the original four.

The Digital Twin should grow gradually, both in the territorial coverage affecting each object and their connections, up to represent the image of the urban fabric, and in the collection of information (by detail of scale and by detail of functions) related to the individual object, thanks also to the increasing incentive and obligation in the actions of central and local government authorities. These actions are directed above all, at central level, to the prevention and management of major risks and consequent emergencies; at local level, to the safety management of the existing private and public building stock. In the private sector, interventions could be characterized by the search for continuity over time and distribution in the urban fabric, which is easier because the resources are private, stimulated and supported also by a non-episodic policy of fiscal facilitations; in the public sector, the search for continuity in interventions could take care of the planning of maintenance with the aim of efficiency as well as safety, but solving the difficulties in finding resources.

The Digital Twin needs the building booklet, it needs a warehouse, open to those who have to work as a team, not only among the four actors mentioned above, which, however, must always be present in the foreground, it needs a guarantee of interoperability, it needs a modularity that depends on interoperability but that must be directed towards the growth over time of the Digital Twin method, which who knows when it will be recognized as a producer of an effective twin. To shorten this time, the four actors must be convinced that the process produces a result that can be immediately used in a framework of complete answers to very precise questions, increasingly numerous and progressively extended to wider sectors in the Digital Twin system.

The maintenance of building and infrastructure structures could become one of the pillars of the smart city if the node of programming in the complex system of the construction process was overcome (Mecca et al., 2020). The "non-programming" depends on the low propensity to spend on maintenance (see SWOT), not intended as a return on investment. There is almost always a tendency to procrastinate maintenance work after the onset of phenomena that interrupt services or eliminate essential safety conditions. The building is not yet perceived as a set of systems, simple or complex, on which it is possible to intervene rigorously for parts, evaluating which risks and dangers may affect the complex of the building as a result of the partial maintenance intervention. But above all, the existing building is not yet designed and built in such a way as to allow maintenance operations for parts or systems (parts = volumetric sectors including systems, finishes, structures; system = complex of elements aimed at achieving an objective connected to one or more functions). In the construction sector it is also necessary to underline the collaboration between the parts substantially oriented to the "optimization of the synthesis", rather than the potential lack of collaboration linked to the "synthesis of optimizations".

The non-acceptance of programming by property is the conditioning obstacle for the chain of actors dependent on the property itself: the administrator, the designer, the executor. To persuade the property of the advantage of planned (and programmable) maintenance goes through the analysis of post-damage maintenance interventions in order to compare costs, including the effects of the damaging event, with technically predictable costs related to maintenance plans. Categories of "systems" or "parts" which acute damage can be associated with could be identified, for which the convenience of maintenance scheduling is evident, concentrating maintenance that can be associated with physiological damage in a "timeless" category. The habit of programming, together with the evident economic convenience, will extend the programmed maintenance over time.

The comparison that can be suggested is the car: in the past fuel and lubricant were used. Today, to keep the warranty alive, there is a list of interventions and periodic inspections, often at predetermined costs. I buy a car; I buy a service. So, why not buy a house and buy a maintenance service linked to an insurance service?

REFERENCES

Agenzia delle Entrate. (2019a). *Gli immobili in Italia. Ricchezza, reddito e fiscalità immobiliare*. Available online at: https://www.agenziaentrate.gov.it/portale/web/guest/gli-immobili-in-italia-2019

Agenzia delle Entrate. (2019b). *Ristrutturazioni edilizie: le agevolazioni fiscali*. Accessed on August 24, 2020: https://www.agenziaentrate.gov.it/portale/documents/20143/233439/Ristrutturazioni+edilizie+it_Guida_Ristrutturazioni_edilizie_Maggio2019.pdf/ed587c35-c2d6-7346-b79f-e2409b6a8c92

Agenzia delle Entrate. (2019c). *Sisma Bonus: le detrazioni per gli interventi antisismici*. Accessed on August 24, 2020: https://www.agenziaentrate.gov.it/portale/documents/20143/233439/Sisma+bonus+le+detrazioni+per+gli+interventi+antisismici_Guida_Sisma_Bonus.pdf/ee5ec719-05ae-0584-897e-f60d34060498

Agenzia delle Entrate. (2020a). *Bonus Facciate*. Accessed on August 24, 2020: https://www.agenziaentrate.gov.it/portale/documents/20143/233439/Guida_Bonus_Facciate.pdf/129df34a-b8b7-5499-a8fb-55d2a32a0b12

Agenzia delle Entrate. (2020b). *Bonus Verde*. Accessed on August 24, 2020: https://www.agenziaentrate.gov.it/portale/web/guest/aree-tematiche/casa/agevolazioni/bonus-verde

Akbarieh, A., Jayasinghe, L. B., Waldmann, D., & Teferle, F. N. (2020). BIM-based end-of-lifecycle decision making and digital deconstruction: Literature review. *Sustainability*, *12*(7), 2670. Advance online publication. doi:10.3390u12072670

Al-Saeed, Y., Edwards, D. J., & Scaysbrook, S. (2020). Automating construction manufacturing procedures using BIM digital objects (BDOs): Case study of knowledge transfer partnership project in UK. *Construction Innovation*, *20*(3), 345–377. doi:10.1108/CI-12-2019-0141

ANCE. (2019). *Osservatorio Congiunturale sull'industria delle costruzioni*. Accessed on January 12, 2021: https://www.ance.it/docs/docDownload.aspx?id=48610

ANCE. (2020). *Osservatorio Congiunturale sull'industria delle costruzioni*. Accessed on August 24, 2020: https://www.ance.it/docs/docDownload.aspx?id=53496

Antonino, M., Nicola, M., Claudio, D. M., Luciano, B., & Fulvio, R. C. (2019). Office building occupancy monitoring through image recognition sensors. *International Journal of Safety and Security Engineering*, *9*(4), 371–380. doi:10.2495/SAFE-V9-N4-371-380

Boje, C., Guerriero, A., Kubicki, S., & Rezgui, Y. (2020). Towards a semantic construction digital twin: Directions for future research. *Automation in Construction*, *114*, 103179. Advance online publication. doi:10.1016/j.autcon.2020.103179

CEPI. (2019). *Joint Report 2019. European and US mutual investments in residential real estate market.* Available online at: http://cepi.eu/uploads/CEPI-OPISAS%20Joint%20Report%202019.pdf

Del Giudice, M., Grosso, R., Mecca, U., Moglia, G., Prizzon, F., & Rebaudengo, M. (2020). The maintenance representation. Research and applications, mixing UAV and digital models. In *Digital Draw Connection*. Springer.

Dodge & Data Analytics. (2018). *SmartMarket Brief: Connecting Owners and Contractors*. Available online at: www.construction.com

DouS. Q.ZhangH. H.ZhaoY. Q.WangA. M.XiongY. T.ZuoJ. M. (2020). Research on construction of spatio-temporal data visualization platform for GIS and BIM fusion. Paper presented at the International Archives of the Photogrammetry. *Remote Sensing and Spatial Information Sciences - ISPRS Archives, 42*(3/W10) 555-563. Retrieved from www.scopus.com doi:10.5194/isprs-archives-XLII-3-W10-555-2020

European Commission. (2019). *Supporting digitalisation of the construction sector and SMEs: Including Building Information Modelling*. Accessed on August 24, 2020: https://ec.europa.eu/docsroom/documents/38281

European Commission. (2020). *EU Buildings Factsheets*. Accessed on August 24, 2020: https://ec.europa.eu/energy/eu-buildings-factsheets_en

European Construction Sector Observatory. (2019a). *Building Information Modelling in the EU construction sector.* European Commission. Accessed on August 24, 2020: https://ec.europa.eu/docsroom/documents/34518

European Construction Sector Observatory. (2019b). *Country Profile Italy*. European Commission. Accessed on August 24, 2020: https://ec.europa.eu/docsroom/documents/39524

European Construction Sector Observatory. (2019c). *Country Profile United Kingdom*. European Commission. Accessed on August 24, 2020: https://ec.europa.eu/docsroom/documents/39527

European Construction Sector Observatory. (2020a). *Country Profile France*. European Commission. Accessed on August 24, 2020: https://ec.europa.eu/docsroom/documents/40289

European Construction Sector Observatory. (2020b). *Country Profile Germany*. European Commission. Accessed on August 24, 2020: https://ec.europa.eu/docsroom/documents/40681

European Construction Sector Observatory. (2020c). *Country Profile Spain*. European Commission. Accessed on August 24, 2020: https://ec.europa.eu/docsroom/documents/40684

European Construction Sector Observatory. (2020d). *Improving the human capital basis*. European Commission. Accessed on August 24, 2020: https://www.clustercollaboration.eu/eu-initiatives/reports/european-construction-sector-observatory-improving-human

Frontera, M. (2019, May 22). *Impennata nei bandi pubblici in Bim: in un anno +223% in numero e +428% in valore, il Sole24Ore*. Available online at: https://www.ilsole24ore.com/art/impennata-bandi-pubblici-bim-un-anno-223percento-numero-e-428percento-valore-ACrVJEG

Gabbiani, B. (2018, February 28). *Architetti e ingegneri in Italia. Troppi o troppo pochi?* Available online at https://www.impresedilinews.it/architetti-e-ingegneri-in-italia-troppi-o-troppo-pochi/

Gerbert, P., Castagnino, S., Rothballer, C., Renz, A., & Filitz, R. (2016). *Digital in Engineering and Construction The Transformative Power of Building Information Modeling.* The Boston Consulting Group. Accessed on August 24, 2020: http://futureofconstruction.org/content/uploads/2016/09/BCG-Digital-in-Engineering-and-Construction-Mar-2016.pdf

Global, F. M. (2018). *Global facilities management market report 2018.* Accessed on August 24, 2020: https://www.hanze.nl/assets/instituut-voor-facility-management/Documents/Public/24315%20Global%20FM%20Market%20Report%202017_0%20%28002%29.pdf

Global Alliance for Buildings and Construction, International Energy Agency and the United Nations Environment Programme. (2019). *2019 global status report for buildings and construction: Towards a zero-emission, efficient and resilient buildings and construction sector.* Accessed on August 24, 2020: https://wedocs.unep.org/bitstream/handle/20.500.11822/30950/2019GSR.pdf?sequence=1&isAllowed=y

Huynh, D., & Nguyen-Ky, S. (2020). Engaging building automation data visualisation using building information modelling and progressive web application. *Open Engineering, 10*(1), 434–442. doi:10.1515/eng-2020-0054

InarCassa (2020). *InarCassa in cifre.* Statistiche su iscritti e pensionati, redditi e volume d'affari.

International Facility Management Association. (n.d.). *Cos'è il Facility Management.* Accessed on August 24, 2020: http://www.ifma.it/index.php?pagina=articolo.php&id_articolo=25&var_id_menu=68&nodata

JouanP.HallotP. (2019). Digital twin: A HBIM-based methodology to support preventive conservation of historic assets through heritage significance awareness. Paper presented at the International Archives of the Photogrammetry. *Remote Sensing and Spatial Information Sciences - ISPRS Archives, 42*(2/W15) 609-615. Retrieved from www.scopus.com doi:10.5194/isprs-archives-XLII-2-W15-609-2019

Kaewunruen, S., & Lian, Q. (2019). Digital twin aided sustainability-based lifecycle management for railway turnout systems. *Journal of Cleaner Production, 228,* 1537–1551. doi:10.1016/j.jclepro.2019.04.156

Kaewunruen, S., Peng, S., & Phil-Ebosie, O. (2020). Digital twin aided sustainability and vulnerability audit for subway stations. *Sustainability, 12*(19), 7873. Advance online publication. doi:10.3390u12197873

Kaewunruen, S., Sresakoolchai, J., & Kerinnonta, L. (2019). Potential reconstruction design of an existing townhouse in washington DC for approaching net zero energy building goal. [Switzerland]. *Sustainability, 11*(23), 6631. Advance online publication. doi:10.3390u11236631

Kaewunruen, S., & Xu, N. (2018). Digital twin for sustainability evaluation of railway station buildings. *Frontiers in Built Environment, 4,* 77. Advance online publication. doi:10.3389/fbuil.2018.00077

Khajavi, S. H., Motlagh, N. H., Jaribion, A., Werner, L. C., & Holmstrom, J. (2019). Digital twin: Vision, benefits, boundaries, and creation for buildings. *IEEE Access: Practical Innovations, Open Solutions, 7,* 147406–147419. doi:10.1109/ACCESS.2019.2946515

Khalil, A., Stravoravdis, S., & Backes, D. (2020). *Categorisation of building data in the digital documentation of heritage buildings.* Applied Geomatics., doi:10.100712518-020-00322-7

Lennartsson, M., Yitmen, I., Movaffaghi, H., & Linderoth, H. (2020). Framework for digital development in industrialized housebuilding. *Advances in Transdisciplinary Engineering, 13*, 335-345. Retrieved from www.scopus.com doi:10.3233/ATDE200171

Liu, Z., Zhang, A., & Wang, W. (2020). A framework for an indoor safety management system based on digital twin. *Sensors (Basel), 20*(20), 1–20. doi:10.339020205771 PMID:33053719

Lu, Q., Xie, X., Heaton, J., Parlikad, A. K., & Schooling, J. (2020). *From BIM towards digital twin: Strategy and future development for smart asset management*. Retrieved from www.scopus.com doi:10.1007/978-3-030-27477-1_30

Lu, R., & Brilakis, I. (2019). Digital twinning of existing reinforced concrete bridges from labelled point clusters. *Automation in Construction, 105*, 102837. Advance online publication. doi:10.1016/j.autcon.2019.102837

Lu, R. D., & Brilakis, I. (2019). Digital twinning of existing bridges from labelled point clusters. *Proceedings of the 36th International Symposium on Automation and Robotics in Construction, ISARC 2019*, 616-623. Retrieved from www.scopus.com doi:10.22260/ISARC2019/0082

Mayer, H. (2018). Digitalization of legacy building data preparation of printed building plans for the BIM process. *Proceedings of the 7th International Conference on Smart Cities and Green ICT Systems*, 304-310. Retrieved from www.scopus.com doi:10.5220/0006783103040310

McGraw Hill Construction. (2014). *The Business Value of BIM for Construction in Major Global Markets*. Accessed on August 24, 2020: https://www.construction.com/toolkit/reports/bim-business-value-construction-global-markets

Mecca, B., & Lami, I. M. (2020). *The appraisal challenge in cultural urban regeneration: An evaluation proposal*. Retrieved from www.scopus.com doi:10.1007/978-3-030-35550-0_5

Mecca, U., Moglia, G., Piantanida, P., Prizzon, F., Rebaudengo, M., & Vottari, A. (2020). How Energy Retrofit Maintenance Affects Residential Buildings Market Value? *Sustainability, 2020*(12), 5213. Retrieved August 24, 2020, from. doi:10.3390u12125213

Mecca, U., Moglia, G., Rebaudengo, M., & Ruffino, P. (2019). BIM and maintenance: real or virtual scenarios? In Ingegno e costruzione nell'epoca della complessità, Colloqui.AT.e 2019, Turin, Italy.

Mecca, U., Piantanida, P., Rebaudengo, M., & Vottari, A. (2019). *Green perspectives for Italian buildings façades. In 19th International Multidisciplinary Scientific GeoConference SGEM 2019 2019, 9*. Section Green Buildings Technologies and Materials. doi:10.5593gem2019V/6.3/S09.038

Ministero delle Infrastrutture e dei Trasporti. (2017). *Ministerial Decree no. 560 of 01/12/2017*. Accessed on August 24, 2020: http://www.mit.gov.it/normativa/decreto-ministeriale-numero-560-del-01122017

Moglia, G., Rebaudengo, M., & Ruffino, P. A. (2019). *"Digital Twin" Utopia? UID per il disegno: 2019; ROMA, Gangemi*. Accessed on August 24, 2020: http://digital.casalini.it/10.36165/1065

Moretti, N., Xie, X., Merino, J., Brazauskas, J., & Parlikad, A. K. (2020). An openbim approach to iot integration with incomplete as-built data. *Applied Sciences (Basel, Switzerland), 10*(22), 1–17. doi:10.3390/app10228287

OICE. (2019). *Rapporto sulle gare BIM 2018 per opere pubbliche: Analisi del mercato e delle gare.* Accessed on August 24, 2020: https://www.oice.it/592206/rapporto-sulle-gare-bim-2018-per-opere-pubbliche

Oreto, P. (2018, April 5). *Codice dei contratti ed i nodi da sciogliere: La qualificazione delle stazioni appaltanti, Rivista Online Lavori pubblici.* Available online at: https://www.lavoripubblici.it/news/2018/04/LAVORI-PUBBLICI/20089/Codice-dei-contratti-ed-i-nodi-da-sciogliere-La-qualificazione-delle-stazioni-appaltanti

Papadonikolaki, E., Leon, M., & Mahamadu, A. M. (2018). Bim solutions for construction lifecycle: A myth or a tangible future? *Paper presented at the EWork and eBusiness in Architecture, Engineering and Construction - Proceedings of the 12th European Conference on Product and Process Modelling, ECPPM 2018,* 321-328. Retrieved from www.scopus.com doi:10.1201/9780429506215-40

Poli, T., Mainini, A. G., Speroni, A., Blanco Cadena, J. D., & Moretti, N. (2020). *The effect of real-time sensing of a window on energy efficiency, comfort, health and user behavior.* Retrieved from www.scopus.com doi:10.1007/978-3-030-33570-0_26

Roxin, A., Abdou, W., Ginhac, D., Derigent, W., Dragomirescu, D., & Montegut, L. (2019). Digital building twins - contributions of the ANR McBIM project. *Proceedings - 15th International Conference on Signal Image Technology and Internet Based Systems, SISITS 2019*, 404-410. Retrieved from www.scopus.com doi:10.1109/SITIS.2019.00071

Schimanski, C. P., Monizza, G. P., Marcher, C., & Matt, D. T. (2019). Pushing digital automation of configure-to-order services in small and medium enterprises of the construction equipment industry: A design science research approach. *Applied Sciences (Basel, Switzerland)*, *9*(18), 3780. Advance online publication. doi:10.3390/app9183780

Scopus Database. (n. d.) Available online at: www.scopus.com

StojanovicV.TrappM.RichterR.DöllnerJ. (2019). A service-oriented indoor point cloud processing pipeline. Paper presented at the International Archives of the Photogrammetry. *Remote Sensing and Spatial Information Sciences - ISPRS Archives*, *42*(2/W17), 339-346. Retrieved from www.scopus.com doi:10.5194/isprs-archives-XLII-2-W17-339-2019

Tahmasebinia, F., Fogerty, D., Wu, L. O., Li, Z., Sepasgozar, S. M. E., Zhang, K., & Marroquin, F. A. (2019). Numerical analysis of the creep and shrinkage experienced in the sydney opera house and the rise of digital twin as future monitoring technology. *Buildings*, *9*(6), 137. Advance online publication. doi:10.3390/buildings9060137

Tchana, Y., Ducellier, G., & Remy, S. (2019). Designing a unique digital twin for linear infrastructures lifecycle management. *Procedia CIRP, 84,* 545-549. Retrieved from www.scopus.com doi:10.1016/j.procir.2019.04.176

United Nations. (2015). *Resolution adopted by the General Assembly on 25 September 2015, Seventieth session, Agenda items 15 and 116.* Accessed on August 24, 2020: https://www.un.org/ga/search/view_doc.asp?symbol=A/RES/70/1&Lang=E

United Nations. (2020). *Sustainable development goals.* Accessed on August 24, 2020: https://www.un.org/sustainabledevelopment/

Vianello, S. (2019, June 20). *La sicurezza dei lavoratori anziani nei cantieri edili.* Available on line at: https://www.unapri.it/news/in-cantiere-non-si-invecchia-la-sicurezza-nell-edilizia-per-i-lavoratori-anziani/

Vivi, Q. L., Parlikad, A. K., Woodall, P., Ranasinghe, G. D., & Heaton, J. (2019). *Developing a dynamic digital twin at a building level: Using cambridge campus as case study.* Paper presented at the International Conference on Smart Infrastructure and Construction 2019, ICSIC 2019: Driving Data-Informed Decision-Making, 67-75. Retrieved from www.scopus.com doi:10.1680/icsic.64669.067

WahbehW.KunzD.HofmannJ.BereuterP. (2020). Digital twinning of the built environment-an interdisciplinary topic for innovation in didactics. Paper presented at the ISPRS Annals of the Photogrammetry. *Remote Sensing and Spatial Information Sciences*, *5*(4) 231-237. Retrieved from www.scopus.com doi:10.5194/isprs-annals-V-4-2020-231-2020

Xie, X., Lu, Q., Rodenas-Herraiz, D., Parlikad, A. K., & Schooling, J. M. (2020). Visualised inspection system for monitoring environmental anomalies during daily operation and maintenance. *Engineering, Construction, and Architectural Management*, *27*(8), 1835–1852. doi:10.1108/ECAM-11-2019-0640

Compilation of References

Abacus. (2020). *Chinese police now have AI helmets for temperature screening*. https://www.scmp.com/abacus/news-bites/article/3052879/chinese-police-now-have-ai-helmets-temperature-screening

Abas, M. A., & Wee, S. T. (2014). Sustainable Solid Waste Management in Malaysia: The Concept of Multi-Stakeholder Governance in Solid Waste Policy Implementation. *Public Policy and Administration Research*, *4*(10), 26–35.

Abbasabadi, N., & Ashayeri, M. (2019). Urban energy use modeling methods and tools: A review and an outlook. *Building and Environment*, *161*, 106270. doi:10.1016/j.buildenv.2019.106270

Abdel-Basset, M., Manogaran, G., Mohamed, M., & Rushdy, E. (2019). Internet of Things in smart education environment: Supportive framework in the decision-making process. *Concurrency and Computation: Practice and Experience*, *31*(10).

Abel, J., Diez, K., & Lennerts, K. (2006). A Standard Database Model for Computer-Aided Facility Management in Hospitals. *Proceedings of the Joint International Conference on Computing and Decision Making in Civil and Building Engineering*, 535–542.

Abergel, T., Dean, B., & Dulac, J. (2017). *Towards a zero-emission, efficient, and resilient buildings and construction sector: Global Status Report 2017*. UN Environment and International Energy Agency.

Abi Ghanem, D., & Mander, S. (2014). Designing consumer engagement with the smart grids of the future: Bringing active demand technology to everyday life. *Technology Analysis and Strategic Management*, *26*(10), 1163–1175. doi:10.1080/09537325.2014.974531

Adu, E. K., & Poo, D. C. (2014). Smart learning: A new paradigm of learning in the smart age. In *Proceedings International Conference Teaching and Learning Higher Education (TLHE)*. Singapore: National Univ. Singapore.

Agenzia delle Entrate. (2019a). *Gli immobili in Italia. Ricchezza, reddito e fiscalità immobiliare*. Available online at: https://www.agenziaentrate.gov.it/portale/web/guest/gli-immobili-in-italia-2019

Agenzia delle Entrate. (2019b). *Ristrutturazioni edilizie: le agevolazioni fiscali*. Accessed on August 24, 2020: https://www.agenziaentrate.gov.it/portale/documents/20143/233439/Ristrutturazioni+edilizie+it_Guida_Ristrutturazioni_edilizie_Maggio2019.pdf/ed587c35-c2d6-7346-b79f-e2409b6a8c92

Agenzia delle Entrate. (2019c). *Sisma Bonus: le detrazioni per gli interventi antisismici*. Accessed on August 24, 2020: https://www.agenziaentrate.gov.it/portale/documents/20143/233439/Sisma+bonus+le+detrazioni+per+gli+interventi+antisismici_Guida_Sisma_Bonus.pdf/ee5ec719-05ae-0584-897e-f60d34060498

Agenzia delle Entrate. (2020a). *Bonus Facciate*. Accessed on August 24, 2020: https://www.agenziaentrate.gov.it/portale/documents/20143/233439/Guida_Bonus_Facciate.pdf/129df34a-b8b7-5499-a8fb-55d2a32a0b12

Agenzia delle Entrate. (2020b). *Bonus Verde*. Accessed on August 24, 2020: https://www.agenziaentrate.gov.it/portale/web/guest/aree-tematiche/casa/agevolazioni/bonus-verde

Agicom. (2018). *Big Data - Interim Report Nell'ambito Dell'Indagine Conoscitiva Di Cui Alla Delibera n. 217/17/CONS*. Rome, Italy: Agicom. Retrieved July 30, 2020, from https://www.agcom.it/documents/10179/10875949/Studio-Ricerca+08-06-2018/c72b5230-354d-444f-9e3f-5467ca450714?version=1.0

AIA. (2008). *Document E202 Building Information Modeling Protocol Exhibit*. https://www.smacna.org/resources/resource/2008/01/04/aia-e202-building-information-modeling-protocol-exhibit

Akbarieh, A., Jayasinghe, L. B., Waldmann, D., & Teferle, F. N. (2020). BIM-based end-of-lifecycle decision making and digital deconstruction: Literature review. *Sustainability*, *12*(7), 2670. Advance online publication. doi:10.3390u12072670

Alam, K. M., & El Saddik, A. (2017). C2PS: A digital twin architecture reference model for the cloud-based cyber-physical systems. *IEEE Access: Practical Innovations, Open Solutions*, *5*, 2050–2062. doi:10.1109/ACCESS.2017.2657006

Albino, V., Berardi, U., & Dangelico, R. M. (2015). Smart Cities: Definitions, Dimensions, Performance and Initiative. *Journal of Urban Technology*, *22*(1), 1, 3–21. doi:10.1080/10630732.2014.942092

Alcott, B. (2008). The sufficiency strategy: Would rich-world frugality lower environmental impact? *Ecological Economics*, *64*(4), 770–786. doi:10.1016/j.ecolecon.2007.04.015

Alexander, K. (1996). *Facilities Management : Theory and Practice*. E & F N Spon. doi:10.4324/9780203475966_chapter_3

Aliberti, A., Bottaccioli, L., Cirrincione, G., Macii, E., Acquaviva, A., & Patti, E. (2018). Forecasting short-term solar radiation for photovoltaic energy predictions. *Proceedings of the 7th International Conference on Smart Cities and Green ICT Systems*, Volume 1: SMARTGREENS, 44-53. 10.5220/0006683600440053

Aliberti, A., Ugliotti, F. M., Bottaccioli, L., Osello, A., Macii, E., Patti, E., & Acquaviva, A. (2018). Indoor air-temperature forecast for energy-efficient management in Smart Buildings. *IEEE Proceedings*, 272-277.

Allam, Z. (2021). *The Rise of Autonomous Smart Cities: Technology, Economic Performance and Climate Resilience*. Palgrave MacMillan. doi:10.1007/978-3-030-59448-0

Allegrini, J., Orehouning, K., Mavromatidis, G., Ruesch, F., Dorer, V., & Evins, R. (2015). A review of modelling approaches and tools for the simulation of district-scale energy systems. *Renewable & Sustainable Energy Reviews*, *52*, 1391–1404. doi:10.1016/j.rser.2015.07.123

Alonso, R., Borras, M., Koppelaar, R. H., & Lodigiani, A. (2019). *SPHERE: BIM Digital Twin Platform*. Proceedings - MDPI. Retrieved from https://sphere-project.eu/download/sphere-digital-twin-definitions-for-buildings/

Al-Saeed, Y., Edwards, D. J., & Scaysbrook, S. (2020). Automating construction manufacturing procedures using BIM digital objects (BDOs): Case study of knowledge transfer partnership project in UK. *Construction Innovation*, *20*(3), 345–377. doi:10.1108/CI-12-2019-0141

Al-Turjman, F., & Imran, M. (Eds.). (2020). IoT Technologies in Smart Cities. From sensors to big data, security and trust. The Institution of Engineering and Technology.

Amaratunga, D., Haigh, R., Sarshar, M., & Baldry, D. (2002). Assessment of facilities management process capability: A NHS facilities case study. *International Journal of Health Care Quality Assurance*, *15*(6), 277–288. doi:10.1108/09526860210442047 PMID:12500653

ANCE. (2019). *Osservatorio Congiunturale sull'industria delle costruzioni*. Accessed on January 12, 2021: https://www.ance.it/docs/docDownload.aspx?id=48610

ANCE. (2020). *Osservatorio Congiunturale sull'industria delle costruzioni*. Accessed on August 24, 2020: https://www.ance.it/docs/docDownload.aspx?id=53496

Andrade, A. (2015). Game engines: A survey. *EAI Endorsed Transactions on Game-Based Learning.*, *2*(6), 150615. doi:10.4108/eai.5-11-2015.150615

Angelidou, M. (2015). Smart cities: A conjuncture of four forces. *Cities (London, England)*, *47*, 95–106. doi:10.1016/j.cities.2015.05.004

Anonymous. (2015). *Python Traveling Salesman Greedy Algorithm (Algorithm)*. Stack Overflow. Available at: https://stackoverflow.com/a/30552710

Anonymous. (2019). *Complete Guide for Ultrasonic Sensor HC-SR04 with Arduino*. Random Nerd Tutorials. Available from: https://randomnerdtutorials.com/complete-guide-for-ultrasonic- sensor-hc-sr04/

Antonino, M., Nicola, M., Claudio, D. M., Luciano, B., & Fulvio, R. C. (2019). Office building occupancy monitoring through image recognition sensors. *International Journal of Safety and Security Engineering*, *9*(4), 371–380. doi:10.2495/SAFE-V9-N4-371-380

Antón, L. A., & Díaz, J. (2014a). Integration of LCA and BIM for Sustainable Construction. *International Journal of Civil and Environmental Engineering*, *8*(5), 1378–1382.

Antón, L. A., & Díaz, J. (2014b). Integration of Life Cycle Assessment in a BIM Environment. *Procedia Engineering*, *85*, 26–32. doi:10.1016/j.proeng.2014.10.525

Anzaldi, G., Corchero, A., Wicaksono, H., McGlinn, K., Gerdelan, A., & Dibley, M. J. (2014). Knoholem: Knowledge-based energy management for public buildings through holistic information modeling and 3d visualization. In I. González Alonso (Ed.), *International Technology Robotics Applications* (pp. 47–56). Springer International Publishing., doi:10.1007/978-3-319-02332-8_5

APE. (2015). https://www.certificato-energetico.it/

Arayici, Y. (2008). Towards building information modelling for existing structures. *Structural Survey, 26*(3), 210-222.

Arcella, G. (2018, May 23). *Open Data e Big Data Pubblici, Cosa Sono e Perché Sono Connessi*. Retrieved July 30, 2020, from https://www.agendadigitale.eu/dati/open-data-e-big-data/

ArcGIS Pro Documentation. (n.d.). *ArcGIS Pro* [Software]. https://pro.arcgis.com/en/pro-app/help/mapping/navigation/infographics.htm

Ariffin, R. N. R., & Zahari, R. K. (2013). Perceptions of the urban walking environments. *Procedia: Social and Behavioral Sciences*, *105*, 589–597. doi:10.1016/j.sbspro.2013.11.062

Armando, A., & Durbiano, G. (2017). *Teoria del progetto architettonico*. Carocci Editore.

Armesto, J., Lubowiecka, I., Ordóñez, C., & Rial, F.I. (2009). FEM modeling of structures based on close range digital photogrammetry. *Automation in Construction, 18*(5), 559-569.

Arun, M. (1999). Smart cities. The Singapore case. *Cities (London, England)*, *16*(1), 13–18. doi:10.1016/S0264-2751(98)00050-X

Ashton, K. (n.d.). *That 'Internet-of-Things' Thing*. Retrieved September 2015, from http://www.rfidjournal.com/articles/view?4986

Assante, D., Caforio, A., Flamini, M., & Romano, E. (2019). Smart education in the context of industry 4.0. *Proceeding IEEE Global Engineering Education Conference (EDUCON),* 1140–1145.

Assobim. (2019). *BIM Report 2019.* Accessed 01/04/2020, available at: https://www.assobim.it/bim-report-2019/

Associazione Torino Internazionale. (2015). *Il terzo Piano Strategico dell'area metropolitana di Torino* [The third Strategic Plan for the Turin metropolitan area]. Author.

Astleithner, F., Hamedinger, A., Holman, N., & Rydin, Y. (2004). Institutions and indicators – The discourse about indicators in the context of sustainability. *Journal of Housing and the Built Environment*, •••, 7–24.

ASviS Alleanza Urbana per lo sviluppo sostenibile. (2018), *L'agenda Urbana per lo sviluppo sostenibile.*

Athanasios, K. P., Gialelis, J. V., Alexakos, C. E., Georgoudakis, M. J., & Koubias, S. (2006). Vertical integration of enterprise industrial systems utilizing web services. *IEEE Transactions on Industrial Informatics*, *2*(2), 120–128. doi:10.1109/TII.2006.875507

Atkin, B., & Brooks, A. (2009). *Total Facilities Management, Facilities* (3rd ed., Vol. 12). Wiley-Blackwell.

Attar, R., Prabhu, V., Glueck, M., & Khan, A. (2010). 210 King Street: A dataset for integrated performance assessment. *Proceedings Spring Simulation Multiconference.*

Attari, S. Z., DeKay, M. L., Davidson, C. I., & De Bruin, W. B. (2010). Public perceptions of energy consumption and savings. *Proceedings of the National Academy of Sciences of the United States of America*, *107*(37), 16054–16059. doi:10.1073/pnas.1001509107 PMID:20713724

Aultman-Hall, L., Roorda, M., & Baetz, B. W. (1997). Using GIS for Evaluation of Neighborhood Pedestrian Accessibility. *Journal of Urban Planning and Development*, *123*(1), 10–17. doi:10.1061/(ASCE)0733-9488(1997)123:1(10)

Aurambout, J., Pettit, C. J., & Lewis, H. (2008). *Virtual Globes: the Next GIS?* Landscape Analysis and Visualisation.

Aurigi, A. (2005). *Making the Digital City: The early shaping of urban internet space*. Ashgate.

Austin, M., Delgoshaei, P., Coelho, M., & Heidarinejad, M. (2020). Architecting Smart City Digital Twins: Combined Semantic Model and Machine Learning Approach. *Journal of Management Engineering*, *36*(4), 1–14. doi:10.1061/(ASCE)ME.1943-5479.0000774

Autodesk Revit. (n.d.). Retrieved November 2015, from http://www.autodesk.com/products/revit-family/overview

Autodesk. (2002). *Building Information Modeling in Practice*. https://images.autodesk.com/apac_grtrchina_main/files/aec_bim.pdf

Aydin, S., & Schnabel, M. A. (2013). Coding Shape Grammars. Hints for generating a parametric design tool for large-scale urban renewal projects. In *Cutting Edge: 47th International Conference of the Architectural Science Association*, (pp. 177-186). The Architectural Science Association (ANZAScA), Australia.

Azhar, S. (2011). Building information modeling (BIM): Trends, benefits, risks, and challenges for the AEC industry. *Leadership and Management in Engineering*, *11*(3), 241–252. doi:10.1061/(ASCE)LM.1943-5630.0000127

Baheti, R., & Gill, H. (2011). Cyber-physical systems. In T. Samad & A. M. Annaswamy (Eds.), *The Impact of Control Technology: Overview, Success Stories, and Research Challenges* (pp. 161–166). IEEE Control Systems Society.

Bahu, J.M., Koch, A., Kremers, E., Murshed, S.M. (2013). *Towards a 3D spatial urban energy modelling approach.* ISPRS Ann. Photogramm. Remote Sens. Spat. Inf. Sci. 2013, II-2/W1.

Baidu. (2020). *How Baidu is bringing AI to the fight against coronavirus*. https://www.technologyreview.com/2020/03/11/905366/how-baidu-is-bringing-ai-to-the-fight-against-coronavirus/

Bainbridge, M., & Finch, E. F. (2009). Getting the attention the facilities manager deserves. *Facilities*, *27*(7/8), 277–290. doi:10.1108/02632770910956139

Bakis, N., Kagioglou, M., & Aouad, G. (2006). Evaluating the business benefits of information systems. In *3rd International SCRI Symposium*. Salford Centre for Research and Innovation, University of Salford.

Ballesty, S., Mitchell, J., Drogemuller, R., Schevers, H., Linning, C., Singh, G., & Marchant, D. (2007). Adopting BIM for facilities management: Solutions for managing the Sydney Opera House. Cooperative Research Centre for Construction Innovation. Report, pp. 1–32.

Balusamy, B., Chilamkurti, N., & Kadry, S. (Eds.). (2021). *Green Computing in Smart Cities: Simulation and Techniques*. Springer Nature AG Switzerland. doi:10.1007/978-3-030-48141-4

Bandyopadhyay, A., Bose, S., & Das, S. (2015). 3D Printing of Biomaterials. *MRS Advances*, *2*, 108–115.

Banu, G. (2018). Masuring innovation using key performance indicators. Procedia Manufacturing, 906-911.

Baranwal, T., & Pateriya, P. K. (2016). Development of IoT based smart security and monitoring devices for agriculture. *Proceedings IEEE 6th International Conference Cloud Systems Big Data Engineering*, 597-602.

Bar-Cohen, Y. (2019). *Advances in Manufacturing and Processing of Materials and Strcutures*. CRC Press. Taylor & Francis Group, LLC.

Barlish, K., & Sullivan, K. (2012). How to measure the benefits of BIM - A case study approach. *Automation in Construction*, *24*, 149–159. doi:10.1016/j.autcon.2012.02.008

Barone, L. (2017). Interoperabilità BIM per l'analisi energetica. In BIM: verso Il catasto del futuro. Conoscere, digitalizzare, condividere. Il caso studio della Città di Torino, (pp. 105-110). Gangemi Editore spa.

Barrett, P. S., & Baldry, D. (2003). *Facilities Management: Towards Best Practice, Facilities Management* (2nd ed., Vol. 2). Wiley-Blackwell.

Barricelli, B. R., Casiraghi, E., & Fogli, D. (2019). A survey on digital twin: Definitions, characteristics, applications, and design implications. *IEEE Access*, *7*(MI), 167653–167671. doi:10.1109/ACCESS.2019.2953499

Barthelat, F. (2007). Biomimetics for next generation materials. *Phil. Trans. R. Soc. A*, 365.

Barthelat, F. (2015) *Architectured materials in engineering and biology: fabrication, structure, mechanics and performance*. Institute of Materials, Minerals and Mining and ASM International.

Based, W. H. E. (2021). *IoT Platforms, Use Cases, Privacy, and Business Models*. doi:10.1007/978-3-030-45316-9

Basu, S., Bale, C. S. E., Wehnert, T., & Topp, K. (2019). A complexity approach to defining urban energy systems. *Cities (London, England)*, *95*, 102358. doi:10.1016/j.cities.2019.05.027

Batty, M. (1995, July). The computable city. In *Fourth international Conference on Computers in Urban Planning and Urban Management*. School of Environmental Studies. University of Melbourne.

Batty, M., Chapman, D., Evans, S., Haklay, M., Kueppers, S., Shiode, N., Smith, A., & Torrens, P. M. (2000). Visualizing the City: Communicating Urban Design to Planners and Decision-Makers. Technical Report Paper 26. Centre for Advanced Spatial Analysis (UCL).

Batty, M. (2005). 2005, Agents, cells, and cities: New representational models for simulating multiscale urban dynamics. *Environment & Planning A*, *37*(8), 1373–1394. doi:10.1068/a3784

Batty, M. (2013). *The New Science of Cities*. The MIT Press. doi:10.7551/mitpress/9399.001.0001

Batty, M., Axhausen, K. W., Giannotti, F., Pozdnoukhov, A., Bazzani, A., Wachowicz, M., Ouzounis, G., & Portugali, Y. (2012). Smart Cities of the Future. *The European Physical Journal. Special Topics*, *214*(1), 481–518. doi:10.1140/epjst/e2012-01703-3

Bauer, F., Recheis, D., & Kaltenböck, M. (2011). Data. Reegle. Info – a new key portal for open energy data. In J. Hřebíček, G. Schimak, & R. Denzer (Eds.), *Environmental Software Systems. Frameworks of eEnvironment* (Vol. 359, pp. 189–194). Springer Berlin Heidelberg. doi:10.1007/978-3-642-22285-6_21

Beesley, P., Chan, M., Gorbet, R., & Kulić, D. (2015). Evolving Systems within Immersive Architectural Environments: New Research by the Living Architecture Systems Group. Next Generation Building, 2(1), 31-56.

BeiDou. (2020). *COVID-19: China's BeiDou Satellites Help Fight Against Epidemic.* POB Point of Beginning. https://www.pobonline.com/articles/102110-covid-19-chinas-beidou-satellites-help-fight-against-epidemic)

Bellini, P., Benigni, M., Billero, R., Nesi, P., & Rauch, N. (2014). Km4City ontology building vs data harvesting and cleaning for smart-city services. *Journal of Visual Languages and Computing*, *25*(6), 827–839. doi:10.1016/j.jvlc.2014.10.023

Belussi, L., Barozzi, B., Bellazzi, A., Danza, L., Devitofrancesco, A., Fanciulli, C., & Scamoni, F. (2019). A review of performance of zero energy buildings and energy efficiency solutions. *Journal of Building Engineering*, *100772*. Advance online publication. doi:10.1016/j.jobe.2019.100772

Bencardino, M., & Greco, I. (2014). The Paradigm of the Modern City: SMART and SENSEable Cities for Smart, Inclusive and Sustainable Growth. In Computational Science and Its Applications – ICCSA 2014: 14th International Conference, Guimarães, Portugal, June 30 – July 3, 2014, *Proceedings, Part II,* (pp. 579-597). Guimarães, Portugal: Academic Press.

Bennacer, L., Amirat, Y., Chibani, A, Mellouk, A., & Ciavaglia, L. (2015). Self diagnosis technique for virtual private networks combining Bayesian networks and case-based reasoning. *IEEE Transactions on Automation Science and Engineering, 12*(1), 354-366.

Bernabei, R. (2015). CAD/CAM and Jewellery Design Education. Making Futures 2015.

Berners-Lee, T., Hendler, J., & Lassila, O. (2001). The Semantic Web: A New Form of Web Content That is Meaningful to Computers Will Unleash a Revolution of New Possibilities. *Scientific American*, *284*(5), 34–43. doi:10.1038cientifi camerican0501-34 PMID:11681174

Bertoldi, P. (2018). *Guidebook 'How to develop a Sustainable Energy and Climate Action Plan (SECAP).* European Commission, Joint Research Centre.

Bhushan, B. (2018). Biomimetics: Bioinspired hierarchical structured surfaces for green science and technology. *Materials Science*, 292.

Biagini, C., Donato, V., & Pellis, D. (2015). Preliminary design through graphs: A tool for automatic layout distribution. *Iconarp International Journal of Architecture and Planning*, *2*(2), 1–13.

Bianco, V., Manca, O., & Nardini, S. (2009). Electricity consumption forecasting in Italy using linear regression models. *Energy*, *34*(9), 1413–1421. doi:10.1016/j.energy.2009.06.034

Bibri, S. E., & Krogstie, J. (2017). Smart sustainable cities of the future: An extensive interdisciplinary literature review. *Sustainable Cities and Society*, *31*, 183–212. doi:10.1016/j.scs.2017.02.016

Bifulco, F., Tregua, M., Amitrano, C. C., & D'Auria, A. (2015). ICT and sustainability in smart cities management. *International Journal of Public Sector Management*, *29*(2), 132–147. doi:10.1108/IJPSM-07-2015-0132

Bilal, M., Oyedele, L. O., Kusimo, H. O., Owolabi, H. A., Akanbi, L. A., Ajayi, A. O., Akinade, O. O., & Delgado, J. M. D. (2019). Investigating profitability performance of construction projects using big data: A project analytics approach. *Journal of Building Engineering*, *26*, 100850. doi:10.1016/j.jobe.2019.100850

Biljecki, F., Ledoux, H., & Stoter, J. (2016). Does a finer level of detail of a 3D city model bring an improvement for estimating shadows? Springer International Publishing.

Biljecki, F., Ledoux, H., & Stoter, J. (2016). *Generation of multi-LOD 3d city models in CityGML with the procedural modelling engine random3dcity*. In ISPRS Annals of the Photogrammetry, Remote Sensing and Spatial Information Sciences, Volume III-4/W1. 1st International Conference on Smart Data and Smart Cities, 30th UDMS, 7–9 September 2016, Split, Croatia.

Biljecki, F., Stoter, J., Ledoux, H., Zlatanova, S., & Çöltekin, A. (2015). Applications of 3D City Models: State of the Art Review. *ISPRS International Journal of Geo-Information*, *4*(4), 4. doi:10.3390/ijgi4042842

BIM Alliance Sweden. (2020). *Vad är BIM?* https://bimalliance.se/vad-aer-bim/

Bisello, A. (2020). Assessing Multiple Benefits of Housing Regeneration and Smart City Development: The European Project SINFONIA. *Sustainability 2020, 12*(19), 1-28.

Blečić, I., Cecchini, A., Fancello, G., Talu, V., & Trunfio, G. A. (2015). Camminabilità e capacità urbane: valutazione e supporto alla decisione e alla pianificazione urbanistica. Territorio Italia, 1, 49 - 65.

Blumsack, S., & Fernandez, A. (2012). Ready or not, here comes the smart grid! *Energy*, *37*(1), 61–68. doi:10.1016/j.energy.2011.07.054

Boje, C., Guerriero, A., Kubicki, S., & Rezgui, Y. (2020). Towards a semantic Construction Digital Twin: Directions for future research. *Automation in Construction*, *114*, 103179. doi:10.1016/j.autcon.2020.103179

Boje, C., Guerriero, A., Kubicki, S., & Rezgui, Y. (2020). Towards a semantic Construction Digital Twin: Directions for future. *Automation in Construction*, •••, 1–16.

Bollinger, L. A., & Evins, R. (2015). HUES: A holistic urban energy simulation platform for effective model integration. In *Proceedings of International Conference CISBAT 2015* (pp. 841-846). Academic Press.

Bollini, L. (2017). Beautiful interfaces. From user experience to user interface design. *The Design Journal, 20*(sup1), S89-S101.

Bolognesi, C., & Galli, A. (2017). The Representation of Socials in Maps: Map of a Great Event in Monza Park. In La Città Creativa; Spazi Pubblici e Luoghi Della Creatività, (pp. 627–35). Rome, Italy: CNAPPC.

Bolton, R., McColl-Kennedy, J. R., Cheung, L., Gallan, A., Orsingher, C., Witell, L., & Zaki, M. (2018). Customer experience challenges: Bringing together digital, physical and social realms. *Journal of Service Management*, *29*(5), 776–808. doi:10.1108/JOSM-04-2018-0113

Bonci, A., Carbonari, A., Cucchiarelli, A., Messi, L., Pirani, M., & Vaccarini, M. (2019). A cyber-physical system approach for building efficiency monitoring. *Automation in Construction*, *102*, 68–85. doi:10.1016/j.autcon.2019.02.010

Bonino, D., Castellina, E., & Corno, F. (2008). The DOG gateway: Enabling ontology-based intelligent domotic environments. *IEEE Transactions on Consumer Electronics*, *54*(4), 1656–1664. doi:10.1109/TCE.2008.4711217

Bonino, D., & De Russis, L. (2018). DogOnt as a viable seed for semantic modeling of AEC/FM. *Semantic Web*, *9*(6), 763–780. doi:10.3233/SW-180295

Bonnemaison, S., & Eisenbach, R. (2009). *Installations by architects: experiments in Building and Design*. Princeton Architectural Press.

Booyse, W., Wilke, D. N., & Heyns, S. (2020). Deep digital twins for detection, diagnostics and prognostics. *Mechanical Systems and Signal Processing*, *140*, 106612. doi:10.1016/j.ymssp.2019.106612

Bordegoni, M., & Rizzi, C. (2011). *Innovation in Product Design – from CAD to Virtual Prototyping*. Springer.

Borelli, G. (2020). La Smart city e la fine della città [The Smart city and the end of the city]. *Sociologia Urbana e Rurale*, *122*(122), 96–122. doi:10.3280/SUR2020-122007

Borga, G. (2013). *City Sensing. Approcci, Metodi e Tecnologie Innovative per La Città Intelligente*. Franco Angeli.

Börner, K., Sanyal, S., & Vespignani, A. (2007). Network Science. In B. Cronin (Ed.), *Annual Review of Information Science & Technology* (Vol. 41, pp. 537–607). Information Today, Inc./American Society for Information Science and Technology.

Borsboom-van Beurden J., Kallaos J., Gindroz B., Costa S. and Riegler J. (2019). *Smart city guidance package*. Norwegian University of Science and Technology/ European Innovation Partnership on Smart Cities and Communities. Faculty of Architecture and Design.

Boschert, S., & Rosen, R. (2016). Digital Twin—The Simulation Aspect. In P. Hehenberger & D. Bradley (Eds.), *Mechatronic Futures*. Springer. doi:10.1007/978-3-319-32156-1_5

Boschert, S., & Rosen, R. (2016). Digital twin—the simulation aspect. In *Mechatronic futures* (pp. 59–74). Springer International Publishing. doi:10.1007/978-3-319-32156-1_1

Bottaccioli, L., Aliberti, A., Ugliotti, F. M., Patti, E., Osello, A., Macii, E., & Acquaviva, A. (2017). Building energy modelling and monitoring by integration of IoT devices and Building Information Models. *Building Digital Autonomy forna Sustainable World (COMPSAC)*, 914-922.

Boudon, R. (1977). *Effets pervers et ordre social*. PUF.

Boulanger, P. M. (2010). Three strategies for sustainable consumption. *SAPI EN. S. Surveys and Perspectives Integrating Environment and Society*, *3*, 2.

Breje, A.-R., Gyorödi, R., Gyorödi, C., Zmaranda, D., & Pecherle, G. (2018). Comparative Study of Data Sending Methods for XML and JSON Models. *International Journal of Advanced Computer Science and Applications*, *9*(12), 198–204. doi:10.14569/IJACSA.2018.091229

Bria, F., Sagarra, O., Rodríguez, J., Balcells, P. (2018). *Deployments of pilots in Barcelona (Decentralized Citizens Owned Data Ecosystem)*, V.1.0, 31.

Brockwell, S., & Horrocks, R. (2020). *BIM, GIS, and Smart Cities: Challenges and Opportunities*. Autodesk University.

Brooks, P. L. (1997). Visualizing data. *DBMS (Redwood City, Calif.)*, *10*(9), 38.

Broto, V. C. (2017). Urban governance and the politics of climate change. *World Development*, *93*, 1–15. doi:10.1016/j.worlddev.2016.12.031

Browne, A. L., Pullinger, M., Medd, W., & Anderson, B. (2014). Patterns of practice: A reflection on the development of quantitative/mixed methodologies capturing everyday life related to water consumption in the UK. *International Journal of Social Research Methodology*, *17*(1), 27–43. doi:10.1080/13645579.2014.854012

Brown, P. (2009). CAD: Do Computers Aid the Design Process After All? *Intersect: The Stanford Journal of Science. Technology in Society*, *2*(1), 52–66.

Brundu, F. G., Patti, E., Del Giudice, M., Osello, A., Macii, E., & Acquaviva, A. (2015). DIMCloud: a distributed framework for district energy simulation and management. In Internet of Things. User-Centric IoT (pp. 331-338). Springer International Publishing. doi:10.1007/978-3-319-19656-5_45

Brundu, F. G., Patti, E., Del Giudice, M., Osello, A., Ramassotto, M., Massara, F., Marchi, F., Musetti, A., Macii, E., & Acquaviva, A. (2017). A scalable middleware-based infrastructure for energy management and visualization in city districts. *EAI Endorsed Trans. Cloud Syst.*, *17*, 1–10. doi:10.4108/eai.28-6-2017.152753

Bruno, S., De Fino, M., & Fatiguso, F. (2018). Historic building information modelling: Performance assessment for diagnosis-aided information modelling and management. *Automation in Construction, 86*, 256-276.

Bruynseels, K., de Sio, F. S., & van den Hoven, J. (2018). Digital Twins in health care: Ethical implications of an emerging engineering paradigm. *Frontiers in Genetics*, *9*(FEB), 1–11. doi:10.3389/fgene.2018.00031 PMID:29487613

Bryde, D., Broquetas, M., & Volm, J. M. (2013). The project benefits of building information modelling (BIM). *International Journal of Project Management., 31*(7), 971-980.

Bryde, D., Broquetas, M., & Volm, J. M. (2013). The project benefits of building information modelling (BIM). *International Journal of Project Management, 31*(7), 971–980. doi:10.1016/j.ijproman.2012.12.001

BSI - British Standards Institute. (2003). *Poor Communication Costing UK Construction Industry*. http://www.bsigroup.com/en-GB/aboutbsi/media-centre/press-releases/2003/9/Poor- Communication-Costing-UK-Construction-Industry/

Bueno, B., Pigeon, G., Norford, L. K., Zibouche, K., & Marchadier, C. (2012). Development and evaluation of a building energy model integrated in the TEB scheme. *Geoscientific Model Development*, *5*(2), 433–448. doi:10.5194/gmd-5-433-2012

Bughin, J., Hazan, E., Labaye, E., Manyika, J., Dahlström, P., Ramaswamy, S., & Cochin de Billy, C. (2016). *Digital Europe: Realizing the continent's potential*. McKinsey Global Institute. https://www.mckinsey.com/business-functions/mckinsey-digital/our-insights/digital-europe-realizing-the-continents-potential

Building Smart International. (2020a). *Enabling an Ecosystem of Digital Twins*. https://www.buildingsmart.org/wp-content/uploads/2020/05/Enabling-Digital-Twins-Positioning-Paper-Final.pdf

buildingSMART International. (2020). https://www.buildingsmart.org/standards/bsi-standards/industry-foundation-classes/

BuildingSMART Website. (n.d.). https://technical.buildingsmart.org/standards/ifc/

BuildingsO. S. M. (n.d.). https://osmbuildings.org/?lat=45.05373&lon=7.65550&zoom=19.6&tilt=30

Bullock, S., Ladley, D., & Kerby, M. (2012). Wasps, termites and waspmites: Distinguishing competence from performance in collective construction. *Artificial Life*, *18*(3), 267–290. Advance online publication. doi:10.1162/artl_a_00065 PMID:22662914

Bultrini, M. (2017). Il walkability index: un indice di camminabilità urbana. In *Qualità dell'ambiente urbano. XIII Rapporto. Edizione 2017. Focus su Mobilità pedonale in città. ISPRA, Stato dell'ambiente 75/2017*. Available on line: https://www.isprambiente.gov.it/files2017/pubblicazioni/stato-ambiente/Focus2017.pdf

Büscher, C., Schippl, J., & Sumpf, P. (2018). *Energy as a Sociotechnical Problem: An Interdisciplinary Perspective on Control, Change, and Action in Energy Transitions*. Routledge. doi:10.4324/9781315186313

Cabeza, L. F., Rincón, L., Vilariño, V., Pérez, G., & Castell, A. (2014). Life cycle assessment (LCA) and life cycle energy analysis (LCEA) of buildings and the building sector: A review. *Renewable & Sustainable Energy Reviews*, *29*, 394–416. doi:10.1016/j.rser.2013.08.037

Cai, Y., Starly, B., Cohen, P., & Lee, Y. S. (2017). Sensor Data and Information Fusion to Construct Digital-twins Virtual Machine Tools for Cyber-physical Manufacturing. *Procedia Manufacturing*, *10*, 1031–1042. doi:10.1016/j.promfg.2017.07.094

Caldera, C., Ostorero, C. L., Manni, V., Galli, A., & Valzano, L. S. (2020). *Cities in Transformation. Computational Urban Planning through Big Data Analytics. Techne, Special Issue, 2*. Firenze University Press.

Callon, M. (1987). Society in the making: the study of technology as a tool for sociological analysis. In W. E. Bijker, T. P. Hughes, & T. Pinch (Eds.), *The social construction of technological systems. New directions in the sociology and history of technology* (pp. 83–106). MIT Press.

Calvano, M., Cirelli, M., & Lo Turco, M. (2020). Display the Invisible. Automated Algorithms to Visualize Complex Phenomena. *Proceedings of the 2nd International and Interdisciplinary Conference on Image and Imagination.*

Camagni, R. (1993). *Principi Di Economia Urbana e Territoriale*. Carocci Editore.

Cândido, G., Colombo, A. W., Barata, J., & Jammes, F. (2011). Service-oriented infrastructure to support the deployment of evolvable production systems. *IEEE Transactions on Industrial Informatics*, *7*(4), 759–767. doi:10.1109/TII.2011.2166779

Caperna, A., Giangrande, A., Mirabelli, P., & Mortola, E. (2013). Partecipazione e ICT. Gangemi.

Caragliu, A., Del Bo, C., & Nijkamp, P. (2011). Smart cities in Europe. *Journal of Urban Technology*, *18*(2), 65–82. do i:10.1080/10630732.2011.601117

Cardoni, A., Noori, A. Z., Marasco, S., Domaneschi, M., Villa, V., Ansari, F., Caldera, C., & Cimellaro, G. P. (2019) Dynamic characterization and vulnerability assessment of a school building in Italy. *Structural Health Monitoring 2019: Enabling Intelligent Life-Cycle Health Management for Industry Internet of Things (IIOT) - Proceedings of the 12th International Workshop on Structural Health Monitoring*, 459-466. 10.12783hm2019/32148

Carli, R., Dotoli, M., & Pellegrino, R. (2018). Multi-criteria decision-making for sustainable metropolitan cities assessment. *Journal of Environmental Management*, *•••*, 46–61.

Carlo Ratti Associati. (2010). *Ka Care*. Retrieved February 30, 2020, from https://carloratti.com/project/ka-care/

Carreón, J. R., & Worrell, E. (2018). Urban energy systems within the transition to sustainable development. A research agenda for urban metabolism. *Resources, Conservation and Recycling*, *132*, 258–266. doi:10.1016/j.resconrec.2017.08.004

Casi, F. (2019 January 28). *Una Nuova Frontiera: La Datizzazione*. Retrieved July 30, 2020, from https://www.consultadibioetica.org/una-nuova-frontiera-la-datizzazione-di-fiorello-casi/

Castells, M. (1996). *The Rise of the Network Society*. Blackwell.

Cavallari Murat, A. (1968). Forma urbana e architettura nella Torino Barocca. Dalle premesse classiche alle conclusioni neoclassiche, Unione tipografico-editrice torinese, Torino.

Cecchini, C., Magrini, A., & Gobbi, L. (2019). A 3d platform for energy data visualization of building assets. In *Proceeding of IOP Conf. Series: Earth and Environmental Science 296*. IOP Publishing. 10.1088/1755-1315/296/1/012035

CEPI. (2019). *Joint Report 2019. European and US mutual investments in residential real estate market*. Available online at: http://cepi.eu/uploads/CEPI-OPISAS%20Joint%20Report%202019.pdf

Cerezo, C., Sokol, J., AlKhaled, S., Reinhart, C., Al-Mumin, A., & Hajiah, A. (2017). Comparison of four building archetype characterization methods in urban building energy modeling (UBEM): A residential case study in Kuwait City. *Energy and Building*, *154*, 321–334. doi:10.1016/j.enbuild.2017.08.029

Cervero, R., & Kockelman, K. (1997). Travel demand and the 3Ds: Density, diversity, and design. *Transportation Research Part D, Transport and Environment*, *2*(3), 199–219. doi:10.1016/S1361-9209(97)00009-6

Champeau, M., Heinze, D., Viana, T., De Souza, E., Chinellatto, A., & Titotto, S. (2020). 4D Printing of Hydrogels: A Review. *Advanced Functional Materials*, *1*(31), 1910606. doi:10.1002/adfm.201910606

Chelleri, L., & Olazabal, M. (2012). *Multidisciplinary perspectives on urban resilience. Workshop Report.* Bilbao, Spain: BC3, Basque Centre for Climate Change.

Chen, Y. Q., Fink, O., & Sansavini, G. (2018). Combined fault location and classification for power transmission lines fault diagnosis with integrated feature extraction. *IEEE Transactions on Industrial Electronics, 65*(1), 561-569.

Cheng, J. C. & Ma, L. Y. (2013). A BIM-based system for demolition and renovation waste estimation and planning. *Waste Management, 33*(6), 1539-1551.

Cheng, J., Zhang, H., Tao, F., & Juang, C. F. (2020). DT-II:Digital twin enhanced Industrial Internet reference framework towards smart manufacturing. *Robotics and Computer-integrated Manufacturing*, *62*, 101881. doi:10.1016/j.rcim.2019.101881

Chen, R. (2011). The development of 3D city model and its applications in urban planning. *Proceedings of the 19th International Conference on Geoinformatics*. 10.1109/GeoInformatics.2011.5981007

Chen, Y., Hong, T., & Piette, M. A. (2017). City-Scale Building Retrofit Analysis: A Case Study using CityBES. In *Proceedings of BS 2017: 15th Conference of IBPSA* (pp. 259-266). International Building Performance Simulation Association.

Chevallier, Z., Finance, B., & Cohen Boulakia, B. (2020). A Reference Architecture for Smart Building Digital Twin. *Proceedings of the International Workshop on Semantic Digital Twins co-located with the 17th Extended Semantic Web Conference (ESWC 2020).*

China Academy of Information and Communications Technology. (2020). *Data and Smart Applications for Epidemic Prevention and Control*. Source: CAICT, Research Report. http://www.caict.ac.cn/english/research/rs/202004/P020200430479193356866.pdf

Chinese Resident Balanced Diet. (2016). Release time: 2016-05-18 声明:关于《中国居民平衡膳食宝塔(2016)》官方版本 发布时间: http://dg.cnsoc.org/article/04/8a2389fd54b964c80154c1d781d90197.html

Choi, J. (2015). 4D Printing Technology: A Review. *3D Printing and Additive Manufacturing*, *2*(4), 159-167. doi:10.1089/3dp.2015.0039

Christaller, W. (1933). *Die Zentralen Orte in Süddeutschland.* Gustav Fischer Verlag.

Chua, C. K., Yeong, W. Y., & An, J. (2017). 3D Printing and Bioprinting in MEMS Technology. *Micromachines*, *8*(12), 229. doi:10.3390/mi8070229 PMID:30400417

Chui, M., Löffier, M., & Roberts, R. (2010). *The Internet of Things* (Vol. 2). McKinsey.

Cimellaro, G. P., Domaneschi, M., De Iuliis, M., Villa, V., Caldera, C., & Cardoni, A. (2019). Fire emergency evacuation in a school building through VR. *COMPDYN Proceedings*, 2046-2055. 10.7712/120119.7057.19951

Cimellaro, G. P., Domaneschi, M., Villa, V., Noori, A. Z., Pietropinto, M. V., Cardoni, A., & Marasco, S. (2019) BIM-based approach for seismic vulnerability analysis: A case study. *Advances in Engineering Materials, Structures and Systems: Innovations, Mechanics and Applications - Proceedings of the 7th International Conference on Structural Engineering, Mechanics and Computation*, 2203-2208. 10.1201/9780429426506-380

Ciribini, A. L. C., Pasini, D., Tagliabue, L. C., Manfren, M., Daniotti, B., Rinaldi, S., & De Angelis, E. (2017). Tracking users' behaviors through real-time information in BIMs: Workflow for interconnection in the Brescia Smart Campus Demonstrator. *Procedia Engineering, 180*, 1484-1494.

Clastres, C. (2011). Smart grids: Another step towards competition, energy security and climate change objectives. *Energy Policy*, *39*(9), 5399–5408. doi:10.1016/j.enpol.2011.05.024

Clifton, K., Ewing, R., Knaap, G. J., & Song, Y. (2008). Quantitative analysis of urban form: A multidisciplinary review. *Journal of Urbanism*, *1*(1), 17–45. doi:10.1080/17549170801903496

Cocchia, A. (2014). Smart and digital city: A systematic literature review. *Smart City*, 13–43.

Cocchia, A. (2014). Smart and Digital City: A Systematic Literature Review. In R. P. Dameri & C. Rosenthal-Sabroux (Eds.), *Smart City: How to Create Public and Economic Value with High Technology in Urban Space* (pp. 13–43). Springer.

Codinhoto, R., & Kiviniemi, A. (2014). Bim for FM: A case support for business life cycle. In S. Fukuda, A. Bernard, B. Gurumoorthy, & A. Bouras (Eds.), *Product Lifecycle Management for a Global Market* (pp. 63–74). Springer-Verlag Berlin Heidelberg. doi:10.1007/978-3-662-45937-9_7

Cognito Project. (2012). Retrieved from Cognito Project EU: https://cordis.europa.eu/project/id/248290

Commissione Europea. (2020). *EUROSTAT Statistiche europee* [EUROSTAT Statistics]. https://ec.europa.eu/info/departments/eurostat-european-statistics_it

Communityviz, (2020). https://communityviz.city-explained.com/

Compton, M., Barnaghi, P., Bermúdez, L., García-Castro, R., Corcho, O., Cox, S., & Huang, V. (2012). The SSN ontology of the W3C semantic sensor network incubator group. *Journal of Web Semantics*, *17*, 25–32. doi:10.1016/j.websem.2012.05.003

Consiglio Comunale di Torino. (2009). Deliberazione 2008 09659/009. *Documento Programmatico finalizzato alla redazione della variante strutturale n. 200 al P.R.G. ai sensi della LR n.56 del 1977 e s.m.i. e della L.R. n.1 del 2007 s.m.i.* [Programme Document for the drafting of structural variant no. 200 to the P.R.G. in accordance with LR no. 56 of 1977 and subsequent amendments and additions and Regional Law no. 1 of 2007 and subsequent amendments and additions]. Author.

Consortium, D. (2016). *Deliverable 1.1 - Districts and buildings selection.* Author.

Coppola, F. F. (Producer), & Reggio, G. (Director). (1982). *Koyaanisqatsi: Life Out of Balance* [Motion Picture]. USA: Island Alive.

Coppola, F. F. (Producer), & Reggio, G. (Director). (1988). *Powaqqatsi: Life in Trasformation.* [Motion Picture]. USA: Island Alive.

Coppola, F. F. (Producer), & Reggio, G. (Director). (2002). *Naqoyqatsi: Life as War.* [Motion Picture]. USA: QATSI Productions, LLC and Miramax Film Corp.

Coraglia, U. M., Simeone, D., Cursi, S., Fioravanti, A., Wurzer, G., & D'Alessandro, D. (2017). A simulation model for logical and operative clash detection. *Sharing Computable Knowl., 2*, 519-534.

Corrado, V., & Ballarini, I. (2013). *Guidelines for Structuring Contextual Data.* SEMANCO Deliverable 3.3. http://www.semanco-project.eu/deliverables.htm

Corrado, V., Ballarini, I., Madrazo, L., & Nemirovskij, G. (2015). Data structuring for the ontological modelling of urban energysystems: The experience of the SEMANCO project. *Sustainable Cities and Society*, *14*, 223–235. doi:10.1016/j.scs.2014.09.006

Corry, E., Coakley, D., O'Donnell, J., Pauwels, P., & Keane, M. (2013). The role of linked data and the semantic web in building operation. *Proceedings of the 13th annual International Conference for Enhanced Building Operations. In 13th annual International Conference for Enhanced Building Operations.*

Costa, G., & Sicilia, Á. (2017). Methodology for data integration using SPARQL Constructs in the AEC industry. In *Proceedings of the 5th Linked Data in Architecture and Construction Workshop* (pp. 13-15). Retrieved from https://linkedbuildingdata.net/ldac2017/files/Presentations/LDAC2017_ARC.pdf

Costa, G., Sicilia, Á., Lilis, G. N., Rovas, D. V., & Izkara, J. (2016). A comprehensive ontologies-based framework to support the retrofitting design of energy-efficient districts. In *eWork and eBusiness in Architecture, Engineering and Construction-Proceedings of the 11th European Conference on Product and Process Modelling* (pp. 673-681). CRC Press, Taylor & Francis Group. 10.1201/9781315386904

Costa, A., Keane, M. M., Torrens, J. I., & Corry, E. (2013). Building operation and energy performance: Monitoring, analysis and optimisation toolkit. *Applied Energy*, *101*(C), 310–316. doi:10.1016/j.apenergy.2011.10.037

Costa, G., & Madrazo, L. (2015). Connecting building component catalogues with BIM models using semantic technologies: An application for precast concrete components. *Automation in Construction*, *57*, 239–248. doi:10.1016/j.autcon.2015.05.007

Costa, G., Sicilia, Á., Oregi, X., Pedrero, J., & Mabe, L. (2020). A catalogue of energy conservation measures (ECM) and a tool for their application in energy simulation models. *Journal of Building Engineering*, *29*, 101102. doi:10.1016/j.jobe.2019.101102

Costello, K., & Omale, G. (2019). *Gartner Survey Reveals Digital Twins Are Entering Mainstream Use.* https://www.gartner.com/en/newsroom/press-releases/2019-02-20-gartner-survey-reveals-digital-twins-are-entering-mai

Court, V. (2018). Energy capture, technological change, and economic growth: an evolutionary perspective. *BioPhysical Economics and Resource Quality, 3*, Article 12.

Cozzi, P. (2015). *Introducing 3D Tiles.* https://cesium.com/blog/2015/08/10/introducing-3d-tiles/

Cozzi, P., Lilley, S., & Getz, G. (2018). *3D Tiles Specification 1.0.* Open Geospatial Consortium (OGC). https://www.ogc.org/standards/3DTiles

Cracknell, D. (2014). *Sector BIM Strategy: Improving BIM Education & Training*. The Higher Education Academy. Available at http://www.heacademy.ac.uk/assets/documents/disciplines/built _environment/BIM/BIM_Skills_CIC_David_Cracknell.pdf

Cui, J. (2020). *Integration of Life Cycle Assessment within Building Information Modeling Environment* [Unpublished master thesis]. KTH, Stockholm, Sweden.

Cullen, G. (1971). *The Concise Townscape*. Architectural Press.

Cullen, G. (1976). *Il paesaggio urbano: morfologia e progettazione*. Calderini Editore.

Cuppini, N. (2020). Cosa si muove sotto lo skyline digitale? Storia e politica della smart city [What moves under the digital skyline? History and politics of the smart city]. *Sociologia Urbana e Rurale, 122*(122), 12–26. doi:10.3280/SUR2020-122002

D'Arcy Thompson, W. (2011). On Growth and Form. New York: Create Space Independent Publishing Platform.

Da Xu, L., He, W., & Li, S. (2014). Internet of Things in industries: A survey. *IEEE Trans. Ind. Informat., 10*(4).

Dan, R., & Tibbits, S. (2014). Active Printed Materials for Complex Self-Evolving Deformations. Scientific Reports, 4. doi:10.1038rep07422

Daniotti, B., Pavan, A., Lupica Spagnolo, S., Caffi, V., Pasini, D., & Mirarchi, C. (2020). Collaborative Working in a BIM Environment (BIM Platform). In BIM-Based Collaborative Building Process Management. Springer.

Darby, S. (2006). *The effectiveness of feedback on energy consumption*. Environmental Change Institute - University of Oxford.

Darong, H., Lanyan, K., Bo, M., Ling, Z., & Guoxi, S. (2018). A new incipient fault diagnosis method combining improved RLS and LMD algorithm for rolling bearings with strong background noise. *IEEE Access: Practical Innovations, Open Solutions, 6*, 26001–26010. doi:10.1109/ACCESS.2018.2829803

Davico, L., Cabodi, C., Guiati, F., Gullino, V., Staricco, L., & Vitale Brovarone, E. (2019). Futuro Rinviato [Posponed Future]. XX RGR. Centro di Ricerca e Documentazione Luigi Einaudi. Torino.

Davis, K. (2020). *Remote Control: Coronavirus Has Millions in China Working From Home*. Sixth Tone. https://www.sixthtone.com/news/1005160/remote-control-coronavirus-has-millions-in-china-working-from-home

Davis, K. (1965). The Urbanization of the Human Population. *Scientific American, 213*(3), 41–53. doi:10.1038cientificamerican0965-40

De Amicis, R., Conti, G., Patti, D., Ford, M., & Elisei, P. (2012). I-Scope–Interoperable Smart City Services through an Open Platform for Urban Ecosystems. In M. Schrenk et al. (Ed.), *Proceedings of REAL CORP 2012* (pp. 1219-1226). Academic Press.

De Ioris, D., Di Pasqua, G., Reginaldi, M., De Pascali, P., Alberti, V., (2015). *L'energia nelle trasformazioni del territorio. Ricerche su governance ed energia nelle trasformazioni del territorio* [Energy in the transformations of the land. Research on governance and energy in the transformation of the land]. Franco Angeli.

De Lange, P., Bähre, B., Finetti-Imhof, C., Klamma, R., Koch, A., & Oppermann, L. (2017). Socio-technical Challenges in the Digital Gap between Building Information Modeling and Industry 4.0. STPIS@ CAiSE, 33-46.

De Luca, D., Del Giudice, M., Dellosta, M., Fonsati, A., Osello, A., & Ugliotti, F. M. (2017). Augmented and virtual reality for smart cities users' awareness. In *INTED 2017, 11th International Technology, Education and Development Conference*, 4333-4342. IATED Academy.

Decreto Ministeriale n. 560, 01/12/2017

DEFRA – Department for Environment, Food and Rural Affairs. (2016). *UK statistics on Waste*. Governmental Report. Available at: https://www.gov.uk/government/uploads/system/uploads/attachment_data/file/593040/UK_statsonwaste_statsnotice_Dec2016_FINALv2_2.pdf

Del Giudice, M., Osello, A., Fonsati, A., De Luca, D., Musetti, A., & Marchi, F. (2017). District data management, modelling and visualization via interoperability. Proceeding of Building Simulation 2017.

Del Giudice, M., De Luca, D., & Osello, A. (2020). HBIM Data Management and Visualization for Healthcare Facilities. In C. M. Bolognesi & C. Santagati (Eds.), *Impact of Industry 4.0 on Architecture and Cultural Heritage* (pp. 239–260). IGI Global., doi:10.4018/978-1-7998-1234-0.ch010

Del Giudice, M., Grosso, R., Mecca, U., Moglia, G., Prizzon, F., & Rebaudengo, M. (2020). The maintenance representation. Research and applications, mixing UAV and digital models. In *Digital Draw Connection*. Springer.

Del Giudice, M., Osello, A., Fonsati, A., De Luca, D., Patti, E., Brundu, F. G., & Acquaviva, A. (2017). District data management, modelling and visualization via interoperability. In *Proceedings of Building Simulations (BS)*. International Building Performance Simulation Association.

Del Giudice, M., & Patti, E. (2014). *BIM and GIS for district modelling*. Research Gate.

Dembski, F., Wössner, U., Letzgus, M., Ruddat, M., & Yamu, C. (2020). *Urban Digital Twins for Smart Cities and Citizens: The Case Study of Herrenberg*. Sustainability - MDPI.

Desogus, G., Di Giuda, G. M., Monni, G., Quaquero, E., Sanna, A., Tagliabue, L. C., & Villa, V. (2018). The building information modeling for the retrofitting of existing buildings. A case study in the University of Cagliari. *International Conference Proceedings of the Seismic and Energy Renovation for Sustainable Cities*, 141-150.

Devlin, B. (1997). *Data Warehouse: From Architecture to Implementation*. Addison-Wesley.

DeWaters, J. E., & Powers, S. E. (2011). Energy literacy of secondary students in New York State (USA): A measure of knowledge, affect, and behavior. *Energy Policy*, *39*(3), 1699–1710. doi:10.1016/j.enpol.2010.12.049

Di Giuda, G. M., Quaquero, E., Villa, V., Tagliabue, L. C., Desogus, G., Sanna, A., & Ciribini, A. L. C. (2018). Towards the cognitive building: information modeling for the energy audit. *TEMA, 4*(2).

Di Giuda, G. M., Villa, V., Paleari, F., & Schievano, M. (2015). *BIM modeling of the existing school heritage for investment planning in Sostenibilita' ambientale e produzione edilizia: La ricerca scientifica sulla Produzione Edilizia nell'era delle nuove sfide ambientali nel Settore delle Costruzioni*. Ingenio.

Dickinson, J., Pardasani, A., Ahamed, S., & Kruithof, S. (2009). A survey of automation technology for realising as-built models of services. *Proceedings 1st International Conference. Improving Construction Use Through Integration Design Solutions*, 365-381.

Digregrorio, G. (2020). *Real-time data and BIM: automated protocol for management and visualization of data in real time* [Unpublished master thesis]. KTH, Stockholm, Sweden.

Dimassi, S., Demoly, F., Cruz, C., Qi, H. J., Kim, K. Y., André, J. C., & Gomes, S. (2021). An ontology-based framework to formalize and represent 4D printing knowledge in design. *Computers in Industry, 126*. doi:10.1016/j.compind.2020.103374

Dimmer, Project Description. (2015). Retrieved from DIMMER: http://dimmer,polito,it/project

Ding, Z., Yuan, C., Peng, X., Wang, T., Qi, H. J., & Dunn, M. L. (2017). Direct 4D printing via active composite materials. *Science Advances*, *3*(4), e1602890. doi:10.1126ciadv.1602890 PMID:28439560

Directive (EU) 2018/844 of the European Parliament and of the Council.

Dodge & Data Analytics. (2018). *SmartMarket Brief: Connecting Owners and Contractors*. Available online at: www.construction.com

Dolk, D. R., & Kirsch, R. A. II. (1987). A relational information resource dictionary system. *Communications of the ACM*, *30*(1), 48–61. doi:10.1145/7885.7888

Domaneschi, M., Villa, V., Cimellaro, G. P., Caldera, C., Zamani Noori, A., Marasco, S., & Ansari, F. (2019). Integrating BIM with on site investigation for seismic vulnerability assessment. *COMPDYN Proceedings*, 544-553. 10.7712/120119.6938.19957

Domìnguez, E., Pérez, B., Rubio, A. L., & Zapata, M. A. (2019). A taxonomy for key performance indicators management. *Computer Standards & Interfaces*, •••, 24–40.

Donato, V. (2014). *Transformability of Existing Building: An approach based on BIM Technologies* (PhD thesis). University of Florence and TU Braunschweig.

Donato, V. (2017). Towards design process validation integrating graph theory into BIM. *Architectural Engineering and Design Management, 13*(1), 22–38. doi:10.1080/17452007.2016.1208602

Dong, B., Lam, K. P., Huang, Y. C., & Dobbs, G. M. (2007). A comparative study of the IFC and gbXML informational infrastructures for data exchange in computational design support environments. In *Proceedings of Building Simulation*. IBPSA.

Dong, B., Lam, K. P., Huang, Y. C., & Dobbs, G. M. (2007). A comparative study of the IFC and gbXML informational infrastructures for data exchange in computational design support environments. *Proceedings: Building Simulation, 2007*, 1530–1537.

Donkers, S., Ledoux, H., Zhao, J., & Stoter, J. E. (2015). Automatic conversion of IFC datasets to geometrically and semantically correct CityGML LOD3 buildings. *Transactions in GIS, 20*(4), 547–569. doi:10.1111/tgis.12162

Donnelly, K. (2010). The technological and human dimensions of residential feedback. An introduction to the broad range of today's feedback strategies. In K. Ehrhardt-Martinez & J. A. Laitner (Eds.), *People-centered initiatives for increasing energy savings* (pp. 13–24). ACEEE.

DouS. Q.ZhangH. H.ZhaoY. Q.WangA. M.XiongY. T.ZuoJ. M. (2020). Research on construction of spatio-temporal data visualization platform for GIS and BIM fusion. Paper presented at the International Archives of the Photogrammetry. *Remote Sensing and Spatial Information Sciences - ISPRS Archives, 42*(3/W10) 555-563. Retrieved from www.scopus.com doi:10.5194/isprs-archives-XLII-3-W10-555-2020

Drygielski, M., Indyka-Piasecka, A., Piwowarczyk, M., Telec, Z., Trawiński, B., & Duong, T. H. (2019, September). Usability testing of data entry patterns implemented according to material design guidelines for the web. In *International Conference on Computational Collective Intelligence* (pp. 697-711). Springer. 10.1007/978-3-030-28377-3_58

Du Plessis, A., Broeckhoven, C., Yadroitsava, I., Yadroitsev, I., Hands, C. H., Kunju, R., & Bhate, D. (2019, March). Beautiful and Functional: A Review of Biomimetic Design in Additive Manufacturing. *Additive Manufacturing, 27*, 408–427. doi:10.1016/j.addma.2019.03.033

Duncan, D. T., Aldstadt, J., Whalen, J., & Melly, S. J. (2012). Validation of Walk Scores and Transit Scores for estimating neighborhood walkability and transit availability: A small-area analysis. *GeoJournal, 78*(2), 407–416. doi:10.100710708-011-9444-4

Dyer, T. D. (2012). *Construction and demolition waste*. Springer. doi:10.1007/978-1-4419-0851-3_118

Eastman, C., Fisher, D., Lafue, G., Lividini, J., Stoker, D., & Yessios, C. (1974). *An Outline of the Building Description System*. Institute of Physical Planning Research Report (No. 50).

Eastman, C., Eastman, C. M., Teicholz, P., & Sacks, R. (2011). *BIM handbook: A guide to building information modeling for owners, managers, designers, engineers and contractors*. John Wiley & Sons.

Eastman, C., Teicholz, P., Sacks, R., & Liston, K. (2011). *BIM Handbook: A Guide to Building Information Modeling for Owners, Managers, Designers, Engineers and Contractors*. Wiley.

Eastman, C., Teicholz, P., Sacks, R., & Liston, K. (2018). *BIM Handbook: A Guide to Building Information Modeling for Owners, Managers, Designers, Engineers, and Contractors*. John Wiley & Sons, Inc.

EC. (2016). *DAREED – Decision support Advisor for innovative business models and useR engagement for smart Energy Efficient Districts*. Retrieved from Cordis: https://cordis.europa.eu/project/id/609082/it

Ecma International. (2020). *The JSON Data Interchange Format, Standard ECMA-404*. http://www.ecma-international.org/publications/standards/Ecma-404.htm

EcoLogic Studio. (2011 January 10). *Ka-Care*. Retrieved March 12, 2020, from http://ecologicstudio.com/v2/project.php?idcat=3&idsubcat=66&idproj=121

EECITIES. (2015). *EECITIES - Energy Efficient Cities. Service platform to support planning of energy efficient cities.* http://www.eecities.com/

Eeckhout, J. (2004). Gibrat's Law for (All) Cities. *The American Economic Review*, *94*(5), 1429–1451. doi:10.1257/0002828043052303

Ehrlich, H. (Ed.). (2017). Extreme Biomimetics. Springer International Publishing AG. doi:10.1007/978-3-319-45340-8

Eid, He, X., Bahr, R., Lin, T.-H., Cui, Y., Adeyeye, A., Tehrani, B., & Tentzeris, M. M. (2020, December). Inkjet-/3D-/4D-Printed Perpetual Electronics and Modules: RF and mm-Wave Devices for 5G+, IoT, Smart Agriculture, and Smart Cities Applications. *IEEE Microwave Magazine*, *21*(12), 87–103. doi:10.1109/MMM.2020.3023310

El Saddik, A. (2018). Digital Twins: The Convergence of Multimedia Technologies. *IEEE MultiMedia*, *25*(2), 87–92. doi:10.1109/MMUL.2018.023121167

Ellegård, K., & Vilhelmson, B. (2004). Home as a pocket of local order: Everyday activities and the friction of distance. *Geografiska Annaler. Series B, Human Geography*, *86*(4), 281–296. doi:10.1111/j.0435-3684.2004.00168.x

Ellen MacArthur Foundation. (2013). *Towards the circular economy: Opportunities for the consumer goods sector*. Ellen Mac Arthur Foundation, Sun, McKinsey Center for Business and Environment.

El-Mekawy, M., Östman, A., & Hijazi, I. (2012). A Unified Building Model for 3D Urban GIS. *ISPRS International Journal of Geo-Information*, *1*(2), 120–145. doi:10.3390/ijgi1020120

Enders, M. R., & Hoßbach, N. (2019). Dimensions of Digital Twin Applications-A Literature Review. *Americas Conference on Information Systems*. Retrieved from https://aisel.aisnet.org/amcis2019/org_transformation_is/org_transformation_is/20/

EPD. The International EPD System. (2020). *What is an EPD?* https://www.environdec.com/What-is-an-EPD/

Erb, R. M., Sander, J. S., Grisch, R., & Studart, A. E. R. (2012). Self-shaping composites with programmable bioinspired microstructures. *Nature Communications*, *4*(1), 1712. doi:10.1038/ncomms2666 PMID:23591879

Eremia, M., Toma, L., & Sanduleac, M. (2017). *The Smart City Concept in the 21th Century*. Science Direct. *Procedia Engineering*, 181.

Eriksson, H. E., Penker, M., Lyons, B., & Fado, D. (2003). *UML 2 Toolkit, CafeScribe* (Vol. 26). John Wiley & Sons.

Ernst & Young. (2016). *Rapporto Smart City Index 2016*. www.comune.bologna.it/iperbole/ piancont/archivionov/tabelle_grafici/EYsmartindex/2016-EY-smart-city-index.pdf

Escribano, C. P., Theologou, N., Likar, M., Tryferidis, A., & Tzovaras, D. (2021). Business Models and Use Cases for the IoT. *IoT Platforms, Use Cases, Privacy, and Business Models*, 51–80. . doi:10.1007/978-3-030-45316-9_3

Esposito de Vita, G. (2009). Le Fonti per l'Urbanistica del XXI Secolo: Luigi Moretti (1907-1973). In Luigi Moretti e La Fondazione Della Rocca Urbanistica e Ricerca Operativa, (pp. 15-76). Rome, Italy: Fondazione Aldo della Rocca.

ESUM. (2017). *Human perception of urban environment*. https://www.esum.arch.ethz.ch

EU Commission - Climate Action. (2020). Retrieved from https://ec.europa.eu/clima/policies/strategies/2050_en

Eugster, P. T., Felber, P. A., Guerraoui, R., & Kermarrec, A.-M. (2003). The many faces of publish/subscribe. *ACM Computing Surveys*, *35*(2), 114–131. doi:10.1145/857076.857078

eur-lex.europa.eu. (2012). Retrieved from https://eur-lex.europa.eu/legal-content/EN/TXT/?uri=celex%3A32012L0027

Eurocities. (2018). *Digital 3D modelling of Rennes Metropole*. Retrieved from eurocities: http://wsdomino.eurocities.eu/eurocities/documents/Digital-3D-modelling-of-Rennes-Metropole-WSPO-A8JDUM#:~:text=Rennes%20Metropole%20has%20developed%20a,on%20buildings%2C%20and%20so%20on

European Commission (EC). (2019). *EU Energy in figures, Statistical Pocketbook 2019*. Publications Office of the European Union.

European Commission. (2007). *A European strategic energy technology plan (SET-plan) - Towards a low carbon future*, COM (2007)723.

European Commission. (2019). *Supporting digitalisation of the construction sector and SMEs: Including Building Information Modelling*. Accessed on August 24, 2020: https://ec.europa.eu/docsroom/documents/38281

European Commission. (2020). *EU Buildings Factsheets*. Accessed on August 24, 2020: https://ec.europa.eu/energy/eu-buildings-factsheets_en

European Commission. (2020). *Horizon 2020 Work Programme 2018 – 2020* (European Commission Decision C(2020) 1862 of 25 March 2020). https://ec.europa.eu/research/participants/data/ref/h2020/wp/2018-2020/main/h2020-wp1820-fet_en.pdf

European Commission. (2020). *Smart Cities*. https://ec.europa.eu/info/eu-regional-and-urban-development/topics/cities-and-urban-development/city-initiatives/smart-cities_en

European Construction Sector Observatory. (2019a). *Building Information Modelling in the EU construction sector*. European Commission. Accessed on August 24, 2020: https://ec.europa.eu/docsroom/documents/34518

European Construction Sector Observatory. (2019b). *Country Profile Italy*. European Commission. Accessed on August 24, 2020: https://ec.europa.eu/docsroom/documents/39524

European Construction Sector Observatory. (2019c). *Country Profile United Kingdom*. European Commission. Accessed on August 24, 2020: https://ec.europa.eu/docsroom/documents/39527

European Construction Sector Observatory. (2020a). *Country Profile France*. European Commission. Accessed on August 24, 2020: https://ec.europa.eu/docsroom/documents/40289

European Construction Sector Observatory. (2020b). *Country Profile Germany*. European Commission. Accessed on August 24, 2020: https://ec.europa.eu/docsroom/documents/40681

European Construction Sector Observatory. (2020c). *Country Profile Spain*. European Commission. Accessed on August 24, 2020: https://ec.europa.eu/docsroom/documents/40684

European Construction Sector Observatory. (2020d). *Improving the human capital basis*. European Commission. Accessed on August 24, 2020: https://www.clustercollaboration.eu/eu-initiatives/reports/european-construction-sector-observatory-improving-human

European Parliament. (2014). *Mapping Smart City in the EU*. https://www.smartcities.at/assets/Publikationen/Weitere-Publikationenzum-Thema/mappingsmartcities.pdf

Eurostat. (2017). *NUTS database*. http://appsso.eurostat.ec.europa.eu/nui/show.do?dataset=apro_cpnh1&lang=en

Evans, S., Savian, C., Burns, A., & Cooper, C. (2019). Digital twins for the built environment - An introduction to the opportunities, benefits, challenges and risks. The Institution of Engineering and Technology (IET).

Falcinelli, R. (2011). *Guardare, pensare, progettare. Neuroscienze per il design*. Stampa Alternativa & Graffiti.

Fan, C., Zhang, C., Yahja, A., & Mostafavi, A. (2019). Disaster City Digital Twin: A vision for integrating artificial and human intelligence for disaster management. *International Journal of Information Management*, *102049*(November). doi:10.1016/j.ijinfomgt.2019.102049

Fanger, P. O. (1970). *Thermal comfort. Analysis and applications in environmental engineering*. Danish Technical Press.

Fannin, R. (2020). *The rush to deploy robots in China amid the coronavirus outbreak*.

Farsi, M., Daneshkhah, A., Hosseinian-Far, A., & Jahankhani, H. (Eds.). (2020). *Digital Twin Technologies and Smart Cities*. Springer Nature Switzerland. doi:10.1007/978-3-030-18732-3

Ferguson, F. (1975). *Architecture, Cities and the Systems Approach*. Braziller.

Fernandez-Anez, V., Fernández-Güellb, M. J., & Giffingerc, R. (2018). Smart City implementation and discourses: An integrated conceptual model. The case of Vienna. *Cities (London, England)*, *78*, 4–16. doi:10.1016/j.cities.2017.12.004

Ferraris, M. (2018). *Scienza Nuova*. Rosenberg & Sellier.

Fetchko, P. (2004). Deployment Optimization of a Boom for Falcon SAT-3 using Elastic Memory Composite Material. *18th Annual AIAA/ USU Conference on Small Satellites*. SSC04-VIII-4.

Fialho, B. C., Codinhoto, R., & Fabricio, M. M. (2019). Trends in BIM and IoT for Reactive Maintenance. *Proceeding of the 36th CIB W78 2019 Conference*.

Fialho, B. C., Codinhoto, R., & Fabricio, M. M. (2020). *BIM and IoT for the AEC Industry: A systematic literature review*. XXIV SIGraDi Congress "Transformative Design" School of Architecture and Design of the Pontific University.

Fielding, R. T. (2000). *Architectural styles and the design of network-based software architectures*. University of California.

Figueira, J., Greco, S., & Ehrogott, M. (2005). *Multiple Criteria Decision Analysis: State of the Art Surveys*. Springer.

Fisteus, J. A., García, N. F., Fernández, L. S., & Fuentes-Lorenzo, D. (2014). Ztreamy: A middleware for publishing semantic streams on the web. *Journal of Web Semantics*, *25*, 16–23. doi:10.1016/j.websem.2013.11.002

Fitoussi, J. P., Sen, A., & Stiglitz, J. (2010). *Mismeasuring our lives: why GDP doesn't add up*. New Press.

FI-WARE. (n.d.). Retrieved September 2015, from https://www.fiware.org/

Flanders-China Chamber of Commerce. (2020). Three clinical Covid-19 vaccine trials approved in China. http://news.flanders-china.be/three-clinical-covid-19-vaccine-trials-approved-in-china

Floridi, L. (2011). *The philosophy of information*. Oxford University Press. doi:10.1093/acprof:oso/9780199232383.001.0001

Fondazione per lo Sviluppo Sostenibile. (2012). *Città intelligenti e sostenibili* [Smart and sustainable cities]. Author.

Fondazione per lo Sviluppo Sostenibile. (2012). *Piacenza Smart City 2020. Il modello smart city per la sostenibilità delle città medie Piacenza Smart City 2020* [The smart city model for the sustainability of medium-sized cities]. Author.

Fondazione Torino Smart City. (2013). *Deliberazione della Giunta Comunale 2013 07373/068. SMILE - Smart mobility.* Inclusion, Life & Health, Energy.

Fonseca, J. A., Nguyen, T., Schlueter, A., & Marechal, F. (2016). City Energy Analyst (CEA): Integrated framework for analysis and optimization of building energy systems in neighborhoods and city districts. *Energy and Building*, *113*, 202–226. doi:10.1016/j.enbuild.2015.11.055

Forrester, J. W., Meadows, D. L., & Meadows, D. H. (1973). *Toward Global Equilibrium.* Wright-Allen Press.

Foulds, C., & Robison, R. (2018). *Advancing Energy Policy: Lessons on the integration of Social Sciences and Humanities.* Springer Nature. doi:10.1007/978-3-319-99097-2

Foundation, N. R. (2018). *Virtual Singapore.* Retrieved from National Research Foundation: https://www.nrf.gov.sg/programmes/virtual-singapore

Fowler, M. (n.d.). *Microservices.* Retrieved September 2015, from https://martinfowler.com/articles/microservices.html

Fox, M., & Kemp, M. (2009). *Interactive Architecture.* Princeton Architectural Press.

Francisco, A., Mohammadi, N., & Taylor, J. E. (2020). Smart City Digital Twin-Enabled Energy Management: Toward Real-Time Urban Building Energy Benchmarking. *Journal of Management Engineering*, *36*(2), 04019045. Advance online publication. doi:10.1061/(ASCE)ME.1943-5479.0000741

Frank, A. G., Mendes, G. H. S., Ayala, N. F., & Ghezzi, A. (2019). Servitization and Industry 4.0 convergence in the digital transformation of product firms: A business model innovation perspective. *Technological Forecasting and Social Change*, *141*, 341–351. doi:10.1016/j.techfore.2019.01.014

Frank, L. D., Saelens, B. E., Powell, K. E., & Chapman, J. E. (2007). Stepping towards causation: Do built environments or neighborhood and travel preferences explain physical activity, driving, and obesity? *Social Science & Medicine*, *65*(9), 1898–1914. doi:10.1016/j.socscimed.2007.05.053 PMID:17644231

Frank, L. D., Sallis, J. F., Conway, T. L., Chapman, J. E., Saelens, B. E., & Bachman, W. (2006). Many pathways from land use to health: Associations between neighborhood walkability and active transportation, body mass index, and air quality. *Journal of the American Planning Association*, *72*(1), 75–87. doi:10.1080/01944360608976725

Frank, L. D., Schmid, T. L., Sallis, J. F., Chapman, J., & Saelens, B. E. (2005). Linking objectively measured physical activity with objectively measured urban form. *American Journal of Preventive Medicine*, *28*(2), 117–125. doi:10.1016/j.amepre.2004.11.001 PMID:15694519

Frantzeskaki, N., Broto, V. C., Coenen, L., & Loorbach, D. (2016). *Urban sustainability transitions.* Routledge.

Fratzl, P. & Burgert, I. (2009). Actuation systems in plants as prototypes for bioinspired devices. *Phil. Trans. R. Soc. A*, *6*, 1541-1557.

Fraunhofer Gesellschaft. (2019). *Morgenstadt - City of the future Initiative.* https://www.morgenstadt.de/en.html.Accessed

Fraunhofer. (2019). *Building from Design to Demolition.* Available: https://www.bau.fraunhofer.de/en/_eldsofresearch/smartbuilding/digital-twin.html

Frazer, J. (1995). *Themes VII: An Evolutionary Architecture.* Architectural Association.

Frontera, M. (2019, May 22). *Impennata nei bandi pubblici in Bim: in un anno +223% in numero e +428% in valore, il Sole24Ore*. Available online at: https://www.ilsole24ore.com/art/impennata-bandi-pubblici-bim-un-anno-223percento-numero-e-428percento-valore-ACrVJEG

Fu, C., Aouad, G., Lee, A., Mashall-Ponting, A., & Wu, S. (2006). IFC model viewer to support nD model application. *Automation in Construction*, *15*(2), 178–185. doi:10.1016/j.autcon.2005.04.002

Fuller, A., Fan, Z., Day, C., & Barlow, C. (2020). Digital Twin: Enabling Technologies, Challenges and Open Research. *IEEE Access: Practical Innovations, Open Solutions*, *8*, 1–1. doi:10.1109/ACCESS.2020.2998358

Fusero, P., Massimiano, L., Tedeschi, A., & Lepidi, S. (2013). *Parametric Urbanism: a New Frontier for Smart Cities. Planum – The Journal of Urbanism, 2(27).*

Gabaix, X. (1999). Zipf's Law for Cities: An Explanation. *The Quarterly Journal of Economics*, *114*(3), 739–767. doi:10.1162/003355399556133

Gabbiani, B. (2018, February 28). *Architetti e ingegneri in Italia. Troppi o troppo pochi?* Available online at https://www.impresedilinews.it/architetti-e-ingegneri-in-italia-troppi-o-troppo-pochi/

Galli, A., & Massimiano, L. (2015). Progettazione Parametrica della Città Attraverso gli Open Data. *Urbanistica Informazioni, Proceedings of IX Giornata di Studi INU, Special Issue September-October 2015*. Rome, Italy: INU Edizioni.

Gangale, F., Mengolini, A., & Onyeji, I. (2013). Consumer engagement: An insight from smart grid projects in Europe. *Energy Policy*, *60*, 621–628. doi:10.1016/j.enpol.2013.05.031

Gao, B., Yang, Q., Zhao, X., Jin, G., Ma, Y., & Xu, F. (2016). 4D printing for biomedical applications. *Trends in Biotechnology*, *9*(9), 746–756. doi:10.1016/j.tibtech.2016.03.004 PMID:27056447

Gartner. (2020). Gartner Special Report "Top 10 Strategic Technology Trends for 2020." *Gartner IT Symposium/Xpo™ 2020.*

Gartner. (2020). *Gartner Top 10 Strategic Technology Trends for 2020.* https://www.gartner.com/smarterwithgartner/gartner-top-10-strategic-technology-trends-for-2020/

Gaudio, F. (2012). *1982-2012. 30 anni di teleriscaldamento a Torino*. IREN Energia.

Ge, Q., Sakhaei, A. H., Lee, H., Dunn, C. K., Fang, N. X., & Dunn, M. L. (2016). Multimaterial 4D Printing with Tailorable Shape Memory Polymers. *Scientific Reports*. doi:10.1038rep31110

Geelenn, D., Reinders, A., & Keyson, D. (2013). Empowering the end-user in smart grids: Recommendations for the design of products and services. *Energy Policy*, *61*, 151–161. doi:10.1016/j.enpol.2013.05.107

Ge, Q., Qi, H. J., & Dunn, M. L. (2013). Active materials by four-dimension printing. *Applied Physics Letters*, *103*(13), 131901. doi:10.1063/1.4819837

Gerbert, P., Castagnino, S., Rothballer, C., Renz, A., & Filitz, R. (2016). *Digital in Engineering and Construction The Transformative Power of Building Information Modeling*. The Boston Consulting Group. Accessed on August 24, 2020: http://futureofconstruction.org/content/uploads/2016/09/BCG-Digital-in-Engineering-and-Construction-Mar-2016.pdf

Gerla, M., Lee, E. K., Pau, G., & Lee, U. (2014). Internet of vehicles: From intelligent grid to autonomous cars and vehicular clouds. *Proceedings IEEE World Forum Internet Things (WF-IoT)*, 241-246. 10.1109/WF-IoT.2014.6803166

Gersch, M., & Goeke, C. (2007). Industry Transformation – Conceptual Considerations from an Evolutionary Perspective. *Journal of Business Market Management*, *1*(2), 151–182. doi:10.100712087-007-0006-5

Giffinger, R., Fertner, C., Kramar, H., Meijers, E., & Pichler-Milhanovic, N. (2007). *Smart Cities: Ranking of European medium-sized cities*. Centre of Regional Science.

Girardin, F., Del Fiore, F., Ratti, C., & Blat, J. (2008). Leveraging Explicitly Disclosed Location Information to Understand Tourist Dynamics: A Case Study. Journal of Location-Based Service, 2(1). doi:10.1080/17489720802261138

Girardin, F., Calabrese, F., Dal Fiore, F., & Ratti, C. (2008, October-December). Digital Footprinting: Uncovering Tourists with User-Generated Content. *IEEE Pervasive Computing*, *7*(4), 36–43. doi:10.1109/MPRV.2008.71

Girardin, F., Vaccari, A., Gerber, A., Birderman, A., & Ratti, C. (2009). Quantifying Urban Attractiveness from the Distribution and Density of Digital Footprints. *International Journal of Spatial Data Infrastructures Research*, 4.

Giusto, D., Iera, A., Morabito, G., & Atzori, L. (2010). *The internet of things: 20th Tyrrhenian workshop on digital communications.* Springer Science & Business Media.

Gladman, A. S., Matsumoto, E. A., Nuzzo, R. G., Mahadevan, L., & Lewis, J. A. (2016a). Biomimetic 4D printing. *Nature Materials*, *15*(4), 413–418. doi:10.1038/nmat4544 PMID:26808461

Gleich, A., Pade, C., Petschow, U., & Pissarskoi, E. (2010) Potentials and Trends in Biomimetics. Nova York: Springer. doi:10.1007/978-3-642-05246-0

Global Alliance for Buildings and Construction (GABC). (2020). *GlobalABC for Building and Construction 2020-2050 Toward a zero emission, efficient and resilient buildings and construction sector*. United Nation Environment Programme. IEA.

Global Alliance for Buildings and Construction, International Energy Agency and the United Nations Environment Programme. (2019). *2019 global status report for buildings and construction: Towards a zero-emission, efficient and resilient buildings and construction sector.* Accessed on August 24, 2020: https://wedocs.unep.org/bitstream/handle/20.500.11822/30950/2019GSR.pdf?sequence=1&isAllowed=y

Global, F. M. (2018). *Global facilities management market report 2018*. Accessed on August 24, 2020: https://www.hanze.nl/assets/instituut-voor-facility-management/Documents/Public/24315%20Global%20FM%20Market%20Report%202017_0%20%28002%29.pdf

Gottschalk, P. (2009). Maturity levels for interoperability in digital government. *Government Information Quarterly*, *26*(1), 75–81. doi:10.1016/j.giq.2008.03.003

Gourlis, G., & Kovacic, I. (2017). Building Information Modelling for analysis of energy efficient industrial buildings – A case study. *Journal of Renewable and Sustainable Energy Reviews*, *68*, 953–963. doi:10.1016/j.rser.2016.02.009

Goy, S., Maréchal, F., & Finn, D. (2020). Data for Urban Scale Building Energy Modelling: Assessing Impacts and Overcoming Availability Challenges. *Energies*, *13*(16), 4244. doi:10.3390/en13164244

Green Building, X. M. L. (gbXML) Schema. (2020). *gbXML*. https://www.gbxml.org/index.html

Greenreport Review. (2013). *Ecco Torino Smart City: Smile, la città intelligente e sostenibile* [Here is Torino Smart City: Smile, the intelligent and sustainable city]. https://www.greenreport.it/news/comunicazione/ecco-torino-smart-city-smile-la-citta-intelligente-e-sostenibile/

Grey, T. (2016), INDICATE Indicator-based Interactive Decision Support and Information Exchange Platform for Smart Cities. *Planning, Climate Change and Energy in Ireland - Irish Planning Institute Guide*, 65-72.

Grey, T. (2016). INDICATE-Indicator-based Interactive Decision Support and Information Exchange Platform for Smart Cities- Project Outline. In Planning, Climate Change and Energy in Ireland - Irish Planning Institute Guide. Academic Press.

Grieves, M. (2014). *Digital Twin : Manufacturing Excellence through Virtual Factory Replication.* White Paper, March. https://www.researchgate.net/publication/275211047_Digital_Twin_Manufacturing_Excellence_through_Virtual_Factory_Replication

Grieves, M. (2014). *Digital Twin: Manufacturing Excellence through Virtual Factory Replication.* Whitepaper.

Grieves, M. W. (2003). PLM-Beyond lean manufacturing. SOC Manufacturing Engineers.

Grieves, M., Vickers, J., Kahlen, F. J., Flumerfelt, S., & Alves, A. (2017). Digital Twin: Mitigating Unpredictable, Undesirable Emergent Behavior in Complex Systems. In F.-J. Kahlen, S. Flumerfelt, & A. Alves (Eds.), Transdisciplinary Perspectives on Complex Systems: New Findings and Approaches (pp. 85–113). Springer International Publishing. doi:10.1007/978-3-319-38756-7_4

Grieves, M., & Vickers, J. (2016). Digital Twin: Mitigating Unpredictable, Undesirable Emergent Behavior in Complex Systems. In *Trans-Disciplinary Perspectives on System Complexity* (pp. 85–114). Springer.

Gröger, G., Kolbe, T. H., & Czerwinski, A. (2006). *Candidate OpenGIS® CityGML Implementation Specification (City Geography Markup Language).* Open Geospatial Consortium (OGC). https://portal.ogc.org/files/?artifact_id=16675

Gröger, G., Kolbe, T., Nagel, C., & Häfele, K. (2019). *OGC City Geography Markup Language (Citygml) Encoding Standard v2. 0. OGC Doc.* http://www.opengeospatial.org/standards/citygml

Gröger, G., Kolbe, T. H., Nagel, C., & Häfele, K. H. (2012). *OGC City Geography Markup Language (CityGML) Encoding Standard.* Open Geospatial Consortium.

Grossi, G., & Pianezzi, D. (2017). Smart cities: Utopia or neoliberal ideologism? *Cities (London, England)*, *69*(9), 79–85. doi:10.1016/j.cities.2017.07.012

Gruber, T. R. (1993). A translation approach to portable ontology specifications. *Knowledge Acquisition*, *5*(2), 199–220. doi:10.1006/knac.1993.1008

Gruber, T. R. (1995). Toward principles for the design of ontologies used for knowledge sharing? *International Journal of Human-Computer Studies*, *43*(5-6), 907–928. doi:10.1006/ijhc.1995.1081

Gudes, O., Kendall, E., Yigitcanlar, T., Pathak, V., & Baum, S. (2010). Rethinking health planning: A framework for organising information to underpin collaborative health planning. *The HIM Journal*, •••, 18–29.

Guerrero, L. A., Maas, G. & Hogland, W. (2013). Solid waste management challenges for cities in developing countries. *Waste Management, 33*(1), 220-232.

Guerriero, A., Kubicki, S., & Reiter, S. (2016) Building information modeling in use: How to evaluate the return on investment? *eWork and eBusiness in Architecture, Engineering and Construction - Proceedings of the 11th European Conference on Product and Process Modelling, ECPPM 2016*, 537–544.

Haarstad, H., & Marikken, W. W. (2019). *Are smart city projects catalyzing urban energy sustainability?* Elsevier. *Energy Policy*, *129*(June), 918–925. doi:10.1016/j.enpol.2019.03.001

Haas, R., & Schipper, L. (1998). Residential energy demand in OECD-countries and the role of irreversible efficiency improvements. *Energy Economics*, *20*(4), 421–442. doi:10.1016/S0140-9883(98)00003-6

Hamidi, B., Bulbul, T., Pearce, A., & Thabet, W. (2014). Potential application of BIM in cost- benefit analysis of demolition waste management. In *Construction Research Congress 2014: Construction in a Global Network.* Construction Institute of ASCE.

Handy & Clifton. (2000). *Evaluating Neighborhood Accessibility: Issues and Methods Using Geographic Information Systems*. The University of Texas - Austin. SWUTC/00/167202-1.

Hannus, M., Samad, A., & Zarli, A. (Eds.). (2010). *ICT Supported Energy Efficiency in Construction. Strategic Research Roadmap and Implementation Recommendations*. REEB Project Consortium.

Hao, W., Yisha, P., & Xiaochun, L. (2019). Integration of BIM and GIS in sustainable built environment: a review and bibliometric analysis. In Automation in Construction. Elsevier.

Hargreaves, T., Nye, M., & Burgess, J. (2010). Making energy visible. A qualitative field study of how householders interact with feedback from smart energy monitors. *Energy Policy*, *38*(10), 6111–6119. doi:10.1016/j.enpol.2010.05.068

Hargreaves, T., Nye, M., & Burgess, J. (2013). Keeping energy visible? Exploring how householders interact with feedback from smart energy monitors in the longer term. *Energy Policy*, *52*, 126–134. doi:10.1016/j.enpol.2012.03.027

Hedegaard, R. E., Kristensen, M. H., Pedersen, T. H., Brun, A., & Petersen, S. (2019). Bottom-up modelling methodology for urban-scale analysis of residential space heating demand response. *Applied Energy*, *242*, 181–204. doi:10.1016/j.apenergy.2019.03.063

Heidi, S. (2018). *Non-Spatial Statistic for Analysing Human Perception of the Built Environment* (Master Thesis). ETH Zurich.

Helsinki, C. o. (2019). *Helsinki's 3D city models*. Retrieved from City of Helsinki: https://www.hel.fi/helsinki/en/administration/information/general/3d/

Hénard, E. (1910). *The Cities of the future, Royal Institute of British Architects. Town Planning Conference, London, UK.*

He, W., & Da Xu, L. (2014). Integration of distributed enterprise applications: A survey. *IEEE Transactions on Industrial Informatics*, *10*(1), 35–42. doi:10.1109/TII.2012.2189221

Hinchy, E. P., O'Dowd, N. P., & McCarthy, C. T. (2019). Using open-source microcontrollers to enable digital twin communication for smart manufacturing. *Procedia Manufacturing*, *38*, 1213–1219. doi:10.1016/j.promfg.2020.01.212

Hollands, R. G. (2008). Will the real smart city please stand up? Intelligent, progressive or entrepreneurial? *City*, *12*(3), 303–320. doi:10.1080/13604810802479126

Hollenstein, D., & Bleisch, S. (2016). *Walkability for different urban granularities*. In The International Archives of the Photogrammetry, Remote Sensing and Spatial Information Sciences, Prague, Czech Republic.

Holló, G. (2015). A new paradigm for animal symmetry. *Interface Focus*, *5*(6), 20150032. Advance online publication. doi:10.1098/rsfs.2015.0032 PMID:26640644

Hong, T., Chen, Y., Luo, X., Luo, N., & Lee, S. H. (2020). Ten questions on urban building energy modeling. *Building and Environment*, *168*, 106508. doi:10.1016/j.buildenv.2019.106508

Howard, E. (1902). *Garden Cities of To-Morrow*. Faber and Faber.

Howell, S. K., Wicaksono, H., Yuce, B., McGlinn, K., & Rezgui, Y. (2019). User centered neuro-fuzzy energy management through semantic-based optimization. *IEEE Transactions on Cybernetics*, *49*(9), 3278–3292. doi:10.1109/TCYB.2018.2839700 PMID:30028719

Hristov, T. (2015 May 8). *Zha Kartal Pendik Masterplan Breakdown*. Retrieved February 4, 2020, from https://tshristov.wordpress.com/2015/05/08/kartal-pendik-masterplan-by-zaha-hadid-architects-work-in-progress/

Hua, X. (2020b). *Intelligent recovery of the old telephone booths.* http://www.xinhuanet.com/english/2020-05/26/c_139089181.htm

Huang, S., Wang, G., Yan, Y., & Fang, X. (2020). Blockchain-based data management for digital twin of product. *Journal of Manufacturing Systems, 54*(August), 361–371. doi:10.1016/j.jmsy.2020.01.009

Huynh, D., & Nguyen-Ky, S. (2020). Engaging building automation data visualisation using building information modelling and progressive web application. *Open Engineering, 10*(1), 434–442. doi:10.1515/eng-2020-0054

Hwang, J. R., Hong, C. H., & Choi, H. S. (2013, May). Implementation of prototype for interoperability between BIM and GIS: Demonstration paper. In *IEEE 7th International Conference on Research Challenges in Information Science (RCIS)* (pp. 1-2). IEEE.

Hwang, J., Jeong, Y., Park, J. M., Lee, K. H., Hong, J. W., & Choi, J. (2015). Biomimetics: Forecasting the future of science engineering, and medicine. *International Journal of Nanomedicine, 10*, 5701–5713. PMID:26388692

IBM Corporation, . (2012). *IBM Smarter Cities. Creating opportunities through leadership and innovation.* Author.

IBM. (2009). *Smarter Thinking for a Smarter Planet.* https://www.ibm.com/smarterplanet/us/en/

IEE Project TABULA and EPISCOPE web site. (n.d.). https://episcope.eu/welcome/

Immortal Project. (2018). Retrieved from Immortal Project EU: http://h2020-immortal.eu/

InarCassa (2020). *InarCassa in cifre.* Statistiche su iscritti e pensionati, redditi e volume d'affari.

INDICATE. (2016). *Project website.* https://cordis.europa.eu/project/id/608775 retrieved at 8/02/2021

Industry 4.0 in Big Data environment. (2013). *German Harting Magazine, 26*, 8–10.

Infoworld. (2020). *What is JSON? A better format for data exchange.* https://www.infoworld.com/article/3222851/what-is-json-a-better-format-for-data-exchange.html

Ingrao, C., Messineo, A., Beltramo, R., Yigitcanlar, T., & Ioppolo, G. (2018). How can life cycle thinking support sustainability of buildings? Investigating life cycle assessment applications for energy efficiency and environmental performance. *Journal of Cleaner Production, 201*, 556–569. doi:10.1016/j.jclepro.2018.08.080

International Facility Management Association. (1980). *What is Facility Management?* https://www.ifma.org/about/what-is-facility-management

International Facility Management Association. (n.d.). *Cos'è il Facility Management.* Accessed on August 24, 2020: http://www.ifma.it/index.php?pagina=articolo.php&id_articolo=25&var_id_menu=68&nodata

Ionov, L. (2012). Bioinspired micro origami by self-folding polymer films. *Macromol. Chem. Phys., 214*, 1178-1183.

ISO/IEC JTC 1. (2014). *Smart cities. Preliminary Report 2014.*

ISTAT. (2011). https://www.istat.it/it/archivio/104317

Istituto Superiore per la Protezione e la Ricerca Ambientale. (2018). *Consumo di suolo, dinamiche territoriali e servizi ecosistemici* [Soil consumption, territorial dynamics and ecosystem services] (Rapporti ISPRA 288). ISPRA.

Ivanic, T., Tadic, Z., & Omazic, M. A. (2015). Biomimicry-an overview. Holistic. Approach. *Environ., 5*(1), 19–36.

Jacobs, J. (1961). *The Death and Life of Great American Cities.* Random House.

Jacobson, I. (1992). *Object-Oriented Software Engineering: A Use Case Driven Approach.* Addison-Wesley.

Jalaei, F., & Jrade, A. (2014). Integrating building information modeling (BIM) and energy analysis tools with green building certification system to conceptually design sustainable buildings. *Journal of Information Technology in Construction, 19*, 494–519.

Jambeck, J. R., Roland Geyer, R., Wilcox, C., Siegler, T. R., Perryman, M., Andrady, A., Narayan, R., & Law, K. L. (2015). Plastic waste inputs from land into the ocean. *Science, 347*(6223), 768–771. doi:10.1126cience.1260352 PMID:25678662

Jáñez Morán, A., Profaizer, P., Herrando Zapater, M., Andérez Valdavida, M., & Zabalza Bribián, I. (2016). Information and Communication Technologies (ICTs) for energy efficiency in buildings: Review and analysis of results from EU pilot projects. *Energy and Building, 127*, 128–137. doi:10.1016/j.enbuild.2016.05.064

Jebamalai, J. M., Marlein, K., & Laverge, J. (2020). Influence of centralized and distributed thermal energy storage on district heating network design. *Energy, 202*, 117689. doi:10.1016/j.energy.2020.117689

Jena, A. (n.d.). Retrieved January 2016, from https://jena.apache.org

Jensen, J. O. (2009). Sustainability Profile for Urban Districts in Copenhagen. In *Proceedings of Sustainable Cities and Regions: Enabling Vision or Empty Talk?* Aalborg University. doi:10.4337/9780857937162

Jia, R., Jin, B., Jin, M., Zhou, Y., Konstantakopoulos, I. C., Zou, H., Kim, J., Li, D., Gu, W., Nuzzo, P., Schiavon, S., Sangiovanni-Vincentelli, A. L., Spanos, C. J., & Arghandeh, R. (2018). Design automation for smart building systems. *Proceedings IEEE, 106*(9), 1680-1699.

Jiang, H., (2013). *A System Dynamics Model for Manpower and Technology Implementation Trade-off and Cost Estimation.* Electronic Theses and Dissertations 202.

Johari, F., Peronato, G., Sadeghian, P., Zhao, X., & Widén, J. (2020). Urban building energy modeling: State of the art and future prospects. *Renewable & Sustainable Energy Reviews, 128*, 109902. doi:10.1016/j.rser.2020.109902

Jones, D., Snider, C., Nassehi, A., Yon, J., & Hicks, B. (2020). Characterising the Digital Twin: A systematic literature review. *CIRP Journal of Manufacturing Science and Technology, 2019*, 36–52. Advance online publication. doi:10.1016/j.cirpj.2020.02.002

Jo, S. K., Park, D. H., Park, H., & Kim, S. H. (2018). Smart Livestock Farms Using Digital Twin: Feasibility Study. *9th International Conference on Information and Communication Technology Convergence: ICT Convergence Powered by Smart Intelligence, ICTC 2018*, 1461–1463. 10.1109/ICTC.2018.8539516

JouanP.HallotP. (2019). Digital twin: A HBIM-based methodology to support preventive conservation of historic assets through heritage significance awareness. Paper presented at the International Archives of the Photogrammetry. *Remote Sensing and Spatial Information Sciences - ISPRS Archives, 42*(2/W15) 609-615. Retrieved from www.scopus.com doi:10.5194/isprs-archives-XLII-2-W15-609-2019

Julin, A., Jaalama, K., Virtanen, J.-P., Pouke, M., Ylipulli, J., Vaaja, M., Hyyppä, J., & Hyyppä, H. (2018). Characterizing 3D City Modeling Projects: Towards a Harmonized Interoperable System. *ISPRS International Journal of Geo-Information, 7*(2), 55. doi:10.3390/ijgi7020055

Jusuf, S. K., Mousseau, B., Godfroid, G., & Soh, J. H. (2017). Path to an Integrated Modelling between IFC and CityGML for Neighborhood Scale Modelling. Urban Science, MDPI, 1-20.

Kaden, R., Kolbe, T.H. (2014). Simulation-based total energy demand estimation of buildings using semantic 3D city models. *Int. J. 3-D Inf. Model, 3*.

Kaewunruen, S., & Lian, Q. (2019). Digital twin aided sustainability-based lifecycle management for railway turnout systems. *Journal of Cleaner Production, 228*, 1537–1551. doi:10.1016/j.jclepro.2019.04.156

Kaewunruen, S., Peng, S., & Phil-Ebosie, O. (2020). Digital twin aided sustainability and vulnerability audit for subway stations. *Sustainability*, *12*(19), 7873. Advance online publication. doi:10.3390u12197873

Kaewunruen, S., Sresakoolchai, J., & Kerinnonta, L. (2019). Potential reconstruction design of an existing townhouse in washington DC for approaching net zero energy building goal. [Switzerland]. *Sustainability*, *11*(23), 6631. Advance online publication. doi:10.3390u11236631

Kaewunruen, S., & Xu, N. (2018). Digital twin for sustainability evaluation of railway station buildings. *Frontiers in Built Environment*, *4*, 77. Advance online publication. doi:10.3389/fbuil.2018.00077

Kamel, A., & Memari, A. M. (2019). Review of BIM's application in energy simulation: Tools, is-sues, and solutions. Automation in Construction, 164-180.

Kamilaris, A., Pitsillides, A., & Trifa, V. (2011). The smart home meets the web of things. *International Journal of Ad Hoc and Ubiquitous Computing*, *7*(3), 145–154. doi:10.1504/IJAHUC.2011.040115

Kamilaris, A., Wohlgemuth, V., Karatzas, K., & Athanasiadis, I. N. (Eds.). (2021). *Advances and New Trends in Environmental Informatics: Digital Twins for Sustainability*. Springer Nature Switzerland AG. doi:10.1007/978-3-030-61969-5

Kang, D. W. (2012 August 13). *Kartal – Pendik Masterplan*. Retrieved February 4, 2020, from https://arcspace.com/feature/kartal-pendik- masterplan/

Karak, T., Bhagat, R. M., & Bhattacharyya, P. (2012). Municipal solid waste generation, composition, and management: The world scenario. *Critical Reviews in Environmental Science and Technology*, *42*(15), 1509–1630. doi:10.1080/10643389.2011.569871

Karami, P., Al-Ayish, N., & Gudmundsson, K. (2015). A comparative study of the environmental impact of Swedish residential buildings with vacuum insulation panels. *Energy and Building*, *109*, 183–194. doi:10.1016/j.enbuild.2015.10.031

Karana, E., Pedgley, O., & Rognoli, V. (Eds.). (2013). *Materials Experience: fundamentals of materials and design*. Butterworth-Heinemann.

Karan, E. P., Irizarry, J., & Haymaker, J. (2015). BIM and GIS Integration and Interoperability Based on Semantic Web Technology. *Journal of Computing in Civil Engineering*, *30*(3), 04015043. doi:10.1061/(ASCE)CP.1943-5487.0000519

Kassem, M., Kelly, G., Dawood, N., Serginson, M., & Lockley, S. (2015). BIM in facilities management applications: A case study of a large university complex. *Built Environment Project and Asset Management*, *5*(3), 261–277. doi:10.1108/BEPAM-02-2014-0011

Kaur, M. J., Mishra, V. P., & Maheshwari, P. (2020). The Convergence of Digital Twin, IoT, and Machine Learning: Transforming Data into Action. In M. Farsi, A. Daneshkhah, A. Hosseinian-Far, & H. Jahankhani (Eds.), *Digital Twin Technologies and Smart Cities. Internet of Things (Technology, Communications and Computing)*. Springer. doi:10.1007/978-3-030-18732-3_1

Kavgic, M., Mavrogianni, A., Mumovic, D., Summerfield, A., Stevanovic, Z., & Djurovic-Petrovic, M. (2010). A review of bottom-up building stock models for energy consumption in the residential sector. *Building and Environment*, *45*(7), 1683–1697. doi:10.1016/j.buildenv.2010.01.021

Keirstead, J., Samsatli, N., & Shah, N. (2010). SynCity: An integrated tool kit for urban energy systems modelling. *Energy efficient cities: Assessment tools and benchmarking practices*, 21-42.

Keirstead, J., Jennings, M., & Sivakumar, A. (2012). A review of urban energy system models: Approaches, challenges and opportunities. *Renewable & Sustainable Energy Reviews*, *16*(6), 3847–3866. doi:10.1016/j.rser.2012.02.047

Keirstead, J., & Shah, N. (Eds.). (2013). *Urban Energy Systems: An Integrated Approach*. Routledge. doi:10.4324/9780203066782

Kelly, G., Serginson, M., Lockley, S., Dawood, N., & Kassem, M. (2013, October). BIM for facility management: a review and a case study investigating the value and challenges. In *Proceedings of the 13th International Conference on Construction Applications of Virtual Reality* (No. 30-31, p. 5). Academic Press.

Kempton, W., & Layne, L. L. (1994). The consumer's energy analysis environment. *Energy Policy*, *22*(10), 857–866. doi:10.1016/0301-4215(94)90145-7

Kezmane, A., Chiaia, B., Kumpyak, O., Maksimov, V., & Placidi, L. (2016). 3D modelling of reinforced concrete slab with yielding supports subject to impact load. *European Journal of Environmental and Civil Engineering*, 1–38. doi:10.1080/19648189.2016.1194330

Khajavi, S. H., Motlagh, N. H., Jaribion, A., Werner, L. C., & Holmström, J. (2019). Digital Twin: Vision, Benefits, Boundaries, and Creation for Buildings. *IEEE Access, 7*, 147406-147419. . doi:10.1109/ACCESS.2019.2946515

Khalil, A., Stravoravdis, S., & Backes, D. (2020). *Categorisation of building data in the digital documentation of heritage buildings*. Applied Geomatics., doi:10.100712518-020-00322-7

Khan, I. A., Liew, C. S., Loy, Y. Y., Lu, W., Myint, P. N., & Soh, K. B. A. (2015). *4D printing with smart materials*. https://www.slideshare.net/Funk9 8/4d-printing-with-smart-materials

Khan, M. A., Algarni, F., & Quasim, M. T. (Eds.). (2021). *Smart Cities: A Data Analytics Perspective*. Springer Nature Switzerland AG. doi:10.1007/978-3-030-60922-1

Kim, Y.-J., Thottan, M., Kolesnikov, V., & Lee, W. (2010). A secure decentralized data-centric information infrastructure for smart grid. *IEEE Communications Magazine*, *48*(11), 58–65. doi:10.1109/MCOM.2010.5621968

Kirihara, S., & Nakata, K. (Eds.). (2021). *Multi-dimensional Additive Manufacturing. The Japan Welding Engineering Society (JWES)*. Springer Nature Singapore Pte Ltd. doi:10.1007/978-981-15-7910-3

Knapp, J., Zeratsky, J., & Kowitz, B. (2016). *Sprint: How to Solve Big Problems and Test New Ideas in Just Five Days*. Simon and Schuster.

Kocher, J., & Lerner, M. (2007). *Walkscore*. Available: https//walkscore.com/

Kolarevic, B. (2003). *Architecture in the Digital Age. Design and Manufacturing*. Routledge.

Kolbe, T. H. (2009). Representing and Exchanging 3D City Models with CityGML. In *Proceedings of the 3rd International Workshop on 3D Geo-Information, Lecture Notes in Geoinformation & Cartography*. Springer Verlag.

Koolhaas, R. (2014). *Are Smart Cities Condemned to Be Stupid?* https://www.archdaily.com/576480/rem-koolhaas-asks-are-smart-cities-condemned-to-be-stupid

Kortman, J., van Ewijk, H., van Konijnenburg, P., Lanting, R., de Groot-van Dam, A., Kleefman, R., & Timmermans, F. (2001). *Duurzaamheidsprofiel van een locatie, Ontwikkeling en est van het DPL instrument versie 1.0*. IVAM Environmental Research, TNO - Building and Construction Research and TNO - Environment, Energy and Process Innovation.

Kovács-Győri, A., Resch, B., & Zeile, P. (2019). Defining and assessing walkability: An concept for an integrated approach using surveys, biosensors and geospatial analysis. *Urban Development Issues*, *62*(1), 5–15. doi:10.2478/udi-2019-0008

Kritzinger, W., Karner, M., Traar, G., Henjes, J., & Sihn, W. (2018). Digital Twin in manufacturing: A categorical literature review and classification. *IFAC-Pap.*, *51*(11), 1016–1022. doi:10.1016/j.ifacol.2018.08.474

Krog, L. (2019). How Municipalities Act under the New Paradigm for Energy Planning. *Sustainable Cities and Society, 47*.

Krüger, A., & Kolbe, T. H. (2012). Building analysis for urban energy planning using key indicators on virtual 3D city models—The energy atlas of Berlin. *The International Archives of the Photogrammetry, Remote Sensing and Spatial Information Sciences*, *39*(B2), 145–150.

Kuechler, B., & Vaishnavi, V. (2008). On theory development in design science research: Anatomy of a research project. *European Journal of Information Systems*, *17*(5), 489–504. doi:10.1057/ejis.2008.40

Kümmel, R. (2011). *The second law of economics: energy, entropy, and the origins of wealth*. Springer Science & Business Media. doi:10.1007/978-1-4419-9365-6

Kupriyanovsky, V., Klimov, A., Voropaev, Y., Pokusaev, O., Dobrynin, A., Ponkin, I., & Lysogorsky, A. (2020). Digital twins based on the development of BIM technologies, related ontologies, 5G, IoT, and mixed reality for use in infrastructure projects and IFRABIM. *International Journal of Open Information Technologies*, *8*(3), 55–74.

Kurakula, V. (2007). *A GIS-Based Approach for 3D Noise Modelling Using 3D City Models* (Master's Thesis). International Institute for Geo-information Science and Earth Observation, Enschede, The Netherlands.

Kuznetsov, S., & Paulos, E. (2010). Rise of the Expert Amateur: DIY Projects, Communities, and Cultures. *Proceedings of the 6th Nordic Conference on HumanComputer Interaction: Extending Boundaries*, 295–304.

Kwon, M. (2004). *One place after another*. MIT Press.

Laaki, H., Miche, Y., & Tammi, K. (2019). Prototyping a Digital Twin for Real Time Remote Control over Mobile Networks: Application of Remote Surgery. *IEEE Access: Practical Innovations, Open Solutions*, *7*, 20235–20336. doi:10.1109/ACCESS.2019.2897018

Laamarti, F., Badawi, H. F., Ding, Y., Arafsha, F., Hafidh, B., & El Saddik, A. (2020). An ISO/IEEE 11073 Standardized Digital Twin Framework for Health and Well-Being in Smart Cities. *IEEE Access: Practical Innovations, Open Solutions*, *8*, 105950–105961. doi:10.1109/ACCESS.2020.2999871

Labs, S. (2019a). *Toronto Tomorrow - A new approach for inclusive growth – Vol. 1 The Plans*. New York: Sidewalk Labs. Retrieved August 20, 2020, from https://sidewalk-toronto-ca.storage.googleapis.com/wp-content/uploads/2019/06/23135619/MIDP_Volume1.pdf

Labs, S. (2019b). Toronto Tomorrow - A new approach for inclusive growth: Vol. 2. *Urban Innovations*. Sidewalk Labs. https://sidewalk-toronto-ca.storage.googleapis.com/wp-content/uploads/2019/06/23135715/MIDP_Volume2.pdf

Ladley, D., & Bullock, S. (2004) Logistic constraints on 3D termite construction. In *Fourth International Workshop on Ant Colony Optimization and Swarm Intelligence* (pp. 178–189). Berlin: Springer-Verlag. 10.1007/978-3-540-28646-2_16

Lai, C. S., Lai, L. L., & Lai, Q. H. (2021). *Smart Grids and Big Data Analytics for Smart Cities*. Springer Nature Switzerland AG. doi:10.1007/978-3-030-52155-4

Lara, A., Costa, E., Furlani, T., & Yigitcanlar, T. (2016). Smartness that matters: Comprehensive and human-centred characterisation of smart cities. *Journal of Open Innovation*, 1-13.

Lasi, H., Fettke, P., Kemper, H.-G., Feld, T., & Hoffmann, M. (2014). Industrie 4.0. *Wirtschaftsinformatik*, *56*(4), 261–264. Advance online publication. doi:10.100711576-014-0424-4

Latosov, E., Kurnitski, J., Thalfeldt, M., & Volkova, A. (2016). Primary Energy Factors for Different District Heating Networks: An Estonian Example. *Energy Procedia*, *96*, 674–684. doi:10.1016/j.egypro.2016.09.126

Latour, B. (2005). *Reassembling the social. An introduction to Actor-Network-Theory*. Oxford University Press.

Laurent, A., Bakas, I., Clavreul, J., Bernstad, A., Niero, M., Gentil, E., Hauschilda, M. Z. & Christensen, T. H. (2014a). Review of LCA studies of solid waste management systems–Part I: Lessons learned and perspectives. *Waste Management, 34*(3), 573-588.

Lazarevic, D., Buclet, N., & Brandt, N. (2012). The application of life cycle thinking in the context of European waste policy. *Journal of Cleaner Production*, *29*, 199–207. doi:10.1016/j.jclepro.2012.01.030

Lazer, D., Pentland, A., Adamic, L., Aral, S., Barabási, A., Brewer, D., Christakis, N., Contractor, N., Fowler, J., Gutmann, M., Jebara, T., King, G., Macy, M., Roy, D., & Van Alstyne, M. (2009). Computational Social Science. *Science, 323*(5915), 721-723. . doi:10.1126cience.1167742

Lee, S. H., Han, J. H., Leem, Y. T., & Yigitcanlar, T. (2008). Towards ubiquitous city: concept, planning, and experiences in the Republic of Korea. In T. Yigitcanlar, K. Velibeyoglu, & S. Baum (Eds.), Knowledge-Based Urban Development: Planning and Applications in the Information Era (pp. 148-169). IGI Global. doi:10.4018/978-1-59904-720-1.ch009

Lee, S. K., Kim, K. R., & Yu, J. H. (2014). BIM and ontology-based approach for building cost estimation. *Automation in Construction, 41*, 96-105.

Lee, A., An, J., & Chua, C. (2017). Two-Way 4D Printing: A Review on the Reversibility of 3D-Printed Shape Memory. *Materials & Design*, *3*(5), 663–674.

Lee, H. W., Oh, H., Kim, Y., & Choi, K. (2015). Quantitative analysis of warnings in building information modeling (BIM). *Automation in Construction*, *51*, 23–31. doi:10.1016/j.autcon.2014.12.007

Lee, J., Azamfar, M., Singh, J., & Siahpour, S. (2020). Integration of digital twin and deep learning in cyber-physical systems: Towards smart manufacturing. *IET Collaborative Intelligent Manufacturing*, *2*(1), 34–36. doi:10.1049/iet-cim.2020.0009

Lee, J., Bagheri, B., & Kao, H. A. (2014). Recent Advances and Trends of Cyber-Physical Systems and Big Data Analysis in Industrial Informatics. *International Conference on Industrial Informatics (INDIN).*

Lee, J., Kao, H. A., & Yang, S. (2014). Service innovation and smart analytics for industry 4.0 and big data environment. *Procedia CIRP*, *16*(1), 3–8. doi:10.1016/j.procir.2014.02.001

Leminen, S., Westerlund, M., Rajahonka, M., & Siuruainen, R. (2012). *Towards IoT ecosystems and business models in Internet of Things, Smart Spaces, and Next Generation Networking*. Springer.

Lennartsson, M., Yitmen, I., Movaffaghi, H., & Linderoth, H. (2020). Framework for digital development in industrialized housebuilding. *Advances in Transdisciplinary Engineering, 13*, 335-345. Retrieved from www.scopus.com doi:10.3233/ATDE200171

Lennerts, K., Abel, J., Pfründer, U., & Sharma, V. (2003). Reducing health care costs through optimised facility management-related processes. *Journal of Facilities Management*, *2*(2), 192–206. doi:10.1108/14725960410808203

Leslie, E., Coffee, N., Frank, L., Owen, N., Bauman, A., & Hugo, G. (2007). Walkability of local communities: using geographic information systems to objectively assess relevant environmental attributes. *Health & Place, 13*(1), 111-22. .healthplace.2005.11.001 doi:10.1016/j

Lewin, K. (1961). *1936, Principles of topological psychology. Trad. it.: Principi di psicologia topologica. Traduzione a cura di A.Ossicini*. Edizioni OS.

Lewis, J. A., & Gladman, A. S. (2017). *Method of 4d printing a hydrogel composite structure*. US Patent App. 14/954,228.

Li, J., Hou, L., Wang, X., Wang, J., Guo, J., Zhang, S., & Jiao, Y. (2014). A Project-Based Quantification of BIM Benefits. *International Journal of Advanced Robotic Systems, 11*(8), 123. doi:10.5772/58448

Lilis, G. N., Giannakis, G., Katsigarakis, K., & Rovas, D. (2018). District-aware Building Energy Performance simulation model generation from GIS and BIM data. In *Proceedings of BSO 2018: 4th building simulation and optimization conference*. BIPSA. Retrieved from https://discovery.ucl.ac.uk/id/eprint/10056593/

Limberger, J. (2014). *Assessing the prospective resource base for enhanced geothermal systems in Europe*. https://www.researchgate.net/publication/320622611_Limberger_et_al_2014_Data-supplement

Lin, L., Zhang, L., & Guo, Y. (2018). Mechanical properties and shape memory effect ofthermal-responsive polymer based on PVA. *Materials Research Express, 5*(1).

Linksmart. (2020). *Linksmart® - Free, open source IoT platform*. Retrieved from http://linksmart.eu/

Linksmart. (n.d.). Retrieved December 2015, from http://linksmart.eu

Li, R., & Klippel, A. (2012). Wayfinding in libraries: Can problems be predicted? *Journal of Map & Geography Libraries, 8*(1), 21–38. doi:10.1080/15420353.2011.622456

Liu, K., & Jiang, L. (2011). Multifunctional Integration: From Biological to Bio-Inspired Materials. Perspective ACS-Nano – American Chemical Society, (5), 6786–6790. doi:10.1021/nn203250y

Liu, Y., van Nederveen, S., & Hertogh, M. (2017). Understanding effects of BIM on collaborative design and construction: An empirical study in China. *International Journal of Project Management, 35*(4), 686-698.

Liu, X., Wang, X., Wright, G., Cheng, J., Li, X., & Liu, R. (2017). A state-of-the-art review on the integration of building information modeling (BIM) and geographic information system (GIS). *ISPRS International Journal of Geo-Information, 6*(2), 53. doi:10.3390/ijgi6020053

Liu, Y., Du, H., Liu, L., & Leng, J. (2014). Shape memory polymers and their composites in aerospace applications: A review. *Smart Materials and Structures, 23*(2), 023001. doi:10.1088/0964-1726/23/2/023001

Liu, Y., Zhang, L., Yang, Y., Zhou, L., Ren, L., Wang, F., Liu, R., Pang, Z., & Deen, M. J. (2019). A Novel Cloud-Based Framework for the Elderly Healthcare Services Using Digital Twin. *IEEE Access: Practical Innovations, Open Solutions, 7*, 49088–49101. doi:10.1109/ACCESS.2019.2909828

Liu, Z., Zhang, A., & Wang, W. (2020). A framework for an indoor safety management system based on digital twin. *Sensors (Basel), 20*(20), 1–20. doi:10.339020205771 PMID:33053719

Li, X., Hijazi, I., Koning, R., Lv, Z., Zhong, C., & Schmitt, G. (2016). Assessing Essential Qualities of Urban Space with Emotional and Visual Data Based on Gis Technique. *ISPRS International Journal of Geo-Information, 5*(11), 218. doi:10.3390/ijgi5110218

Lo Turco, M. (2015). Rappresentare e gestire patrimoni immobiliari: Il BIM per il Facility Management. *Territorio Italia, 2*, 33–38.

Lorek, S., & Spangenberg, J. H. (2019). Energy sufficiency through social innovation in housing. *Energy Policy, 126*, 287–294. doi:10.1016/j.enpol.2018.11.026

Lösch, A. (1954). The Economics of Location: A Pioneer Book in the Relations Between Economic Goods and Geography. Yale University Press.

Lu, Q., Xie, X., Heaton, J., Parlikad, A. K., & Schooling, J. (2020). *From BIM towards digital twin: Strategy and future development for smart asset management*. Retrieved from www.scopus.com doi:10.1007/978-3-030-27477-1_30

Lu, R. D., & Brilakis, I. (2019). Digital twinning of existing bridges from labelled point clusters. *Proceedings of the 36th International Symposium on Automation and Robotics in Construction, ISARC 2019*, 616-623. Retrieved from www.scopus.com doi:10.22260/ISARC2019/0082

Lu, R., & Brilakis, I. (2019). Digital twinning of existing reinforced concrete bridges from labelled point clusters. *Automation in Construction, 105*, 102837. Advance online publication. doi:10.1016/j.autcon.2019.102837

Lurija, A. (1973). *The working brain. Trad. it.: Come lavora il cervello. Introduzione alla neuropsicologia*. Edizione Il Mulino.

Lutzenhiser, L. (2014). Through the energy efficiency looking glass. *Energy Research & Social Science, 1*, 141–151. doi:10.1016/j.erss.2014.03.011

Lu, W., Li, Y., Cheng, Y., Meng, D., Liang, B., & Zhou, P. (2018). Early fault detection approach with deep architectures. *IEEE Transactions on Instrumentation and Measurement, 67*(7), 1679–1689. doi:10.1109/TIM.2018.2800978

Lynch, K. (1960). L'immagine della città. Versione italiana a cura di Paolo Ceccarelli, Marsilio Editore, 2001-2006, Venezia.

Lynch, K. (1990). Wasting Away. Sierra Club Books.

Lynch, K., & Appleyard, D. (1964). The View from the Road. MIT Press.

Lynch, K., & Rivkin, M. (1990). A Walk around the Block. In T. Banerjee & M. Southworth (Eds.), *City Sense and City Design, Writings and Projects of Kevin Lynch* (pp. 185–205). MIT Press.

Ma, C., Chou, D. C., & Yen, D. C. (2000). Data warehousing, technology assessment and management. *Industrial Management & Data Systems, 100*(3), 125–135. doi:10.1108/02635570010323193

Madakam, S., Ramaswamy, R., & Tripathi, S. (2015). Internet of Things (IoT): A Literature Review. *Journal of Computer and Communications, 3*(05), 164–173. doi:10.4236/jcc.2015.35021

Madrazo, L., Nemirovski, G., & Sicilia, Á. (2014a). Shared Vocabularies to Support the Creation of Energy Urban Systems Models. In European Commission (Ed.), ICT for a Low Carbon Economy - EEBuilding Data Models - Energy Efficiency Vocabularies & Ontologies (pp. 130-150). Publications Office of the European Union.

Madrazo, L., Sicilia, Á., Ortiz, D., & Pleguezuelos, J. (2014b). *Operative integrated platform.* SEMANCO Deliverable 5.6. http://www.semanco-project.eu/deliverables.htm

Madrazo, L., Sicilia, Á., Ortiz, D., Oliveras, J., Ronn, T., Lynch, D., Crosbie, T., & Carpenter, M. (2015). *Energy service platform web portal / Energy service platform web portal contents.* SEMANCO Deliverables 5.7 / 7.6. http://www.semanco-project.eu/deliverables.htm

Madrazo, L., Massetti, M., Sicilia, Á., Wadel, G., & Ianni, M. (2015). SEíS: A semantic-based system for integrating building energy data. *Informes de la Construcción, 67*(537), e060. Advance online publication. doi:10.3989/ic.13.048

Madrazo, L., Nemirovski, G., & Sicilia, Á. (2013, September). Shared vocabularies to support the creation of energy urban systems models. *Proceedings of 4th ICT for Sustainable Places Conference, EEBuilding Data Models, Energy Efficiency Vocabularies and Ontologies. Nice, France.*

Madrazo, L., Sicilia, Á., & Gamboa, G. (2012). SEMANCO: Semantic tools for carbon reduction in urban planning. In *Proceedings of the 9th European Conference on Product and Process Modelling*. CRC Press. 10.1201/b12516-143

Madrazo, L., Sicilia, Á., Massetti, M., López Plazas, F., & Ortet, E. (2018). Enhancing energy performance certificates with energy related data to support decision making for building retrofitting. *Thermal Science*, *22*(3, Suppl. 3), 957–969. doi:10.2298/TSCI171005028M

Mandon, C., Blum, L., & Marquette, C. (2017). 3D–4D Printed Objects: New Bioactive Material Opportunities. *Micromachines*, *8*(12), 102. doi:10.3390/mi8040102

Mantis Project. (2018). Retrieved from Mantis Project EU: http://www.mantis-project.eu/

Manville, C., Cochrane, G., Cave, J., Millard, J., Pederson, J. K., Thaarup, R. K., Liebe, A., Wissner, M., Massink, R., & Kotterink, B. (2014), Mapping smart cities in the EU. European Parliament. Policy Department A: Economic and Scientific Policy, Bruxelles.

Marinakis, V., Doukas, H., Tsapelas, J., Mouzakitis, S., Sicilia, Á., Madrazo, L., & Sgouridis, S. (2020). From big data to smart energy services: An application for intelligent energy management. *Future Generation Computer Systems*, *110*, 572–586. doi:10.1016/j.future.2018.04.062

Markard, J. (2018). The next phase of the energy transition and its implications for research and policy. *Nature Energy*, *3*(8), 628–633. doi:10.103841560-018-0171-7

Market Research Future. (2017). *Global Digital Twin Market is Estimated to Grow at a Cagr of 37% From 2017 to 2023.* Available: https://www.marketresearchfuture.com/reports/digital-twin-market-4504

Marr, D. (1982). Vision. A Computational Investigation into the Human Representation and Processing of Visual Information. Freeman & Company.

Marr, D., & Nishihara, H. K. (1978, February 23). Representation and Recognition of the Spatial Organization of Three-Dimensional Shapes. *Proceedings of the Royal Society of London. Series B, Biological Sciences*, *200*(1140), 269–294. doi:10.1098/rspb.1978.0020 PMID:24223

Martin, D. (2020). *China retreats online to weather coronavirus storm.* Agence France-Presse.

Martín, S., Serna, V., Álvarez, S., García, M. A., Hernández, G., & Sicilia, Á. (2019). OptEEmAL: it-supported design tool for the generation of optimised energy retrofitting scenarios at district level. *European Conference on Computing in Construction*. 10.35490/EC3.2019.169

Massara, F. (2016). Trying to visualize GIS & BIM information on the web: a solution using Leaflet and Cesium. *FOSS4G Bonn 2016.*

Massara, F., Provera, F., & Ramassotto, A. (2019) Streaming and rendering the Turin 3D geospatial content through GIS and BIM integration. *FOSS4G Bucharest 2019.*

Matarneh, S. T., Danso-Amoako, M., Al-Bizri, S., Gaterell, M., & Matarneh, R. (2019). Building information modeling for facilities management: A literature review and future research directions. *Journal of Building Engineering*, *24*, 100755. doi:10.1016/j.jobe.2019.100755

Matt, D. T., Marcher, C., Malacarne, G., & Brozzi, R. (2018). *BIM Report Alto Adige 2018. Sondaggio sulla conoscenza e sull'uso del BIM tra gli operatori del settore delle costruzioni.* Accessed 01/04/2020, available at: https://www.fraunhofer.it/content/dam/italia/it/documents/per_il_sito_BIM%20report%20Alto%20Adige_GT_Versione%20per%20PDF.pdf

Mattei, M. (2013). Quando la città diventa smart. In C. Ratti (Ed.), *Smart city, smart citizen* (pp. 9–17). Egea.

Maya-Drysdale, D., Krog Jensen, L., & Vad Mathiesen, B. (2020). Energy Vision Strategies for the EU Green New Deal: A Case Study of European Cities. *Energies*, •••, 13.

Mayer, H. (2018). Digitalization of legacy building data preparation of printed building plans for the BIM process. *Proceedings of the 7th International Conference on Smart Cities and Green ICT Systems*, 304-310. Retrieved from www.scopus.com doi:10.5220/0006783103040310

Mayer-Schöenberger, V., & Cukier, K. (2013). *Big data: a revolution that will transform how we live, work and think.* John Murray.

Ma, Z., & Ren, Y. (2017). *Integrated applications of BIM and GIS: an overview*. Procedia Engineering, Science Direct.

McAleer, R. S., Kogut, P., & Raes, L. (2020). Digital Urban European Twins. Change the way you see the city. Project white Paper.

McCoy, K., Oliver, J. J., Borden, D. S., & Cohn, S. I. (2018). Nudging waste diversion at Western State Colorado University: Application of behavioural insights. *International Journal of Sustainability in Higher Education, 19*(3, I3), 608–621. doi:10.1108/IJSHE-05-2017-0063

McGraw Hill Construction. (2014). *The Business Value of BIM for Construction in Major Global Markets.* Accessed on August 24, 2020: https://www.construction.com/toolkit/reports/bim-business-value-construction-global-markets

McMillan, T. E. (2005). Urban Form and a Child's Trip to School: The Current Literature and a Framework for Future Research. *Journal of Planning Literature, 19*(4), 440–456. doi:10.1177/0885412204274173

Meadows, D.H., Meadows, D.L., Renders, J., & Berhens, W.W. (1972). *The Limits to Growth.* Universe Book.

Meadows, D. H., Meadows, D. L., & Randers, J. (2004). *Limits to Growth. The 30-Year Update*. Earthscan.

Mecca, B., & Lami, I. M. (2020). *The appraisal challenge in cultural urban regeneration: An evaluation proposal.* Retrieved from www.scopus.com doi:10.1007/978-3-030-35550-0_5

Mecca, U., Moglia, G., Rebaudengo, M., & Ruffino, P. (2019). BIM and maintenance: real or virtual scenarios? In Ingegno e costruzione nell'epoca della complessità, Colloqui.AT.e 2019, Turin, Italy.

Mecca, U., Moglia, G., Piantanida, P., Prizzon, F., Rebaudengo, M., & Vottari, A. (2020). How Energy Retrofit Maintenance Affects Residential Buildings Market Value? *Sustainability, 2020*(12), 5213. Retrieved August 24, 2020, from. doi:10.3390u12125213

Mecca, U., Piantanida, P., Rebaudengo, M., & Vottari, A. (2019). *Green perspectives for Italian buildings façades. In 19th International Multidisciplinary Scientific GeoConference SGEM 2019 2019, 9.* Section Green Buildings Technologies and Materials. doi:10.5593gem2019V/6.3/S09.038

Meex, E., Hollberg, A., Knapen, E., Hildebrand, L., & Verbeeck, G. (2018). Requirements for applying LCA-based environmental impact assessment tools in the early stages of building design. *Building and Environment, 133*, 228–236. doi:10.1016/j.buildenv.2018.02.016

Mehta, V. (2008). Walkable Streets: Pedestrian Behavior, Perceptions and Attitudes. *Journal of Urbanism, 1*(3), 217–245. doi:10.1080/17549170802529480

Mehta, V. (2013). Evaluating Public Space. *Journal of Urban Design, 19*(1), 53–88. doi:10.1080/13574809.2013.854698

Meijer, A., & Bolivar, M. P. R. (2015). Governing the smart city: A review of the literature on smart urban governance. *International Review of Administrative Sciences, 82*(2), 392–408. doi:10.1177/0020852314564308

Melia, A., Nolan, E., & Kerrigan, R. (2015). INDICATE: towards the development of a virtual city model, using a 3D model of Dundalk city. In *CISBAT 2015 International Conference "Future Buildings and Districts - Sustainability from Nano to Urban Scale"* (pp. 925-930). Lausanne, Switzerland: Lausanne, EPFL Solar Energy and Building Physics Laboratory.

Melia, A., Nolan, E., & Kerrigan, R. (2015). INDICATE: towards the development of a virtual city model, using a 3d model of Dundalk city. *Proceedings of International Conference CISBAT 2015 "Future Buildings and Districts – Sustainability from Nano to Urban Scale.*

Menassa, C. C., & Ortiz-Vega, W. (2013). Uncertainty in Refurbishment Investment. In F. Pacheco Torgal, M. Mistretta, A. Kaklauskas, C. Granqvist, & L. Cabeza (Eds.), *Nearly Zero Energy Building Refurbishment* (pp. 143–175). Springer. doi:10.1007/978-1-4471-5523-2_6

Mendoza, A. J. (2015). *Legal and Social Implications of the 3D Printing Revolution.* CMC Senior Theses. Paper 1032. https://scholarship.claremont.edu/cmc_theses/1032

Mengolini, A., & Vasiljevska, J. (2013). *The social dimension of Smart Grids. Consumer, community, society.* EU Commission, JRC Scientific and Policy Reports, 31.

Meredith, S. (2018 May 17). *Two-Thirds of World Population Will Live in Cities by 2050 - UN Says.* Retrieved February 4, 2020, from https://www.theguardian.com/world/2018/may/17/two-thirds-of-world-population-will-live-in-cities-by-2050-says-un

Merton, R. K. (1936). The unanticipated consequences of purposive social action. *American Sociological Review, 1*(6), 894–904. doi:10.2307/2084615

Miafodzyeva, S., & Brandt, N. (2013). Recycling behaviour among householders: Synthesising determinants via a meta-analysis. *Waste and Biomass Valorization, 4*(2), 221–235. doi:10.100712649-012-9144-4

Microsoft. (2020). *What is. NET?* https://dotnet.microsoft.com/learn/dotnet/what-is-dotnet

Miller, C. A., Iles, A., & Jones, C. F. (2013). The social dimensions of energy transitions. *Science as Culture, 22*(2), 135–148. doi:10.1080/09505431.2013.786989

Milton, M., De La, C. O., Ginn, H. L., & Benigni, A. (2020). Controller-Embeddable Probabilistic Real-Time Digital Twins for Power Electronic Converter Diagnostics. *IEEE Transactions on Power Electronics, 35*(9), 9852–9866. doi:10.1109/TPEL.2020.2971775

Ministero delle Infrastrutture e dei Trasporti. (2017). *Ministerial Decree no. 560 of 01/12/2017.* Accessed on August 24, 2020: http://www.mit.gov.it/normativa/decreto-ministeriale-numero-560-del-01122017

Minoli, D., Sohraby, K., & Occhiogrosso, B. (2017). IoT considerations, requirements, and architectures for smart buildings in Energy optimization and next-generation building management systems. *IEEE Internet of Things Journal, 4*(1), 269–283. doi:10.1109/JIOT.2017.2647881

Mitchell, W. J. (2001). *Smart Cities.* https://smartcities.media.mit.edu/

Mitrofanova, Y. S., Sherstobitova, A. A., & Filippova, O. A. (2019). *Modeling smart learning processes based on educational data mining tools in Smart Education and e-Learning 2019.* Springer.

Miyauchi, A., & Shimomura, S. (2019). *Industrial Biomimetics.* Taylor & Francis Group. doi:10.1201/9780429058837

Moglia, G., Rebaudengo, M., & Ruffino, P. A. (2019). *"Digital Twin" Utopia? UID per il disegno: 2019; ROMA, Gangemi.* Accessed on August 24, 2020: http://digital.casalini.it/10.36165/1065

Mohanty, S., Choppali, U., & Kougianos, E. (2016). *Everything You wanted to Know about Smart Cities.* Research Gate.

Molitor, C., Groß, S., Zeitz, J., & Monti, A. (2014). MESCOS - A multienergy system cosimulator for city district energy systems. *IEEE Transactions on Industrial Informatics, 10*(4), 2247–2256. doi:10.1109/TII.2014.2334058

Mome.ni, Hassani, Liu, & Ni. (2017). A review of 4D printing. *Materials & Design*, *122*, 42–79. doi:10.1016/j.matdes.2017.02.068

Moretti, L. (1954). *Structure Comme Forme*. United States Lines Paris Review.

Moretti, M. (2016). *Senso e paesaggio. Analisi percettive e cartografie tematiche in ambiente GIS*. Franco Angeli.

Moretti, N., Xie, X., Merino, J., Brazauskas, J., & Parlikad, A. K. (2020). An openbim approach to iot integration with incomplete as-built data. *Applied Sciences (Basel, Switzerland)*, *10*(22), 1–17. doi:10.3390/app10228287

Morgenstern, P., Lowe, R., & Chiu, L. F. (2015). Heat metering: Socio-technical challenges in district-heated social housing. *Building Research and Information*, *43*(2), 197–209. doi:10.1080/09613218.2014.932639

Mosannenzadeh, F., Bisello, A., Vaccaro, R., D'Alonzo, V., Hunter, G., & Vettorato, D. (2017). Smart Energy City. *Cities (London, England)*, *64*, 54–65.

Motamedi, A., Wang, Z., Yabuki, N., Fukuda, T., & Michikawa, T. (2017). Signage visibility analysis and optimization system using BIM-enabled virtual reality (VR) environments. *Advanced Engineering Informatics*, *32*, 248–262. doi:10.1016/j.aei.2017.03.005

Motlagh, N. H., Khajavi, S. H., Jaribion, A., & Holmström, J. (2018). An IoT based automation system for older homes: A use case for lighting system. *Proceedings IEEE 11th International Conference Service-Oriented Computer Application (SOCA),* 1-6.

Motlagh, N. H., Taleb, T., & Arouk, O. (2016). Low-altitude unmanned aerial vehicles-based Internet of Things services: Comprehensive survey and future perspectives. *IEEE Internet of Things Journal*, *3*(6), 899–922.

MQTT. (n.d.). Retrieved November 2015, from http://mqtt.org

Mukherjee, T., & DebRoy, T. (2019). A digital twin for rapid qualification of 3D printed metallic components. *Applied Materials Today*, *14*, 59–65. doi:10.1016/j.apmt.2018.11.003

Munda, G. (2005). Multiple criteria decision analysis an sustainable development. In *Multi Criteria Decision Analysis, State of the Art Surveys* (pp. 954–986). Springer.

Munier, N. (2005). *Introduction to Sustainability: road to a better future*. Springer.

Murgante, B., & Borruso, G. (2013). Smart cities: Un'analisi critica delle opportunità e dei rischi [Smart cities: a critical analysis of opportunities and risks]. *GEOmedia*, *3*, 6–10.

Musetti, A. (2016). *Deliverable 7.8. Third Dissemination Event*. INDICATE Project.

Nagar, S. V., Chandrashekar, A. C., & Suvarna, M. (2020). Optimized Additive Manufacturing Technology Using Digital Twins and Cyber Physical Systems. In *Cyber-physical Systems and Digital Twins: Proceedings of the 16th International Conference on Remote Engineering and Virtual Instrumentation*. Springer Nature Switzerland AG.

Nageler, P., Koch, A., Mauthner, F., Leusbrock, I., Mach, T., Hochenauer, C., & Heimrath, R. (2018). Comparison of dynamic urban building energy models (UBEM): Sigmoid energy signature and physical modelling approach. *Energy and Building*, *179*, 333–343. doi:10.1016/j.enbuild.2018.09.034

Naghshbandi, S. N. (2017). BIM for facility management: Challenges and research gaps. *Civil Engineering Journal*, *2*(12), 679–684. doi:10.28991/cej-2016-00000067

Naisbitt, J. (1982). *Megatrends: Ten New Directions Transforming Our Lives*. Warner Books Inc.

Nasaruddin, N., Ito, T., & Tuan, T. B. (2018). Digital twin approach to building information management. *Proceedings Manufacturing Systems Division Conference*, 304.

Negri, E., Fumagalli, L., & Macchi, M. (2017). A Review of the Roles of Digital Twin in CPS-based Production Systems. *Procedia Manufacturing*, *11*(June), 939–948. doi:10.1016/j.promfg.2017.07.198

Negroponte, N. (1995). *Being Digital*. Alfred A. Knopf Inc.

Nemirovski, G., Nolle, A., Sicilia, Á., Ballarini, I., & Corrado, V. (2013). Data Integration Driven Ontology Design, Case Study Smart City. In *Proceedings of 3rd International Conference on Web Intelligence, Mining and Semantics, WIMS'13*. Association for Computing Machinery (ACM). 10.1145/2479787.2479830

Nettstraeter, A. (2012). Architectural reference model for IoT. *EC FP7 IoT-A (257521) D, 1*, 2.

Networkworkworld. (2020). *Internet of Things*. https://www.networkworld.com/article/3280225/what-is-digital-twin-technology-and-why-it-matters.html

Neves, D. T., Santos, M., & Pinto, M. (2015). ReActOR: A middleware as a service to interact with objects remotely. In *Proc. ICIT 2015* (pp. 2433-2439). IEEE. 10.1109/ICIT.2015.7125456

Ng, I., Scharf, K., Pogrebna, G., & Maull, R. (2015). Contextual variety, Internet-of-Things and the choice of tailoring over platform: Mass customisation strategy in supply chain management. *International Journal of Production Economics*, *159*, 76–87. doi:10.1016/j.ijpe.2014.09.007

Niknam, M., & Karshenas, S. (2017). A shared ontology approach to semantic representation of BIM data. *Automation in Construction*, *80*, 22–36. doi:10.1016/j.autcon.2017.03.013

Niles, I., & Pease, A. (2001). Towards a standard upper ontology. In *Proceedings of the international conference on Formal Ontology in Information Systems* - Volume 2001, (pp. 2-9). 10.1145/505168.505170

Noble, D., & Kaminski, C. (2015). Introduction for bioinspiration. *Interface Focus*, *5*(4), 20150052. doi:10.1098/rsfs.2015.0052

Nolle, A., & Nemirovski, G. (2013). ELITE: An Entailment-Based Federated Query Engine for Complete and Transparent Semantic Data Integration. In *Description Logics* (pp. 854-867). Retrieved from http://ceur-ws.org/Vol-1014/paper_50.pdf

Nouvel, R., Brassel, K. H., Bruse, M., Duminil, E., Coors, V., Eicker, U., & Robinson, D. (2015). SimStadt, a new workflow-driven urban energy simulation platform for CityGML city models. In *Proceedings of International Conference CISBAT 2015* (pp. 889-894). Academic Press.

Nouvel, R., Schulte, C., Eicker, U., Pietruschka, D., & Coors, V. (2013). CITYGML-based 3D city model for energy diagnostics and urban energy policy support. *Proceedings of BS2013: 13th Conference of International Building Performance Simulation Association*, 218-225.

NRC. (2012). *NRC Publications Archive (NPArC) Archives des publications du CNRC (NPArC)*. Author.

NYC. (2020). *NYC 3-D Building Model*. Retrieved from The Official Website of the City of New York: https://www1.nyc.gov/site/doitt/initiatives/3d-building.page

O'Donnell, J., See, R., Rose, C., Maile, T., Bazjanac, V., & Haves, P. (2011). SimModel: A domain data model for whole building energy simulation. In *Proceedings of Building Simulation 2011: 12th Conference of International Building Performance Simulation Association*, (Vol. 12). Retrieved from http://www.ibpsa.org/proceedings/BS2011/P_1223.pdf

O'Dwyer, E., Pan, I., Acha, S., & Shah, N. (2019). Smart energy systems for sustainable smart cities: Current developments, trends and future directions. *Applied Energy*, *237*, 581–597. doi:10.1016/j.apenergy.2019.01.024

Odling-Smee, J., & Turner, J. S. (2011). Niche Construction Theory and Human Architecture. *Biological Theory*, *6*(3), 283–289. doi:10.100713752-012-0029-3

OGC. (2020). https://www.ogc.org/standards/3DTiles

Ogrodnik, K. (2020). Multi-criteria analysis of smart cities in Poland. *Geographia Polonica*, •••, 163–181.

OICE. (2019). *Rapporto sulle gare BIM 2018 per opere pubbliche: Analisi del mercato e delle gare*. Accessed on August 24, 2020: https://www.oice.it/592206/rapporto-sulle-gare-bim-2018-per-opere-pubbliche

OneM2M. (n.d.). Retrieved September 2015, from https://onem2m.org/

ONU. (n.d.). *Department of Economic and Social Affair - Population dynamic*. Retrieved from https://population.un.org/wup/Download/

Oracle. (2020). https://www.oracle.com/technetwork/java/javaee/tech/index.html

Oreto, P. (2018, April 5). *Codice dei contratti ed i nodi da sciogliere: La qualificazione delle stazioni appaltanti, Rivista Online Lavori pubblici*. Available online at: https://www.lavoripubblici.it/news/2018/04/LAVORI-PUBBLICI/20089/Codice-dei-contratti-ed-i-nodi-da-sciogliere-La-qualificazione-delle-stazioni-appaltanti

Osello, A. (2016). Deliverable 1.5 - Districts and Buildings energy typologies. Turin.

Osello, A. (2016). Deliverable 3.1.3 - District Information Modeling: implementation and standard definition. Turin.

Osello, A. (2012). *The Future of Drawing with BIM for Engineers and Architects*. Dario Flaccovio Editore.

Osello, A., Acquaviva, A., Del Giudice, M., Patti, E., & Rapetti, N. (2016). District Information Models. The DIMMER project: BIM tools for the urban scale. In R. Pagani & G. Chiesa (Eds.), *Urban data. Tecnologie e metodi per la città algoritmica* (pp. 244–275). FRANCOANGELI.

Osello, A., Davardoust, S., Semeraro, F., Lucibello, G., Barone, L., Dellosta, M., ... Ugliotti, F. M. (2016). The centrality of Representation with BIM. *DISEGNARECON*, *9*(16), 5–1.

Osello, A., & Ugliotti, F. M. (2015). *BIM, GIS, AR per il facility management*. Dario Flaccovio Editore.

Osman, A., Yahya, A. A., & Kamal, M. B. (2018). A benchmark collection for mapping program educational objectives to ABET student outcomes: Accreditation. In *Proceedings 5th International Symposium on Data Mining Applications*. Springer.

Osservatorio Nazionale Smart City. (2018). *Vademecum per la cittá intelligente*. www.osservatoriosmartcity.it

Otto, F. (2011). Occupyind and Connecting - Thoughs on Territories and Spheres of Influence with Particular Reference to Human Settlement. Edition Axel Menges GmbH.

Otto, F., & Rasch, B. (1996). *Gestalt Finden. Auf Dem Weg Zu Einer Baukunst Des Minimalen*. Edition Axel Menges GmbH.

Paelke, V. (2014). Augmented reality in the smart factory: Supporting workers in an industry 4.0. environment. *Proceedings of the 2014 IEEE Emerging Technology and Factory Automation (ETFA),* 1-4. 10.1109/ETFA.2014.7005252

Pagani, R., & Chiesa, G. (2017). *Urban Data. Tecnologie e Metodi per La Città algoritmica*. FRANCOANGELI.

Papadonikolaki, E., Leon, M., & Mahamadu, A. M. (2018). Bim solutions for construction lifecycle: A myth or a tangible future? *Paper presented at the EWork and eBusiness in Architecture, Engineering and Construction - Proceedings of the 12th European Conference on Product and Process Modelling, ECPPM 2018,* 321-328. Retrieved from www.scopus.com doi:10.1201/9780429506215-40

Papastamatiou, I., Doukas, H., Spiliotis, E., & Psarras, J. (2016). How "OPTIMUS" is a city in terms of energy optimization? E-SCEAF: A web based decision support tool for local authorities. *Information Fusion*, *29*, 149–161. doi:10.1016/j.inffus.2015.10.002

Papastamatiou, I., Marinakis, V., Doukas, H., & Psarras, J. (2017). A decision support framework for smart cities energy assessment and optimization. *Energy Procedia*, *111*, 800–809. doi:10.1016/j.egypro.2017.03.242

Parliament, E. (2018). Directive (EU) 2018/844. European Parliament and of the Council of 30 May.

Parr, J. B. (1978). Models of the Central Place System: A More General Approach. *Urban Studies (Edinburgh, Scotland)*, *15*(1), 33–50. doi:10.1080/713702286

Parr, J. B. (2015). Exploring the Urban System of von Thünen's Isolated State. *Papers in Regional Science*, *94*(1), 1161–1175.

Parr, J. B. (2017). Central Place Theory: An Evolution. *Review of Urban and Regional Development Studies*, *29*(3), 151–162.

Parsanezhad, P. (2019). *Towards a BIM-enabled Facility Management: Promises, Obstacles and Requirements* (Doctoral dissertation). KTH Royal Institute of Technology.

Partnership for a Walkable America, Pedestrian and Bicycle Information Center. (2005). *Walkability Checklist: How walkable is your community?* https://cachampionsforchange.cdph.ca.gov/en/Documents/Let%27sWalk.pdf

PAS 1192-2:2013 (2013). Specification for information management for the capital/delivery phase of construction projects using building information modelling

Pasichnyi, O., Levihn, F., Shahrokni, H., Wallin, J., & Kordas, O. (2019). Data-driven strategic planning of building energy retrofitting: The case of Stockholm. *Journal of Cleaner Production*, *233*, 546–560. doi:10.1016/j.jclepro.2019.05.373

Pasini, D., Mastrolembo Ventura, S., Rinaldi, S., Bellagente, P., Flammini, A., & Ciribini, A. L. C. (2016). Exploiting Internet of Things and building information modeling framework for management of cognitive buildings. In *Smart Cities Conference (ISC2) in IEEE International* (pp. 1-6). IEEE. 10.1109/ISC2.2016.7580817

Patti, E., Acquaviva, A., Jahn, M., Pramudianto, F., Tomasi, R., Rabourdin, D., Virgone, J., & Macii, E. (2016, September). Event-driven user-centric middleware for energy-efficient buildings and public spaces. *IEEE Systems Journal*, *10*(3), 1137–1146. Advance online publication. doi:10.1109/JSYST.2014.2302750

Patti, E., Acquaviva, A., & Macii, E. (2013). Enable sensor networks interoperability in smart public spaces through a service oriented approach. *IEEE International Workshop on Advances in Sensors and Interfaces (IWASI)*, 2-7. 10.1109/IWASI.2013.6576081

Patti, E., Syrri, A. L., Jahn, M., Mancarella, P., Acquaviva, A., & Macii, E. (2016, March). Distributed Software Infrastructure for General Purpose Services in Smart Grid. *IEEE Transactions on Smart Grid*, *7*(2), 1156–1163. Advance online publication. doi:10.1109/TSG.2014.2375197

Patton, D. (2020). *China's 2019 pork output plunges to 16-year low as disease culls herd.* Reuters. https://www.reuters.com/article/us-china-economy-output-pork/chinas-2019-pork-output-plunges-to-16-year-low-as-disease-culls-herd-idUSKBN1ZG08H

Pauwels, P. (2015). *IFCtoRDF converter*. Available online: https://github.com/pipauwel/IFCtoRDF

Pauwels, P., Corry, E., & O'Donnell, J. (2014). Representing SimModel in the web ontology language. In Computing in Civil and Building Engineering (pp. 2271-2278). doi:10.1061/9780784413616.282

Pauwels, P., Sijie Zhang, S., & Lee, Y.-C. (2017). Semantic web technologies in AEC industry: A literature overview. *Automation in Construction*, *73*, 145–165. doi:10.1016/j.autcon.2016.10.003

Pauwels, P., & Terkaj, W. (2016). EXPRESS to OWL for construction industry: Towards a recommendable and usable ifcOWL ontology. *Automation in Construction*, *63*, 100–133. doi:10.1016/j.autcon.2015.12.003

Pei, E. (2014). 4D Printing-revolution or fad? *Assembly Automation*, *34*(2), 123–127. doi:10.1108/AA-02-2014-014

Peng, H., Ma, S., & Spector, J. M. (2019). *Personalized adaptive learning: An emerging pedagogical approach enabled by a smart learning environment in Foundations and Trends in Smart Learning*. Springer.

Pereira, G. V., Cunhab, M. A., Lampoltshammera, T. J., Paryceka, P., & Testac, M. G. (2017). Increasing collaboration and participation in smart city governance: A cross-case analysis of smart city initiatives. *Information Technology for Development*, *23*(3), 526–553. doi:10.1080/02681102.2017.1353946

Persiani, S. (2019). *Biomimetics of Motion. Nature-Inspired Parameters and Schemes for Kinetic Design*. Springer International Publishing AG.

Peterson, J. L. (1981). *Petri net theory and the modeling of systems*. Prentice Hall PTR.

Phang, T., Chen, C., & Tiong, R. (2019). New Model for Identifying Critical Success Factors Influencing BIM Adoption from Precast Concrete Manufacturers' View. *Journal of Construction Engineering and Management*, *146*(4). Advance online publication. doi:10.1061/(ASCE)CO.1943-7862.0001773

PiemonteC. S. I. (2019). https://github.com/csipiemonte/ogc-citygml-utils

Piro, M., Ballarini, I., & Corrado, V. (2020). Use of Energy Performance Certificates data repositories in Urban Building Energy Models. *AiCARR Journal*, *64*(5), 44–48.

PMI - Project Management Institute. (2013). *The high cost of low performance: The essential roles of communications*. https://www.pmi.org/-/media/pmi/documents/public/pdf/learning/thought-leadership/pulse/the-essential-role-of-communications.pdf

Poggi, A., Lembo, D., Calvanese, D., De Giacomo, G., Lenzerini, M., & Rosati, R. (2008). Linking data to ontologies. In *Journal on data semantics X* (pp. 133–173). Springer., doi:10.1007/978-3-540-77688-8_5

Poli, T., Mainini, A. G., Speroni, A., Blanco Cadena, J. D., & Moretti, N. (2020). *The effect of real-time sensing of a window on energy efficiency, comfort, health and user behavior*. Retrieved from www.scopus.com doi:10.1007/978-3-030-33570-0_26

Polly, B., Kutscher, C., Macumber, D., Schott, M., Pless, S., Livingood, B., & Van Geet, O. (2016). From zero energy buildings to zero energy districts. ACEEE Summer Study on Energy Efficiency in Buildings.

Pont, U. J., Ghiassi, N., Fenz, S., Heurix, J., & Mahdavi, A. (2015). SEMERGY: Application of semantic web technologies in performance-guided building design optimization. *Journal of Information Technology in Construction*, *20*(8), 107–120. Retrieved August 21, 2020, from https://itcon.org/papers/2015_8.content.00001.pdf

Powells, G., Bulkeley, H., Bell, S., & Judson, E. (2014). Peak electricity demand and the flexibility of everyday life. *Geoforum*, *55*, 43–52. doi:10.1016/j.geoforum.2014.04.014

Primrose, S. B. (2020). *Biomimetics – Nature-inspired design and innovation*. Wiley Blackwell. doi:10.1002/9781119683360

Princen, T. (2005). *The logic of sufficiency*. MIT Press.

Provera, F. (2017). *Implementation of BIM methodology for district data visualization*. https://webthesis.biblio.polito.it/6408/1/tesi.pdf

Provera, F. (2017). *Implementation of BIM methodology for district data visualization*. Master Thesis in Building Engineering, Politecnico di Torino.

Provincia di Torino, (2002). *Scheda della Fondazione di Torino Smart City per lo Sviluppo Sostenibile* [Torino Smart City Foundation for Sustainable Development sheet]. Relazione Annuale, Allegato1.

Provincia di Torino. (2011). *Relazione illustrativa del Piano Territoriale di Coordinamento Provinciale – Ptc 2* [Illustrative report of the Provincial Coordination Territorial Plan - Ptc2]. Città Metropolitana di Torino.

Qi, Q., & Tao, F. (2018). Digital Twin and big data towards smart manufacturing and industry 4.0: 360 degree comparison. *IEEE Access, 6*, 3585-3593.

Qi, Q., Tao, F., Zuo, Y., & Zhao, D. (2018). Digital Twin service towards smart manufacturing. *Proceedings CIRP, 72*, 237-242.

Qin, Q., & Tao, F. (2018). Digital Twin and big data towards smart manufacturing and industry 4.0: 360 degree comparison. *IEEE Access, 6*, 3585-3593.

Qi, Q., & Tao, F. (2018). Digital Twin and Big Data Towards Smart Manufacturing and Industry 4.0: 360 Degree Comparison. *IEEE Access: Practical Innovations, Open Solutions*, *6*, 3585–3593. doi:10.1109/ACCESS.2018.2793265

Rafiemanzelat, R., Emadi, M. I., & Kamali, A. J. (2017). City sustainability: The influence of walkability on built environments. *Transportation Research Procedia*, ●●●, 24.

Ramchurn, S. D., Vytelingum, P., Rogers, A., & Jennings, N. R. (2012). Putting the 'smarts' into the smart grid: A grand challenge for artificial intelligence. *Communications of the ACM*, *55*(4), 86–97. doi:10.1145/2133806.2133825

Randall, T. A., & Baetz, B. W. (2001). Evaluating Pedestrian Connectivity for Suburban Sustainability. *Journal of Urban Planning and Development*, *127*(1), 1–15. doi:10.1061/(ASCE)0733-9488(2001)127:1(1)

Rasheed, A., San, O., & Kvamsdal, T. (2020). Digital twin: Values, challenges and enablers from a modeling perspective. *IEEE Access: Practical Innovations, Open Solutions*, *8*, 21980–22012. doi:10.1109/ACCESS.2020.2970143

Rasheed, A., San, O., & Kvamsdal, T. (2020). Digital Twin: Values. *Challenges and Enablers from a Modeling Perspective in IEEE Access*, *8*, 21980–22012.

Ratajczak, J., Riedl, M., & Matt, D. T. (2019). BIM-based and AR application combined with location-based management system for the improvement of the construction performance. *Buildings*, *9*(5), 118. doi:10.3390/buildings9050118

Ratti, C. (2013). *Smart city, smart citizen*. Egea.

Ratti, C., & Claudel, M. (2017). *La città di domani, Come le reti stanno cambiando il futuro urbano*. Einaudi.

REEIMAN Project. (2017). Retrieved from REEIMAN Project EU: http://www.reemain.eu/

Reeves, L. (1995). Data mining. *Data Management Review*, *5*(7), 12–19.

Reinhart, C. F., & Cerezo Davila, C. (2016). Urban building energy modeling – A review of a nascent field. *Building and Environment*, *97*, 196–202. doi:10.1016/j.buildenv.2015.12.001

Reinhart, C. F., Dogan, T., Jakubiec, J. A., Rakha, T., & Sang, A. (2013). UMI - An urban simulation environment for building energy use, daylighting and walkability. In *Proceedings of BS 2013: 13th Conference of IBPSA* (pp. 476-483). International Building Performance Simulation Association.

Reiss, J. (1999). *From 'Margin to Center: The Spaces of Installation Art*. MIT Press.

Remixstudio. (2012). *Overlapped City*. Retrieved February 30, 2020, from http://www.remixstudio.org/research/design/overlapped-city

Remmen, P., Lauster, M., Mans, M., Fuchs, M., Osterhage, T., & Müller, D. (2018). TEASER: An open tool for urban energy modelling of building stocks. *Journal of Building Performance Simulation*, *11*(1), 84–98. doi:10.1080/19401493.2017.1283539

Renda, A., & Cellai, C. (2016). *Strategia Europa 2020: cronaca di una revisione annunciata* [Europe Strategy 2020: chronicle of an announced revision]. Seminario CINSEDO-Centro Interregionale Studi e Documentazione.

Rennie, J. (1869). Insect Architecture. Bell and Daldy.

Riddhi, D., Siddhi, D., & Hemal, T. (2020). Digital Twins: Current problems in Smart City and Recommendations for future technology. *International Research Journal of Engineering and Technology, 7*(5).

Rišová, K. (2020). Walkability research: Concept, methods and a critical review of post-socialist studies. *Geograficky Casopis*, *72*(3), 219–242.

Ritchie, H., & Roser, M. (2018). *Urbanization, Our World in Data*. Oxford, UK: Oxford University Press. Retrieved July 30, 2020, from https://ourworldindata.org/urbanization

RitchieH.RoserM. (2018). *Urbanization*. https://ourworldindata.org/urbanization

Robinson, D., Haldi, F., Kämpf, J., Leroux, P., Perez, D., Rasheed, A., & Wilke, U. (2009). CitySim: Comprehensive micro-simulation of resource flows for sustainable urban planning. In *Proceedings of BS 2009: 11th Conference of IBPSA* (pp. 1083-1090). International Building Performance Simulation Association.

Rockart, J. F. (1979). Chief executives define their own data needs. *Harvard Business Review*, *57*(2), 81–93. PMID:10297607

Rogers, R. (1998). Cities for a Small Planet. Westview Press.

Romero, A., Egusquiza, A., & Izkara, J. L. (2014). Integrated decision support tool in energy retrofitting projects for sustainable urban districts. In *Proceedings of the SB14 World Sustainable Buildings*, Vol. 3, (pp. 268-274). Retrieved from https://gbce.es/archivos/ckfinderfiles/WSB14/CreatingNewResources_volume3.pdf

Romero, A., Izkara, J., Mediavilla, A., Prieto, I., & Pérez, J. (2016). Multiscale building modelling and energy simulation support tools. *ECPPM 2016 - 11th European Conference on Product & Process Modelling*, 315-322.

Romero, A., Izkara, J. L., Mediavilla, A., Prieto, I., & Pérez, J. (2016). Multiscale building modelling and energy simulation support tools. *Proceedings of the 11th ECPPM*, 315-322.

Rosen, R., von Wichert, G., Lo, G., & Bettenhausen, K. D. (2015). About the importance of autonomy and Digital Twins for the future of manufacturing. *IFAC-PapersOnLine, 48*(3), 567-572.

Rosen, R., Von Wichert, G., Lo, G., & Bettenhausen, K. D. (2015). About the importance of autonomy and digital twins for the future of manufacturing. *IFAC-PapersOnLine*, *28*(3), 567–572. doi:10.1016/j.ifacol.2015.06.141

Ross, L. (2010). *Virtual 3D City Models in Urban Land Management—Technologies and Applications* (Ph.D. Thesis). Technische Universität Berlin, Berlin, Germany.

Rosslyn, I. (2018 July 24). Action on Overpopulation Still Urgent. *Victorian Humanist, 57*(6). Retrieved February 4, 2020, from https://vichumanist.org.au/environment/action-overpopulation-urgent/

Roxin, A., Abdou, W., Ginhac, D., Derigent, W., Dragomirescu, D., & Montegut, L. (2019). Digital building twins - contributions of the ANR McBIM project. *Proceedings - 15th International Conference on Signal Image Technology and Internet Based Systems, SISITS 2019*, 404-410. Retrieved from www.scopus.com doi:10.1109/SITIS.2019.00071

Rüßmann, M., Lorenz, M., Gerbert, P., & Waldner, M. (2015). Industry 4.0. *The Future of Productivity and Growth in Manufacturing Industries, Boston Consulting Group, 9*, 1–14.

Saaty, T. (1980). *The analytic hierarchy process*. McGraw Hill.

Sabir, A., Abbasi, N. A., & Islam, N. (2018). *An electronic data management and analysis application for abet accreditation*. arXiv:1901.05845.

Sacks, R., Eastman, C., Lee, G., & Teicholz, P. (2018). BIM Handbook: A Guide to Building Information Modeling for Owners, Designers, Engineers, Contractors, and Facility Managers. Wiley.

Sacks, O. (1985). The Man Who Mistook His Wife For a Hat. Trad. it.: L'uomo che scambiò sua moglie per un cappello (trad. di Morena, C.). *Adelphi Edizioni, Milano, 2001*, 318.

Sacks, R., Eastman, C., Lee, G., & Teicholz, P. (2018). *BIM Handbook: A Guide to Building Information Modeling for Owners, Designers, Engineers, Contractors, and Facility Managers*. Wiley. doi:10.1002/9781119287568

Saddik, A. E. (2018). Digital Twins: The convergence of multimedia technologies. *IEEE Multimedia, 25*(2), 87-92.

Saelens, B. E., Sallis, J. F., & Frank, L. D. (2003). Environmental correlates of walking and cycling: Findings from the transportation, urban design, and planning literatures. *Annals of Behavioral Medicine, 25*, 80–91.

Sampaio, A. Z., & Simoes, D. (2017). Maintenance of buildings supported on BIM methodology. *Proceedings of XIV DBMC-Durability of Building Materials and Components, Ghent, Belgian*, 83-84.

Saracco, R., De Kerckhove, D., & Nisiotis, L. (2020). *The Role of Personal Digital Twins in Control of Epidemics.* An IEEE Digital Reality White Paper.

Saran, S., Oberai, K., & Wate, P. (2018). Utilities of Virtual 3D City Models Based on CityGML: Various Use Cases. *Journal of the Indian Society of Remote Sensing.*

Saunders, H. D. (2015). Recent evidence for large rebound: Elucidating the drivers and their implications for climate change models. *Energy Journal, 36*(1), 23–48. doi:10.5547/01956574.36.1.2

Schaeben, C. (2017). *People's Perception of Urban and Architectural Features*. Retrieved from: https://www.research-collection.ethz.ch/handle/20.500.11850/266634

Schick, L., & Winthereik, B. R. (2013). Innovating relations–or why smart grid is not too complex for the public. *Science & Technology Studies, 26*(3), 82–102. doi:10.23987ts.55289

Schimanski, C. P., Monizza, G. P., Marcher, C., & Matt, D. T. (2019). Pushing digital automation of configure-to-order services in small and medium enterprises of the construction equipment industry: A design science research approach. *Applied Sciences (Basel, Switzerland), 9*(18), 3780. Advance online publication. doi:10.3390/app9183780

Schleich, B., Anwer, N., Mathieu, L., & Wartzack, S. (2017). Shaping the digital twin for design and production engineering. *CIRP Annals - Manufacturing Technology, 66*(1), 141–144. doi:10.1016/j.cirp.2017.04.040

Schleich, B., Anwer, N., Mathieu, L., & Wartzack, S. (2017). Shaping the Digital Twin for design and production engineering. *CIRP Annual, 66*(1), 141-144.

Schluse, M., Atorf, L., & Rossmann, J. (2017). Experimentable Digital Twins for model-based systems engineering and simulation-based development. *Proceedings Annual IEEE International System Conference*, 628– 635.

Schluse, M., Priggemeyer, M., Atorf, L., & Rossmann, J. (2018). Experimentable Digital Twins: Streamlining simulation-based systems engineering for industry 4.0. IEEE Transactions on Industrial Informatics, 14(4), 1722-1731.

Schluse, M., Priggemeyer, M., Atorf, L., & Rossmann, J. (2018). Experimentable Digital Twins-Streamlining Simulation-Based Systems Engineering for Industry 4.0. *IEEE Transactions on Industrial Informatics*, *14*(4), 1722–1731. doi:10.1109/TII.2018.2804917

Schweigkofler, A., Romagnoli, K., Steiner, D., Riedl, M., & Matt, T. D. (2020) *Methodology and operating tool for urban renovation: the case study of the Italian city of Meran.* SSPCR 3rd International Conference on Smart and Sustainable Planning for Cities and Regions.

Sciacovelli, A., Verda, V., & Borchiellini, R. (2013). *Numerical design of thermal systems.* Clut.

Scopus Database. (n. d.) Available online at: www.scopus.com

Seepersad, C. C. (2014). Challenges and Opportunities in Design for Additive Manufacturing. 3D Printing and Additive Manufacturing, 1(1), 10-13. doi:10.1089/3dp.2013.0006

SEMANCO. (2015). *Semantic Tools for Carbon Reduction in Urban Planning*. http://www.semanco-project.eu/

SemsorGrid4Env. (n.d.). Retrieved September 2015, from https://www.semsorgrid4env.eu/

SENML Draft. (n.d.). Retrieved from https://tools.ietf.org/html/draft-jennings-senml-10

Shahinmoghadam, M., & Motamedi, A. (2019). Review of BIM-centred IoT deployment: state of the art, opportunities, and challenges. In *ISARC. Proceedings of the International Symposium on Automation and Robotics in Construction* (Vol. 36, pp. 1268-1275). IAARC Publications.

Shah, N. (2013). Modelling urban energy systems. In J. Keirstead & N. Shah (Eds.), *Urban Energy Systems. An Integrated Approach.* Routledge.

Shah, S. A., Seker, D. Z., Rathore, M. M., Hameed, S., Ben Yahia, S., & Draheim, D. (2019). Towards Disaster Resilient Smart Cities: Can Internet of Things and Big Data Analytics Be the Game Changers? *IEEE Access: Practical Innovations, Open Solutions*, *7*, 91885–91903. doi:10.1109/ACCESS.2019.2928233

Shanghai Air Pollution. Real-time Air Quality Index. (2020). [Data set], https://aqicn.org/city/shanghai/

Sharma, S. B., & Tabak, V. (2008). Rapid Agent Based Simulation of People Flow for Design of Spaces. In H. J. P. Timmermans & B. de Vries (Eds.), *Design & Decision Support Systems in Architecture and Urban Planning*. University of Technology Eindhoven.

Sherrington, C. S. (1906). The Integrative Action of the Nervous System. New Haven: Yale University Press.

Shin, D. G., Kim, T. H., & Kim, D. E. (2017). Review of 4D printing materials and their properties. *International Journal of Precision Engineering and Manufacturing-Green Technology*, *4*(3), 349–357. doi:10.100740684-017-0040-z

Shmelev, S., & Shmeleva, I. (2019). Multidimensional sustainability benchmarking for smart megacities. *Cities (London, England)*, •••, 134–163.

Shove, E., Pantzar, M., & Watson, M. (2012). The dynamics of social practice. Everyday life and how it changes. *Sage (Atlanta, Ga.)*. Advance online publication. doi:10.4135/9781446250655

Shove, E., & Warde, A. (2002). Inconspicuous consumption: the sociology of consumption, lifestyles, and the environment. In T. Benton (Ed.), *Sociological theory and the environment: classical foundations, contemporary insights* (pp. 230–251). Rowman & Littlefield.

Sicilia, Á., & Costa, G. (2017). Energy-Related Data Integration Using Semantic Data Models for Energy Efficient Retrofitting Projects. In Multidisciplinary Digital Publishing Institute Proceedings, (Vol. 1, No. 7, 1099, pp. 1-5). doi:10.3390/proceedings1071099

Sicilia, Á., Costa, G., Corrado, V., Gorrino, A., & Corno, F. (2015). A Semantic Decision Support System to optimize the energy use of public buildings. In *CIB W78 Conference* (pp. 676-685). Retrieved from https://itc.scix.net/pdfs/w78-2015-paper-070.pdf

Sicilia, Á., Madrazo, L., & Costa, G. (2015). *Building a semantic-based decision support system to optimize the energy use in public buildings: the OPTIMUS project. Sustainable Places 2015 event.* Savona.

Sicilia, Á., Madrazo, L., & Pleguezuelos, J. (2014). Integrating multiple data sources, domains and tools in urban energy models using semantic technologies. In *Proceedings of the 10th European Conference on Product and Process Modelling* (pp. 837-844). 10.1201/b17396-134

Sicilia, Á., Nemirovski, G., & Nolle, A. (2017). Map-On: A web-based editor for visual ontology mapping. *Semantic Web*, *8*(6), 969–980. doi:10.3233/SW-160246

Silva, M., Gaviao, L. O., Gomes, C., & Lima, G. (2017). A proposal for the application of multicriteria analysis to rank countries according to innovation using the indicators provided by the World World Intellectual Property Organization. *RAI Revista de Administração e Inovação*, 188-198.

Silva, J., Mojica, J., Pineres, A., Rojas, A., Acosta, S., Giuliany, J. G., & Sanabria, E. S. (2020). Algorithms for the Control of Key Performance Indicators for Smart Cities. *Procedia Computer Science*, •••, 971–976.

Silvast, A. (2013). *Anticipating interruptions. Security and risk in a liberalized electricity infrastructure*. University of Helsinki – Department of Social Research.

Simpson, J., & Simpson, K. (2018). Visual engagement with urban street edges: Insights using mobile eye-tracking. *Journal of Urbanism*.

Simulation of Systems in Materials Handling. Logistics and Production–Fundamentals, VDI-Standard 3633 Part 1, 2016. Available: www.vdi.eu/3633

Sitte, C. (1953). L'arte di costruire le città, Officine Grafiche dell'Editore Antonio Vallardi, Milano.

Skjølsvold, T. M., & Lindkvist, C. (2015). Ambivalence, designing users and user imaginaries in the European smart grid: Insights from an interdisciplinary demonstration project. *Energy Research & Social Science*, *9*, 43–50. doi:10.1016/j.erss.2015.08.026

Smart Built Environment. (2020). *Projekt DigSam - Digital Samhällsbyggnadsprocess*. https://www.smartbuilt.se/digsam

Smart Cities Information System. (2020). Retrieved from EC-Funded Projects Tracked by the Smart Cities Information System: https://smartcities-infosystem.eu/sites-projects/projects

Smil, V. (2010). *Energy transitions: history, requirements, prospects*. ABC-CLIO.

Sola, G. (2009). *Natural Forms and Optimum Models*. Retrieved February 30, 2020, from https://www.domusweb.it/en/reviews/2009/03/05/natural-form-and-optimum-models.html

Sola, A., Corchero, C., Salom, J., & Sanmarti, M. (2020). Multi-domain urban-scale energy modelling tools: A review. *Sustainable Cities and Society*, *54*, 101872. doi:10.1016/j.scs.2019.101872

Sonetti, G., Arrobbio, O., Lombardi, P., Lami, I. M., & Monaci, S. (2020). "Only social scientists laughed": Reflections on Social Sciences and Humanities integration in European energy projects. *Energy Research & Social Science*, *61*, 101342. doi:10.1016/j.erss.2019.101342

Southworth, M. (1997). Walkable suburbs? An evaluation of neotraditional communities at the urban edge. *Journal of the American Planning Association*, *63*(1), 28–44. doi:10.1080/01944369708975722

Southworth, M. (2005). Designing the walkable city. *Journal of Urban Planning and Development*, *131*(4), 246–257. doi:10.1061/(ASCE)0733-9488(2005)131:4(246)

Southworth, M., & Ben-Joseph, E. (1995). Street standards and the shaping of suburbia. *Journal of the American Planning Association*, *61*(1), 65–81. doi:10.1080/01944369508975620

Southworth, M., & Owens, P. M. (1993). The evolving metropolis studies of community, street form at the urban edge. *Journal of the American Planning Association*, *59*(3), 271–287. doi:10.1080/01944369308975880

Sovacool, B. K. (2016). How long will it take? Conceptualizing the temporal dynamics of energy transitions. *Energy Research & Social Science*, *13*, 202–215. doi:10.1016/j.erss.2015.12.020

Spuybroek, L. (2005). *The Structure of Vagueness. Textile The Journal of Cloth and Culture, 3(1).*

Srewil, Y., Ismail, A., & Scherer, R. J. (2016). A method to integrate virtual-physical construction environment in framework of CPS approach. *Smart SysTech 2016; European Conference on Smart Objects, Systems and Technologies, VDE*, 1-8.

Srewil, Y., & Scherer, R. J. (2017). Construction objects recognition in framework of CPS. *2017 Winter Simulation Conference (WSC),* 2472-2483. 10.1109/WSC.2017.8247976

Staniaszek, D., Rapf, O., Faber, M., Nolte, I., Bruel, R., Fong, P., Lees, E., & Verheyen, L. (2013). A Guide to Developing Strategies for Building Energy Renovation–Delivering Article 4 of the Energy Efficiency Directive. *The Buildings Performance Institute Europe (BPIE).* Retrieved from http://www.bpie.eu/documents/BPIE/Developing_Building_Renovation_Strategies.pdf

Stern, D. I. (2017). How accurate are energy intensity projections? *Climatic Change*, *143*(3–4), 537–545. doi:10.100710584-017-2003-3

StojanovicV.TrappM.RichterR.DöllnerJ. (2019). A service-oriented indoor point cloud processing pipeline. Paper presented at the International Archives of the Photogrammetry. *Remote Sensing and Spatial Information Sciences - ISPRS Archives*, *42*(2/W17), 339-346. Retrieved from www.scopus.com doi:10.5194/isprs-archives-XLII-2-W17-339-2019

Stoychev, G., Razavi, M., Wang, X., & Ionov, L. (2017). 4D Origami by Smart Embroidery. *Macromolecular Rapid Communications*, *38*(18), 1700213. doi:10.1002/marc.201700213 PMID:28759164

Straubel, M., Block, J., Sinapius, M., & Hühne, C. (2011). Deployable Composite Booms for Various Gossamer Space Structures. *52nd AIAA/ASME/ASCE/AHS/ASC Structures, Structural Dynamics and Materials Conference.* 10.2514/6.2011-2023

Studio, A. (2010). *Ka Care*. Retrieved July 30, 2020, from http://www.atmosstudio.com/KA-CARE

Suárez-Figueroa, M. C., Gómez-Pérez, A., & Fernández-López, M. (2012). The NeOn methodology for ontology engineering. In *Ontology engineering in a networked world* (pp. 9–34). Springer. doi:10.1007/978-3-642-24794-1_2

Suárez-Figueroa, M. C., Gómez-Pérez, A., Motta, E., & Gangemi, A. (2012). *Ontology Engineering in a Networked World.* Springer. doi:10.1007/978-3-642-24794-1

Succar, B. (2009). Building information modelling framework: A research and delivery foundation for industry stakeholders. *Automation in Construction, 18*(3), 357-375.

Sun, H. (2020). *Alibaba says AI can identify coronavirus infections with 96% accuracy.* Nikkei Asian Review. https://asia.nikkei.com/Spotlight/Coronavirus/Alibaba-says-AI-can-identify-coronavirus-infections-with-96-accuracy

Swan, L. G., & Ugursal, V. I. (2009). Modeling of end-use energy consumption in the residential sector: A review of modeling techniques. *Renewable & Sustainable Energy Reviews, 13*(8), 1819–1835. doi:10.1016/j.rser.2008.09.033

Swan, L. G., Ugursal, V. I., & Beausoleil-Morrison, I. (2009). Implementation of a Canadian residential energy end-use model for assessing new technology impacts. In *Proceedings of BS 2009: 11th Conference of IBPSA* (pp. 1429-1436). International Building Performance Simulation Association.

Swyngedouw, E. (2005). Governance innovation and the citizen: The Janus face of governance beyond the state. *Urban Studies (Edinburgh, Scotland), 42*(11), 1991–2006. doi:10.1080/00420980500279869

Tagliabue, L. C., Consalez, L., Mastrolembo Ventura, S., & Ciribini, A. L. C. (2016). The Educational Building in the Circular Economy: an Extended Concept of Asset Quality. In Convegno Internazionale ISTeA 2016, BACK TO 4.0: Rethinking the Digital Construction Industry, Naples.

Tagliabue, L. C., Cecconi, F. R., Maltese, S., Rinaldi, S., Ciribini, A. L. C., & Flammini, A. (2021). Leveraging Digital Twin for Sustainability Assessment of an Educational Building. *Sustainability, 13*(2), 480. doi:10.3390u13020480

Tagliabue, L. C., & Villa, V. (2017). *Il BIM per le scuole. Analisi del patrimonio scolastico e strategie di intervento.* Hoepli.

Tahmasebinia, F., Fogerty, D., Wu, L. O., Li, Z., Sepasgozar, S. M. E., Zhang, K., & Marroquin, F. A. (2019). Numerical analysis of the creep and shrinkage experienced in the sydney opera house and the rise of digital twin as future monitoring technology. *Buildings, 9*(6), 137. Advance online publication. doi:10.3390/buildings9060137

Talamo, C. (2014). *La gestione integrata delle informazioni nei processi manutentivi.* Dall'anagrafica degli edifici ai sistemi BIM.

Talen, E., & Koschinsky, J. (2013). The Walkable Neighborhood: A Literature Review. *International Journal of Sustainable Land Use and Urban Planning,* (1), 42—63.

Talen, E. (2003). Neighborhoods as Service Providers: A Methodology for Evaluating Pedestrian Access. *Environment and Planning. B, Planning & Design, 30*(2), 181–200. doi:10.1068/b12977

Tang, P., Huber, D., Akinci, B., Lipman, R., & Lytle, A. (2010). Automatic reconstruction of as-built building information models from laser-scanned point clouds: A review of related techniques. *Automation in Construction, 19*(7), 829-843.

Tang, S., Shelden, D. R., Eastman, C., Pishdad-Bozorgi, P., & Gao, X. (2019). A review of building information modeling (BIM) and the Internet of Things (IoT) devices integration: Present status and future trends. *Automation in Construction, 101*, 127-139.

Tao, F., Cheng, J., Qi, Q., Zhang, M., Zhang, H., & Sui, F. (2018). Digital twin-driven product design, manufacturing and service with big data. *International Journal Advanced Manufacturing Technology, 94*(9-12), 3563-3576.

Tao, F., Cheng, Y., Xu, L. D., Zhang, L., & Li, B. H. (2014). CCIoT-CMfg: Cloud computing and Internet of Things-based cloud manufacturing service system. *IEEE Transactions on Industrial Informatics, 10*(2), 1435-1442.

Tao, F., Sui, F., Liu, A., Qi, Q., Zhang, M., Song, B., Guo, Z., Lu, S. C. Y., & Nee, A. (2019). Digital twin-driven product design framework. *International Journal of Production Research, 57*(12), 3935-3953.

Tao, F., Cheng, J., Qi, Q., Zhang, M., Zhang, H., & Sui, F. (2018). Digital twin-driven product design, manufacturing and service with big data. *International Journal of Advanced Manufacturing Technology*, *94*(9–12), 3563–3576. doi:10.100700170-017-0233-1

Tao, F., Sui, F., Liu, A., Qi, Q., Zhang, M., Song, B., Guo, Z., Lu, S. C. Y., & Nee, A. Y. C. (2019). Digital twin-driven product design framework. *International Journal of Production Research*, *57*(12), 3935–3953. doi:10.1080/00207543.2018.1443229

Tao, F., Zhang, H., Liu, A., & Nee, A. Y. C. (2019). Digital Twin in Industry: State-of-the-Art. *IEEE Transactions on Industrial Informatics*, *15*(4), 2405–2415. doi:10.1109/TII.2018.2873186

Tarkhov, D., & Vasilyev, A. (2020). *Semi-empirical neural network modeling and digital twins development*. Elsevier.

Tavakoli, J., & Tang, Y. (2017). Hydrogel based sensors for biomedical applications: An updated review. *Polym.*, *8*(12), 364. doi:10.3390/polym9080364 PMID:30971040

Taylor, M. C. (2002). *The Moment of Complexity: Emerging Network Culture*. University of Chicago Press.

Tchana, Y., Ducellier, G., & Remy, S. (2019). Designing a unique digital twin for linear infrastructures lifecycle management. *Procedia CIRP, 84,* 545-549. Retrieved from www.scopus.com doi:10.1016/j.procir.2019.04.176

Teicholz, P. (2013). *BIM for Facility Managers*. John Wiley & Sons. doi:10.1002/9781119572633

Telensa. (2020). *Urban Data Project*. Retrieved from Urban Data Project: https://urbandataproject.org/

Teoh, J. E. M., An, J., Chua, C. K., Lv, M., Krishnasamy, V., & Liu, Y. (2016). Hierarchically self-morphing structure through 4D printing. *Virtual and Physical Prototyping*. Advance online publication. doi:10.1080/17452759.2016.1272174

Testoni, C. (2016). *Towards Smart City*. FRANCOANGELI.

Tezel, A., & Aziz, Z. (2017). From conventional to IT based visual management: a conceptual discussion for lean construction. *ITcon, 22*, 220-246. https://www.itcon.org/2017/12

The British Standards Institution. (2014). Smart cities – Vocabulary. *PAS*, *180*, 2014.

The British Standards Institution. (2018). *Smart city framework – Guide to establishing strategies for smart cities and communities.* https://www.bsigroup.com/en-GB/smart-cities/Smart-Cities-Standards-and-Publication/PAS-181-smart-cities-framework/

The Engineering Toolbox. (n.d.). https://www.engineeringtoolbox.com/density-solids-d_1265.html

The Norwegian EPD Foundation. (2020). *EPD-Norge Digi*. https://digi.epd-norge.no/

The Prince's Foundation. (2020). *Walkability and Mixed-Use. Making Valuable and Healthy Communities.* https://content.knightfrank.com/research/2139/documents/en/walkability-and-mixed-use-making-valuable-and-healthy-communities-7667.pdf?fbclid=IwAR161EMGOJ4kszJPO-DgWJ2wLUWWKVwG2ufSKaCs1BoL6H7jT73e5C77hEg

The PVGIS Database. (2017). https://re.jrc.ec.europa.eu/PVGIS5-release.html

Think: Act Magazine. (2019). *The Smart City Breakaway*. Roland Berger GMBH.

Throndsen, W. (2017). What do experts talk about when they talk about users? Expectations and imagined users in the smart grid. *Energy Efficiency*, *10*(2), 283–297. doi:10.100712053-016-9456-5

Thürer, M., Pan, Y. H., Qu, T., Luo, H., Li, C. D., & Huang, G. Q. (2016). Internet of Things (IoT) driven system for reverse logistics: Solid waste collection. *Journal of Intelligent Manufacturing*, *2016*, 1–10.

Tibbits. (2014). 4D printing and universal transformation. *Acadia 2014 - Design Agency*, 539-548.

Tibbits, S. (2014, January/February). Multi-material Shape Change. *Architectural Design.*, *84*(1), 116–121. doi:10.1002/ad.1710

Tibbits, S., & Cheung, K. (2012). Programmable materials for architectural assembly and automation. *Assembly Automation*, *32*(3), 216–225. doi:10.1108/01445151211244348

Titotto, S. (2013). *Branching morphogenesis: from growth patterns of vascular networks to biomimicry devices* (PhD Thesis). University of Sao Paulo & Politecnico di Torino.

Titotto, S. L. M. C. (2014). Models and prototypes of biomimetic devices to architectural purposes. *SCIRES-IT - SCIentific RESearch and Information Technology, 4,* 55-62. http://caspur-ciberpublishing.it/index.php/scires-it/article/view/11117

Titotto, S. L. M. C., & Dobelis, M. (2017). Teaching CAD/CAM: exchanging experiences from Brazilian and Latvian courses In *Engineering Graphics BALTGRAF-14 Proceedings of the Fourteenth International Conference BALTGRAF Tallinn 2017.* Tallinn: TTU - Tallinn University of Technology.

Titotto, S. L. M. C., Oliveira, D. R., & Ferrante, A. (2015). Space and energy: relationships among architects from nature. *SCIRES-IT - SCIentific RESearch and Information Technolog*y, *5*, 115-120. http://caspur-ciberpublishing.it/index.php/scires-it/article/viewFile/11425/10618

Tocci, G. (2020). La Governance delle città intelligenti. Tecnologia, innovazione, smartness [Governance of smart cities. Technology, innovation, smartness]. *Sociologia Urbana e Rurale*, *122*(122), 44–61. doi:10.3280/SUR2020-122004

Tomlinson, R. F. (1968). A geographic information system for regional planning. In Symposium on Land Evaluation, Commonwealth Scientific and Industrial Research Organization. Melbourne: MacMillan of Australia.

Torabi Moghadam, S., Delmastro, C., Corgnati, S. P., & Lombardi, P. (2017). Urban energy planning procedure for sustainable development in the built environment: A review of available spatial approaches. *Journal of Cleaner Production*, *165*, 811–827. doi:10.1016/j.jclepro.2017.07.142

Torabi Moghadam, S., & Lombardi, P. (2019). An interactive multi-criteria spatial decision support system for energy retrofitting of building stocks using CommuntiyVIZ to support urban energy planning. *Building and Environment*, *163*, 106233. doi:10.1016/j.buildenv.2019.106233

Torabi Moghadam, S., Lombardi, P., & Mutani, G. (2017). A mixed methodology for defining a new spatial decision analysis towards low carbon cities. *Procedia Engineering*, *198*, 375–385. doi:10.1016/j.proeng.2017.07.093

Torabi Moghdam, S., Lombardi, P., Ugliotti, F. M., Osello, A., & Mutani, G. (2016). BIM-GIS Modelling for Sustainable Urban Development. *NEWDIST*, *2016*(Special Issue), 339–350.

Torricelli, G. P. (2003a). *Introduzione Alla Geografia Economica - Lezione 2*. Retrieved February 30, 2020, from http://www.gpt.adhoc.ch/Geoeconomica/Modulo2/Lezione2.pdf

Torricelli, G. P. (2003b). *Introduzione Alla Geografia Economica - Lezione 3*. Retrieved February 30, 2020, from http://www.gpt.adhoc.ch/Geoeconomica/Modulo2/Lezione3.pdf

Trindade, E., Hinnig, M. P., Costa, E. M., Marques, J. S., Bastos, R. C., & Yigitcanlar, T. (2017). Sustainable development of smart cities: a systematic review of the literature. *Journal of Open Innovation: Technology, Market and Complexity.*

Tsikos, M., & Negendahl, K. (2017). *Sustainable design with respect to LCA using parametric design and BIM tools* [Paper presentation]. World Sustainable Built Environment Conference, Wan Chai, Hong Kong. https://backend.orbit.dtu.dk/ws/portalfiles/portal/133787517/Sustainable_Design_with_Respect_to_LCA_Using_Parametric_Design_and_BIM_Tools.pdf

Tsiompras, A. B., & Photis, Y. N. (2017). What matters when it comes to "Walk and the city"? Defining a weighted GIS-based walkability index. *Transportation Research Procedia, 24*, 523–530. doi:10.1016/j.trpro.2017.06.001

Turk, Z. (2017). Ten questions concerning building information modelling. *Building and Environment, 107*, 274–284. doi:10.1016/j.buildenv.2016.08.001

Turner, J. (2009). *The Extended Organism: The Physiology of Animal-Built Structures*. Harvard Univ Press.

Turner, J. (2011). Termites as models of swarm cognition. *Swarm Intelligence, 5*(1), 19–43. doi:10.100711721-010-0049-1

Turner, J. S. (2000). Architecture and morphogenesis in the mound of Macrotermes michaelseni (Sjöstedt) (Isoptera: Termitidae, Macrotermitinae) in northern Namibia. *Cimbebasia, 16*, 143–175.

Turner, J. S., & Soar, R. C. (2008) Beyond biomimicry: What termites can tell us about realizing the living building. *Proceedings of the 1st International Conference on Industrialised, Integrated, Intelligent Construction (I3CON)*, 221.

Ugliotti, F. M. (2017). Utilizzi del modello BIM per il FM. In BIM: verso Il catasto del futuro. Conoscere, digitalizzare, condividere. Il caso studio della Città di Torino, (pp. 56-73). Gangemi Editore spa.

Ugliotti, F. M., Dellosta, M., & Osello, A. (2016). A BIM-based energy analysis using Edilclima EC770 plug-in. Case study: Archimede Library, EEB Project. *Procedia Engineering. WMCAUS, 6*(161), 3–8. doi:10.1016/j.proeng.2016.08.489

Uhlemann, T. H. J., Lehmann, C., & Steinhilper, R. (2017). The Digital Twin: Realizing the Cyber-Physical Production System for Industry 4.0. *Procedia CIRP, 61*, 335–340. doi:10.1016/j.procir.2016.11.152

UNI 11337 (2017). Edilizia e opere di ingegneria civile - Gestione digitale dei processi informativi delle costruzioni.

UNI EN ISO 19650 -1(2019). Organization and digitization of information about buildings and civil engineering works, including building information modelling (BIM) — Information management using building information modelling. Parte 1: Concetti e principi.

UNI EN ISO 19650 -2 (2019). Organization and digitization of information about buildings and civil engineering works, including building information modelling (BIM) — Information management using building information modelling. Parte 2: Fase di consegna dei cespiti immobili.

UNI. (2017). UNI 11337-4:2017. Edilizia e opere di ingegneria civile - Gestione digitale dei processi informativi delle costruzioni - Parte 4: Evoluzione e sviluppo informativo di modelli, elaborati e oggetti.

United Nations Organization DESA. (2014). *World Urbanization Prospects*. Retrieved February 10, 2020, from https://population.un.org/wpp/

United Nations Organization. (2018). *The World's Cities in 2018*. Retrieved February 10, 2020, from https://www.un.org/en/events/citiesday/assets/pdf/the_worlds_cities_in_2018_data_booklet.pdf

United Nations. (1992, June 14). *Sustainable Development*. Retrieved from Agenda 21: https://sustainabledevelopment,un,org/content/documents/Agenda21,pdf

United Nations. (2015). *Resolution adopted by the General Assembly on 25 September 2015, Seventieth session, Agenda items 15 and 116.* Accessed on August 24, 2020: https://www.un.org/ga/search/view_doc.asp?symbol=A/RES/70/1&Lang=E

United Nations. (2020). *Sustainable development goals.* Accessed on August 24, 2020: https://www.un.org/sustainabledevelopment/

Unreal Engine 4 Documentation. (n.d.). *Unreal Datasmith* [Software]. https://docs.unrealengine.com/en-US/Engine/Content/Importing/Datasmith/index.html

Uschold, M., & King, M. (1995). Towards a methodology for building ontologies. *Workshop on Basic Ontological Issues in Knowledge Sharing, held in conjunction with IJCAI-95.*

Ustundag, A., & Cevikcan, E. (2018). *Industry 4.0: Managing The Digital Transformation.* Springer. doi:10.1007/978-3-319-57870-5

Uwiringiyimana, M. M., Nandagopal, G., Guan, Y., Vinkovič, S., Kölsch, J., & Heinz, C. (2021). IoT Platforms. *IoT Platforms, Use Cases, Privacy, and Business Models*, 21–49. . doi:10.1007/978-3-030-45316-9_2

Vachálek, J., Bartalský, L., Rovný, O., Šišmišová, D., Morháč, M., & Lokšík, M. (2017). The digital twin of an industrial production line within the industry 4.0 concept. *2017 21st International Conference on Process Control (PC), Strbske Pleso*, 258-262. 10.1109/PC.2017.7976223

Vamvakidis, G. (2017). Fabricating Gradient Transparency: A Bio-Inspired Digital Design Model and Prototyping Methods. 3D Printing and Additive Manufacturing, 4(1), 41-52. doi:10.1089/3dp.2016.0050

Vandenberg, L. (2020). *How Can China Fulfill Its Rising Meat Demand?* The Diplomat. https://thediplomat.com/2020/04/how-can-china-fulfill-its-rising-meat-demand/

Vanolo, A. (2015). Smart city e sviluppo urbano: alcune note per un'agenda critica [Smart city and urban development: some notes for a critical agenda]. *Scienze del Territorio 3. Ricostruire le città,* 111-118. . doi:10.13128/Scienze_Territorio-16256

Vassileva, I., Dahlquist, E., & Campillo, J. (2016). The citizens' role in energy smart city development. *Energy Procedia, 88*, 88. doi:10.1016/j.egypro.2016.06.055

Venkov, D. (2017). *Shortest Path through Points (connect the dots).* Available from: https://forum.dynamobim.com/t/shortest-path-through-points-connect-the-dots/17610/4

Verbeke, S., Ma, Y., Van Tichelen, P., Bogaert, S., & Gómez Oñate, V. (2018). Support for setting up a Smart Readiness Indicator for buildings and related impact assessment - Final report. Brussels: Flemish Institute for Technological Research NV ("VITO").

Verda, V., Chirio, M., & Ponta, A. (2006). A Thermoeconomic Based Tool for the Real-time Management of a Cogenerative Combined Cycle. *ASME 2006 International Mechanical Engineering Congress and Exposition.*

Verda, V., & Baccino, G. (2014). Primary energy reductions in district heating networks through variation of the thermal load profile of the users. *Proceedings of ECOS*, 15-19.

Verda, V., & Colella, F. (2011). Primary energy savings through thermal storage in district heating networks. *Energy, 36*(7), 4278–4286. doi:10.1016/j.energy.2011.04.015

Verma, U. (2018). *What are Digital Twins in Smart Buildings?* Available: https://inbuildingtech. com/bms/digital-twin-commercial-office-building/

Vianello, S. (2019, June 20). *La sicurezza dei lavoratori anziani nei cantieri edili.* Available on line at: https://www.unapri.it/news/in-cantiere-non-si-invecchia-la-sicurezza-nell-edilizia-per-i-lavoratori-anziani/

Vignais, N., Miezal, M., Bleser, G., Mura, K., Gorecky, D., & Marin, F. (2012). Innovative system for real-time ergonomic feedback in industrial manufacturing; Ltd and The Ergonomics Society. *Elsevier, 44*(4), 566-574.

Viitanen, J., & Kingston, R. (2014). Smart cities and green growth: Outsourcing democratic and environmental resilience to the global technology sector. *Environment & Planning A*, *46*(4), 803–819. doi:10.1068/a46242

Villa, V., Domaneschi, M., Cimellaro, G., & Caldera, C. (2018). BIM-Based Approach for Seismic Risk Analysis. In Streamlining Information Transfer between Construction and Structural Engineering. AAE-06-1-AAE-06-6. doi:10.14455/ISEC.res.2018.145

Villa, V., & Di Giuda, G. M. (2016). *Collaborative Design and Collaborative Contracting a Defiant Challenge in Digitizing the analogical thoughts in architecture: a menace or a promise?* Ingenio.

Vinagre Díaz, J. J., Wilby, M. R., Rodríguez González, A. B., & García Muñoz, J. (2013). EEOnt: An ontological model for a unified representation of energy efficiency in buildings. *Energy and Building*, *60*, 20–27. doi:10.1016/j.enbuild.2013.01.012

Vincent, J. (2001). Stealing ideas from nature, Deployable Structures, 51-58.

Vincent, J. (2006). Biomimetics: Its practice and theory. *Journal of the Royal Society, Interface*, 3. PMID:16849244

Vinco, S., Bottaccioli, L., Patti, E., Acquaviva, A., Macii, E., & Poncino, M. (2018). *GIS-Based Optimal Photovoltaic Panel Floorplanning for Residential Installations.* Paper presented at 2018 Design, Automation & Test in Europe (DATE). 10.23919/DATE.2018.8342049

Vis, D., & van den Berg. (2010). *Best practices and methods Handbook.* http://www.uvm.edu/~cdl/2012workshop/BEE%20Biomass%20Methods%20Handbook%202010.pdf

Vivi, Q. L., Parlikad, A. K., Woodall, P., Ranasinghe, G. D., & Heaton, J. (2019). *Developing a dynamic digital twin at a building level: Using cambridge campus as case study.* Paper presented at the International Conference on Smart Infrastructure and Construction 2019, ICSIC 2019: Driving Data-Informed Decision-Making, 67-75. Retrieved from www.scopus.com doi:10.1680/icsic.64669.067

Volk, R., Stengel, J., & Schultmann, F. (2014). Building Information Modeling (BIM) for existing buildings: Literature review and future needs. *Automation in Construction, 38*, 109-127.

von Thünen, J. H. (1842). Der Isolierte Staat in Beziehung Auf Landwirtschaft Und Nationalökonomie, 2018th ed. London, UK: Forgotten Books.

W3C. (2020a). *XML Essentials*. https://www.w3.org/standards/xml/core

W3C. (2020b). *XML Path Language (XPath).* https://www.w3.org/TR/1999/REC-xpath-19991116/#location-paths

WahbehW.KunzD.HofmannJ.BereuterP. (2020). Digital twinning of the built environment-an interdisciplinary topic for innovation in didactics. Paper presented at the ISPRS Annals of the Photogrammetry. *Remote Sensing and Spatial Information Sciences*, *5*(4) 231-237. Retrieved from www.scopus.com doi:10.5194/isprs-annals-V-4-2020-231-2020

Wan, C., Cheung, R., & Qiping Shen, G. (2012). Recycling attitude and behaviour in university campus: A case study in Hong Kong. *Facilities*, *30*(13/14), 630–646. doi:10.1108/02632771211270595

Wang, M., Zhang, G., Zhang, C., Zhang, J., & Li, C. (2013). An IoT-based appliance control system for smart homes. *Proceedings IEEE 4th International Conference Intelligence Control Information Processing and Cloud Computing Process*, 744-747.

Wang, J., Ye, L., Gao, R. X., Li, C., & Zhang, L. (2019). Digital Twin for rotating machinery fault diagnosis in smart manufacturing. *International Journal of Production Research*, *57*(12), 3920–3934. doi:10.1080/00207543.2018.1552032

Wang, Y., Wang, X., Wang, J., Yung, P., & Jun, G. (2013). Engagement of facilities management in design stage through BIM: Framework and a case study. *Advances in Civil Engineering*, *2013*, 1–8. doi:10.1155/2013/189105

Watróbski, J., Jankowski, J., Ziemba, P., Karczmarczyk, A., & Zioło, M. (2019). Generalised framework for multi-criteria method selection. *Omega*, *86*, 107–124. doi:10.1016/j.omega.2018.07.004 PMID:30671511

Watson, J. (Ed.). (2020). *Biosurveillance Market. 07 mai 2020.* Reports and Data. https://www.globenewswire.com/fr/news-release/2020/05/07/2030018/0/en/Biosurveillance-Market-To-Reach-USD-5-62-Billion-By-2027-Reports-and-Data.html

Wei, T., Zhou, J., & Zhang, H. (2019). Rebound effect of energy intensity reduction on energy consumption. *Resources, Conservation and Recycling*, *144*, 233–239. doi:10.1016/j.resconrec.2019.01.012

White, G., Zink, A., Codecá, L., & Clarke, S. (2021, March). A digital twin smart city for citizen feedback. *Cities (London, England)*, *110*, 103064. doi:10.1016/j.cities.2020.103064

Whitlock, K., Abanda, F. H., Manjia, M. B., Pettang, C., & Nkeng, G. E. (2018). BIM for construction site logistics management. *Journal of Engineering, Project, and Production Management*, *8*(1), 47.

Wilson, D. C., Rodic, L., Scheinberg, A., Velis, C. A., & Alabaster, G. (2012). Comparative analysis of solid waste management in 20 cities. *Waste Management & Research*, *30*(3), 237–254. doi:10.1177/0734242X12437569 PMID:22407700

Wilson, L., Danforth, J., Cerezo Davila, C., & Harvey, D. (2019). How to Generate a Thousand Master Plans: A Framework for Computational Urban Design. In *Proceedings of SimAUD 2019, Society for Modeling & Simulation International (SCS)*. Society for Modeling & Simulation International (SCS).

Wolman, A. (1965). The Metabolism of Cities. *Scientific American*, *213*(3), 179–190. doi:10.1038cientificamerican0965-178 PMID:14338520

Wolsink, M. (2018). Social acceptance revisited: Gaps, questionable trends, and an auspicious perspective. *Energy Research & Social Science*, *46*, 287–295. doi:10.1016/j.erss.2018.07.034

Wolters, M., Nemirovski, G., & Nolle, A. (2013). ClickOn_A: An Editor for DL-Lite_A Based Ontology Design. In *Description Logics* (pp. 1000-1010). Retrieved from https://www.semanticscholar.org/paper/ClickOn_A%3A-An-Editor-for-DL-Lite_A-Based-Ontology-Wolters-Nemirovski/b449eea3c1d57ecdf55f1d71d9dc13b844e553c2

Wolters, M., Nemirovski, G., & Nolle, A. (2013). ClickOnA: An Editor for DL-LiteA based Ontology Design. *Proceedings of the 26th International Workshop on Description Logics.*

Wong, K., & Fan, Q. (2013). Building information modelling (BIM) for sustainable building design. *Facilities, 31*(3-4), 138-157.

World Bank. (2018). Tool for Rapid Assessment of City Energy (TRACE 2.0): A Manual for Experts and City Officials. Energy Sector Management Assistance Program. World Bank.

World Health Organisation. (2016a). *Shanghai Consensus on Healthy Cities.* https://www.who.int/healthpromotion/conferences/9gchp/9gchp-mayors-consensus-healthy-cities.pdf?ua=1

World Health Organisation. (2016b). *9th Global Conference on Health Promotion, Shanghai 2016.* https://www.who.int/healthpromotion/conferences/9gchp/programme/en/

World Health Organization. (n.d.). *Urban population growth.* Retrieved February 2016, from https://www.who.int/gho/urban_health/situation_trends/urban_population_growth_text/en/

Wortmann, J. C. (1983). *A Classification Scheme for Master Production Schedule. Efficiency of Manufacturing Systems.* Plenum Press., doi:10.1007/978-1-4684-4475-9_10

Wright, L., & Davidson, S. (2020). How to tell the difference between a model and a digital twin. *Advanced Modeling and Simulation in Engineering Sciences*, *7*(1), 13. doi:10.118640323-020-00147-4

Wu, H., He, Z., & Gong, J. (2010). A virtual globe-based 3D visualization and interactive framework for public participation in urban planning processes. *Computers, Environment and Urban Systems.*

Würstle, P., Santhanavanich, T., Padsala, R., & Coors, V. (2020). The Conception of an Urban Energy Dashboard using 3D City Models. In *e-Energy '20: Proceedings of the Eleventh ACM International Conference on Future Energy Systems* (pp. 523–527). Virtual Event Australia: Association for Computing Machinery. 10.1145/3396851.3402650

Wüstenhagen, R., Wolsink, M., & Bürer, M. J. (2007). Social acceptance of renewable energy innovation: An introduction to the concept. *Energy Policy*, *35*(5), 2683–2691. doi:10.1016/j.enpol.2006.12.001

Wynes, S., Zhao, J., & Donner, S. D. (2020). How well do people understand the climate impact of individual actions? *Climatic Change*, 1–14.

Xia, H. (2020a). *Shanghai to invest 270 bln yuan in new infrastructure in next 3 years.* http://www.xinhuanet.com/english/2020-05/07/c_139038827.htm

Xie, X., Lu, Q., Rodenas-Herraiz, D., Parlikad, A. K., & Schooling, J. M. (2020). Visualised inspection system for monitoring environmental anomalies during daily operation and maintenance. *Engineering, Construction, and Architectural Management*, *27*(8), 1835–1852. doi:10.1108/ECAM-11-2019-0640

Xu, Y., Sun, Y., Liu, X., & Zheng, Y. (2019). A Digital-Twin-Assisted Fault Diagnosis Using Deep Transfer Learning. IEEE Access, 7, 19990-19999. doi:10.1109/ACCESS.2018.2890566

Xu, L., Francisco, A., Taylor, J. E., & Mohammadi, N. (2021). Urban Energy Data Visualization and Management: Evaluating Community-Scale Eco-Feedback Approaches. *Journal of Management Engineering*, *37*(2), 04020111. doi:10.1061/(ASCE)ME.1943-5479.0000879

Yalcinkaya, M., & Singh, V. (2014, July). Building information modeling (BIM) for facilities management–literature review and future needs. In *IFIP International Conference on Product Lifecycle Management* (pp. 1-10). Springer. 10.1007/978-3-662-45937-9_1

Yamamura, S., Fan, L., & Suzuki, Y. (2017). Assessment of urban energy performance through integration of BIM and GIS for smart city planning. *Energy Procedia*, *180*, 1462–1472. doi:10.1016/j.proeng.2017.04.309

Yang, C., Wang, B., Li, D., & Tian, X. (2017). Modelling and characterization for the responsive performance of CF/PLA and CF/PEEK smart materials fabricated by 4D printing. *Virtual and Physical Prototyping*, *12*(1), 69–76. doi:10.1080/17452759.2016.1265992

Ye, Z., Yin, M., Tang, L., & Jiang, H. (2018). Cup-of-Water theory: A review on the interaction of BIM, IoT and blockchain during the whole building lifecycle. In *ISARC. Proceedings of the International Symposium on Automation and Robotics in Construction* (*Vol. 35*, pp. 1-9). IAARC Publications.

Ye, X., & Lin, H. L. (Eds.). (2021). *Spatial Synthesis: Computational Social Science and Humanities*. Springer.

Yin, Z., & Hou, J. (2016). Recent advances on SVM based fault diagnosis and process monitoring in complicated industrial processes. *Neurocomputing, 174*, 643-650.

Yinfen, Z. (2019). *Innovation Drives Development, Technology Leads the Future: G60 Science & Technology Innovation Valley*. Shanghai International Studies University, SESTA. http://en.shisu.edu.cn/resources/features/G60-Science-Technology-Innovation-Valley

Yu, H., Miao, C., Leung, C., & White, T. J. (2017). Towards AI-powered personalization in MOOC learning in npj Sci. *Learn.*, *2*, 15. PMID:30631461

Zaha Hadid Architects. (2006). *Kartal Masterplan*. Retrieved February 30, 2020, from https://www.zaha-hadid.com/masterplans/kartal-pendik-masterplan/

Zhang, Y. Y., Kang, K., Lin, J. R., Zhang, J. P., & Zhang, Y. (2020). Building information modeling–based cyber-physical platform for building performance monitoring. *International Journal of Distributed Sensor Networks*, *16*(2), 1–21. doi:10.1177/1550147720908170

Zhiliang, M., & Yuan, R. (2019). *Integrated application of BIM-GIS: An overview*, Science Direct. *Procedia Engineering*, 196.

Zhou, M., Yan, J., & Feng, D. (2019). Digital twin and its application to power grid online analysis. *CSEE Journal of Power and Energy Systems*, *5*(3), 391–398. doi:10.17775/CSEEJPES.2018.01460

Zhou, Y., Ding, L., Rao, Y., Luo, H., Medjdoub, B., & Zhong, H. (2017). Formulating project-level building information modeling evaluation framework from the perspectives of organizations: A review. *Automation in Construction*, *81*, 44–55. doi:10.1016/j.autcon.2017.05.004

Zuehlke, D. (2010). SmartFactory—Towards a factory-of-things. *Annual Reviews in Control*, *34*(1), 129–138. doi:10.1016/j.arcontrol.2010.02.008

About the Contributors

Osman Arrobbio is a Research Fellow at the Department of Culture, Politics and Society. He holds a PhD in Sociology with a background in Political Sciences and Sustainable Development from the University of Torino. He has mainly carried out research activities in projects funded by the European Union on issues related to the energy transition. Specifically on the involvement of stakeholders and citizens, on energy feedback systems, on the integration of Social and Human Sciences, on collective actions in the energy field. He is currently engaged in the COMETS project. His other research interests concern climate warming, mitigation strategies, theory of practices and sociology of technique.

Ilaria Ballarini graduated in Architecture at Politecnico di Torino in 2007, has achieved her PhD in "Technological Innovation for Built Environment" in 2011. She is an Associate Professor in building physics and building energy systems at Politecnico di Torino, where she teaches in bachelor's and master of science's degree courses of Architecture, Building Engineering and Energy Engineering. Her main research activity concerns the topics of building physics, energy modelling of buildings and building stocks, procedures for energy performance assessment and cost-optimisation, building energy audit and certification.

Fabrizio Barbero has a degree in Computer Science and he is specialized in Artificial Intelligence, Expert systems and Machine Learning technologies. He is currently Director of Platform & Integration Division at CSI-Piemonte, a full service provider company working for Piedmont Regional Public Administrations, with responsibility for integration at Enterprise level of service and data cross platform, web and mobile solutions and identity management technology. He has in charge new initiatives in Big Data management and architecture for Internet of Things. Fabrizio is experienced in redesigning development organization focusing on project management, engineering and SOA best practices. He has set up Software Development Factories and worked intensively in the standardization of architecture, application development processes and techniques, using object oriented, agile methodologies, dev-ops practices and Java development. Under Fabrizio's leadership, CSI-Piemonte has successfully achieved the delivery of SOA and ALM platforms, Open data and Big Data technology for enterprise information analysis and coordinated open source technologies adoption. He led the implementation of CSI-Ecosystem, a federation of Information System based on a newborn CSI-Piemonte Private Cloud. He is currently working on the delivery of "Yucca" Smart Data Platform, the Piedmont data hub for Internet of Things, People and Systems, exploring infrastructure for big data and advanced analytics projects. He is member of Confidustria Digitale IoT Group.

Ambra Barbini is working as scientific collaborator within the research group Process Engineering and Construction of the Fraunhofer Italia Research Institute in Bolzano, since December 2019. After graduating in Building Engineering-Architecture at the University of Trento in July 2019, with a Master thesis developed in collaboration with the Fraunhofer Italia Research Institute, Ms. Barbini worked on the research project "BIM Methodologies for new industrialization of the energy requalification of the existing building heritage" as a post-graduate scholarship-holder at the University of Trento.

Olivia Becher is a DPhil student at the University of Oxford specialising in water system risks and adaptation. Olivia holds an undergraduate degree in Civil Engineering from the University of Bath and an MSc in Water Science Policy and Management at the University of Oxford.

Maurizio Bocconcino is an engineer for the environment and the territory and PhD in representation and relief for the protection of the and territorial heritage, he is Associate Professor of Drawing and Surveying at DISEG - Department of Structural, Structural, Building and Geotechnical Engineering of Politecnico di Torino (Italy). Member of the Urban and Social Regeneration Group of Fondazione Sviluppo e Crescita CRT, he deals with information systems for the study and representation of the territory and the city.

Lorenzo Bottaccioli is currently Assistant Professor in the EDA group. He received his B.Sc. in Mechanical Engineering in 2011 at University of Perugia. During the B.Sc. he took part in the draft, and in the early stage analysis, of the student transportation survey. In 2014 he received his M.Sc. in Energetic and Nuclear Engineering at Politecnico di Torino, with a thesis on sustainable Energy Communities. Since November 2014 he joined the Department of Computer and Control Engineering at Politecnico di Torino first as Ph.D. student and then as Postdoctoral Research Fellow. In 2018, he received the Ph.D. in Computer and Control Engineering. He participates in the ministerial project "Zero Energy Buildings in Smart Urban Districts" of the Italian Technology Cluster for Smart Community. As well He is involved in several European Funded Projects focused on Smart Cities and Smart Communities. His main research interests concern on Energy Efficiency and Smart Cities/Communities with a particular attention on the spatial representation of Energy data and on the development of software solution for the planning, analysis and optimization of smart energy systems.

Ezio Nicolas Bruno Urbina is an engineer in charge of the technical office of the Gaslini Istituto IRCCS public pediatric hospital in Genoa since 2020. Previously he was in the technical office of the Galliera public hospital in Genoa since 2006 dealing with the maintenance and renovation of hospital buildings. He was Project Manager of the large renovation project called "New Galliera Hospital". He led the Galliera team during the development of the INDICATE project. In 2020 he obtained a master in InfraBIManagement at the Politecnico di Torino.

Carlo Caldera graduated in Civil Engineering at the Politecnico di Torino. He is former Full Professor of Building Technology and Integrated Building Design at Department ISEG of Politecnico di Torino and member of the Scientific Board of the Responsible Risk Resilience Centre - R3C. His studies addressed to construction processes, according to an approach related to sustainability, and process control in building design and maintenance. His researches focused on traditional building practices, sustainable integration of technology for building services and construction techniques, with the integration of environmental assessment and design.

Bernardino Chiaia is a Full Professor of Structural Engineering at the department of Structural, Geotechnical and Building Engineering of the Polytecnic of Turin. Since 2008 Dean of the Faculty of Engineering, International Telematic University Uninettuno, Member of the Board - Italian Institute of Geophysics and Volcanology (INGV - Rome). Author of more than 190 scientific publications on subjects of Structural Engineering, Materials Engineering and Fracture Mechanics. In particular, 85 papers have been published on the most important Journals of the field. Since 2018 scientific coordinator of SISCON "Safety of Infrastructures and Constructions".

Ricardo Codinhoto's expertise is BIM for life-cycle assessment. His research focus on understanding how to anticipate the implications that design decisions have on use, maintenance, recycling and decommissioning of buildings. He led the 2 stage commissioned research project for the Manchester Town Hall Complex Project (first UK facility to operate with BIM support). Regarding large and complex buildings involving vulnerable user groups and wellbeing, he led and participated on research funded by EPSRC, NIHR, MRC and also working closely to the NHS Estates. He contributed to CPD initiatives led by CIOB, RIBA, RICS and CIAT. He also developed commissioned research for the Government Office for Science that supported policy-making. Current research also focus on Design for Disassembly and Recycling.

Vincenzo Corrado, civil engineer, is a Full Professor of building physics and building energy systems at Politecnico di Torino, where he coordinates a unit of TEBE (Technology Energy Building Environment) Research Group focused on: building physics, building energy modelling, procedures for energy audit and certification, indoor environmental comfort, legislation, and technical standards. He is former President of IBPSA-Italy (Italian chapter of the International Building Performance Simulation Association), Italian delegate of CEN/TC 89 (Thermal performance of buildings and building components) and of ISO/TC 163 (Thermal performance and energy use in the built environment).

Daniela De Luca, born in Turin in 1991, architect and PhD student at Politecnico di Torino. Since 2015, she studies and applies the use of the BIM methodology for different contexts. In particolar way, the use of BIM methodology per energy saving, at the European Project: DIMMER (District Infromation Modelling and Management for Energy Reduction). In addition, she has studying new applications of the BIM method in Industry 4.0, for maintenance, training, security and storage using Virtual and Augmented Reality. Since 2016, she has been part of the research group coordinated by Prof. Osello at the drawingTOthefuture and VR@polito laboratory.

Monica Dettori born in 1994 in Ozieri (SS), building engineer at the Polytechnic University of Turin. Since 2019 she is a research fellow at the DrawingTOthefuture research laboratory and she is investigating the use of Building Information Modeling (BIM) methodology for energy efficiency and facility management. She is specifically involved in the analysis of the BIMtoBEM process for performing energy simulations in the industrial environment and she is also working in the field of integration between BIM and Integrated workplace management system (IWMS) platforms.

Umberto di Staso has been working in Territorium Online since 2017 as senior analyst and developer. From 2012 to 2017 he worked for Fondazione Graphitech, a computer graphics research centre located in Trento, Italy. In 2012, he received a master degree in Computer Science from the University of Trento. During his working experience, Mr. Di Staso has been involved in several European and National funded research projects, mainly under the ICT-PSP, FP7 and FESR2014-2020 programmes. In this context, his responsibilities ranged from analysis, design and development of ICT services to project management and coordination.

Giuseppe Digregorio (1996), after a bachelor degree in Civil and Environmental Engineering at the Politecnico di Bari, is currently enrolled at the Politecnico di Turin for a master in Structural Engineering and he spent one year at KTH in Stockholm. He is currently conducting researches in the field of BIM with particular attention on real time data and visual programming.

Vincenzo Donato is an Architectural Engineer graduate from the University of Florence and received a double PhD degree from the University of Florence (Italy) and the Technical University of Brainwashing (Germany). Currently is lecturer at the University of Florence. From 2016 till 2018 he was appointed as lecturer at Polytechnic of Turin and he works at the Department of Structural, Geotechnical and Building Engineering (DISEG). He carries out research and teaching activities focusing on various aspects related to "Building Information Modelling". Vincenzo has practising experience from working as a BIM manager and as an Architect. He developed capability to work on international architectural studies, developing ideas and concepts as well standards and procedures for supporting the design of hospitals and schools using BIM. He has also been involved in the retrofitting and the renovation of existing buildings in which BIM was required. Currently, his work is focused on applying BIM and diagnostic techniques for protecting historical and monumental heritage.

Arianna Fonsati was born in Turin (TO) on 20.02.1992. She received a graduate Architecture degree at Politecnico di Torino in 2015. She was Research Fellowship at Politecnico di Torino and she is now a PhD Student in Urban and Regional Development. Her research activity is related to BIM Standards development and implementation, working both at DrawingTOtheFuture lab and Lombardi Ingegneria.

Gian Vincenzo Fracastoro is emeritus professor of Building Physics at Politecnico di Torino, Energy Department.

Andrea Galli graduated in Architectural Engineering at the Politecnico di Torino and Management, Innovation and Service Engineering at Scuola Superiore Sant'Anna in Pisa, he is Director of Strategy and development at Accurat, a data-driven innovation and design firm based in Milan and New York and teaches computational design as Adjunct Professor at Politecnico di Milano. Formerly project leader at Carlo Ratti Associati, over the last ten years he has been testing the impact of a computational design approach to projects bridging between the physical and the digital environment.

Francesco Gavino Brundu received his Ph.D. from Politecnico di Torino in 2017. He works as a Postdoctoral Research Scientist at Columbia University in the city of New York, where his main research focus is on the application of computational and statistical analysis to biological data.

Sara Giaveno is a PhD Candidate in Architecture. History and Project at Politecnico di Torino and currently she is developing the PhD thesis in collaboration with CRH, a global leader in building materials. She held a II Level Master in BIM Manager at Politecnico di Milano in 2017, and in 2016 she obtained the master degree in Architecture Construction City at Politecnico di Torino with a thesis developed in collaboration with the University of Pennsylvania (Penn Design). Her work is focused on managing urban transformation with digital tools and her field of investigation includes the sociotechnical issues involved in digital revolution.

Kjartan Gudmundsson is an associate professor at the division of Sustainable Buildings at KTH in Stockholm. The principal aim of research is new theoretical methods and building technology that can bring about a sustainable society with certain emphasis on building physics modelling as well as environmental sustainability. Recent work relates to the development of superinsulation materials for buildings while current efforts are concerned with the use of digital tools for the design and analysis of buildings with focus on BIM, visual programming, data capture and the use of APIs. Mr. Gudmundsson's teaching activities include courses in Architectural Engineering, Building Technology and Building Physics.

Elisa Guelpa is researcher at the Energy Department of Politecnico di Torino. She received her master degree in Energy Engineering in 2012 and the Ph.D. in Energy Engineering in 2016, both from the Politecnico di Torino. The research activity she works on is related with thermodynamics and heat transfer problems. The main topics she is working on are district heating systems (specifically for peak shaving, optimal network expansion, best failure management, network modelling and demand side management), thermal storage (optimal design, localization and heat transfer enhancement) and modelling of large/complex energy systems (district multi-energy systems, fire systems, multi-source energy systems, renewable energy plants). Dr Guelpa is author of more than 20 papers published in international journals and more than 20 conference proceedings. According with Scholar, he has received 475 citations, with an H-Index of 18. She is co-author of a book about thermodynamic methods for reduction of thermal energy consumption in industries and power plants. She collaborated on various R&D projects financed by governmental agencies, both Italian and European, and by companies. She is guest editor of Applied Science, for the special issue on Modelling and Optimization of Thermal Energy Storage Systems. She is member of the scientific committee and topic leader at international conferences.

Ezz Eldin Hemdan has published much work in the area of Internet of Things, Internet of Nano Things and Industrial Internet of Things and Big Data analytics. Dr.Ezz has also done much work in applying machine learning and big data analytics approaches in crucial domains including Finance, Healthcare, and Cyber Security.

Jonathan Heron is a Civil Engineer graduated at the University of Bath. He is currently a structural engineer designer at Robert West Civil Engineering Consultancy. He is responsible for developing structural solutions for mainly permanent and temporary rail projects. His research and practice interests are Building Information Modelling and how it can be used to improve the construction industry.

Marco Jahn is Senior Researcher at Fraunhofer FIT. He obtained his diploma in computer science from Ulm University in 2006 and his PhD from RWTH Aachen in 2016. He worked as software developer at denkwerk GmbH before moving to Fraunhofer FIT. There he works as researcher and (technical) project manager in various European research projects in the areas of IoT, Smart Buildings, and Smart Cities. Furthermore, he leads the IoT Platforms team and coordinated the IoT Large-Scale Pilot Project MONICA.

Enrico Macii was born in Torino, Italy, on February 7, 1966. He is a Full Professor of Computer Engineering at Politecnico di Torino. Prior to that, he was an Associate Professor (from 1998 to 2001) and an Assistant Professor (from 1993 to 1998) at the same institution. From 1991 to 1997 he was also an Adjunct Faculty at the University of Colorado at Boulder. He holds a Laurea Degree in Electrical Engineering from Politecnico di Torino (1990), a Laurea Degree in Computer Science from Università di Torino (1991) and a PhD degree in Computer Engineering from Politecnico di Torino (1995). From 2009 to 2016 he was the Vice Rector for Research at Politecnico di Torino; he was also the Vice Rector for European Affairs (2007-2009), the Vice Rector for Technology Transfer (2009-2015) and the Rector's Delegate for International Affairs (2012-2015).His research interests are in the design of electronic digital circuits and systems, with particular emphasis on low power consumption aspects. In the last decade, he has extended his research activities to areas such as bioinformatics, energy efficiency in buildings, districts and cities, sustainable urban mobility, clean and intelligent manufacturing.

Amged Sayed Abdelmageed Mahmoud is an assistant professor in the department of Industrial Electronics and Control Engineering, Faculty of Electronic Engineering, Menoufia University, Menof Egypt. His research is focused in the field of Artificial Intelligence Algorithms (Fuzzy system, type-2 Fuzzy, Adaptive fuzzy, Neural Network, learning algorithms), Artificial intelligence with machine vision, Biomedical system, Embedded systems using microcontroller, Industrial IoT, Cyber-Physical System, Networked control and Optimal control.

Giada Malacarne graduated in Building Engineering and Architecture with a specialization in BIM Manager at the Polytechnic of Milan. Since 2015 she has been working at Fraunhofer Italy in the research group "Process Engineering in Construction". Her research focuses on the implementation of innovative information and communication technologies for a more effective and efficient design and construction process. Giada mainly collaborates with local public authorities and is an active member of EUnet4DBP - the European network for the digitalization of the building permit process.

Valentino Manni graduated in Architecture at Politecnico di Torino and got a Ph.D. Degree at Politecnico di Milano in Architectural and Environmental Technologies. Since 1996, he has taught both at Politecnico di Milano and at Politecnico di Torino. At present he works at the Department of Architecture and Design of Politecnico di Torino as Adjuncted Professor in Architectural Technology, holding the course "Culture and Fundamentals of Architectural Technology".

Sara Mannoni is Head of the GIS and Cartography Center of Excellence of CSI Piemonte. Many years of experience relating to issues related to Land Planning and Governance, Civil Protection and Land Defense which, over time, resulted in the management and coordination of Areas that develop and manage GIS (Geographic Information System) Services and Tools for Public Administration of Piedmont Region, Italy.

Fabrizio Massara is software architect of Geographic Information Systems (GIS), Geospatial and Smart Cities Applications, recently developed interest in Building Information Modeling (BIM) interaction and interoperability with the geospatial field. After a significative experience as free-lance senior software engineer of GIS and Open Source Applications, currently employed as Solution Designer for CSI Piemonte, Italy.

Dominik Matt holds the Chair for Production Systems and Technologies and heads the research department "Industrial Engineering and Automation (IEA)" at the Faculty of Science and Technology at the Free University of Bozen-Bolzano. Moreover, Prof. Matt is the director of Fraunhofer Italia in Bolzano. His research focuses on the areas of Industry 4.0 and Smart Factory, Applications of AI in Manufacturing, Lean and Agile Production, on the planning and optimization of assembly processes and systems, as well as on organizational and technical aspects of in-house logistics.

Umberto Mecca, born in Turin on 07.08.1988, graduated in Building Systems Engineering (LM-24) at Politecnico di Torino in 2016, qualified to practice his profession and enrolled in the Register of Engineers of the Province of Turin (Section A - Civil and Environmental Sector) in 2017, holds a Research Fellowship at the Department of Structural Building and Geotechnical Engineering (DISEG) at Politecnico di Torino.

Caterina Mele, Ph.D, Associate Professor, member of the Board of Directors of the Politecnico di Torino (Italy); member of the International Biomimicry Network ; appointed Investigador Externo Invitado of the Group of Technological Studies of the ITDHAu Institute of the Faculty of Architecture and Urban Planning of the Universidad Nacional del Nordeste, Resistencia, Chaco (Argentina). Her studies and publications mainly concern the building technology and sustainability.

Giuseppe Moglia (Torino, 1950), engineer (Politecnico di Torino, 1974, civil, building section), is professor for subjects ICAR 17, Drawing, in Politecnico di Torino. He works in Department of Structural, Geotechnical and Building Engineering, where he was supervisor of the Quality Assessment. He is author of about 80 publications on books, periodicals and in conventions. Scientific work is focused on regulatory framework for urban and regional planning and on survey, especially at urban scale. Studies are to mention on historic urban fabric of Turin, in preparation for Master Plan of 1995. Teaching in Politecnico School of Engineering was conducted especially in Drawing for Civil and Building Engineering Programmes, on preparing students for work in design. This activity has extended to distance learning. Since 2010-2011 he is tenured of Survey and Urban Design at MSc in Building Engineering. He has also taught courses for Università di Torino Interschool Programme in Strategic Sciences, held at Italian Army School of Applied Military Science and Institute of Military Studies. In Politecnico di Torino since 1999 he is involved in Educational Quality Assessment and from 2001 to 2007 he was Supervisor of Degree Programme in Building Engineering, part of Professional Degree Reinforcement Programme funded by European Union. From 2007 to 2012 he was Assistant Dean of I School of Engineering, at first responsible for master's degree and new educational projects, and from 2009 as President of Education Area in Building Engineering. From 2012 to 2015 he was Chairman of Building Engineering Collegio. From 2019 he is a member of Politecnico di Torino Quality Assessment Presidium. He is Supervisor of EUR-ACE Accreditation (EURopean ACcreditation of Engineering programmes) of ENAEE (European

Network for Accreditation of Engineering Education) of Civil Engineering Programmes. He is ANVUR Reviewer, system expert (National Agency for the Evaluation of Universities and Research Institutes).

Carlo Ostorero graduated at Politecnico di Torino in Building Engineering, Ph.D. in Architectural Design, has spent periods of studies and research in several Countries and Universities such as Faroe Island (DK), Portugal, Montenegro, Switzerland and Holland. In the last 25 years he taught at Faculty of Architecture of University of Parma and at Faculty of Building Engineering of Politecnico di Torino as Adjuncted Professor of "Restoration and Building Refurbishment". In Turin he also works as architect in Studio Dedalo Architettura (studiodedaloarchitettura.it).

Dario Padovan is Associate Professor of Sociology at the University of Turin and he is member of the Department of Culture, Politics and Society. He works since long time on history of social sciences, environmental sociology, social metabolism, sustainable consumption, sociology of risk, environmental conflicts. He is coordinator of the Unesco Chair on Sustainable Development at the University of Torino. He is member of Editorial Committee of Theomai Journal (Buenos Aires), Chosmos & History (Melbourne, Australia). He is member of Esa RN 12 on "Environment and Sociology", and of RN 6 on "Sociology of Consumption". Recently is doing research on household metabolism, environmental accounting, urban metabolism regulation, sociology of climate change, energy transition, environmental conflicts, socio-technical systems. Currently he is Principal Investigator of the project called "Comets - Collective action models for Energy Transition and Social Innovation", funded by European Commission on H2020-LC-SC3-2018-2019-2020 call - Grant Agreement No. 837722.

Roberto Pagani is professor of Architectural Technology at Politecnico di Torino. Currently, he is Scientific Attaché of the Italian Ministry of Foreign Affairs and International Cooperation at the Consulate General of Italy in Shanghai.

Gabriele Pasetti Monizza (ORCID ID: https://orcid.org/0000-0003-2197-6267) is the scientific coordinator for the team Digital Design & Engineering in the PEC team at Fraunhofer Italia. He leads the research focus group which aims at developing methods and approaches for the digital transformation of product development process towards smart manufacturing systems. He is adjunct professor at the Free University of Bolzano for digital transformation of building industry within the Professional Bachelor in Wood Engineering. He has developed his expertise in applied research within the last 10 years at Fraunhofer Italia. He has been the project manager and the principal investigator of several research projects, such as the BIM Simulation Lab project. He is author of several publications in digital transformation of the building industry. His primary research focus is automation and computational design in building product development.

Edoardo Patti received both M.Sc. and the Ph.D. degrees in Computer Engineering at Politecnico di Torino in 2010 and 2014, respectively. During the Academic Year 2014/2015, he was Academic Visiting at the University of Manchester. He is currently Assistant Professor in the EDA group. He is also involved in various European funded projects focused on Smart City. His research interests concern: i) Ubiquitous Computing; ii) Internet of Things; iii) Smart Systems and Cities; iv) Software architectures with particular emphasis on infrastructure for Ambient Intelligence; v) Software solutions for simulating and optimizing energy demand response; vi) Software solutions for energy data visualization for increasing user awareness.

Marco Piovano is a researcher in the PEC team at Fraunhofer Italia. He received his degree in Architecture at the Politecnico of Turin with a thesis about H-BIM for the Facility Management (FM), with a particular focus on interoperability, data extraction and management. During his studies he worked in an architectural firm supporting the design of residential buildings. He joined Fraunhofer Italia in 2017 as a scientific collaborator within the research group Process Engineering in Construction. He is the project manager of AR4FM project. His research focuses on the implementation of innovative digital technologies for the real estate and, in particular, on the implementation and test of BIM, AR technologies for Facility Management in pilot projects.

Matteo Piro graduated in March 2020 at Politecnico di Torino in Building Engineering, with a thesis focused on the methodologies and tools to model a building stock from an energy and environmental point of view. One of the main objectives was to develop guidelines to fill the uncertainty in finding the data necessary to create a UBEM model and apply them to a case study.

Francesco Prizzon is Full Professor at Politecnico di Torino, DIST Interuniversity Department of Regional and Urban Studies and Planning, he is teaching Feasibility and economic evaluation of projects (Master Degree in Building Engineering). From November 2006 to 2012, he was Head of the Fondazione Compagnia di San Paolo Programma Housing Technical Evaluation Committee, launched to prepare, finance and monitor all Social Housing interventions of the Foundation. From 2004 to 2014 member of the Technical Steering Committee of SiTI, Istituto Superiore dei Sistemi Territoriali per l'Innovazione (Higher Institute of Territorial Systems for Innovation), founded jointly by Politecnico di Torino and Fiondazione Compagnia di San Paolo. From 2010 to 2019 member of the Advisory Committee of FASP - Fondo Abitare Sostenibile Piemonte; from 2014 to 2019 member of the Fondo Housing Sociale Liguria Advisory Committee, Social Housing Real Estate Funds as independent expert. Since 2005 founding member of the Scientific Committee of SIEV, Italian Society of Appraisal and Valuation.

Anna Rabbia graduated in Architecture at the Polytechnic of Turin (Italy), she is a Member of the Urban and Social Regeneration Urban and Social Regeneration Group of Fondazione Sviluppo e Crescita CRT. She analyzes issues related to urban redevelopment and quality quality, through indicators for the social and economic impact of investments, all with the support of Geographic Information Systems.

Adelaide Ramassotto is Program Manager for CSI Piemonte with responsibilities of sustainable development of local public authorities. After significative experience in Geographic Information Systems (GIS), environmental, land and open data information systems, the main occupation now is in the field of Energetic and Facility Management. The union of professional skills with creativity allowed to create the Dressmap brand in 2013.

Manuela Rebaudengo, Building Engineer (2003), Ph.D. (2007) in Project Appraisal, has been Assistant Professor since 2011, teaching assets and investments appraisal. Since 2004 she has been carrying out research and advisory activities for public and private institutions on real estate market, public works and public-private partnerships, economic and financial sustainability of projects. Member of

R3C, Responsible Risk Resilience Centre of Politecnico di Torino. Since 2017, Representative of the Politecnico di Torino in the Working Group of Public Contracts Regional Observatory, as expert on public contracts. Since 2018, she has been included in the "experts long list for teaching assignments on public contracts by ITACA and coordinates, at the Politecnico di Torino, permanent training courses on the assessment and management topics of public contracts. Since 2019, she has been teaching "Assessment of construction and maintenance costs" in the 2nd level Master's course "Architecture of wooden constructions" at the Politecnico di Torino. The research activity has developed mainly around real estate market values and procedures for real estate developers (feasibility and preliminary investments assessment, hope value, HBU, …); then, economic impact of abnormally low bids in public tenders and public works evaluation (trends, procedures and related costs). Now, as R3C member, the research deals with territorial resilience assessment through maintenance regulation; design of resilient scenarios for urban brownfield sites towards agenda 2030; buildings and urban maintenance conditions assessment; urban and building strategies to reduce the overall impact of the built environment; construction risks assessment, estimating risks and technical-economic impacts for preventive risk mitigation actions.

Michael Riedl obtained his Ph.D. degree at the Technical University of Munich in mechanical engineering with a specialization in systematic product development, automotive engineering and simulation technologies. Subsequently he worked at AUDI AG in the department for technical development with a focus on concept development. At Fraunhofer Italia, he is deputy director and team leader of the research group "Automation and Mechatronics Engineering". His current research is on automation technologies as well as the development of mechatronic and cyber physical systems and their application in smart buildings, precision agriculture and the factory of the future.

Laura Rietto, PhD in Energy and Architectural Engineer, is actually working at the Politecnico di Torino and as freelance professional engineer. She obtained her PhD title in 2017 at the Politecnico di Torino, working on the integration of systems from building scale to urban scale, specializing in energy dynamic simulations of buildings and systems and in the forecasting of the thermal energy demand by means of neural networks. In her M.Sc. studies she focused on NZEBs and energy efficiency solutions for buildings and systems and she is active in these fields of research.

Katrien Romagnoli works at Fraunhofer Italia as scientific researcher. After obtaining her MSc. in Environmental Engineering at the University of Ghent (Belgium) and a Master Thesis from the University of Manitoba (Canada), she worked for 3 years as a sustainable energy use consultant at DNV GL. She is specialized in energy (efficiency) management (ISO 50001, EN 16247, energy culture etc.) and establishing decarbonization strategies for governmental organizations (EU, federal and regional).

Christoph Schimanski is a Civil Engineer; Doctoral student in the field of BIM-supported Construction Management; Scientific Collaborator at Fraunhofer Italia Research in the research group of "Process Engineering in Construction"

Alice Schweigkofler works as a researcher in the Process Engineering in Construction (PEC) team at Fraunhofer Italia. She obtained her Master´s degree in Building Engineering and Architecture at the University of Bologna and during her studies she spent one year at the KIT, Karlsruher Institut of Technology. She has recently concluded a specialization course in the field of sustainable urban regeneration at the Politecnico di Milano and currently focuses on the digitalization of processes for the renovation of buildings and the application of smart city concepts.

Alessandro Sciullo is a Research Fellow at the Department of Culture, Politics and Society. Holds a PhD in Economic Sociology and a Master in Public Policy Analysis from the University of Torino and his research experience are policy evaluation, organizational analysis, innovation diffusion and energy transition. He is currently involved in three H2020 projects: in GEMEX project he deals with strategies for improving public acceptance of geothermal energy; in COMETS he is project manager and responsible for the DCEP implementation; in eCREW he is leader of 2 tasks related to public engagement and designing energy related performance indicators. He is also Member of the management board of the EERA e3s Joint Programme.

Elisa Spallarossa is graduated in civil engineering, She is BIM Manager and RICS expert in Project Management(MRICS). From June 2015 to June 2018, she held the position of Executive Engineer and BIM Manager at Struttura Complessa R.U.P. Nuovo Galliera, E.O. Ospedali Galliera in Genoa. As BIM manager she is in charge of the drafting and management of contract documents and technical documentation including: documents, contents and methods of the Bim process applied to design, construction and Asset Management (EIR: AIR, IDP, etc.). Since April 2017, she is part of the UNI/CT 033/GL 05 commission "Codification of products and construction processes in construction" for the drafting of the voluntary standard UNI11337. Since February 2017 she has acquired the ICMQ Certification in Expert in Building Information Modeling - BIM MANAGER and renewed it according to the provisions of the UNI 11337-4 standard in April 2019. She is an expert in integrated systems of planned maintenance for infrastructure networks and real estate assets, she also held the role of AASHTO representative in Italy for the implementation of Asset Management systems.

Dieter Steiner works as a thematic coordinator and researcher in the Automation and Mechatronics (AME) team at the Fraunhofer Italia Innovation Engineering Center. He got his Master's degree in computer engineering from the Vienna University of Technology. He has over a decade of experience in applied research in the fields of computer vision, smart sensors and data fusion, with applications in automated assembly-, as well as human-centered production systems. He has participated in various FP7 and H2020 projects, as well as national and regionally funded research projects. His current research focuses on machine vision and -learning, cooperative robotics and cyber-physical systems in the context of Industry 4.0.

Silvia Titotto is Architect and Urban Planner (2005), Master (2008) and PhD (2013) from University of Sao Paulo via the Design and Architecture Program as well as PhD (2013) from Politecnico di Torino (Italy) via the Technological Innovation for the Built Environment Program at DISEG Department. She is currently a permanent professor at the Graduate Program in Engineering and Innovation Management and she has obtained her tenure at the Center for Engineering, Modelling and Applied Social Sciences at the Federal University of ABC. She served in military service as an officer of the Brazilian Navy as part of the Corp of Naval Engineers. She was a researcher at the Computational Mechanics Laboratory of the USP Polytechnic School. She taught courses on Bio-Inspired Design and Prototyping at UNESP (Bauru), Politecnico di Torino (Italy), Università di Bologna (Italy), Silesian University of Technology (Poland), Riga Technical University (Latvia) and Fraunhofer IWM (Germany). She is the principal investigator/leader of the 4D Printing and Biomimetics (4DB) Research Group in the CNPq directory and collaborates with interdisciplinary research groups abroad. She is fluent in Portuguese / English / Italian and communicates in French / Spanish / German. Since 2001, she has been exhibiting responsive installation artworks and kinetic prototypes in Brazil and the European Union, with institutional support and industry sponsorships. Since 2002, she has been researching the interdisciplinary field of Bioinspired Engineering, seeking technological innovations inspired by solutions and strategies presented by biological structures and their functions. She is interested in the following subjects: biomimetics/biomimicry, digital fabrication, 4D and 5D printing, installation art and lightweight structures.

Francesca Maria Ugliotti was born in Borgosesia (Vercelli, Italy) on 19.03.1987 She is an Assistant Professor RTD-A at the Department of Structural Engineering, Geotechnics and Construction of the Politecnico di Torino. She received a graduate Building Engineering degree at Politecnico di Torino in 2011 and a first-level Specializing Master in Real Estate Management at Università degli Studi di Torino in 2012. Her research is focused on the intelligent use of data that characterize existing assets, exploiting the potential offered by the Building Information Modelling methodology. Through the PhD in Urban and Regional Development, she broadened the topic from the building to the urban scale, pursuing the objective of providing managers of real estate assets, both public and private, with innovative tools for data collection, use and visualization.

Luca Saverio Valzano got a M.Sc.Degree both in Building Engineering and in Architecture at Politecnico di Torino, where he further got a Second Level M.Sc. Degree in "Design and Building of Tall Buildings". At present he works as Research Fellow at the Department of Architecture and Design of Politecnico di Torino, addressing to the Architectural Technology. He also works as professional in Studio LVA in Turin.

Valentina Villa is an M.Eng and PhD in Construction Management at Polytecnic of Milan. Now she's Assistant Professor at the department of Structural, Geotechnical and Building Engineering of the Polytecnic of Turin. Her research is situated in the field of Technology & Innovation, with a special focus on digitalization of processes and methodology in AEC Industry. She's the scientific coordinator of Polytecnic of Turin for the project: "A Distributed Digital Collaboration Framework for Small and Medium-Sized Engineering and Construction Enterprises, (2019-2022). Since 2016 she deals with Building Information Modeling and Management. She has over 100 papers and articles, including proceedings in major international conferences.

Mariapaola Vozzola, Engineer and PhD, Fellow at DISEG - Department of Structural, Geotechnical and Building Engineering of the Politecnico di Torino (Italy). She conducts research activities in the field of survey and BIM modelling with a focus on urban resilience design and graphic analysis and design representation codes.

Index

D

E

F

G

H

I

J

K

L

M

N

O

P

R

S

T

U

V

W

X

www.ingramcontent.com/pod-product-compliance
Lightning Source LLC
LaVergne TN
LVHW080617300926
849162LV00005B/1188